Industrial Crops and Uses

Industrial Crops and Uses

Edited by

Bharat P. Singh

Fort Valley State University, Fort Valley, Georgia, USA

www.cabi.org

CABI is a trading name of CAB International

CABI Head Office
Nosworthy Way
Wallingford
Oxfordshire OX10 8DE
UK

Tel: +44 (0)1491 832111
Fax: +44 (0)1491 833508
E-mail: cabi@cabi.org
Website: www.cabi.org

CABI North American Office
875 Massachusetts Avenue
7th Floor
Cambridge, MA 02139
USA

Tel: +1 617 395 4056
Fax: +1 617 354 6875
E-mail: cabi-nao@cabi.org

A catalogue record for this book is available from the British Library, London, UK.

Library of Congress Cataloging-in-Publication Data

Industrial crops and uses / edited by Bharat P. Singh.
p. cm.
Includes bibliographical references and index.
ISBN 978-1-84593-616-7 (alk. paper)
1. Field crops–Utilization. 2. Field crops–Industrial applications.
I. Singh, Bharat P.

SB185.I53 2010
633.8--dc22

2009048875

ISBN-13: 978 1 84593 616 7

Commissioning Editor: Meredith Carroll
Production Editor: Shankari Wilford

Typeset by SPi, Pondicherry, India.
Printed and bound in the UK by CPI Antony Rowe, Chippenham, UK.

Contents

Contributors

Lorenzo Aleman, Institute for Agricultural Biosciences, Oklahoma State University, Ardmore, Oklahoma, USA. E-mail: alemans@okstate.edu

Randy D. Allen, Institute for Agricultural Biosciences, Oklahoma State University, Ardmore, Oklahoma, USA. E-mail: randy.allen@okstate.edu

William Anderson, USDA-ARS Crop Genetics and Breeding Research Unit, Tifton, Georgia, USA. E-mail: Bill.Anderson@ ARS.USDA.GOV

Monica Båga, Department of Plant Sciences, College of Agriculture and Bioresources, University of Saskatchewan, Saskatoon, Canada. E-mail: Monica.baga@usask.ca

Scott Bean, USDA-ARS, Grain Marketing and Production Research Center, Manhattan, Kansas, USA. E-mail: Scott.Bean@ ARS.USDA.GOV

B. Todd Campbell, Coastal Plains Soil, Water and Plant Research Center, USDA-ARS, Florence, South Carolina, USA. E-mail: Todd.Campbell@ARS.USDA.GOV

Enrico Casadei, Former Senior Officer of the Food and Agriculture Organization of the United Nations and member of the Network for Natural Gums and Resins in Africa (NGARA), Nairobi, Kenya. E-mail: enrico.casadei@fao.org

Jonathan Y. Chen, Texas Material Institute Material Science and Engineering Program, University of Texas at Austin, Austin, Texas, USA. E-mail: jychen2@austin.utexas.edu

Paul Chen, Centre for Biorefining and Bioproducts and Biosystems Engineering Department, University of Minnesota, St Paul, Minnesota, USA. E-mail: chenx088@umn.edu

Yanling Cheng, Centre for Biorefining and Bioproducts and Biosystems Engineering Department, University of Minnesota, St Paul, Minnesota, USA and Biochemical Engineering College, Beijing Union University, Beijing, China.

Ravindra N. Chibbar, Department of Plant Sciences, College of Agriculture and Bioresources, University of Saskatchewan, Saskatoon, Canada. E-mail: ravi.chibbar@usask.ca

Ben Chikamai, Executive Secretary of the Network for Natural Gums and Resins in Africa (NGARA), Nairobi, Kenya. E-mail: benchikamai@ngara.org

Terry A. Coffelt, New Crops, Plant Physiology and Genetics Research Unit, US Arid-Land Agricultural Research Center, USDA-ARS, Maricopa, Arizona, USA. E-mail: Terry.Coffelt@ARS.USDA.GOV

Gillian Eggleston, Commodity Utilization Research Unit, USDA-ARS-SRRC, New Orleans, Louisiana, USA. E-mail: Eggleston@ARS.USDA.GOV

Holger Fischer, Faserinstitut Bremen e.V. – FIBRE –, Am Biologischen Garten 2, Bremen, Germany. E-mail: fischer@faserinstitut.de

Michael A. Foster, Texas Agricultural Experiment Station, Texas A&M University, Pecos, Texas, USA. E-mail: ma-foster@tamu.edu

Seedhabadee Ganeshan, Department of Plant Sciences, College of Agriculture and Bioresources, University of Saskatchewan, Saskatoon, Canada. E-mail: Pooba.ganeshan@usask.ca

Lori Hinze, Southern Plains Agricultural Research Center, USDA-ARS College Station, Texas, USA. E-mail: Lori.Hinze@ARS.USDA.GOV

Lou Honary, National Ag-Based Lubricant Center, University of Northern Iowa, Cedar Falls, Iowa, USA. E-mail: lou.honary@uni-nabl.org

Murray B. Isman, Faculty of Land and Food Systems, University of British Columbia, Vancouver, Canada. E-mail: murray.isman@ubc.ca

Sarita Jaiswal, Department of Plant Sciences, College of Agriculture and Bioresources, University of Saskatchewan, Saskatoon, Canada. E-mail: saj105@mail.usask.ca

Jay-lin Jane, Department of Food Science and Human Nutrition, Iowa State University, Ames, Iowa, USA. E-mail: jjane@iastate.edu

Thomas Klasson, Commodity Utilization Research Unit, USDA-ARS-SRRC, New Orleans, Louisiana, USA. E-mail: Thomas.klasson@ARS.USDA.GOV

Peter M. Kopittke, School of Land, Crop and Food Sciences, The University of Queensland, St Lucia, Queensland, Australia and Cooperative Research Center for Contamination Assessment and Remediation of the Environment (CRC-CARE), Salisbury, South Australia, Australia. E-mail: p.kopittke@uq.edu.au

Xiangyang Lin, Centre for Biorefining and Bioproducts and Biosystems Engineering Department, University of Minnesota, St Paul, Minnesota, USA; MOE Biomass Engineering Center, Nanchang University, Jiangxi, China and College of Bioscience and Biotechnology, Fuzhou University, Fujian, China.

Frank Liu, Veterinary Diagnostic Laboratory, College of Veterinary Medicine, University of Minnesota, St Paul, Minnesota, USA. E-mail: liuxx063@umn.edu

Yuhuan Liu, Centre for Biorefining and Bioproducts and Biosystems Engineering Department, University of Minnesota, St Paul, Minnesota, USA and MOE Biomass Engineering Center, Nanchang University, Jiangxi, China.

Enzo Lombi, Cooperative Research Center for Contamination Assessment and Remediation of the Environment (CRC-CARE), Salisbury, South Australia, Australia and Center for Environmental Risk Assessment and Remediation, University of South Australia, Mawson Lakes, South Australia, Australia. E-mail: Enzo.Lombi@unisa.edu.au

Colleen McMahan, Western Regional Research Center, USDA-ARS, PWA, WRRC-CIU, Albany, California, USA. E-mail: Colleen.McMahan@ARS.USDA.GOV

Clodualdo C. Maningat, MGP Ingredients Inc, Atchison, Kansas, USA. E-mail: Ody.Maningat@mgpingredients.com

Neal W. Menzies, School of Land, Crop and Food Sciences, The University of Queensland, St Lucia, Queensland, Australia and Cooperative Research Center for Contamination Assessment and Remediation of the Environment (CRC-CARE), Salisbury, South Australia, Australia. E-mail: n.menzies@uq.edu.au

Denis J. Murphy, Biotechnology Unit, University of Glamorgan, Trefforest, Wales, UK. E-mail: dmurphy2@glam.ac.uk

Jörg Müssig, Hochschule Bremen, Department for Biomimetics, University of Applied Sciences, Bremen, Germany. E-mail: jmuessig@bionik.hs-bremen.de

Ravi Naidu, Center for Environmental Risk Assessment and Remediation, University of South Australia, Mawson Lakes, South Australia, Australia. E-mail: Ravi.Naidu@unisa.edu.au

Priya Padmanabhan, Department of Biology, Western Kentucky University, Bowling Green, Kentucky, USA. E-mail: priya.padmanabhan@wku.edu

Lee Panella, Crops Research Laboratory USDA-ARS-NPA, Fort Collins, Colorado, USA. E-mail: Lee.Panella@ARS.USDA.GOV

Dennis T. Ray, Division of Horticultural and Crop Sciences, University of Arizona, Tucson, Arizona, USA. E-mail: dtray@email.arizona.edu

Roger Ruan, Centre for Biorefining and Bioproducts and Biosystems Engineering Department, University of Minnesota, St Paul, Minnesota, USA and MOE Biomass Engineering Center, Nanchang University, Jiangxi, China. E-mail: ruanx001@umn.edu

Shivendra V. Sahi, Department of Biology, Western Kentucky University, Bowling Green, Kentucky, USA. E-mail: shiv.sahi@wku.edu

Dev Shrestha, Biological and Agricultural Engineering Department, University of Idaho, Moscow, Idaho, USA. E-mail: devs@uidaho.edu

Ralph E.H. Sims, Center for Energy Research, Massey University, New Zealand. E-mail: Ralph.SIMS@iea.org

Bharat P. Singh, Agricultural Research Station, Fort Valley State University, Fort Valley, Georgia, USA. E-mail: singhb@fvsu.edu

Hari P. Singh, Agricultural Research Station, Fort Valley State University, Fort Valley, Georgia, USA. E-mail: singhh@fvsu.edu

R. Siva, School of Biotechnology, Chemical and Biomedical Engineering, VIT University, Vellore, India. E-mail: rsiva77in@rediffmail.com

Kazuhiro Takamizawa, Department of Applied Biological Sciences, Gifu University, Yanagido, Gifu, Japan. E-mail: tak2003@gifu-u.ac.jp

Thomas Tew, Sugarcane Research Unit, SRRC-USDA-ARS, Houma, Louisiana, USA. E-mail: Thomas.Tew@ARS.USDA.GOV

Jon Van Gerpen, Biological and Agricultural Engineering Department, University of Idaho, Moscow, Idaho, USA. E-mail: jonvg@uidaho.edu

Yiqin Wan, Centre for Biorefining and Bioproducts and Biosystems Engineering Department, University of Minnesota, St Paul, Minnesota, USA and MOE Biomass Engineering Center, Nanchang University, Jiangxi, China.

Donghai Wang, Department of Biological and Agricultural Engineering, Kansas State University, Manhattan, Kansas, USA. E-mail: dwang@ksu.edu

Xiaoquan Wang, Centre for Biorefining and Bioproducts and Biosystems Engineering Department, University of Minnesota, St Paul, Minnesota, USA.

Rungtiwa Wongsagonsup, Department of Food Science and Human Nutrition, Iowa State University, Ames, Iowa, USA. E-mail: rungtiwa@iastate.edu

Xiaorong Wu, Department of Biological and Agricultural Engineering, Kansas State University, Manhattan, Kansas, USA. E-mail: xwu3786@k-state.edu

Renyong Zhao, Hunan University of Technology, Zhengzhou, China. E-mail: zry8600@hotmail.com

Preface

The 21st century promises to be the period of a new industrial revolution in which the criteria for the selection of raw materials will be decided by their renewable nature and environmentally friendly accessibility, in contrast to the preceding century in which petroleum, a depleting raw material source and cause for multiple health and environmental concerns, dominated the industrial scene. The industries of the future will rely increasingly on resources such as the wind and the sun, which previously have been untapped, and old sources of raw materials such as plants, which were abandoned during the last century in favour of petroleum. Plants, because of their potential to yield a wide range of products, their renewable nature and their positive environmental impacts, are destined to play a key role in this new industrial economy. These products will include liquid fuels, chemicals, lubricants, plastics, paints, building materials, insecticides and pharmaceuticals, to name a few. In addition, plants will play a crucial role in the reclamation of soil, water and air polluted by the petroleum-based industries. How fast and by how much industries based on plant raw material will develop depends on a multitude of factors, such as research into crop production and processing, infrastructure development for plant-based industries, the cost competitiveness of bio-based products against petroleum-derived products, consumer demand and political support.

In recent years, the pace of research and development in the production and processing of different crops for various categories of industrial products has accelerated. The literature, however, awaits review and consolidation in a systematic manner so as to be a useful reference in teaching and research. This fact became apparent to me when I was approached recently by the dean of a university in South-east Asia to assist in developing a graduate study concentrating on industrial crops. We tried to find a book that could guide us in developing courses for this purpose. The book that we were looking for would have covered crop species of importance, the current status of research and development in their production, harvesting, handling and processing, and current and prospective end uses for these crops. We were unable to find such a comprehensive book. My colleagues and friends encouraged and coaxed me to take the lead in compiling such a text. I developed this book proposal with this specific purpose in mind. It was a difficult exercise to identify appropriate authors for the various chapters, considering that they required vastly different expertise, although with the common denominator of industrial crops. I am indebted to all the contributors to the book from around the world who are noted experts in their fields. Chapter authors were asked to include information on industrial crops suitable for a textbook for graduate-level students and, in addition, to include current research and developments of interest to researchers and professionals devoted to the industrial utilization of plants. The controversies, scopes and prospects, sustainability and planning and policy issues associated with different plant-based industries were also to be addressed where needed for the benefit of political and industrial planners.

The book has been divided into eight parts. The first part overviews industrial crops. It is essentially the editor's opinion on the various issues related to industrial crops, covered in detail in place of a lengthy preface. The second part is devoted to the crops and processes involved in generating energy in the form of heat and electricity, as well as liquid fuel. Ethanol production from sugar crops, grain crops and lignocellulosic crops has been covered under separate chapters to explain the crops fully and the processes and by-products produced in each case. At the end of this section, biodiesel production from oilseed crops has been described. The third part is devoted to the narration of industrial oil types and their uses separate from biofuel, and also contains an in-depth deliberation of the research aimed at improving oilseed crops for industrial uses. The fourth part is on industrial starch and similarly consists of two chapters, the first describing the characteristics of different crop starches and their appropriate uses for various applications other than ethanol production, followed by a review of the research towards improving the quantity and quality of starch composition in plants. The fifth part describes fibre and dye crops. It contains five chapters, two each on cotton and bast fibres and one on dye crops. The discussion on cotton includes the production and processing of the crop, as well as research accomplishments in improving cotton fibre quality. Bast fibre coverage consists of important feedstock crops, the manufacturing process from plant to products, and various end uses for these fibres. The final chapter of this section describes the plants that supply natural fibre dyes and the process of fibre dyeing. The sixth part concerns crops that are the sources of rubber, gums, resins and waxes. The two important rubber crops, the rubber tree and guayule, are covered in separate chapters to include adequate deliberation on their culture, breeding and quantitative and qualitative aspects of the latex produced by these two crops. The chapter on gums, resins and waxes describes these compounds and their sources, as well as their trade. The seventh part focuses on insecticides, covering the important insecticidal crops and their active ingredients, as well as the prospects for using natural insecticides in pest control. The eighth and final part of the book shifts emphasis from the utilization of industrial crops for the generation of products to their deployment for such applications as the remediation of contaminated soils. The two chapters in this section cover the principles of plant-based remediation, the traits that make certain crops suited for such use and a description of these remedial crops.

Every chapter of this book covers several disciplines such as agronomy, plant breeding, biotechnology, biochemistry, process engineering, etc. The chapter authors had the daunting task of interconnecting these disciplines in order to explain the production and processing aspects of crops leading to their end use. The book clearly shows that plant biomass is rich in chemicals and materials and can be used in its original form or converted into other chemical forms or products. The development of crops for industrial uses is just beginning, but the combination of conventional breeding methods with modern molecular breeding tools makes accelerated progress possible. Similarly, there is much scope to incorporate green biochemical technologies to make biomass processing to products environmentally friendly and to expand the list of products and chemicals that can be derived from a particular plant feedstock. Thus, the challenges and opportunities in the exploitation of crops for various industrial purposes are enormous.

Finally, I would like to thank my colleague, Hari P. Singh, who was prepared to help whenever asked. I dedicate this book to my 2-year-old granddaughter, Naya, whose limitless energy has been inspirational in keeping me focused when feeling tired and exhausted from constant editorial work.

Bharat P. Singh
Fort Valley, Georgia, USA
October 2009

1

Overview of Industrial Crops

Bharat P. Singh

Introduction

Crops are commonly categorized as food, feed or industrial, based on their use. Wheat, rice, sugarcane, sugarbeet, Irish potato, groundnut, beans and peas, etc., are called food crops because of their primary use for human consumption. Similarly, crops like maize, sorghum, oats, barley, hay and silage, which are grown mainly for animal consumption, are labelled as feed crops. Finally are a group of crops that although grown commercially for products of economic value, have limited value for food or feed. Examples of industrial crops are cotton, flax and tobacco. This definition, however, is very narrow and does not reflect correctly the current and potential utilization of different crops. For example, while maize is considered a feed crop, it is the human staple food for Central America. Maize is also an important feedstock for several industrial products, including bioplastic and biofuel. In a similar manner, classification of sorghum as a feed crop is ironic considering the millions in Africa who depend on it for survival. Tobacco, a crop grown in the past only for the manufacturing of cigarettes, holds promise of becoming a factory of made-to-order drugs. Also, the utilization of sugarcane has expanded from being mainly for sugar to biofuel and a host of co-products. With new uses constantly being discovered for the commonly cultivated crops, it has become increasingly difficult to allocate a crop to a single category; some crops can belong to different groups, depending on the purpose of their production.

Industry is defined as the production of goods and services. Therefore, the term 'industrial crop' has been used in this book to describe goods that can be produced from plants (e.g. energy, biochemicals, biocomposites) and services that different plants can provide (e.g. phytoremediation to detoxify soil, biochar to sequester carbon).

Different criteria such as use, chemical properties and raw material volume are used to separate feedstocks into different categories. Often, these categories overlap. Based on use, feedstocks may be grouped for the production of liquid fuel, lubricant, solvent, fibre, rubber, pharmaceuticals, insecticide, plastics, etc. On the basis of chemical properties, they may be separated into classes such as oil, starch, sugar, terpenoid, etc. Feedstocks are also separated by the quantity of raw materials needed to produce a unit of industrial product and its value. Accordingly, they can be for the industrial processing of low-value, high-volume products such as liquid fuel or high-value, low-volume products such as pharmaceuticals. However, the same pine trees that are the feedstock for turpentine, a low-value, high-volume product, become a high-value, low-volume feedstock when turpentine is processed further to isolate pharmaceuticals, flavourings and fragrances. The categorization of plant feedstocks is fluid and will become of lesser value as scientists and engineers are able to exploit the full potential of complex macromolecules to produce a wide array of products from the same feedstock.

The history of industrial crops can be traced back to primitive humans' earliest encounters with fire. The subsequent knowledge that fire could also be created artificially by friction led to its utilization for such purposes as keeping the body warm, cooking food, warding off predators and making spears, bowls and digging sticks. In due course, this primitive discovery became the basis of the establishment of the first industry in the form of energy. The value of wood as a natural resource for fire was known from the very beginning as humans watched fire caused by lightening kept alive by forest wood until extinguished by rain. Increasing the efficiency of energy produced from wood by converting it into charcoal through pyrolysis was among the early pioneering industrial inventions. It was used in the smelting of a variety of metals such as aluminum, copper and iron, until replaced by coke. Charcoal production remains alive to this day and continues to be an important cooking fuel in many of the world's rural areas (Fig. 1.1). With time, humans learned that not only was the cellulosic biomass of the plant a reservoir of energy, but the seeds were also. Seeds of certain plant species were found to contain liquid that could be extracted easily and when burned, produced light. The ancient Egyptians burned castor oil to light lamps. In some parts of the world, lamps using plant oil are still used on religious occasions.

The 18th century revolution in agriculture and industry laid the foundation for the accelerated use of plant feedstocks for multiple industrial uses. Cotton was the leading industrial crop during this period and Britain led the league of industrialized nations. Cotton material during this period made up half Britain's exports. This industrial success of Britain was made possible by the importation of cheap cotton from its colony in the USA, where at that time approximately 80% of the world's cotton crop was produced. Industrial crops like jute also gained prominence because of their utility in the protective wrapping of cotton and other products during shipment.

Feedstocks for the synthetic industry were, for the most part, supplied by plants

Fig. 1.1. Primitive method of charcoal production in Ratnakiri, Cambodia.

until approximately 100 years ago. Plants were the source of inks, paints, dyes, adhesives, glues and other industrial chemicals. Cotton, wool and some minor crops provided yarn for weaving clothes. The source of the first plastic introduced in the 1880s was cotton and in the 1920s, plastics used in manufacturing car radios were derived from wood pulp. The first synthetic fibre, rayon, was derived from cotton. One-third of the cultivated land in Western Europe was devoted to producing energy and industrial raw materials as recently as the 1950s (Kjoller, 1998).

While plants were gaining importance as the raw material source for other industries, they were losing ground to coal as the source of energy. For thousands of years, humans had known that coal was a good source of energy but were not able to harness it until the beginning of the 19th century because of insurmountable quarrying problems. Engineering advances made the industrial mining of coal feasible. The discovery of technology to convert coal into kerosene on an industrial scale in the mid-19th century led to a gradual replacement of plant oil by kerosene oil, because it was cheaper and more readily available. Kerosene also laid the foundation of the modern chemical industry. During the period preceding World War I, it became possible to derive a host of chemicals from coal. The next wave of energy source came in the form of petroleum. Similar to coal, humans' knowledge of petroleum went back thousands of years, but the lack of technology to drill wells to pump it out of the earth and refine it to usable products was a stumbling block to its industrial utilization. By the beginning of the 20th century, these hurdles had been overcome. Vast deposits of petroleum in the upper strata of the earth crust within confined geographical boundaries were discovered during this period, lending itself to inexpensive extraction and transportation in bulk. As petroleum contained a high energy density and was easily adaptable for internal combustion engines, it rapidly became the choice energy source worldwide.

In the 1920s, the industrial economy based on living plants began shifting towards a fossil fuel-based economy. Petroleum became the dominant source for synthetic products after World War II, as chemists were able to convert large quantities of gaseous and liquid by-products during the production of gasoline for a new category of synthetics named petrochemicals. These chemicals replaced those produced from plant-based feedstock comprised of starch, vegetable oil and cellulose during the early industrial period. As a consequence, the quantity of plant matter consumed for industrial products declined drastically, except for papermaking. Glass, metal and paper were replaced with plastics in a wide range of products. Today, petroleum-derived products have entered all facets of our life, from clothing to food dyes and vinegar. Petrochemicals are used to make tens of thousands of final products.

Lately, the advantage in the concentration of petroleum deposits in a few countries has turned into a definitive hindrance to the reliable supply of this vital commodity. Several of these countries are beset by political and economic problems that threaten their stability, and petroleum pipelines are often sabotaged as a political statement. Petroleum-exporting countries have organized a cartel in order to coordinate production to their advantage, interfering with the governance of price under a free market system. Oil embargos have also been used as a political tool. Threats of unpredictable and artificially created shortages of petroleum essential for energy and synthetic products have become a matter of urgent concern for oil-importing countries. There is not much chance of increasing petroleum production in western countries because of organized opposition to drilling for new oil discoveries governed by the fear of oil spills and other ecological damage. There is also growing apprehension that the world may run out of easily extractable petroleum in the not too distant future. The next generation's sources of petroleum will be based on mining rather than drilling, and the cost of procuring and processing into usable forms will be much higher. In addition, these sources may pose an environmental threat on several fronts. Mining operations may alter the existing landscape and vegetation, even under the strictest legislation, put pressure on water resources needed for agriculture, human and animal use and disturb existing ecological wildlife habitats. There are various projections that within the next 100 years, the world will run out of fossil fuel. So, there is rethinking on the heavy dependence on petroleum and a sincere search for alternatives has begun. The new emphasis is on finding resources that are inexhaustible because of their renewable nature. In this context, the plant kingdom has again come into focus as an industrial raw material, because it can be replenished.

Comparative Analyses of Petroleum and Plant Feedstocks

Both plant and petroleum feedstocks are hydrocarbon. The preference for petroleum feedstock emanates from its several advantages over plant feedstocks. First and foremost, the technologies for the industrial uses for petroleum are already in existence; thus, the considerable investment required in developing new technologies, with an uncertain outcome, has already taken place. It is much cheaper to extract petroleum out of full reserve wells compared to the recurring cost of producing crops. Petroleum is also easier to handle because larger mass can be accommodated per volume in liquid form compared to the solid form of plants. An estimate of the extractable oil reservoirs of wells can be made with accuracy to ensure continued supply against the dependence on year-to-year climate for biomass production. In addition, once the crude oil is cracked to extract petrol, the liquid and gaseous by-products are processed easily into petrochemicals for the manufacturing of thousands of products, while the technology to use the generated by-products of plant feedstocks is still in the evolutionary stage. Because of these advantages, petrochemical-derived products at present cost much less than those from plant sources.

However, there are several drawbacks to heavy petroleum dependence. Petroleum is a non-renewable resource and is thus subject to depletion. It is no longer a dependable feedstock and its price fluctuates widely, making it difficult to fix a floor to the price of products utilizing it as a raw material. The new and stricter health and environmental regulations are adding considerably to the manufacturing cost of petroleum products. Crude oil is carried long distances in giant oil tankers to refineries in different parts of the world from a few countries through narrow sea straits, making them vulnerable to attack and hijack. Accidental oil spills out of these tankers are of such a mammoth proportion that they cause long-term harm to wide swathes of ocean ecosystems. The spill of about 40 million l of crude oil from the Exxon Valdez tanker in the Prince William Sound in Alaska on 24 March 1989, the ensuing endangerment of habitat for a large number of marine species and the continued contamination of the seashore after years of extensive clean-up remain fresh in the minds of people worldwide.

The biggest advantage of using plant feedstocks is that they are a renewable resource and thereby can last indefinitely. Feedstocks for biochemicals are products of agriculture worldwide, which diversifies the supply source and thus lowers the risk of disruption. Transportation risks to the environment due to spillage are also lower with biochemicals because the manufacturing distance between different points of supply are greatly reduced compared to petrochemicals.

Recent advances in biological sciences, coupled with progress in process engineering have increased the competitiveness of plant-based products. Research in crop improvement and production systems has resulted in substantial increases in the yield of crops needed as stocks for various industries. The new cultivars are modified physiologically for higher photosynthetic efficiency and are genetically altered to fend against competition and invasion. Improved production systems have reduced year-to-year variation in crop yield due to climate, thereby assuring an uninterrupted supply of raw materials. Relatively recent tools in molecular breeding and genetic engineering hold promise of making alterations to the genetic make-up of plants suited for various bio-based industrial products more feasible, thus increasing their appeal for industrial use. It is expected that the price gap between biochemicals and petrochemicals will narrow as more and more co-products are derived from the processing of plant feedstocks. The price of plant feedstocks will decrease as more farmers begin growing needed crops. The biochemical industry is in the very early technological stage; there is tremendous room for improvement compared to a mature petrochemical. The fast pace of genetic advance brings forth the possibility of plants being factories of made-to-order chemicals. In this new cycle of competition between petrochemicals and biochemicals, the competitiveness of agricultural raw materials for different industries has improved substantially compared to that in the World War II era.

Laser *et al.* (2009) have provided a model-based analysis of the cost of production from biorefineries as if they were fully developed similar to the petroleum refineries of today. They considered the two emerging processing approaches – biological and thermochemical – for the production of fuels, power and/or animal protein. According to their calculations, the cost of production amounted to US$0.36–0.56/l (US$1.37–2.16/gallon) petrol equivalent for a plant with the capacity of 4535 dry t feedstock/day, considering 12% internal rate of return, 35% debt fraction and 7% loan rate. However, when the biological production of ethanol was combined with the thermochemical production of fuels and/or power on the same scale and financial structure, the cost came down to US$0.25–0.33/l (US$0.96–1.24/gallon) petrol equivalent.

Reaching the level of maturity in biological processing comparable to oil refinery is a formidable task and will require huge private and public investment. A substantial portion of expenditure will be incurred for research and development, where the return to investment is uncertain. Setting up these refineries will also require raising a large amount of capital for an industry with an unproven track record. Certain disadvantages of plant feedstocks are due to limitations imposed by their biological properties. The energy density of ethanol is 68% on a mass basis and 71% on a volume basis of petrol (Davis, 2006). Plant feedstocks are made up of a large number of species, each varying in chemical composition, which makes the fractionation technology plant specific, needing modification whenever one feedstock is replaced with another to meet raw material needs.

Environmental and health concerns

Transportation fuel is the largest end use of petroleum. It is blamed for 25% of greenhouse gas emission and that share is rising because of the dramatic increase in cars in countries with emerging economies like China and India (Worldwatch Institute, 2007). The processing of crude oils into chemicals often creates pollution problems and raises issues concerning the health and safety of workers. Petrochemicals such as cleaning solvents are preferred because they are cheap and possess excellent cleaning properties and the ability to dissolve a wide variety of contaminants. But, they impact the environment negatively and pose a hazard to the workers producing them and the people who use them. Environmental standards are being tightened continually for the manufacturers and end-users of chemicals throughout the world. The US Clean Air Act Amendment of 1990 (EPA, 2008) has identified 189 petrochemicals as hazardous air pollutants. These chemicals are subject to strict emission standards and manufacturers are required to implement prohibitively expensive maximum available control technologies to lower emissions. Violators are subject to heavy penalties. The Clean Air Act also regulates the emission of organic volatile chemicals, which evaporate and can undergo chemical reactions in the lower atmosphere and produce ground-level smog, contributing to ozone gas.

Biochemicals obtained from plant feedstocks can be an alternate raw material or replacement for petrochemicals. Phenol, a chemical used heavily in the plastic and textile industries, can be manufactured both from petroleum and plant feedstock. As the physical and chemical properties of phenols remains the same irrespective of the raw material, the associated health hazards are also identical. On the other hand, the petroleum-derived cleaning solvent 1,1,1 trichloroethane, can be replaced by D-limonene extracted from citrus fruit, deemed a safer chemical. Botanically derived chemicals generally are less toxic, flammable and corrosive than petroleum-derived chemicals. Manufacturers using botanical feedstock generally avoid the costly permits required in petrochemical-based processing, save on hazardous waste disposal costs and run low risk of compliance penalties. Biochemicals also degrade easily and can be disposed of safely and inexpensively by end-users. In addition, the processes for transforming biological raw materials into final products often require less energy and chemicals than the corresponding petrochemical processes (Kjoller, 1998).

The CO_2 emitted by petrochemicals is an added greenhouse gas burden on the atmosphere because it does not return to the soil to make additional petrochemicals. Although plant feedstocks also emit CO_2 during processing, it should not add to the atmospheric CO_2 because the CO_2 released when the feedstocks degrade can be reabsorbed by replacement crops. As the cycle of feedstock cultivation and use will continue indefinitely, the equilibrium of consumption and emission will be stable, making the whole process close to carbon neutral. Farrell *et al.* (2006) reported ethanol, compared with petroleum, produced from lignocellulosic biomass reduced greenhouse gas emission by 88%, while a 94% reduction in greenhouse gas emission has been reported for switchgrass–ethanol (Schmer *et al.*, 2008).

Strict emission laws have narrowed considerably the emission advantages of biofuel over petrol. On the other hand, nitrogen oxide and volatile organic compound emission can be problematic in engines not properly calibrated to run on biofuels. There are dangers also of deforestation and the accompanying loss of biodiversity and threats of extinction to endangered species in order to expand the cultivable land area for plant feedstocks. The clearing of forest for planting sugarcane in Brazil and oil palm in Malaysia and Indonesia justify such concerns.

A chemical derived from plant feedstock cannot automatically be assumed to be non-hazardous to health. For example, turpentine is an organic solvent obtained by the distillation of resin obtained from trees, mainly pine trees, and it is a health as well as a fire hazard. Turpentine is a skin, eye, mucous membrane and upper respiratory tract irritant in humans. It may also cause skin sensitization and central nervous system, gastrointestinal and urinary tract problems. The lowest estimated oral dose reported to be lethal in humans is 441 mg/kg (RTECS, 1989). The National Fire Protection Association has assigned a flammability rating of 3 (severe fire hazard) to turpentine. Contact of turpentine with oxidation catalysts or with strong oxidizing agents (especially chlorine) may cause fires and explosions. Such fire releases toxic gases and vapours such as carbon monoxide and the partial oxidation products of terpenes (HSDB, 1989).

Consumer perception

Consumer acceptance of petrochemical-derived products was driven by the cheap price, large selection and the general perception that their use was safe. Several synthetic products such as asbestos fibres, plasticizers and surface-active ingredients originally considered safe have been shown subsequently to pose health hazards. Regular media reports of possible harmful and allergenic residues in petrochemical-derived products of daily use such as textiles, detergents, cosmetics and packaging materials have created concern and anxiety. Consumers have become reluctant to use paints, lacquers, wood preservatives and inks that contain organic solvents. Recently, a new breed of consumer labelled 'green' has emerged. It consists largely of people with the ability to pay higher prices for commodities produced using environmentally responsible raw materials and easily degradable products. This has given plant-derived raw materials a distinct advantage. Further, consumer preference has been rising for products made out of plant raw materials that are grown organically without using any commercial fertilizers, herbicides or pesticides. Garments made from organic cotton fabrics and clothes dyed with natural dyes have become trendy and occupy racks of high-class fashion boutiques. This trend has also stimulated markets for insecticides, repellents and deterrents of botanical origin.

The Plant-based Industry's Impact on the Rural Economy and Agriculture Sustainability

Agriculture sustainability is tied to rural livelihood sustainability. Farmers are being squeezed by the rising prices of external inputs like fuel, fertilizer, labour, etc., while their ability to pass on added expenses are constantly declining due to the globalization of the economy. The price structure and cost of production of soybean, a crop that has multiple uses, can illustrate this point vividly. The price of soybean has not changed between 1981 and 2005; US$223 and US$208/t in 1981 and 2005, respectively. As a consequence of price stagnation, the total cost of producing a hectare of soybean for American farmers in 2005 was US$663 against the gross value of US$654/ha of the harvest (SoyStats, 2007). The situation is the same worldwide. Rural communities heavily reliant on income generated from farming throughout the world are being emptied due to poor economics.

Creating an alternate demand for currently produced crops or finding alternate crops with market demand will generate new income for farmers. Diversification of enterprises will also give farmers the ability to switch between markets that offer the best prices for the produce and to select enterprises requiring lesser inputs in years of adverse climatic conditions. Some of the new crops for industrial use can be grown on marginal land that is generating little or no income at present, thereby increasing farm output.

World peace and prosperity is tied inherently to sound and stable rural economies. Without jobs in villages, the present trend of mass migration to cities and its accompanied urban congestion and chaos will worsen with time. The establishment of rural resource-based industries offers new hope for rural economies. Biomass needed for biochemical industries depends on agriculture and forestry located in rural areas. Biomass feedstocks are much more voluminous than petroleum feedstock; thus, it is economical for biochemical industries to be situated adjacent to the raw material source rather than in the proximity of urban areas. By generating greater demand for agricultural commodities, biochemical programmes have the potential to increase employment significantly in rural areas. Job growth will be spread across the spectrum from the growing and harvesting of feedstock to employment in the industries involved in the conversion and processing of feedstocks to different products. A study conducted in Georgia, USA (Shumaker *et al.*, 2007), found that the construction of a 375 million l maize-based ethanol plant created a one-time economic output impact of US$130 million to the Georgia economy. Economic activity related to construction generated US$51.7 million in labour income for 1203 jobs. The production of ethanol created annual economic output impacts of US$335.8 million. Plant operations accounted for 50.2% of the total output impact, while 49.8% was attributed to maize produced in Georgia. Ethanol production generated US$37.6 million in labour income to the Georgia economy from 1030 jobs. It also contributed US$3.8 million and US$3.1 million in taxes to the state and local governments, respectively. The Worldwatch Institute (2007) has compiled the following job-related vital statistics comparing petroleum and biofuel industries: (i) biofuel industries require about 100 times more workers per joule of energy produced than petroleum industries; (ii) in Germany, the biodiesel industry generates roughly 50 times more jobs per tonne of raw oil than diesel production; (iii) job creation cost in Brazil is 25 times less for the ethanol industry than for the petroleum industry; (iv) in sub-Saharan Africa, a region-wide blend of biofuels to the tune of 10% for gasoline and 5% for diesel would provide jobs for between 700,000 and 1.1 million people; and (v) a jatropha farm in India could yield 313 person days/ha in the first year of the plantation and 50 person days/ha over the next 30–40 years.

Although the overall impact of the expansion of agriculture into industrial feedstock will be positive for the rural economy, some agro-industries may be affected adversely. Diverting grains for biochemicals may raise its price for cattle producers. Since most oil feedstocks currently are edible oils, not only the price of edible oil but also food industries such as margarine, salad dressing and mayonnaise production will experience hikes in their raw material cost. Due to the diversion of grains to industry, the quantity of grain needed to be stored long term will decline and grain storage operations will shrink.

The sustainability of agriculture as a system should improve from the expansion of agriculture into industrial raw material production. Most food crops are annual and their production requires yearly land tillage and planting that disturbs soil structure and texture and soil ecosystems, thereby creating soil erosion problems. The monoculture of commodity crops has also created increased pest and disease pressure. In temperate climates, farmland often sits idle during winter months, generating no farm income, as most commodity crops are grown during summer. A substantial number of feedstocks are derived from perennial plants requiring negligible tillage once planted. Soil microorganisms in no-tillage are undisturbed, creating habitat stability, promoting species interaction and above- and belowground biological diversity (Altieri, 1995). Tillage decreases soil organic C and N, while they remain intact in no-tillage (Cambardella and Elliott, 1992). Sainju and Singh (1997) observe that only 60–70% of N applied to field crops is taken by plants. The remainder leaches to the groundwater or is lost by runoff to aboveground water reservoirs, creating pollution problems and raising nitrate concentration above the health hazard threshold (Nielson and Lee, 1987). Perennial crops often serve as catchment for these chemicals and prevent runoff. It is easier to establish legume perennial cover, companion or intercrop in conjunction with a perennial main crop, thereby reducing or eliminating the input of commercial nitrogen fertilizer. Legume cover crops develop large populations of mycorrhizal fungi that are efficient in extracting P from deficient soils in exchange for photosynthate from legumes (Clark, 2007).

Concerns of soil depletion have been voiced over the advocated use of maize stover for biofuel production. Maize stover protects the soil against erosion in the absence of crop cover and supplies organic matter to the soil in order to keep it healthy. Maskina *et al.* (1993) found that the removal of maize stover resulted in grain and biomass yield reduction for several succeeding years, even if the practice was discontinued in the following years. Using the carbon content values (Johnson *et al.*, 2006) based on the US national grain-yield database, Wilhelm *et al.* (2007) concluded that the carbon amount in the maize residual biomass was more than that needed to meet soil needs and could be otherwise used. Several questions, including the economics of stover-based biofuel production and the method of determining the sustainably harvestable stover amount, still remain unanswered.

Common Crops for Food and Non-food Uses

Throughout history, humans have used some 10,000 plant species for food, but today's diet is based on just over 100 species, a dozen of which represent 70% of human consumption. The crop statistics of the USA for 2006 revealed that 82% of the land was devoted to only three crops, maize, soybean and wheat, planted on 31, 29 and 22% of the total area, respectively. This was followed by cotton on 6%, sorghum on 3%, oats on 2% and sunflower, rice and barley each on 1% of land. All other crops were planted on only 4% of cultivated land (SoyStats, 2007). No wonder, then, that the same handful of crops figure in meeting food and non-food needs.

The crops occupying a prominent place in agricultural production are so placed because of their relative advantage in production and utility. These plants have a higher yielding ability, can be grown under wider climatic conditions and produce food products that are preferred by consumers. Farmers have experience producing, harvesting and storing these crops and have made investments in machinery for their planting through to harvesting. As the current agricultural production system is geared for the production of food crops, recently developed plant-based industries also look to the same supply source to meet their raw material needs.

The trends in the sharing of crops between food and industrial uses can be illustrated using soybeans. The present proportion of soybeans used as industrial feedstocks is miniscule in proportion to that used for human and animal dietary purposes. Only 1.11 million t out of 39.38 million t of the 2006 US production were used in industries. However, the gradual annual increase in soybean use for industries is obvious. While the increase in US soybean consumption between 2001 and 2006 by animals and humans was to the tune of 0.67 and 0.37%, respectively, the increase was 26.57% by industries (United Soybean Board, 2007). The use of soybean oil for biodiesel in the USA is primarily legislation driven. The US Congress passed legislation in 1999 that allowed public vehicle fleets to earn credits for using a blend of 20% biodiesel and 80% petroleum diesel, commonly called B-20. US commercial production of biodiesel in 1999 amounted to 1.9 million l. By 2004, soy-biodiesel production had increased to 94.6 million l and after passing of the second piece of legislation under the Energy Policy Act 2005 for Biodiesel Tax Incentives, production climbed to 851.7 million l in 2006.

What in the past was a food versus feed debate has now become a food versus fuel debate. The use of maize for ethanol in the USA, rapeseed oil for biodiesel in Europe and sugar for ethanol in Brazil has raised the price of these commodities on the world market. However, when analysed rationally, these competing uses have been good for bringing depressed agricultural commodity prices to a healthy level. There is no evidence that increased use of plants for feedstock has led to food commodity shortages in the marketplace. Gains in agricultural productivity, together with the subsidies provided to farmers in western countries, have resulted in the production of food grains far exceeding the demand for food. The developed nations have used the developing nations as a dumping ground for their surplus grains, putting pressure on subsistence farmers in those countries. Poor countries are not able to protect the livelihood of their farmers on globalized markets because they are bound by international laws or prevented through treaties. A good example of this is NAFTA (North American Free Trade Agreement), a trilateral treaty between the USA, Canada and Mexico, which lowered food prices significantly in Mexico. The agreement depressed the income of Mexican farmers to the extend that they were forced out of the only livelihood they had known for generations and were compelled to migrate to Mexican urban centres and into the USA in search of a livelihood. Those people working on farms and in low-skilled jobs in the USA are looked upon as illegal immigrants, but many among them view themselves as NAFTA refugees. The suicide of farmers in India due to high debt and low income is another example of the plight of farmers resulting from the globalization of agriculture. Increased demand in the rich countries for agricultural commodities for industrial purposes will offset some of these imbalances and hopefully will again make farming a sustainable livelihood in poor countries. The importance of agriculture to the rich versus the poor countries can be seen by the fact that agriculture provides less than 2% of income and employment in the former, but it is the source of 35% of the gross domestic product of the latter (Watkins and von Braun, 2003). Greater demand by agro-industries for certain crops will increase their production in the poor countries, benefiting subsistence farmers and improving their standard of living. At present, these farmers number around 800 million (IFAD, 2002). Undoubtedly though, increased prices will have an adverse effect on the urban poor, who will have to spend more of their income on food purchases. A safety network for the urban poor will be needed in order to avoid hunger. However, caring for them should be easier through ration cards or other subsidy mechanisms because they are densely clustered in urban centres, whereas the rural poor are difficult to reach as they are sparsely scattered. Protests are also expected from the middle classes, who have become accustomed to allocating a proportionately low share of their budget for food than for housing, clothing and transportation.

For those critics against the very concept of using agriculture for any purpose other than for food, a reminder of the history of agriculture itself would be in order. Agriculture always has been a source of both food and non-food products. In the early days of the industrial revolution, farming and forestry were the main sources of raw materials for industries. Both the food and non-food sectors were important sources of farm income. One of the root causes of the demise of small and family-owned farms can be traced to decreased outlets for farm products. Advancement in agriculture led to surplus production beyond the demand for food and feed and so depressed commodity prices. At present, food scarcity is seldom a production problem but rather the absence of effective distribution channels. Moreover, an increase in non-food production is not synonymous with a corresponding reduction in food production as most farm crops contain raw material components for both food and non-food utilization. For example, oilseed rape can be used for the following economic food and non-food purposes: straw chips for composite industries, the paper and pulp industry, the packaging industry and energy; straw meal for energy; seed oil for the food industry, biodiesel, biodegradable lubricants and the paper and lacquer industry; seed protein for the feed industry, the fermentation industry and the food industry; syrup for the feed industry and the food industry; and hulls for energy. Increases in food grain prices are expected to stabilize in the future as plant feedstock-based industries are restructured using next-generation technologies that utilize all grain and oilseed components and produce both food and non-food products, and non-food plant feedstocks replace food feedstocks. While market supply and demand should be left to determine the market food grain price, worldwide monitoring to ensure that feedstock demand does not create food grain shortage would seem appropriate.

Complexity of Selecting Feedstock Crops

Agriculture as an industrial feedstock source did not attract much attention until biofuel production started on a measurable scale during the past few years. Now it is facing scrutiny similar to petroleum with respect to suitability, sustainability, efficiency, profitability, environmental impact and appropriateness of use. The use of maize for ethanol production in the USA is a suitable example to illustrate this point. The first and foremost critics object to diverting a major crop for human food and animal feed to any subsidiary use. Long-term viability of ethanol is also questioned on economic grounds. The ethanol industry receives governmental subsidy at the farm as well as at the processing level. The US Agricultural Adjustment Act of 1933, and subsequently amended to the US Agricultural Act of 1949, put into place subsidies to farm crops to lower national risk, to offer security for domestic food producers and to instil lower food prices in the marketplace to benefit all consumers. In the USA, there was a 15% increase in the hectares planted under maize in 2006 over 2005, mainly to produce maize-based ethanol (NRC, 2008). However, maize farmers received subsidies irrespective of the purpose of production. Ethanol production is further supported by federal subsidy.

Grain crops require high inputs of fertilizers and pesticides that create a possible spectre of pollution of below and above groundwater from percolation or runoff of residual inputs. The environmental costs of increased land area under grain extend beyond individual farms and on to local streams, watersheds and coastal areas. The mid-western states which drain into the Mississippi river produce most of the maize grown in the USA. Alexander *et al.* (2008) linked seasonal hypoxia in the northern Gulf of Mexico as being due mainly to fluxes of N, and to a lesser degree P. Maize and soybean cultivation are responsible for approximately 52% of the N and 25% of the P load of the Mississippi river. Oxygen-deficient water threatens the survival of marine life and productivity in the Gulf of Mexico, which is the source of a major percentage of US harvested shrimp and oyster and a significant percentage of commercial fish. The Energy Independence and Security Act (EISA) passed by the US Congress in December 2007, which sets a high goal of biofuel production, raises the possibility of bringing substantially more farmland under maize production, thereby further increasing the nutrient load of this water body.

According to Pimentel and Patzek (2005), the net energy balance of ethanol from maize starch, switchgrass and wood was −29, −50 and −57%, respectively, and −27% of biodiesel from soybean. They pointed out that when all the costs such as seed, fuel for tractors, power for irrigation and fertilizer were taken into account, the input–output ratio for maize to ethanol became economically questionable. Wesseler (2007) disputed the above study on the grounds that it used too wide energy input boundaries, thus inflating energy requirements. Instead, he proposed to base the balance energy differential by separating the consumption attributed to biofuels from that attributed to other purposes. Done this way, the energy balance for food crops turned positive. Schmer *et al.* (2008) calculated that switchgrass produced 540% more renewable energy than non-renewable energy consumed to generate ethanol. Farrell *et al.* (2006) found the Pimentel and Patzek study flawed on the cost–return calculations because of outdated input costs and lack of proper accounting of co-product values. Ulgiati (2001) showed that when maize was dry milled for ethanol production, the input–output ratio was in the range of 0.59–1.36, depending on whether the value of maize stover (approximately 70% of the plant) for energy generation for the ethanol plant and the revenue from dry distillers grains with solubles (DDGS) had been taken into consideration. The production cost of ethanol from maize and the energy required for processing will decline further once the emerging membrane technology called pervaporation becomes available to substitute the distillation step partially or completely (Vane, 2005). It can be argued that subsidies and price premiums for biofuel are appropriate considering the hidden cost of oil dependence– oil-related defence expenditure, the loss of domestic investments and government revenues, and the cost of disruption in the oil supply.

In 2004, 11% of the US maize harvest yielded ethanol to meet roughly 1.7% of the 2004 fuel demand. If all the maize grown in the USA was converted to ethanol, it would satisfy about 15% of the transportation needs, while the US Energy Policy Act of 2005 mandated replacing 30% of the nation's liquid fuel from fossil to renewable sources by 2030. Achieving this national mandate is clearly not attainable through maize. So it is clear that, at best, maize can serve as a transitional feedstock and, in the long-term, comprise only a small portion of the sustainable feedstock pool. The pace of refinery construction for maize feedstock is already showing signs of slowing down. The new emphasis is on inducting a wide range of potential non-food crops to create a crop pool that will enable the country to reach the national biofuel goals without major encroachment on prime farmland or a large increment in new agricultural land. The future feedstocks may comprise some food crops, including maize, sorghum, soybean and sugarcane; off-season crops; energy crops like switchgrass, energycane, napiergrass and jatropha; agricultural residues of stover and straw; forest residues such as tree thinning, logging scraps, sawdust; wastes including recycled grease, garbage and manure; algae and others. Some 16 commercial-scale cellulosic ethanol production plants are expected to be operable in the USA by 2010 (NRC, 2008). Turning forest residues and wastes into energy are environmentally sound and attractive options. Using wood that builds up on the forest floor will reduce the potential of forest fires and create healthier forests. Similarly, converting waste into energy provides a mode of sustainable resolution to the landfill problem.

Potentials and Challenges of New Crops for Industrial Uses

The so-called new crops, for the most part, have been known to humankind for thousands of years, but have been neglected because of low value or utility. They consist of plant species that are either low yielding, unfit for mechanized production or in low demand for human or animal food. But some of the little-noticed plants are now emerging as crops of value for industrial purposes. For example, jatropha, jojoba, crambe, milkweed and meadowfoam, which had little or no impact 10–15 years ago, are beginning to boost farm return and help keep farmers afloat. These crops do not compete with the commodity crops for market share or displace traditional commodity crops because they are for different markets. Not allowed to grow on fertile land because of competition with food crops, these plant species have undergone natural selection for survival in poor and low-water soils. India has set aside 40.5 million ha for jatropha and expects the oil from it to account for 20% of its diesel consumption by 2011 (Padgett and Myers, 2009). A goal that is perhaps too ambitious and challenging. Australia, China, Brazil and Kenya also plan to plant a large acreage of jatropha.

Most new crops have not been cultivated on a commercial scale and as a result have not undergone selection for higher yield and industry-desired traits. Years of research may be needed to amend a plant species to suit industrial feedstock criteria. For example, cuphea is an annual plant that produces small seeds rich in saturated medium-chain fatty acids. The seed contains up to 35% crude oil and the defatted meal contains as much as 33% crude protein (Evangelista *et al.*, 2006). The United States Department of Agriculture (USDA) Agricultural Research Service (ARS) identified cuphea as a potential industrial oil crop as far back as the 1960s, and the Research Center at Peoria, Illinois, characterized the oil extracted from a number of cuphea species at that time. Some critical improvements needed for the successful marketing of the crop for industry have been achieved, but others remain to be accomplished. By crossing *Cuphea viscosissima* × *C. lanceolata*, it was possible to enhance the concentration of capric (C-10) acid used in the formation of lubricants, soaps and detergents. Cuphea has an indeterminate growth habit and flowers throughout the growing season. The early pods shatter before harvest, reducing the yield. Selection and breeding for resistance to shattering is needed. There are also problems associated with mechanically harvesting the plant green (at moisture levels above 50%) in an effort to prevent shattering and in post-harvest handling of such seed. Cuphea oil as an industrial feedstock also has shortcomings of containing high levels of chlorophyll, which may act as a pro-oxidant and poison the catalyst used in the bleaching step of oil refining.

Some crops with proven industrial value and with an established method of production are prohibited from large-scale production due to regulatory restrictions. Castor seed contains approximately 50% oil, which is composed of 80–90% ricinoleic acid, used in a number of processes to create unique chemicals and polymers (Lowery *et al.*, 2007). However, castor also contains a toxin protein, ricin, which can be deployed in chemical warfare and bioterrorism. Until the research under way to develop a castor plant with low ricin is fully successful, castor production will remain limited to a few countries only. Hemp is another crop whose industrial potential exploitation has been hampered by the presence of undesirable chemical compounds.

When introducing new crops from another part of the world for industrial use, their invasive potential has to be studied carefully. An introduced crop may grow more vigorously in a new environment than in its natural habitat because of the absence of natural enemies and may encroach upon land reserved for the production of other crops. Kudzu was introduced to the USA from Asia as an ornamental vine with fragrant violet purple flowers and in the 1940s was planted on 1.21 million ha of farmland, supported by government payment. An additional 73 million seedlings were planted along roadsides to control soil erosion. However, this aggressive growth habit species overtook and choked all vegetation in the vicinity of its planting. As a result, kudzu was officially labelled a 'weed' in 1970 and upgraded to 'noxious weed' in 1997.

The path to encouraging sufficient numbers of farmers to produce new crops in order to achieve the feedstock production level required for industrial use, establishing industrial units to process them into commercial products and marketing the new product successfully is laden with hurdles. Such an endeavour requires large investments in research and development, a long-term commitment, and it is risky.

Sustainable Supply Chain

The great unknown in agriculture is the weather. Crop yields can fluctuate from year to year, depending on the weather. However, industrial supply must be reliable for industry to meet market demand. Economics does not permit industries to build huge stockpiles of raw materials; thus, it falls on suppliers to be consistent in their delivery. The need of industries for raw materials fluctuates with market demand, which is governed by the fluctuating buying power of the consumer based on the local, national and world economy. Thus, farmers entering into industrial crop production will have to be more agile than those in the production of food crops. The conventional agriculture supply chain has been built on a production model that relies on the predicted demand for food needs of a known number of the human population in the world. Most national governments build in a security mechanism against shortfall by stockpiling during years of excess production to ward against famine during years of adverse weather. Industries will also have to adjust to the agricultural supply chain which requires constant raw material innovation and is made up of a large number of suppliers of the same raw material as against one-time raw material innovation and a large single supplier of synthetically produced raw material. Plants are constantly under threat from insects and pests and the methods of production have to be adjusted depending on soil conditions, even on a single farm. Therefore, plant breeders become the first link in the industrial feedstock supply chain. Seed producers make the next link to supply seeds for planting the crop. Growers, in turn, produce the crops to be used by industries. As feedstocks have to be collected from multiple farms and delivered to the industrial establishment, collection and delivery in itself becomes a supply chain link. This task has to be carried out either by a middleman or by farmers forming cooperatives.

The backbone of the agricultural-based industries is the willingness of farmers to produce crops as per industry demand. The profitability in production has to be lucrative enough to shift from or add to the familiar model of producing crops for human and animal consumption, as this will involve redeployment of land and other fixed resources (labour and machinery). It takes a number of years for farmers to become fully versed in the production of a new crop and to achieve the full yield potential of the crop. If the market size does not expand with the increased production of feedstock, the demand side of the supply–demand equation will diminish, putting the pricing pressure, and consequent attractiveness of the crop to the farmer.

Primary Product and Co-products of Value

One of the main challenges the nascent plant product industries face while trying to compete with the petroleum-based industries is that while the prime biochemical generates the bulk of its revenue, the income of the petroleum industry is leveraged by the fractionation of the raw material for multiple commercial biochemicals, along with the primary biochemical. Thus, addressing the issues of by-product utilization and environmentally friendly disposal of residual solid and liquid biomass and gaseous emission with the ultimate aim of no waste are essential for sustainable agro-based industries. The mere fact that a large mass of agricultural feedstocks is required for product manufacturing also generates a large leftover mass which requires channelling for further processing into co-products, routing for animal feed at minimal hauling cost and reducing the disposable solid volume for landfill to a minimum or to zero. Alternative routing of processing leftover biomass unsuited for animal feed for useful purposes such as the generation of power can minimize disposal problem while creating additional value. Finding commercial uses for gaseous emissions during manufacturing can increase the competitiveness of agro-industries further. For example, in the process of fermenting sugars to ethanol, about 50% of feedstock mass and energy is lost to CO_2. Although CO_2 if captured can be used for several industrial applications such as refrigeration and cooling, inert gas in chemical processes, fire extinguisher filler, shield gas in welding, and as a raw material for several chemicals, the total market size is relatively small and unable to absorb a new supply influx. Thus, new innovative ways of utilizing large volumes of CO_2 are needed. An algae biofuel enterprise in the vicinity of an ethanol production facility should be able to use substantial amounts of CO_2 and help establish a renewable biofuel hub. Greenhouse operations can also be started in the vicinity to utilize CO_2 for increased yields of vegetables and ornamentals. Research is in progress for the anaerobic conversion using *Escherichia coli* of glycerol and CO_2 generated during biodiesel production into succinic acid, an important industrial chemical (Buchanan, 2007). In these scenarios, the economic return from CO_2 can be calculated in figuring the economics of ethanol production.

Turning a low-value by-product into a higher value product is another strategy important for increased return from plant feedstocks. In dry-grind maize processing, ethanol is produced from the starch portion of the kernel and the remaining material is dried to produce low-value DDGS. In order to add value, Srinivasan *et al.* (2008) devised a method to separate the light fibre fraction of the DDGS from the non-fibre material using air fractionation. The fibre fraction of this separation was then used to prepare a valuable product called maize fibre gum. This food-grade gum is a complex carbohydrate that can be used to stabilize natural flavours in soft drinks or as an emulsifier in other food or industrial products. Jojoba is a semi-arid perennial shrub whose seed contains 50–60% oil. Discovering a use for the main biochemical in multiple products and fractionation of raw material into components of value and their production on a commercial scale need to go hand-in-hand in order to gain the most value out of a plant feedstock. Jojoba oil is similar to that of sperm whale oil in skin-softening properties. As a result, many skin lotions in the market contain jojoba oil. But attempts are under way to expand the use of the oil further in suntan lotions by modifying it to give it the ability to absorb ultraviolet radiation (Harry O Kuru *et al.*, 2004). Simmondsin analogues have been isolated from jojoba meal (Van Boven *et al.*, 1994, 1996; Erhan *et al.*, 1997) that show application for reducing body weight, fat and blood lipid levels. Detoxification and altering the bitterness of jojoba meal have been attempted to make it suitable for animal feed, and some success has been reported (Medina and Trejo-Gonzalez, 1990; Abbott *et al.*, 1991; Lanzani, 1991).

Research Needs and Accomplishments

Feedstock yield and quality

There are multiple opportunities to improve crops to meet industrial ends. The focus of these improvements could be developing production systems that apply across crops, improving the overall efficiency of a crop, or changing the characteristics of plants to meet a particular objective. The need to improve water use efficiency in crop production is well recognized. This requires technological advancement both in the handling of water and its efficient use by plants. Rainwater harvest and storage and its subsequent delivery in measured amounts based on the plants' needs in the ever-changing environmental conditions will make it possible to use this scarce resource available for crop production on a long-term basis and increase the land under irrigation. On the plant front, water use efficiency can be achieved by adapting cultural practices such as planting date, tillage methods, etc., or improving the plant's physiological processes to conserve water. Plant water use efficiency is also tied to photosynthetic efficiency. By utilizing the C-4 pathway, tropical crops such as maize and sugarcane need only half of the amount of water compared to temperate cereals that use the C-3 pathway to fix CO_2 into dry matter (Miflin, 1998). Nutrient use efficiency is an important area requiring across-the-crop improvement. Proper management of N to prevent its seepage to groundwater or runoff to aboveground water reservoirs is important for safeguarding human health and the environment. Enhancing the storage of organic N in the soil profile will decrease the need to apply N fertilizer to produce crops. This can be achieved by developing proper crop rotation and residue management practices. Substantial near-term crop N requirement can be met by growing legume cover or companion crops. The adoption of precision agriculture will enhance producers' ability to base input decisions according to best management practices and cost–benefit considerations. Improvement in photosynthetic efficiency is another research area that needs attention. Riggs *et al.* (1981) noted that the total biomass accumulated during a season had remained unchanged in barley varieties released in the UK over the past 100 years. Evans (1993) reported that gains in crop yields had largely been attained primarily by improving the partitioning of photosynthate to the harvested portion of the crop without changing the quantity of total biomass. Increases in total biomass will be particularly useful for industries engaged in the generation of electricity and biofuel.

Research interest in feedstock crops is expected to accelerate in the coming years. Improvements in plant physiology and genetics to enhance the biomass-producing abilities of crops belonging to the grass family such as switchgrass, miscanthus, napiergrass and high-fibre sugarcane, and fast-growing trees such as poplar hybrids and bamboo will gain momentum. Conventional breeding will remain an important tool in improving plants for industrial uses and will be used alone or most probably in combination with recent genomic tools. As the genomes of more crops are sequenced and the genetic control of important crop traits is better understood, the ability of breeders to manage variations present in their crops will improve. The pace and feasibility for meeting breeding objectives will advance as more and more breeders use genetic marker technology for trait recognition and incorporation.

A significant breeding achievement in developing a high biomass crop for biofuel was achieved in 2007, when scientists at the USDA's Agricultural Research Services Sugarcane Research Laboratory, New Orleans, the Louisiana State University's Agricultural Center's Agricultural Experiment Station and the American Sugar Cane League jointly released three high-fibre sugarcane varieties (L 79-1002, HoCP 91-552 and Ho 00-961) as candidate feedstocks for the biofuels industry. Breeding efforts and agronomic studies are under way to develop even higher biomass-yielding sugarcane cultivars that possess greater levels of cold tolerance so as to allow a longer harvest season. In these attempts, crosses are being made between energycane cultivars and its near relatives, *Miscanthus* spp. and *Erianthus* spp.

Transformation technology has made it possible to make changes in crops for enhanced value. One good example of such a change is the fatty acid composition in soybean oil, with implications both for food and industrial uses. Soybean oil is composed of palmitic, stearic, oleic, linoleic and linolenic fatty acids. These fatty acids are different from one another in all-important characteristics such as melting points, oxidative stabilities and chemical functionalities, which make the full utilization of the soybean oil for one purpose difficult. Researchers at DuPont using genetic modification (GM) have developed soybean seeds with an oleic acid content of approximately 80% of the total oil (Kinney, 1997), as compared to 25% of the total oil in the conventional soybean seeds. This was accomplished by downregulating the expression of the *FAD2* gene, which encoded the enzyme that converted the monounsaturated oleic acid to the polyunsaturated linoleic acid. Later, downregulation of genes that controlled the expression of palmitic acid (Fat B genes) were added to produce soybean seed oil with greater than 85% oleic acid content and saturated fatty acid levels less than 6% (Buhr *et al.*, 2002). High oleic acid content is a much-desired property for lubricant oils because it provides oxidative stability. In contrast, low oxidative stability is favoured in oils used in coating applications, such as paints, inks and varnishes, because they dry fast. The low oxidizing linolenic acid content of soybean seed oil has been enhanced from 10% to in excess of 50% of total oil by increasing the expression of the *FAD3* gene, which encodes the enzyme that converts linoleic acid to linolenic acid (Cahoon, 2003). These types of oils may also complement ongoing research aimed at improving the industrial value of soybean oil through 'green chemistry'. One of the notable examples of success of such research has been the discovery of methods for producing polyols from soybean oil (Crandall, 2002). Soybean-derived polyols can be used in a number of industrial applications, including the production of polyurethanes.

Genetic engineering has made possible the import of suitable genes not otherwise available within a crop species. Gene introgression has been exploited commercially for improving insect resistance and herbicide tolerance in important crops like soybean, maize, cotton and oilseed rape. However, the potential exists to modify a large array of traits using traits across plant species and other living organisms. For example, Cahoon *et al.* (2001) reported that expression of a gene from pot marigold encoding an enzyme that introduced conjugated double bonds into polyunsaturated fatty acids resulted in the accumulation of calendic acid, a novel conjugated polyunsaturated fatty acid, to amounts of 20–25% of the reported total soybean seed oil. Calendic acid is even more oxidatively unstable than linolenic acid, thus improving the drying properties of coating applications. However, the level of calendic acid concentration in soybean at 20–25% remains much lower than the 55% concentration found in marigold.

Induction of genes that accelerate lignin degradation can be valuable for ethanol production from lignocellulosic biomass. Laccase enzymes in fungi take part in lignin degradation (Thurston, 1994). Hood *et al.* (2003) generated transgenic maize plants for a fungal laccase gene using an *Agrobacterium*-mediated system. The molecular parameters that induced the highest expression were the maize embryo-preferred globulin 1 promoter and targeting of the protein to the cell wall. Plants derived from a single transgenic event varied in expression level, and the variation in expression levels was heritable. Within the seed, expression in these plants was primarily in the embryo and was associated with seed browning and limited germination. High oil germplasm was used to increase germination, as well as to assist in increasing expression 20-fold in five generations through breeding and selection.

Research investment on a crop depends on market size and profitability. It is very expensive to develop products such as pesticides and thus requires a considerable market size to recoup the investment. Seed companies are more willing to invest in developing hybrid seeds that create repeat markets than inbred seeds, which after purchasing once, farmers can multiply themselves. In which crops marker technology and transformation techniques will be applied will depend on the objective of improvement, the value added to the crop by the incorporation of the trait and the size of the market for the crop. It has been estimated that the costs for generating the necessary data and information for regulatory agencies is in the range of at least US$1–10 million for a given transgenic trait (Parrot and Clemente, 2004). Transgenic traits incorporated for a particular industrial purpose will be aimed at improving the concentration of a specific biochemical and will require high-volume market demand to interest commercial research and development investments. The promise of genetic engineering at the present cost structure may be realized only for traits with high value.

Postharvest and processing

Plant raw materials harvested at the farm level have to be transported to the manufacturing site or stored for time-bound delivery. Research in technology that cuts the transportation cost or develops a production–processing hub will improve the competitiveness of the final industrial product. Plant materials undergo changes after harvest that cause spoilage and reduction in mass. Tuberous feedstocks need curing for long storage life. Optimum temperature and humidity during storage vary with the crop. Each crop has a unique set of postharvest insects and diseases and requires appropriate control procedures to control the infestation. Certain feedstocks are susceptible to rodent damage and can cause potential quality degradation of the finished products. These problems call for developing economical and efficient storage procedures and the design of facilities suited to store different types of feedstocks. To be turned into industrial raw materials, feedstocks may need to undergo fractionation, combination with other feedstocks or upgrading using various mechanical, chemical and biochemical means. All these steps are subject to improvement through research.

Microbial digestion of feedstock eliminates chemicals used in processing. It can result in reduced costs of production and have environmental benefits. Several methods have been developed to obtain oil from maize germ and other oil-rich plant materials using aqueous enzymatic methods (Moreau *et al.*, 2007). Unlike traditional oil extraction methods, these new bioprocesses are performed without the use of presses and without organic solvents. Three types of enzymes (cellulases, proteases or pectinases) have been found to be effective for aqueous enzymatic oil extraction. Sheridan (2005) reported genetically altered heat-stable α-amylase to degrade maize for ethanol production. The high cost of enzymes has been a major hurdle to the commercialization of aqueous enzymatic oil extraction methods. Recent advances in enzyme production technology provide hope of a substantial reduction in enzyme cost.

Much effort will be needed in developing technologies for the full utilization of the by-products generated by biochemical industries and to keep the balance in the supply–demand equation to ward against price collapse due to oversupply. For example, glycerol is used in the soap, cosmetics, ink, lubrication and preservative industries. However, the demands for glycerol for current uses are not expected to grow significantly in the future, while the amount generated during biofuel production will increase considerably as its use becomes more common. Furthermore, glycerol leaving the transesterfication reactor has many impurities, including methanol, which mandates it to be treated as a hazardous waste. Therefore, it is important to develop technologies that use low-grade crude glycerol. Three research approaches – aqueous-phase reforming (APR), chemical conversion and bioconversion – are currently under way to utilize crude glycerol. Bioconversion will be used here to show the impressive array of possibilities that exist for new uses of glycerol. Dharmadi *et al.* (2006) achieved a 93% yield of glycerol to succinic acid by *E. coli*. Papanikolaou *et al.* (2002) found that *Yarrowia lipolytica* yeast could convert raw glycerol to citric acid. Suthers and Cameron (2005) obtained a US patent for producing HP (3-hydroxypropionic acid) using a genetically engineered bacterial strain hosting two enzymes, glycerol dehydratase and aldehyde dehydrogenase. Arbige (2005) developed a novel thermostable dehydratase enzyme catalysing the conversion of glycerol to HPA (3-hydroxypropionaldehyde). Glycerol also provided a carbon source for the successful fermentative production of butanol by *Clostridium pasteurianum*, PHAs (polyhydroxyalkanoic acid) by *Pseudomonas oleovorans* and *P. corrugate*, propionic acid by *Propionibacterium acidipropionici* and PD (1,3 propanediol) by *C. butyricum* (Biebl, 2001; Ashby, 2005; Barbirato *et al.*, 1997; Papanikolaou *et al.*, 2000, respectively).

Harnessing the full potential of biomass requires utilizing every component of complex plant macromolecules. Plant tissue is composed of carbohydrate, protein, oil, lignin and a host of other chemicals in small quantities, depending on the plant species. Separating individual components economically using environmentally friendly procedures is a difficult challenge, but also has promise of high dividends. Such technologies will permit the development of products that are high volume but have a low market value at one end, such as biofuel and biodegradable plastic, products of intermediate volume with an intermediate market value in the middle, such as succinic acid and lactic acid, and products of low volume but a high value at the other end, such as pharmaceuticals.

New product development and commercialization

The shift from synthetic towards natural products is expected to be gradual, but the trend is unmistakably visible. A number of examples from different industrial sectors can be cited to elucidate this trend: car manufacturers have begun the transition to natural fibres for inside door coverings, instrument panels and ceilings; the electronics industry is shifting to biopolymers for making print plates and cabinets; the detergent industry is relying increasingly on surfactants made of biological raw materials; and a number of industries including paint and lacquer, printing ink and wood preservative have started substituting fossil-fuel solvents with biological solvents.

The new product development process begins with identifying crops that produce feedstock possessing the ingredients needed for the potential products. Transforming the idea for a new product is a complex, multistage process. The research cost of developing a new product is substantial, but according to some estimates, the cost of the technology transfer of research findings, forgetting the cost of the commercialization of a new product or process, is at least nine times the cost of doing original research (Hall, 1993). The US Congress has attempted to ease the technology transfer process by implementing the Federal

Technology Transfer Act of 1986 (Public Law 99-502). This law permits technology transfer agreement between a commercial firm and the USDA. This gives the commercial firm the first right to exclusive licences on patented discoveries and improved access to USDA scientists and facilities. Technology transfer does not guarantee that a product eventually will arrive at the market and that there will be sufficient buyers of the new product. Subsequent activities needed could include assembling further financial resources for ensuing research and development, conducting marketing studies, pilot-plant evaluations, scale-up of production and distribution activities.

Summary

The consumer preference for plant-derived products over petroleum-derived products is increasing steadily. Plant feedstocks are favoured as they are a renewable resource and are perceived to cause less environmental damage. However, at present, the industry is in its infancy and is miniscule compared to the size of the petroleum industry. The growth rate of plant feedstock industries has been slow because of the entrepreneurial risks and market price competitive disadvantage to petroleum-based products. Up until now, most crop improvements have been aimed at improving traits associated with food and feed uses. Industries have their unique needs to concentrate certain plant constituents of industrial values requiring special research attention. Plant feedstocks will become more attractive to entrepreneurs when it becomes possible to process them at the biorefinery level. Technologies that can process plant feedstock at that level are still at the developmental stage. Use of food crops for biofuel has generated concerns of food shortage, but these will dissipate as industries shift to biomass-based feedstocks. The future prospects for plant-based industry appear bright with proper capital and scientific investment.

References

Abbott, T.P., Nakamura, K.L., Buchlz, G., Wolf, J.W., Palmer, M.D., Gasdorf, J.H., Nelson, C.T. and Kleiman, R. (1991) Process of making animal feed and protein isolates from jojoba meal. *Journal of Agricultural and Food Chemistry* 39, 1488–1493.

Alexander, R.B., Smith, R.A., Schwarz, G.E., Boyer, E.W., Nolan, J.V. and Brakebill, J.W. (2008) Phosphorus and nitrogen delivery to The Gulf of Mexico from differences in the Mississippi River Basin. *Environmental Science and Technology* 42, 822–830.

Altieri, M.A. (1995) *Agroecology: The Science of Sustainable Agriculture*. Westview Press, Boulder, Colorado.

Arbige, M.V. (2005) Fermentation process for making 1,3-propanediol. *Industrial Bioprocessing* 27, 2.

Ashby, R.D. (2005) Synthesis of short-/medium-chain-length poly(hydroxyalkanoate) blends by mixed culture fermentation of glycerol. *Biomacromolecules* 6, 2106–2112.

Barbirato, F., Chedaille, D. and Bories, A. (1997) Propionic acid fermentation from glycerol: comparison with conventional substrates. *Applied Microbiology and Biotechnology* 47, 441–446.

Biebl, H. (2001) Fermentation of glycerol by *Clostridium pasteurianum* – batch and continuous culture studies. *Journal of Industrial Microbiology and Biotechnology* 27, 18–26.

Buchanan, G. (2007) New direction for research in bioenergy and biobased products. In: Janick, J. and Whipkey, A. (eds) *Issues in New Crops and New Uses*. ASHS Press, Alexandria, Virginia.

Buhr, T., Sato, S., Ebrahim, F., Xing, A., Zhou, Y., Mathiesen, M., Schweiger, B., Kinney, A., Staswick, P. and Clemente, T. (2002) Ribozyme termination of RNA transcripts down-regulate seed fatty acid genes in transgenic soybean. *Plant Journal* 30, 155–163.

Cahoon, E.B. (2003) Genetic enhancement of soybean oil for industrial uses: prospects and challenges. *AgBioForum* 6, 11–13.

Cahoon, E.B., Ripp, K.G., Hall, S.E. and Kinney, A.J. (2001) Formation of conjugated delta8, delta10 double bonds by delta12-oleic acid desaturase related enzymes: biosynthetic origin of calendic acid. *Journal of Biological Chemistry* 276, 2637–2643.

Cambardella, C.A. and Elliott, E.T. (1992) Particulate soil organic matter changes across grassland cultivation sequence. *Soil Science Society of America Journal* 56, 777–783.

Clark, A. (2007) *Managing Cover Crop Profitability*, 3rd edn. Sustainable Agriculture Network, Beltsville, Maryland.
Crandall, L. (2002) Bioplastics: a burgeoning industry. *Inform* 13, 626–630.
Davis, G.W. (2006) Using E85 in vehicles. In: Minteer S. (ed.) *Alcohol Fuels*. Taylor and Francis, Boca Raton, Florida, pp. 137–152.
Dharmadi, Y., Murarka, A. and González, R. (2006) Anaerobic fermentation of glycerol by *Escherichia coli*: a new platform for metabolic engineering. *Biotechnology and Bioengineering* 94, 821–829.
EPA (US Environment Protection Agency) (2008) Clean Air Act (http:www.epa.gov./air/caa.htm, accessed 22 July 2009).
Erhan, S.M., Abbott, T.P., Nabetani, H. and Purcell, H. (1997) Simmondsin concentrate from defatted jojoba meal. *Industrial Crops and Products* 6, 147–154.
Evangelista, R.L., Wu, Y.V. and Hojilla-Evangelista, M.V. (2006) Characterization of proteins in cuphea (PSR23) seeds. *Journal of American Oil Chemists' Society* 83,785–790.
Evans, L.T. (1993) *Crop Evolution, Adaptation and Yield*. Cambridge University Press, Cambridge, UK.
Farrell, A.E., Plevin, R.J., Turner, B.T., Jones, A.D., O'Hare, M. and Kammen, D.M. (2006) Ethanol can contribute to energy and environmental goals. *Science* 311, 506–508.
Hall, J.T. (1993) Cooperative research and development agreements under the Technology Transfer Act of 1986. In: Janick, J. and Simon, J.E. (eds) *New Crops*. Wiley, New York, pp. 674–676.
Harry O Kuru, R.E., Mohamed, R.E.A. and Abbott, T.P. (2004) Synthesis of novel ultraviolet filters from new crop (jojoba) oil. *Association for the Advancement of Industrial Crops Conference*, Minneapolis, Minnesota, 19–22 September 2004 (www.aaic.org/04programe.htm, accessed 22 July 2009).
HSDB (Hazardous Substances Data Bank) (1989) *Turpentine*. National Library of Medicine, Bethesda, Maryland.
Hood, E.E., Bailey, M.R., Beifuss, K., Magallanes-Lundback, M., Horn, M.E., Callaway, E., Drees, C., Delaney, D.E., Clough, R. and Howard, J.A. (2003) Criteria for high level expression of a fungal laccase gene in transgenic maize. *Plant Biotechnology Journal* 1, 129–140.
IFAD (International Fund for Agricultural Development) (2002) *IFAD Strategic Framework for 2002–2006*. IFAD, Rome.
Johnson, J.M.F., Allmaras, R.R. and Reicosky, D.C. (2006) Estimating source carbon from crop residues, roots and rhizodeposits using the national grain-yield database. *Agronomy Journal* 98, 622–636.
Kinney, A.J. (1997) Genetic engineering of oilseeds for desired traits. In: Setlow, J.K. (ed.) *Genetic Engineering Volume 19*. Plenum Press, New York, pp. 149–166.
Kjoller, C. (1998) The biorefinery: a quality investment opportunity. In: Shewry, P.R., Napier, J.A. and Davis, P.J. (eds) *Engineering Crop Plants for Industrial Uses*. Portland Press, London, pp. 205–216.
Lanzani, A. (1991) A wet process technology applied to jojoba seed to obtain oil and detoxified protein meal. *Journal of the American Oil Chemists Society* 60, 722–774.
Laser, M., Larson, E., Dale, B., Wang, M., Greene, N. and Lynd, L.R. (2009) Comparative analysis of efficiency, environmental impact, and process economics for mature biomass refining scenarios. *Biofuels, Bioproducts, and Biorefining* 3, 247–270.
Lowery, C., Alud, C., Rolfe, R., McKeon, T. and Goodrum, J. (2007) Barriers to commercialization of a castor cultivar with reduced concentration of ricin. In: Janick, J. and Whipkey, A. (eds) *Issues in New Crops and New Uses*. ASHS Press, Alexandria, Virginia, pp. 97–100.
Maskina, M., Power, J., Doran, J. and Wilhelm, W. (1993) Residual effects of non-till crop residues on corn yield and nitrogen uptake. *Soil Science Society America Journal* 57, 1555–1560.
Medina, L.A. and Trejo-Gonzalez, A. (1990) Detoxified and debittered jojoba meal: biological evaluation and physical-chemical characterization. *Cereal Chemistry* 67, 476–479.
Miflin, B. (1998) Crops for the 21st century. In: Shewry, P.R., Napier, J.A. and Davis, P.J. (eds) *Engineering Crop Plants for Industrial Uses*. Portland Press, London, pp. 1–9.
Moreau, R.A., Johnston, D., Dickey, L.C., Parris, N. and Hicks, K.B. (2007) Aqueous enzymatic oil extraction: a 'green' bioprocess to obtain oil from corn germ and other oil-rich plant materials. In: Eggleston, G. and Vercellotti, J.R. (eds) *The Industrial Application of Enzymes on Carbohydrate-Based Materials*. American Oil Chemists' Society Press, Champaign, Illinois, pp. 101–120.
Nielson, E.G. and Lee, L.K. (1987) *The Management and Costs of Groundwater Contamination from Agricultural Chemicals: A National Perspective. Agricultural Economics Report No. 576*. Economic Research Service, USDA, Washington, DC.
NRC (National Research Council) (2008) *Transition to Sustainability Through Research and Development on Ecosystem Services and Biofuels*. The National Academies Press, Washington, DC.

Padgett, T. and Myers, F. (2009) The next big biofuel? *Time Magazine*, Time Inc, 29 January 2009.

Papanikolaou, S., Ruiz-Sanchez, P., Pariset, B., Blanchard, F. and Fick, M. (2000) High production of 1,3-propanediol from industrial glycerol by newly isolated *Clostridium butyricums* strain. *Journal of Biotechnology* 77, 191–208.

Papanikolaou, S., Muniglial, L., Chevalot, I., Aggelis, G. and Marc, I. (2002) *Yarrowia lipolytica* as a potential producer of citric acid from raw glycerol. *Journal of Applied Microbiology* 92, 737–744.

Parrot, W.A. and Clemente, T.A. (2004) Transgenic soybean. In: Specht, J. and Boerma, R. (eds) *Soybeans: Improvement, Production, and Uses, Agronomy Monograph No. 16*, 3rd edn. American Society of Agronomy, Madison, Wisconsin, pp. 265–302.

Pimentel, D. and Patzek, T.W. (2005) Ethanol production using corn, switchgrass and wood; biodiesel production using soybean and sunflower. *Natural Resources Research* 14, 65–76.

RTECS (Registry of Toxic Effects of Chemical Substances) (1989) *Turpentine*. National Library of Medicine, Bethesda, Maryland.

Riggs, T.J., Hanson, P.R., Start, N.D., Milles, D.M., Morgan, C.L. and Ford, M.A. (1981) Comparison of spring barley varieties grown in England and Wales between 1880 and 1980. *Journal of Agricultural Science* 97, 599–610.

Sainju, U.M. and Singh, B.P. (1997) Winter cover crops for sustainable agricultural systems: influence on soil properties, water quality, and crop yields. *HortScience* 32, 21–28.

Schmer, M.R., Vogel, K.P., Mitchell, R.B. and Perrin, R.K. (2008) Net energy of cellulosic ethanol from switchgrass. *Proceedings of the National Academy of Sciences, USA* 105, 464–469.

Sheridan, C. (2005) It came from beneath the sea. *Nature Biotechnology* 23, 1199–1201.

Shumaker, G.A., Luke-Morgan, A., McKissick, J.C. and Flanders, A. (2007) *The Economics of Ethanol Production in Georgia*. Publication No. FR-07-02, The University of Georgia, Athens, Georgia.

SoyStats (2007) *Soy Stats Guide*. American Soybean Association, St Louis, Missouri.

Srinivasan, R., Yadav, M.P., Belyea, R.L., Rausch, K.D., Pruiett, L.E., Johnston, D., Tumbleson, M.E. and Singh, V. (2008) Fiber separation from distillers dried grains with solubles using a larger elutriation apparatus and use of fiber as a feedstock for corn fiber gum production. *Biological Engineering* 1, 39–49.

Suthers, P.F. and Cameron, D.C. (2005) Production of 3-hydroxypropionic acid in recombinant organisms. United States Patent Number 6852517.

Thurston, C.F. (1994) The structure and function of fungal laccases. *Microbiology* 140, 19–26.

Ulgiati, S. (2001) A comprehensive energy and economic assessment of biofuels: when green is not enough. *Critical Review in Plant Sciences* 20, 71–106.

United Soybean Board (2007) *Soybean Market View Database*. United Soybean Board, Chesterfield, Missouri.

Van Boven, M., Daenens, P., Cokelaere, M. and Janssen, G. (1994) Isolation and structure elucidation of the major simmondsin analogs in jojoba meal by 2-dimensional nmr-spectroscopy. *Journal of Agricultural and Food Chemistry* 42, 2684–2687.

Van Boven, M., Daenens, P., Tyga, J. and Cokelaere, M. (1996) Determination of simmondsins and simmondsin ferulates in jojoba meal and feed by high-performance liquid chromotography. *Journal of Agricultural and Food Chemistry* 44, 2239–2243.

Vane, L.M. (2005) A review of pervaporation for product recovery from biomass fermentation process. *Journal of Chemical Technology and Biotechnology* 80, 603–629.

Watkins, K. and von Braun, J. (2003) *Time to Stop Dumping on the World Poor, 2002–2003. IFPRI Annual Report*. International Food Policy Research Institute, Washington, DC.

Wesseler, J. (2007) Opportunities ('costs') matter: a comment on Pimentel and Patzek 'Ethanol production using corn, switchgrass, and wood, biodiesel production using soybean and sunflower'. *Energy Policy* 35, 1414–1416.

Wilhelm, W.W., Johnson, J.M.F., Karlen, D.L. and Lightle, D.T. (2007) Corn stover to sustain soil organic carbon further constrains biomass supply. *Agronomy Journal* 99, 1665–1667.

Worldwatch Institute (2007) Implications for agriculture and rural development. In: *Biofuels for Transport*. Earthscan, London, pp. 120–137.

2

Bioenergy Industry Status and Prospects

PAUL CHEN, YIQIN WAN, XIAOQUAN WANG, YANLING CHENG, YUHUAN LIU, XIANGYANG LIN AND ROGER RUAN

Introduction

World energy demands are rising and fossil resources are declining. While we should improve energy efficiency and encourage energy conservation, alternative energy resources are required for sustainable development of the world's economy. Bioenergy is renewable energy made from plant-derived organic matter, collectively termed 'biomass'. Biomass is widely available and can be converted through a wide range of technologies to different forms of energy, chemicals and materials that conventionally are derived from fossil resources (Fig. 2.1). Many industrial crops potentially could make a significant contribution to bioenergy production.

The bioenergy industry holds great promise for our economy, environment and society. Sustainable development of the bioenergy industry will not only reduce global reliance on fossil energy but also bring new economic opportunities to many developing countries and the rural areas of developed countries. Value will be added to otherwise unused, underused or improperly used biomass resources; non-arable lands may be used to grow energy crops; rural communities will be able to access local energy supplies; and related industries will be established nearby. All these will offer new growth opportunities and higher incomes to agricultural and forest producers, bioenergy producers and related equipment manufacturers and energy distributors. This will also improve the standard of living and create new employment opportunities (Fig. 2.2).

The net greenhouse gas emissions from bioenergy sources are less than those from petroleum sources (Fig. 2.3).

The environmental benefits of bioenergy are at least threefold: (i) it reduces the use of environmentally unfriendly fossil resources; (ii) biomass sequesters CO_2 and therefore is more carbon neutral than fossil energy; and (iii) it helps waste management. There are increasing concerns about the undesirable impacts of bioenergy activities on our ecosystem and environment. These concerns include land and water resource use and competition with food supplies, soil erosion due to excessive removal of biomass and decreased biodiversity due to monoculture and short-rotation crops. These concerns may be addressed through informed planning that keeps ecological and environmental factors in mind, government policy that offers incentives to good agricultural practice, adequate management systems that control biomass collection and technological development which facilitates biomass production using less land and water, non-arable land and high-yield crops. Further life-cycle studies of key parameters are indeed necessary to clarify the social, economical and environmental impacts of the bioenergy industry.

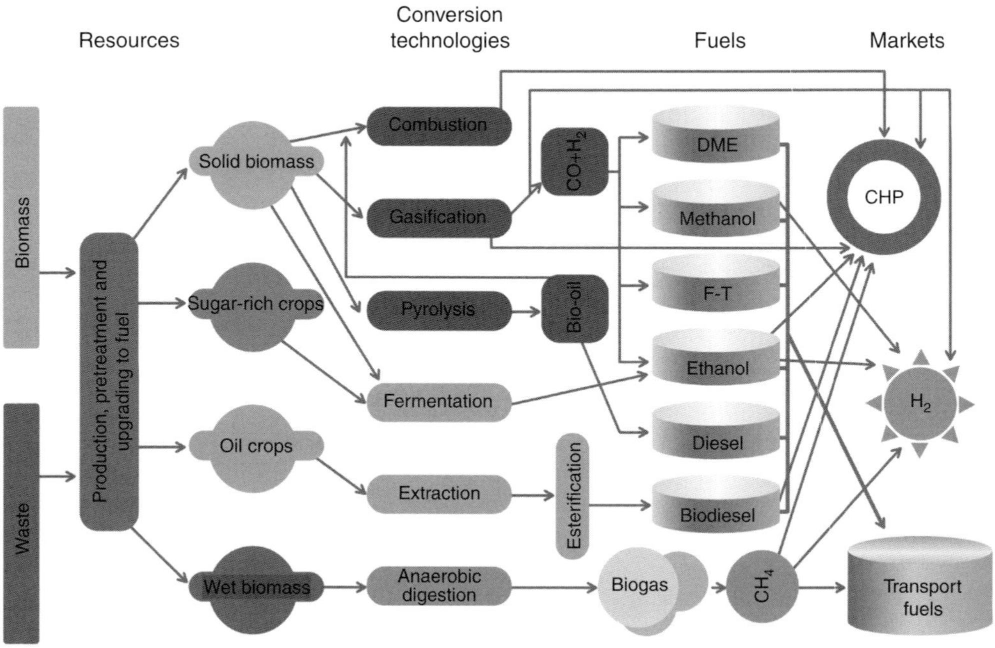

Fig. 2.1. Bioenergy from biomass through different pathways (European Commission, 2006).

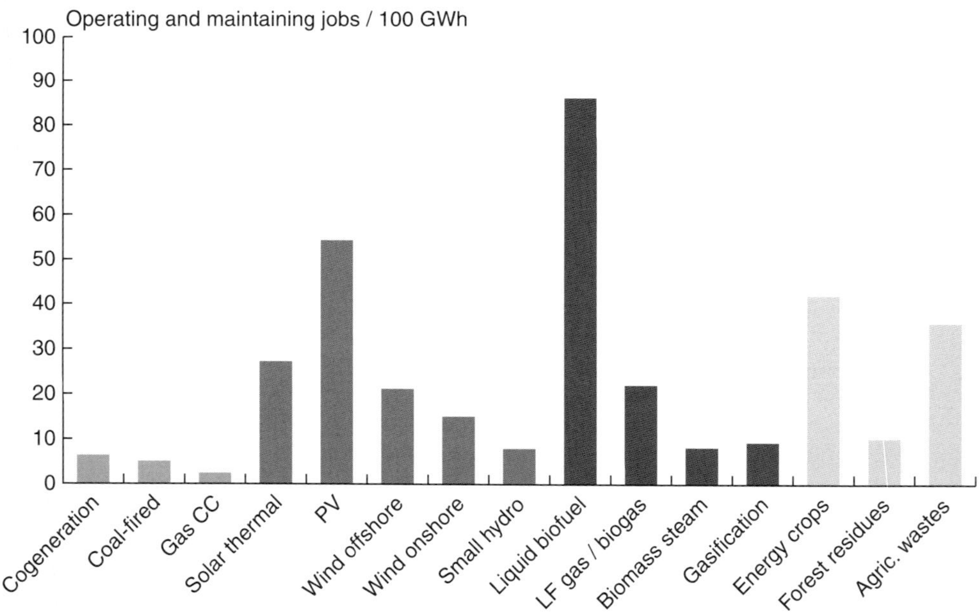

Fig. 2.2. Employment requirements for energy projects. Labour is required for operating and maintaining various renewable energy projects after their construction, with bioenergy projects also requiring additional labour to produce and deliver the biomass to the plant (International Energy Agency, 2007 © OECD/IEA, 2007).

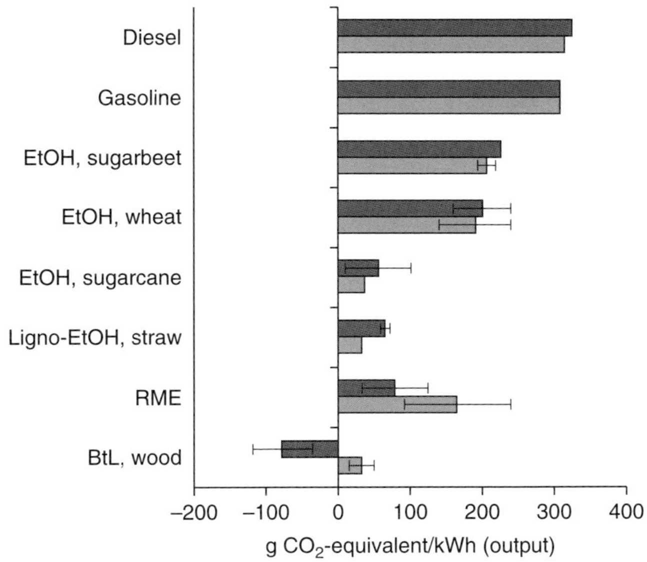

Fig. 2.3. The net life cycle greenhouse gas emissions of fossil fuels and various biofuels. EtOH (ethanol), Ligno-EtOH (ethanol from lignocellulosics), RME (rapeseed methyl ester) and BtL (biomass to liquids). (Edwards *et al.*, 2007, with permission).

Developed sustainably and used efficiently, bioenergy can induce growth in developing countries, reduce oil demand and address environmental problems. The potential benefits include: reduction of greenhouse gases, recuperation of soil productivity and degraded land, and economic benefits from adding value to agricultural activities and improving access to and quality of energy services. The production of bioenergy involves a range of technologies, including solid combustion, gasification and fermentation. These technologies produce energy from a diverse set of biological resources – traditional crops, crop residues, energy-dedicated crops, manure and the organic component of urban waste. The results are bioenergy products that provide multiple energy services: cooking fuel, heat, electricity and transportation fuels (Fig. 2.1). It is this very diversity that holds the potential of a win–win situation for the environment and for social and economic development. Bioenergy has to be viewed not as a replacement for oil but as one element of a portfolio of renewable sources of energy. Coherent and mutually supportive environmental and economic policies may be needed to encourage the emergence of a globally dispersed bioenergy industry that will pursue a path of sustainable development.

Bioenergy Consumption and Future Outlook

The world used about 488 exajoules (EJ) of commercial energy in 2005 (Fig. 2.4). The demand for energy is growing rapidly and is expected to double or perhaps triple during this century.

Current global energy resources consist of fossil oil, coal, natural gas, nuclear power and renewables (Fig. 2.5).

Biomass provides an average of about 10% (45 ± 10 EJ) of the total energy supply, making it by far the most important renewable energy source in use. However, in developing countries, the biomass supply proportion is as high as 20–30% or up to 50–90% of the total energy demand in a number of countries. Biomass combustion is responsible for over 90% of the current production of secondary energy carriers from biomass. Commercial bioenergy for industry, power generation or transport fuels makes up a very small portion of total biomass-derived energy (about 7 EJ/year in 2000),

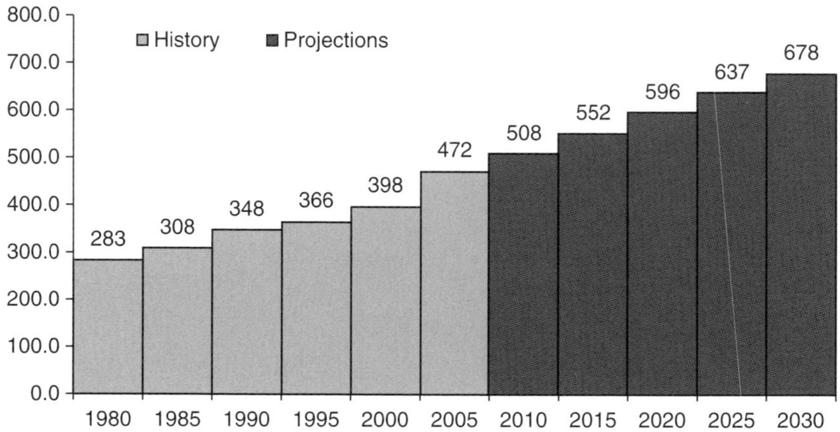

Fig. 2.4. World marketed energy consumption and projections in exajoules (EJ), 1980–2030 (US Energy Information Administration (EIA), 2009).

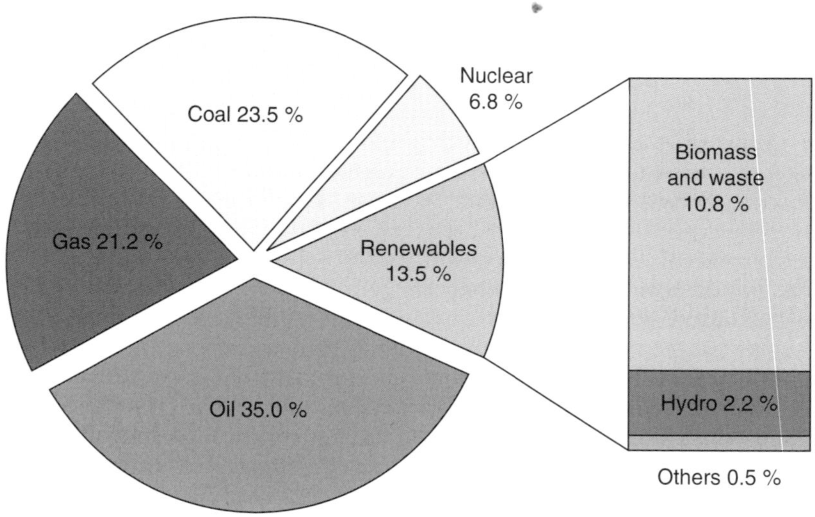

Fig. 2.5. Share of biomass in world total primary energy supply (IEA-Bioenergy, 2007).

but is expected to grow rapidly (Fig. 2.6). The current use of modern bioenergy may be constrained by a number of barriers but primarily biomass feedstock-related issues. Maize and soybean, current primary biomass feedstocks for transportation fuel production in the USA, have to compete with food production for arable land use and could drive a possible rise in both biofuel and food prices, which could have a huge negative impact on the large population groups in developing countries. Harvesting, collection and transport of lignocellulosic biomass remain costly and a logistics barrier that limits the size of ethanol production plants and economies of scale. Many conversion technologies are still at the developing and demonstration stage, and their techno-economics has not been well studied for different regions in the world with various levels of natural resources and economic development. However, with technological advancements and change in world energy supplies and prices, it is estimated that a range from 200 to 400 EJ/year in biomass harvested for energy production may be expected during this century. The

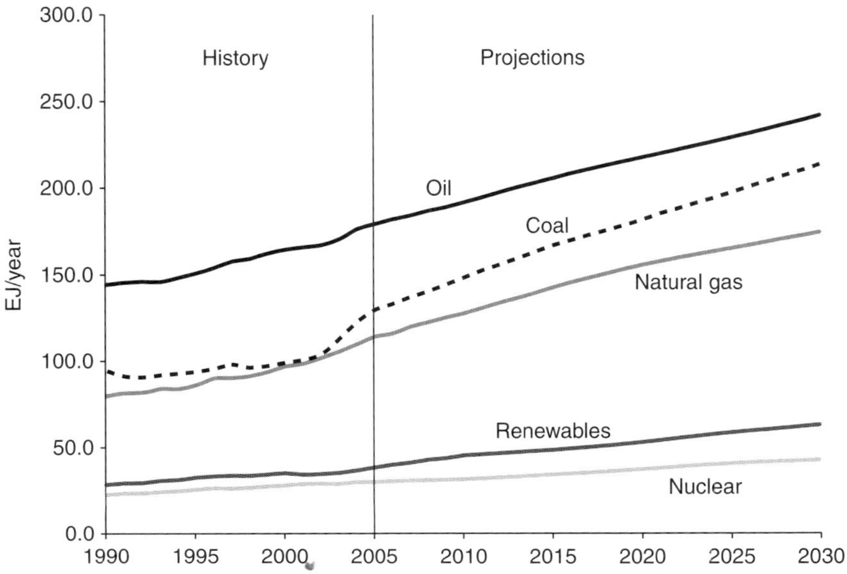

Fig. 2.6. World marketed use of energy by fuel type (EIA, 2009).

European Union (EU) and the USA have made bold projections of bioenergy development in their renewable energy roadmap (Table 2.1).

Biomass Resources and Potential

The total world annual biomass production is estimated at 2740 quads (a quad of energy, or quadrillion British thermal units, is equivalent to 170 million barrels of oil), which is about eight times the total annual world consumption of energy from all sources (about 340 quads) (BRIC, 2000). At present, only about 7% of the annual production of biomass is used. In the USA, biomass energy resources account for 3% of the total energy consumption, or 47% of the total renewable energy consumption. A USDOE (US Department of Energy) and USDA (United States Department of Agriculture) funded study reports that there are more than 1.3 billion dry t of biomass produced annually from forest land and agricultural land alone, which could be converted into fuels to meet one-third of the current demand for transportation fuels. With other types of biomass and increasing interest in growing energy speciality crops, biomass resources will have a great potential to meet a large portion of US energy needs.

Table 2.1. The US vision goals for bioenergy and bioproducts (BRDI, 2006).

	Units	2000	2004	2010	2015	2020	2030 (% increase from 2004)
Biofuels	Market share (%)	0.7	1.2	4.0	6.0	10.0	20.0 (156,667%)
	Consumption (billion gasoline-equivalent litres)	4.2	7.9	30.3	48.8	85.9	193.1 (232,857%)
Biopower	Market share (%)	3.0	3.0	4.0	5.5	7.0	7.0 (13,333%)
	Consumption (EJ)	2.1	2.2	3.3	3.4	3.6	4.0 (8,095%)
Bioproducts	Production (kg)	5.8	8.0	10.8	12.0	16.1	25.1 (21,420%)

Type and availability

The US Biomass Research and Development Board classified biomass into the following three categories, based on the maturity of their production processes:

- First generation feedstocks include maize for ethanol and soybeans for biodiesel. These feedstocks are produced using mature production processes and are currently in commercial use. Future cost savings due to technique refinements are likely to be marginal.
- Second generation feedstocks consist of the residues or 'leftovers' from crop and forest harvests without a food use. Technologies for production and processing of these feedstocks are emerging, with significant potential for reducing production and processing costs.
- Third generation feedstocks are speciality energy crops, representing long-term significance for sustainable development of the biofuels industry. These feedstocks include perennial grasses, fast-growing trees and algae. Further research and development are necessary.

Biomass feedstock can also be divided into the following major categories based on their plant origins.

Forestland biomass

Forestland biomass includes logging residues, other removal residues, thinning from timberland and other forestland, mill residues, urban wood waste and conventionally sourced wood. Traditionally, woody biomass is used as fuelwood (direct burning), charcoal and black liquor (the spent pulping chemicals and the lignin component of wood after chemical pulping). Today, with increasing interest in cellulosic ethanol, forest biomass has become a major feedstock for ethanol fermentation.

The US forestland produces a total of about 20 billion dry t of biomass, among which there are about 368 million dry t of residues and wastes, including the recovered residues generated by traditional logging activities and residues generated from forest cultural operations or clearing of timberlands; the recovered residues generated from fuel treatment operations on timberland and other forestland; the direct conversion of roundwood to energy (fuelwood) in the residential, commercial and electric utility sectors; and forest products industry residues and urban wood residues. How much of these residue and waste resources are available for bioenergy production depends on how they can be recovered, which is a function of tree form, technology and timing of the removal of the biomass from the forests.

Agricultural biomass

Agricultural biomass is the largest available feedstock resource including starch, sugar and oil crops, crop residues and animal manure. Maize and soybean are the major feedstock for ethanol and biodiesel fuel production in the USA. In 1980, less than 1% of US maize was used for ethanol production. Total maize production increased from 241.3 million t (9.5 billion bushels) in 2001 to 330.1 million t (13 billion bushels) in 2007 and the share of maize used for ethanol has jumped from 7 to 24% since 2001. The total world production of ethanol in 2006 was estimated at over 46 hm^3 and projected to be over 82 hm^3 by 2012. Biodiesel production from major producing countries was 7495 t in 2007, which is much smaller than ethanol fuels. In the USA, biodiesel use is now mandated to grow from 1890 million l (500 million gallons) in 2009 to 3.78 billion l (1 billion gallons) in 2012. The EU mandates for 5.75% biofuels in the transportation sector by 2010 are expected to drive rapid growth of biodiesel in the major EU economies. Increasing demands for biodiesel will certainly put enormous pressure on vegetable oil supply.

Other grains (oats, sorghum, wheat, rice and barley) and sugar crops (sugarcane, sugarbeet) are used for ethanol production (Kim and Dale, 2004). In the USA, eight ethanol plants use grain sorghum (milo) as a feedstock for ethanol production, consuming approximately 15% of US grain sorghum (12.83 million t [505 million bushels]). Grain sorghum produces a similar amount of ethanol per bushel as maize, but the sorghum yield is lower than for maize. In Brazil, the second largest ethanol producer in the world, ethanol is produced mainly from sugarcane (see Chapter 4, this volume).

With increasing demand for ethanol and competition with food and feed supplies, agricultural crop residues (lignocellulosic biomass) become very attractive. Researchers have been developing technologies to convert cellulosic biomass to ethanol (see Chapter 6). Agricultural crop residues are the biomass that remains in the field after harvest. The most common residues include maize stover (stalks, leaves and/or cobs) and straw associated with wheat, rice, barley or oat production. The world produces about 1.5 billion t of cellulosic biomass, which could produce up to 442 Gl/year of ethanol (Kim and Dale, 2004). The USA is projected to produce 446 million dry t of agricultural crop residues annually. Among these crop residues is maize stover, with a potential annual production of 256 million dry t.

Speciality energy crops

Energy crops are fast-growing speciality crops that are grown for the purpose of producing energy. These include annual perennial crops such as forage sorghum and switchgrass, fast-growing trees such as willow and poplar, and microalgae. Forage sorghum grows from 1.8 to 3.7 m tall and produces more dry matter tonnage than grain sorghum. Herbaceous crops such as switchgrass, lucerne and miscanthus are receiving considerable attention. Switchgrass yields on average about 9.9–24.7 dry t/ha/year. Woody trees such as hybrid poplar and willow yield on average 12.4–29.6 dry t/ha/year. These yields are much higher than most of the grains and maize stover (BRDR, 2008).

There is increasing interest in growing other starch and oil crops for biofuels. For example, oilseed crops such as Indian mustard, rapeseed and many nut crops such as *Jatropha curcas* and oil palm are potential vegetable oil sources for biodiesel. The lipid profiles of oils from these crops are different from soybean oils. Therefore, modification of the manufacturing processes is necessary.

Another potential alternative feedstock under active development is microalgae. Microalgae grow 10–20 times faster than agricultural crop plants and certain microalgal species can accumulate large amounts of lipids or oil (30–60% of dry weight). Microalgae could help avoid a number of the sustainability issues associated with land use, freshwater use, deforestation and food production. Microalgae production also provides significant environmental benefits through effective CO_2 sequestration. Despite intensive studies on microalgae in the past 40 years, there is no economically viable microalgae production system commercially available on the market. Large-scale microalgae production is facing the following key challenges: (i) high costs of nutrients and CO_2; (ii) poor performance of microalgae species/strains; (iii) making enclosed photobioreactors cost-effective; (iv) high costs of artificial lighting and pumping; and (v) high costs of harvesting and oil extraction.

Animal manures, food processing residues and wastes and municipal solid wastes

These wastes and residues are considered secondary and tertiary biomass feedstocks. Animal manures typically are applied as organic fertilizers on farms. Excess manures can be converted to bioenergy, which would reduce nutrient runoff and contamination of surface water and groundwater resources (He *et al.*, 2000; Perlack *et al.*, 2005; Kyoung *et al.*, 2007). Food processing residues are generated in the manufacture and distribution of food for a number of reasons, including spoilage, removal of unusable portions, discarding of substandard products and packaging failure. Municipal solid wastes (MSWs) consist of food scraps, papers, yard wastes, cardboard, plastics, woods, etc. A major concern and challenge for both food processing residues and MSWs as bioenergy feedstocks is their inconsistency.

Collection and delivered costs of biomass feedstock

Most biomass feedstocks are widely distributed in loose form and need to be collected, packaged, stored and shipped to conversion facilities. Technologies and equipment currently exist for biomass harvest, collection, baling and storage. New methods may be needed for non-commonly collected and new biomass crops. Nevertheless, the following factors must be considered in planning biomass collection and transportation (Sokhansanj *et al.*, 2003):

- low bulk density;
- spoilage due to high moisture;
- variability in physical and chemical characteristics;
- geographical and seasonal variations in biomass;
- conflicting demands on labour and machines;
- combustibility;
- competition from soil nutrients;
- local regulations on storage and transport;
- sensitivity to price structure for companion products and farm commodities.

Although cellulosic residues in the field are rather inexpensive (Perlack and Turhollow, 2003), getting the residues to the processing plants and converting them to fermentable sugars is very costly (Wyman, 2007). A study funded by the US Department of Energy shows that the delivered costs for maize stover range from US$43.10/dry t for a 500 dry t/day facility (4662 km^2 collection area, 35.4 km average one-way hauling distance) to about US$51.60/dry t for a 4000 dry t/day facility (36,260 km^2 collection area, 100 km average one-way hauling distance) (Perlack and Turhollow, 2003). The difference in delivered costs between facility sizes reflects transport costs, which account for 33% of total delivered costs for a 500 dry t/day facility and 40% for a 4000 dry t/day facility. Research has found that the financial advantage provided by large processing capacity may be offset by the high delivery costs of feedstock and suggests that biomass industry development should include smaller-scale facilities to be economically viable (English, 2003). Furthermore, compared with maize ethanol production, additional processing costs are needed to convert cellulosic feedstock to fermentable sugars, which would raise feedstock-associated costs to as high as 70–80% of the final product cost (Wooley *et al.*, 1999). Thus, cellulosic ethanol faces challenges to reduce costs in feedstock transport and processing.

The above analysis leads us to believe that future economically viable alternative biomass processing systems must cut down feedstock-related costs significantly by reducing transport costs and developing more efficient processing technologies. Bio-oil's BTU (British thermal unit) is 7.5 times of that of biomass. If biomass feedstock can be processed into bio-oils on the farm and the bio-oils can be used directly as a boiler fuel or transported to a central biorefinery for further processing, significant cost savings can be realized. To achieve this, alternative biomass energy production systems must be developed.

The Distributed Biomass Energy Production System (DBEPS) concept relies on scalable technologies that can be implemented on average-size farms where crop residues are converted to bio-oils with minimal transportation (Ruan *et al.*, 2008). The bio-oils produced can be used as home heating oil or transported to a central biorefinery where upgrading and manufacture of other products can be carried out. Any DBEPS must meet the following criteria: (i) affordable capital cost; (ii) low transport costs; (iii) easy to operate (turnkey) technology; and (iv) economic and social benefits for the rural community. A 151–189 million l cellulosic ethanol plant costs about US$300 million to build, while the cost for building an on-farm DBEPS facility would be lower than US$200,000. Feedstock may be collected from one farm or neighbouring farms with minimal transport costs. Operating a DBEPS facility should not require special experience or expertise. Farms can use the bio-oils or sell for profit.

Biomass feedstock outlook

Considering the benefits and concerns discussed in the beginning of this chapter, future biomass feedstock supplies depend on a number of complex factors, particularly those related to sustainability, which may give rise to a number of uncertainties of biomass supplies. Table 2.2 shows the energy potential in biomass with key assumptions.

Table 2.2. Overview of the global potential of biomass for energy (EJ/year) to 2050 for a number of categories and the main preconditions and assumptions that determine these potentials.[a]

Biomass category	Energy potential in biomass up to 2050	Main assumptions and remarks
Energy farming on current agricultural land	0–700 EJ (more average development: 100–300 EJ)	Potential land surplus: 0–4 Gha (average: 1–2 Gha). A large surplus requires structural adaptation towards more efficient agricultural production systems. When this is not feasible, the bioenergy potential could be reduced to zero. On average, higher yields are likely because of better soil quality: 8–12 dry t/ha/year[a] is assumed.
Biomass production on marginal lands	60–110 EJ	On a global scale, a maximum land surface of 1.7 Gha could be involved. Low productivity of 2–5 dry t/ha/year.[b] The net supplies could be low due to poor economics or competition with food production.
Residues from agriculture	15–70 EJ	Potential depends on yield/product ratios and the total agricultural land area, as well as type of production system. Extensive production systems require reuse of residues for maintaining soil fertility. Intensive systems allow for higher utilization rates of residues.
Forest residues	30–150 EJ	The sustainable energy potential of the world's forests is unclear – some natural forests are protected. Low value: includes limitations with respect to logistics and strict standards for removal of forest material. High value: technical potential. Figures include processing residues.
Manures	5–55 EJ	Use of dried dung. Low estimate based on global current use. High estimate: technical potential. Utilization (collection) in the longer term is uncertain.
Organic wastes	5–50 EJ	Estimate on basis of literature values. Strongly dependent on economic development, consumption and the use of biomaterials. Figures include the organic fraction of MSW and waste wood. Higher values possible by more intensive use of biomaterials.
Combined potential	40–1100 EJ (200–400 EJ)	Most pessimistic scenario: no land available for energy farming; only utilization of residues. Most optimistic scenario: intensive agriculture concentrated on the better quality soils. In parentheses: average potential in a world aiming for large-scale deployment of bioenergy.

[a] Source: Hooper and Li, 1996; Berndes *et al.*, 2001; Smeets and Faaij, 2007.
[b] Heating value: 19 GJ/t dry matter.

Bioenergy Technologies

Biomass processing technologies are widely grouped into first generation and second generation. First generation technologies are well established. These include transesterification of plant oils, fermentation of plant sugars and starch for liquid biofuel production, anaerobic fermentation of organic residues to generate biogas, combustion of organic materials for heat recovery or combined heat and power (CHP) systems for the production of both heat and electrical power (Table 2.3). Second generation or advanced technologies often refer

Table 2.3. Major bioenergy technologies, feedstocks used and energy produced.

Technology	Conversion process	Major biomass feedstock	Energy or fuel produced
Densification	Physical	Agricultural residues	Solid fuels (briquette, pallets)
Direct combustion	Thermochemical	Wood, agricultural and municipal solid waste, residential fuels	Heat, steam, electricity
Gasification	Thermochemical	Wood, agricultural waste, municipal solid waste	Low- or medium-energy intensity producer gas
Pyrolysis	Thermochemical	Wood, agricultural and municipal solid waste	Synthetic fuel oil (biocrude), charcoal
Anaerobic digestion	Biochemical (anaerobic)	Animal manure and agricultural waste, landfills' wastewater	Medium intensity gas (methane)
Ethanol production	Biochemical (aerobic)	Sugar or starch crops, wood waste, pulp sludge, grass straw	Ethanol
Biodiesel production	Chemical	Rapeseed, soybeans, waste vegetable oil, animal fats	Biodiesel
Methanol production	Thermochemical	Wood, agricultural and municipal solid waste	Methanol

to the conversion of lignocellulosic materials to fuels. These technologies comprise a range of alternatives such as enzymatic production of lignocellulosic ethanol, synthetic gas-based fuels, pyrolysis oil-based biofuels, gasification and others that are not yet economically viable and the technical aspects are still under development.

Physical conversion

Several physical processes have been used to transform biomass to energy products or intermediate feedstocks. These processes include dewatering, drying, size reduction and densification. Water removal in the form of liquid (dewatering) or vapour (drying) is often used as a pretreatment for other conversion processes or to meet the moisture content requirement for certain solid fuels. Dewatering may be accomplished through filtration, centrifugation, pressing or extrusion. Wet biomass may be dried in natural air, sun or artificial heat. Size reduction is used as a pretreatment for other conversions or in preparation of biomass for direct fuel use. Shredding, cutting, grinding/milling, pulping and steam explosion are often employed in size reduction. Densification of loose biomass is called biomass briquetting or palletizing. It facilitates easy transportation and better handling and storage, besides being efficient in use as an alternative fuel to coal and firewood. The high temperature developed during the high-pressure densification process assists the inherent lignin present in the biomass to bind the biomass and form densified fuel briquettes or pallets.

Direct combustion

Direct combustion is the burning of biomass in air to convert biomass into heat, mechanical power or electricity using various equipment; e.g. stoves, furnaces, boilers, steam turbines and turbogenerators. Wet biomass needs to be dried to below 50% moisture content for direct combustion. The net bioenergy conversion efficiencies for direct biomass combustion power plants range from 20 to 40%. Direct combustion is simple and has the advantage that it employs well-developed, commercially available technology. The disadvantages include thermal penalties associated with burning high-moisture fuels, agglomeration and ash fouling due to alkali compounds in biomass and relatively low thermodynamic efficiencies for steam power plants.

Biochemical conversion

Two biological conversion processes are commonly used in bioenergy production, namely ethanol fermentation and anaerobic digestion. Ethanol is produced mostly through yeast fermentation of sugars, which may be derived from sugar crops (see Chapter 4), grains (see Chapter 5) and lignocellulosics (see Chapter 6). Sugars (glucose, fructose and sucrose) can be fermented directly to ethanol by yeasts, while grain starch and lignocelluosics must be hydrolysed to fermentable sugars first. The ethanol yields from different biomass feedstocks are listed in Table 2.4.

There are large-scale commercial plants for both sugar-based and starch-based ethanol production. The conversion of lignocellulosic biomass to ethanol is much more complex. Enzymatic hydrolysis of lignocellulosics, which are composed mainly of cellulose, hemicellulose and lignin, has been proven difficult and costly. Pretreatments are usually required to increase the surface area of lignocellulosic materials and thus make polysaccharides more susceptible to enzymatic attack (hydrolysis). Pretreatment is one of the most costly steps in converting lignocellulose to sugars, accounting for about 20–30% of the total processing costs.

Anaerobic digestion is often employed to convert organics in animal wastes and municipal sludge to biogas (mainly methane and carbon dioxide) by bacteria under anaerobic conditions. Anaerobic digestion is a mature technology and is widely used. Biogas can be used directly in gas turbines to produce electricity.

Table 2.4. Ethanol yields from different biomass feedstocks.

Crop	Ethanol l/ha	US gallon/acre
Miscanthus	14,031	1,500
Switchgrass	10,757	1,150
Sweet potatoes	10,000	1,069
Poplar wood (hybrid)	9,354	1,000
Sweet sorghum	8,419	900
Sugarbeet	6,679	714
Sugarcane	6,192	662
Maize	3,461	370
Cassava	3,835	410
Wheat	2,591	277

Thermochemical conversion

In gasification conversion, lignocellulosic feedstocks are converted to a combustible gas mixture called 'synthesis gas' (syngas) or 'producer gas' through partial oxidation reactions at high temperature, ranging typically from 700 to 1100°C. Syngas may vary in composition with type and moisture content of feedstock, type of gasifier, gasification conditions, etc. Syngas can be burned to produce heat or used in gas engines or gas turbines to produce electricity. Gasification units are commercially available. Syngas clean-up and conditioning has been identified as a key technical barrier to the commercialization of biomass gasification technologies and has the greatest impact on the cost of clean syngas. Catalytic reforming and fermentation of syngas to other chemicals such as short-chain fatty acids, methanol, ethanol, other mixed alcohols, hydrogen, aldehydes, olefins and polyhydroxyalkanoates (PHA) are being investigated.

Pyrolysis is another important thermochemical conversion process in which biomass is degraded to bio-oil, syngas and chars at medium-high temperature (300–700°C) in the absence of oxygen. Biomass is heated usually through a heated surface or sands. A new type of pyrolysis process using microwave heating is being developed. The technical advantages of microwave-assisted pyrolysis (MAP) over conventional pyrolysis include:

1. Microwave heating is uniform and easy to control.
2. It does not require a high degree of feedstock grinding (e.g. large chunks of wood logs can be used) and can handle mixed feedstock (e.g. MSWs).
3. The conversion products (pyrolytic gas and bio-oils) are cleaner than those from gasification and conventional pyrolysis because it does not have to use biomass powder and does not require agitation and fluidization.

4. The syngas produced has a higher heating value since it is not diluted by the carrying gas for fluidizing the biomass materials.
5. Microwave heating is a mature technology and development of microwave heating systems for biomass pyrolysis is of low cost. Wood wastes, sludge, slaughter wastes and MSWs have been tested with microwave pyrolysis (Aubin and Roy, 1980; Elliott, 1994; Diebold and Czernik, 1997).

The bio-oil produced may be refined to liquid fuels or converted to other chemicals. The product fraction ratio of bio-oil:solid char:syngas varies primarily with heating rate and biomass composition. The ratios for gasification, slow pyrolysis and fast pyrolysis are 5:10:85, 30:35:35 and 75:12:13, respectively. Biomass pyrolysis has not been broadly commercialized. Complexity and instability of bio-oil is the key barrier to the commercialization of biomass pyrolysis. There is an ongoing focused effort to stabilize bio-oils from biomass pyrolysis. Another new development in pyrolysis is catalytic pyrolysis in which catalysts are premixed with biomass feedstock prior to thermal treatment. By using carefully selected catalysts, the thermochemical degradation reactions are directed to produce bio-oil with desirable chemical profiles.

Chemical conversion

Biodiesel processing is a mature commercial technology. Biodiesel is essentially methyl esters of fatty acids, made through a chemical process called transesterification in which reactions between vegetable oil and alcohol (methanol or ethanol) are catalysed by alkali (KOH or NaOH). The reactions produce two products – methyl esters and glycerin. The glycerin as a by-product is removed from the methyl esters (biodiesel). Biodiesel can be used in compression-ignition (diesel) engines with little or no modifications. Biodiesel is simple to use, biodegradable, non-toxic and essentially free from sulfur and aromatics.

Biorefining

Biorefining is a concept derived from petroleum refining. A biorefinery uses biomass as feedstock as opposed to fossil resources used in a petroleum refinery. The goal of biorefining is to produce a wide range of products such as fuels, materials, chemicals, etc., from biological resources, much like those made from fossil resources. Because biomass is not a uniform feedstock, several biorefinery platforms such as biological platforms and

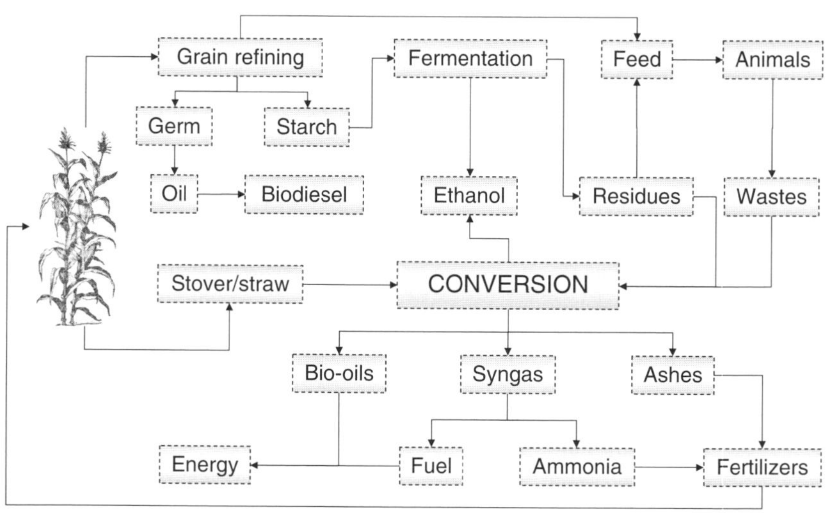

Fig. 2.7. Schematic diagram of a maize-based biorefinery.

thermochemical platforms have been proposed based on the unique characteristics of the biomass feedstocks used. A biorefinery uses a portfolio of conversion and refining technologies and may be integrated with biomass feedstock production. Figure 2.7 illustrates a biorefinery where the maize crop is fully utilized to produce fuels, biomaterials, feeds, foods, fertilizers, etc. An integrated biorefinery is capable of producing multiple product streams and thus multiple income streams from a single biomass feedstock and, therefore, is more economically viable than single product-based production schemes.

References

Aubin, H. and Roy, C. (1980) Study on corrosiveness of wood pyrolysis oils. *Fuel Science and Technology International* 8, 77–86.

Berndes, G., Azar, C., Kaberger, T. and Abrahamson, D. (2001) The feasibility of large-scale lignocellulose-based bioenergy production. *Biomass and Bioenergy* 20, 371–383.

BRDI (Biomass Research and Development Initiative) (2006) Vision for bioenergy and bio-based products in the US (http://www.climatevision.gov/sectors/electricpower/pdfs/bioenergy_vision.pdf, accessed 14 April 2009).

BRDR (Biomass Research and Development Board) (2008) Increasing feedstock production for biofuels: economic drivers, environmental implications, and the role of research (http://www.brdisolutions.com/default.aspx, accessed 16 April 2009).

BRIC (Biomass Resource Information Clearinghouse) (2000) What is biomass? Biomass Resource Information Clearinghouse. National Renewable Energy Laboratory, Golden, Colorado.

Diebold, J.P. and Czernik, S. (1997) Additives to lower and stabilize the viscosity of pyrolysis oils during storage. *Energy Fuels* 11, 1081–1091.

Edwards, R., Larivé, J.-F., Mahieu, V. and Rouveirolles, P. (2007) Well-to-wheels analysis of future automotive fuels and power trains in the European context, Version 2c. European Commission, Ispra, Italy.

EIA (Energy Information Administration) (2009) Annual Energy Outlook 2009 (http://www.eia.doe.gov/oiaf/aeo/, accessed 20 April 2009).

Elliott, D.C. (1994) Water, alkali, and char in flash pyrolysis oils. *Biomass Bioenergy* 7, 179–186.

English, B. (2003) The importance of user facility size and biomass feedstock distribution in determining economic viability. Bio-Energy Analysis Group, Department of Agricultural Economics, University of Tennessee, Knoxville, Tennessee.

European Commission (2006) The state and prospects of European energy research: comparison of commission, member and non-member states' R&D portfolios. EUR 22397 – The State and Prospects of European Energy Research. European Commission, Brussels, Belgium.

He, B., Zhang, Y., Yin, Y., Funk, T.L. and Riskowski, G.L. (2000) Conversion of swine manure: an alternative process of waste treatment and renewable energy production. *Transactions of the ASAE. American Society of Agricultural Engineers* 43, 1827–1833.

Hooper, R.J. and Li, J. (1996) Summary of the factors critical to the commercial application of bioenergy technologies. *Biomass and Bioenergy* 11, 469–474.

IEA-Bioenergy (2007) Benefits of bioenergy. IEA Bioenergy, Rotorua, New Zealand.

Kim, S. and Dale, B.E. (2004) Global potential bioethanol production from wasted crops and crop residues. *Biomass and Bioenergy* 26, 361–375.

Kyoung, S.R., Cantrell, K., Douglas, E. and Hunt, P.G. (2007) Catalytic wet gasification of municipal and animal wastes. *Industrial and Engineering Chemistry Research* 46, 8839–8845.

Perlack, R.D. and Turhollow, A.F. (2003) Feedstock cost analysis of corn stover residues for further processing. *Energy* 28, 1395–1403.

Perlack, R.D., Wright, L.L., Turhollow, A.F., Graham, R.L., Stokes, B.J. and Erbach, D.C. (2005) *Biomass as Feedstock for A Bioenergy and Bioproducts Industry: The Technical Feasibility of a Billion-Ton Annual Supply.* Oak Ridge National Laboratory, Oak Ridge, Tennessee.

Ruan, R., Chen, P., Hemmingsen, R., Morey, V. and Tiffany, D. (2008) Size matters: small distributed biomass energy production systems for economic viability. *International Journal of Agricultural and Biological Engineering* 1, 64–67.

Smeets, E.M.W. and Faaij, A.P.C. (2007) Bioenergy potentials from forestry in 2050. *Climatic Change* 82, 353–390.

Sokhansanj, S., Cushman, J. and Wright, L. (2003) Collection and delivery of feedstock biomass for fuel and power production. *Agricultural Engineering International: The CIGR Journal of Scientific Research and Development*, 5 (http://www.ecommons.cornell.edu/bitstream/1813/10360/1/Sokhansanj%20invited%202Feb2003.pdf, accessed March 12, 2009).

Wooley, R., Ruth, M., Glassner, D. and Sheehan, J. (1999) Process design and costing of bioethanol technology: a tool for determining the status and direction of research and development. *Biotechnology Progress* 15, 794–803.

Wyman, C.E. (2007) What is (and is not) vital to advancing cellulosic ethanol. *Trends in Biotechnology* 25, 153–157.

3

Heat and Power Generation by Gasification and Combustion

Ralph E.H. Sims

Introduction

This chapter examines the potential for industrial energy crops used as feedstocks to produce heat and power in a range of conversion plant technologies and scales. The potential for both crop residues and forest residues is included as their availability will affect the future demand for industrial crops grown for energy purposes. Some emphasis is placed on using forest biomass in heat and power plants, as there is relatively little experience or information available on the use of industrial crops.

Biomass Resources for Heat and Power Generation

Biomass continues to be the largest renewable energy contributor to global primary energy. It differs from other renewable energy resources in that it can be a substitute for all fossil fuel-based products by being converted into heat and electricity, as well as into liquid or gaseous fuels used for transport. Competing uses for non-food biomass resources also include biomaterials and biochemicals, as produced in biorefineries. For example, woodchips and pellets from Canada have been shipped to Scandinavia for use in the combined heat and power (CHP) plants common there.

Biomass currently contributes over 10% of primary energy to meet world annual demand. This is mainly in the form of traditional, non-commercial biomass such as scavenged fuelwood, dried animal dung and crop residues used on open fires and in crude, low-efficiency stoves to provide heat for basic cooking and household heating requirements. Sources of commercial biomass include crop residues, forest arisings and charcoal, livestock and human wastes, food and fibre processing wastes, municipal solid organic wastes and a range of dedicated energy crops including short-rotation forests, vegetative grasses, oilseed crops, sugarcane and cereals. This wide range of feedstocks can be processed into bioenergy carriers in the form of heat, electricity, solid fuels (such as woodchips, pellets, briquettes, logs), liquid biofuels (methanol, ethanol, butanol, esters, oils) and gaseous biofuels (synthesis gas, biomethane, hydrogen) suitable for transport applications. The solid, liquid or gaseous fuels can be combusted to provide the energy services desired, or stored for future use. The competing uses for non-food biomass resources also include biomaterials and biochemicals produced in complex biorefineries. Woody biomass and straw can be used initially to provide materials (for buildings, furniture, paper, etc.) and then recycled for energy purposes at the end of the product life.

Global trade in biomass in the easily transportable forms of pellets, woodchips and biofuels could help compensate for regional differences in the availability, accessibility and demand for biomass resources. Of the approximately 45 EJ (approximately 4500 Mt) of biomass consumed worldwide in 2006, around 7% was used to fuel combustion/steam turbine plants that generated 240 TWh of electricity at a conversion efficiency of around 20% on average (Fig. 3.1). Approximately 25% of this biomass was consumed in electricity-only generation plants. The remainder was used to generate electricity in CHP plants where the heat was either used on-site (as in pulp and paper manufacturing plants) or sold and taken off-site, often for use in district heating schemes to supply useful heat to buildings and industry throughout the locality.

Most of the data on current biomass demand are very uncertain. For example, the commercial use of charcoal and firewood in households in industrialized countries, and by industry in developing countries, has very limited data available but is possibly around 9 ± 6 EJ/year. Volumes of commercial solid biomass delivered to large CHP plants and district heating schemes are usually recorded and the heat and power produced metered. But at locations where the biomass is collected and used on-site, such as sugarcane bagasse, rice husks and wood process residues, there is often no measurement taken of the volumes used or of the heat and power produced. In small direct-heat plants in factories and buildings, metering is rarely used at all. The volumes and types of traditional biomass consumed by rural communities for cooking and heating in developing countries is even less certain, with data based on just a few surveys. So, the data shown in Fig. 3.1 are only indicative, with uncertainty ranges around ± 10–20%.

Projections of future biomass supply arising from industrial crops are also uncertain. These depend on assumptions of future land use competition, rate of continuing increase in crop yields, water availability, climate change impacts, etc. By how much traditional inefficient open fire combustion of biomass for heating and cooking will have been replaced by more modern biomass systems or other fuels (such as liquid dimethyl ether, DME) over the next few decades is impossible to predict. Competition for land will depend on world population, type of diet chosen, potential crop yield increases, uptake of genetically modified crops, water availability, soil nutrient conservation, deforestation protection, climate change effects of droughts and extreme weather events, etc. Estimates for modern bioenergy applications are therefore wide-ranging, since prediction is very difficult.

Present supply chains to transport, store, handle and process large volumes of biomass are costly and inefficient, so how these will develop in the future is also uncertain. Improved efficiency of energy conversion in

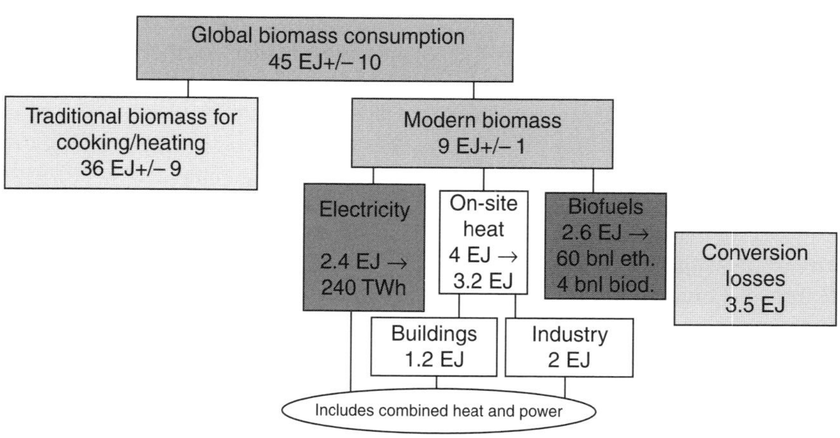

Fig. 3.1. Contribution of biomass to global primary and consumer energy supplies in 2006 including non-commercial traditional biomass used for cooking and heating in rural communities (based on IEA data).

mature bioenergy plants (such as combustion for heat or CHP using steam turbines or anaerobic digestion) can be expected, along with the greater uptake of gasification and pyrolysis if these processes become cheaper and more reliable. But how rapidly these improvements will occur is uncertain. The many co-benefits arising from well-managed modern bioenergy systems (including improved health, reduced air pollution, employment and support for rural communities) could create more rapid bioenergy deployment than predicted if they are integrated successfully into the policy development process. Competition for the limited biomass resource is already occurring in some countries, so exactly what shares of the available biomass will become available for the production of heat, power, CHP, transport biofuels, biomaterials, biochemicals, soil conditioning and maintenance of soil nutrients in future cannot be determined. It can only be said that, overall, there appears to be good potential for a significant increase in the share of total primary energy coming from biomass by 2050.

In the literature (see, for example, Hoogwijk *et al.*, 2003; Parikka, 2004; Haberl *et al.*, 2007), it is shown that by 2050 the technical potential for biomass supply could range between approximately 100 and 700 EJ/year, but the economic potential is likely to be much lower, though depending in part on future carbon and fossil fuel prices. The main reason for the differences in biomass potentials in the literature is that the two most crucial parameters in energy crop production, land availability and crop yield levels, are both very uncertain and therefore subject to widely different estimates, as are projections about the future availability of woody biomass forests and residues from agriculture and forestry, environmental protection requirements and sustainable management of soils and water reserves. The IEA *World Energy Outlook 2009* (IEA, 2009a) shows that by 2030 biomass will remain at around 10% of primary energy, taking into account population changes. However, much of this biomass resource will be used more efficiently than at present to provide more energy services. Analysis in the IPCC (2007) 4th Assessment Report showed there should be enough biomass supply available to meet the growing global demand for bioenergy products by 2030. It was assumed in the IEA Energy Technology Perspectives (2008) report that up to 180 EJ/year of primary biomass could become available for energy purposes in 2050.

Other projections show that by 2050 around 25% of world primary energy will come from biomass using available farm, forest and urban residues and by growing perennial energy crops, but these have more ambitious assumptions for future land use. Of the 13.2 billion ha (Gha) of the world's total land area, 1.5 Gha are currently used to produce arable crops and 3.5 Gha are under pasture for meat, milk and wool production from grazing animals. Crops grown specifically for biofuels currently use only 0.025 Gha. In Brazil, for example, where over 40% of total gasoline demand is provided by ethanol and more is exported, the ethanol is produced from sugarcane grown on just 1% of the 320 mega ha (Mha) of all arable and pasture land in that country. Some of the 1 Gha of marginal and degraded lands unsuitable for food production (such as from rising salinity levels on millions of hectares in Australia) could be reclaimed for productive use by growing selected energy crops.

Regional analysis by the VTT Technical Research Centre of Finland gave more moderate assessments than most other estimations. Agricultural and forestry residues and wastes were estimated to be the most available types of biomass (Table 3.1) (and hence probably the most cost-competitive, depending on collection and storage costs). The basic data used included agricultural, arable and permanent cropland areas, agricultural by-products (rice husks, bagasse, straw), forest biomass by-products from pulp and paper mills and the timber industry, municipal solid waste (MSW) and firewood. Their estimation was based on a techno-ecological analysis of biomass potential using current technology and taking future limited water resources into account. It was assumed that 30% of forest residues could be utilized as energy fuel. Around 30% of the total potential can arise from growing specialist energy crops that need a relatively small amount of the arable land now used to grow conventional food and fibre crops.

Table 3.1. Regional estimates of available biomass resources (EJ/year) in 2050, as analysed by VTT, Finland.

Region	Bark	Sawdust	Forest residues	Black liquor	Bagasse	Rice husk	Straw	Energy crops	Waste	Fuelwood	Total
Africa	0.05	0.05	0.14	–	0.67	0.01	2.80	3.50	0.18	4.96	*12.37*
Australia/NZ	0.04	0.01	0.10	–	0.27	–	–0.72	0.88	0.03	0.06	*2.10*
Canada	0.16	0.13	0.41	–	–	–	0.66	0.41	0.03	0.02	*1.83*
S/Central America	0.13	0.05	0.33	–	4.10	0.09	1.70	2.13	0.28	1.89	*10.69*
China	0.08	0.01	0.20	–	0.66	0.65	2.05	2.57	0.93	1.56	*8.72*
Eastern Europe	0.07	0.02	0.17	–	–	–	0.65	0.81	0.14	0.15	*2.02*
Former USSR	0.13	0.05	0.34	–	–	–	2.93	3.66	0.19	0.49	*7.77*
India	0.02	0.02	0.04	–	1.70	0.46	2.30	2.88	0.50	2.46	*10.38*
Japan	0.01	0.01	0.03	–	0.01	0.04	0.06	0.08	0.19	–	*0.44*
Middle East	0.01	0.01	0.03	–	–	–	0.80	0.50	0.16	0.05	*1.55*
Mexico	0.01	–	0.02	–	0.32	–	0.36	0.45	0.07	0.31	*1.53*
S Korea	–	–	–	–	–	0.03	0.02	0.03	0.05	0.02	*0.16*
USA	0.34	0.18	0.87	–	0.19	0.04	2.50	3.13	0.47	0.35	*8.07*
Western Europe	0.20	0.11	0.52	–	–	0.01	1.08	1.28	0.50	0.27	*3.98*
Other Asia	0.07	–	0.18	–	1.51	0.78	1.59	1.99	0.44	2.33	*8.89*
TOTAL	*1.31*	*0.66*	*3.39*	*2.33*[a]	*9.44*	*2.10*	*20.23*	*24.29*	*4.15*	*14.94*	*82.83*

[a]Not disaggregated into regions.

The analysis shows that all regions have some biomass resource growth potential, particularly Africa, South America, India and China, thus highlighting the opportunity for increased bioenergy uptake in developing countries. Other analyses show greater biomass contributions could come from energy crops, including from arid lands, although production costs, sustainable production methods, required land use change, competition for land for food and fibre production and possible impacts on yields from droughts and floods due to climate change all remain very uncertain.

The International Energy Agency's Bioenergy Implementing Agreement estimated 100–300 EJ/year of biomass from energy crops grown on present arable and pasture land could become available in 2050 without jeopardizing the future food and fibre supply (IEA Bioenergy, 2007a). Most of this could be produced for up to US$4/GJ, assuming perennial crops are grown. The remainder would come from biomass grown on marginal and degraded lands. Residues from agriculture could produce a further 15–70 EJ/year. However, the integration of food, fibre and energy production will depend partly on higher land use efficiencies, particularly in some developing countries where there are potentials for improvement.

Priorities for the use of limited biomass resources are difficult to predict with any degree of confidence. The amount of biomass available from both residues and energy crops by 2030 will be dependent on the efficiency of the world agricultural system. The future opportunity to increase the productivity of crops and the patterns of dietary change will effect the future competition for arable land. Average yields per hectare of agricultural crops have continued to increase since the 'green revolution' began in the 1960s. Yields of wheat and maize, for example, have increased by 250%/ha as a result of better management, breeding and selection of improved varieties and hybrids, irrigation and drainage, increased mechanization, improved storage methods and more intensive inputs of fertilizers and agri-chemicals. Rice and cotton yields have increased by 200%. This trend could continue by using better land management practices, having more control over chemical inputs using precision farming techniques, and with the wider deployment of genetically modified crops. Hence, the area of land needed for future food and fibre production could be less than projected.

Direct land use for biomass production in competition with food production, and indirect land use from the displacement of other crops to new areas, remains the subject of much debate. Agricultural commodity price fluctuations, for the most part, can be attributed to factors unrelated to the competing production of biomass. These are: increasing demands for food and animal fodder; speculation on international food markets; changing diets to more meat and milk products; poor harvests due to extreme weather events, including droughts and floods; and higher energy prices increasing the costs of production (including fertilizers), food processing and transport.

Low productivity in agriculture in many regions has resulted from unsustainable land use, erosion and loss of soils, deforestation and poverty. Increased productivity over time as a result of better farm management, new technologies, improved crop varieties, energy-related capital investment and capacity building will increase the intensity of land use gradually so that sufficient land could become available for the growing demands of food, fodder, fibre and fuel production. Needing to be accounted for in analyses is the fact that many crops used for energy also provide co-products and services such as protein, animal fodder and soil conservation benefits.

Commercial biomass markets could become a major factor in raising the economic viability of rural enterprises, especially in developing countries. Increased investment in infrastructure for biomass processing, distribution and transport would also result. At least some of this infrastructure will also contribute to the overall development of the agricultural sector. The deployment of new and improved technologies for biomass combustion at both the domestic and industrial scales should lead to more efficient fuel use and reduced health problems. Some countries such as Nepal are dependent on traditional biomass to meet up to 90% of their total energy demand. The efficiency of small stoves and fireplaces used for cooking could be improved significantly by introducing improved designs, which would also reduce the air emissions and improve the health of the users by avoiding smoke and carbon monoxide inhalation.

Bioenergy from agricultural biomass typically has been limited to local heat production from the combustion of straw, bagasse, rice husks, coconut shells, etc., although small to medium CHP plants are becoming increasingly common at sugar mills and other plants. In addition, biogas produced from animal manure, green crops and other forms of organic wastes can be used for heat and power generation (as well as for transport fuels if scrubbed to remove the CO_2 and H_2S contaminants). Residues can have significant economic potential, particularly where there are costs associated with their disposal, for example when generated at a forest or food processing plant.

Biomass may also provide farmers with additional income. In Denmark, Spain and Romania, cereal straw and other farm residues already play a significant role in bioenergy production, and in Brazil, Australia, South Africa and elsewhere, sugarcane bagasse is commonly used for power generation, both for use at the mill and, where the mill is located close to power lines, for export to the grid as a revenue earner. Vegetative grass crops such as *Miscanthus* and reed canarygrass can be grown for combustion in commercial grate boilers for heat production, and there are fluidized bed boilers in which straw and coal are co-fired. Small-scale power generation technologies ($< 500\,kW_e$) exist based on the steam cycle, but have relatively low efficiency with relatively high power production costs. In all cases, one key issue to resolve is how best to store the biomass in order to be able to operate the bioenergy plant for as long a season as possible in order to spread the investment costs.

The main barriers include fuel logistics, fuel quality fluctuations (due to variable rainfall or varieties for example) and price fluctuations due to varying interest by farmers to supply the plant and the economic variables for farmers and biomass companies to deliver the fuel. Baling, storage and transport technologies are available in various size classes; however, technical improvements in harvesting, storage, transport, fuel preparation and other measures are still needed for virtually all biomass feedstocks.

Different biomass resources suit different applications. Specific objectives for using biomass are affected by the quantity, quality and cost of available feedstocks; the relative location of the biomass resource in relation to its consumers; the conversion technologies available; the type and value of the energy services required; and the specific co-products and benefits offered. The following sections of this chapter outline the technologies available for using the biomass resource for heat and power generation at the larger scale. The supply chain issues are covered briefly and new concepts and designs of conversion plants are discussed. The text does not cover cooking stoves and other appliances used in the domestic sector, nor does it cover liquid biofuels.

Harvesting, Logistics and Pretreatment of Biomass

Most biomass feedstocks have a lower energy density on a per volume or per mass basis compared with equivalent fossil fuels. This makes their collection, transport, storage and handling more costly per unit of energy. Delivered costs to the bioenergy conversion plant can be minimized if the biomass can be sourced from a location where it is already concentrated, but this is not always possible.

Well-designed fuel supply chains are therefore critical to supply reliable and competitive biomass all year round, especially to large-scale heat and power plants. The biomass feedstock is bulky, deteriorates over time and is not easy to store and handle. Experiences from traditional agriculture and forest harvesting and transport methods can be utilized, and the integration of conventional agricultural and silvicultural operations with the procurement of the biomass can help to decrease costs. Harvesting of biomass with low specific density, high moisture content and modest storage properties often requires new types of machinery, logistics and procurement practices to be developed.

In a fuel chain supplying just one large bioenergy plant, hundreds of farmers and forest owners and numerous machinery contractors could be involved in the supply chain. At biomass feedstock production sites, pretreatment of the biomass can help eventually to provide a homogeneous product delivered to the conversion plant with a narrow particle size range and uniform moisture content.

Procuring the biomass usually consists of a sequence of individual operations performed to harvest and process feedstocks for delivery from source to conversion plant. Every link in the production chain has to be optimized to improve profitability without compromising the quality or security of supply. Typical phases are harvesting, baling or shredding, off-road transport to roadside, measurement, transport from roadside to intermediate store or conversion plant, and receiving, handling and storage at the plant. Control and management of the process system therefore needs good organization and smart logistical analysis.

For short-rotation forest crops, the position of the chipper or shredder in the supply chain largely determines the state of biomass during transport and, consequently, whether subsequent machines are dependent on each other. Chipping may take place in the field at a central site, at the roadside, at an intermediate terminal or at the plant where the chips are to be finally used. For vegetative grasses, the chipping would be replaced by baling operations.

- Chipping at a central landing site is the traditional option. The biomass can be taken by the harvester to the landing, where it can be stacked into storage piles perhaps 4–5 m high, ready for the chipping/shredding operation to be performed.
- Alternatively, chipping in the field can be more costly as it requires a highly mobile chipper suitable for cross-country operations and often equipped with a 15–20 m^3 chip tipping container.
- Chipping at the plant or intermediate terminal enables the chipper to become independent of the chip trucks. The technical and operative availability of the equipment increases, control of the procurement process is facilitated, demand for labour is decreased and the control of fuel quality is improved. However, since the investment cost is high, only large plants can afford a stationary chipper/crusher. The low bulk density of the unchipped biomass is often the weak link in this system, because truck transportation takes place in the form of relatively loose residues.
- Baling or bundling the material as an intermediate step, rather than chipping on-site, makes the system less vulnerable, with waiting times between machines eliminated, storage options facilitated and the entire system easier to control. Due to the costs, however, current baling technology is only suitable for large-scale operations with a suitable shredder sited at the plant. Hence, it remains more common for forest biomass to arrive at a conversion plant as woodchips and short-rotation crops as bundles or loose billets.

In recent years, a new truck transportation system of baled forest residues up to around 100 mm maximum diameter has been developed. In this Timberjack/John Deere process, logging residues are collected from the forest after stemwood extraction or whole stems of short-rotation crops cut. These are then compressed and tied into 70 cm diameter, 3.2 m long bales or composite residue 'logs'. A single bale of green residues at around 50% moisture content (wet basis) weighs around 500 kg and has an energy content of about 3.6 GJ. Bales are transported to the roadside using a conventional forwarder and then taken to the plant using a conventional truck, or over longer distances by train. With many indirect cost savings and better supply security, bundling has proved to be an effective solution for large-scale woodchip production in Finland.

Costs of delivered biomass vary with country and region due to variations in crop yields, terrain, labour costs and machinery costs (Table 3.2). The costs of delivered forest biomass in Sweden, as an example, are around €4/GJ, with forwarding 18% of the total, chipping 33%, transport 27% and the stumpage fee paid to the forest owner accounting for

Table 3.2. Variation in the delivered costs of woody biomass in different European countries relate to the wide cost fluctuations for each operation (€/hour) due to varying machinery and labour charges. Data from Asikainen *et al.*, 2007.

Country	Forwarder €/h	Chipper €/h	Transport €/h	Loading/unloading €/h
Czech Republic	56	124	77	35
Finland	75	129	106	63
France	84	155	111	67
Hungary	55	124	77	34
Poland	54	120	74	33
Slovakia	55	126	77	32
Spain	65	128	86	47
UK	82	172	116	60

most of the remainder (Junginger *et al.*, 2005). The delivered cost has declined by around a third over the past two decades, due to lowered costs in all activities because of increased field experience, better system management and improved machinery designs.

The state of feedstock production development for forest residues from different countries differs much more than for agricultural residues and crops. Lack of suitable and cost-effective harvesting technology exists in many countries; for example, when harvesting forest residues in steep terrain. In some places on rugged terrain unsuitable for machinery, cable haulers are used to extract whole trees after manual felling to a landing where the stemwood is separated and the remaining arisings are available for chipping on-site or bundling.

Non-technical barriers can sometimes become a barrier in efficient biomass production and handling. For example, increasing the availability of low cost biomass in the short term may involve policies to subsidize desired land use change by encouraging farmers to grow new crops. In the longer term, it could result from investment in research to obtain higher crop yields with lower inputs by developing genetically modified crops.

For energy crops, where an average crop yield of around 10–12 t dry matter/ha/year is possible, an energy yield would be typically 180–220 GJ/ha. There are often significant differences between the potential crop yields measured from small-scale experimental plots and the yields actually reached in practice at the commercial scale. This reflects the various climatic conditions, soil types, average grower competence, losses at harvest, storage and transport, etc.

Harvesting methods depend on the crop type. If existing farm harvesting equipment can be used, the cost is far cheaper than if specialist harvesters will need to be developed (as in the case of short-rotation coppice *Salix*). For vegetative grass-type energy crops, harvest methods to give low losses and high bale densities are essential to reduce the overall costs. Long-distance transport of reed canarygrass, for example, can form a significant share of production and delivery costs for, say, a large commercial processing plant of 400,000 t/year. This would require feedstock brought in from 100 km radius or more to ensure operation all year round and 24 h a day. At present, the full payload capacity of a truck usually cannot be met due to the light bulk matter of the relatively dry material. Denser bales are being developed but, ideally, their production needs to avoid higher energy inputs. The interaction between truck payload, truck volume and the moisture content of the biomass affects the cost per GJ of energy delivered by the biomass (Fig. 3.2). More detailed analysis of the biomass supply chain can be found in other publications (see, for example, Sims, 2002; IEA, 2007a).

Bioenergy Heat and Power Combustion Technologies

A wide range of technologies exist for the combustion of biomass to provide useful heat that can also be utilized for power generation.

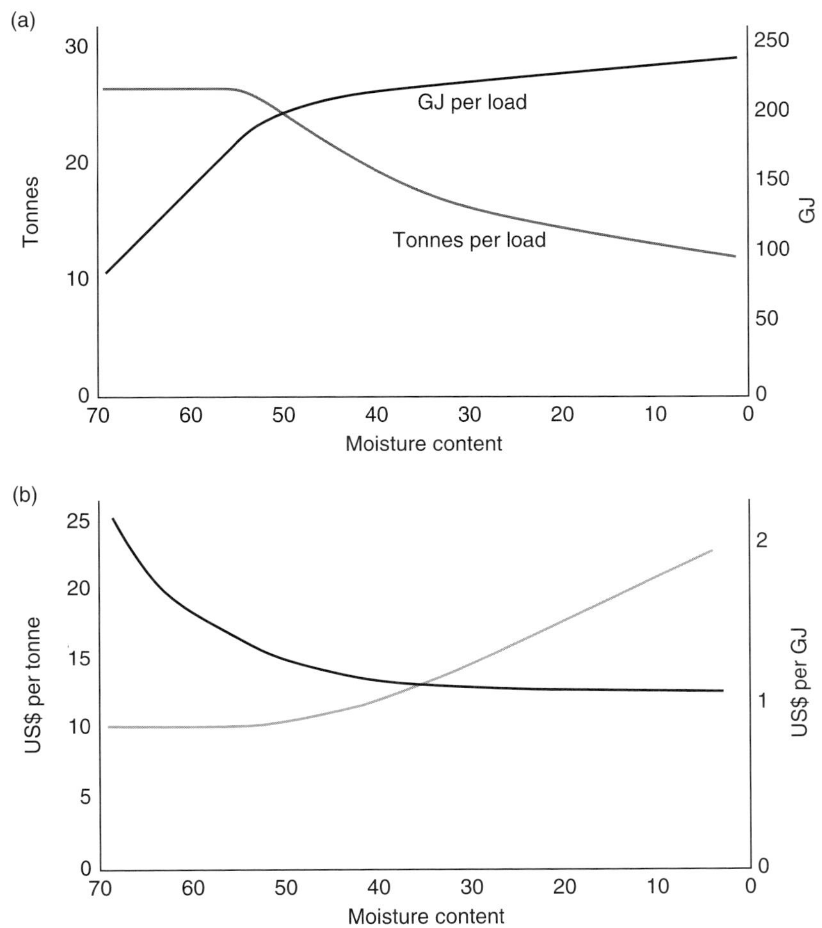

Fig. 3.2. The interaction between biomass moisture content and transport costs is exemplified for a 36 m^3 truck with a 26 t payload that is weight-limited when carrying wet biomass but volume-limited for drier loads (a). More energy is carried when the loads are drier. However, the cost per tonne carried increases for dry biomass loads (b), but the more important cost per GJ of energy delivered is optimum when the biomass is around 30–40% moisture content, wet basis.

The interactions are varied and complex. Traditionally, the combustion of dry biomass has been used to produce heat. Some of this heat, in turn, has been used to raise steam and power steam turbines to generate electricity. The more efficient gasification of solid biomass to produce power from gas engines or gas turbines has been achieved at both small and large scales – although with varying degrees of success. Pyrolysis to produce transportable crude bio-oils in order to be able to separate the biomass source from the energy demand remains under development. Anaerobic digestion using bacteria to produce methane biogas that can then be combusted to produce heat or power is an alternative conversion process, particularly useful for biomass with high moisture content. In most cases, the combined production of useful heat and power (CHP) is a more efficient overall process.

The characteristics of the biomass fuel supplied need to match the type, size and design of burner. For example, woodchips need not be pulverized or dried before firing into fluidized bed boilers. It is sufficient to crush the biomass fuel mechanically to facilitate its feeding into the combustion bed. For gasifiers, however, a uniform particle size with low moisture content is critical. Fluidized bed combustors (see below) can tolerate quite a wide particle size distribution and high moisture content because of the stabilizing effect of the bed, but some fuels with high alkali contents can cause the bed particles to aggregate, leading to reduced performance. Particle size of the biomass feedstock is limited for rotary fuel feeders. When preprocessing and handling reactive fuels, the risks of explosive dust formation and fires are normally reduced if the fuel moisture content is above 40%.

Dry woody biomass, peat and bark are well-understood fuels, having been combusted for heat for decades. Their net heat value varies between 8 and 14 MJ/kg of fuel, depending on moisture content and, to a lesser degree, the plant species. Many standard designs of burners/boilers of varying capacities exist in the market. Some challenges remain with designing burners for some woody biomass fuels in relation to their varying moisture contents and high alkali levels, which can cause problems in the burner. Woody biomass, vegetative grasses and short-rotation forest crops tend to be low in potassium, sodium, calcium and sulfur, which can cause slagging, fouling and corrosion in the boiler. Crop residues are acceptable feedstocks if they are reasonably dry, free of soil and relatively uniform. However, wheat and rice straw, rice husks, sorghum and bagasse can all have high mineral contents, leading to the need for more careful plant operation and higher levels of maintenance. High chlorine content in some feedstocks can also lead to higher dioxin levels. Demolition wood can have additional problems if it has been treated with preservative chemicals, and some processed wood products incorporating resins and glues can produce other problems.

The organic fractions of MSW and refuse-derived fuels (RDF) tend to have a lower net heat value (5–10 MJ/kg), which can give many challenges to the burner designer and operator in order to minimize and monitor any toxic emissions and contain them below acceptable low levels. Similarly, ash disposal needs to be undertaken in an acceptable manner due to the presence of heavy metals. Wet biomass fuels and green crops may be preferable as feedstock for anaerobic digestion to produce biogas.

Overall, the potential for bioenergy to make a significantly large contribution to the energy supply mix is subject to the sustainable production of biomass, improved efficiency of the supply chain, successful development and deployment of new thermochemical technologies and improved biochemical conversions such as anaerobic digestion. Most of the biomass fuel will come from crop and forest residues including industrial black liquor, with the remainder from industrial crops purpose-grown for energy. Increased biomass conversion will likely occur in systems that co-fire the biomass with gas or coal, or in CHP plants where more of the heat is utilized. More efficient integrated gasification, combined cycle (IGCC) plants and pyrolysis plants are still under the early stages of development, the latter having the potential to separate the biomass production and utilization locations by making transport of the energy-dense bio-oil more competitive. The potential development and deployment of such immature technologies is difficult to predict (IEA, 2008).

Bioenergy heat generation

The heat services required by people and processes in homes, commercial buildings and factories make up 40–50% of world final energy demand. Approximately 400 GW of modern biomass-fired heating equipment is currently in operation worldwide, consuming around 600 Mt of biomass/year to provide around 40 EJ of useful heat (assuming a 70% conversion efficiency).

Little commercial data exist for domestic heating from biomass. Even if the capacity of an installed burner is known (e.g. 10 kW), the amount of biomass consumed and heat produced depends on whether it is used for 100, 1000 or 8000 h/year. However, it is thought that in energy terms, bioenergy used annually for direct heating (excluding traditional use of biomass) is two to three times all the renewable electricity generated (excluding that from large hydropower plants) and two to three times all the liquid biofuels production for transport (IEA, 2007b). The current heating share from modern biomass is 2–3% of total heat demand. When traditional biomass is included, the contribution is probably around 10%, although global heat data statistics are very uncertain (see the section headed 'Biomass Resources for Heat and Power Generation'). Figure 3.3 shows the current and projected future of biomass used for direct heating.

Commercial solid biomass combusted in boilers and stoves to provide space and process heat in the building and industry sectors includes district heat and CHP plants in Scandinavia and elsewhere. For example, Sweden supplied over 60% of its district heat using a range of biomass fuels in 2006. Biomass is projected to increase its share of global primary energy through increased CHP installations in industrial facilities and increased growth demand for the heating of buildings. In developing countries, increased use of modern biomass technologies is also projected, leading to more efficient and sustainable use of the local biomass resource.

The costs of bioenergy heating technologies vary according to local conditions. Where the biomass is abundant and conventional energy sources are expensive, they can be cost-competitive with conventional fossil fuel heating systems and therefore require few supporting policies (IEA, 2007b). Bioheat is a relatively cheap option in China, for example, particularly in rural areas where gas and electricity are not widely reticulated. There is potential for the installed costs of

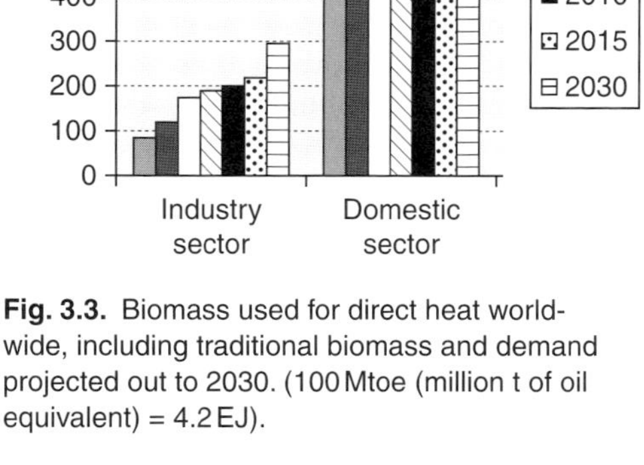

Fig. 3.3. Biomass used for direct heat worldwide, including traditional biomass and demand projected out to 2030. (100 Mtoe (million t of oil equivalent) = 4.2 EJ).

bioenergy heating technologies to be reduced considerably due to further learning experience, training, mass production and improved performance efficiency.

Some small-scale boiler designs enable several types of biomass including straw, woodchips and organic solid waste to be used as fuels. Typical capacities of small-scale combustion technologies are 1–30 MW for mechanical grates, 2–10 MW for a fluidized bed and 7–20 MW for a circulating fluidized bed. Use of cereal straw as feedstock for small farm-scale heat and CHP plants started more than 25 years ago in Denmark, with the current annual volume combusted being around 0.5 Mt and representing about 15% of the total amount of straw produced. A further 0.8 Mt/year is used in straw-fired district heating plants ranging from 0.6 to 9 MW capacity with a typical maximum boiler temperature of 120°C and a maximum pressure of 6 bar, giving conversion efficiencies up to 85%.

At the farm-scale, batch-fired boilers are always installed in combination with a storage tank that can absorb the heat energy from one firing of 1–4 large round or square straw bales. The energy content of the straw is better utilized because the boiler can operate at full load. Boiler efficiency has increased from 35 to 40% in 1980 to over 80% today, due to better control of the air supply. Continuous boilers are fed by a conveyor that is loaded with straw bales once or twice a day, which are then shredded and fed into the boiler automatically at a rate linked with the varying heat demand. Danish straw-fired CHP plants range from 2 to 30 MW_e and 7 to 70 MW_{th}, giving overall efficiencies of 85–90%. Similar technologies would suit feedstocks from vegetative grasses.

Woody biomass in the form of logs or pellets is commonly used for heating houses and small industries, though poorly designed or maintained biomass heating systems, and the use of wet biomass, can cause significant amounts of particulates and other impurities to be emitted up the flue. In traditional heat-retaining fireplaces in Scandinavia, some of the heat is stored in the stone structure for later even release into the room over a longer period of time, giving an overall efficiency of about 80–85%. Enclosed stoves of 'double burning' designs can provide heat quickly and with high outputs. Initially, the log, once fed into the stove, gasifies over the grate, with the gas flowing to the after-burning space under or over the grate, where it burns efficiently at high temperatures (up to 1000°C). The efficiency is usually between 70 and 80% and a water heating system can be incorporated.

Pellet stoves are fed automatically and are growing in popularity. Electricity input is needed for the combustion air fans, hot air circulation and fuel feed system. A compact unit has a small, inbuilt chamber for the pellets that is filled automatically by a screw auger from a large external pellet storage tank. This is filled as needed by a delivery tanker truck using a pneumatic hose. A small dwelling in a cold region needs about 4 t pellets/year, so a store of around 8 m^3 would hold a year's supply. A smaller store with more frequent fills may be less costly to construct, although the pellets can be cheaper if bought in larger quantities.

Pellet production costs in Austria and Sweden were investigated for nine plants with production capacities ranging between 430 and 79,000 t/year (Thek and Obernberger, 2004). Production costs in Austria, including capital investment and plant maintenance costs, were around €10/GJ, or €6/GJ if dry biomass was available and drying could be avoided. In Sweden, production costs were slightly lower, though it was not clear why.

Since compressed pellets can only be manufactured using dry sawdust or straw at < 15% moisture content (wet basis), the combustion equipment is clean and gives around 80% efficiency as it is designed for a homogeneous and dry fuel with high energy value and low ash content in stable conditions. Since the pellets are dry with a relatively high energy value, it takes little time for the burner to reach a high temperature in the combustion zone as relatively little moisture has to be driven off. A temperature of around 500°C can be retained easily throughout the whole combustion process. As the reaction time is short during down-combustion, it is possible to let the pellet burner function on an 'on/off' basis as for an oil burner, but without increasing the emissions. Maintenance is required to clean the burner, remove the ash and sweep the boiler, taking about 10–15 min every 1–2 weeks.

Bioenergy power generation

Over 40 GW of biomass-fired power generation capacity is installed worldwide, which in 2006, assuming a capacity factor of around 55%, generated around 240 TWh of electricity. The plants with an average conversion efficiency of 25% consumed around 250 Mt of biomass. Typically, 1 ha of fertile land in the temperate regions could support enough sustainably grown biomass to generate around 5–10 MWh/year of electricity. Virtually all this biomass could come from crop and forest residues, but if it were to come from energy crops, approximately 25 Mha of land would be needed.

Combustion is the common conversion system, with gasification remaining largely at the demonstration stage. Designs are basically grate, fluidized bed or circulating fluidized bed designs.

Grate boiler combustion

The simplest and most common boiler combustion design is where the biomass is fired on to a fixed or moving grate. The fuel system aims to feed the fuel evenly over the whole grate width, as there is little lateral movement, even if the grate is sloping or mechanically reciprocating. If the fuel is not distributed evenly, some of the primary air blown up from below will escape through the grate where the air resistance is lowest. For grate boilers above 3 MW_e, fuel feeding is usually by gravity, whereas in underfeed grates the fuel is fed by a screw on to the middle of the grate, from where it spreads to all parts. Maximum capacity of grate boilers is usually below 30 MW_e.

Combustion air is typically fed in two or three phases. Primary air is fed from under the grate, with secondary and tertiary air above the grate to combust the volatile combustible compounds. In addition, efficient combustion to produce low NO_x emissions requires a sophisticated secondary air system and special furnace design with two combustion zones. Fuel feed systems have to be designed so that uncontrolled air cannot return back through to the boiler to avoid the risk of fire in unstable conditions if a back-fire gets from the boiler to the fuel feed system.

The heat output of a grate boiler varies by the fuel and grate design, but is typically between 300 and 1000 kW/m^2. The anticipated moisture content of the available biomass fuel affects the output, and consequently the boiler design and grate dimensions. The largest part of the grate is used for the drying of the biomass particles after they enter. The drying time, and hence heat output capacity, can be shortened using preheated air, by adjusting the geometry of the boiler or by decreasing the individual particle size of the fuel to increase the evaporation area. After entry, the fuel is dried initially on the grate, then pyrolysed by driving off the gaseous volatile compounds. Gasification of the volatile compounds begins at 150–200°C to give carbon monoxide, hydrogen and different hydrocarbons. It reaches a peak between 250 and 500°C, when a large amount of tar and volatile compounds are also formed. The pyrolysis process ends at about 800°C after producing combustible and liquid tar compounds, which burn well in the flame when sufficient oxygen is available.

The remaining solid char is burned from the fuel surface without a flame when a high enough temperature is reached and sufficient oxygen is available. Char combustion is relatively slow and requires more grate area than pyrolysis, but the time can be reduced by decreasing the original particle size of the fuel or by increasing the temperature by heating the air input. This could cause problems with ash melting and therefore affect the reliability and maintenance of the plant.

The variable melting temperature of the ash depends on the biomass fuel used. With woody biomass, the ash melts at over 1400°C. Ash melting problems can be decreased by using mechanically water-cooled grates linked with the water circulation of the boiler or by using cold air in the final combustion zone. Most of the ash will leave the system as grate ash, which can be recycled on to nearby land to return some of the nutrients and trace elements in the biomass.

Grate boiler costs per kW of electricity capacity vary widely with size, design and location, but are relatively costly for power generation alone. For example, one 1.14 MW$_e$ plant in Austria built in 1995 then upgraded in 2000 has been quoted as having a relatively high total capital cost of €4.5 M. Therefore, it is more usual to invest in CHP plants where the heat can also be utilized (see below).

Typical conversion efficiencies for combustion grate CHP plants are around 90%, but over 100% efficiency can be calculated for heating plants with a flue gas condenser because the lower heating value of the biomass is used to calculate the fuel input. The moisture in the fuel lowers the heating value because part of the energy content is used to evaporate the moisture. The latent energy stored in this water vapour can be recovered in a flue gas condenser and therefore is not lost in the process. If in theory the biomass fuel were to be oven dried to remove all moisture, then the efficiency would be less than 100%.

Fluidized bed combustion

Bubbling fluidized bed (BFB) and circulating fluidized bed (CFB) combustion technologies first became commercially viable in the 1970s. Today, many new solid fuel-fired boilers with a total capacity of over 5 MW are fluidized bed boiler designs, particularly in Scandinavia. Elsewhere, grate designs are preferred, particularly where pellets, straw, wood process residues and MSW are used as fuel. However, since fluidized bed boilers have several benefits when using biomass as fuel, special materials and construction can be undertaken for straw, bagasse and rice husk fuels. These have all been combusted successfully. The temperature of a fluidized bed is typically 800–900°C; the lower limit is due to the combustion reactivity of the biomass fuels and the upper limit is to avoid fuel ash sintering.

BFB combustion is similar to grate combustion, but the burners are especially suitable for variable quality biomass fuels since they have better temperature control because the material of the bed boosts mixing and hence heat transfer. Typically, they consist of a 0.5–1.5 m deep bed of sand, ash, fuel, dolomite and limestone of typical particle size distribution between 0.5–1.5 mm. This dense bed of about 1000 kg/m^3 sits on a porous distribution plate through which air is blown at a velocity of about 1 m/s, sufficient to fluidize the bed. The particles remain in the bed, although some small particles can be carried out with the fluidizing gas flow and larger particles sink on to the distribution plate.

CFB combustion resembles pulverized fuel combustion, but the better temperature control of the furnace ensures ignition of the biomass fuel without the need for a high temperature flame. In CFBs, the bed material particle size is smaller than BFBs at 0.1–0.6 mm and the fluidizing velocity is higher at 4–6 m/s. This changes the fluidizing conditions so that the bed particles are blown up with the fluidizing gas above the bed. Part of the bed material is carried out through the furnace to the second pass of the boiler. These particles are separated from the flue gas by a cyclone, electrostatic precipitator, fabric filter or U beams, then circulated back to the bed after separation and recycling of any unburned biomass particles. The high share of circulating inert material helps smooth the temperature profile throughout the furnace.

The boiler tubes in the bubbling bed area need to be protected by fireproof ceramic coatings to avoid the heat transfer surfaces corroding and eroding. Vertical heat transfer surfaces located in the oxidizing zone, such as membrane walls of the furnace, are less prone to wear.

The biomass particles dry and pyrolyse instantly when the fuel is fed into the hot bed. Around 30–40% of combustion air is used as the fluidizing air and the rest is utilized for the combustion of the pyrolysis gases above the bubbling bed. Most of the small fuel particles burn in this region, where the combustion temperatures can be up to 1100–1200°C. Other gas, oil and coal burners can be located here and operated simultaneously with the bed as co-firing systems.

The selection between grate and fluidized bed boilers depends on the quality of the ash and the share of physical impurities in the biomass fuel. Fuels with high alkali content and a low ash melting point cannot be burned in a fluidized bed because the fluidization will be disturbed. Heavy physical impurities such as metal particles in MSW fuels cannot be fluidized, so they sink on to the air distribution plate, disturb the fluidization and are difficult to remove from the furnace. However, new solutions for keeping the bed operational with MSW fuels have been developed (IEA, 2009b).

Bioenergy combined heat and power generation

Only around 25–45% of the chemical energy in the biomass fuel (measured as the lower heating value) can be converted into electricity at a stand-alone bioenergy power plant. The rest is lost as low-temperature waste heat into the air and cooling water. One of the main options for increasing the efficiency of power generation and the competitiveness of bioenergy has been the trend away from building separate plants to provide heat and power towards CHP production.

Where there is heat demand for hot water or process steam, CHP is the most profitable choice for power production using biomass. Pulp and paper mills, sawmills, sugarcane plants, rice mills, etc. traditionally have used most of their biomass residues for steam and electricity production on-site. Municipality-owned plants generally use the heat for district heating. The capital investment is normally higher to build a CHP plant than separate power and heating plants, but it is cheaper to operate as less fuel is required (Fig. 3.4).

Cogeneration plants have a typical overall efficiency of 80–90%, over double the efficiency of condensing power plants that generate electricity only and greater than the 60–65% combined efficiency achieved when linked with a separate heat plant. Therefore, the heat and power production costs of CHP are lower. The owner of a CHP plant may consume all the power and heat on-site, or sell all or part to other customers. The return on investment depends on local heat demand, the period of demand for the heat load, competing grid electricity tariffs, the purchase price of available biomass and the capital investment costs.

Grate firing based on various designs was the dominant technology until the 1980s, when fluidized bed technology became popular since it allowed larger variations of fuel particle size, moisture content and heat values. It can use higher steam values to give power–heat ratios

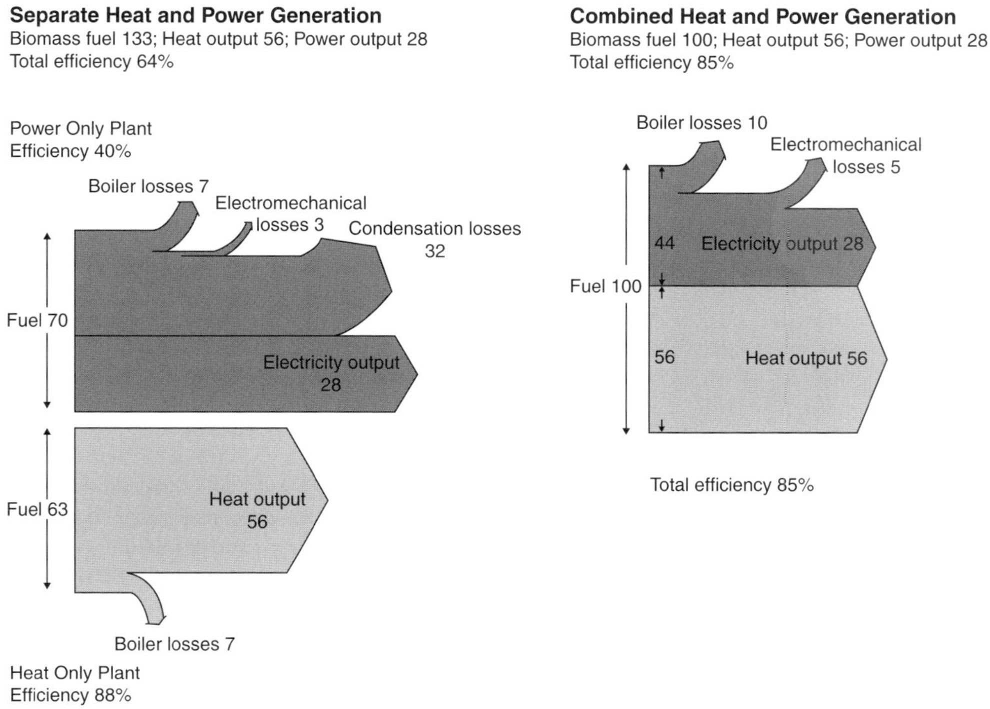

Fig. 3.4. Primary biomass energy consumption in a CHP generation plant compared with generating the same amount of heat and power in separate plants (based on Alakangas and Flyktman, 2001).

up to 0.5. The technology is also feasible for co-firing of biomass with recycled fuels and coal. Emission limits can be controlled for reasonable costs, but handling of the ash can be a problem in co-firing applications. Savolainen *et al.* (2001) analysed ten Swedish biomass CHP plants, five fluidized bed and one grate used for industry applications and four fluidized beds with the heat used for district heating. The plant capacities ranged between 1 and 60 MW_e, the efficiencies were between 84 and 88% and economies of scale for the investment costs were evident (Fig. 3.5).

Technically, all condensing power plant designs can be modified for cogeneration of heat and power, but this is only warranted where there is a useful application for the heat. The rejected heat can be utilized by raising the temperature of circulating water to 120–200°C for industrial processes or to 70–120°C for space heating where there is a large enough district heating system, but at the cost of reduced power generation. In all cases, cogeneration saves fuel in comparison to the separate generation of heat and power and is normally a good investment where the heat can be used or sold.

In Finland, biomass fuels are used almost entirely for heat. The country has nearly 100 CHP plants with a total capacity over 1500 MW_e. Most pulp mills using unbarked logs have installed CHP plants. The Alholmens Kraft CHP plant in Pietarsaari is the largest biomass-fuelled power plant in the world, producing 100 MW_{th} process steam for use in the adjacent paper mill, 240 MW_e for use both on-site and sale to a utility and 60 MW_{th} heat output for district heating.

The improvement of power–heat ratios in biomass CHP plants is based on the increase in the superheated steam temperature and pressure of steam boilers. The largest plants are designed for 165 bar/545°C, but supercritical values will be introduced in future fluidized bed boiler designs. The biomass fuel consumption of large CHP plants (possibly up to 500 MW_{e+th}) could be up to 3000 t/day, requiring 100 truck deliveries. This scale could therefore benefit from the co-firing of fossil fuels

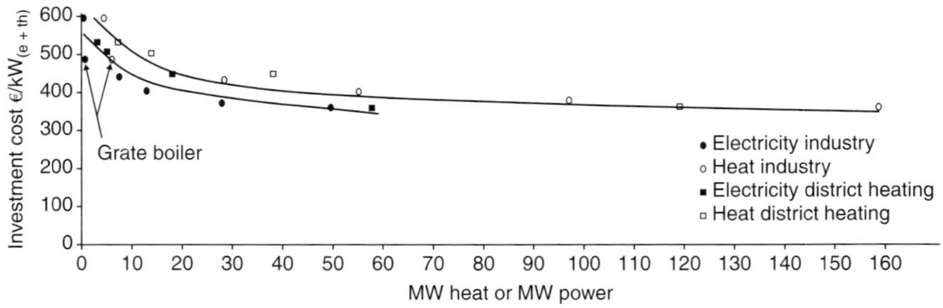

Fig. 3.5. Specific investment costs for woody biomass CHP plants in Sweden shown for separate heat and power outputs for one grate-fired and five fluidized bed plants owned by industry and four fluidized bed plants with the heat used in district heating schemes, showing economies of scale for both heat and electricity* (based on Savolainen *et al.*, 2001). *Each CHP plant is evident by the horizontal pairs of solid and open circles (industry) and squares (district heat).

due to the large transport distances involved to collect sufficient biomass. CFB combustion is optimal for co-firing, with good possibilities of reaching low sulfur and nitrogen oxide emissions without flue gas cleaning.

The capital costs of eight grate boilers, eight BFB and five CFB biomass CHP plants located in Sweden, Finland, Denmark, Austria, Germany and the Netherlands up to 30 years old were analysed by VTT in Finland. A correlation exists between plant capacity and capital investment costs (Fig. 3.6). Grate boilers built between 1994 and 2004 tended to be smaller than BFB plants (originally built between 1969 and 2002 but with the older plants reconstructed in the mid-1990s) and CFB plants (built between 1990 and 2002). Costs varied with heat capacity, power capacity and design. A strong relationship for economies of scale is evident when comparing specific costs for both heat and power capacity in terms of €/kW_{e+th} (Fig. 3.7). The higher costs of the smaller-scale grate boilers are part of the reason for fluidized bed plants gaining in popularity in recent years.

Small-scale CHP technologies below 10 MW_e are commonly designed around combustion/steam boilers, but other options such as gasification/combustion/steam, gasification/internal combustion engine, pyrolysis bio-oil/diesel engine, heat use with Stirling engines, heat use in an organic Rankine cycle and landfill or biogas production/gas engines have all been demonstrated (IEA, 2008). Heating systems can now be controlled automatically to decrease operation and maintenance inputs and reduce emissions. Highest efficiencies and lowest emissions are reached when burning homogeneous and dry fuels (such as sawdust pellets or woodchips) and by continuous operation of equipment to meet constant heat loads. In the future, flue gas cleaning with particle separation and catalysts will be equipped in smaller units to reach air quality standards and maximum particulate levels, especially when operated in densely populated areas. They will also be able to follow the heat load.

Small CHP plants < 5 MW_e using solid biomass are most often grate-fired designs, with steam pressure around 50–60 bar and low power–heat ratios typically at around 0.2. Small fluidized bed CHP steam boilers tend to be employed to give higher power–heat ratios around 0.3–0.35 under full load. In plants of less than 1 MW_e, the steam could be produced in a grate boiler and electricity generated from a steam engine. The power–heat ratio is lower, but specific investment costs can also be lower at around €500/kW_{th}. Future efforts should be put into the development of small-scale (1–5 MW_e) plants with lower specific investment costs and better power–heat ratios.

Co-firing

Biomass mixed in with coal or natural gas as a dual fuel to displace some CO_2 from the

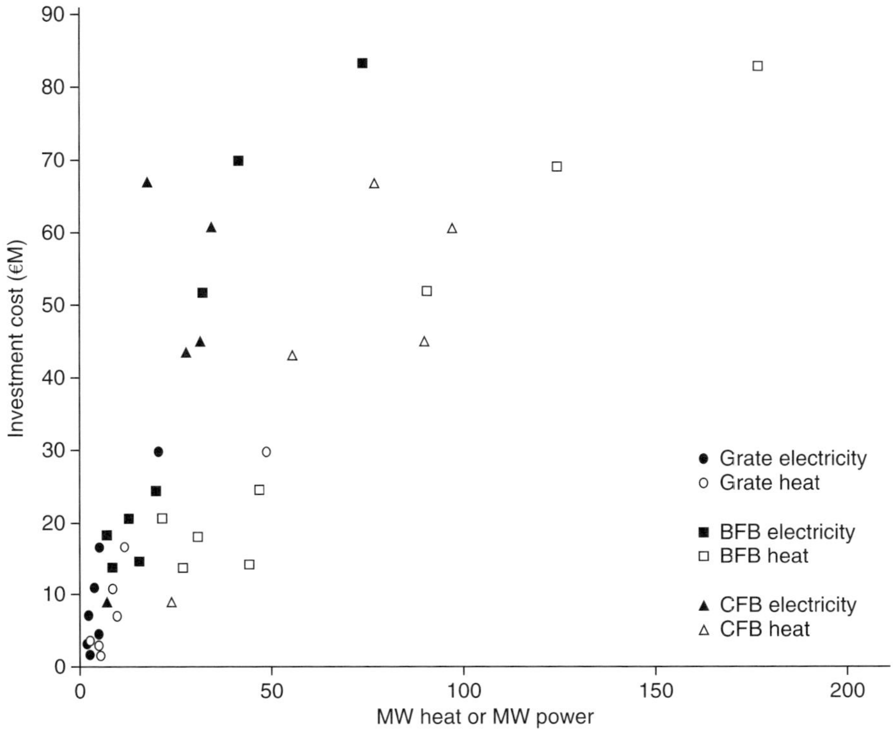

Fig. 3.6. Capital investment costs of several existing biomass-fired, European combustion grate boiler, bubbling fluidized bed (BFB) and circulating fluidized bed (CFB) plants for CHP generation correlated with the plant heat or power capacity (based on VTT analysis in an unpublished 2007 report to the IEA).

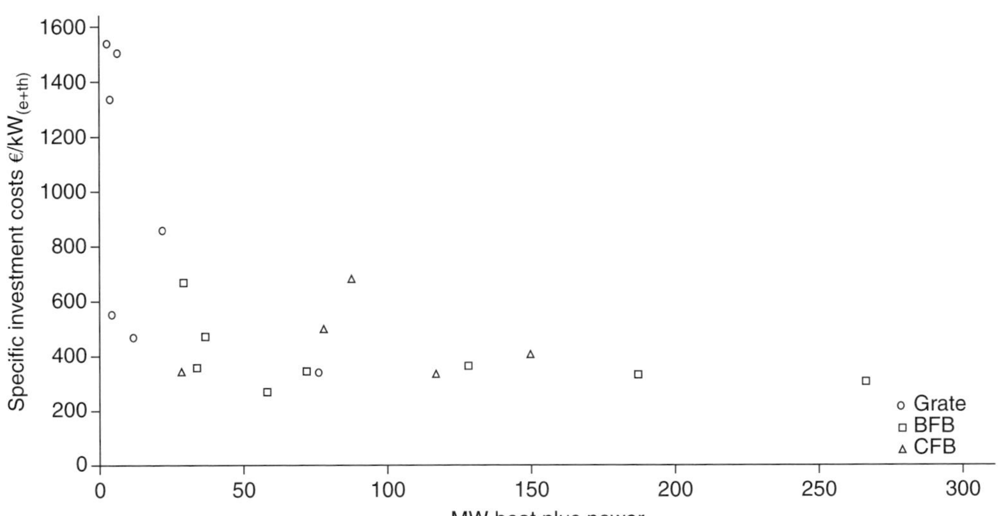

Fig. 3.7. Specific capital investment costs of several existing biomass-fired, European combustion grate boiler, bubbling fluidized bed (BFB) and circulating fluidized bed (CFB) plants for CHP generation correlated with the plant heat and power capacity (based on VTT analysis in an unpublished 2007 report to the IEA).

fossil fuels has good potential. Around 40% of global electricity is currently generated in coal-fired power stations. Displacing one percentage point with biomass could reduce global emissions by 60 Mt CO_2/year (IEA, 2008). Specific investment costs for biomass implementation in existing coal-fired boilers at between €50–150 kW_{th} are around half that of building new bioenergy plants of a capacity to use a similar amount of biomass. If the biomass is combusted independently, the energy conversion efficiency is often lower than when combusted in larger coal-fired plants and the S and NO_x emissions possibly can be less economically controlled. Where the availability of biomass varies, or greater flexibility is sought to purchase biomass according to current price levels, the reliability is less critical with co-firing as the coal or gas can be used alone.

The biomass can be co-fired in pulverized coal boilers as pellets or by direct feeding of finely crushed biomass, though both these fuels have an economic and energy cost in their production. In BFB and CFB boilers, the coal and biomass are usually fed into the boiler using similar handling systems, with biomass volumes restricted depending on the burner design and required heat output. Boiler output and efficiencies decrease when biomass is added because the biomass energy density is lower than that of coal and is due to the limited capacity of existing grinding equipment and fans. Co-firing is usually feasible up to 5–10% biomass by volume of the total fuel input, with little loss of total output capacity. This volume also often corresponds to the amounts of biomass available within a reasonable economic collection radius for a large-scale plant. Higher biomass energy shares are possible by employing separate biomass gasifiers or boilers adjacent to the coal plant. Large once-through supercritical fluidized bed boilers could become feasible in future to co-fire biomass and produce electricity with an efficiency of up to 50%.

To obtain an acceptable outcome when co-firing biomass and coal, the particle size of the biomass needs to be small enough to guarantee a long enough retention time in the boiler for complete combustion. The combustion temperature in pulverized coal boilers is around 1000–1250°C, so to prevent slagging, the biomass fuel used needs a reasonably high ash melting point. Biomass will change the ash composition and this could cause problems in the utilization of coal ash for various applications, such as cement manufacture.

Recovery boilers at pulp mills

In Kraft pulp mills, the liquor remaining after sulfate cooking of the pulp is a mixture of chemicals and dissolved wood material, mainly lignin. Recovery of the chemicals for reuse is based on the combustion of this spent 'black liquor' in a recovery boiler. The total efficiency of power and heat production in the recovery boiler has improved steadily over time because of increases in the dry-solids content of the black liquor. A solids content of 65% was normal 20 years ago, but 80–85% is now more common in new facilities. Hence, the steam-generating capacity of the recovery boiler has increased by 10% and SO_2 emissions have been virtually eliminated.

Recent emphasis to improve conversion efficiency is to increase the power–heat ratio of the recovery boiler's energy process by improving steam parameters with temperatures above 500°C (limited by the risk of corrosion) and preheating the feed water and combustion air with recirculated steam from the turbine. Preheating of air by flue gases is not undertaken due to the risk of corrosion associated with moisture condensation. Processes for gasifying the black liquor are currently under development.

Gasification of Biomass for Heat and Power

Gasification technologies have been developed in order to produce fuel gas from biomass feedstocks for use in gas engines and gas turbines for CHP applications and co-firing in gas boilers for heat. Large-scale plants use gas turbines for power generation, whereas small-scale systems (< 10 MW_e) use gas engines driving a generator. Atmospheric fluidized and fixed bed gasifiers have been well demonstrated and are commercially available for co-firing and heat plant applications, but their market share compared to combustion has remained low because of their reputation for unreliability and their competitive advantage being limited to niche markets. Several demonstration plants in Sweden, the UK and USA have been shut down after varying lengths of operation, some after thousands of hours. The problem is that both capital and operating costs remain relatively high compared with less efficient biomass combustion systems.

During gasification, the dry biomass feedstock (below 20% moisture content wet basis) is partially oxidized at high temperatures by using restricted amounts of oxygen to form a synthesis gas mixture of CO, H_2 and CH_4 together with higher hydrocarbons suitable for combustion. Higher moisture content fuels will lower the heating value of the gas. The oxygen carrier can be air, oxygen or steam (IEA, 2008). Less oxygen has to be introduced than for full stoichiometric combustion of the biomass. The pyrolysis process is followed by gasification of the residual carbon and its combustion to provide the heat to drive the process. Carbon monoxide and hydrogen formation is an endothermic reaction requiring heat from partially combusting the residual carbon or by introducing external heat. Formation of tars and condensates should be minimized as cleaning the gas to avoid problems in later use is costly and challenging. For example, when using the raw gas in a fire-tube boiler equipped with a gas burner designed to suit the synthesis gas, any tar-like substances present need to be combusted in the boiler to avoid fouling the convection part of the system. Similar fouling can occur if the uncleaned gas is combusted in an internal combustion engine. The continuing interest in thermochemical liquid biofuel production is based mainly on combining the synthesis gas products chemically after gasification to form liquids particularly suitable for aviation, heavy truck or marine applications.

Various designs of gasifier work under different conditions that affect the content of the synthesis gas mixture, the carbonized residues, ash, fuel-based impurities, CO_2 and the amounts of condensing liquids and tars formed during pyrolysis.

- *Fixed bed gasifiers* depend on a slow downwards flow of biomass and gas through the fuel bed requiring a relatively even quality and particle size of biomass fuel such as woodchips. The particle size should be a few centimetres across to allow the gas to pass through. These designs are more competitive at the small scale below 10 MW_{th}. Counter-current designs feed the biomass fuel into the reactor from the top, from where it flows slowly downwards through the drying, pyrolysis, gasification and combustion zones, where operating temperatures are typically around 200–600°C. Ash is removed from the bottom of the reactor. Air, oxygen or steam travel up inside the reactor through a grid at the bottom that supports the fuel. The gases from the pyrolysis and drying phases produce the synthesis gas with relatively high hydrocarbon and tar concentrations. Co-current designs, where the biomass and gas both flow in the same direction, split the hydrocarbons and tars produced from the pyrolysis phase into more simple compounds when they flow through the hotter combustion and gasification phases. Therefore, the problematic tar concentrations are typically smaller than in counter-current designs.
- *Fluidized bed gasifiers* rely on fuel particles about 10 mm diameter to form a bed which floats in the flow of the oxygen carrier (air, steam or oxygen) coming from below. The gases produced leave the reactor at the top. This design of gasifier has good mass transfer between the oxygen carrier and biomass particles because of the effective mixing of the particles. This provides a more even temperature distribution in the reactor and a higher gas temperature. Any incompletely reacted fuel particles in the production gas leaving the reactor are separated and returned to the gasifier. The tar concentration in the production gas is lower than in a fixed bed gasifier. Compared with the BFB design, the CFB

has a faster flow speed of the oxygen carrier and an increased gas output per area of the reactor bed. The higher flow velocity increases the mass of biomass leaving the reactor that is returned to the process by a cyclone. At higher flow speeds, the retention time of the fuel particles in the reactor are decreased. This design therefore best suits biomass fuels with more easily volatile compounds. Fluidized bed gasifiers are more cost-effective at the larger scale, with BFBs in the 20–60 MW_{th} range and CFBs over 60 MW_{th}.

Gasification can be carried out either pressurized or under atmospheric pressure. Cost savings should be gained with pressurized reactors as the equipment can be smaller, giving particular benefits in large-scale plants using gas cleaning equipment. However, the materials need to be resilient to corrosion. Highly pressurized components are significantly more expensive, but the costs can be offset if the pressure can also be utilized in the energy production process. Pressurized gasification of biomass and gas cleaning technologies needed prior to gas turbine combustion have not yet been successfully demonstrated as economically viable in large-scale biomass integrated gasification combined-cycle (BIGCC) plants, in spite of large investments in several demonstration plants. The conversion efficiency is higher than for the combustion/steam turbine route, but the capital costs remain too high, requiring them to be spread over a high number of annual operation hours, such as in paper and pulp mills, in order for the plant to be come economically viable. Significant efforts to develop more advanced technology could provide new concepts that improve the economy of IGCC plants.

Costs and efficiencies are widely varying due to the immaturity of the technology, although small-scale systems are commonly used in countries such as India, where labour costs to maintain and clean the plants are very low. An analysis of several gasifier plants using woodchips conducted by European Bio-CHP in 2006 gave capital investment costs ranging from €0.21 M for a 0.1 MW_e/0.32 MW_{th} (€512/kW_{e+th}) in Northern Ireland to €5.1 M for a 1.45 MW_e/2.8 MW_{th} (€1200/kW_{e+th}) plant in Denmark. An Austrian 1.45 MW_e plant not utilizing the heat had an investment cost of over €5000/kW_e.

Black liquor gasification

At present, most mills dry and combust the black liquor as fuel in their recovery boilers to provide process heat and power (see the section headed 'Bioenergy Heat and Power Combustion Technologies'). Alternative gasification processes have been proposed to utilize the synthesis gas to produce a variety of liquid fuels for transport (IEA Bioenergy, 2007b). These include the Chemrec process based on a high temperature molten-phase entrained-flow gasifier and the Manufacturing and Technology Conversion International (MTCI) process using a low temperature fluidized bed gasifier.

Currently, for each tonne of pulp produced in a Kraft pulp mill, around 1.7 t dry weight of black liquor is produced containing around half the organic material by weight that was originally in the wood in the form of lignin. A typical mill therefore has an available energy source of 250–500 MW, more than is needed to power the plant. Globally, around 170 Mt dry matter/year of black liquor is produced, with a total energy content of approximately 2 EJ/year. It has already been partially processed during the pulping operation into a form suitable for pumping. In the proposed gasification process, additional biomass, perhaps from industrial crops or forest residues, could be added so that the plant can remain self-sufficient in heat and power, as well as produce liquid biofuels from the synthesis gas. This process is being demonstrated in Sweden where a 20 t dry matter/day, pressurized 27 bar, high temperature pilot plant has been operated for more than 3000 h since 2005 (Chemrec, 2007). The next stage being planned is to produce 4–5 t/day of DME (dimethyl ether) fuels and then to scale-up the process to a 500 t/day plant to be constructed in the USA.

Technology Comparisons, Costs and Potential

A wide range of bioenergy conversion technologies is under continuous development to produce cheaper bioenergy carriers for both small-scale residential and large-scale industrial applications. Based on the technologies discussed above and on IEA Bioenergy (2007a) analysis, a comparison of plant capacities, efficiencies and costs for plants operating in OECD countries gives an indication of the opportunities for bioenergy heat and power systems. The data cover a wide range due to variations in plant scale, maturity of technology, detailed design variables, type and quality of biomass feedstocks and regional differences (Table 3.3). Plants operating in non-OECD countries may benefit from cheaper labour costs and biomass fuels.

Power and heat generation costs vary widely with fuel type, scale, technology and location. Annual operating and maintenance costs are around US$100–170/$kW_e$, though they are lower for co-firing and almost double for MSW plants. Capacity factors can vary between 35 and 75%, but tend to be around 70–80% for MSW plants. All plants have a life of around 20–25 years and a 1–3 year planning and construction time (though may be up to 5 years for MSW plants).

The main variable is the delivered cost of the biomass fuel. When the biomass feedstock is harvested and collected from arable land, stored and delivered to the plant, it can range from US$7–16/GJ. For comparison, the price for energy straw delivered to a heating plant in Denmark is about US$8/GJ. When using biomass already available on-site, it can cost US$0–5GJ or even be negative where alternative disposal costs are avoided. MSW is a special case where in some instances a charge is levied to dispose of the waste. Hence, the fuel feedstock costs can range from US$ −2 to +1/GJ. Heat and power generation costs from bioenergy are likely to decline in the future due to improved technology and greater project experience.

Future Opportunities

Total global primary energy assumptions from fossil fuels, renewable energy and nuclear in 2050 range from around 600–1400 EJ. The potential contribution from biomass under such wide-ranging projections

Table 3.3. Typical plant size, efficiency and capital cost for a range of bioenergy conversion plant technologies (IEA Bioenergy, 2007a; IEA, 2008).

Conversion type	Typical capacity	Net efficiency	Investment costs
Anaerobic digestion	< 10 MW_e	10–15% electrical 60–70% heat	US$2300–4000/$kW_e$
Landfill gas	< 200 kW_e–2 MW_e	10–15% electrical	US$1500–2500/$kW_e$
Combustion for heat	0.1–5 kW 5–50 kW_{th} residential 1–5 MW_{th} industrial	Open fires 10–20% Stoves 40–50% Furnaces 70–90%	Very low ~ US$25/$kW_{th}$ US$350–950/$kW_{th}$
MSW combustion power	10–100 MW_e	15–30%	US$7000–9000/$kW_e$
Combustion for power	10–100 MW_e	20–40%	US$2000–3000/$kW_e$
Combustion for CHP	Small < 10 MW_e Large 10–50 MW_e	60–90% overall 80–100% overall	US$3500–4500/$kW_e$ US$3000–4000/$kW_e$
Co-firing with coal	Existing 5–100 MW_e New 100–300+ MW_e	25–40% 35–45%	US$130–1300/$kW_e$ + power station costs
Gasification for heat	50–500 kW_{th}	80–90%	US$850–1000?/$kW_{th}$
BIGCC for power	Demos 5–10 MW_e Future 30–200 MW_e	40–50% plus 45–55%?	US$4500–6000/$kW_e$ US$1200–2500?/$kW_e$
Gasification for CHP using gas engines	0.1–1 MW_e	60–80% overall	US$1200–3500/$kW_e$
Pyrolysis for bio-oil	Demos 10 t/h Future 100 t/h?	60–70% ~ 85%? if using char	US$850?/$kW_{th}$ for 10 MW_{th}

is difficult to assess without making some key assumptions.

- Bioenergy used for heat and power generation will become more competitive with natural gas, oil and coal if their global prices rise and carbon charges are widely applied.
- Policies relating to energy security, climate change, health, trade, employment, rural industry support, sustainable development, transport, etc., also often support industrial crop production and bioenergy deployment.
- If policy makers and decision makers better recognize the several co-benefits that bioenergy offers, this could impact positively on its deployment over the next few decades.
- The wood processing and sugar industry sectors will continue to expand their use of biomass for heat and power in certain regions.
- Average bioenergy plant size is increasing continually due to economies of scale and the improved availability, reliability and delivery of biomass resources.
- Plant capacity tends to be limited by the local availability of biomass which, unlike coal or gas, has to be collected from a widely disbursed area. The extra cost of biomass transport, especially when it is traded internationally, can be outweighed by lower investment costs per kW and increased conversion efficiencies of larger-scale commercial plants. The costs will vary with local conditions. Certainly, sugarcane mills have long used efficient delivery systems for 300,000 t or more of biomass/year, using specialist trucks and even mini-rail infrastructure systems to transport the freshly harvested crop into the plant. There is no reason why bioenergy heat and power plants could not follow a similar model.
- Advanced combustion technologies (BFB and CFB boilers, for example) and co-firing of biomass with coal are continuing growth areas.
- The potential growth of gasification technology is more difficult to predict but could be significant. BIGCC systems offer higher efficiencies and reduced costs. If linked with CFB combustion, they could also be used for co-firing with coal.
- Electricity generation from combustion/steam turbines in plants up to 100 MW_e capacity could cost between €40–60/MWh, or possibly less when co-fired. BIGCC plants, once they become commercially viable, could in theory reduce these costs by 25–50% due to their higher efficiency.
- For off-grid applications, decentralized biomass-fired power plants may have a reasonably high potential, especially in regions where electricity grids or extensions are not being planned. If there is also a demand for heat (as could be the case for district heating in Canada, Russia, China, etc.), then CHP systems would be particularly attractive.
- Exactly how traditional biomass users in developing countries might access more efficient bioenergy services over the next few decades is not clear. Inefficient traditional biomass use may decline due to fewer people living in rural areas in 2050. The greater uptake of more efficient stoves, small-scale biogas systems and even biomass-based cooking fuels such as DME or ethanol gels could have an impact on biomass demand and the health of users.
- Using existing biomass supplies for fuel in modern bioenergy plants rather than on open fires could provide the necessary energy services more efficiently by producing process heat, power and liquid fuels. A conversion to 'modern' bioenergy would certainly assist some developing countries to better meet their sustainable development goals.

Competing markets for biomass

Biomass conversion to heat and power is mature with CHP, co-firing and various combustion technologies proving to be reliable, efficient and clean when they are designed and assembled properly and when biomass with acceptable and consistent properties is used. However, with a predicted growth in demand for heat, power, CHP, transport fuels, biomaterials and biochemicals, coupled with growing concerns over soil and water conservation and soil quality (that could decline if less residual organic matter was incorporated after harvest of the crop in order to increase the biomass volumes extracted for combustion), it is difficult to assess the priority use for future biomass.

Assuming that an international cap on carbon will be imposed in future, biomass to generate heat as a substitute for producing heat from fossil fuel combustion is probably the most cost-effective use in terms of US$/t CO_2 avoided. Electricity generation from biomass feedstocks competes with nuclear and other renewables, especially where greenhouse gas (GHG) mitigation is a major objective. It can be argued, therefore, that biomass use for this purpose is not warranted since other options also exist. Carbon dioxide capture and storage (CCS) could become feasible for large coal- and gas-fired heat and CHP plants, but since these are normally sited close to the heat demand, the cost of transporting the captured CO_2 to suitable reservoirs could be relatively high.

For transport fuels, there are also alternatives to biofuels, including compressed natural gas (cng), liquid propane gas (lpg) and liquid fuels produced from a range of non-conventional oil sources such as tar sands and coal–liquids. Many of these fuels, however (with the exception of cng and lpg), result in higher CO_2 emissions per km travelled than when using gasoline or diesel. The use of biofuels is therefore seen to be beneficial – assuming that life-cycle analyses endorse sustainable production methods of the biomass and that there are real GHG mitigation potentials, though this is not necessarily the case for all biofuels produced today.

The versatility of biomass in its many forms will ensure future wide-ranging markets for energy carriers, biochemicals and biomaterials. Although heat and power generation dominate its current use, these may not be the most relevant or productive future markets. Some extreme theoretical cases are considered here to exemplify this point.

- Assuming second-generation biofuels become technically and economically feasible, then if all the 180 EJ of biomass projected to be available in 2050 were to be converted to liquid fuels at 65% efficiency, the volumes produced would exceed the current world demand for transport fuels.
- If, however, all the biomass was used solely for power generation, half of it in efficient BIGCC plants and half in less efficient combustion/steam systems to give an average conversion efficiency of 40%, then around three-quarters of the projected total power demand in 2030 of around 30,000 TWh theoretically could be met.
- By 2050, biomass consumed to produce bioplastics could have an energy content of around 50 EJ, equivalent to the energy in the oil displaced to make the same volume of plastics (Hoogwijk *et al.*, 2003). The development of biorefineries to meet this demand and also to produce a range of energy products has been proposed and is under evaluation.
- Biomass used for construction materials, including timber and fibreboard, may also increase in the future when the embedded carbon is valued more when compared with the GHG emissions of similar structures made from steel, aluminium and concrete. The additional demand by 2050, plus that assumed for paper, could be higher than today's total world biomass demand (Hoogwijk *et al.*, 2003). Much of the timber used in construction, as well as for paper and cardboard production, eventually ends up in waste streams and hence ultimately becomes a low-value biomass feedstock.

Selection of future markets for competing biomass products will depend on the priority objectives of a government. For example, if avoiding CO_2 emissions is the major concern, then co-firing biomass to displace some coal will have higher avoided emission reductions per unit of biomass than might converting it to transport biofuels to displace gasoline or diesel. However, this option could give greater energy security.

Net avoided emissions depend on the application that the biomass is used for, as well as on the supply chain efficiency. Should the electricity sector become more decarbonized over time due to the introduction of more nuclear, renewables, CCS and fuel switching from coal to gas-fired power plants in the supply mix, biomass use for power generation will become less attractive.

How individual countries will develop strategies and policies to assess the best use of their land and biomass resources in future is impossible to determine, but any suggested allocation away from food production or from the present efficient use of biomass in heat and CHP markets should be considered carefully (IEA Bioenergy, 2007a). The option for biomass to displace oil in transport fuels and plastics manufacture is not possible using other forms of renewable energy and this option may therefore have added national benefits for energy supply security.

Conclusions

Whether used locally or traded as a future commodity, biomass in the future could provide useful revenue for producers, market opportunities for exporters, a lever for sustainable rural development and enhanced agricultural production methods in developing countries, as well as possibly providing more stability in the fluctuating energy markets that result from fossil fuel price fluctuations. However, concerns at competing uses for land and sustainability of production are widely expressed.

IEA Bioenergy (2007a) estimated that by 2050, 100 EJ/year could be supplied from agricultural residues and wastes, 75 EJ/year from regeneration of 60 Mha of degraded and marginal lands and 125 EJ/year from growing crops on 400 Mha of current agricultural and pasture lands. This could be achieved without food or fibre shortages due to better crop yields and improved land management. Taking up 7% of the global land surface and 20% of current farmland for this form of energy production, however, is seen by many as being very ambitious, but too conservative by others. A projection of 180 EJ/year of total biomass supply in 2050 (IEA, 2008) errs on the more conservative side as it assumes that certification schemes will limit the volumes of industrial crops for biomass that might otherwise be produced.

The opportunity exists for biomass to increase its current contribution to primary energy by up to fourfold in the next 3–4 decades, though this is subject to its sustainable production, improved efficiency of the supply chain, development of new thermochemical technologies, including BIGCC conversion, and improved biochemical conversions, for anaerobic digestion plants for example.

Business opportunities are increasingly evident with the current trend towards ethical investments for environmental benefits, equity and rural development. The constraints of developing a bioenergy project are well understood (IEA, 2007a) and should become easier in future as a greater understanding develops. The costs of bioenergy are anticipated to decline over time from both learning experiences and economies of scale in larger commercial plants. The current electricity generation costs of around €50–150/MWh could reduce to €40–100/MWh by 2050; biofuels at €8–25/GJ could lower to €6–10/GJ and heat produced at between €3–15/GJ today will remain similar (IEA Bioenergy, 2007a).

Overall, there is high potential for biomass use to expand by 2050 and to reduce GHG emission by displacing some fossil fuel uses. Multi-fuel heat, power and CHP plants that can use biomass or coal or natural gas as a back-up are low risk and can also be cost-competitive, particularly where there is a price imposed for CO_2 emissions (biomass being classed as carbon neutral). The IPCC (2007) consider the demand for biomass in 2030 will be less than the potential supply. By 2050, this could still be the case but it will be dependent on the many uncertainties as outlined above. The contribution that industrial crops might play in the future primary energy supply will be limited by competition for land and water, but they are expected to make a significant input to providing heat and power over the next few decades.

References

Alakangas, E. and Flyktman, M. (2001) Biomass CHP technologies. *VTT Energy Reports 7/2001*. VTT Technical Research Centre of Finland, Jyväskylä, Finland, 62 pp.

Asikainen, A., Liiri, H., Laitila, J. and Peltola, S. (2007) Cost of forest biomass supply and harvesting technology markets in EU 27. *Proceedings of Bioenergy 2007: 3rd International Bioenergy Conference and Exhibition*, 3–6 September, Jyväskylä, Finland. FINBIO – Bioenergy Association of Finland, Jyväskylä, Finland.

Chemrec (2007) Processes for energy and chemicals recovery from black liquor in pulp mills (http://www.chemrec.se/Technology.aspx, accessed 5 June 2010).

Haberl, H., Erb, K.H. and Krausmann, F. (2007) Human appropriation of net primary production. *Internet Encyclopaedia of Ecological Economics*. Institute of Social Ecology, Vienna.

Hoogwijk, M., Faaij, A., van den Broek, R., Berndes, G., Gielen, D. and Turkenburg, W. (2003) Exploration of the ranges of the global potential of biomass for energy. *Biomass and Bioenergy* 25, 119–133.

IEA (2007a) Good practice guidelines – bioenergy project development and biomass supply, International Energy Agency, *OECD/IEA*, Paris (http://www.iea.org/publications/free_new_Desc.asp?PUBS_ID=1933, accessed 5 June 2010).

IEA (2007b) Renewables for heating and cooling – untapped potential. International Energy Agency IEA/OECD, Paris (http://www.ied.org/publications/free_new_Desc.asp?PUBS_ID=1975, accessed 5 June 2010).

IEA (2008) *Energy Technology Perspectives*. Chapter 9 Biomass and Bioenergy. International Energy Agency, IEA/OECD, Paris (http://www.iea.org, accessed 5 June 2010).

IEA (2009a) *World Energy Outlook 2009*. International Energy Agency, IEA/OECD, Paris (http://www.iea.org).

IEA (2009b) Bioenergy task on 'Integrating energy recovery into solid waste management systems' (http://www.ieabioenergytask36.org, accessed on 5 June 2010).

IEA Bioenergy (2007a) Potential contribution of bioenergy to the world's future energy demand. International Energy Agency, Bioenergy Implementing Agreement (http://www.ieabioenergy.com/LibItem.aspx?id=5584, accessed 5 June 2010).

IEA Bioenergy (2007b) Black liquor gasification; summary and conclusions from the 2004 workshop (http://www.ieabioenergy.com/LibItem.aspx?id=5611, accessed 5 June 2010).

IPCC (2007) Chapter 4, Energy Supply. *The Fourth Assessment Report on Climate Change – Mitigation*. Cambridge Press, UK (http://www.ipcc.ch, accessed 5 June 2010).

Junginger, M., Faaij, A., Björheden, R. and Turkenburg, W.C. (2005) Technological learning and cost reductions in wood fuel supply chains in Sweden. *Biomass and Bioenergy* 29, 399–418.

Parikka, M. (2004) Global biomass fuel resources. *Biomass and Bioenergy* 27, 613–620.

Savolainen, I., Tuhkanen, S. and Lehtilä, A. (2001) Technology and Mitigation of Greenhouse Gas Emissions – Background Study for the Finnish Climate Change Action Plan. Publication 1/2001. Ministry of Trade and Industry, Helsinki, 198 pp. (in Finnish with English summary).

Sims, R.E.H. (2002) *The Business of Bioenergy – in Business and in Practice*. Earthscan, London.

Thek, G. and Obernberger, I. (2004) Wood pellet production costs under Austrian and in comparison to Swedish framework conditions. *Biomass and Bioenergy* 27, 671–693.

4

Ethanol from Sugar Crops

Gillian Eggleston, Thomas Tew, Lee Panella and Thomas Klasson

Introduction

There are a number of driving forces in favour of the current worldwide expanded production of bioenergy including fuel ethanol, namely, uncertainty in global energy prices and availability of crude petroleum oil, environmental concerns and national security. The worldwide impetus to produce alternatives to petroleum-based fuels and the relatively low profit for sugar are putting pressure on the sugar industry to diversify for sustainability. Sugar crops, mainly sugarcane, sugarbeet and sweet sorghum, fit well into the emerging concept of a renewable carbohydrate feedstock because of their availability, and they are among the plants giving the highest yields of carbohydrates per hectare. As of January 2009, approximately 50% of the world's fuel ethanol production was from sugar crops (mostly the fermentation of either sugarcane juice or molasses), with the remaining 50% from grain crops (Table 4.1). Currently in the USA, the dominant feedstock for ethanol production is maize grain, but most other ethanol producing countries use sugarcane and sugarbeet as their primary feedstocks. Both Brazil and India have large-scale sugarcane-based fuel ethanol production.

Sugar crops have the advantage over grain crops because they can be grown in a much larger area of the world (Fig. 4.1). Another unique advantage sugar crops have over grain or even cellulosic crops is that they require less processing because of being directly fermentable. Starch-based grain ethanol is also expected to provide only a fraction of the fuel ethanol needed in the long term because (i) there is a finite amount of grain production and (ii) grain is also used as a feed and food source (Rooney *et al.*, 2007). Furthermore, crop management knowledge and postharvest technologies for sugar crops are well known and already available, particularly for sugarcane and sugarbeet.

Expansion of ethanol production is expected to rely on starchy grains and sugar products from sugar crops, and cellulosic biomass including agricultural and forestry residues. Companies and government agencies in several countries are currently sponsoring research into the development of energy sugarcanes and sugarbeets. Processes to convert energy sugarcanes and high fibre content sugarbeets (as well as traditional and cellulosic by-products of sugarcane and sugarbeet) into fuel ethanol are also under intense investigation (Brumbley *et al.*, 2007; Eggleston *et al.*, 2007). Thus, the sugar industry is currently faced with the reality that sugar, molasses and bagasse or pulp can no longer be

Table 4.1. World fuel ethanol production by country (million litres 2008). Data from F.O. Lichts 2009 Yearbook.

Country	Million litres/year	World production (%)
USA (maize)	35,315	44.8
Brazil (sugarcane juice and molasses)	26,790	34.0
China (maize and wheat)	4,100	5.2
India (sugarcane molasses)	1,963	2.5
France (sugarbeets)	1,550	2.0
World total	*78,885*	*100.0*

regarded as the final products of a factory or refinery. Instead, the sugar industry should be regarded as a biomass-based industry that is not only equipped to manufacture products for the food sector, but also value-added biofuels and energy.

The future use of sugar crops for energy will depend on the demand for energy and the ability of these crops to be a competitive resource within the national and international policy environments. An efficient low-cost sugar crop feedstock needs to have the following characteristics: high in sugar (sucrose) content, high yielding, remunerative for growers and have low cultivation costs. As energy crops, sugarcane, sugarbeet and sweet sorghum can be converted to liquid fuel (ethanol), heat and electricity. This chapter focuses on sugarcane, sugarbeet and sweet sorghum for fuel ethanol production as most research has been undertaken on this energy source and because it has the highest potential to succeed in the short term. However, for fuel ethanol to succeed, it will need to compete effectively with alternative energy sources and be produced in a sustainable, environmentally acceptable manner. Commercialization of fuel ethanol will depend mostly on economic factors, such as government subsidies.

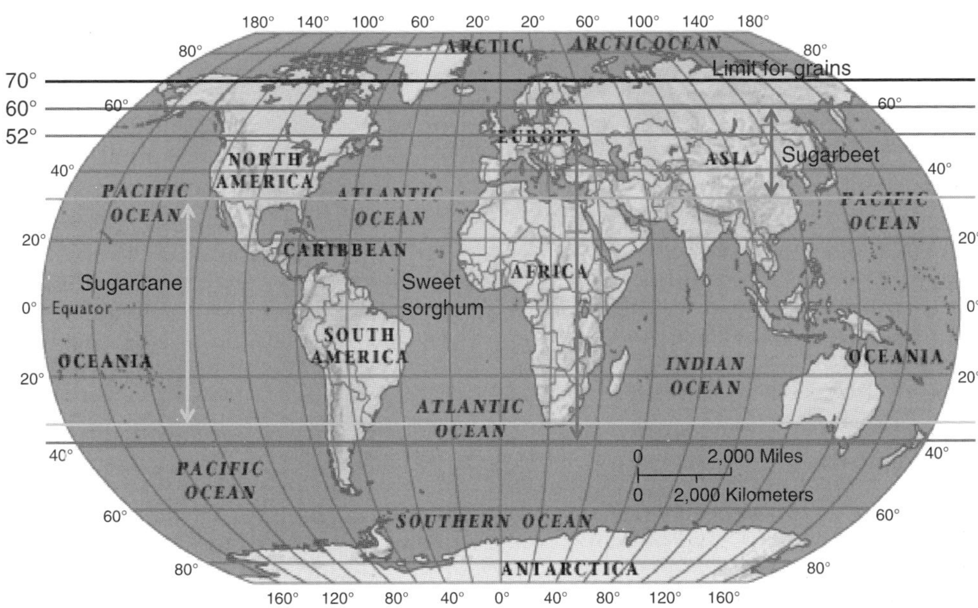

Fig. 4.1. World map depicting approximate areas where sugarcane, sugarbeet, and sweet sorghum can be grown (adapted from Debor, 2009).

The Chemistry of Sugar Conversion to Ethanol

The conversion of sugars into ethanol (CH_3CH_2OH) occurs through fermentation – one of the oldest chemical processes known to humans. Fermentation is defined as an enzymatically anaerobic controlled transformation of an organic compound. With respect to ethanol production from sugar, fermentation refers to the anaerobic conversion of sugars to ethanol by yeasts and some types of bacteria. The equation for the fermentation of glucose in the presence of yeast is:

$$C_6H_{12}O_6 \dashrightarrow 2CH_3CH_2OH + 2CO_2$$

The dashed arrow indicates the stepwise transformation of glucose to ethanol through intermediates, pyruvate and acetaldehyde (Embden–Meyerhof pathway).

During fermentation, the yeast cells gain energy from the breakdown of the sugar. The CO_2 bubbles through the liquid and dissipates into the air. The ethanol remains in the liquid but the increased production of ethanol inhibits and can eventually kill the yeast. The initial fermentation mixture can contain up to ~ 15% ethanol (per cent depends on the yeast strain used); higher concentrations of ethanol cannot be achieved by fermentation because the yeast becomes inactivated/killed. To generate higher ethanol concentrations, distillation is required. The fermentation process has other limiting factors such as temperature; > 27°C kills the yeast and < 15°C results in yeast activity that is too slow. The ideal conditions for yeast fermentation are: (i) temperature 18–24°C; (ii) sugar (not more than 1 kg/4.5l); (iii) acid pH 3.5–5.5; (iv) nutrients (di-ammonium source of nitrogen); (v) tannin (correct amount); and (vi) oxygen (the first part of fermentation requires oxygen for yeast replication) (Anon., 2009a). For example, in the industrial fermentation of sugarcane blackstrap molasses (30–40 wt% sucrose; 10–20 wt% invert sugars [glucose and fructose]; 28–35% wt% non-sugars), the molasses is diluted to a mash containing 10–20% wt% sugar. After the pH of the mash is adjusted to ~ pH 4.0–5.0 with mineral acid, it is inoculated with yeast and the fermentation is undertaken non-aseptically at 20–32°C for ~ 1–3 days. The ferment produced typically contains ~ 6–10% wt% ethanol, which is then purified.

Yeasts owe their inverting (breakdown of sucrose into glucose and fructose with invertase) and fermentative properties to enzymes. Yeast enzymes include invertase, zymase, maltase, lactase, hexosephophatase, reductase, carboxylase, melibiase and endotryptase, as well as proteolytic enzymes. Yeast enzymes vary with yeast species; hence, various yeasts behave differently towards various sugars. However, most yeasts can invert and then ferment sucrose because invertase is of common occurrence in yeasts. The yeasts of primary interest to industrial fermentation of ethanol include *Saccharomyces cerevisiae*, *S. uvarum*, *Schizosaccharomyces pombe* and *Kluyveromyces* spp.

A great number of bacteria, for example, *Zymomonas mobilis*, are also capable of fermenting sugars into ethanol. Many of these bacteria, however, generate multiple end products in addition to ethanol, such as other alcohols (e.g. butanol), organic acids, polyols, ketones and various gases (methane, carbon dioxide and hydrogen). *Z. mobilis* possesses advantages over *S. cerevisiae* yeast with respect to ethanol productivity and tolerance. The industrial bottlenecks in *Z. mobilis* are (i) its inability to convert complex carbohydrate polymers like cellulose, hemicelluloses and starch into ethanol, (ii) its resulting by-products and (iii) formation of extracellular levan.

The production of cellulosic ethanol from lignocellulosic material such as bagasse is much more difficult than from sugars. This is because cellulose and plant cell wall materials must first be converted to fermentable sugars. Once simple sugars (glucose and xylose) are formed, enzymes from yeast or bacteria can ferment them readily into ethanol (Liu and Cotta, 2008). The complete breakdown of lignocellulose biomass materials is accomplished via a combination of physical, chemical and enzymatic pretreatments that usually generates hydrolysates containing disaccharides (cellubiose, xylobiose) and monosaccharides including five- (xylose, arabinose) and six-carbon (glucose, mannose) sugars. This is discussed further in the section on sugarcane harvesting. These hydrolysates also can contain organic acids, sugar degradation products and polyphenolics which may inhibit fermentation.

Sugarcane Production

Sugarcane is a large stature perennial grass that is cultivated in approximately 80 countries in tropical, semi-tropical and subtropical regions of the world (Fig. 4.1), primarily for its ability to store high concentrations of sucrose in the stalk. It grows relatively rapidly and produces high yields. Traditionally, sugarcane is used for its sugar content to produce table sugar (sucrose), and approximately 70% of the world's sugar supply originates from sugarcane. Sugarcane now has increased stature as a potential energy crop because it is an inexpensive, abundant and renewable source of sugars for the new biorefineries that are producing fuel ethanols and electrical energy (cogeneration). The fermentable sugars are available as sucrose, invert sugars (glucose and fructose) and sugars that are 'released' from the remaining fibre after processing. The reasons for this interest in sugarcane are due to the fact that it is a C-4 crop with a relatively high photosynthetic capacity and yield, and also is the second most prolific tropical grass (Table 4.2). It possesses the ability to partition carbon into sucrose at up to 42% of the dry weight of the stalk and contains a mobile pool of hexose sugars through most of its life cycle. Sugarcane is vegetatively propagated, has multiple ratoons, and collection and transportation systems for sugarcane are already in place.

Sugarcane planting and rationing

Sugarcane is a vegetatively propagated crop. Stem sections (seed pieces) or topped whole stems (whole stalks) containing several lateral buds are planted in wide rows (1.5–2.0 m). Propagation ratios from a seed cane field to a commercially planted field can range widely, but are generally between 1:3 and 1:10, depending on seed-cane cutting and planting methodology (hand versus machine) and desired planting density (single stem sections end-to-end to several stem sections side-by-side within the row).

Sugarcane is managed almost universally as a perennial crop, the plant crop being followed by multiple ratoon crops. Often, the plant crop is allowed a longer growth cycle (16–18 months) than the ratoon crops, which are harvested at 12-month intervals. Also, ratoon crops usually are harvested in the drier, winter months when the climate favours a natural shift in energy partitioning from growth toward sucrose storage. In a few areas of the world, sugarcane is cropped for 24 months, including Hawaii and the upland regions of Costa Rica, Peru and South Africa. Conversely, in the most temperate of cane-growing areas, such as in Louisiana, USA, the physiological age of the crop at harvest is well below 12 months because the crop is largely dormant from late autumn (October–

Table 4.2. Yields and photosynthetic capture efficiency of various crops (adapted from Klass, 1994).

Location	Crop	Yield t/(hm^2/year)	Solar energy capture efficiency (%)
Minnesota, USA	Maize	24.0	0.79
New Zealand	Temperate grass	29.1	1.02
New South Wales, Australia	Rice	35.0	1.04
Columbia	Pangola grass	50.2	1.50
West Indies	Tropical forest, mixed ages	59.0	1.55
Hawaii, USA	Sugarcane	74.9	2.24
Java	Sugarcane	86.8	2.59
Puerto Rico, USA	Napiergrass	106.0	2.78

December) until springtime (March–April), to lessen the effect of freeze damage.

The number of ratoons varies greatly from location to location. Yields generally trend downward from one ratoon to the next, as a function of climate, soil and soilborne diseases, harvest conditions and the differential innate ratooning ability of varieties. The high costs associated with planting must be weighed against anticipated reduction in yields per additional ratoon to determine the optimum number of ratoon crops.

Sugarcane is subject to multiple biotic and abiotic stresses, which can affect total yield adversely. Ratoon stunting disease (*Leifsonia xyli* subsp. *xyli*), smut (*Ustilago scitaminea*), rust (*Puccinia* spp.), leaf scald (*Xanthomonas albilineans*) and mosaic (sugarcane and sorghum mosaic virus strains) are among the major diseases that affect sugarcane productivity adversely throughout the world. Several sugarcane borers are capable of impacting yield adversely by damaging the stalks of sugarcane. Most important in Louisiana and Florida, where most of the sugarcane in the USA is grown, are the stem borer (*Diatraea saccharalis*) and the Mexican rice borer. As with other crops, there are many abiotic stresses that can affect yield adversely. Sugarcane is limited in where it can be grown successfully because of its long crop cycle and low tolerance to cold temperatures. It requires more water than sugarbeets and is more sensitive to droughts than is sorghum. In spite of this, sugarcane has demonstrated a remarkable ability to grow under a wide range of soil and climatic conditions, exhibiting relatively stable yields in locations where other crops have not consistently succeeded.

Sugarcane harvesting

In many countries around the world, particularly developing countries, sugarcane is still hand-cut. The field sugarcane already may have been burnt, mainly to remove trash (leaves and tops) – a practice that is decreasing gradually, mostly because of pressure from environmental groups. Hand cutting of sugarcane into whole stalks is very laborious and developed countries cannot afford the labour. As a consequence, mechanical harvesting has been introduced in the past 50 years. Mechanical harvesting includes soldier harvesting of sugarcane into whole stalks or combine harvesting into billeted (~ 23 cm stalk pieces) sugarcane. Once the sugarcane has been cut, it has been injured and is susceptible to deterioration. Thus, cut-to-processing (crush) delays are critical and dependent on the environment, particularly temperature and humidity. As billeted sugarcane has more injured sites, it is more susceptible to deterioration than whole-stalk cane. Generally, the deterioration of sugarcane harvest methods follows the following decreasing order: burnt billets > green billets >> burnt whole stalk > green whole stalk. Eggleston *et al.* (2008) have published a more detailed review on sugarcane deterioration.

Sugarcane as Feedstock for Fuel Ethanol Production

Sugar to fuel ethanol

Traditionally, sugarcane is used commercially as a combination food and fuel crop, producing sugar and fibrous bagasse for steam to run the factories, respectively. About one half of the organic matter in sugarcane is sugar (mainly sucrose with glucose and fructose) and the other half is fibre. Sugarcane processing for sucrose production often occurs in two stages. Firstly, extracted juice (sucrose yields range between 10 and 15% wet weight of sugarcane) is processed into raw sugar (golden brown/yellow crystals; > 98.5% purity) at factories in tropical and subtropical areas where sugarcane is grown (Eggleston, 2008). Unit processes in raw sugar manufacture include extraction of the juice by tandem milling or diffusion, which leaves the fibrous by-product bagasse, clarification of the extracted juice, multi-stage evaporation, vacuum crystallization and finally centrifugation to produce the raw sugar and by-product molasses. Secondly, raw sugar is transported to a refinery where it is refined (using very similar unit processes as those used in raw sugar manufacture) to white refined sugar (> 99.9% purity). In those countries that produce ethanol at the sugarcane factory, juice and/or molasses are diverted for ethanol production in a distillery located next to the factory.

Brazil has large-scale sugarcane-based ethanol production from the juice or molasses. A typical flowchart of the manufacture of both sugar and fuel ethanol in Brazil is illustrated in Fig. 4.2. The success of the PROALCOHOL programme in Brazil can be attributed to the production of sugarcane fuel ethanol at the lowest costs of any other country due to favourable climate and economic conditions, i.e. low labour costs and government support (Henniges, 2007). Over 30 years of research in Brazil has also meant great improvements in efficiency and ethanol yields (Amorim, 2007). Furthermore, Brazil has over 30 years of experience in bioethanol production and a large potential for the expansion of sugarcane production, making it a long-term dominant competitor in the worldwide sugar ethanol market (Henniges, 2007).

In a recent feasibility study for the US Department of Agriculture (USDA) on ethanol production from sugarcane and sugarbeet in the USA, only the economics for ethanol production from molasses compared favourably to maize (Shapouri *et al.*, 2006). However, this could change, i.e. through government subsidies and higher gasoline and maize prices. The theoretical yield of ethanol from sucrose was estimated conservatively to be ~ 680 l/t of sucrose, but in practice is expected to be only 86.6% of the theoretical yield, i.e. 534 l/t of sucrose (Shapouri *et al.*, 2006). Based on average sugar recovery rates, 1 t of sugarcane would, therefore, be expected to yield 81.4 l of ethanol. The cost of converting sugarcane into ethanol was estimated to be ~ US$9.10/l, and 62% was feedstock cost (Shapouri *et al.*, 2006).

Although fuel ethanol production from sugarcane molasses may currently be feasible in the USA, similar to other crops, great enhancement of the cost structure of fuel ethanol production from sugarcane will only occur if cellulosic waste by-products are incorporated. Furthermore, cellulosic fuel ethanol production would reduce greenhouse gases (biofuels are carbon-neutral sources of energy). For these reasons, there is emphasis now on the development and commercialization of cellulosic conversion processes (Ragauskas *et al.*, 2007). For sugarcane,

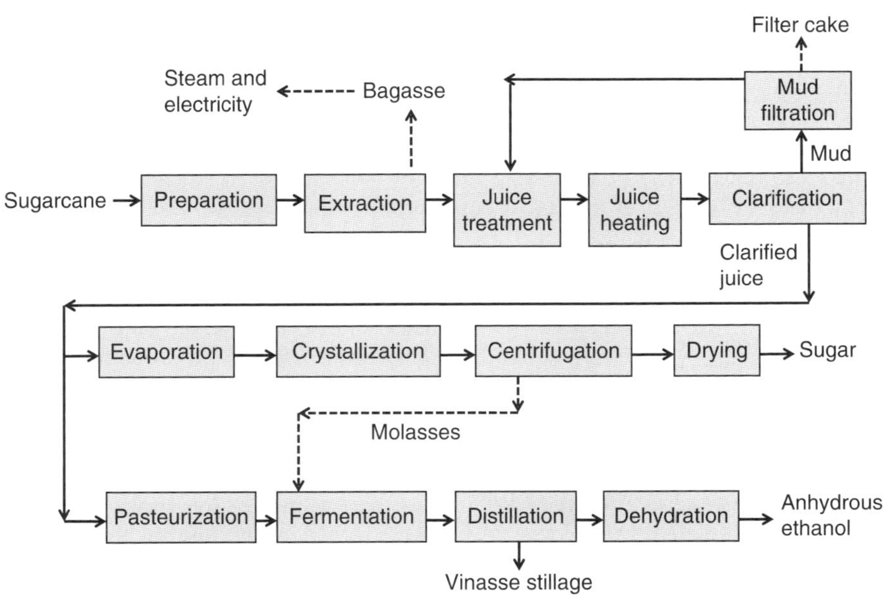

Fig. 4.2. Typical flow chart of the manufacture of sugar and fuel ethanol from sugarcane in Brazil. Use of clarified juice or molasses for ethanol production depends on the current price of these two commodities (adapted from Oliverio, 2008). Dashed lines represent by-products.

fibrous bagasse, trash (leaves and tops) and energy canes could be used.

Fibre (cellulosic biomass) to fuel ethanol

Sugarcane contains two sources of fibre – trash (green leaves + brown leaves + tops) from the field plant and bagasse left over after extraction of the juice. On a wet weight basis, sugarcane total trash comprises 16.5–20% of the field plant (12-month sugarcane) and up to 30% in 24-month sugarcane (Muir *et al.*, 2009). On a similar basis, bagasse constitutes 11–18% of sugarcane biomass. The bagasse itself typically contains 40% water, 56.5% fibre and 3.5% soluble components. The actual amounts are dependent on the variety, environmental conditions and location.

Bagasse

Sugarcane bagasse composition is illustrated in Fig. 4.3. Much research attention is focused on the production of cellulosic ethanol from bagasse worldwide, particularly in Australia (Brumbley *et al.*, 2007), Brazil (Oliverio and Hilst, 2005) and the USA (Rein, 2007). However, which of the developing technologies will turn out to be commercial winners or losers remains to be seen. As bagasse is an important steam and electricity source in the sugar industry, an integrated factory approach to use bagasse for cellulosic ethanol production should be considered. Martin *et al.* (2007) have already put forward such an approach, which includes the production of fuel ethanol from the carbohydrate fraction and the use of the remaining lignin fraction as a solid fuel for generating energy for the factory and distillery. Lora *et al.* (2006) recently described the thermodynamic limits for the production of fuel ethanol and electricity from sugarcane.

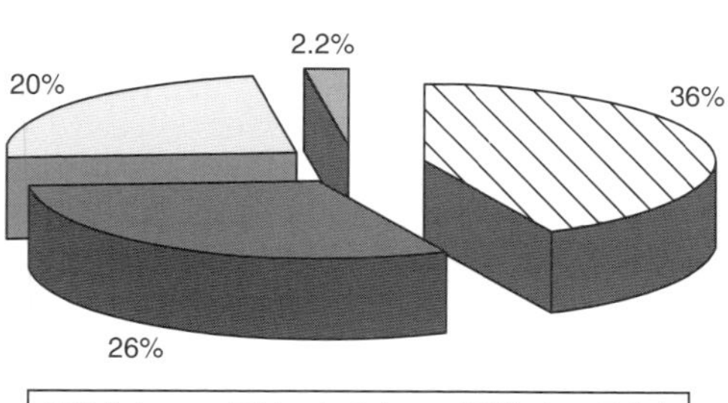

Fig. 4.3. Approximate dry weight composition of sugarcane bagasse.

A current impediment to the production of cellulosic ethanol from bagasse in the USA is that a demonstrated economically viable technology is still not available, but is expected. With the expected high throughput of cellulose, hemicelluloses and lignin, new tools, techniques and analytical methods are urgently required, for example, in the assessment of bacterial contamination in fuel ethanol production (Eggleston *et al.*, 2007). Regardless of which processing technologies end up being incorporated into the new commercial-scale sugarcane biorefineries, almost all will generate some waste products that cannot be converted to value-added materials. The disposal of such waste or stillage (vinasse) with high BOD (biological oxygen demand) serves as a real problem in many countries that have stringent environmental standards. Therefore, unlike in Brazil where stillage is spread on fields, the USA and other countries will have to find other disposal mechanisms. Some researchers (Sricharoenchaikul *et al.*, 2002) have already suggested that such stillage is an ideal candidate for thermochemical conversion to syngas.

Availability of bagasse remains a large problem in the USA, but not in Brazil. For example, in Louisiana most bagasse is presently being burned in only moderately efficient boilers to produce steam energy to fuel the factory. Co-feedstocks of bagasse and sugarcane trash (leaves and tops; wet and dry) may be feasible for cellulosic ethanol production (but not steam production due to the formation of corrosive acids). Another co-cellulosic feedstock with bagasse could be, for example, rice straw, maize stover, switchgrass, etc. Currently, there are billions of tonnes of other cellulosic residues not being utilized in the USA, as well as much underutilized land. Many of these other cellulosic feedstocks have lower lignin contents than bagasse (Eggleston *et al.*, 2007) and are degraded more easily. However, a drawback to their use as a co-feedstock with bagasse would be the necessary collection and transportation costs.

Sugarcane trash (leaves + tops)

Sugarcane trash (green and brown, dried leaves plus growing point region [apical internodes or top]) has advantages over bagasse as a feedstock source for cellulosic ethanol production. Sugarcane trash is also currently being underutilized in numerous countries. Compared to bagasse, sugarcane trash contains approximately the same or slightly less lignin and is, therefore, as easily degraded. Singh *et al.* (2008) recently reported the effect of biological treatments on sugarcane trash for its conversion to fermentable sugars.

The use of sugarcane trash as a biomass source for ethanol production is dependent on the amount of dry mass available. Typical per cent tissue weights on a dry mass basis for US commercial sugarcane varieties grown mid-season in Louisiana were reported recently (Eggleston *et al.*, 2009) and are listed in Table 4.3. Although varietal variation occurred; on average 64% of the sugarcane dry biomass were stalks and 36% total trash, which was similar to results reported for South African sugarcane by Purchase *et al.* (1990) and Muir *et al.* (2009). Thus, over one-third of the total dry mass from sugarcane is from trash, with green leaves delivering most dry mass of all the trash tissues. This also means that over one-third of sugarcane dry mass is trash which is currently not being utilized or is underutilized as a biomass source. Self-shucking varieties (brown, dried leaves fall off the field cane before harvesting) such as L 99-233 in Louisiana deliver less total trash on a dry mass basis than the other varieties, but have greater dry mass in the stalk.

Often, sugarcane trash is burned in the field or left as a cover in the field after harvesting to contribute as an organic soil fertilizer or is delivered to the factory, where it affects upstream and downstream processing detrimentally (Muir *et al.*, 2009). Leaving excess trash on the field can reduce subsequent ratoon yields (Viator *et al.*, 2005). Although some trash should be utilized as a soil fertilizer, there is still plenty available for use as biomass. Furthermore, the worldwide shift away from the harvesting of burnt to unburnt (green) sugarcane means even more trash is becoming available to collect in the field or at the factory. Its use as a biomass source simultaneously would lift a great burden from sugarcane processors (Eggleston *et al.*, 2009). For trash to work as a biomass source, research into economical ways to collect and transport excess field trash, preferably after solar drying in the field to create greater dry mass (Purchase *et al.*, 2008), is needed. Trash that is harvested and delivered with the stalks could also be separated at the factory; trash separation technologies at the factory are available (Schembri *et al.*, 2002), including dry cleaning before the sugarcane is shredded. However, questions still remain on how efficient trash separation technologies perform

Table 4.3. Average tissue weights in per cent on a dry mass basis (potential biomass) of field sugarcane varieties (from Eggleston *et al.*, 2009, with permission).

	Tissue on a dry weight basis[a,b] (%)					
Tissue	LCP 85-384	HoCP 96-540	L 97-128	L 99-226	L 99-233	Average ± SD
S (stalk)	58.8	63.7	61.1	64.0	71.0	63.7 ± 4.6
GPR	5.1	4.8	6.1	4.5	4.8	5.1 ± 0.6
GL	19.9	17.2	20.0	20.7	16.2	18.8 ± 2.0
BL	16.3	14.3	12.8	10.7	8.1	12.4 ± 3.2
Total trash: GPR + GL + BL	*41.3*	*35.3*	*39.9*	*35.9*	*29.1*	*36.3 ± 4.8*

[a]$N = 4$.
[b]dry weight was calculated as 100% moisture content. Average per cent moisture contents for all five varieties were: S (stalk) = 83.3 ± 1.2; GPR (growing point region) = 79.62 ± 1.9; GL (green leaves) = 59.7 ± 2.3; BL (brown, dried leaves) = 12.3 ± 3.2.

while not removing valuable sucrose from the stalks (Viator *et al.*, 2006; Eggleston *et al.*, 2009). Furthermore, excessively large piles of trash could be created at the factory that will have to be used quickly. Trash shredders can reduce trash to bagasse-like consistency (Schembri *et al.*, 2002).

Energy sugarcane

Another promising alternative solution to the lack of bagasse availability could be high-fibre 'energy' or 'biomass' sugarcanes. Energy crops are currently part of a grand plan for biomass production and biorefineries (Ragauskas *et al.*, 2007). The challenge is to develop energy crops with a suite of desirable physical and chemical traits, while increasing biomass yields by a factor of two or more (Ragauskas *et al.*, 2007). Energy crops such as energy sugarcanes can be grown on marginal lands that previously have been underutilized, and they often require less fertilizer and water inputs. Little work has been accomplished on the breeding and cultivation of most grasses, including sugarcane, for increased biomass yields. Furthermore, there will be no conflict between energy crops and those crops grown for food production, as most US crops are being fed to animals (Dale, 2007). Therefore, the time is ripe for intensive breeding of energy sugarcane varieties.

Alexander (1985) argued that biomass yield of energy canes purposely grown to maximize total caloric output could be in the range of two- to threefold that of conventional sugarcanes by: (i) using semi-wild high-fibre hybrids; (ii) using the whole plant, including tops and leaves; and (iii) promoting growth throughout the life cycle of the crop.

In sugarcane, more rapid genetic gain (faster breeding) should occur for total biomass yield than for sugar yield. First, growth would not be restricted intentionally during the life cycle of the crop, as occurs under cropping systems designed to impose stress as a means of maximizing sugar content. Second, a wider array of germplasm of potential value would be available to the breeder once stringent standards for sucrose and fibre levels are relaxed. Finally, high-fibre canes tend to remain erect later into the season, allowing the breeder to conduct selection closer to harvest time (Ming *et al.*, 2006).

Sugarcane varieties for the Louisiana industry have already been developed in a cooperative effort between the USDA's Agricultural Research Service's (ARS) Sugarcane Research Unit, the Louisiana State University's (LSU) Agricultural Center and the American Sugar Cane League (Fig. 4.4). During the 13-year selection process for varietal development, the sugar yield potentials of candidate varieties are compared to commercial standards. Often, varieties are discarded because their fibre levels exceed 16%, a level which raw sugar manufacturers consider unacceptable for processing. Some of these discarded varieties continue to be used as parents in the breeding programmes conducted by ARS and LSU because of their positive attributes. With the expanding interest in higher fibre, three of these high-fibre sugarcane varieties (L 79-1002, HoCP 91-552 and Ho 00-961) were released for commercial planting in 2007 (Anon., 2007a). All of these varieties produced dry biomass yields in excess of 25 t/ha when averaged over three successive fall (autumn) harvests.

At 29–30° north latitude, the sugar belt in Louisiana represents one of the most temperate areas of sugarcane production. Cold tolerance is a highly desirable trait in sugarcane grown in Louisiana and a major emphasis of the breeding programmes. Researchers at ARS's Sugarcane Research Unit in Houma, Louisiana, are introgressing cold tolerance from wild species of *Saccharum* (*S. spontaneum*) into their parental clones used to develop commercial sugarcane varieties. These early-generation hybrids have greater biomass yields than either parent due to hybrid vigour and exhibit levels of cold tolerance that are greater than the commercial parent used in the cross and Brix (% dissolved solids) levels that are only slightly lower than the commercial parent. The most vigorous of these hybrids are subjected to further backcrossing to enhance sugar and reduce fibre levels. During the backcrossing process, much of the vigour and cold tolerance is

Fig. 4.4. Evaluating potential energy canes at first clonal stage in Louisiana. Individual clones are established in 2 m length plots. Energy canes are much taller than normal canes.

lost (Legendre and Burner, 1995). However, where fuel ethanol and not sugar production is the primary focus, these cold-tolerant, high biomass-producing early-generation hybrids could be excellent candidates as varieties for the biofuels industry.

Researchers have identified five ARS-bred high-fibre sugarcane clones that have demonstrated post-winter regrowth ability in geographic locations more northerly than Louisiana. These five clones are currently being evaluated to determine their geographic range of adaptability at several state-level universities in a multistate trial (Brian Baldwin, Mississippi State University, personal communication).

Technology for large-scale cellulosic ethanol production from sugarcane fibre

There are two technological prerequisites for large-scale cellulosic ethanol production from sugarcane fibre:

1. Effective and economical pretreatment to increase the accessibility/digestibility of cellulose and hemicellulose.
2. Complete utilization of all biomass components: carbohydrates, lignin, lipids, minerals, organic acids, etc. (Dale, 2007).

Many advances have been made in these two prerequisite areas in the past few years and readers are referred to Gray (2007), Pu *et al.* (2008), and Taherzadeh and Karimi (2008) for more details. Prerequisite 1 is the description of a biochemical process but, alternatively, the fibre can be broken down into its elemental components using heat and chemical synthesis (thermochemical process). The primary thermochemical processes include pyrolysis and gasification, which are followed by a chemical synthesis process (Erickson, 2007). Erickson (2007) reported recently that the most prevalent emerging thermochemical technology was gasification, followed by catalytic conversion of the synthesis gas (syngas) to ethanol, and gave an overview of the available technologies.

The reason for the prerequisite biochemical process to produce fuel ethanol from bagasse is that cellulose is very difficult to break down, and bagasse contains a higher amount of lignin than other cellulosic feedstocks. As a consequence, the pretreatment of the cellulosic material is necessary to degrade the protective layer of lignin and hemicelluloses and make it accessible for hydrolysis (Fleischer and Senn, 2006).

Generally, pretreatment of cellulosic material occurs at high pressures and temperatures and represents the most intensive capital investment for a biomass to ethanol process. There are several different pretreatment chemistries available (Gray, 2007). The five most common pretreatment methods are: hot water extraction, dilute acid with sulfuric acid or SO_2 (steam explosion), ammonia fibre explosion (AFEX), ammonia recycle percolation (ARP) and lime. These treatments can be coupled to solvent extraction to separate the various components. Each pretreatment has its own reactor and chemical requirements that affect its performance and economics, as well as those of the subsequent enzymatic hydrolysis and fermentation processes. Acidic pretreatments tend to hydrolyse hemicelluloses compared to the alkaline methods and do not require enzymes to hydrolyse them. As a consequence, there is a trade-off in cost between pretreatment and enzyme requirements.

Eggeman and Elander (2005) found that each pretreatment was capital-intensive. Low-cost pretreatment reactors in some pretreatment processes are often counterbalanced by higher costs associated with the pretreatment catalyst recovery or higher costs for ethanol recovery. Optimization of enzyme blends for each pretreatment as well as improved conditioning of hydrolysates at process sugar concentrations were projected to differentiate further the economics of the different pretreatments. Removal of fermentation inhibitors formed during hydrolysis would also be of benefit (Martin *et al.*, 2007). However, cellulase enzyme costs are no longer considered the main economic barrier to the commercialization of cellulose–fuel ethanol processes (Eggleston *et al.*, 2007), although further enzyme cost reductions are still required (Sharon Shoemaker, University of California, Davis, California, USA, personal communication). In 2005, Novozymes Corporation reported (Anon., 2005) that, with research grants from the US Department of Energy and the National Renewable Energy Laboratory, enzyme production costs had reduced 30-fold since 2001 from US\$1.45/l ethanol to US\$0.03–0.05. This was accomplished through an integration of biotechnological techniques to give significant improvements in enzyme yield activity, and thermostability.

Scale-up of cellulosic ethanol production

As much as technology for lignocellulosic ethanol has advanced, only commercial fuel ethanol facilities based on grain starch or sugar exist today in the major fuel ethanol producing countries. A review of existing and planned cellulosic ethanol facilities by Solomon *et al.* (2007) has been published recently. Cellulosic ethanol pilot plants with capacities of < 30 hm^3/year were constructed between 1985 and 2006, giving way for demonstration-scale plants starting in 2004. Eleven commercial plants were planned for start-up in 2007–2008 for conversion of wood, straw, stover, bagasse and garbage; however, of these, only one is currently operational.

Denini S/A Indústrias de Base (Brazil), the largest sugarcane distillery producer in Brazil, was the first company that built a devoted pilot-scale bagasse-based ethanol facility (Solomon *et al.*, 2007). While scale-up plans to larger facility have been reported later (Oliverio and Hilst, 2004; Oliverio, 2006), commercialization has yet to take place. Hawaii, which passed legislation that went into effect in 2006 to mandate ethanol/gasoline blends (Reyes, 2006), still has no fuel ethanol facilities; and one planned facility for conversion of sugarcane bagasse to ethanol (Solomon *et al.*, 2007) has given way in the short term to a sugar-based ethanol platform (Voegele, 2009).

While ethanol from cellulosic biomass via pretreatment, hydrolysis, fermentation and distillation is the most discussed method of conversion; the US Department of Energy (USDOE) is supporting industrial efforts to develop pilot-scale facilities for alternative routes. In a 2007 February news release (USDOE, 2007a), the USDOE announced an investment of up to US$385 million in six new near-commercial biorefinery projects, which would be developed over 4 years. The selected companies and their technologies were: (i) Abengoa Bioenergy Biomass of Kansas, for biochemical production of fuel ethanol and thermochemical production of synthesis gas for power, and possibly ethanol in the future (430 hm^3 t/year; wheat straw, grain sorghum stubble, switchgrass); (ii) ALICO, Inc, for thermochemical processing to syngas, followed by fermentation of gas to produce ethanol (530 hm^3/year; yard, wood and citrus peel waste, and eventually energy cane feedstocks); (iii) BlueFire Ethanol, Inc, for concentrated acid hydrolysis, followed by fermentation to ethanol (720 hm^3/year; green and wood waste); (iv) Broin Companies (now POET), for weak acid and enzymatic hydrolysis, followed by fermentation to ethanol (1180 hm^3/year; maize fibre, cobs and stalks); (v) Iogen Biorefinery Partners, for weak acid and enzymatic hydrolysis followed by fermentation to ethanol (680 hm^3/year; wheat, barley and rice straw, maize stover and switchgrass); and (vi) Range Fuels Thermochemical, for catalytic conversion to ethanol and methanol (1510 hm^3/year; wood residues and wood-based energy crops). Since these awards were announced, ALICO (La Belle, Florida) has withdrawn as lead and a start-up company, New Planet Energy (Los Angeles, California), has taken over this project (Salsibury, 2008). Reportedly, Iogen (Ottowa, Canada) has also withdrawn from the DOE project, focusing instead on developing their technology in Canada (Anon., 2008a; Curtis, 2008). Range Fuels have broken ground for their facility in Soperton, Georgia (USDOE, 2007b), and the company has estimated that production will commence in 2010. POET (formerly Broin, Sioux Falls, South Dakota) recently secured a US$14.27 million contract and production is anticipated in 2011 (Shea, 2009). BlueFire Ethanol (Irvine, California) has experienced permitting delays and has yet to break ground on their Lancaster, California, cellulosic ethanol plant (Anon., 2008b). Abengoa Bioenergy has not begun construction of their Kansas facility, which will be co-located with a maize–ethanol plant, but groundbreaking is anticipated to begin in 2010.

The first company to produce cellulosic ethanol on a commercial basis outside the major ethanol producing companies was BioEthanol Japan, which began operation in early 2007 with an operating capacity of 14 hm^3 with waste wood as its raw material (Anon., 2007b). While small in size, its production was anticipated to increase to 40 hm^3 in 2008. This collaborative venture between Verenium (USA), Marubeni Corp (Japan) and Tsukishima Kikai Co (Japan) uses a weak-acid biomass pretreatment technology purchased by Verenium from SunOpta (Brampton, Canada) and a fermentation process based on a technology developed at the University of Florida and commercialized by Celunol Corp (USA). The fermentation process has two trains with five-carbon sugars fermented separately from the main six-carbon sugar fermentation, which also provides for cellulose hydrolysis. Celunol and Diversa Corp (USA) merged in 2006 to form Verenium and the company has continued its collaboration with Marubeni. The technology is now developed for commercial conversion of sugarcane bagasse in Thailand (Anon., 2008c). Verenium also commissioned its 53 hm^3 demonstration plant for conversion of non-food biomass (including sugarcane bagasse) in Jennings, Louisiana, in May 2008 (Ehrlich, 2008a) and has announced its intention of partnering with BP to break ground on a commercial-scale ethanol plant in Florida, with an estimated start in 2012 and an operating capacity of 1360 hm^3/year.

In Canada, Iogen announced its first commercial delivery of ethanol from its pilot plant to Shell in September 2008 (Ehrlich, 2008b). The ethanol production costs from (operational or planned) biochemically-based facilities is not known, but the latest estimates suggest that the cost could be in the range of US$0.60/l, not including tax or subsidies (Solomon *et al.*, 2007), but is predicted to decrease in the next 10 years (Curtis, 2008).

Sugarbeet Production

Sugarbeets are currently harvested and processed in numerous temperate and subtropical countries (Fig. 4.1). Sugarbeet is a relatively new crop plant, although the domestication of the beet took place in prehistoric times (Ford-Lloyd *et al.*, 1975; Ford-Lloyd, 2005). It was the development of the methodology necessary to measure the concentration of sucrose in solution and the economic necessity of the British blockade of France in the early 19th century that spurred the creation of a sugar industry in France and Germany. Since then, the industry has spread across Europe to Asia, North and South America and North Africa (Francis, 2006). Sugarbeet has been cultivated in southern Spain and Italy, Greece, Morocco, Tunisia, Egypt, Algeria, Afghanistan, Iran, Iraq, Pakistan and Syria (Esteban, 1999). FAO (Anon., 2009b) lists 57 countries where sugarbeet is grown and a world production of 247.9 million t in 2007.

Sugarbeet is primarily a spring planted crop of the northern temperate zones, between 30° and 60° (Draycott, 2006a), although it is grown in Chile, Venezuela and Uruguay in South America, and experimental efforts to test cultivation have been undertaken in South Africa and Australia. Sugarbeets, however, can be cultivated in subtropical and tropical arid areas where irrigation is available. In these areas, it is an autumn planted crop with a harvest in the spring and summer. Sugarbeet is remarkably adaptable to soil conditions as long as the pH is above about 6.5. Soil types range from peat (San Joaquin Delta, California) to low organic, mineral desert soils with a pH greater than 8.0 (R.T. Lewellen, USDA-ARS, Salinas, California, personal communication). In arid tropical areas, with sufficient irrigation, it will withstand temperatures upwards of 40°C. In humid tropical and subtropical areas, at high temperature, disease becomes a factor limiting cultivation. Disease can be caused by a number of fungal soil pathogens such as *Sclerotium rolfsii* (Galletti *et al.*, 2006), by *Cercospora* leaf spot (caused by *C. beticola*) and other foliar pathogens. If chemical control is not available, these diseases can limit yield severely (Sadeghian and Sharifi, 1999).

Sugarbeet is a biennial plant, which during the first year produces an enlarged root and hypocotyl to store sucrose. After a period of vernalization (usually 90–120 days at temperatures just above freezing), using the sucrose stored in the root, the plant produces a flower stalk and sets seed. When the crop is grown for sucrose (versus seed), the agricultural product of interest is the root, which is produced in that first year, and is highlighted in this chapter.

Planting

Sugarbeet planted in temperate zones is a spring planted crop, which develops throughout the summer and is harvested throughout the autumn. Soil preparation depends on the cropping system, soil type, climatic conditions and farmers' available equipment. Tillage may consist of one tillage operation in a no-till system to six or seven in other systems. For consistent germination, a good seedbed is necessary because the sugarbeet seed is relatively small and usually planted to a depth of 2.0–2.5 cm (Henriksson and Håkansson, 1993; Smith, 2001; Håkansson *et al.*, 2006).

Sucrose yield is correlated directly to the amount of sunlight intercepted by the sugarbeet leaves when they reach a maximum leaf area index (LAI) (Elliott and Weston, 1993; Milford, 2006). Time of planting is an attempt to maximize the intercepted solar radiation and is dependent mainly on temperature, moisture and the time and amount of land a farmer has to plant. In studies at the University of Nebraska, 50% germination required 30 soil degree days (based on a minimum of 4°C) (Yonts *et al.*, 2001). However the sugarbeet seedling is sensitive to cold and will not survive prolonged temperatures below –2.5°C. Sufficient moisture for germination is necessary, regardless of whether it comes from rainfall or irrigation.

Most of the sugarbeet seed planted is hybrid seed to maximize heterosis and assure monogermity. In North America and Europe, seed is planted to stand with at least 80,000–100,000 plants/ha. Quite often, the seed is pelleted with fungicide or insecticide; this has the added advantage of producing a uniformly round seed to put through precision planters. Throughout the growing season, weeds are controlled using mechanical tillage, herbicide application or hand hoeing. Sugarbeet that has been genetically enhanced for resistance to the herbicide glyphosate is grown in North America but not Europe at this time. Nutrition, cultural practices and planting dates vary widely for sugarbeets, which are reviewed comprehensively in Cooke and Scott (1993) and Draycott (2006b).

Winter sugarbeets

When sugarbeet is treated as an autumn planted crop (in subtropical and tropical arid areas) with a harvest in the spring and summer, it is named a winter beet or a tropical-type of beet. There are both advantages and disadvantages to growing beets as an autumn planted winter crop. Because of the longer growing period (9 months instead of 6 or 7), yields are higher. Water use efficiency is improved in Mediterranean climates, due to cooler temperatures and frequent rainfall in the winter season; and disease problems may be less, dependent on the environment. Difficulties include the necessity to breed for extreme resistance to flowering (due to colder winter temperatures vernalizing the crop). There are different pest problems, and resistance to these pests must be incorporated into the hybrids; because of the frequent rainfall, weed control can be difficult; and the logistics of harvest are much more complicated (Esteban, 1999; Jonsson, 1999; Sadeghian and Sharifi, 1999). Many of the specific practices are reviewed in Cooke and Scott (1993) and Draycott (2006b). Irrigation of winter beet has been reviewed by Morillo-Velarde *et al.* (2001) and Morillo-Velarde and Ober (2006).

Harvest

The time of sugarbeet harvest depends on temperature, weather and factory capacity. Sugarbeet does not have to die to be harvested, nor does it undergo senescence in the autumn – it will grow as long as the environment allows (i.e. no frost kill of the leaves which halts photosynthesis). Most processors want a sucrose concentration of at least 14% (fresh weight of root) before harvest begins. It is critical that the beets be lifted (normally with a mechanical harvester) before the beet is frozen. Once a sugarbeet has been frozen solid (and the root killed), its quality declines rapidly if thawed and bacterial and fungal pathogens multiply in severe cases, which makes those roots stored with them unsuitable for processing.

In milder European climates, sugarbeet may be left in the ground until mid-December without threat of severe freezing. In most areas of North America, the sugarbeet harvest is finished by the end of October and the sugarbeets are piled for storage. In Europe, this is done most often on farm; however, in the USA, the beets are received by the factory at harvest and stored, either on the factory grounds or on the processor-owned piling grounds.

Winter beets are harvested starting in the spring and into the heat of the summer. In the Imperial Valley of California, they are harvested in April or May through until late July or early August. Because of the extremely high summer temperatures (+40°C), the harvested beets cannot be stored in piles as are autumn harvested beets. The factory maintains a 24–36 h supply to be sliced, which is a tremendous logistical undertaking. None the less, yields of refined white sugar per area are the largest in the world in this location.

Constraints to sugarbeet crop yields

Sugarbeet is subject to multiple biotic and abiotic stresses, which affect total yield. Sugar yield is made up of three components: root yield (weight), the per cent sucrose of the beet and the impurities in the juice extracted from the beet (mostly sodium and potassium cations, amino acids and betaine), which interfere with the extraction of the sucrose from that juice. In addition to increasing one or more of the yield components, much breeding work is for resistance to diseases, which vary from region to region and can have devastating effects on all yield components (reviewed in Bosemark, 2006; Panella and Lewellen, 2007). Major diseases include rhizomania (caused by Beet necrotic yellow vein virus), *Cercospora* leaf spot (caused by *C. beticola*) and a number of root rotting fungi, nematodes and insect pests (Whitney and Duffus, 1986). In rainfed areas, intermittent drought also can affect the yield of sugarbeet. Diseases and pests are managed with a combination of chemical control, cultural practices and resistant cultivars. None the less, pests and diseases can devastate a crop when environmental conditions are permissive or proper management practices have not been followed.

Sugarbeet as a Feedstock for Fuel Ethanol Production

Sugarbeets are refined at factories and, unlike sugarcane, are produced directly into white, refined sugar (see Eggleston, 2008, for a full description of the process) (Fig. 4.5). This makes the processing of sugarbeet ethanol more economical than for sugarcane (Anon., 2009c). The major by-products from sugarbeet processing are the pulp, which is the fibrous root material left over after the extraction process, and molasses, which consists of the soluble impurities left after sucrose extraction (Harland *et al.*, 2006). Both are used as animal feed. Sugarbeets generally contain more sucrose than sugarcane (Shapouri *et al.*, 2006). The sucrose content of sugarbeets ranges from 16–18% (wet weight), whereas sucrose in sugarcane varies from 10–15%. In a recent study, it was calculated that the yield of ethanol from the sucrose in a sugarbeet was 118.5 l/t of root (wet weight) (Shapouri *et al.*, 2006). This calculation was based on a refined sucrose recovery of 15.5% (of wet weight) and a yield of 20 kg of sucrose from 1 t of beet molasses.

The dry mass in the sugarbeet is dependent on the environment in which it is grown (Milford, 2006). In 13 years yield data from irrigated beets grown at the Broom's Barn Experimental Station, Suffolk, UK, per cent dry mass varied from 30–38% of the fresh wet weight, with most of the variation in the beet tops (5–14%). The root dry mass remained relatively constant at ~ 25% of the wet weight (23–26%) (Scott and Jaggard, 1993). The portion of sucrose in the dry mass

Fig. 4.5. Sugarbeet roots grown in the Imperial Valley of California are being delivered for processing. With a high percentage of sucrose and low lignin content, sugarbeets have tremendous potential as a feedstock for fuel ethanol.

of the root ranged from 72–78% (Milford, 2006).

Pulp as a source of fermentable sugars

The pulp represents the 22–28% of the dry mass of the sugarbeet root that is not solubilized during the sugarbeet extraction process. A large portion of the pulp might provide additional feedstock for the production of cellulosic biofuels if the sugars can be released from the biomass (Table 4.4). Atlantic Biomass Conversions (USA) reported that approximately 50–60% of the theoretically available sugars could be solubilized using enzyme digestion (Kozak and Laufer, 2009). A high-protein pellet (~ 35% crude protein) is the co-product of this process, which potentially could be sold as animal feed (Kozak and Laufer, 2009). Sugarbeet pulp solubilized in this manner could provide additional feedstock for ethanol conversion not included in the 118.5 l/t figure calculated by Shapouri *et al.* (2006).

A 30% conversion rate would yield ~ 356 l ethanol/dry t pulp and a 50% conversion rate would yield ~ 589 l/dry t. If the root is 25% dry weight of the sugarbeet plant, and sucrose ~ 75% of the dry root weight, then 1 t of sugarbeet would yield about 6.25% pulp by dry weight. If this is added to the 118.5 l/t from the sucrose and molasses, then there is a potential total of 140.7 or 155.2 l ethanol/t root wet weight at 30 and 50% conversion of the pulp, respectively. There will be, however, a financial cost for the enzymatic conversion of the pulp.

Table 4.4. Approximate percentage composition of the various components of beet pulp. These would be expected to vary with growing environment and cultivar (data from Harland *et al.*, 2006, and Kozak and Laufer, 2009).

Beet pulp components	Beet pulp composition[a] (%)
Methanol and acetic acid	5
Pectin	19
Xylose/hemicellulose	16
Arabinose/hemicellulose	16
Total hemicellulose	32
Cellulose	20
Total available – cellulose/ hemicellulose/pectin, (C-5/C-6) feedstock	71

[a]The rest of the beet pulp is made up of lignin, crude protein and encrusted mineral substances.

Sugarbeet and ethanol yields

The yield of sugarbeet is dependent on many factors, including whether it is irrigated or rainfed, disease pressured or there are other abiotic stresses present. All things being equal, the amount of sucrose in the beet is directly proportional to the amount of solar radiation intercepted by a full canopy (Milford, 2006). The growing areas of the USA have very different root yields. A 3-year average (2006–2008) gives 56.2 t/ha in the Great Lakes region (Michigan), 55.1 t/ha in the upper Midwest (North Dakota, Minnesota), 54.2 t/ha in the Great Plains (Montana, Wyoming, Nebraska, Colorado), 80.0 t/ha in the Intermountain region (Oregon, Idaho) and 82.7 t/ha in California (Anon., 2009c). Sucrose content also varies by location and would play an important part in ethanol yield if the ethanol were to be produced from the fermentation of sucrose. For ethanol produced solely from pulp and molasses, the yield would range between 7 and 11 l/t wet weight of sugarbeet assuming a 30 and 50% conversion of pulp to feedstock, respectively. US sugarbeet production in 2008 was approximately 21,719,411 t, giving a potential range of ethanol production of 431,402,000–714,105,000 l using only sugarbeet co-products. Regional production potential using co-products (and assuming a 40% pulp conversion rate), would give an ethanol production yield range from ~ 2170 l/ha in the Great Plains or upper Midwest to ~ 3310 l/h in California.

Shapouri *et al.* (2006) calculated the economics of the conversion of sugarbeet to ethanol and remarked that the economic feasibility depended on both the cost of production of ethanol and the price of oil, which determines the price of gasoline. Since then, we have seen the cost of gasoline go up to well over US$120/barrel and plummet to US$40/barrel, making it difficult to predict this economic feasibility. Ethanol plants using sugarbeet as feedstock are operating or being planned in some EU countries and in other parts of the world.

Developing an energy sugarbeet

If sugarbeet were to be developed as an ethanol feedstock in the USA, most likely biomass would be the driving factor and sucrose content secondary. Higher biomass yields are possible using fodder beet germplasm as a parent in hybrids with sugarbeet (Doney and Theurer, 1984; Theurer *et al.*, 1987). Biomass yield potential is dependent on the interception of solar radiation (Kenter *et al.*, 2006), which gives the autumn sown sugarbeet, with its longer growing season, a decided advantage (Hoffman, 2008). In the Imperial Valley of California, the average yields in 2006 and 2007 have been 82.9 and 85.3 t/ha, respectively. However, because the beets are harvested from April through August, there is almost a doubling of yield between the fields harvested in April and those harvested in August (Kaffka, 2009). The yield potential for winter beet is tremendous in an irrigated cropping system. As an example, in 2004 a 40 ha field in the Imperial Valley of California gave average yields of 142.4 t/ha, with an average sucrose yield of 8.2 t/ha (Kaffka, 2009). This is a potential ethanol yield of about 20,000 l/ha using both co-products and sucrose from that crop.

Overall, there is tremendous potential for ethanol production from sugarbeet, especially with winter beet, which has a longer growing season and, therefore, a much higher yield potential. In the current economic situation, most US growers would want beets that could be grown for either sucrose or ethanol, ensuring flexibility. Sugarbeet pulp and molasses are also potentially excellent feedstocks for ethanol. It makes economic sense to co-locate ethanol plants, or at least enzymatic digestion facilities, next to sugarbeet factories where the pulp is produced. As with all potential feedstocks, economics will determine the feasibility of developing the sugarbeet crop as an ethanol feedstock.

Sweet Sorghum Production

Sweet sorghum has advantages over other sugar crops and is expected to play an important role as a fuel ethanol feedstock. It is a member of the grass (*Poaceae*) family, which also includes grain and fibre sorghum, and shares some features with sugarcane. Sweet sorghum is a C-4 crop characterized by a high photosynthetic efficiency and a stalk that is rich in fermentable sugars (sucrose, glucose and fructose) for syrup or ethanol production. It has one of the highest intercepted radiation use efficiencies (RUEs) of any plant species, on a par with sugarcane, allowing it to grow rapidly under optimum conditions. Although sweet sorghum is native to the tropics, it is well adapted to temperate (particularly summer) climates (Gnansounou *et al.*, 2005). It is an annual plant sensitive to photoperiod and temperature and has a short duration (3–5 months to maturity depending on the climate, which is much shorter than the typical 12–18 months for sugarcane) and, therefore, in tropical and subtropical climates can yield up to three crops a year (Table 4.5). Minimum temperatures for germination are 7–10°C and 15°C for growth (Fernandez and Dolores Curt, 2005); optimum temperatures for growth are 27–30°C. Sweet sorghum is easy to cultivate from seed and requires less water and fertilizer than sugarcane and has a greater tolerance of climatic and soil conditions. Indeed, sweet sorghum is considered one of the most drought-resistant agricultural crops and has the capability of remaining dormant during the driest periods (Woods, 2000). Moreover, its real potential lies in its growth under suboptimal conditions where the combinations of high RUE and high water and nutrient use efficiencies allow it to continue producing a sugar- and fibre-rich stalk when other crops would struggle (Woods, 2000).

Sweet sorghum yields will vary with varieties/hybrids used, location (soil, water, environment, pests and diseases), inputs and production practices (Vermerris *et al.*, 2008). Generally, biomass yields of sweet sorghums in Florida can range from 18 to 107 t/ha and juice content range from 65 to 80%. The combined sugar content of the juice

Table 4.5. Advantages and disadvantages of sweet sorghum as a fuel ethanol feedstock crop compared with sugarcane and maize[a] (modified from Yusupov, 2009).

Parameter	Sweet sorghum	Sugarcane	Maize
Crop duration (months)	3–5	12–18[b]	4
Relative energy efficiency	1:8	1:9	1:1.6
Water requirement/crop (cm)	33	36	81
Grain yield (t/ha)	4.9	0	3.6
Stalk yield (t/ha)	74	76	45
Ethanol from grain (l/ha)	0	0	3832
Ethanol from juice (l/ha)	5748	8319	0
Relative stalk deterioration	Fast	Medium fast	n/a
Juice purity (%)[c]	~ 75	82–88	n/a

[a]Average values given.
[b]12 months is most common.
[c]Purity = (sucrose/Brix) × 100.

varies between 9–15%. Sugar yields vary from 3.6–15.5 t/ha. The bagasse represents ~ two-thirds of the dry mass.

Sweet Sorghum as a Feedstock for Fuel Ethanol Production

Most current applications of sweet sorghum around the world are still for small-scale syrup and forage production, although the juice, grain and bagasse from sweet sorghum have manifold uses. More recently, there has been a dramatic increase in interest of sweet sorghum for large-scale fuel ethanol production from both the juice and bagasse (Tew and Cobill, 2006, 2008). Interest in sweet sorghum as an ethanol feedstock started in the late 1970s and early 1980s, but has revived recently for the double concerns about petroleum and the climate. Furthermore, this interest has been researched for the past 40 years in many continents and countries and continues to be evaluated in North America (Rooney *et al.*, 2007), Africa (Matsika and Yamba, 2006), Europe (Fernandez and Dolores Curt, 2005), China (Gnansounou *et al.*, 2005), India (Uppal, 2007) and Australia (Webster *et al.*, 2004). In 2009, there were a few commercial projects including the Ecology Mir group projects in Central Asia, where sweet sorghum is one of many feedstocks (Yusupov, 2009). Two companies in the USA are currently developing large commercial projects. Louisiana Green Fuels LLC is currently adding on to a sugarcane syrup facility to produce ethanol from both sweet sorghum and sugarcane syrup (R. Hulett, Louisiana Green Fuels LLC, 2008, personal communication). Highlands Envirofuels Company is currently constructing a 76 million l sweet sorghum–ethanol plant north-west of Lake Okeechobee, Florida, after receiving a US$7 million state grant in 2008 (Anon., 2008d).

Problems associated with processing of sweet sorghum to fuel ethanol

The sugars content of typical sweet sorghum juice consists of 85% sucrose, 9% glucose and 6% fructose. The purity (sucrose/Brix × 100) of sweet sorghum is, on average, ~ 75% (this will alter with variety and environment), which is markedly lower than for sugarcane and sugarbeets (~ 85%) (Table 4.5). This makes the clarification and subsequent crystallization of sweet sorghum juice more difficult and explains why most sweet sorghum ethanol facilities will produce ethanol from syrup. The impurities, aconitic acid and starch, in particular, have been reported to be relatively high in sweet sorghum compared to sugarcane and are the main causes of reduced crystallization. Juice extraction yields of sweet sorghum across tandem mills have been reported to be ~ 87% (Gnansounou *et al.*, 2005) and will be higher if diffusers are used instead. Because of the relatively high fibre content of sweet sorghum compared to sugarcane, higher extraction yields are not expected (Woods, 2000).

Another technological drawback of sweet sorghum is that the sugar will deteriorate in the stalk quickly after harvesting (the time will depend on the environmental conditions, particularly temperature and humidity). Therefore, the juice must be extracted soon after cutting and then processed quickly to prevent sugar losses into non-fermentable acids. Harvest–process delay times still have to be optimized for different climates. Although, combine and forage harvesters to produce billeted and chopped sweet sorghum, respectively, have been advocated, two research groups (Eiland *et al.*, 1982; Webster *et al.*, 2004) have shown that forage chopped sweet sorghum loses fermentable sugars rapidly in a warm climate. This can be attributed to the fact that every time the stalk is cut, it is injured and therefore made more susceptible to microbial activity, which uses sucrose. As a consequence, the length of the billet may also have to be adjusted depending on the average temperatures and relative humidity. Practices to improve sugar and ethanol yields could include the identification of varieties with different maturing dates and deterioration rates for harvesting and factory delivery at different times of the season.

Breeding sweet sorghum for fuel ethanol

Successful use of sweet sorghum as an industrial feedstock for ethanol production will depend on the varieties chosen and if technical processing challenges can be surmounted. A good variety will resist lodging (falling down in the field), be vigorous, a good height and yield, resistant to major diseases, have good drought tolerance and high yield and fermentable sugar contents. A combination of short-, medium- and long-season varieties of sweet sorghum may be needed to maintain high levels of sugars in the stalks over the entire harvesting season.

Breeding programmes were initiated in the mid-1970s to develop high-yielding sweet sorghum specifically for ethanol production, which takes intensive effort (Rooney *et al.*, 2007). These breeding efforts were dropped in the late 1980s with the reduced demand for alternative fuels but, with the renewed interest in bioethanol, there has been increased research in sweet sorghum, including the breeding of hybrid sweet sorghum (Rooney *et al.*, 2007). Sweet sorghum hybrids are expected to result in modest yield increases (Clark, 1981) but make the logistics of seed production much more feasible. Sweet sorghum hybrids are particularly suited for production in tropical areas where drought or crop rotation restrictions limit sugarcane production (Rooney *et al.*, 2007). High-yielding sweet sorghum hybrids have been tested and are now being used for ethanol production in India, and are being evaluated in other areas too (Rooney *et al.*, 2007). Breeding and development programmes for sweet sorghum germplasm and hybrids are also being conducted in the USA (Texas A&M and Purdue Universities) (Rooney *et al.*, 2007).

Integrated sweet sorghum–sugarcane systems

There has been recent interest and momentum (Webster *et al.*, 2004) for integrating sweet sorghum with sugarcane to produce fuel ethanol from sugar by constructing an ethanol distillery adjacent to a sugarcane factory. Thus, existing sugarcane harvesting and factory facilities can be utilized, causing an increased efficiency of production, land, personnel and other resources by lengthening the processing season. The ideal scenario reported by Ferraris (1988) was to process sweet sorghum either immediately before or after the sugarcane processing season, to avoid major maintenance works and reducing any possible disruptions in harvesting the sugarcane crop. Integrating sweet sorghum with sugarcane for ethanol production will increase both the efficiency and duration with which bioenergy can be produced (Woods, 2000). Woods (2000) stated, however, that 'care is needed in implementing a sweet sorghum–sugarcane integrated system because the logistics of doing so are complicated and the range of applicable technologies wide'. The viability of the integrated sweet sorghum–sugarcane system will be dependent on maintaining high sugarcane yields and achieving sustainable and high sweet sorghum yields.

Sweet sorghum, compared to sugarcane and sugarbeet, can grow in more parts of the world (Fig. 4.1) and is easier to grow, harvest and transport than sugarcane, at about one-third of typical cultivation costs and using significantly less water (Matsika and Yamba, 2006). The by-products of leaves and bagasse are higher in value as animal feeds than the equivalent products from sugarcane. The length of period during which sweet sorghum will be available for processing in an integrated sweet sorghum–sugarcane system will depend on the planting date, land area available for sweet sorghum cultivation, season length of variety planted, crop management (including any feedback loops between sorghum and sugarcane), minimum biomass quality parameters acceptable to the factory, the processing rate of the factory (will depend on its capacity, which is usually limited by the size of the evaporator station), the market for products, the impact of new technologies if introduced and government policies towards energy and environmental issues (Woods, 2000). Sugarcane factories stand 'idle' for significant periods of time during the year (up to 9 months in Louisiana), which provides an ideal opportunity for sweet sorghum to be processed for at least part of that period. Both sweet sorghum and sugarcane syrup can be stored to enable more economic processing at the distillery all year round, but research is still needed to optimize storage, e.g. optimum Brix, under different environmental conditions. Because sweet sorghum is adapted to semi-arid areas and makes optimum use of scarce resources such as water and nutrients, its use should result in net improvements in the resource use efficiency for bioenergy production on sugar estates (Woods, 2000). Sweet sorghum bagasse can also be used to produce steam, energy and electricity, but research is still needed in these areas.

Disclaimer

Mention of trade names or commercial products in this article is solely for the purpose of providing specific information and does not imply recommendation or endorsement by the US Department of Agriculture.

References

Alexander, A.G. (1985) The energy cane alternative. *Sugar Series*, Volume 6. Elsevier, Amsterdam.

Amorim, H. (2007) Ethanol production in Brazil: a successful history. *Proceedings of the Sugar Processing Research Conference*. SPRI, New Orleans, Louisiana, pp. 44–47.

Anon. (2005) http://www.novozymes.com.

Anon. (2007a) Notice of release of high-fiber sugarcane varieties L 79-1002, Ho 00-961, and HoCP 91-552. *Sugar Bulletin* 85, 21–26.

Anon. (2007b) BioEthanol Japan begins production of cellulosic ethanol from wood scraps; uses Celunol technology. *Green Car Congress*, 16 January.

Anon. (2008a) Iogen enters due diligence phase for cellulosic venture in Saskatchewan. *Bioenergy Canada*, June/July.

Anon. (2008b) BlueFire issues annual letter to the shareholders. *Marketwire*, Los Angeles, CA, 23 December.

Anon. (2008c) Verenium and Marubeni advance the development of cellulosic ethanol facilities in Asia. *Reuters*, 15 July.

Anon. (2008d) New ethanol project in Florida using sweet sorghum. *Tampa Bay News*, 22 September.

Anon. (2009a) Alcoholic fermentation by yeast cells (http://www.yobrew.co.uk/fermentation.php, accessed 24 July 2009).

Anon. (2009a) Alcoholic fermentation by yeast cells (http://www.yobrew.co.uk/fermentation.php, accessed 24 July 2009).

Anon. (2009b) Sugar and sweeteners: recommended data. ERS Briefing Room (http://www.ers.usda.gov/briefing/sugar/data.htm, accessed 24 July 2009).
Anon. (2009c) Sugar beet ethanol. Ethanol-Still-Plans.com (http://www.ethanol-still-plans.com/sugar_beet_ethanol.htm, accessed 24 July 2009).
Bosemark, N.O. (2006) Genetics and breeding. In: Draycott, A.P. (ed.) *Sugar Beet*. Blackwell Publishing Ltd, Oxford, UK, pp. 50–88.
Brumbley, S.M., Purnell, M.P., Petrasovits, L.A., Nielsen, L.K. and Twine, P.H. (2007) Developing the sugarcane biofactory for high-value biomaterials. *International Sugar Journal* 109, 5–15.
Clark, J. (1981) The inheritance of fermentable carbohydrates in stems of *Sorghum bicolor* (L.) Moench. PhD thesis, Texas A&M University, College Station, Texas.
Cooke, D.A. and Scott, R.K. (1993) *The Sugar Beet Crop: Science Into Practice*. Chapman and Hall, London, UK.
Curtis, B. (2008) US ethanol industry: the next inflection point. BCurtis Energies and Resource Group, San Francisco, California.
Dale, B. (2007) Cellulosic ethanol and sustainability: there is no 'food vs. fuel' conflict. *American Chemical Society 233rd National Meeting*. ACS, Washington, DC, No. 33.
Debor, H. (2009) Building energy independent ethanol plants in sub-tropical climates using sorghum ethanol. *Proceedings of the Sweet Sorghum Ethanol Association Meeting*, Orlando, Florida (http://www.sseassociation.org, accessed 2 March 2009).
Doney, D.L. and Theurer, J.C. (1984) Potential of breeding for ethanol fuel in sugarbeet. *Crop Science* 24, 255–257.
Draycott, A.P. (2006a) Introduction. In: Draycott, A.P. (ed.) *Sugar Beet*. Blackwell Publishing Ltd, Oxford, UK pp. 1–8.
Draycott, A.P. (2006b) *Sugar Beet*. Blackwell Publishing Ltd, Oxford, UK.
Eggeman, T. and Elander, R.T. (2005) Process and economic analysis of pretreatment technologies. *Bioresource Technology* 96, 2019–2025.
Eggleston, G. (2008) Sucrose and related oligosaccharides. In: Fraser-Reid, B., Tatsuta, K. and Thiem, J. (eds) *Glycoscience*. Springer-Verlag, Berlin Heidelberg, Germany, pp. 1163–1182.
Eggleston, G., Salassi, M., Richard, E. and Birkett, H. (2007) Sustainability of the sugar industry: future value addition from sugarcane. *International Sugar Journal* 109, 415–432.
Eggleston, G., Morel du Boil, P.G. and Walford, S.N. (2008) A review of sugarcane deterioration in the United States and South Africa. *Proceedings of the South African Sugar Technologists Association* 81, 72–85.
Eggleston, G., Grisham, M., Tew, T., Triche, R. and Antoine, A. (2009) Potential biomass quantity and processing quality of trash tissues by different US sugarcane varieties. *International Sugar Journal* 111, 108–118.
Ehrlich, D. (2008a) Verenium starts up demonstration cellulosic facility. *Cleantech Group*, 28 May.
Ehrlich, D. (2008b) Iogen ships cellulosic ethanol to Shell. *Cleantech Group*, 26 September.
Eiland, B.R., Clayton, J.E. and Bryan, W.L. (1982) Losses of fermentable sugars in sweet sorghum during storage. ASAE Paper No. 83-3109. American Society of Agricultural and Biological Engineers, St Joseph, Michigan, pp. 10.
Elliott, M.C. and Weston, G.D. (1993) Biology and physiology of the sugar-beet plant. In: Cooke, D.A. and Scott, R.K. (eds) *The Sugar Beet Crop: Science Into Practice*. Chapman and Hall, London, UK, pp. 37–65.
Erickson, J.C. (2007) Overview of thermochemical biorefinery technologies. *International Sugar Journal* 109, 163–173.
Esteban, J.A. (1999) Caractérisation de la betterave de semis d'automne. *Proceedings of the 62nd International Institute of Beet Research Congress*. IIRB, Brussels, pp. 6–7.
Fernandez, J. and Dolores Curt, M. (2005) New energy crops for bioethanol production in the Mediterranean region. *International Sugar Journal* 107, 622–627.
Ferraris, R. (1988) Sweet sorghum: agronomic evaluation and potential application for Australian agro-industry. Research Report No. 5. CSIRO Division of Tropical Crops and Pastures publication, Brisbane, Australia.
Fleischer, S. and Senn, T. (2006) Cellulosic and starch-based raw materials in ethanol production. *Zuckerindustrie* 131, 766–769.
Ford-Lloyd, B.V. (2005) Sources of genetic variation, genus *Beta*. In: Biancardi, E., Campbell, L.G., Skaracis, G.N. and De Biaggi, M. (eds) *Genetics and Breeding of Sugar Beet*. Science Publishers, Inc, Enfield, New Hampshire, pp. 25–33.

Ford-Lloyd, B.V., Williams, A.L. and Williams, J.T. (1975) A revision of *Beta* section *Vulgares* (*Chenopodiaceae*), with new light on the origin of cultivated beets. *Botanical Journal of the Linnean Society* 71, 89–102.
Francis, S.A. (2006) Development of sugar beet. In: Draycott, A.P. (ed.) *Sugar Beet*. Blackwell Publishing Ltd, Oxford, UK, pp. 9–29.
Galletti, S., Burzi, P.L., Sala, E., Marinello, S. and Cerato, C. (2006) Reducing *Sclerotium* root rot incidence in sugar beet by integration of different biological approaches. *Proceedings of the 69th International Institute of Beet Research Congress*. IIRB, Brussels, pp. 325–329.
Gnansounou, E., Dauriat, A. and Wyman, C.E. (2005) Refining sweet sorghum to ethanol and sugar: economic trade-offs in the context of North China. *Bioresource Technology* 96, 985–1002.
Gray, K.A. (2007) Cellulosic ethanol-state of the technology. *International Sugar Journal* 109, 145–151.
Håkansson, I., Henriksson, L. and Blomquist, J.E. (2006) Soil tillage and crop establishment. In: Draycott, A.P. (ed.) *Sugar Beet*. Blackwell Publishing Ltd, Oxford, UK, pp. 114–133.
Harland, J.I., Jones, C.K. and Hufford, C. (2006) Co-products. In: Draycott, A.P. (ed.) *Sugar Beet*. Blackwell Publishing, Ltd, Oxford, UK, pp. 443–463.
Henniges, O. (2007) Bioethanol production in Germany – is the framework sufficient? *Zuckerindustrie* 132, 231–240.
Henriksson, L. and Håkansson, I. (1993) Soil management and crop establishment. In: Cooke, D.A. and Scott, R.K. (eds) *The Sugar Beet Crop: Science Into Practice*. Chapman and Hall, London, pp. 156–177.
Hoffman, C.M. (2008) Bioenergy from sugar beet – physiological aspects of yield formation. *Proceedings of the 71st International Institute of Beet Research Congress*. IIRB, Brussels, pp. 117–124.
Jonsson, B.O. (1999) Development of a winter beet alternative for S. Europe. *Proceedings of the 62nd International Institute of Beet Research Congress*. IIRB, Brussels, pp. 69–76.
Kaffka, S.R. (2009) Fertilizer N effects on yield and root quality for high-yielding, fall-planted sugarbeets in the Imperial Valley. *Proceedings of the America Society of Sugar Beet Technologists*. ASSBT, Denver, Colorado, pp. 37.
Kenter, C., Hoffmann, C.M. and Märländer, B. (2006) Effects of weather variables on sugar beet yield development (*Beta vulgaris L.). European Journal of Agronomy* 24, 62–69.
Klass, D.L. (1994) Fuels from biomass. In: Kroschwitz, J.I. and Howe-Grant, M. (eds) *Kirk-Othmer Encyclopedia of Chemical Technology*. John Wiley and Sons, New York, pp. 16–110.
Kozak, R. and Laufer, C.S. (2009) Addition of a thermostabilized pectin methylesterase significantly enhances the rate of saccharification of sugar beet pulp by the commercial pectinase preparation Pectinex® Ultra SPL. *Proceedings of the 35th Meeting of the American Society of Sugar Beet Technologists*. ASSBT, Denver, Colorado, pp. 24.
Legendre, B.L. and Burner, D.M. (1995) Biomass production of sugarcane cultivars and arly-generation hybrids. *Biomass and Bioenergy* 8, 55–61.
Liu, S. and Cotta, M.A. (2008) Gram-positive bacteria as biocatalysts to convert biomass derived sugars into biofuel and chemicals. In: Hou, C.T. and Shaw, J.F. (eds) *Biocatalysis and Bioenergy*. John Wiley and Sons, New York, pp. 249–264.
Lora, E.E.S., Zampieri, M., Nogueira, L.A., Leal, M.R. and Cobas, V.M. (2006) Thermodynamic limits for the production of ethanol and electricity from sugarcane. *Zuckerindustrie* 131, 759–765.
Martin, C., Almazan, O., Marcet, M. and Jönsson, L.J. (2007) A study of three strategies for improving the fermentability of sugarcane bagasse hydrolysates for fuel ethanol production. *International Sugar Journal* 109, 33–39.
Matsika, E. and Yamba, F.D. (2006) Harnessing sweet sorghum for bioenergy in Zambia. *Renewable Energy for Development* 19, 6–8.
Milford, G.F. (2006) Plant structure and crop physiology. In: Draycott, A.P. (ed.) *Sugar Beet*. Blackwell Publishing Ltd, Oxford, UK, pp. 30–49.
Ming, R., Moore, P.H., Wu, K.K., D'Hont, A., Tew, T.L., Mirkov, T.E., Da Silva, J., Schnell, R.J., Brumbley, S.M., Lakshmanan, P., Jifon, J., Rai, M., Comstock, J.C., Glaszmann, J.C. and Paterson, A.H. (2006) Sugarcane improvement through breeding and biotechnology. *Plant Breeding Reviews* 27, 17–118.
Morillo-Velarde, R. and Ober, E.S. (2006) Water use and irrigation. In: Draycott, A.P. (ed.) *Sugar Beet*. Blackwell Publishing Ltd, Oxford, UK, pp. 221–255.
Morillo-Velarde, R., Cavazza, L., Cariolle, M. and Beckers, R. (eds) (2001) Irrigation de la betterave scurière en zone méditerranéenne. *Advances in Sugar Beet Research*, Volume 3. IIRB, Brussels.
Muir, B., Eggleston, G. and Barker, B. (2009) Factory trials to determine the effect of green trash on downstream processing. *Proceedings of the International Society of Sugar Cane Technologists Processing Workshop*. ISSCT, Quatre-Bornes, Mauritius, pp. 22.

Oliverio, J.L. (2006) Technological evolution of the Brazilian sugar and alcohol sector: Dedini's contribution. *International Sugar Journal* 108, 120–129.
Oliverio, J.L. and Hilst, A.G.P. (2004) DHR-DEDINI hidrolise rapida (DEDINI rapid hydrolysis) – revolutionary process for producing alcohol from sugar cane bagasse. *International Sugar Journal* 106, 168–172.
Oliverio, J.L. and Hilst, A.P. (2005) DHR (Dedini rapid hydrolysis): revolutionary process producing alcohol from sugarcane bagasse. *Proceedings International Society of Sugar Cane Technologists* 25, 320–327.
Panella, L. and Lewellen, R.T. (2007) Broadening the genetic base of sugar beet: introgression from wild relatives. *Euphytica* 154, 382–400.
Pu, Y., Zhang, D., Singh, P.M. and Ragauskas, A.J. (2008) The new forestry biofuels sector. *Biofuels, Bioproducts, and Biorefining* 2, 58–73.
Purchase, B.S., Lionnet, G.R.E., Reid, M.J., Wienese, A. and De Beer, A.G. (1990) Options for, and implications of, increasing the supply of bagasse by including tops and trash with cane. *Proceedings of the Sugar Processing Research Conference*. Sugar Processing Research Institute, New Orleans, Louisiana, pp. 229–243.
Purchase, B.S., Wynne, A.T., Meyer, E. and Van Antwerpen, R. (2008) Is there profit in cane trash? – Another dimension to the assessment of trashing versus burning. *Proceedings of the South African Sugar Technologists Association* 81, 86–99.
Ragauskas, A.J., Williams, C.K., Davison, B.H., Britovsek, G., Cairney, J., Eckert, C.A., Frederick, W.J. Jr, Hallett, J.P., Leak, D.J., Liotta, C.L., Mielenz, J.R., Murphy, R., Templer, R. and Tschaplinkski, T. (2007) The path forward for biofuels and biomaterials. *Science* 311, 484–489.
Rein, P. (2007) Biofuels research at the Audubon Sugar Institute. *Sugar Journal* 69, 9–10.
Reyes, B.J. (2006) Isle ethanol producers still year away. *Honolulu Star Bulletin* Honolulu, Hawaii, 23 January.
Rooney, W.L., Blumenthan, J., Bean, B. and Mullet, J.E. (2007) Designing sorghum as a dedicated bioenergy feedstock. *Biofuels, Bioproducts, and Biorefining* 1, 147–157.
Sadeghian, S.Y. and Sharifi, H. (1999) Improvement of sugar beet for combined resistance to bolting and *Cercospora* leaf spot. *Proceedings of the 62nd International Institute of Beet Research Congress*. IIRB, Brussels, pp. 61–67.
Salisbury, S. (2008) Start-up takes over abandoned ethanol project. *Palm Beach Post*, Palm Beach, Florida, 3 June.
Schembri, M.G., Hobson, P.A. and Paddock, R. (2002) The development of a prototype factory-based trash separation plant. *Proceedings of the Australian Sugar Cane Technologists* 24, 12–18.
Scott, R.K. and Jaggard, K.W. (1993) Crop physiology and agronomy. In: Cooke, D.A. (ed.) *The Sugar Beet Crop: Science Into Practice*. Chapman and Hall, London, UK, pp. 179–237.
Shapouri, H., Salassi, M. and Fairbanks, J.N. (2006) The economic feasibility of ethanol production from sugar in the United States. Joint publication of OEPNU, OCE, USDA, and LSU, pp. 69 (http://www.usda.gov/oce/reports/energy/EthanolSugarFeasibilityReport3.pdf, accessed 24 July 2009).
Shea, B. (2009) POET gets $14 M boost: cellulosic ethanol plant gets state funds. *The Messenger*, Fort Dodge, Iowa, 27 February.
Singh, P., Suman, A., Tiwari, P., Arya, N., Gaur, A. and Shrivatava, A.K. (2008) Biological pretreatment of sugarcane trash for its conversion to fermentable sugars. *World Journal of Microbiology and Biotechnology* 24, 667–673.
Smith, J.A. (2001) Tillage and seedbed preparation. In: Wilson, R.G., Smith, J.A. and Miller, S.D. (eds) *Sugarbeet Production Guide*. University of Nebraska, Lincoln, Nebraska, pp. 23–36.
Solomon, B.D., Barnes, J.R. and Halvorsen, K.E. (2007) Grain and cellulosic ethanol: history, economics, and energy policy. *Biomass and Bioenergy* 21, 416–425.
Sricharoenchaikul, V., Agrawal, P., Frederick, W.J. Jr (2002) Black liquor gasification characteristics. 1. Formation and conversion of carbon-containing product gases. *Industrial and Engineering Chemistry Research* 41, 5640–5649.
Taherzadeh, M.J. and Karimi, K. (2008) Pretreatment of lignocellulosic wastes to improve ethanol and biogas production: a review. *International Journal of Molecular Sciences* 9, 1621–1651.
Tew, T.L. and Cobill, R. (2006) Evaluation of sweet sorghum as a complementary bioenergy crop to sugarcane in Louisiana. *Sugar Journal* 69, 27.
Tew, T.L. and Cobill, R. (2008) Genetic improvement of sugarcane (*Saccharum* spp.) as an energy crop. In: Vermerris, W. (ed.) *Genetic Improvement of Bioenergy Crops*. Springer, New York, pp. 249–272.

Theurer, J.C., Doney, D.L., Smith, G.A., Lewellen, R.T., Hogaboam, G.J., Bugbee, W.M. and Gallian, J.J. (1987) Potential ethanol production from sugar beet and fodder beet. *Crop Science* 27, 1034–1040.
Uppal, J. (2007) Indian ethanol program – first baby step but great expectations. *International Sugar Journal* 109, 189–194.
USDOE (US Department of Energy) (2007a) DOE selects six cellulosic ethanol plants for up to $385 million in federal funding. *Press Release*, 28 February.
USDOE (US Department of Energy) (2007b) Secretary Bodman touts importance of cellulosic ethanol at Georgia biorefinery groundbreaking. *Press Release*, 6 November.
Vermerris, W., Rainbolt, C., Wright, D. and Newman, Y. (2008) Production of biofuel crops in Florida: sweet sorghum. University of Florida IFAS Extension Publication No. SS AGR 293, pp. 1–7.
Viator, R.P., Johnson, R.M. and Richard, E.P. Jr (2005) Challenges of post-harvest residue retention management in the Louisiana sugarcane industry. *Proceedings of the International Society of Sugar Cane Technologists* 25, 238–244.
Viator, R.P., Richard, E.P. Jr, Viator, B.J., Jackson, W., Waguespack, H. and Birkett, H. (2006) Sugar cane chopper harvester extractor fan and ground speed effects on sugar yield, cane quality, and field losses. *Applied Engineering in Agriculture* 23, 31–34.
Voegele, E. (2009) Sugarcane economics. *Ethanol Producer Magazine*, March.
Webster, A.J., Hoare, C.P., Sutherland, R.F. and Keating, B.A. (2004) Observations of the harvesting, transporting and trial crushing of sweet sorghum in a sugar mill. *Proceedings of the Australian Society of Sugarcane Technologists*. ASSCT, Australia, pp. 26.
Whitney, E.D. and Duffus, J.E. (1986) *Compendium of Beet Diseases and Insects*. APS Press, St Paul, Minnesota.
Woods, J. (2000) Integrating sweet sorghum and sugarcane for bioenergy: modelling the potential for electricity and ethanol production in SE Zimbabwe. PhD thesis, King's College, London, UK.
Yonts, C.D., Smith, J.A. and Wilson, R.G. (2001) Planting. In: Wilson, R.G, Smith, J.A. and Miller, S.D. (eds) *Sugarbeet Production Guide*. University of Nebraska, Lincoln, Nebraska, pp. 43–74.
Yusupov, A. (2009) Agro-industrial complex for Central Asia using sweet sorghum. *Proceedings of the Sweet Sorghum Ethanol Association Meeting*, Orlando, Florida (http://www.sseassociation.org, accessed 24 July 2009).

5

Ethanol from Grain Crops

Renyong Zhao, Xiaorong Wu, Scott Bean and Donghai Wang

Introduction

Ethanol as a transportation fuel and fuel additive

Ethanol as a transportation fuel in the USA is not new. The first fuel-flexible car using ethanol or ethanol-blended gasoline was the Ford Model T in 1908. During the First World War and in the 1930s, use of ethanol as transportation fuel for cars was common. Because of the discovery of cheap crude oil and the rise of the petroleum industry, interest in fuel ethanol waned and its production declined drastically (Solomon *et al.*, 2007; Kamm *et al.*, 2008). The oil crisis in the 1970s revived the fuel ethanol industry in the USA, but production was very low. Fuel ethanol production in the USA increased steadily in the 1980s with the phasing out of leaded gasoline by the EPA. Annual US production reached a historical 3.79 billion l (one billion gallons) in 1992. The real momentum for rapid growth started with the ban of methyl tertiary butyl ether (MTBE) in the most populated states in 2002 and the passage of the 2005 Energy Policy Act and Renewable Fuel Standards. In 2006, the US annual fuel ethanol production reached 18.36 billion l (4.85 billion gallons) and surpassed the Brazil ethanol production of 17.72 billion l (4.68 billion gallons) (Fig. 5.1). According to the Renewable Fuels Association (RFA), the USA has been the world's top fuel ethanol producer since then and produced 34.83 billion l (9.2 billion gallons) in 2008. Approximately 90% of the world's fuel ethanol is produced by the USA and Brazil (Fig. 5.2). Ethanol production in the EU, China and other countries has also grown rapidly. There are two main causes for this rapid increase in fuel ethanol production around the world: (i) to increase energy security for oil-deficient countries (Arunachalam and Fleischer, 2008); and (ii) to lower carbon dioxide emissions, a major contributor to the climate changes in the past two centuries (Johnson *et al.*, 2007).

The large-scale production and use of fuel ethanol not only can reduce significantly the amount of crude oil imported from politically instable regions and improve national energy security, but also can have enormous environmental benefits. The use of fuel ethanol can greatly reduce harmful gases (CO, CO_2, volatile organic compounds and SO_2) and particulate matter emission and improve air quality (RFA, http://www.ethanolrfa.org/resource/facts/environment/). The production and use of fuel ethanol in the USA alone in 2008 reduced CO_2 emissions by 14 Mt, which is equal to removing 2.1 million cars from America's roads (Dinneen, 2009).

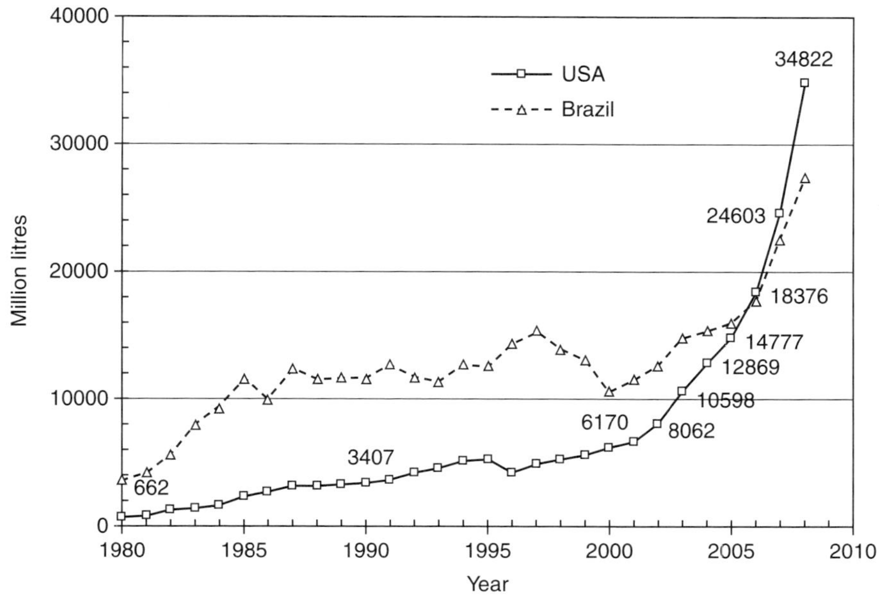

Fig. 5.1. Historical fuel ethanol production in the USA and Brazil (Goldemberg, 1996; Unica website at http://english.unica.com.br/dadoscotacao/estatistica/; and RFA, 2009). 1 gallon = 3.785 l.

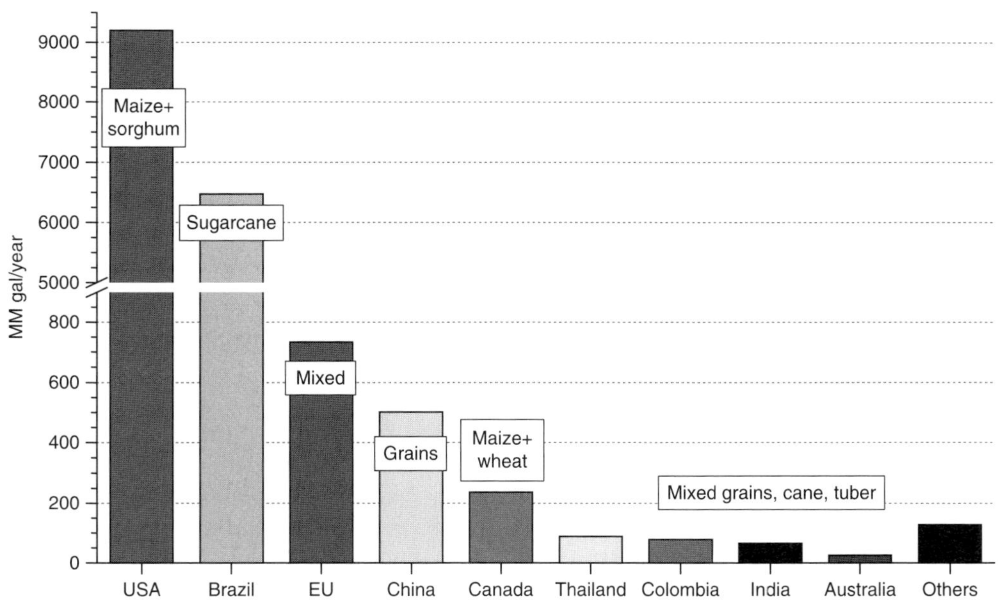

Fig. 5.2. World fuel ethanol production in 2008 (RFA, 2009). 1 gallon = 3.785 l.

Ethanol from grain crops

With the exception of Brazil, which uses sugarcane as feedstock for fuel ethanol production, most of the world's fuel ethanol is produced from various kinds of cereal grains. In the USA, approximately 95% of its fuel ethanol is produced from maize, about 4% from grain sorghum and around 1% from other feedstock such as wheat, potato, whey, etc. In 2008, the fuel ethanol industry in the USA consumed approximately 23% of the 2008 maize and sorghum crops (RFA, 2009). In Canada, the annual fuel ethanol production in 2008 reached 901 million l (238 million gallons), with approximately 70% from maize, 25% from wheat and the rest from barley and agricultural and forest waste (USDA-FAS, 2007a).

Fuel ethanol production in China began officially in 2004 and reached 1.9 billion l (502 million gallons) in 2008 (USDA-FAS, 2007b; RFA, 2009). About 50% of China's current fuel ethanol is produced from grains (mostly maize, wheat, rice and sorghum); the other 50% is from cassava and sweet potatoes.

The annual fuel ethanol production in the 27 EU states was approximately 2.78 billion l (733 million gallons) in 2008 (Fig. 5.2). The feedstocks for fuel ethanol production in the EU are more diverse than other regions because of the differences in climate and geographic conditions among the member states. The major feedstocks include wheat (39%), rye (17%), barley (6%), wine alcohol (16%) and sugarbeet (USDA-FAS, 2007c).

Kim and Dale (2004) carried out a thorough review on potential grains for fuel ethanol production in different areas of the world, which showed that nearly 13 billion gallons of fuel ethanol could be produced from the world's waste crops.

Chemistry of Ethanol Production

The basic principles in modern ethanol production from cereal grains involve hydrolysis of starches to glucose using starch hydrolysing enzymes (thermostable α-amylase and amyloglucosidase) and conversion of glucose and other fermentable sugars (maltose, sucrose and fructose) into ethanol using yeast (Fig. 5.3). Every gram of starch theoretically may be converted into 1.11 g of glucose, and each gram of glucose may generate 0.511 g of ethanol.

Ethanol Production Process from Grain Crops

Depending on the processing characteristics, ethanol production from cereal and other starch-rich materials may be defined as either a wet milling process or a dry grind process. Two major differences between the dry grind and wet milling processes are the way in which major components (co-products) other than starch in cereal grains are handled and the initial building cost. The initial building cost for dry grind ethanol plants is between US\$0.317 and US\$0.793/l (US\$1.20 and US\$3.00/gallon) capacity, which is approximately two to four times less expensive than the cost for wet milling facilities (Bothast and Schlicher, 2005; Shapouri and Gallagher, 2005). Conventional dry grind ethanol plants only generate one co-product, dried distillers grain with solubles (DDGS), whereas wet milling ethanol plants can produce a variety of co-products depending on the feedstocks used and market demand. As newer processing technologies enter the dry grind ethanol industry, differences between dry grind and wet milling become less distinct in terms of co-product varieties and ethanol production process (Rausch and Belyea, 2006). Before 2000, a major portion of the US fuel ethanol (6.1 billion l or 1.6 billion gallons in 2000) was produced with the wet milling process in ethanol plants owned by large corporations. Most ethanol plants built since then have been dry grind ethanol plants, and the portion of fuel ethanol

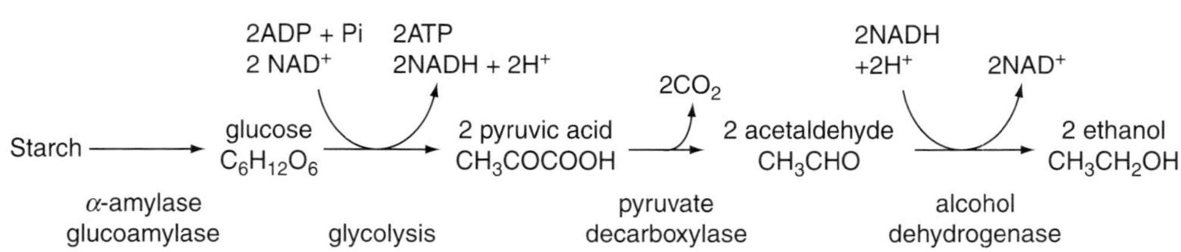

Fig. 5.3. The process of converting starch to ethanol (adapted from Nichols and Bothast, 2008).

produced by the dry grind process has grown steadily and rapidly. By 2007, more than 80% of the 24.6 billion l (6.5 billion gallons) of fuel ethanol produced in the USA was from dry grind ethanol plants (Graybosch *et al.*, 2009). Data from the Renewable Fuels Association indicate the growth trend is still strong. Fuel ethanol production in the USA in 2008 was approximately 34.1 billion l (9.0 billion gallons) and the total annual capacity had reached 46.9 billion l (12.4 billion gallons) by 193 nameplate ethanol plants as of February 2009. Of this amount, 39.4 billion l (10.4 billion gallons) were operating capacity and 7.8 billion l (2.07 billion gallons) were under construction/expansion capacity (RFA, http://www.ethanolrfa.org/industry/locations/, accessed February 2009).

Ethanol production from dry grind process

In a conventional dry grind ethanol plant, cereal grains or other starch-rich feedstocks go through five stages of processing to be converted to fuel ethanol: grinding, liquefaction, saccharification and fermentation, distillation and dehydration (Fig. 5.4).

Grinding

At this stage, kernels of cereal grains (maize, wheat, sorghum, barley, rye or oats) are ground into a meal with hammer mills or roller mills. The process is simple, but the efficiency of grinding has a great effect on the final ethanol yield. Inefficient grinding results in coarser particle sizes in the ground meal. Coarser meal could reduce the final ethanol yield by 5–10% (Kelsall and Lyons, 2003), or even up to 20% (Naidu *et al.*, 2007) depending on particle size. Therefore, routinely checking the particle size distribution of ground meal and making timely adjustments to mill settings are important and essential practices for any dry grind ethanol facility.

Liquefaction

The finely ground cereal meal is mixed with process water (40–60°C) and a small portion

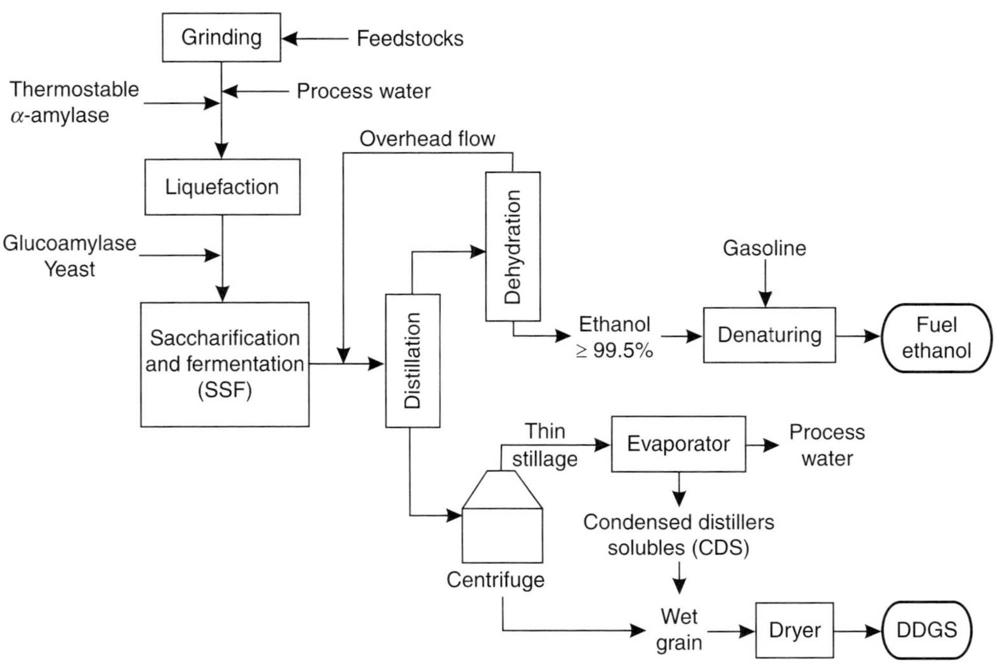

Fig. 5.4. Schematic outline of conventional dry grind process for fuel ethanol production (modified from Kelsall and Lyons, 2003; Singh, 2008).

of thermostable α-amylase (~ 0.02%) to form a slurry. The slurry is then heated to 90–120°C with a jet cooker to gelatinize starches in the slurry. The gelatinized slurry is kept in a liquefaction tank at ≈ 90°C and the major dose of thermostable α-amylase (~ 0.04–0.06%) is added until the starches are totally liquefied (in 1–2 h).

Saccharification and fermentation

Most dry grind fuel ethanol plants use the simultaneous saccharification and fermentation (SSF) method. The liquefied mash is cooled to 30–32°C through a heat exchanger, transferred to a fermentation tank and adjusted to pH 4–5 with sulfuric acid. Glucoamylase (0.06–0.12%) is added to break down dextrins and oligosaccharides in the liquefied mash into glucose and maltose. At the same time, yeast is inoculated into the mash at ~ 1×10^7 cell/ml. Some ethanol plants may also add another amylase, rhizozyme, which has some cellulase and pectinase activity in addition to α-amylase activity, and therefore could release fermentable sugars from some non-starch polysaccharides and increase ethanol yield to a certain extent. In addition to ethanol, large quantities of CO_2 are generated during fermentation and may be recovered for other industries (e.g. carbonating beverages, making dry ice). The fermentation process normally finishes in 40–60 h, with 12–14% (v/v) ethanol concentration in the finished beer (Nichols *et al.*, 2008). Some ethanol concentration may even reach 19–20% (v/v) (Kelsall and Lyons, 2003).

Distillation

The finished beer with residual solids is pumped through distillation columns to separate ethanol from the solids and water. The highest ethanol concentration achievable by distillation is about 96%. The residues after distillation, called stillage, are separated by centrifugation into solid (grain residue) and liquid (thin stillage) fractions. The thin stillage is condensed by evaporation into syrup, which may be mixed back with the grain residue to form wet distillers grain (WDG). The WDG can be sold directly as animal feed or dried to produce DDGS.

Dehydration

The small amount of water in the azeotropic ethanol is removed with molecular sieves. As the 96% pure azeotropic ethanol passes through the molecular sieves, water molecules are trapped in the sieves, but ethanol molecules are too large to enter the sieves. The output from the dehydration column is pure ethanol (200 proof), which is then denatured with a small amount of gasoline (~ 3–5%) into fuel ethanol.

As described previously, WDG or DDGS are the only co-products of the conventional dry grind ethanol process (CO_2 may also be a co-product if recovered, but its market is very limited). Market values of WDG and DDGS are very important to a dry grind ethanol plant because these products account for up to 20% of a plant's annual revenue. Because of the fast growth of the dry grind ethanol industry, market values for WDG and DDGS are not very promising. Processing WDG or DDGS into other value-added products or expanding the variety of co-products produced is an urgent task for the dry grind ethanol industry.

Ethanol production from wet milling process

The wet milling ethanol process is much more complex than the dry grind process and has more diverse co-products. Theoretically, any cereal grain may be processed with wet milling, but only maize and wheat are used commercially. The actual wet milling process for maize is very different from that for wheat because of the different characteristics of gluten proteins in each grain. Figure 5.5 is a flow chart of the maize wet milling ethanol process, which can be distinguished from the dry grind process by the steeping, degerm, defibre and starch/gluten separation processes. Rausch and Belyea (2006) provide a detailed description of co-products from the maize wet milling process.

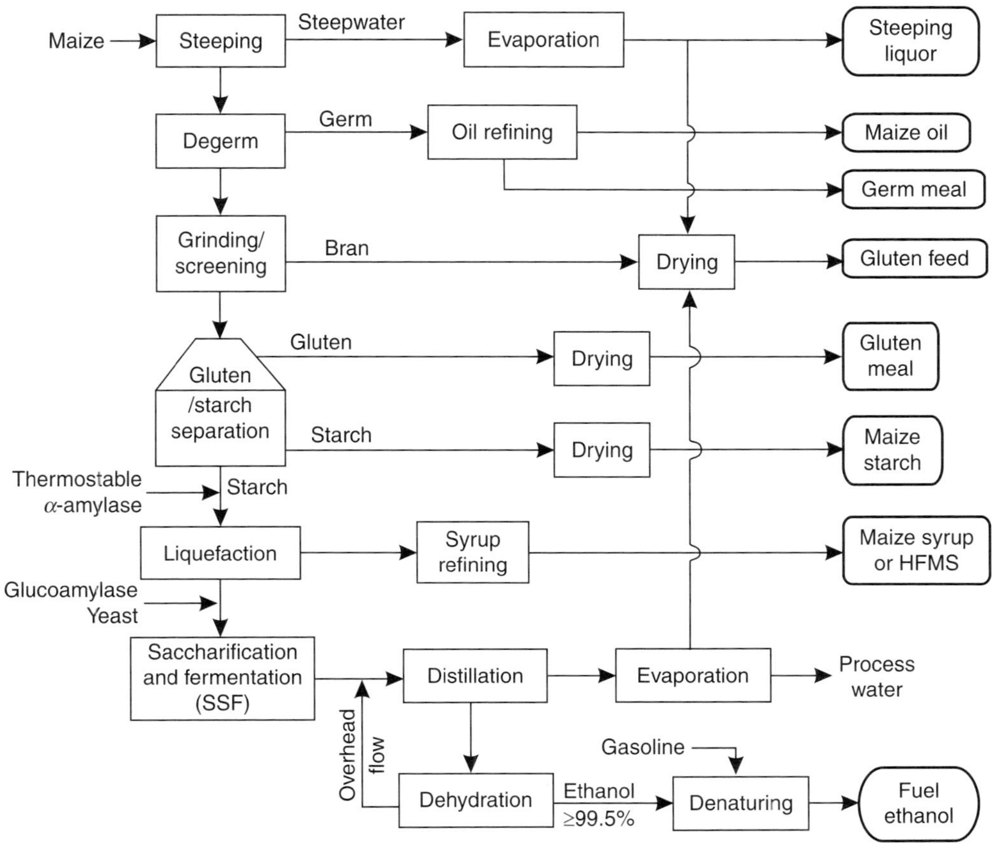

Fig. 5.5. Schematic chart of the ethanol wet milling process (modified from Galitsky *et al.*, 2003; Bothast and Schlicher, 2005).

Steeping

Steeping, a crucial process in wet milling, involves a biochemical, chemical and mechanical process in which maize kernels are conditioned to achieve optimal separation of germ, fibre and gluten from starch in downstream processing. Maize kernels are soaked in a hot (50–52°C), dilute sulfurous acid solution (0.12–0.20%) for 24–48 h, which hydrates and softens the maize kernels and leaches solubles from the germ. Sulfurous acid inhibits the growth of other undesirable bacteria in the steeping water and the activity of lactic acid-producing bacteria helps maintain an acidic environment, which facilitates disintegration of the endosperm protein matrix surrounding the starch granules by SO_2 (Singh and Eckhoff, 1996b). The steep water is concentrated to 30–55% solids by evaporation. The resulting steeping liquor, or heavy steep water, containing 35–45% proteins, could be sold as fermentation nutrients or mixed into other milling residues to produce gluten feed.

Degerm

The steeped maize kernels are torn apart in a degermination mill and a mixture of maize germ and starchy slurry is formed. The oil-rich germ is separated from the main starch and gluten slurry with hydrocyclones because the density of germ is lower than that of the starchy slurry. The removed germ is subsequently washed, dewatered and dried. After the oil is extracted and refined into maize oil, the spent germ is processed into germ meal or used in maize gluten feed (MGF).

Defibre

The slurry from the degermination mill passes through a series of washing, grinding and screening operations. Fibrous material (pericarp and endosperm cell wall fibres) is separated from the starch and gluten slurry by staying on top of the screens (50 μm opening) and subsequently is dried as MGF. Maize fibre may be further processed into maize fibre oil or maize fibre gum (Singh *et al.*, 2000a,b).

Starch/gluten separation

The germ- and fibre-free starch/gluten slurry is separated by centrifugation into a top light stream (gluten) and bottom heavy stream (starch) according to the density differences between gluten (1.1 g/cm^3) and starch (1.5 g/cm^3). The heavier starch stream is then washed continuously with water to remove residual gluten and other solubles in a series of hydrocyclones to the setting purity. The lighter gluten stream containing 1.0–1.5% solids (60–70% of the solids are protein) is then concentrated by centrifugation, dewatered and dried to form a high-protein co-product, maize gluten meal.

After the starches are separated from gluten and purified, they can be used to produce maize starch, modified maize starches, maize syrup, high fructose maize syrup (HFMS) or ethanol. For fuel ethanol production, the starch slurry will go through liquefaction, SSF, distillation and dehydration processes, as described for the dry grind ethanol process. The major advantage of the wet milling ethanol process is its diverse, high-value co-products, such as maize oil, germ meal, maize gluten meal (MGM), MGF, maize starch, modified maize starch, maize syrup and HFMS. Another advantage of the wet milling process is its ability to divert its resources easily to various products to meet the ever-changing market demand.

Wet milling of wheat usually starts with wheat flour as feedstock (or wheat is first separated into germ, bran and wheat flour in a dry milling process) because wheat gluten from the whole-kernel wet milling process loses its functionality in bread making. Sayaslan (2004) and Van Der Borght *et al.* (2005) provided thorough, in-depth reviews of the wheat wet milling process. Four different processes – the Martin, Alf-Laval/Raisio, hydrocyclone and high-pressure disintegration (HD) – have been used in commercial wheat wet milling to separate and purify wheat gluten from wheat starch on the basis of differences in water solubility, density and particle size (Sayaslan, 2004). Recoveries for both starch and gluten from wheat flour are around 80% in vital wheat gluten and wheat starch (Van Der Borght *et al.*, 2005). The protein stream is purified and processed into high-value, food-grade vital wheat gluten (~ 80% protein) and the starch stream can be processed into high quality starch (≈ 98% starch) or other products, as described for the maize wet milling process. Residual fibres, pentosans and some water solubles separated from starch and gluten purifying processes are concentrated, dewatered and dried into wheat gluten feed.

Improvement and recent advances in the ethanol industry

To be economically viable, an ethanol plant has to reduce production costs by improving production and energy efficiency and at the same time expanding co-product varieties and maximizing the market values of co-products (Bothast and Schlicher, 2005). Great progress has been made during the past decade in all areas related to fuel ethanol production, such as feedstock variety breeding, processing technologies, equipment and process efficiency. From 2001 to 2007, ethanol yield per unit weight increased by 6.4% and energy and water consumption decreased by 21.8% and 26.6%, respectively, in the dry grind industry (Wu, 2008).

New cereal hybrids

New cereal hybrids with high ethanol yields for dry grind ethanol production or high extractable starch for the wet milling industry have been developed and marketed. Hybrids with high total fermentables (HTF) or high fermentable maize (HFM) hybrids from

Mycogen, Monsanto, Pioneer, Syngenta and other companies could increase ethanol yield by 3–5%. These new varieties increase ethanol yield from the normal 409 to 432 l/t (2.75 gallons/bushel to 2.90 gallons/bushel), which could lower the production cost of ethanol significantly (Rendleman and Shapouri, 2007). Many hybrids possess both HTF and high extractable starch traits, which will perform well in both wet milling and dry grind processes. New waxy wheat and sorghum hybrids with high ethanol fermentation efficiency are also being introduced into the market. Waxy hybrids have shown better processing quality (low viscosity and quick liquefaction) and a higher starch–ethanol conversion rate (less residual starch in DDGS) than their normal counterparts in laboratory studies of the dry grind process (Wu *et al.*, 2006b).

Processing technologies

The most energy-intensive processes in ethanol plants are the cooking and distillation processes, which account for up to 25% of the total ethanol production cost (Shapouri and Gallagher, 2005). Processing with raw starch-hydrolysing enzymes not only greatly reduces energy consumption but also significantly lowers the content of residual sugar, organic acids and glycerol in the spent grains. This lowers production costs and improves the quality of DDGS (Lewis, 2006, 2007). POET (used to be Broin, Sioux Falls, South Dakota) integrated fractionation technology successfully (the BFRAC [Broin Fractionation] process) with raw starch hydrolysis technology (the BPX [Broin Project X] process, using an acid fungal amylase from Novozyme) in several of its dry grind ethanol plants. Plants with such integrated technologies will have higher ethanol yield, increased nutrient quality and flowability of DDGS, lower plant emissions and reduced energy costs of up to 15%. Integrated technologies have been proven commercially viable (Bryan, 2005; Lewis, 2007).

The quick-germ (QG), quick-germ/quick-fibre (QGQF) and enzymatic-milling (E-milling) technologies developed at the University of Illinois are other examples of upstream fractionation technology (Singh and Eckhoff, 1996a; Singh *et al.*, 1999, 2005). Although not yet commercialized, these technologies have the potential to lower ethanol production costs by generating high-value co-products (maize oil, germ meal, maize fibre and high-protein DDGS), reducing energy consumption and increasing the capacity of the original dry grind equipment by 7–20% (Eckhoff, 2001; Singh *et al.*, 2005; Ramirez *et al.*, 2009). Decortication of cereal grains may be another way to improve dry grind ethanol production. Corredor *et al.* (2006) reported that decortication could improve the fermentation performance of grain sorghum significantly in the dry grind process, as well as reduce the amount but increase the protein content of DDGS. Similar results have been reported for other cereals (Wang *et al.*, 1997a).

Fermentation technology

The dry grind ethanol process usually operates on a batch basis. Some wet milling plants run on a continuous basis, which can last for several months to a year. Continuous processing has the advantages of lower equipment and maintenance costs, less labour, less yeast usage, no peak utility period and better ethanol yield. Arifeen *et al.* (2007) provides additional details about features of continuous ethanol fermentation.

Development of new fermentation microorganisms with specific features, such as tolerance to high ethanol concentrations (up to 18–23% v/v) and fairly high temperatures (up to 60°C), has greatly improved fermentation rates and efficiency in the past decade.

Because fibre in DDGS is considered a restriction factor for its application in non-ruminant animal feed, many ethanol producers plan to convert the fibres in cereal or DDGS into ethanol, which can increase ethanol yield and the value of DDGS as a protein supplement. Some genetically engineered yeasts (Ho *et al.*, 1998; Liu *et al.*, 2008) and bacteria (Alterthum and Ingram, 1989; Zhang *et al.*, 1995; Luli *et al.*, 2008) have been constructed to co-ferment both C-5 and C-6 carbohydrates from cellulose hydrolysates and have achieved very promising results. Pilot-scale and demonstration tests have been conducted on these engineered organisms.

Improvements are still needed to increase the conversion efficiency of cellulosic biomass and ethanol concentration in the finished beer before these organisms are used in commercial ethanol production.

Another approach to increase ethanol yield is to reduce the formation of glycerol during ethanol fermentation by yeast. One molecule of ethanol could have been produced for every molecule of glycerol produced during ethanol fermentation by yeast. Because glycerol concentration in finished beer is approximately 1%, 0.5% ethanol could be gained without any extra cost if the glycerol level was reduced by half of its current level. Some industry researchers have worked on this subject, but no exciting progress has been reported.

Ethanol recovery technology

The revolutionary technology improvement in fuel ethanol production in the 1980s was the replacement of azeotropic distillation with molecular sieve dehydration (MSD) to process 96% pure ethanol into anhydrous fuel-grade ethanol. This not only lowered the original capital investment (US$250,000 per distillation column) and energy consumption significantly (saving ~ 20% energy), but also improved the working environment greatly (no exposure to hazardous chemicals) (Rendleman and Shapouri, 2007). However, the MSD process has a complex control system, needs intensive maintenance and generates a purge stream of 60–80% ethanol that must be redistilled.

A new membrane-based dehydration technology has been developed by Vaperma (Saint-Romuald, Canada) and succeeded in 7.57 million l/year scale test. This technology, called Siftek, separates ethanol water vapours in polymeric hollow fibres. When ethanol and water vapours pass these hollow fibres (0.2 mm-thick walls and 2000 Å-thick skin), ethanol is dehydrated while water vapour permeates the membrane and is desorbed at the outer surface of the membrane under low pressure (low vacuum). This membrane dehydration process could turn a stream of 40–60% ethanol continuously into 99% fuel-grade ethanol. If coupled with the current MSD dehydration process, the Siftek membrane dehydration process could save approximately 40% of the dehydration and reboiling energy of the normal MSD process and increase the capacity of the MSD units by 20–25% (Plante *et al.*, 2008).

Biological and engineering constraints for processing mixed feedstocks (simultaneously or sequentially)

Modern ethanol plants process several thousand tonnes of feedstock daily. It is impossible to acquire the same kind of feedstock year-round. Because different kinds of feedstock vary greatly in their physical (shape, size, hardness, etc.), chemical (moisture, starch, protein, oil, fibre, etc.) and biological (enzyme inhibitors, mycotoxin contamination) properties, mixed feedstock presents a constant challenge in all critical process steps for dry grind and wet milling ethanol production. Possible problems related to inconsistent feedstocks at various stages of the processing are:

- Feedstock receiving: quality control for heterogeneous or mixed feedstocks is difficult.
- Grinding: equipment must be adjusted and grinding results must be checked more frequently.
- Steeping: achieving uniform steeping effects on feedstocks with varying sizes and shapes is challenging, which inevitably influences the efficiency of degerm and defibre processes.
- Liquefaction: the different contents of tannin, pentosans and other non-starch polysaccharides may cause problems if not known in advance.
- SSF process: elements unfavourable to the liquefaction process similarly may affect the saccharification process, but nutrients from different feedstocks may be complimentary and beneficial to the fermentation process.
- Co-product quality: inconsistent or mixed feedstocks may change co-product quality favourably or unfavourably to some degree.

DDGS as co-product from dry grind process

More than 80% of the fuel ethanol currently produced in the USA is produced with the conventional dry grind process; this means DDGS is the largest product stream in an ethanol plant. Sales from DDGS account for 15–20% of the annual revenue of a maize dry grind ethanol plant. Thus, quality and market value of DDGS are just as important as ethanol yield to the success or failure of an ethanol plant.

The US ethanol industry generated 22 million metric tons of DDGS in 2008; approximately 20% of the 2008 DDGS was exported (Schill, 2009). Because of its high fibre content, DDGS is used mainly (84%) in the cattle and dairy industries (42% each); swine and poultry industries account for only 11 and 5%, respectively (Christiansen, 2009). The DDGS from newer, modified dry grind processes (e.g. BFRAC, BPX, QGQF) has much higher protein and lower oil and fibre contents because the germ and fibre portions are removed upfront; this product is called high-protein distillers dried grains (HPDDG). Application of HPDDG will not be limited to the dairy and cattle industries and recent tests have showed that HPDDG can be a high quality protein source for both swine (Widmer *et al.*, 2008) and poultry (Applegate *et al.*, 2009).

In the conventional dry grind ethanol process, about two-thirds of the feedstock is starch, which is converted into ethanol, and the other one-third is processed into DDGS. The contents of protein, oil and fibre, even mycotoxins if they exist, in the DDGS will be about three times more concentrated than those in the original feedstock (Table 5.1).

Generally, DDGS from feedstocks with high protein content, such as wheat, will have higher crude protein content, and DDGS from feedstocks with high fibre (oats) or oil (maize) content will have higher crude fibre or oil contents. Certainly, starch content also plays a role in the chemical composition of DDGS. Higher starch content in the starting material

Table 5.1. Major components of DDGS from different feedstocks and dry grind processes (%, dry basis [db]).

DDGS	Crude protein[a]	Starch	Crude fat	Crude fibre	Ash	Sources
Barley	28.7–29.5	n.a.	4.45–10.4	12.1–15.4	2.04–6.50	Wu, 1986; Kim *et al.*, 1989
Maize	28.1–33.3	4.7–5.9	8.2–12.6	7.1–10.6	4.3–6.7	Spiehs *et al.*, 2002; Belyea *et al.*, 2004
Millet (pearl)	30.7	3.5	19.2	4.3	5.2	Wu *et al.*, 2006a
Oats	13.6	n.a.	9.70	28.8	3.81	Kim *et al.*, 1989
Rye	24.5–25.0	1.1	2.29–11.2	14.5–33 (TDF)	2.36–5.32	Kim *et al.*, 1989; Wang *et al.*, 1997b
Sorghum (grain)	29.0	n.a.	7.30	19.33	2.57	Kim *et al.*, 1989
	36.6	4.7	9.2	8.7	5.5	Wu and Sexson, 1984
Sorghum (sweet)	36.8	4.36	9.80	6.54	4.89	Wu, 1987
Triticale	33.5–34.2	1.0–2.0	3.55–4.10	34.0–35.8 (TDF)	4.7–4.9	Wang *et al.*, 1997b
Wheat (hard)	36.5	n.a.	3.09	10.4	4.80	Wu *et al.*, 1984
Wheat (soft)	30.3	n.a.	4.02	10.3	4.78	Wu *et al.*, 1984
Wheat (durum)	42.7	0.8	5.6	n.a.	5.3	Lee *et al.*, 1991
Wheat (CPS)	41.9	1.5	3.9	n.a.	5.9	Lee *et al.*, 1991
Wheat:maize (30:70)	33.5	1.1	12.1	n.a.	5.3	Lee *et al.*, 1991
Wheat:maize (70:30)	35.9	1.8	9.9	n.a.	4.9	Lee *et al.*, 1991

Continued

Table 5.1. Continued.

DDGS	Crude protein[a]	Starch	Crude fat	Crude fibre	Ash	Sources
Sorghum						
10% decorticated	46.6–53.1	5.4–5.5	11.4–13	5.8–6.9	3.6–4.0	Corredor *et al.*, 2006
20% decorticated	51.6–56.8	5.6–5.7	11.8–14.2	3.4–4.7	3.9–3.9	
HPDDG (maize)	44.6	n.a.	4.18	20.4 ADF	1.90	Tedeschi *et al.*, 2009
BPX DDGS (maize)	30.8	n.a.	11.2	12.8 ADF	4.19	
QG	35.9	n.a.	4.83	8.22 ADF	4.05	Singh *et al.*, 2005
QGQF	49.3	n.a.	3.85	6.80 ADF	4.13	Singh *et al.*, 2005
E-mill	58.9	n.a.	4.53	2.03 ADF	3.24	Singh *et al.*, 2005

[a]Protein contents were calculated by multiplying N% with 5.7 for wheat and 6.25 for others. n.a. = not available; CPS = Canadian Prairie Spring; TDF = total dietary fibre; ADF = acid detergent fibre.

(e.g. decorticated sorghum or fractionalized maize) will result in less spent grain and higher contents of residual components such as protein, fat, fibre and ash.

As the fuel ethanol industry booms, DDGS will become a multibillion dollar market. But there are no well-accepted standard methods to monitor DDGS quality. Standard methods applied to other commodities often give very different results when the same items are analysed in different laboratories. The American Feed Industry Association (AFIA), the National Maize Growers Association (NCGA) and the RFA funded a multiparty collaborative research project on selecting approximate analysis methods determining moisture, crude protein, crude fat and crude fibre content in DDGS. Five of the 13 methods evaluated were selected as recommended methods for proximate composition analyses of DDGS (Table 5.2) and released as AFIA's final report and guidelines (AFIA, 2007).

In addition to the nutrient values of DDGS, special attention should be paid to some contaminants, especially mycotoxins in feedstocks. Mycotoxins in feedstocks will not be destroyed during the ethanol production process; instead, they will be concentrated multiple times in various co-products (~ 3 times in DDGS). Ignoring the possible existence of mycotoxins in DDGS could cause catastrophic economic loss to an ethanol plant because regulatory bodies in the USA and EU, the major importer of US DDGS, have already set limits or guidelines for common mycotoxins such as aflatoxins, deoxynivalenol, zearalenone, ochratoxin A, fumonisins, T-2 and HT-2 toxin (Kyprianou, 2006; Wu and Munkvold, 2008).

Table 5.2. AFIA recommended methods for DDGS composition analysis.

Analysis	Method
Moisture	Lab dry matter (105°C/3 h) (NFTA 2.2.2.5)
Crude protein	Protein (crude) in animal feed – combustion (AOAC 990.03) or Protein (crude) in animal feed and pet food (copper catalyst) (AOAC 2001.11)
Crude fat	Oil in cereal adjuncts (petroleum ether) (AOAC 945.16)
Crude fibre	Fibre (crude) in animal feed and pet food (F.G. crucible) (AOAC 978.10)

Assessing Grain Crop Quality for Ethanol Production

According to the estimation by Schultze *et al.* (2005), feedstocks represent between 55 and 70% of bioethanol processing costs, which leaves little doubt that grain crop quality (e.g. variations in the fermentation performance of grain crops in terms of ethanol yield, conversion efficiency, ease of mashing, fermentation rate and the yield and quality of DDGS) affects ethanol producers' profit margins. Ethanol yield and conversion efficiency are the major quality traits of cereal grains used as feedstocks to produce fuel ethanol. Grains with higher ethanol yield per unit will generate more revenue.

Laboratory fermentation tests

Laboratory fermentation is the most direct and reliable method of evaluating the fermentation qualities of grain crops. Many laboratories have reported dry grind procedures, most of which are grouped into three types: (i) Ingledew and associates (Thomas and Ingledew, 1990; Ingledew *et al.*, 1995, 1999; Thomas *et al.*, 1995; Wang *et al.*, 1997b, 1999); (ii) Singh and co-workers (Singh and Graeber, 2005; Singh *et al.*, 2006; Naidu *et al.*, 2007); and (iii) Wang and colleagues (Zhan *et al.*, 2003, 2006; Wu *et al.*, 2006a,b, 2007). For Type 1 and Type 3 procedures, a weighed quantity of meal is first mixed into a vessel with a known quantity of water and α-amylase is added. Then, slurries are cooked at 90–98°C for 45–60 min. After the slurry temperature is reduced to 80°C, a second dose of α-amylase is added and liquefaction proceeds for an additional 30 min at 80°C. For the Type 2 procedure, all required α-amylase is added to the slurries in one step and the slurries are cooked at 85–90°C for 90 min. Saccharification is conducted at 30–40°C for 30 min in the Type 1 procedure, at 55–60°C for 2 h in the Type 2 procedure, and at 60°C for 30 min in the Type 3 procedure. There is no pH adjustment in Type 1. The pH of mashes is adjusted to 4.1–4.2 using 1M H_2SO_4 before saccharification in Type 2 and adjusted to 4.2–4.3 using 2M HCl before inoculation in Type 3. In Types 1 and 2, industrial strains of active dry yeast are used and activated at 38°C for 20–25 min before inoculation; in Type 3, a laboratory strain of *Saccharomyces cerevisiae* (ATCC 24860) is used and pre-cultured at 30°C for 48 h before inoculation. Lee *et al.* (2000) and Mojovic *et al.* (2006) reported procedures similar to these three types, and Devantier *et al.* (2005) reported a procedure that used jet cooking during mash preparation.

The traditional procedure described by Wu *et al.* (2007) was tedious and laborious, especially the steps of mash preparation, pH adjustment and yeast pre-culture preparation. To save time, increase experiment repeatability and mimic fuel ethanol production in the dry grind industry, that procedure was modified by Zhao *et al.* (2009a) as follows: liquefaction was simplified from two steps (95°C for 45 min and 80°C for 30 min) to one step (86°C for 90 min), yeast pre-culture preparation was substituted with commercially available active dry yeast and the SSF procedure was applied. There was a strong linear relationship in ethanol yields of 18 sorghum hybrids between the traditional and the SSF procedure. Ethanol yield improved significantly with SSF.

Because starch is converted to ethanol in a dry grind process, it seems logical to assume that the amount of starch is related to ethanol yield. However, even with the same laboratory protocol tailored to simulate a commercial production process, controversial observations about the relationship between starch content and alcohol yield exist. Swanston *et al.* (2007) reported that starch content did not correlate significantly with ethanol yield; Kindreda *et al.* (2008) found a positive relationship between alcohol yield and starch concentration, but starch content could explain only 37% of the variance in alcohol yield; and in a report reviewed by Smith *et al.* (2006), there was a much better correlation between starch and alcohol yield (R^2 = 0.78). Recently, using a dry grind procedure, Lacerenza *et al.* (2008) reported that starch content in spring wheat was correlated highly with ethanol yield (R^2 = 0.60). The inconsistent results were due to the inherent variability and difficulty in the starch measurements (Smith *et al.*, 2006; Kindreda *et al.*, 2008). Another explanation may be related to fermentation procedures. Using sorghum grain, Zhao *et al.* (2009a) found a positive correlation between total starch and ethanol yield (R^2 = 0.86) in SSF that was stronger than that in traditional fermentation (R^2 = 0.78). Protein content has been reported to correlate negatively with ethanol yield and predicts ethanol yield with more precision than starch content (Swanston *et al.*, 2005, 2007; Smith *et al.*, 2006; Kindreda *et al.*, 2008). Notably, a recent study

(Zhao *et al.*, 2009b) showed highly significant correlations between ethanol yield and both total starch and protein content. For all 30 wheat cultivars with a broad range of genotypes, up to 95% of variation in ethanol yield could be explained by total starch and up to 84% by protein.

Crop and grain quality effects on ethanol production (maize, sorghum, wheat and others)

In North America, ethanol is produced almost entirely from maize with one of two methods, the wet milling (18%) or dry grind (82%) processes (RFA, 2007). Ethanol production has become an important and expanding market for maize grain. In 2007, conversion to ethanol accounted for 58.5 million t (2.3 billion bushels) of maize, nearly 18% of the 333 million t US maize crop (13.1 billion bushels) (RFA, 2008). Starch is the major carbohydrate storage component in maize kernels, comprising 70–72% of the kernel weight on a dry-weight basis (Bothast and Schlicher, 2005). Current technologies allow for 408 ± 191 of ethanol/t (2.74 ± 0.13 or 2.51–2.87 gallons/bushel) (wet milling process) to 418 ± 131 of ethanol/t of maize (2.81 ± 0.09 or 2.62–2.96 gallons/bushel) (dry grind process) (Wu *et al.*, 2007). On the basis of the extensive database published annually for hybrid maize varieties in Illinois, Patzek (2006) obtained detailed statistical estimates of the mean extractable maize starch (66.18 ± 1.13%) and mean theoretical ethanol yield (393 ± 71/t or 2.64 ± 0.05 gallons/wet bushel) and concluded that the US ethanol industry had inflated its ethanol yields consistently by counting 5% by volume of gasoline denaturant as ethanol.

Maize quality is associated with two products: ethanol and DDGS. Quality tests commonly performed on incoming maize in a dry grind ethanol plant include moisture content, broken maize and foreign material, test weight and mycotoxins. Laboratory and commercial testing has demonstrated that maize hybrids differ in ethanol yield potential (Haefele *et al.*, 2004). Three maize quality factors (kernel composition, endosperm hardness and planting location) affected or controlled by genetics and environment affect dry grind ethanol yields. Approximately 23% of observed variability in ethanol yield was due to hybrid and planting locations, and maize hybrids with increasing true density gave higher final ethanol yields (Singh *et al.*, 2004). Predicting ethanol yield from total starch has both practical and theoretical limits (Haefele *et al.*, 2004). Research from Pioneer, a seed company, showed that the HTF trait of maize was a more accurate indicator of dry grind ethanol production than total starch (Bryan, 2003). Additionally, Pioneer developed a point-of-sale assay using whole grain near-infrared (NIR) technology that allowed ethanol plants to predict the value of maize for ethanol production by identifying HTF grain arriving at the plant.

Because starch is the principal component of maize, other cereal grains including sorghum, wheat, millet, rice and barley are obvious ethanol feedstocks in areas where maize production is limited.

Researchers and ethanol producers have shown that grain sorghum is a reasonable feedstock (technically acceptable, fits the infrastructure and can be economically viable) for ethanol and could make a larger contribution to the nation's fuel ethanol requirements. Interest in using grain sorghum for bioindustrial applications is now growing in the USA (Farrell *et al.*, 2006). Because of climate diversity and continuing decline of water resources, it is crucially important to use available dry land to grow grain sorghum in a way that ensures sustainable economic development and rational economic distribution.

Variety, ratio of amylose to amylopectin, protein–starch interaction, tannin level, mash viscosity, formation of amylose–lipid complexes and particle size of ground sorghum meal had significant effects on ethanol yield and conversion efficiency from grain sorghum (Zhan *et al.*, 2003; Wu *et al.*, 2007). Starch content in grain sorghum could be 64–74%, which could result in a 22% difference in ethanol yield for the same amount of grain. Ethanol yields from sorghums with similar starch contents varied as much as 7.4%, which indicated that not all the starch contributed equally to ethanol yield (Wu *et al.*,

2007). Waxy and heterowaxy varieties generally have higher fermentation efficiencies than non-waxy varieties because amylose is likely to form amylose–lipid complexes in seeds or during mashing that are resistant to enzymatic hydrolysis (Wu *et al.*, 2006b). As in other cereal grains, protein content in grain sorghum is inversely proportional to starch content and thus has a negative effect on ethanol yield (Zhan *et al.*, 2003; Wu *et al.*, 2007; Zhao *et al.*, 2008a). There could be as much as an 8% difference in ethanol yield among sorghum varieties with similar protein content. The most probable reason for the adverse effects of protein on ethanol fermentation is formation of a web-like protein matrix by cross-linking of sorghum proteins during mashing or cooking that prevent starch granules in the enmeshed matrix from gelatinization, limit their accessibility to enzyme hydrolysis and consequently lower the digestibility of sorghum starch (Zhang and Hamaker, 1998; Wu *et al.*, 2007; Zhao *et al.*, 2008a,b). The degree of protein cross-linking varied among sorghum cultivars; the lower the degree of protein cross-linking, the easier the enzyme access to starch granules, gelatinized starch or oligosaccharides held by the cross-linked protein matrix and the better the fermentation performance (Zhao *et al.*, 2008a,b). Adverse effects of tannin on digestibility of sorghum protein and starch have long been recognized. Tannins are well known for their adverse effect on starch digestibility, which is due to their ability to interact with proteins (including hydrolytic enzymes), metal ions and polysaccharides (Davis and Hoseney, 1979; Deshpande and Salunkhe, 1982; Rooney and Pflugfelder, 1986; Schofield *et al.*, 2001). Wu *et al.* (2007) reported that liquefaction of starch in tannin sorghums was more difficult and slower than in normal and waxy sorghums. Tannins in sorghum retarded the hydrolysis process and resulted in viscous mash, a problem that could not be solved simply by increasing amylase usage during mashing. There was a remarkable difference in mashing properties among representative grains with the normal dosage of α-amylase, and tannin content was correlated highly with mashing properties (Zhao *et al.*, 2008c).

Wheat is the most widely cultivated cereal and a staple food for the world's population. Wheat flour is used in the composition of breads, noodles, cereals and many other food products. In Canada and Europe, wheat has been used to produce potable and fuel ethanol (Thomas and Ingledew, 1990, 1992; Sosulski and Sosulski, 1994; Thomas *et al.*, 1996; Loyce and Meynard, 1997; Wang *et al.*, 1997a,b; Freeze and Peters, 1999; Swanston *et al.*, 2005, 2007; Agu *et al.*, 2006; Kindreda *et al.*, 2008). Because bioethanol is used worldwide as a renewable component of fuels, wheat is considered a main energy crop in Europe (Loyce *et al.*, 2002; Smith *et al.*, 2006; Rigler *et al.*, 2007). Wheat varieties producing 'feed class' grain with high starch content and thus relatively low protein content have been highlighted as the preferred ideotype for ethanol production (Sosulski and Sosulski, 1994; Smith *et al.*, 2006; Kindreda *et al.*, 2008). Smith *et al.* (2006) concluded that ethanol yield from the best varieties grown under ideal UK conditions were likely to exceed 4000 l/ha, which was comparable to maize-based biofuel production systems in the USA.

Wheat markets in the USA traditionally have been for milling (principally for the baking industry) and export. Therefore, most research efforts with respect to wheat quality traits have been targeted toward protein quantity, composition, structure, genetic basis and functionality desirable for food use. Few US ethanol plants currently use wheat as a feedstock (RFA, 2008). In addition, existing ethanol plants that use wheat as a feedstock use a wet milling process to produce gluten, and the isolated starch can then be used for ethanol production if desired. Thus, high starch content of the incoming wheat is not critical because ethanol is only one of a number of valuable products. Starch content of US wheat varieties has been reported to be 63–72% (Lineback and Rasper, 1988). The opportunity to use wheat as a feedstock affords a choice to ethanol facilities being constructed in some agricultural areas outside major maize growing regions where climatic and economic conditions are favourable for wheat production. The great advantage of and motivation for using wheat in the fuel ethanol industry is the opportunity to choose high-yielding, locally adapted grains, which will result in reduced transportation costs and other local benefits (Agu *et al.*, 2006). In addition, poor quality (e.g. weather damaged, immature) wheat grain less suitable for either human or livestock use may be used for ethanol production. To date, there has been little effort in breeding wheat varieties specifically for fuel ethanol production. Compared with maize, factors affecting ethanol yield from wheat are not well understood. Little information is available on the fermentation performance of wheat varieties in a dry grind process. Lacerenza *et al.* (2008) recently reported that different classes of spring wheat were equally suitable for ethanol production in terms of ethanol conversion efficiency and ethanol yield and pointed out that traditional selection for milling and baking quality was not consistent with maximal ethanol yield per hectare.

Considering that yeast growth normally accounted for approximately 8% of the sugars available for fermentation, Smith *et al.* (2006) predicted conversion efficiency nearer to 92% for UK wheat. In a recent study, Zhao *et al.* (2009b) observed the DDGS of non-waxy wheat cultivars contained only 0.7–1.6% of unconverted starch, which indicated a high conversion rate of starch. In contrast, there was 5–6% residual starch in commercial maize (Kim *et al.*, 2008) and sorghum (Corredor *et al.*, 2006) DDGS. These results, in conjunction with fermentation tests, indicate that conversion efficiency of wheat is generally superior to that of maize and sorghum and comparable to that of pearl millet (Zhao *et al.*, 2009b). There was no difference between hard and soft wheat in terms of conversion efficiency (i.e. variation in starch quality related to ethanol fermentation was not observed). For non-waxy wheat, selection of wheat cultivars as a feedstock for production of fuel ethanol would become simpler than selection of wheat for bread-making because grain hardness did not influence the extent of starch conversion and starch and protein content were the most significant factors in determining ethanol yield as well as yield and quality of DDGS (Zhao *et al.*, 2009b). Obviously, a wheat cultivar with higher starch content is desirable because it will produce more ethanol and less DDGS per tonne of grain, resulting in less residual material and greater energy savings during DDGS drying. The most effective way to increase ethanol yield from wheat is to increase the amounts of starch and sugar in the grain (Smith *et al.*, 2006). Soft wheat is superior to hard wheat in fermentation performance because of its higher starch content. Soft wheat generally yields far more than hard wheat (Economic Research Service/USDA, 2008; Lacerenza *et al.*, 2008). At present, the challenge for the US fuel ethanol industry is that wheat has a low agronomic yield compared with maize. Wheat for biofuel production should give higher grain yields, making it more financially viable for growers. For UK wheat, Kindreda *et al.* (2008) predicted a grain yield of 9.6 t/ha at the economically optimum fertilizer nitrogen rate. Thus, soft wheat is potentially a technically and economically attractive crop for fuel ethanol production.

Compared with non-waxy samples, waxy wheat exhibited some performance advantages during fermentation, such as novel mashing property, rate of fermentation, reduced nitrogen food requirement and conversion efficiency (Zhao *et al.*, 2009b). Waxy cultivars had extremely low peak viscosity during liquefaction. The dry grind industry could increase the solids content in a mash, lower α-amylase dosages, or decrease energy requirements for stirring systems when waxy wheat is used as a feedstock. With nitrogen food supplemented into the mash, waxy cultivars had faster rates of fermentation than their non-waxy counterparts. A shorter batch fermentation time would result in greater ethanol output, more savings in facilities' energy consumption and less risk of bad production that must be recycled. Without nitrogen food supplemented to the mash, fermentation rates for waxy cultivars were comparable to those for non-waxy cultivars with nitrogen food. This should be very beneficial to ethanol producers because the cost of exogenous nitrogen food could be avoided without loss of production rate. Regardless of higher levels of free sugars in grain, waxy cultivars have overall higher conversion efficiency than non-waxy cultivars.

Biologically and economically, pearl millet is a feasible supplemental feedstock for dry grind maize–ethanol facilities in south-eastern USA (Wilson *et al.*, 2007). Research to support profitable cropping systems that supply an adequate supply of feedstock is needed. Regionally grown pearl millet should benefit rural economies in the south-east that are planning to import feedstock for ethanol production.

References

AFIA (2007) Evaluation of analytical methods for analysis of dried distillers grains with solubles. *AFIA Sub-Working Group Final Report and Recommendations*. AFIA, Arlington, Virginia, pp. 1–20.

Agu, R.C., Bringhurst, T.A. and Brosnan, J.M. (2006) Production of grain whisky and ethanol from wheat, maize, and other cereals. *Journal of the Institute of Brewing* 112, 314–323.

Alterthum, F. and Ingram, L.O. (1989) Efficient ethanol production from glucose, lactose, and xylose by recombinant *Escherichia coli*. *Applied and Environmental Microbiology* 55, 1943–1948.

Applegate, T.J., Troche, C., Jiang, Z. and Johnson, T. (2009) The nutritional value of high-protein maize distillers dried grains for broiler chickens and its effect on nutrient excretion. *Poultry Science* 88, 354–359.

Arifeen, N., Wang, R., Kookos, I.K., Webb, C. and Koutinas, A.A. (2007) Process design and optimization of novel wheat-based continuous bioethanol production system. *Biotechnology Progress* 23, 1394–1403.

Arunachalam, V.S. and Fleischer, E.L. (2008) The global energy landscape and materials innovation. *MRS Bulletin* 33, 264–276.

Belyea, R.L., Rausch, K.D. and Tumbleson, M.E. (2004) Composition of maize and distillers dried grains with solubles from dry grind ethanol processing. *Bioresource Technology* 94, 293–298.

Bothast, R.J. and Schlicher, M.A. (2005) Biotechnological processes for conversion of maize into ethanol. *Applied Microbiology and Biotechnology* 67, 19–25.

Bryan, T. (2003) Pioneer 'rapid assay' identifies hybrids for above average ethanol production potential. *Ethanol Producer Magazine* September, 36–38.

Bryan, T. (2005) Changing the game. *Ethanol Producer Magazine* August, 2.

Christiansen, R. (2009) DDGS: supplying demand. *Ethanol Producer Magazine* March, 3.

Corredor, D.Y., Bean, S.R., Schober, T. and Wang, D. (2006) Effect of decorticating sorghum on ethanol production and composition of DDGS. *Cereal Chemistry* 83, 17–21.

Davis, B. and Hoseney, R.C. (1979) Grain sorghum condensed tannins. I. Isolation, estimation, and selective adsorption by starch. *Cereal Chemistry* 56, 310–314.

Deshpande, S.S. and Salunkhe, D.K. (1982) Interactions of tannic acid and catechin with legume starches. *Journal of Food Science* 47, 2080–2083.

Devantier, R., Pedersen, S. and Olsson, L. (2005) Characterization of very high gravity ethanol fermentation of maize mash. Effect of glucoamylase dosage, pre-saccharification and yeast strain. *Applied Microbiology and Biotechnology* 68, 622–629.

Dinneen, B. (2009) 2009 state of the ethanol industry address (http://www.ethanolrfa.org/objects/documents//2009_state_of_the_industry.pdf, accessed 15 March 2009).

Eckhoff, S. (2001) Use the quick germ method – lower your ethanol production costs. *Seed World* 139, 14–16.

Economic Research Service/USDA (2008) Wheat: planted acreage, harvested acreage, production, yield, and farm price (http://www.ers.usda.gov/Data/Wheat/Yearbook/WheatYearbookTable01.pdf, accessed 30 April 2009).

Farrell, A.E., Plevin, R.J., Turner, B.T., Jones, A.D., O'Hare, M. and Kammen, D.M. (2006) Ethanol can contribute to energy and environmental goals. *Science* 311, 506–508.

Freeze, B.S. and Peters, T. (1999) A note on the profitability of wheat–ethanol–feedlot production in Alberta. *Canadian Journal of Agricultural Economics* 47, 67–78.

Galitsky, C., Worrell, E. and Ruth, M. (2003) Energy efficiency improvement and cost saving opportunities for the maize wet milling industry: an ENERGY STAR® guide for energy and plant managers. A report by Ernest Orlando Lawrence, Berkeley National Laboratory, University of California, Berkeley, California (http://repositories.cdlib.org/lbnl/LBNL-52307/, accessed 2 February 2009).

Goldemberg, J. (1996) The evolution of ethanol price in Brazil. *Energy Policy* 24, 1127–1128.

Graybosch, R.A., Liu, R.H., Madl, R.L., Shi, Y.-C., Wang, D. and Wu, X. (2009) New uses for wheat and modified wheat products. In: Caver, B.F. (ed.) *Wheat: Science and Trade (World Agriculture Series).* Wiley-Blackwell, Ames, Iowa.

Haefele, D., Owens, F., O'Bryan, K. and Sevenich, D. (2004) Selection and optimization of maize hybrids for fuel ethanol production. In: *Proceedings of ASTA 59th Annual Maize and Sorghum Research Conference.* American Seed Trade Association, Alexandria, Virginia, pp. 21.

Ho, N.W.Y., Chen, Z. and Brainard, A.P. (1998) Genetically engineered *Saccharomyces* yeast capable of effective co-fermentation of glucose and xylose. *Applied Environmental Microbiology* 64, 1852–1859.

Ingledew, W.M., Jones, A.M., Bhatty, R.S. and Rossnagel, B.G. (1995) Fuel alcohol production from hull-less barley. *Cereal Chemistry* 72, 147–150.

Ingledew, W.M., Thomas, K.C., Hynes, S.H. and McLeod, J.G. (1999) Viscosity concerns with rye mashes used for ethanol production. *Cereal Chemistry* 76, 459–464.

Johnson, J.M.F., Coleman, M.D., Gesch, R., Jaradat, A., Mitchell, R., Reicosky, D. and Wilhelm, W.W. (2007) Biomass–bioenergy crops in the United States: a changing paradigm. *American Journal of Plant Science and Biotechnology* 1, 1–28.

Kamm, B., Kamm, M., Gruber, P.R. and Kromus, S. (2008) Biorefinery systems– an overview. In: Kamm, D.B., Gruber, D.P.R. and Kamm, M. (eds) *Biorefineries – Industrial Processes and Products.* Wiley-vch Verlag GmbH and Co, Weinheim, Germany, pp. 3–38.

Kelsall, D.R. and Lyons, T.P. (2003) Grain dry milling and cooking procedures: extracting sugars in preparation for fermentation. In: Jacques, K.A., Lyons, T.P. and Kelsall, D.R. (eds) *The Alcohol Textbook: A Reference for the Beverage, Fuel and Industrial Alcohol Industries.* Nottingham University Press, Bath, UK, pp. 9–31.

Kim, C.H., Maga, J.A. and Martin, J.T. (1989) Properties of extruded dried distiller grains (DDG) and flour blends. *Journal of Food Processing and Preservation* 13, 219–231.

Kim, S. and Dale, B.E. (2004) Global potential bioethanol production from wasted crops and crop residues. *Biomass and Bioenergy* 26, 361–375.

Kim, Y., Mosier, N.S., Hendrickson, R., Ezeji, T., Blaschek, H., Dien, B., Cotta, M., Dale, B. and Ladisch, M.R. (2008) Composition of maize dry-grind ethanol by-products: DDGS, wet cake, and thin stillage. *Bioresource Technology* 99, 5165–5176.

Kindreda, D.R., Verhoevena, T.M.O., Weightman, R.M., Swanston, J.S., Agu, R.C., Brosnan, J.M. and Sylvester-Bradley, R. (2008) Effects of variety and fertiliser nitrogen on alcohol yield, grain yield, starch and protein content, and protein composition of winter wheat. *Journal of Cereal Science* 48, 46–57.

Kyprianou, M. (2006) Commission recommendation of 17 August 2006 on the presence of deoxynivalenol, zearalenone, ochratoxin A, T-2 and HT-2 and fumonisins in products intended for animal feeding. *Official Journal of European Union* L 229/7, 1–3.

Lacerenza, J.A., Martin, J.M., Talbert, L.E., Lanning, S.P. and Giroux, M.J. (2008) Relationship of ethanol yield to agronomic and seed quality characteristics of small grains. *Cereal Chemistry* 85, 322–328.

Lee, W.J., Sosulski, F.W. and Sokhansanj, S. (1991) Yield and composition of soluble and insoluble fractions from maize and wheat stillages. *Cereal Chemistry* 68, 559–562.

Lee, W.J., Yoon, J.R., Park, K.J. and Chung, K.M. (2000) Fermentation of maize and wheat with supplementation of inactive dry brewer's yeast. *Journal of American Society of Brewing Chemists* 58, 155–159.

Lewis, S.M. (2006) VHG, BPXTM™ and BFRAC™: innovation in biorefining. In: Tumbleson, M.E. (ed.) *Proceedings of the Corn Utilization and Technology Conference.* NCGA, St Louis, Missouri, pp. 10–14.

Lewis, S.M. (2007) Method for producing ethanol using raw starch. *United States Patent Application Publication* Number US 2007/0196907 A1, pp.1–47.

Lineback, D.R. and Rasper, V.F. (1988) Wheat carbohydrates. In: Pomeranz, Y. (ed.) *Wheat Chemistry and Technology,* Volume I, 3rd edn. American Association of Cereal Chemists, St Paul, Minnesota, pp. 277–372.

Liu, Z.L., Saha, B.C. and Slininger, P.J. (2008) Lignocellulosic biomass conversion to ethanol by *Saccharomyces.* In: Wall, J.D., Harwood, C.S. and Demain, A. (eds) *Bioenergy.* ASM Press, Washington, DC, pp. 17–36.

Loyce, C. and Meynard, J.M. (1997) Low input wheat management techniques are more efficient in ethanol production. *Industrial Crops and Products* 6, 271–283.

Loyce, C., Rellier, J.P. and Meyard, J.M. (2002) Management planning for winter wheat with multiple objectives (2): ethanol–wheat production. *Agricultural Systems* 72, 33–57.

Luli, G.W., Jarboe, L. and Ingram, L.O. (2008) The development of ethanologenic bacteria for fuel ethanol production. In: Wall, J.D., Harwood, C.S. and Demain, A. (eds) *Bioenergy*. ASM Press, Washington, DC, pp. 129–137.
Mojovic, L., Nikolic, S., Rakin, M. and Vukasinovic, M. (2006) Production of bioethanol from maize meal hydrolyzates. *Fuel* 85, 1750–1755.
Naidu, K., Singh, V., Johnston, D.B., Rausch, K.D. and Tumbleson, M.E. (2007) Effects of ground maize particle size on ethanol yield and thin stillage soluble solids. *Cereal Chemistry* 84, 6–9.
Nichols, N.N. and Bothast, R.J. (2008) Production of ethanol from grain. In: Vermerris, W. (ed.) *Genetic Improvement of Bioenergy Crops*. Springer, New York, pp. 75–88.
Nichols, N.N., Monceaux, D.A., Dien, B.S. and Bothast, R.J. (2008) Production of ethanol from maize and sugarcane. In: Wall, J.D., Harwood, C.S. and Demain, A. (eds) *Bioenergy*. ASM Press, Washington, DC, pp. 3–15.
Patzek, T.W. (2006) A statistical analysis of the theoretical yield of ethanol from maize starch. *Natural Resources Research* 15, 205–212.
Plante, P., De Caumia, B., Roy, C. and Noel, G. (2008) Ethanol processing with vapour separation membranes. *United States Patent Application Publication* Number US2008/0207959 A1, pp. 1–13.
Ramirez, E., Johnston, D., McAloon, A. and Singh, V. (2009) Enzymatic maize wet milling: engineering process and cost model. *Biotechnology for Biofuels* 2, 1–9.
Rausch, K. and Belyea, R. (2006) The future of co-products from maize processing. *Applied Biochemistry Biotechnology* 128, 47–86.
Rendleman, C.M. and Shapouri, H. (2007) New technologies in ethanol production. *USDA Agricultural Economic Report* AER–842. Office of Energy Policy and New Uses, USDA, Washington, DC, pp. 1–30.
RFA (2007) Building new horizons. Ethanol industry outlook 2007 (http://www.ethanolrfa.org/media/pdf/outlook_2007.pdf, accessed 15 March 2009).
RFA (2008) Changing the climate. Ethanol industry outlook 2008 (http://www.ethanolrfa.org/media/pdf/outlook_2008.pdf, accessed 15 March 2009).
RFA (2009) Growing innovation. Ethanol industry outlook 2009, pp. 1–36 (http://www.ethanolrfa.org/objects/pdf/outlook/RFA_Outlook_2009.pdf, accessed 15 March 2009).
Rigler, Z., Kurthy, G. and Szollosi-Nagy, D. (2007) Final conclusions report of the project 'Central-eastern European Biofuels Sustainable Day' (http://www.etanol.info.hu/hun/letoltesek/Publishable_result_oriented_report.doc, accessed 15 March 2009).
Rooney, L.W. and Pflugfelder, R.L. (1986) Factors affecting starch digestibility with special emphasis on sorghum and maize. *Journal of Animal Science* 63, 1607–1623.
Sayaslan A. (2004) Wet-milling of wheat flour: industrial processes and small-scale test methods. *Lebensmittel-Wissenschaft und-Technologie* 37, 499–515.
Schill, S.R. (2009) Distillers grains exports nearly double in 2008. *Ethanol Producer Magazine* March issue (http://www.ethanolproducer.com/article.jsp?article_id=5398, accessed 15 March 2009).
Schofield, P., Mbugua, D.M. and Pell, A.N. (2001) Analysis of condensed tannins: a review. *Animal Feed Science Technology* 91, 21–40.
Schultze, T., Krenn, W. and Potschacher, P. (2005) Basic considerations and concepts in the planning and construction of grain-based bioethanol plants. *Zuckerindustrie* 130, 707–711.
Shapouri, H. and Gallagher, P. (2005) USDA's 2002 ethanol cost-of-production survey. Agricultural Economic Report 841. Office of Chief Economist and Office of Energy Policy and New Uses, USDA, Washington, DC, pp. 1–19.
Singh, N. and Echhoff, S.R. (1996a) Wet milling of maize – a review of laboratory-scale and pilot plant-scale procedures. *Cereal Chemistry* 73, 659–667.
Singh, V. (2008) Emerging technologies in dry grind ethanol production. In: Hou, C.T. and Shaw, J. F. (eds) *Biocatalysis and Bioenergy*. John Wiley and Sons, Hoboken, New Jersey, pp. 239–247.
Singh, V. and Eckhoff, S.R. (1996b) Effect of soak time, soak temperature, and lactic acid on germ recovery parameters. *Cereal Chemistry* 73, 716–720.
Singh, V. and Graeber, J.V. (2005) Effect of maize hybrid variability and planting location on ethanol production. *Transactions of ASAE* 48, 709–714.
Singh, V., Moreau, R.A., Doner, L.W., Eckhoff, S.R. and Hicks, K.B. (1999) Recovery of fiber in the maize dry-grind ethanol process: a feedstock for valuable coproducts. *Cereal Chemistry* 76, 868–872.
Singh, V., Doner, L.W., Johnston, D.B., Hicks, K.B. and Eckhoff, S.R. (2000a). Comparison of coarse and fine maize fiber for maize fiber gum yields and sugar profiles. *Cereal Chemistry* 77, 560–561.

Singh, V., Moreau, R.A., Haken, A.E., Hicks, K.B. and Eckhoff, S.R. (2000b) Effect of various acids and sulfites in steep solution on yields and composition of maize fiber and maize fiber oil. *Cereal Chemistry* 77, 665–668.

Singh, V., Murthy, G.S., Graeber, J.V. and Tumbleson, M.E. (2004) Grain quality issues related to maize dry grind processing. In: *Proceedings of the 2004 International Quality Grains Conference*. Purdue University, Indianapolis, Indiana, pp. 1–6.

Singh, V., Johnston, D.B., Naidu, K., Rausch, K.D., Belyea, R.L. and Tumbleson, M.E. (2005) Comparison of modified dry-grind maize processes for fermentation characteristics and DDGS composition. *Cereal Chemistry* 82, 187–190.

Singh, V., Batie, C.J., Aux, G.W., Rausch, K.D. and Miller, C. (2006) Dry-grind processing of maize with endogenous liquefaction enzymes. *Cereal Chemistry* 83, 317–320.

Smith, T.C., Kindred, D.R., Brosnan, J.M., Weightman, R.M., Shepherd, M. and Sylvester-Bradley, R. (2006) Wheat as a feedstock for alcohol production. *Research Review No. 61*. The Home-Grown Cereals Authority (HGCA), London.

Solomon, B.D., Barnes, J.R. and Halvorsen, K.E. (2007) Grain and cellulosic ethanol: history, economics, and energy policy. *Biomass and Bioenergy* 31, 416–425.

Sosulski, K. and Sosulski, F. (1994) Wheat as a feedstock for fuel ethanol. *Applied Biochemistry and Biotechnology* 45/46, 169–180.

Spiehs, M.J., Whitney, M.H. and Shurson, G.C. (2002) Nutrient database for distiller's dried grains with solubles produced from new ethanol plants in Minnesota and South Dakota. *Journal of Animal Science* 80, 2639–2645.

Swanston, J.S., Newton, A.C., Brosnan, J.M., Fotheringham, A. and Glasgow, E. (2005) Determining the spirit yield of wheat varieties and variety mixtures. *Journal Cereal Science* 42, 127–134.

Swanston, J.S., Smith, P.L., Gillespie, T.L., Brosnan, J.M., Bringhurst, T.A. and Agu, R.C. (2007) Associations between grain characteristics and alcohol yield among soft wheat varieties. *Journal of the Science of Food and Agriculture* 87, 676–683.

Tedeschi, L.O., Kononoff, P.J., Karges, K. and Gibson, M.L. (2009) Effects of chemical composition variation on the dynamics of ruminal fermentation and biological value of maize milling (co)products. *Journal of Dairy Science* 92, 401–413.

Thomas, K.C. and Ingledew, W.M. (1990) Fuel alcohol production: effects of free amino nitrogen on fermentation of very-high-gravity wheat mashes. *Applied and Environmental Microbiology* 56, 2046–2050.

Thomas, K.C. and Ingledew, W.M. (1992) Relationship of low lysine and high arginine concentrations to efficient ethanolic fermentation of wheat mash. *Canadian Journal of Microbiology* 38, 626–634.

Thomas, K.C., Dhas, A., Rossnagel, B.G. and Ingledew, W.M. (1995) Production of fuel alcohol from hull-less barley by very high gravity technology. *Cereal Chemistry* 72, 360–364.

Thomas, K.C., Hynes, S.H. and Ingledew, W.M. (1996) Practical and theoretical consideration in the production of high concentrations of alcohol by fermentation. *Process Biochemistry* 31, 321–331.

USDA-FAS (2007a) GAIN Report – Bio-fuels Canada 2007. GAIN Report CA7041 (http://www.fas.usda.gov/gainfiles/200708/146292128.pdf, accessed 15 March 2009).

USDA-FAS (2007b) GAIN Report – China Bio-fuels Annual 2007. GAIN Report CH7039 (http://www.fas.usda.gov/gainfiles/200706/146291348.pdf, accessed 15 March 2009).

USDA-FAS (2007c) GAIN Report – EU-27 Bio-fuels Annual 2007. GAIN Report E47051 (http://www.fas.usda.gov/gainfiles/200707/146291616.pdf, accessed 15 March 2009).

Van Der Borght, A., Goesaert, H., Veraverbeke, W.S. and Delcour, J.A. (2005) Fractionation of wheat and wheat flour into starch and gluten: overview of the main processes and the factors involved. *Journal of Cereal Science* 41, 221–237.

Wang, S., Sosulski, K., Sosulski, F. and Ingledew, M. (1997a) Effect of sequential abrasion on starch composition of five cereals for ethanol fermentation. *Food Research International* 30, 603–609.

Wang, S., Thomas, K.C., Ingledew, W.M., Sosulski, K. and Sosulski, F.W. (1997b) Rye and triticale as feedstock for fuel ethanol production. *Cereal Chemistry* 74, 621–625.

Wang, S., Ingledew, W.M., Thomas, K.C., Sousulski, K. and Sosulski, F.W. (1999) Optimization of fermentation temperature and mash specific gravity for fuel alcohol production. *Cereal Chemistry* 76, 82–86.

Widmer, M.R., McGinnis, L.M., Wulf, D.M. and Stein, H.H. (2008) Effects of feeding distillers dried grains with solubles, high-protein distillers dried grains, and maize germ to growing-finishing pigs on pig performance, carcass quality, and the palatability of pork. *Journal of Animal Science* 86, 1819–1831.

Wilson, J.P, McAloon, A.J., Yee, W., McKinney, J., Wang, D. and Bean, S.R. (2007) Biological and economic feasibility of pearl millet as a feedstock for ethanol production. In: Janick, J. and Whipkey, A. (eds) *Issues In New Crops and New Uses*. ASHS Press, Alexandria, Virginia, pp. 56–59.

Wu, F. and Munkvold, G.P. (2008) Mycotoxins in ethanol co-products: modeling economic impacts on the livestock industry and management strategies. *Journal of Agricultural and Food Chemistry* 56, 3900–3911.

Wu, M. (2008) Analysis of the efficiency of the US ethanol industry 2007. Center for Transportation Research, Argonne National Laboratory, US Department of Energy, pp. 7.

Wu, X., Wang, D., Bean, S.R. and Wilson, J.P. (2006a). Ethanol production from pearl millet using *Saccharomyces cerevisiae*. *Cereal Chemistry* 83, 127–131.

Wu, X., Zhao, R., Wang, D., Bean, S.R., Seib, P.A., Tuinstra, M.R., Campbell, M. and O'Brien, A. (2006b) Effects of amylose, maize protein, and maize fiber contents on production of ethanol from starch-rich media. *Cereal Chemistry* 83, 569–575.

Wu, X., Zhao, R., Bean, S.R., Seib, P.A., McLaren, J.S., Madl, R.L., Tuinstra, M., Lenz, M.C. and Wang, D. (2007) Factors impacting ethanol production from grain sorghum in the dry-grind process. *Cereal Chemistry* 84, 130–136.

Wu, Y.V. (1986) Fractionation and characterization of protein-rich material from barley after alcohol distillation. *Cereal Chemistry* 63, 142–145.

Wu, Y.V. (1987) Recovery of protein-rich by-products from sweet sorghum grain stillage after alcohol distillation. *Cereal Chemistry* 64, 244–247.

Wu, Y.V. and Sexson, K.R. (1984) Fractionation and characterization of protein-rich material from sorghum alcohol distillation. *Cereal Chemistry* 61, 388–391.

Wu, Y.V., Sexson, K.R. and Lagoda, A.A. (1984) Protein-rich residue from wheat alcohol distillation: fractionation and characterization. *Cereal Chemistry* 61, 423–427.

Zhan, X., Wang, D., Tuinstra, M.R., Bean, S.R., Seib, P.A. and Sun, X.S. (2003) Ethanol and lactic acid production as affected by sorghum genotype and location. *Industrial Crops and Products* 18, 245–255.

Zhan, X., Wang, D., Bean, S.R., Mo, X., Sun, X.S. and Boyle, D. (2006) Ethanol production from supercritical-fluid-extrusion cooked sorghum. *Industrial Crops and Products* 23, 304–310.

Zhang, G. and Hamaker, B.R. (1998) Low alpha-amylase digestibility of cooked sorghum flours and the effect of protein. *Cereal Chemistry* 75, 710–713.

Zhang, M., Eddy, C., Deanda, K., Finkelstein, M. and Picataggio, S. (1995) Metabolic engineering of a pentose metabolism pathway in ethanologenic *Zymomonas mobilis*. *Science* 267, 240–243.

Zhao, R., Bean, S.R., Ioerger, B.P. and Wang, D. (2008a) Impact of mashing on sorghum proteins and its relationship to ethanol fermentation. *Journal of Agricultural and Food Chemistry* 56, 946–953.

Zhao, R., Bean, S.R. and Wang, D. (2008b) Sorghum protein extraction by sonication and its relationship to ethanol fermentation. *Cereal Chemistry* 85, 837–842.

Zhao, R., Bean, S.R., Wu, X. and Wang, D. (2008c). Assessing fermentation quality of grain sorghum for fuel ethanol production using Rapid Visco-Analyzer. *Cereal Chemistry* 85, 830–836.

Zhao, R., Bean, S.R., Wang, D., Park, S.H., Schober, T.J. and Wilson, J.D. (2009a) Small-scale mashing procedure for predicting ethanol yield of sorghum grain. *Journal of Cereal Science* 49, 230–238.

Zhao, R., Wu, X., Seabourn, B.W., Bean, S.R., Guan, L., Shi, Y.-C., Wilson, J.D., Madl, R.L. and Wang, D. (2009b) Comparison of waxy versus non-waxy wheat in fuel ethanol fermentation. *Cereal Chemistry* 86, 145–156.

6

Ethanol from Lignocellulosic Crops

Kazuhiro Takamizawa, William Anderson and Hari P. Singh

Introduction

Wood, grasses and most of the plant litter represent the major part of the biomass in nature and collectively are called lignocellulose. Regardless of the source, lignocellulosic materials are composed mainly of cellulose, hemicellulose and lignin. Over 150 billion t of organic substances are photosynthesized annually which consist of the above three major constituents, with an average proportion of 4:3:3 (Lachke, 2002). Typical compositions of cellulose, hemicellulose and lignin in lignocellulosic plant materials and in plant residues are presented in Tables 6.1 and 6.2, respectively. Fengel and Wegener (1989) and Argyropoulos and Menachem (1997) estimate that there is 2.5–4 × 10^{11} t of cellulose and 2–3 × 10^{11} t of lignin on the earth, representing 40 and 30% of organic matter carbon, respectively, with other polysaccharides comprising 26%. Because cellulose, hemicellulose and lignin are closely associated in plants, isolating these compounds to a pure state is virtually impossible. They also are not distributed uniformly in the plant cell wall; the S2-layer of the secondary wall has the highest percentage of cellulose, and the middle lamella has the highest percentage of lignin, but all three compounds are present in every cell wall layer (Sjöström, 1993; Kuhad *et al.*, 1997). Their distribution in the different parts of the plant is also not uniform. Gramineous plants have more variation than woody plants. In addition, some grasses contain considerable amounts of pectin in the middle lamella, whereas wood contains only small quantities of extractives, inorganic compounds and pectin compounds (Fengel and Wegener, 1989; McDougall *et al.*, 1993; Kuhad *et al.*, 1997). Based on weight percentage, cellulose and hemicelluloses are higher in hardwoods compared to softwoods and wheat straw, while softwoods have higher lignin content.

Historically, lignocellulosic biomass is used in direct combustion to produce heat or anaerobically to produce charcoal. However, biomass can be converted into a number of other forms of energy. Conversion to electricity can be accomplished by burning or co-firing with coal at power plants. Recently, emphasis has also focused on biomass as a feedstock for the production of biofuels for transportation or synthesis of industrial chemicals. There are two primary avenues for conversion of lignocellulosic materials. Biochemical conversion uses enzymes from living organisms to degrade plant cell wall components, except for lignin, to sugars that can then be fermented to alcohols. This is similar to the means that microbes from ruminant animals use to degrade and supply nutrients to the host from forage grasses

Table 6.1. Cellulose, hemicellulose and lignin in lignocellulosic biomass.

	Cellulose	Hemicellulose	Lignin	
		(%)		References
Softwood				
Pine	40.1	29.3	27.8	Timell, 1967
Spruce	46.1	24.6	26.3	Cowling and Kirk, 1976
Hardwood				
Eucalyptus	41.7	15.1	27.2	Ferrara and Kling, 1987
Birch	44.9	32.7	19.3	Cowling and Kirk, 1976
Poplar	47.6	27.4	19.8	Wayman and Parekh, 1990
Monocotyledons				
Wheat straw	42.0	32.0	10.0	Jackson, 1977
Barley straw	44.0	27.0	7.0	Theander and Åman, 1978
Oat straw	41.0	16.0	11.0	Jackson, 1977
Paddy straw	33.0	26.0	7.0	Jackson, 1977
Maize straw	53.3	15.0	16.2	Wayman and Parekh, 1990
Sugarcane bagasse	43.7	30.6	11.8	Castro and Machado, 1989

Table 6.2. Chemical composition of agricultural wastes.

Kinds	Hemi	Cell	Lig	Xyl	Ara	Glc	Gal
Pistachio shells	48.9	39.5	11.6	36.0	0.8	2.2	3.7
Walnut shells	33.5	41.4	25.1	18.7	3.7	0.2	0.2
Sunflower seed peel	28.1	46.7	25.3	18.3	0.1	0.3	0.8
Barley bran	30.4	44.1	25.5	14.4	4.4	0.5	0.1
Chestnut shells	30.1	45.5	24.4	12.6	6.0	4.6	ND
Rice bran	38.3	35.1	26.6	10.7	0.9	0.2	ND
Ginkgo shells	45.1	32.4	22.5	8.7	2.9	0.5	0.3
Rice straw	31.3	38.9	29.8	8.5	2.3	3.5	0.4
Peanut shells	22.9	50.7	26.4	8.4	1.2	0.7	ND
Barley tea dregs	29.8	10.3	59.9	4.8	1.5	20.6	0.1
Japanese tea dregs	16.4	56.5	27.1	1.5	0.3	1.6	ND
Banana peel	8.2	49.9	41.9	1.4	1.5	2.6	ND
Orange peel	4.2	32.3	63.5	1.1	0.8	0.9	0.6
Tea dregs	21.8	51.4	26.8	0.8	0.8	0.5	1.2
Carrot dregs	7.2	37.6	55.2	0.4	0.9	0.1	1.1
Coffee dregs	10.8	64.1	25.1	0.2	0.3	1.4	1.1

Hemi, hemicellulose; Cell, cellulose; Lig, lignin; Xyl, D-xylose; Ara, arabinose; Glc, glucose; Gal, galactose; ND, less than 0.1 g. Hydrolysis was performed with 0.5% H_2SO_4 for 1 h.

and legumes. Thermochemical conversion encompasses an array of methods that combines varying levels of heat and pressure to convert biomass to bio-oils, syngas or biochemicals. These products can then, in turn, be processed to fuels such as ethanol. The residues from thermochemical processes are char and ash.

Biomass Feedstocks

The US Department of Energy (DOE) first became interested in cellulosic biomass as a renewable fuel in 1978 and in subsequent years attempted to determine the most promising plant species for use as bioenergy feedstocks (McLaughlin and Ksvos, 2005). A joint

US Department of Agriculture (USDA) and DOE report estimated 1.3 billion t of potential dry biomass available annually in the USA to help offset 30% of the total gasoline use in the country by the year 2030 (Perlack *et al.*, 2005). Furthermore, the National Resource Defense Council (NRDC) determined that biofuels could meet 50% of the total oil use in the US transportation sector by 2050 (Greene *et al.*, 2004). Table 6.3 provides a summary comparison of different feedstocks, and individual species are further described below.

Crop residues

Of the row crops, most of the current emphasis has been on characterization and assessment of maize stover. It is estimated that maize stover could supply 75 million t of dry biomass annually (Perlack *et al.*, 2005). Stover is made up of the stalk (50–56%), leaf (20–30%), husk (< 10%) and cob (15%) (Shinners and Binversie, 2007). Varvel and Wilhelm (2008a) found that cob biomass composed approximately 20% of grain biomass over different hybrids, cropping systems and N fertilization. Cell wall components of the different portions of the stover were analysed by Akin *et al.* (2006). The total aromatics including lignin were three- to fourfold higher in the stem pith and rind compared to the leaf sheath and blade. Subsequently, when these fractions were treated with ferulic acid esterase and cellulase that break down ester-bonded phenolic compounds and cellulose, respectively, stems produced higher levels of phenolic compounds and lower levels of fermentable sugar. The authors conclude that leaf components of the stover are much more amenable to conversion to fermentable sugars (Akin *et al.*, 2006). Research is being performed to determine the amount and portions of the stover that can be harvested without having adverse effect on soil fertility. Graham *et al.* (2007) identified regions in the Midwest USA that could support large biorefineries with 1 million t/year feedstock demands while factoring in wind erosion and nutrient replacement costs. Varvel and Wilhelm (2008b) concluded that soil organic carbon levels could be maintained with modest removal of maize stover.

Until very recently, all of the genetic improvement in maize has been for kernel or silage yields and quality (de Leon and Coors, 2008). Of the overall 1.4%/year forage yield increase in maize over the past 80 years, most of the increase has been due to grain yield increases (2.4%/year) versus only 0.7%/year for the stover (Lauer *et al.*, 2001). During this time, stover neutral detergent fibre (NDF) and *in vitro* dry matter digestibility (IVDMD) have remained the same. Increasing the biomass production in maize will likely come as a result of increasing stover yields through

Table 6.3. Potential cellulosic feedstock species and reported range of dry matter yields.

Feedstock	Species	Range in DM yields (t/ha/year)	References
Maize stover	*Zea mays*	7–14	Johnson *et al.*, 2007; Shinners and Binversie, 2007
Pine pulpwood	*Pinus taeda* and *P. elliotti*	3–7	Fox *et al.*, 2004; Allen *et al.*, 2005
Willow	*Salix* spp.	4–8	Kszos *et al.*, 2001
Poplar	*Populus* spp.	4–8	Kszos *et al.*, 2001
Wildrye	*Leymus* spp.	4–10	Anderson *et al.*, 2008a
Reed canarygrass	*Phalaris arundinacea*	10–15	Anderson *et al.*, 2008a
Sorghum	*Sorghum bicolor*	25–30	Rooney *et al.*, 2007
Sugarcane/energy cane	*Saccharum* spp.	17–70	Tew and Cobill, 2008
Miscanthus	*Miscanthus* spp.	30–45	Clifton-Brown *et al.*, 2008
Switchgrass	*Panicum virgatum*	9–23	McLaughlin and Ksvos, 2005
Bermudagrass	*Cynodon* spp.	7–25	Anderson *et al.*, 2008a
Napiergrass	*Pennisetum purpureum*	10–40	Anderson *et al.*, 2008a

genes such as *Leafy* (*Lfy1*), which generate extra nodes and leaves, and *grassy tiller1* (*gt1*), which alter lateral branching, or traits that confer regrowth (de Leon and Coors, 2008).

Improving traits that confer stover degradation to fermentable sugars should also improve stover as a bioenergy feedstock source. Although most silage quality improvement has relied on selection for grain components, some improvements in stover digestibility have been achieved. Genes including *brown midrib* are available within maize germplasm for genetic improvement (Cherney *et al.*, 1991; de Leon and Coors, 2008). Compositional make-up of maize stover has been investigated (Sheehan *et al.*, 2001) to determine better which stover components and which specific cell wall traits to target for genetic improvement. To improve maize stover for bioenergy, efforts will be focused primarily on increasing maize biomass by modifying plant morphology, improving drought tolerance and photosynthetic ability (de Leon and Coors, 2008).

Forest products

Traditionally, the marketplace for forest products has been primarily for construction or paper products. However, due to the abundance of forests and economic difficulty that many portions of the industry have endured, emphasis is now being place on the use of forest products for conversion to energy. A number of initiatives are in place for conversion of wood products to electricity (Zerbe, 2006). Wood pellets supply homes and small industries with the feedstock to replace or reduce dependence on electricity from the grid (Cassidy and Ashton, 2007). Also, since wood is high in cellulose, conversion to ethanol is another avenue for use in the USA and elsewhere (Frederick *et al.*, 2008). The southern USA, in particular, has abundant wood resources, providing 60% of the nation's wood supply. The pulpwood market that uses small diameter, high quality trees has declined in recent years, resulting in an overabundance of small diameter trees that can provide the initial biomass feedstock to biofuel production plants (Hubbard *et al.*, 2007).

Woody biomass consists of logging residues and thinnings, branches, tops, stumps and other woody debris from commercial harvesting operations left behind at the harvest site, as well as milling residues and small diameter trees. In the European Union (EU), 46% of the harvested timber was used for fuel, which accounted for about 3.5% of their energy needs (Hubbard *et al.*, 2007). There is approximately 87 million ha of forestland in southern USA from which, on average, 147 million green t are harvested for pulpwood each year (Wear and Greis, 2002). Between 25 and 45% of the total biomass of trees is left at the harvest site (Hubbard *et al.*, 2007). It has been estimated that the cost of producing chips from tops and limbs can be as low as US$11/t. Chipping tops, limbs and understory vegetation increased production costs by approximately US$1/t (Foster and Mayfield, 2007). Conventional, small-scale woody biomass harvesting systems (Ashton and Jackson, 2007) or pre-processing and drying the biomass (Jackson, *et al.*, 2007) have been proposed for collecting the biomass to be used for bioenergy. Hubbard *et al.* (2007) concluded that the most cost-effective harvesting system was in-woods chipping as part of a conventional logging or thinning operation and bulk vans as the most cost-efficient mode of transporting preprocessed woody biomass.

Wood processing residues such as black liquor, sawdust and bark are an additional source of feedstock. In 2002, over 80 million t of processing residues were available and 70% was used to manufacture materials such as particleboard, while the rest was used to power the processing plants (McKeever, 2002). It is estimated that 1.5×10^{-12} MJ/year of energy is generated from burning pulp mill residues mainly from black liquor (Peter, 2008). This material could be used more extensively for biofuel production, despite the fact that it currently has value in the marketplace.

Genetic improvement of pines can be an important aspect to developing managed forests for multiple uses. Loblolly pine and slash pine comprise half of the world's industrial forest plantations (Siry *et al.*, 2006), though longleaf and shortleaf pines are also important in south-eastern USA (Peter, 2008).

Loblolly is a more aggressive competitor and the preferred species in plantations that have well-drained soils with good fertility, while slash pine grows better on marginal soils or sites with high groundwater (Peter, 2008). Slash pine has higher nitrogen use efficiency than loblolly (Roth *et al.*, 2007). Slash and longleaf pines have more tolerance to disease and fire, respectively. The southern pines have shown great genetic and phenotypic diversity (Cornelius, 1994; Brown *et al.*, 2004; Peter, 2008), which has been used for mass selection for improvement in growth rates and rust resistance (Vergara *et al.*, 2007) in both loblolly and slash pines. Clonal propagation by root cuttings and somatic embryogenesis has been developed (Nehra *et al.*, 2005) for pines, which allows for rapid multiplication and selection of superior lines and reduces time from initial crosses to commercial production. Breeders have also devised methods to detect fast-growing genotypes at very early stages of development (Peter, 2008). Genetic engineering tools have also been developed using expressed sequence tags (ESTs), single nucleotide polymorphisms (SNP) and genetic transformation (Peter, 2008). These have resulted in the development of marker-assisted breeding tools (Neale, 2007) and transgenic trees with reduced lignin (Peter, 2008).

Management and harvesting techniques for pulpwood and timber is well established. Some research has been done on pine managed as short-rotation woody crops (SRWC) (Rockwood and Dippon, 1989; Peter, 2008). However, pine cannot be regenerated by coppicing, which is a disadvantage compared to other wood species. In general, pines are started in a nursery and transplanted into the field when the seedlings are about 1 year old. Pines can tolerate drought, poor soils and other extreme conditions. Peter (2008) summarizes the efforts to use pine in biorefineries. At the time of this publication, pilot plants for conversion of cellulose to ethanol are being constructed in south-eastern USA and they are proposing to use predominantly available pine as a feedstock. One advantage that pine has over angiosperms and grasses is that the hemicellulose is composed of hexoses, which can be fermented more readily (Peter, 2008).

Coppice tree species

A number of hardwood species have been considered as woody perennial energy crops (WPEC). *Eucalyptus* species, sycamore, sweet gum and others have been studied as biomass sources (Kszos *et al.*, 2001). Among those studied, willow and poplar are considered the most promising as possible dedicated bioenergy crops. Both are coppiceable, grow well in temperate climates, and improving yield and other traits through breeding has been achieved (Smart and Cameron, 2008). They have been grown extensively in Europe as energy crops and have drawn greater interest as a biomass feedstock in the USA.

Willow

Willow can grow in almost all climates, but the predominant testing of the genus for biofuels has occurred in northern USA and the northern European countries (Smart and Cameron, 2008). Several species, primarily within the subgenus *Vetrix*, have been tested and improved through breeding. The major species of study and genetic improvement in Europe has been *Salix viminalis* and in the USA, *S. eriocephala*. A recent study indicated a great deal of heterosis between *S. eriocephala* lines for both yield components and disease resistance (Cameron *et al.*, 2008). Interspecific hybrids have also shown significantly improved biomass yield and rust resistance (Smart and Cameron, 2008). The use of molecular genetic techniques for willow breeding has begun only recently and amplified fragment length polymorphism (AFLP) analysis was used to determine genetic distances among the species within *Salix* (Trybush *et al.*, 2008). The genus is perennial and deciduous, is shade intolerant but grows well in riparian and wetland areas. The major disease of willow is rust (*Melampsora epitea*), while numerous insect pests include chrysomelid beetles, aphids and leafhoppers (Smart and Cameron, 2008). Interest has grown in willow as a bioenergy crop because it could be grown in many arable locations in the north-eastern USA that were once farmed in row crops during the 19th century.

Willow plantations are established by planting dormant stems in the soil, which will root and generate shoots from dormant buds. Various planting equipment has been developed and is used in many areas of Europe (Smart and Cameroon, 2008). Plants are coppiced after the first year's growth only, to stimulate multiple stem formation the following year. Though most of the nutrients are recycled to the soil, additional nutrients will enhance growth (Adegbidi *et al*., 2003). Nutrients from wastewater streams or other wastes had been suggested which would add benefit in reducing environmental pollutants. Biomass is then collected every 3–4 years, using a variety of harvest methods. Harvesting is done when trees are dormant, leaves have fallen and minerals have been translocated back to the root. Harvest equipment can range from forage harvesters with adapted heads to sugarcane harvesters, and biomass can be removed from the field as chipped, billeted or whole material (Smart and Cameroon, 2008). The predominant use of willow for bioenergy has been in Europe and Sweden in particular, which has 15,000 ha of willow in cultivation. The willow chips and wood residues are either burned in heating plants that can also generate electricity or co-fired with coal for electricity (Smart and Cameron, 2008).

Poplar

Populus is a very diverse genus with some very important members such as eastern cottonwood, European aspen, bigtooth aspen and trembling aspen. Most of the species are dioecious, allowing for controlled crossing. Breeding is normally performed on detached branches of mature trees that can be rooted in the greenhouse. Poplars have the same advantages and limitations as the willows. Poplars have the advantage of being perennial and can be coppiced for many years with few inputs. However, poplar species do not tolerate shade and have similar disease and pest problems as willow. Poplar has been chosen as the model tree species for study (Bradshaw *et al*., 2000). Breeding methods and trait studies on poplar are summarized by Davis (2008). The elucidation of the genetic sequence of the poplar cultivar 'Nisqually-1' (*P. trichocarpa*) will aid in annotating genes involved with biomass quantity and quality (Tuskan *et al*., 2006). Quantitative trait loci (QTL) for many above- and belowground traits (Wullschleger *et al*., 2005) and total aboveground biomass (Rae *et al*., 2008) have been identified. These tools, as well as transgenic manipulation of poplar for studies of gene expression (Davis, 2008), give plant physiologists and plant breeders the means to determine the genetic control of specific traits that can be coupled with specific conversion technologies.

Though seeds are produced readily, the majority of the poplar plantations are propagated vegetatively from dormant 'whips' (shoots from coppiced trees), which are rooted under mist chambers (Davis, 2008). The management of popular plantations for harvesting biomass for energy is very similar to willow, which has been described earlier. Yields of poplar range between 7.7 and 12.5 t/ha/year for diverse areas of the USA (Davis, 2008). The feedstock can be directed in either woodchips or pellets for combustion or to conversion to ethanol, as described for willow.

Dedicated herbaceous feedstock crops

Most of the commonly proposed dedicated cellulosic biomass feedstock crops other than wood are native or non-native grasses. There are numerous ways to separate the species of grasses for discussion. The diverse monocots are annual or perennial, native to the USA or non-native, and they vary according to their adapted environment. Cool-season grasses are commonly defined by having the C-3 carbon photosynthetic pathway, which is common for the majority of plants on earth. Warm-season grasses have the C-4 pathway, which is defined as plants that produce the 4-carbon molecules, malate and aspartate, as the first products of carbon fixation. The warm-season grasses are much more efficient due to the Kranz leaf anatomy that reduces the loss of carbon which occurs during photorespiration (Moser *et al*., 2004). The theoretical maximum efficiency in converting sunlight into biomass at 30°C and CO_2 concentrations at 380 ppm is 4.6% for C-3 plants and 6.0% for C-4 plants (Carroll and Somerville, 2009). However, C-4 plants are more efficient only when CO_2 concentrations are low, temperatures high and water scarce (Ainsworth *et al.*, 2008), and it has been estimated that C-3 plants will have equal efficiency in the future when CO_2 concentrations reach 700 ppm (Zhu *et al.*, 2008).

Warm-season grasses thrive in tropical and semi-tropical climates, with optimum temperature for growth at 35–38°C and occur generally between 30°N and 30°S latitude. Cool-season (thus the name) species survive predominantly in latitudes above 30° with shorter growing seasons and cooler climates. Annual and perennial species exist for both warm- and cool-season grasses. The C-4 grass species generally require approximately half or less water and nitrogen of C-3 grasses for comparable yields. Thus, the warm-season grasses have a higher water use efficiency (WUE) and nitrogen use efficiency (NUE) than the cool-season species (Wedin, 2004). These characteristics allow exploitation of C-4 species in central and southern USA for use as forage or biomass feedstock for bioenergy.

Cool-season grasses and legumes

Temperate north and western pastures of the USA consist primarily of a mixture of cool-season (C-3) grasses and legumes. Though the primary use has been for forage or hay, a few have drawn attention as possible bioenergy feedstocks. Reed canarygrass is a persistent forage grass with a wide range of habitats that produces between 10–15 t/ha/year in Midwestern USA. It is propagated by seed and is tolerant to drought and waterlogging (Wright, 1990). There is considerable genetic variability within the species for traits important for improvement in production for biomass (Anderson *et al.*, 2008a).

WILDRYE. Wildrye is a C-3 grass particularly well adapted to the mountain and inter-mountain areas of western USA. The genus does well in riparian areas in silt and clay soils, tolerates cold climates and has good salt tolerance (Wasser, 1982). Yields have ranged from 3–10 t/ha/year with minimum inputs and unimproved varieties. Genetic maps are being developed and relevant traits are being determined with QTL markers (Larson *et al.*, 2006). The wildryes are propagated by seed and harvested as hay, similar to all forages.

LUCERNE. Lucerne is important forage for livestock due primarily to the high N content of the leaves (Arinze *et al.*, 2003). Most of the production of lucerne occurs in temperate areas, with California, South Dakota and Wisconsin as the leading producers in the USA. The majority of breeding work on lucerne has concentrated on disease resistance, cold tolerance and autumn (fall) dormancy. Subsequently, very little improvement in yield has occurred. Higher yields are attainable with higher percentage of stem (Samac *et al.*, 2006). However, the stem of lucerne contains high amounts of fibre, which is less digestible and reduces the quality of lucerne used as hay. Leaves can be separated easily from the stems and the stems in term used for bioenergy (Arinze *et al.*, 2003). The stems would be considered the co-product for biofuels, with the leaves used as feed. Recent research indicated that high biomass genotypes produced 37% higher stem polysaccharides than traditional genotypes, which theoretically could double ethanol yields (Lamb *et al.*, 2007). The composition of lucerne stems has been studied over maturities and environments. It has been concluded that cell wall components can be manipulated genetically for reduction in lignin or increase in cellulose (Samac *et al.*, 2006). Lucerne stems are more recalcitrant to cell wall decomposition than grasses (Dien *et al.*, 2006), but in turn give greater syngas yields than grasses during pyrolysis (Boateng *et al.*, 2006).

Warm-season grasses

SORGHUM. Sorghum is a very important grain, sugar and forage crop worldwide. The species is a potential dedicated bioenergy feedstock based on its great yield potential and superior tolerance to drought, marginal soil conditions and wide range of climates (Pederson and Rooney, 2004). Grain sorghum can be a supplement to maize for feed or conversion to ethanol. The sweet sorghums produce high-sugar juice in the stem comparable to sugarcane. Forage or biomass types, which can include sudangrass (*Sorghum* × *drummondii*) or sorghum/sudangrass hybrids, can reach 4m in height and can produce large amounts of cellulosic biomass.

The fibrous root system of sorghum, which can extend to 1.5m (Pederson and Rooney, 2004), allows the crop to maximize nutrient and water uptake in the soil. Seed should be planted at a depth of between 6 and 40mm at rates of 17–28kg/ha. Nitrogen rates of 40–160kg/ha have been recommended (Saballos, 2008), but rotations with legumes may almost be sufficient for good production (Taylor, 1988). Potassium may be required, especially as it may effect sugar production (Saballos, 2008). The drought tolerance of sorghum is superior to most crops, except for pearl millet. Sorghum requires approximately half of the water needed by maize for optimal production (Pedersen and Rooney, 2004).

Yields of forage or cellulosic sorghums up to 30t/ha (Rooney *et al.*, 2007) have been reported. Sweet sorghum varieties allow for the use of both the high-sugar sap and the remaining cellulosic bagasse. The most promising scenario for harvest and use is to extract the juice in the field, followed by harvest of bagasse by baling or chopping after field drying (Li *et al.*, 2004; Kundiyana *et al.*, 2006).

Since sorghum is self-pollinating, pure line pedigree selection is a common method of genetic improvement in the species (Saballos, 2008). This method can be used to produce either open-pollinated cultivars or inbred parents for hybrid production. Male sterility (ms_3) genes are also used for mass selection and development of improved male sterile parental lines (A-lines) for later restoration with R-lines to produce fertile hybrids (Smith and Frederiksen, 2000).

The primary trait of interest for sorghum is biomass yield, as it is for most bioenergy crops. Plant height, photoperiod sensitivity, stem thickness, number of stems per plant and maturity are among the morphological traits that effect yield in sorghum. Morgan *et al.* (2002) documented how many of these traits were interrelated. For example, four major genes have been determined to control plant height by affecting internode length and are designated *Dwarfing* (Dw) 1–4 (Quinby and Karper, 1954). The partial dominance of the *Dw* genes are additive in nature and plants range in height from approximately 0.5m (all recessive dwarf genes) to over 4m (all dominant). As many as six genes for maturity (*Ma*) have also been discovered (Quinby, 1967; Rooney and Aydin, 1999) and plants with the recessive allele (ma_1) are photoperiod insensitive to allow flowering in the long days of the temperate regions of the world. Depending on the environment and production practices, tiller number per plant may be very important for biomass production in sorghum. Tillering has been mapped using QTL (Patterson *et al.*, 1995; Hart *et al*, 2001). Another important trait is reduced susceptibility to stem lodging. Besides plant height, morphological traits such as strength of stem rind (Hondroyianni *et al.*, 2000) and the stay-green trait (Haussmann *et al.*, 2002) may be important in reduced lodging.

The overall superior stress tolerance of sorghum is an advantage over many other potential biomass feedstocks. Drought resistance has been associated with the stay-green trait and three QTLs have been mapped (Subudi *et al.*, 2000). Heat tolerance genes have also been described (Khizzah *et al.*, 1993). Few studies have been performed on other abiotic stress tolerances in sorghum and more germplasm screening is necessary. If biomass crops are to be grown on marginal soils to avoid conflict with food row crops, then tolerance to acidic, alkaline or saline soils, as well as excessive or low phosphorus availability, become priority breeding objectives. Aluminum tolerance has been found and identified as a single locus with multiple alleles (Troeh and Thompson, 2005; Caniato *et al.*, 2007). The two most important disease problems of sorghum are anthracnose (*Colletotrichum graminicola*) and downy mildew (*Peronosclerospora sorghi*). Resistance to these diseases is present in broad germplasm and genetic improvement has been accomplished through breeding and selection (Smith and Frederiksen, 2000).

Sorghum has become a very important bioenergy feedstock. Increases in starch from the grain, fermentable free sugars and lignocellulose can be achieved genetically through breeding and selection. Recently, a significant amount of emphasis has been put on the genomics of sorghum, which has been summarized by Paterson (2008). Complete sequencing of an inbred line has been accomplished. Sorghum is genetically related to other important bioenergy feedstocks such as maize and sugarcane, but it has a relatively small genome size and is diploid. For that reason, it has become a genomic model plant for C-4 grasses and a subject of comparative genomics (Paterson, 2008). Further, targeting-induced local lesion in genomes (TILLING) has been applied successfully to sorghum (Xin *et al.*, 2008).

SUGARCANE/ENERGY CANE. Sugarcane is a tall perennial grass grown primarily for the production of processed sugar (sucrose) in nearly 80 countries in tropical or semi-tropical regions and is a major crop in Brazil as well as Louisiana and Florida in the USA. An excellent review of the taxonomy, anatomy, cultivation and use of sugarcane is presented by Tew and Cobill (2008). Prior to around 1900, the original genotypes grown for sugar were *Saccharum officinarum*. More genetically diverse material was later used to avoid susceptibility to various diseases such as sugarcane mosaic virus (SCMV). More recently, breeding for lignocellulosic bioenergy conversion has been performed to incorporate higher fibre content for cold tolerance and less lodging.

A more complete description of sugarcane production has been presented earlier (see Chapter 4, this volume). Sugarcane is propagated vegetatively by planting stems from which axillary buds emerge. The stems are planted in rows from 1.5 to 3 m apart. Usually, the cane is harvested annually, with ratoon crops lasting a few years (Tew and Cobill, 2008). Much of the management and harvest timing is directed toward obtaining the highest production of sucrose. This is usually obtained with long, warm growing seasons followed by a cooler season that allows the plant to partition more energy toward sucrose storage in the stem.

The fibrous residue of sugarcane (bagasse) is often used to fuel boilers that provide necessary energy to make sugarcane mills self-sufficient. Excess energy from the burning of the bagasse can be converted to electricity and sold to utilities, as has been done in Hawaii (Payne, 1991) and Florida (Tew and Cobill, 2008). Sugarcane bagasse contains comparable to slightly higher amounts of energy (18–19 MJ/kg) to switchgrass (Jenkins *et al.*, 1998; McKendry, 2002; Tew and Cobill, 2008). The relatively high energy content is due to the high lignin content (23–32%) in sugarcane (Tew and Cobill, 2008). If bagasse is burned or used in thermochemical processes, the high lignin would be a desirable trait.

Breeding programmes divide sugarcane into three physiological types in relation to conversion to bioenergy. *Saccharum* spp. used for sugar has already been discussed and makes up the majority of cultivars grown currently for sugar production (Chapter 4). Type 1 energy cane maintains a high level of free sugar (> 13%) but also has a much higher level of fibre (> 17%) than traditional sugarcane (12%). Type 2 energy cane is referred to as canes with low sugar (< 5%) and much higher levels of fibre (30%) (Tew and Cobill, 2008). Type 1 energy cane is bred to maximize both sugar and fibre and allows for the incorporation of wild *Saccharum* germplasm such as *S. spontaneum* and increases the genetic base of the crop (Ming *et al.*, 2006). Type 2 canes are currently being bred to increase fibre content only and the sole attention is lignocellulosic. In breeding for fibre only, fibre content and stalk number have been found to be correlated positively (Jackson, 1994). The increased fibre content appears to improve tolerance to more temperate climates (Tew and Cobill, 2008). Intergeneric hybridization with *Erianthus arundinaceus* (Cai *et al.*, 2005) and *Miscanthus* spp. (Lo *et al.*, 1986) has been accomplished. Genes for high yields from *E. arundinaceus* (up to 127 t/ha/ year) (Matsuo *et al.*, 2006) and *Miscanthus* spp., as well as drought, flood, cold and disease tolerance, may be brought into these hybrids. Hybrids with *Miscanthus* have yielded up to 175 t/ha/year (Lo *et al.*, 1986).

MISCANTHUS SPP. The genus *Miscanthus* is closely related to sugarcane and native to South-east Asia. This species is comprised of about 12 species (Clifton-Brown *et al.*, 2008), of which *M. sacchariflоris*, *M. sinensis* and hybrids have shown the greatest potential for bioenergy. Dry matter yields as high as 45 t/ha/year have been observed. *Miscanthus* is unusual in that it is highly tolerant of cool weather and produces high yields in temperate areas of the USA such as Illinois, despite being a C-4 perennial (Heaton *et al.*, 2008). Most research and production has been centred in Europe and has concentrated on a single clone of *M.* × *giganteus*, which is a sterile triploid hybrid clone of *M. sacchariflоris* and *M. sinensis* (Jones and Walsh, 2001). In the USA, field tests in Illinois have produced yields of 38 t/ha/year compared to 12 t/ha/year for switchgrass (*Panicum virgatum*) (Heaton *et al.*, 2008). The authors claim that *M.* × *giganteus* could provide 260% more ethanol per hectare than maize ethanol.

Besides the tremendous yields obtained in even temperate climates, germplasm of *Miscanthus* appears to have very good drought tolerance, frost and low temperate tolerance, high disease and pest tolerance, as well as variation in lignin, cellulose and hemicellulose content (Clifton-Brown *et al.*, 2008). At present, the major weakness to the use of *Miscanthus* for biofuels is the cost of establishment. Rhizomes are dug from established fields and planted in furrows, which requires special equipment. However, progress has been reported in direct seeding and may be used in the future (Clifton-Brown *et al.*, 2008).

SWITCHGRASS. Switchgrass was identified by the DOE as the initial dedicated energy crop because it was native to the USA, had high biomass yields and possessed the ability to grow on marginal lands with low inputs (McLaughlin *et al.*, 1999). One of the advantages of switchgrass over other C-4 grasses mentioned in this chapter is that it can be established easily by seed, if herbicides are used to control the weeds (Vogel, 2004). Switchgrass is an important pasture grass in central and eastern USA, and breeding programmes have concentrated on increased yields and improved rumen digestibility.

There are lowland and upland ecotypes and tetraploid and octaploid cytotypes of switchgrass (Vogel, 2004). The lowland types are found in low-lying areas such as flood plains and are also the predominant ecotypes for the milder southern climates. The upland types are shorter, more adapted to northern climates and are mostly octaploids. These ecotypes have been further subdivided into more regional areas (Casler *et al.*, 2004). The majority of improved varieties has come from mid-west breeding programmes and are upland types, but the lowland ecotypes have shown greater yield potential (Bouton, 2008).

Switchgrass tolerates a range of soil conditions, from sandy to heavy clay and pH from 5–7 (Moser and Vogel, 1995). Relatively low fertilization rates are required for production, though it responds well to increased fertility (Bouton, 2008). The recommended seeding rate is approximately 5–6 kg pure live seed/ha (Ball *et al.*, 2007).

Switchgrass produces the greatest yields when harvested once or twice a year (Vogel, 2004). Reported dry matter yields are in the range of 9.5–23.0 t/ha/year over multiple year dry land production systems (McLaughlin and Ksvos, 2005). Pests such as grasshoppers and diseases such as *Helminthosporium sativa* have become problems when switchgrass is grown in monoculture (Vogel, 2004). Harvesting of switchgrass can be accomplished using current hay equipment, which is an advantage over other thicker-stemmed tall biomass crops that require more heavy-duty equipment. On-farm production cost estimates for the upper mid-west were estimated to be approximately US$50/t (Perrin *et al.*, 2008). This information, along with models such as the Agricultural Land Management Alternatives with Numerical Assessment Criteria (ALMANAC) (Kiniry *et al.*, 2008) can be used by growers to assess the economic feasibility of growing switchgrass for biofuels on establishing a viable market.

Recurrent selection has been the normal method for cultivar development and has proven successful. Polycrossing of superior genotypes and selections generated progeny with 33% greater yields than the parents in Georgia, USA (Bouton, 2008). Though increased dry matter yield has been the primary trait of interest in breeding programmes, selection for altered cell wall components has produced genotypes with varying levels of forage digestibility and lignin content. A study by Sarath *et al.* (2008) indicated that the degree of forage digestibility was correlated highly with stem lignin content. However, in their study, there was a great deal of variation within both the high and low lignin lines for the amount of phenolic acids and sugars released after pretreatment with esterases and cellulases, indicating that genotypes with altered lignin and DMD could arise from several different combinations of genes. Molecular mapping techniques are being initiated for switchgrass and a genetic map from a cross of varieties 'Alamo' and 'Summer' has been published (Missaoui *et al.*, 2005).

BERMUDAGRASS. Bermudagrass is the most widely used perennial forage in southern USA. Though there are numerous *Cynodon* species, forage bermudagrass is normally classified as *C. dactylon* and the highly productive coarse stargrass as *C. nlemfuënsis* (Taliaferro *et al.*, 2004). They are perennial in nature and go through a winter dormancy period. The greatest genetic improvement in forage yield has been through selection of hydrids that require vegetative propagation (Burton, 1956). Sprigs that consist of rhizomes, stolons, roots and stems are used to establish fields. Aboveground mature stem cuttings can also be propagated successfully. Seeded varieties also exist but, to date, have not produced dry matter yields comparable to the sprigged cultivars.

Once differences were found among bermudagrass germplasm for rumen digestibility, methods were devised to measure this important quality trait efficiently in the laboratory (Tilley and Terry, 1963; Monson *et al.*, 1969). These methods, along with breeding and selection among full-sib controlled crosses, were performed to develop high-yielding forage bermudagrass lines with improved quality (Burton, 1972; Burton *et al.*, 1993). Histochemical techniques revealed that one of these improved lines, Coastcross 1, had fewer ester linkages, while Tifton 85 appeared to have a greater proportion of ester versus recalcitrant ether linkages within the cell walls (Mandebvu *et al.*, 1999). It has been proposed that multiple mechanisms and genes for reduced recalcitrance are involved and that further genetic improvement can be achieved (Anderson and Akin, 2008). Significant differences in sugars and ethanol were observed between Coastal and the much less recalcitrant Tifton 85 when subjected to hydrolysis with commercial esterase, and cellulase enzymes followed by fermentation indicated that sugar production and subsequent fermentation to ethanol may be correlated with rumen digestibility (Anderson *et al.*, 2008b). However, when bermudagrass was subjected to pyrolysis, there was no difference in syngas production (Boateng *et al.*, 2007), likely due to the similar concentrations of cell wall components.

Production and harvesting of bermudagrass for dry hay is well established and the hay undergoes enzymatic degradation easily. The major disadvantages of bermudagrass are cost of production, mainly from fertilizer, and the competitive prices commanded by the livestock industry. However, old or low quality hay reserves could provide a source for bioenergy feedstocks (Anderson *et al.*, 2008b).

NAPIERGRASS. Napiergrass is another tall, high-yielding C-4 grass that is a potential feedstock. It is a tall, leafy grass and is an important forage crop for much of the tropics. Napiergrass grown at three locations in Georgia over 6 years consistently produced almost double the dry matter (27 t/ha/year) compared to Alamo switchgrass (15 t/ha/year) (Bouton, 2002). It is photoperiod-sensitive and will not flower until day length becomes 11 h or less and will stop growth at 10°C (Hanna *et al.*, 2004). Napiergrass is propagated clonally from stems or rhizomes because genotypes produce small amounts of seed and they are highly heterozygous. Stems can be planted in trenches similar to sugarcane, no deeper than 10 cm (Anderson *et al.*, 2008a). Harvesting can be performed with silage choppers or sugarcane equipment and should be performed after senescence but prior to loss of appreciable leaves. As with other perennial crops, nutrient and stem moisture declines after senescence as they are translocated to the roots. Yields tend to increase as harvest interval increases, with maximum dry matter being attained after 24 weeks (Woodard and Prine, 1991).

Breeding of napiergrass during the 1980s in Georgia and Florida created high-producing germplasm and cultivars (Burton, 1989; Schank *et al.*, 1989). Napiergrass has also been intercrossed successfully with other *Pennisetum* species such as the apomictic species *P. squamulatum* and pearl millet (Hanna *et al.*, 1984, 2004). A great deal of genetic variability is available and plant introductions have been evaluated for fibre and rumen digestibility (Anderson *et al.*, 2008a). In general, the stem is much more recalcitrant due to higher lignin and non-degradable fibre (Anderson *et al.*, 2005, 2008b). Breeding efforts will target genes for yield and either increased or decreased lignin, depending on the requirement of the process for feedstock conversion. The tall, stemmy bunch grasses with higher lignin content, such as napiergrass and energy cane, should be more amenable to thermochemical conversion.

Miscellaneous grasses

Among some of the other C-4 grasses that have been considered as biomass feedstocks are the native species big bluestem, indiangrass and eastern gamagrass. These species are planted by seed. However, they are all very slow in germinating, establishing and producing substantial biomass. Big bluestem had greater *in vitro* fermentability compared to other warm-season grasses (Weimer and Springer, 2007) and produced greater ethanol yields from SSF than switchgrass (Jung and Vogel, 1992). When pyrolysis was used to produce bio-oil, big bluestem not only produced more but also gave a higher return on investment (Anderson *et al.*, 2008a).

Eastern gamagrass is another native warm-season grass that is slow to establish but has produced greater yields than either big bluestem or sand bluestem (Weimer and Springer, 2007). There have been some advances in breeding and selection of eastern gamagrass with the development of triploids, and tetraploids that are apomictic as well as cultivars with greater rumen digestibility (Anderson *et al.*, 2008a). In nature, the native prairie grasses did not exist in monoculture but were in mixed stands. Research has discovered that mixtures may out-yield monoculture of these grasses. The native habitat on marginal lands could benefit from the mixtures and also provide biomass to the bioenergy industry (Mulkey *et al.*, 2008).

Chemistry of Cellulose and Hemicelluloses

Cellulose

Cellulose is a linear polymer chain which is formed by joining the anhydroglucose units into glucan chains. The chemical structure is shown in Fig. 6.1. Basically, cellulose is classified into crystal and non-crystal structures of polymers of D-glucose with β-1,4 bonds. These anhydroglucose units are bound together by β-(1,4)-glycosidic linkages. Due to its linkage, cellobiose is established as the repeat unit for cellulose chains. The degree of polymerization (DP; i.e. molecular weight of one glucose unit) of native cellulose is in the range of 7000–15,000.

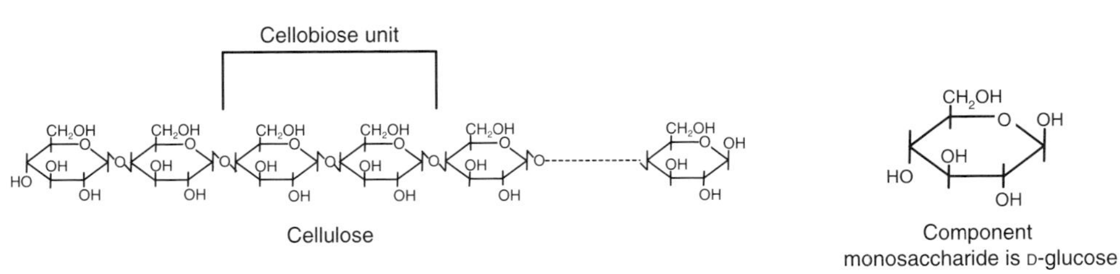

Fig. 6.1. Chemical structure of cellulose.

By forming intramolecular and intermolecular hydrogen bonds between OH-groups within the same cellulose chain and the surrounding cellulose chains, the chains tend to arrange in parallel and form a crystalline supermolecular structure. Then, bundles of linear cellulose chains (in the longitudinal direction) form a microfibril, which is oriented in the cell wall structure.

Hemicelluloses

Unlike cellulose, hemicelluloses show structural variability consisting of different monosacharide units. In addition, the polymer chains of hemicelluloses have short branches and are amorphous. Because of the amorphous morphology, hemicelluloses are partially soluble or swellable in water. The backbone of the chains of hemicelluloses can be a homopolymer (generally consisting of single sugar repeat unit) or a heteropolymer (mixture of different sugars). Hemicelluloses by definition are the short branched chain heteropolysaccharides of mixed hexosans and pentosans. D-Xylose and L-arabinose are the major constituents of pentosans, while D-glucose, D-galactose and D-mannose are the constituents of hexosans. The types of hemicelluloses are often classified according to the sugar residues present. Commonly existing hemicelluloses are D-xylan, L-arabino-D-xylan, L-arabino-D-glucurono-D-xylan, D-gluco-D-mannan and D-galacto-D-gluco-D-mannan. The types and amounts of hemicelluloses vary widely, depending on plant materials, stage of growth, growth environment, physiological conditions, etc. (Horwath and Colonna, 1984; Hoq *et al.*, 1992; Horitsu *et al.*, 1992). Among the most important sugars of the hemicellulose components is xylose. The conceptional chemical structure of hemicellulose is illustrated in Fig. 6.2.

In hardwood xylan, the backbone chain consists of xylose units which are linked by *β*-(1,4)-glycosidic bonds and branched by *α*-(1,2)-glycosidic bonds with 4-*O*-methylglucuronic acid groups. In addition, *O*-acetyl groups sometimes replace the OH-groups in positions C2 and C3 (Fig. 6.3). For softwood xylan, the acetyl groups are fewer in the backbone chain. However, softwood xylan has additional branches consisting of arabinofuranose units linked by *α*-(1,3)-glycosidic bonds to the backbone (Fengel and Wegener, 1989).

The close association of hemicelluloses with cellulose and lignin confer rigidity to the plant cell wall. As compared to cellulose, hemicelluloses are low molecular weight polymers with a degree of polymerization of around 200. Native hemicellulose like xylan is composed of 85–90% D-xylose and a small amount of L-arabinose and traces of glucuronic acid. Hemicelluloses have more branches and are less crystalline than cellulose. The glycosidic linkages between the anhydro-D-xylose residues in xylan are more susceptible to acid or enzymatic hydrolysis than linkages between anhydro-D-glucose residues in cellulose. As a result, pentose sugars can be obtained readily in hydrolysates with sufficient yields from hemicellulose.

Xylan is the most abundant polysaccharide present in the plant cell wall (Fig. 6.3). It is a heteropolymer consisting of a backbone of *β*-1,4-linked D-xylose residues which can be modified by various substituents: 1,2-linked *α*-D-glucuronic acid or 4-*O*-methyl-*α*-D-glucuronic acid residues can be present, as well as 1,2- and 1,3-linked *α*-L-arabinose residues. In some cases, these L-arabinose

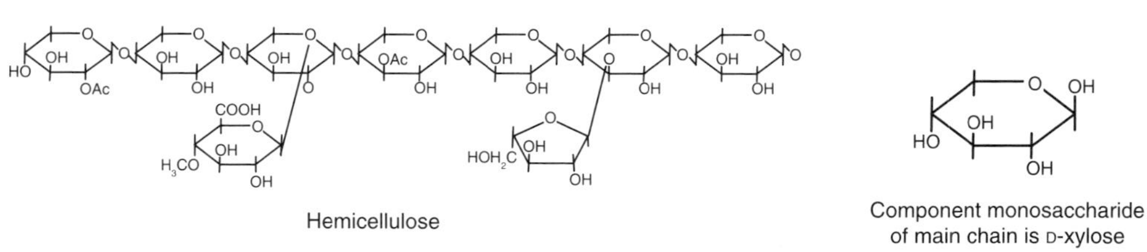

Fig. 6.2. Conceptional chemical structure of hemicelluloses.

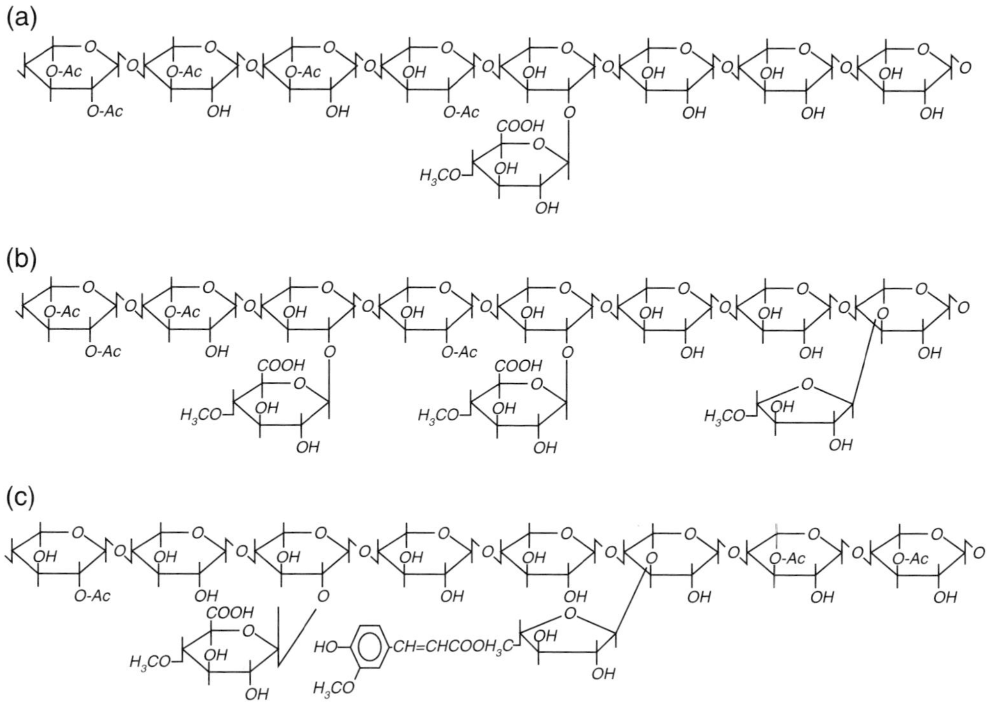

(a) a typical hardwood O-acetyl-4-O-methylglucuronoxylan
(b) a typical softwood arabino-4-O-methylglucuronoxylan
(c) a typical grass arabino-4-O-methyglucuronoxylan

Fig. 6.3. Hypothetical structure of various xylans.

residues are esterified with ferulic and *p*-coumaric acid, enabling cross-linking of the xylan to the lignin matrix. Depending on the source of the plant material, the D-xylose residues in the backbone can be modified by acetylation at the C-2 or C-3 position (Wilkie, 1979). Complete degradation of this heteropolymer requires the synergistic action of a spectrum of enzymes (to be discussed later).

Biochemical Conversion of Cellulosic Biomass

Pretreatment methods for ethanol production

Conversion of cellulose crops to bioethanol is divided into two processes: saccharification (or liquefaction) and ethanol fermentation. It is difficult for yeasts to ferment cellulose crops directly without breaking down the hard structure of plants chemically into monosaccharide. The term 'saccharification' or 'liquefaction' means decomposition of cellulose crops, mainly their individual cellulose and hemicellulose to sugars by using appropriate methods suited for individual crops. Before saccharification, reducing the molecular size of cellulose is desirable to improve decomposition efficiency. Typical physical methods for pretreatment of cellulose crops are mechanical treatment, steam explosion and semi-supercritical treatment. Mechanical treatment methods include cutting, grinding and milling, which use shearing and impact forces to yield a fine or super-fine substrate having a low crystallinity, which would lead to a reduction in the subsequent reactor volume and reaction period for obtaining bioethanol.

Milling

Milling helps in the distribution of reactants throughout the plant material (Colberg, 1988). Ball milling is an effective means of pretreatment (Cote, 1981). In addition to reducing particle size, ball milling disrupts the crystalline structure and breaks down the chemical bonding of long-chain molecules, increasing the accessibility of enzymes for decomposition of the substrate.

Chemical hydrolysis

Chemical treatment methods are effective for structural modification and/or decomposition of cellulose crops, but by-products generated by chemical reactions are difficult to recycle or dispose. Acids such as sulfuric, hydrochloric, nitric, acetic and phosphoric are generally used to prepare cellulose crop hydrolysate. Under acidic conditions, after the hydrolysis of cellulose and hemicelluloses, decomposition of the resulting monosaccharides takes place and D-xylose is decomposed five times faster than D-glucose (Kubata *et al.*, 1992, 1994). D-Xylose is decomposed into furfural, while D-glucose is chemically transformed to hydroxymethyl furfural. Both types of products are harmful for microbial fermentation (Kubata *et al.*, 1997). Although steaming can be used to volatilize and remove furfural, the best results can be obtained by passing the hydrolysate through an ion exchange column (Kubata *et al.*, 1995; Kuhad *et al.*, 1997). Cho *et al.* (2000) have demonstrated that neutralization, filtration, anion- and cation-resin treatments are necessary (Fig. 6.4) for xylitol fermentation using acid hydrolysed groundnut shell as a medium for *Candida tropicalis* (Fig. 6.5).

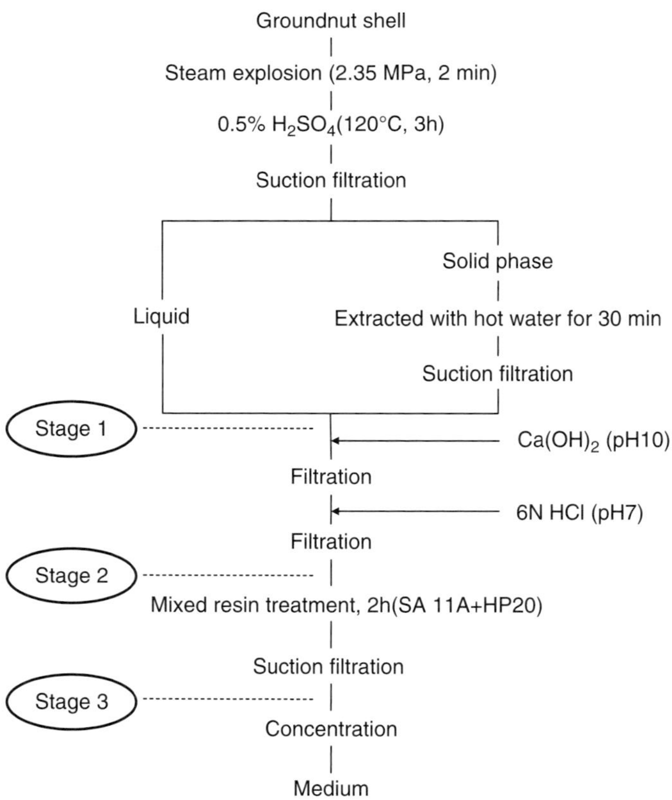

Fig. 6.4. Proposed method for making medium from groundnut shell for xylitol production by *Candida tropicalis* IFO 0618.

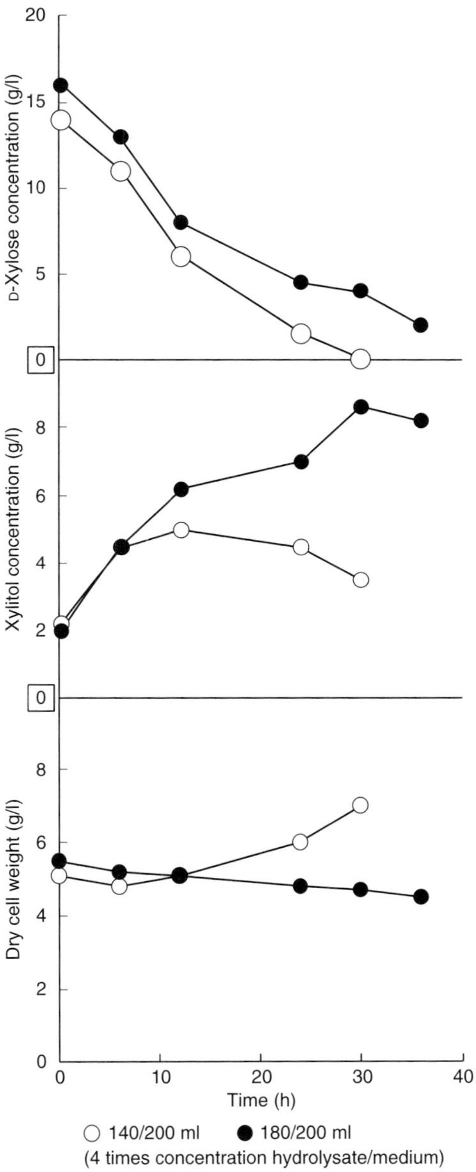

Fig. 6.5. Effect of higher inoculation amount of *Candida tropicalis* IFO 0618 on xylitol production using newly developed medium from groundnut shell.

Enzymatic hydrolysis: cellulase, xylanase, arabinofuranosidase, esterase, glucosidase and xylosidase

Lignocellulose is a very complex structure polymer with many binding types between constituted molecules and materials. Many enzymes possessing different activity to substrates are required for the decomposition of lignocellulose. One enzyme may catalyse one type of reaction (reaction specificity) and another may use only a particular substrate or substrate of specific chemical structure (substrate specificity). Important enzymes for the decomposition of lignocelluloses are cellulase for cellulose and xylanase for hemicellulose. Shellfish, plants, protozoa and microorganisms secrete cellulase and hemicellulase or xylanase in natural environments. Insects such as cockroaches and termites, which eat wood, ruminants such as cows and sheep and herbivores such as horses all harbour microorganisms which secrete cellulase in their alimentary canal and digest cellulose to D-glucose (Castro and Machado, 1989).

Cellulases are divided into three categories; celobiohydrase (β-1,4-D-glucan glucohydrolase; EC 3.2.1.91), endoglucanase (endo-β-1,4-glucanase; EC 3.2.1.4) and β-glucosidase; EC 3.2.1.21). Celobiohydrase is an exo-type reaction enzyme and hydrolyses crystal-cellulose to D-glucose. Partially hydrolysed cellulose, free β-1,4-glucan, is hydrolysed randomly by endoglucanase and the resulting oligosaccharides are decomposed to D-glucose by the reaction of β-glucosidase (Fig. 6.6).

Commercially available cellulases are derived from microorganisms and are usually a mixture of cellulotic and hemicellulotic enzymes. *Clostridium* spp., *Bacillus* spp., *Penicillium* spp., *Humicola gresea*, *Acremonium cellulolyticus*, *Tricoderma viride* and *T. reesei* are typical cellulase-producing microorganisms. Among them, *T. reesei* secretes at least three kinds of celobiohydrase and two kinds of endoglucanase and is a well-known excellent cellulase producer. *Aspergillus niger* generates a higher amount of β-glucosidase.

Hydrolysis of cellulose is expressed by the following reaction equation:

$$(C_6H_{10}O_5)n + nH_2O \longrightarrow nC_6H_{12}O_6$$

cellulose water D-glucose

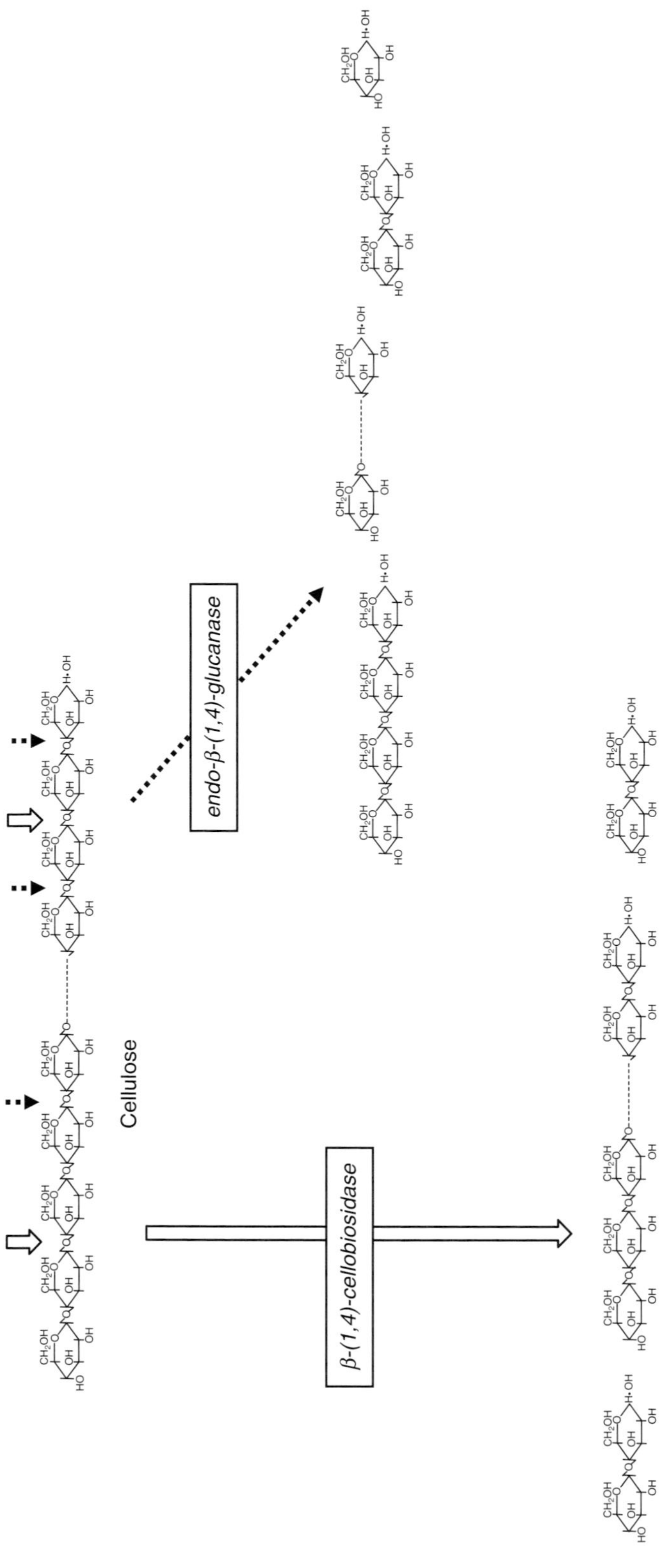

Fig. 6.6. Enzymatic hydrolysis of cellulose.

The final product of enzymatic hydrolysis of cellulose is D-glucose.

Enzymes involved in xylan degradation are endoxylanase (*β*-1,4-D-xylan xylohydrolase; EC 3.2.1.8), *β*-xylosidase (EC 3.2.1.37), acetylxylan esterase (EC 3.1.1.72), L-arabinose, releasing enzymes such as *α*-L-arabinofuranosidase (EC 3.2.1.55), and arabinoxylan arabinofuranohydrolase, *α*-glucuronidase (EC 3.2.1.139), feruloyl esterase and *p*-coumaroyl esterase (Biely, 1985). Enzymes have been found that have a high specificity towards arabinoxylan or arabinoxylan-derived oligosaccharides and which are active against L-arabinofuranosyl groups linked to double-substituted D-xylopyranosyl residues (Laere *et al*., 1997).

Endoxylanases randomly attack xylan at the D-xylose backbone, thus lowering the degree of polymerization rapidly. *Clostridium* spp., *Bacillus* spp., *Aspergillus* spp., *Penicillium* spp., *Streptomyces* spp., *Tricoderma* spp., *Rhizomucor* spp. and *Aeromonas caviae* are typical hemicellulase-producing microorganisms. The xylo-oligosaccharides are further hydrolysed by *β*-xylosidase. These degradations are expressed in Fig. 6.7.

Most xylanases are strongly inhibited in their activity by side groups attached to the D-xylose backbone, e.g. acetyl groups, L-arabinose or D-glucuronic acid residues. Accessory enzymes, like *α*-L-arabinofuranosidase, acetylxylan esterase or *α*-glucuronidase, which remove these side groups, have been shown to have strong synergistic activity (enhancement of activity over the added activities of the individual enzymes) with endoxylanases, thus facilitating the complete degradation of the heteropolymer (Kormelink *et al*., 1993).

Hydrolysis of hemicellulose is expressed by the following reaction equation:

$$(C_5H_8O_4)n + nH_2O \longrightarrow nC_5H_{10}O_5$$

hemecellulose water D-xylose

Hemicellulose is decomposed to the monosaccharide, D-xylose.

It is important to note that many cellulases have the activities of hemicellulose decomposition as well, e.g. *β*-glucosidase possesses activity as well as *β*-xylosidase. It must also be recognized that the decomposition ability of celluloses by cellulases depends on the source of enzyme. The synergy effect of applying two or more different enzymes increases decomposition efficiency and specific decomposition rate.

ORDER AND PERIOD OF ENZYME APPLICATION. Hydrolysis of plant residues can be influenced by the sequence of enzyme application. An example of oat spelt xylan hydrolysis for extraction of D-xylose is provided. The enzymes used were partially purified supernatant of culture broth of *Penicillium* spp. AHT-1 and *Rhizomucor pusillus* HHT-1 (Cho *et al*., 2002; Rahman *et al*., 2003). Activities of xylanase, *β*-D-xylosidase, *α*-L-arabinofuranosidase and CMCase of *Penicillium* spp. strain AHT-1 were 60.0, 0, 0.01 and 0.50 U/ml, respectively, and those of *R. pusillus* HHT-1 were 0.69, 0.16, 0.42 and 0.09 U/ml. The reaction conditions for hydrolysis were: xylanase 28 U/ml, *β*-D-xylosidase 0.2 U/ml and *α*-L-arabinofuranosidase 19 U/ml at 40°C for 72 h. The result is shown in Fig. 6.8. Step 1 is the result after xylanase application showing peaks of xylobiose, xylotriose and oligoxylose in high performance liquid chromatography (HPLC). It indicates that xylanase produced by *Penicillium* spp. AHT-1 has an endo-type reaction mechanism. After treatment of *β*-D-xylosidase and *α*-L-arabinofuranosidase solution, a peak of D-xylose was produced but generation of L-arabinose did not occur (Step 2). When *α*-L-arabinofuranosidase was reacted first, it resulted in the production of L-arabinose (Step 3), followed by xylobiose, xylotriose and related oligoxylose generated after serial reaction by xylanase (Step 4), and then D-xylose by reacting with *β*-D-xylosidase (Step 5). Comparing the peak height of D-xylose in Step 5 with that in Step 2, it is clear that the extracted D-xylose amount in Step 5 is significantly greater than that in Step 2. It means that application of *α*-L-arabinofuranosidase, xylan side chain degrader prior to the treatment of xylanase will help the reaction between xylanase and xylan, resulting in efficient hydrolysis of xylan.

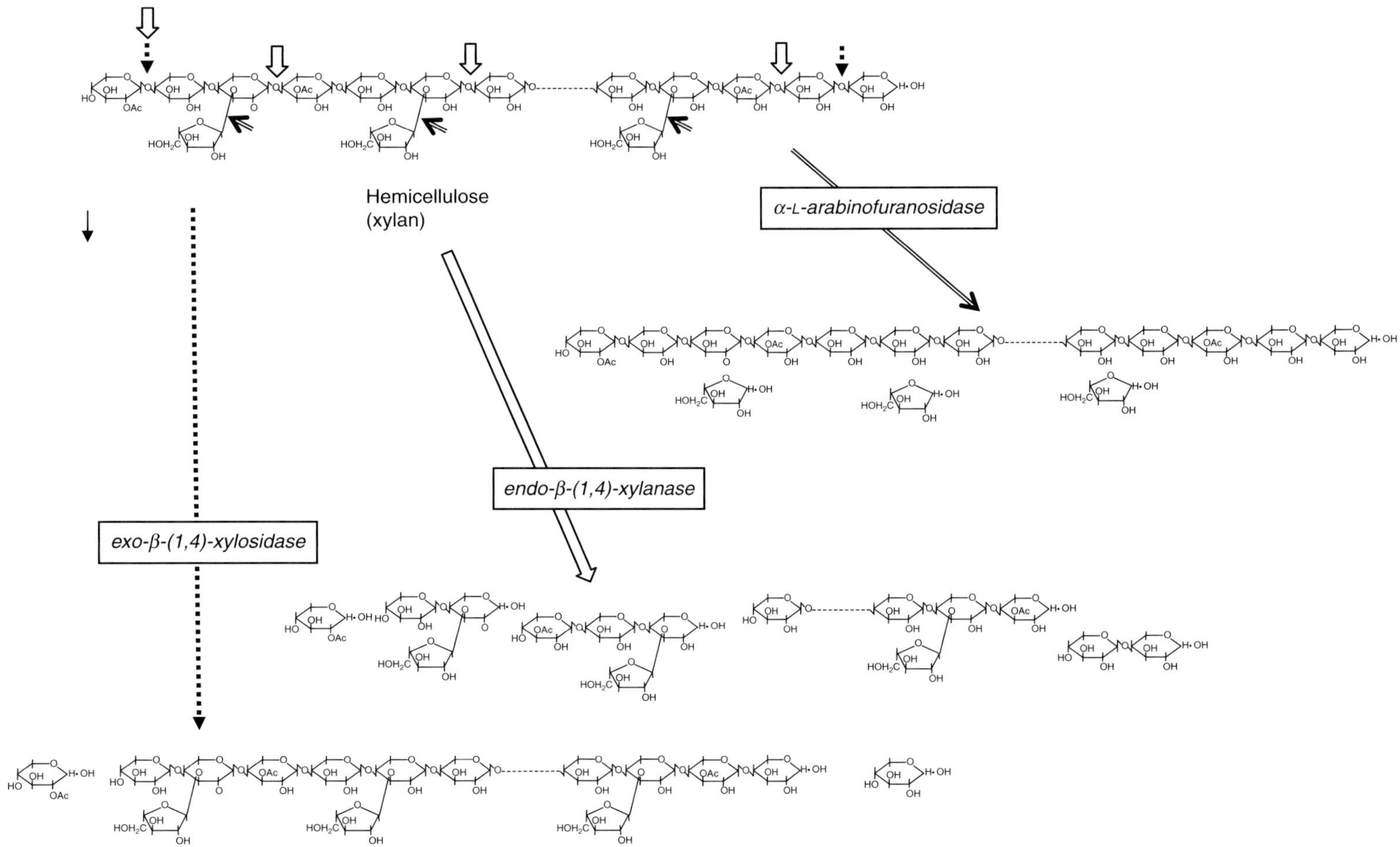

Fig. 6.7. Enzymatic hydrolysis of hemicellulose.

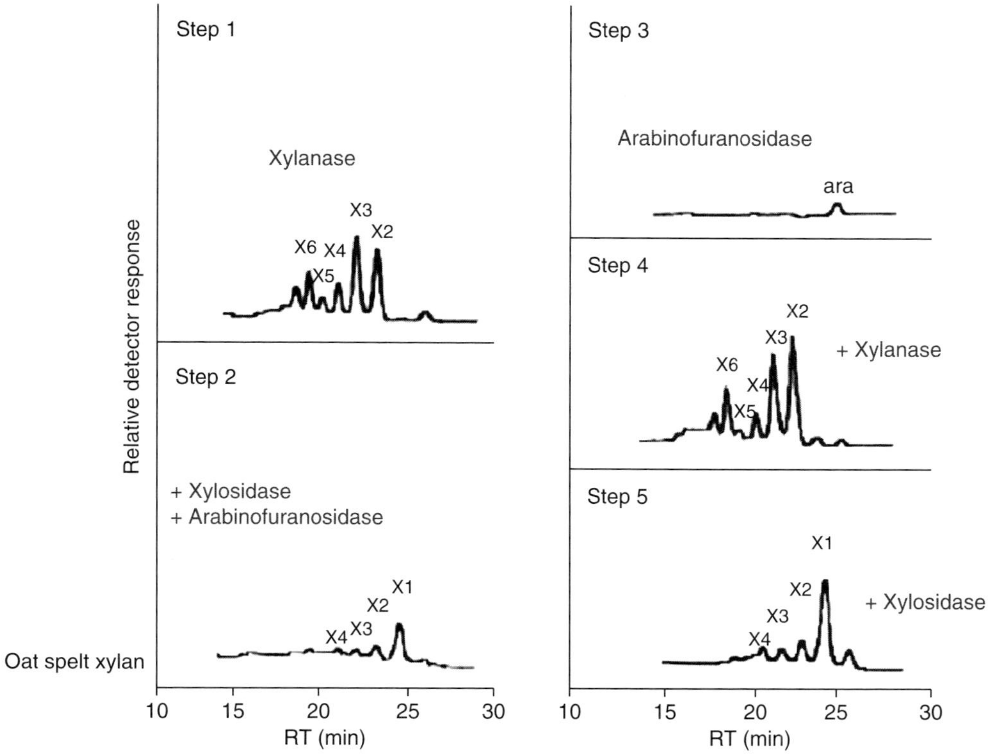

Fig. 6.8. Hydrolysis of oat spelt xylan using holocellulase.

Figure 6.9 shows the hydrolysis reaction kinetics of plant residues by hemicellulase. Average particle size of each material was 90 μm or less. D-Xylose concentrations were increased with reaction time and in the case of weeds, the value turned to a constant after a 40 h incubation. The maximum concentration of D-xylose was 11.5 g/100 g material dry base. In the cases of beechwood sawdust, Japanese cedar sawdust and bamboo sawdust, concentrations of D-xylose extracted by hemicelluloses treatment were stable after 72 h reaction; 12.1, 10.3 and 9.1 g/100 g material dry base, respectively. Recovery yields of D-xylose based on chemical analysis of each material were 82% for weeds, 72% for beechwood sawdust, 93% for Japanese cedar sawdust and 70% for bamboo sawdust, respectively.

ADVANTAGE OF ENZYMATIC HYDROLYSIS FOR CASCADE UTILIZATION OF CELLULOSE CROPS. Compared with the chemical and physical pretreatment of cellulose crops, the advantage of enzymatic pretreatment is that it produces less side-products such as furfural, 5-hydroxymethyl furfural, phenolic compounds and acetic acid. These inhibit the activity of microorganisms for ethanol fermentation. Furfural, an overdecomposition product of D-xylose, is converted to furalic alcoholic compounds by yeast during alcohol fermentation. This reaction proceeds faster than ethanol production from D-glucose and D-xylose, obstructing ethanol fermentation (Delgenes *et al.*, 1996; Mart'in *et al.*, 2002, 2007). Acetic acid and phenolic compounds derived from lignin are also inhibitors for microorganism growth and activity, resulting in less fermentation yields (Delgenes *et al.*, 1996).

The reactions of enzymes are selective for cellulose and hemicelluloses leaving non-reactive lignin residue. Lignin is a potential resource for polyurethane, carbon fibre, phenolic resins and also pellets for furnaces. Cellulose is a source for D-glucose;

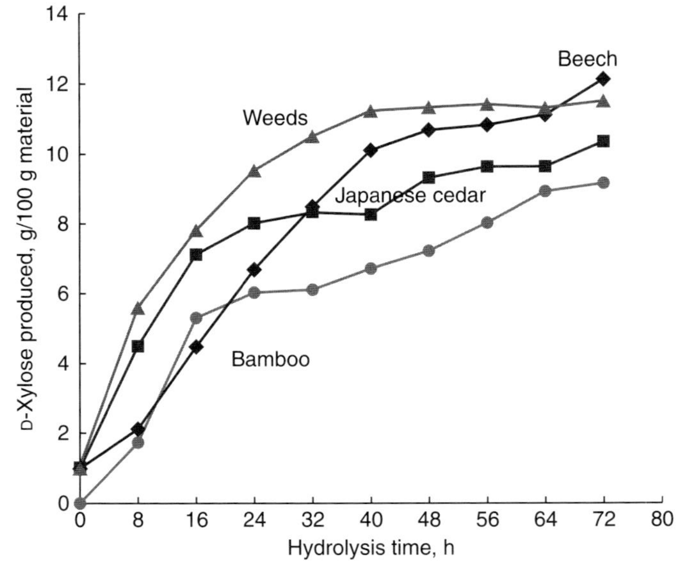

Fig. 6.9. Hydrolysis reaction kinetics of plant residues by hemicellulase.

hemicellulose is for D-xylose. Enzymatic hydrolysis helps for cascade utilization of cellulose crops.

Conversion of monosaccharides to ethanol

Biochemical pathway

After obtaining monosaccharides, D-glucose and D-xylose, alcohol fermentation by microorganisms proceeds as follows. Alcohol fermentation from D-glucose by microorganisms is performed mainly by the yeast, *Saccharomyces cerevisiae*. The yeast is widely distributed in nature. The peel of every type of ripe fruit and flowers harbours wild *S. cerevisiae*. Wine is a typical product fermented from grape by yeast.

The biochemical common pathway from D-glucose to ethanol is glycolysis, also called the Embden–Meyerhof–Parnas (EMP) pathway after its major discoverers, and is shown in Fig. 6.10. The pathway is an anaerobic process and can be divided into three stages, each involving a series of enzymatic reactions. Stage I is conversion of D-glucose to fructose 6-phosphate and comprises preparatory reactions of phosphorylation and isomeration. These are not redox reactions and do not release energy but lead to the production of two molecules of a key intermediate, glyceraldehyde 3-phosphate. In Stage II, redox reactions occur and energy is conserved in the form of adenosine triphosphate (ATP) and two molecules of pyruvate are produced. In Stage III, redox reactions again occur and fermentation products, mainly ethanol, are formed.

Another important glycolytic pathway, called the Entner–Doudoroff pathway (Fig. 6.11) was first discovered in an isolated bacterium, *Zymomonas mobilis* from an agave/maguey. It is now known to be a widely found pathway for sugar catabolism in bacteria, especially among the species of the pseudomonad group.

The initial D-xylose metabolism is important to comprehend ethanol fermentation from D-xylose and is shown in Fig. 6.12. D-Xylose is reduced to xylitol under anaerobic/anoxic conditions using nicotineamide adenine dinucleotide (phosphate) reduced form (NAD(P)H) as a co-factor, followed by oxidation to D-xylulose using NAD(P). It is essential to control oxidation and the reduction in the potential to convert D-xylose to

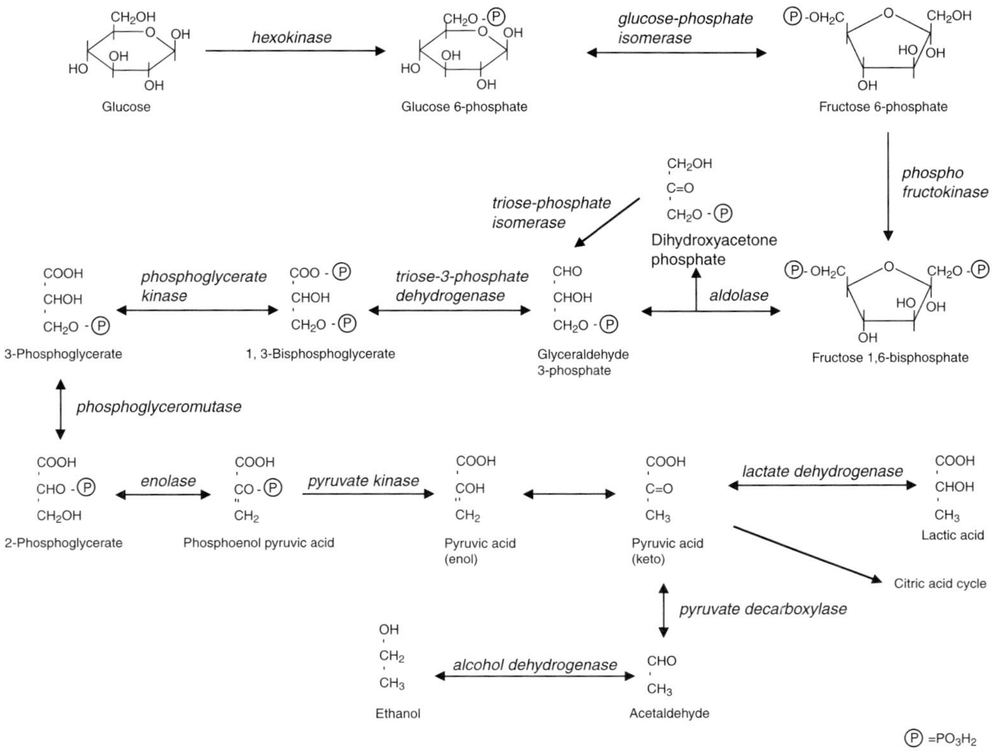

Fig. 6.10. Embden–Meyerhof–Parnas pathway (EMP).

D-xylulose. Conversion of D-xylose directly to D-xylulose is carried out by bacteria only. D-Xylulose is the starting material for the pentose phosphate pathway (Fig. 6.13) to the EMP pathway. The pentose phosphate pathway is thus the gateway of D-xylose to glycolysis, followed by ethanol production.

Reaction stoichiometry

S. cerevisiae and *Z. mobilis* can ferment D-glucose to ethanol using different pathways. The overall reaction from glucose to ethanol by microorganisms is expressed as the Gay–Lussac equation:

$$C_6H_{12}O_6 \longrightarrow 2C_2H_5OH + 2CO_2$$

D-glucose ethanol carbon dioxide

Microorganisms produce 51.4g of ethanol from 100g of D-glucose. The yield based on weight is reduced to approximately half, but 97.61% energy (based on standard generation enthalpy) of D-glucose can be conserved as ethanol. In this sense, ethanol fermentation is very attractive as an energy conservative method but, in reality, we must recognize that approximately 49g of greenhouse gas, equivalent to approximately 13g of carbon, are released by fermentation.

D-Xylose fermentation is performed by yeasts having the fermentation ability of five carbon sugars; *Pichia stipitis*, *Pachysolen tannophilus*, *Candida* spp. The reaction equation is as follows:

$$3C_5H_{10}O_5 \longrightarrow 5C_2H_5OH + 5CO_2$$

D-xylose ethanol carbon dioxide

The theoretical fermentation yield from D-xylose to ethanol is 0.511g ethanol/g D-xylose and is close to the yield from

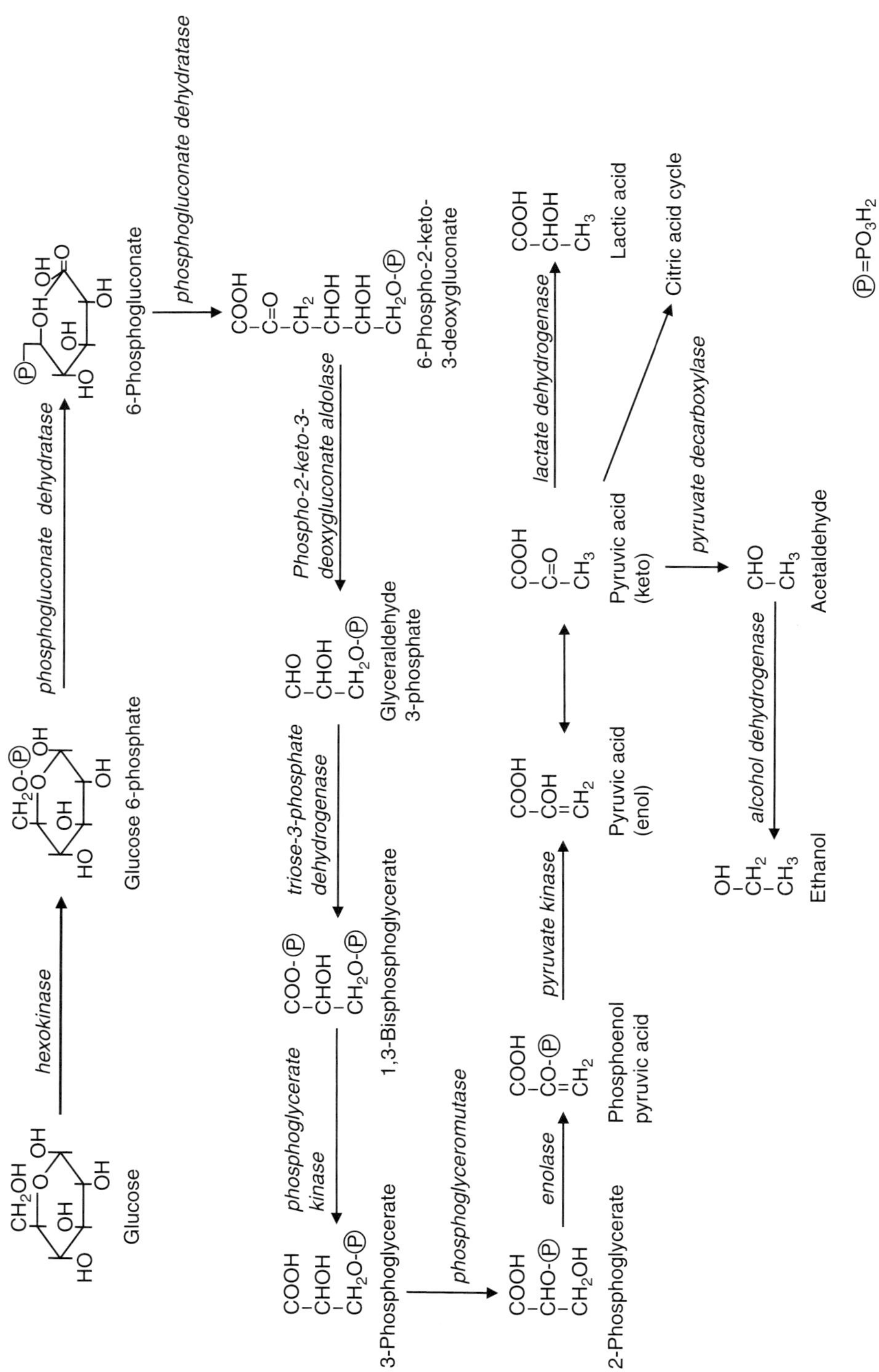

Fig. 6.11. Entner–Doudoroff pathway.

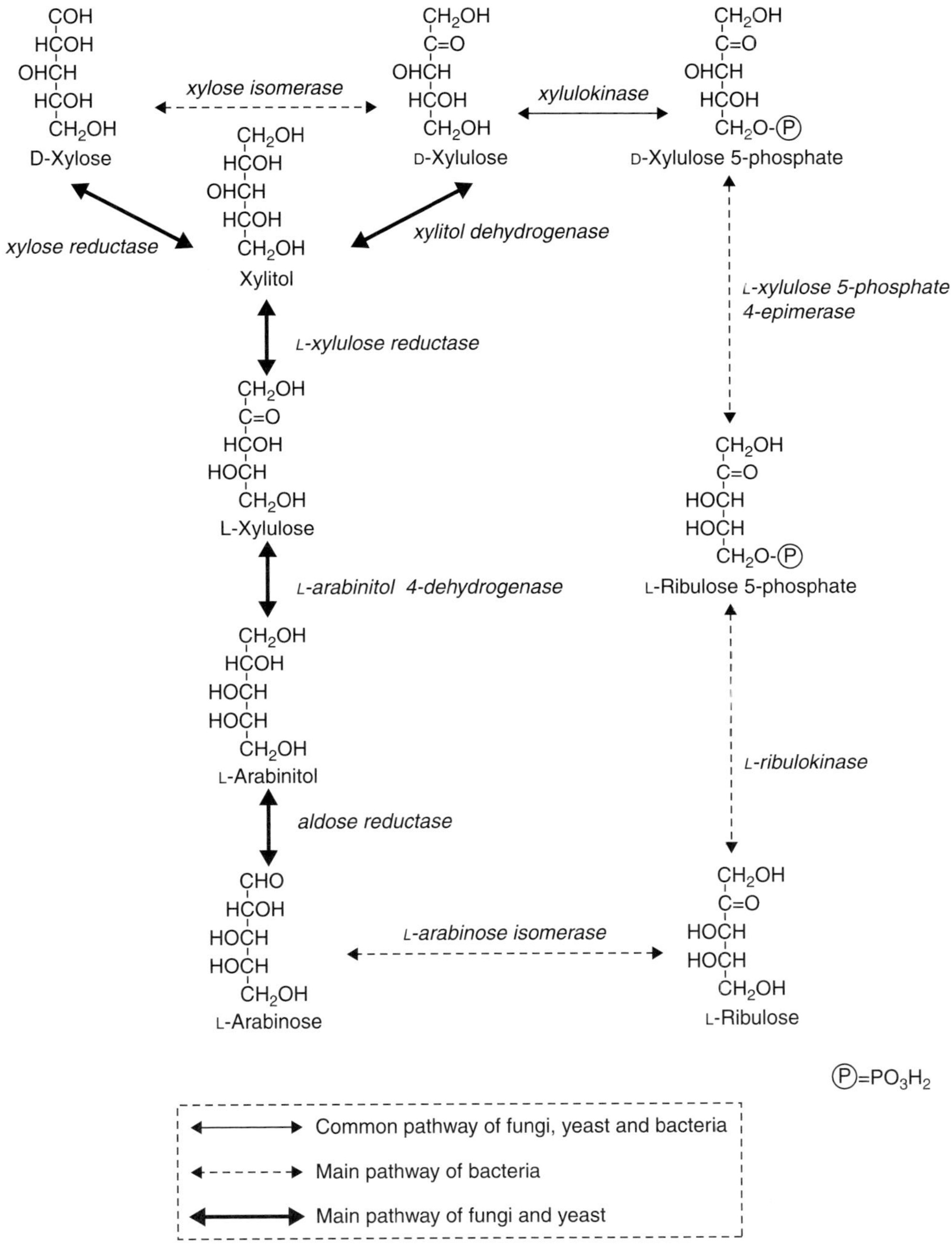

Fig. 6.12. Initial metabolism of D-xylose by microorganisms.

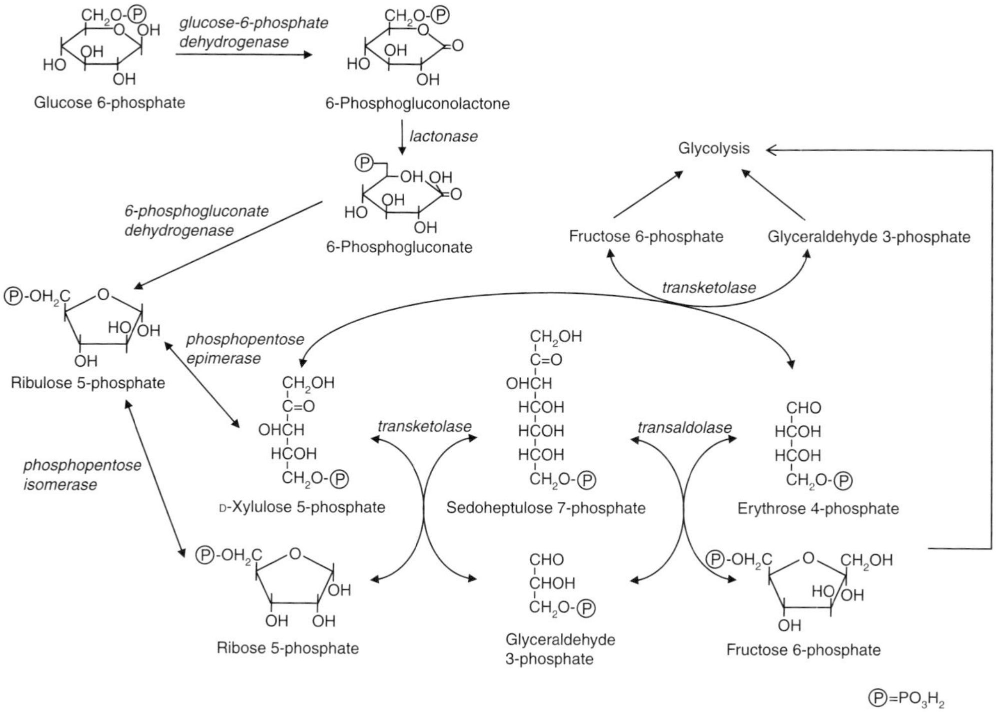

Fig. 6.13. Pentose phosphate pathway

D-glucose. These yeasts also have activity to use D-glucose.

Traditional microorganisms and genetically engineered microorganisms

Typical and traditional D-glucose fermentation yeast is *S. cerevisiae*. The strain is widely distributed in nature and has been used since ancient times to make bread, wine, beer, Japanese sake and other alcoholic beverages. It ferments D-glucose, D-fructose and sucrose to ethanol and is tolerant to produced alcohol. It contains invertase, which is an enzyme capable of hydrolysing sucrose to D-glucose and D-fructose. *S. cerevisiae* is also tolerant to lower acidic pH, and ethanol fermentation can be performed continuously for more than 87h under pH 4.5 conditions without contamination of other microorganisms (Nigam, 2000).

Microorganisms able to ferment D-xylose belong to yeast: *P. stipitis*, *C. shehatae*, *C. tenius*, *P. tannophilus* and *Brettanomyces noardenensis* (Toivola *et al.*, 1984). *Z. mobilis*, *Zymobacter palmae* and *Corynebacter glutamicum* are typical fermentation bacteria of D-glucose.

S. cerevisiae is an excellent yeast for ethanol production from D-glucose but it is impossible for it to metabolize pentose. There have been many attempts to transfer genes related to D-xylose metabolism of *P. stipitis* to *S. cerevisiae* but a clone worthy for industrial use has not yet been obtained (Byron and Chu, 2007).

Ethanol production is associated with ATP production. In the case of D-glucose fermentation, *S. cerevisiae* generates two molecules of ethanol and two molecules of ATP from D-glucose through the Embden–Meyerhof–Parnas pathway (Fig. 6.10). In the same condition, *Z. mobilis* uses a different metabolic pathway, the Entner–Doudoroff shunt (Fig. 6.11), and produces two molecules of ethanol and one molecule of ATP. For getting energy from D-glucose, the Embden–Meyerhof–Parnas pathway is more efficient than the Entner–Doudoroff shunt. But, for supplementation of energy shortage, growth is suppressed and the glucose uptake rate is faster than that of yeast. As a result, the ethanol production rate of *Z. mobilis* (0.47) is slightly higher than the value of *S. cerevisiae* (0.44) (Cowling and Kirk, 1976).

Z. mobilis can use D-glucose, sucrose and D-fructose as substrates, but cannot ferment D-xylose. Many researchers are trying to add D-xylose fermentation ability to *Z. mobilis*. Zhang *et al.* have obtained a transformant *Z. mobilis* ZM4 (pZB5) harbouring an operon encoding D-xylose assimilation and pentose phosphate pathway and have succeeded in ethanol fermentation from mixer of D-glucose and D-xylose (Zhang *et al.*, 1995). The strain also fermented rice straw hydrolysed by sulfuric acid and an ethanol yield of 0.47 was obtained (Davis *et al.*, 2005).

Z. palmae can use a wide variety of substrates, D-glucose, glucobiose, glucotriose, hexose and sugar alcohol, but cannot utilize D-xylose (Okamoto *et al.*, 1993). Important genomes related to D-xylose metabolisms, transaldorase, transketolase, xylose-isomerase and xylulokinase of *E. coli*, were introduced using promoters of *Z. mobilis* into *Z. palmae*, resulting in the transformant that ferments both D-glucose and D-xylose (Yanase *et al.*, 2007). However, genetic stability of the transformant was not very high. Another transformant of *Z. palmae* expressing β-glucosidase fermented cellobiose with 95% ethanol yield (Yanase *et al.*, 2005).

Transformants of *E. coli* harbouring genomes related to ethanol fermentation of *Z. mobilis* have been obtained to ferment D-xylose and L-arabinose. *E. coli* KO11, harbouring D-xylose fermentation ability, is not so good at fermenting D-glucose and other hexoses. Co-cultivation with *S. cerevisiae* led to increased ethanol yield up to 81% and to the reduction of nutrients in broth and an initial amount of inoculum (Okuda *et al.*, 2008). *E. coli* FBR5 produced 90% ethanol yield from hydrolysate of maize cob bagasse (Dien *et al.*, 2000).

Arming yeast, displaying endoglucanase of *Trichoderma reesei* and β-glucosidase of *A. aculeatus* on the cell surface of *S. cerevisiae*, can perform saccharification and fermentation of cellulose at a time (Kondo and Ueda, 2004). The ethanol yield from cellulose was 0.48 and is equivalent to 93% of the theoretical yield.

Process for ethanol production

The overall system for bioethanol production is shown in Fig. 6.14. Starting from a stockyard of cellulose crops for storing and drying, major processes are: pretreatment, saccharification/liquefaction, ethanol fermentation, concentration and distillation, and waste treatment. Depending on the characteristics of the cellulose crop, cutting, milling, fine-milling or heat treatment are necessary as a pretreatment before saccharification/liquefaction. Enzymatic hydrolysis of cellulose crops requires adequate buffer, solution of inorganic minerals for keeping the structure of enzymes, and neutralization after acidic hydrolysis. In the case of sulfuric acid hydrolysis, calcium is conventionally used as a neutralizer and gypsum (hydrated calcium sulfate) is produced as a by-product. Enzymes are more expensive than chemicals and how to reduce the amount of enzyme used for saccharification is a major concern industrially. To develop more active hydrolytic enzymes could be one way of providing a solution and another will be to develop methods that can prolong the activity of enzymes.

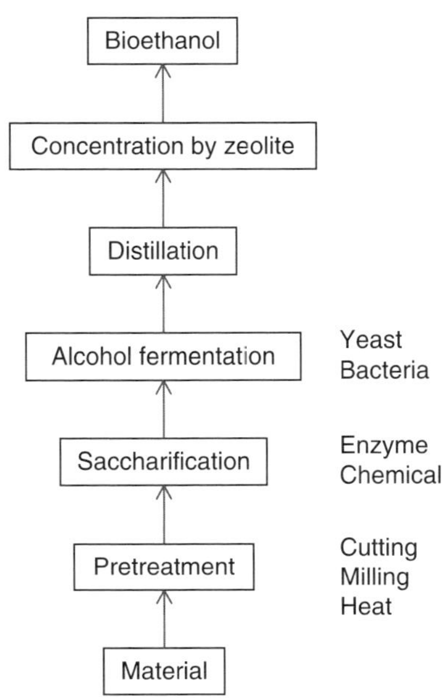

Fig. 6.14. Principal process flow of bioethanol production.

Batch operation is conventionally used for ethanol fermentation from saccharified cellulose crops. The reactor used for fermentation is basically a regular fermentor with modification, depending on the property of the material. A propeller-type mixing paddle is suitable rather than a plain paddle, because of the viscosity of the saccharified broth. Sand drain is also necessary. Except for using genetically modified microorganisms which ferment both hexose and pentose, *S. cerevisiae* or other hexose-assimilating microorganisms are solely used because the major material for ethanol fermentation is D-glucose. To increase the yield of fermentation, pentose-assimilating microorganisms such as *P. stipitis* and *P. tannophilus* must be introduced. Simultaneous fermentation of hexose and pentose, serial fermentation of hexose followed by pentose or vice versa are applicable (Taniguchi *et al.*, 1997; Rouhollah *et al.*, 2007). Figure 6.15 presents the result of simultaneous fermentation of saccharified lawn using *P. stipitis* and *S. cerevisiae* or *Z. mobilis*, carried out in our laboratory. Comparing the results of *P. stipitis–S. cerevisiae* with that of *P. stipitis–Z. mobilis* maximum ethanol concentration, 7.6 g/l, is the same, but the latter has a higher yield value of 82%. Probably, the combination of hexose fermenter and pentose fermenter will be changeable, depending on the characteristics of materials.

One attractive method would be to combine saccharification and fermentation processes together. Japanese sake production traditionally uses the combined method and its advantage is the higher concentration of ethanol that is obtained. *A. oryzae* for saccharification and *S. cerevisiae* for alcohol fermentation are used. Ethanol concentration of beer and wine are around 5 and 12%, respectively, but that of sake is up to 15%, without the supply of any D-glucose.

Multiple stage distillation is used for concentration of ethanol from broth, followed by zeolite resin concentration, resulting in 95% or more purity of ethanol. We must recognize that concentrations of fermented product from sugarcane bagasse, wheat straw, woodchips, lawn, weeds and shredder paper are approximately 18 (Mart'in *et al.*, 2002), 24 (Saha *et al.*, 2005), 30, 8, 4.4 and 24%, respectively, and more than half of cellulose crops remain as unused waste.

Waste produced after fermentation is filter-pressed solids and liquid waste after distillation. Conventionally, solid wastes are considered as materials for composting, and fuel pellet and liquid wastes can be utilized as a fertilizer. Making compressed board (biomass board) from solid waste will be an attractive and effective use of solid waste.

In our study, we have found that a typical golf course having 18 holes in Japan produces approximately 1200 m^3 (equivalent to approximately 600 t) of cut grass for course maintenance. Moisture content of cut grass is around 90% and contains 30% D-glucose and 16% D-xylose on a dry basis. Assuming saccharification yield, fermentation yield and distillation and concentration yield are 95,

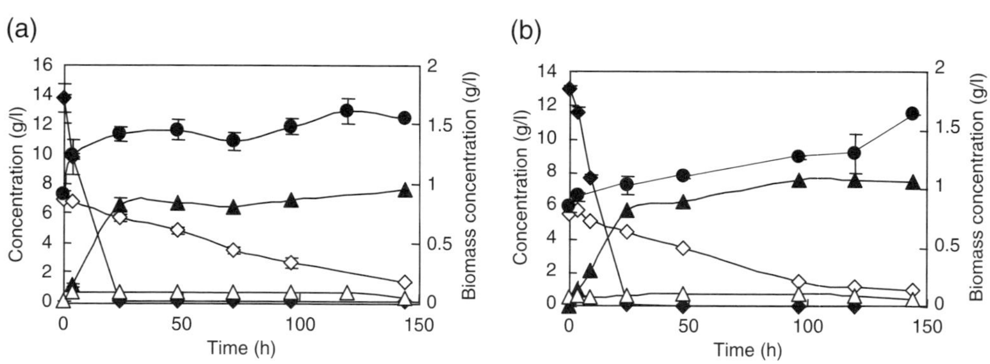

Fig. 6.15. Fermentation of Korean Lawn (KL) saccharified with acclimated *Pichia stipitis* and (a) *Saccharomyces cerevisiae* or (b) *Zymomonas mobilis*. Concentrations of ethanol (close triangle), D-glucose (close diamond), D-xylose (open diamond), acetate (open triangle) and biomass (close circle) were measured.

90 and 95%, respectively, 13.8t ethanol can be obtained from a golf course in 1 year. The amount is equivalent to 11.3kl of gasoline in calorie base. The annual amount of gasoline for truck and vehicle maintenance for a golf course is between 6.5 and 20kl, and 17–50% of gasoline can be obtained from cut grass. In another calculation, five cars can be driven by the ethanol produced.

Thermochemical Conversion of Cellulosic Biomass

Generally, thermochemical conversion is classified as pyrolysis or gasification. Gasification requires oxygen for full conversion, while pyrolysis is performed primarily in the absence of oxygen. The products of pyrolysis are dependent on temperature and residence time. 'Slow' pyrolysis (400–500°C) has been used for centuries to produce charcoal. Of more recent interest, a charcoal-like substance called 'biochar' from slow pyrolysis is being investigated as a means to sequester carbon in the soil and as a tool to reduce global warming. The Amazonian Indians first used this as a way of enhancing soil productivity. A new organization has been formed recently called the International Biochar Initiative (IBI) to support its production and use (http://www.biochar-international.org/abouttheibi/aboutibi.html).

For fast pyrolysis, the biomass is heated much quicker and at more controlled temperatures to produce bio-oils (60–75%) and char (15–25%) (Mohan *et al.*, 2006). Bio-oil is a very diverse mixture of compounds from cellulose, hemicellulose and lignin, which varies by feedstock and thermochemical conditions. The low pH (2–3) of bio-oil is due to hydroxyacetaldehyde (< 10wt%), acetic acid (5%) and formic acids (3%), which are the compounds with highest concentrations. Bio-oil has less value than other hydrocarbons due to its high water and oxygen content, but can be converted to green diesel by either direct catalytic cracking or passing through a gas phase. Also, bio-oil can be fermented to produce ethanol or alkanes (Lercher, 2009).

As temperature and pressure is increased, more biomass is converted to gas or syngas. The gas, which is composed primarily of carbon monoxide and hydrogen gases with varying levels of carbon dioxide, low organic carbon molecules and tar, can then be converted to alcohols by catalysis from organic or inorganic sources. The Fischer–Tropsch process was first invented in the 1920s in Germany. The process uses inorganic catalysts such as iron, cobalt, nickel or ruthernium, palladium, platinum or rhodium (Inderwildi *et al.*, 2008) to convert the primary gasification products, carbon monoxide and hydrogen syngases, to alkanes or alcohols at moderate heat (150–300°C). The optimum H_2:CO ratio varies by catalysts. Biomass tends to produce a lower ratio which favours iron as the best catalyst. All catalysts are sensitive to high sulfur-containing compounds, which can be an important quality factor in potential feedstocks. The industry has developed a number of proprietary techniques to target specific products and to improve efficiency.

Various microbes have been discovered that are effective biocatalysts for the Fischer–Tropsch synthesis of fuels and chemicals. Madhukar *et al.* (1996) described three rod-shaped bacteria that converted CO into acetate, ethanol and methanol. *Clostridium ljungdahlii* has been the major Gram-positive bacteria studied for ethanol production from syngas (Klasson *et al.*, 1992; Najafpour and Younesi, 2006). *Clostridium autoethanogenum* also produces acetate and ethanol (Cotter *et al.*, 2009). Acetate production was from 5 to 20 times greater than ethanol in the first experiments in actively growing cell cultures. However, with manipulation of reaction parameters, more favourable ethanol/acetate ratios were obtained (Klasson *et al.*, 1992), but at the expense of decreased *Clostridium* cell growth. Attempts at developing resting cell cultures to produce higher levels of ethanol generally have been unsuccessful to date (Cotter *et al.*, 2009).

Acknowledgement

The contribution of Professor Siaw Onwana Agyeman of Tokyo University of Agriculture and Technology in the critical reading of the manuscript is acknowledged.

References

Adegbidi, H.G., Briggs, R.D., Volk, T.A., White, E.H. and Abrahamson, L.P. (2003) Effect of organic amendments and slow-release nitrogen fertilizer on willow biomass production and soil chemical characteristics. *Biomass and Bioenergy* 25, 389–398.

Ainsworth, E.A., Leakey, A.D., Ort, D.R. and Long, S.P. (2008) FACE-ing the facts: inconsistencies and interdependence among field, chamber and modeling studies of elevated CO_2 impacts on crop yield and food supply. *New Phytologist* 179, 5–9.

Akin, D.E., Morrison, H.W. III, Rigsby, L.L., Barton, F.E. II, Himmelsbach, D.S. and Hicks, K.B. (2006) Corn stover fractions and bioenergy. *Applied Biochemistry and Biotechnology* 129–132, 104–116.

Allen, H.L., Thomas, R. and Campbell, R.G. (2005) What is ahead for intensive pine plantation silviculture in the South? *Southern Journal of Applied Forestry* 29, 62–69.

Anderson, W.F. and Akin, D.E. (2008) Structural and chemical properties of grass lignocelluloses related to conversion for biofuels. *Journal of Industrial Microbiology and Biotechnology* 35, 355–366.

Anderson, W.F., Peterson, J.D., Akin, D.E. and Morrison, W.H. III (2005) Enzyme pretreatment of grass lignocellulose for potential high-value co-products and an improved fermentable substrate. *Applied Biochemistry and Biotechnology* 121–124, 303–310.

Anderson, W.F., Casler, M.D. and Baldwin, B.S. (2008a) Improvement of perennial forage species as feedstock for bioenergy. In: Vermeris, W. (ed.) *Genetic Improvement of Bioenergy Crops*. Springer, New York, pp. 309–345.

Anderson, W.F., Dien, B.S., Brandon, S.K. and Peterson, J.D. (2008b) Assessment of bermudagrass and bunch grasses as feedstock for conversion to ethanol. *Applied Biochemistry and Biotechnology* 145, 13–21.

Argyropoulos, D.S. and Menachem, S.B. (1997) Lignin. In: Eriksson K.E.L. (ed.) *Advances in Biochemical Engineering/Biotechnology*. Springer-Verlag, Germany, pp. 127–158.

Arinze, E.A., Schoenau, G.J., Sokhansanj, S. and Adapa, P. (2003) Aerodynamic separation and fractional drying of alfalfa leaves and stems – a review and new concept. *Drying Technology* 21, 1669–1698.

Ashton, S. and Jackson, B. (2007) Small-scale woody biomass harvesting systems. In: Hubbard, W., Biles, L., Mayfield, C. and Ashton, S. (eds) *Sustainable Forestry for Bioenergy and Bio-based Products: Trainers Curriculum Notebook*. Southern Forest Research Partnership, Inc, Athens, Georgia, pp. 135–137.

Ball, D.M., Hoveland, C.S. and Lacefield, G.D. (2007) *Southern Forages, 4th edition*. International Plant Nutrition Institute, Norcross, Georgia.

Biely, P. (1985) Microbial xylanolytic systems. *Trends in Biotechnology* 3, 286–290.

Boateng, A.A., Jung, H.G. and Adler, P.R. (2006) Pyrolysis of energy crops including alfalfa stems, reed canarygrass, and eastern gamagrass. *Fuel* 85, 2450–2457.

Boateng, A.A., Anderson, W.F. and Phillips, J.G. (2007) Bermudagrass for biofuels: effect of two genotypes on pyrolysis product yield. *Energy and Fuels* 21, 1183–1187.

Bouton, J. (2008) Improvement of switchgrass as a bioenergy crop. In: Vermerris, W. (ed.) *Genetic Improvement of Bioenergy Crops*. Springer, New York, pp. 295–306.

Bouton, J.H. (2002) Bioenergy crop breeding and production research in the southeast – final report for 1996 to 2001. Oak Ridge National Laboratory Report ORNL/SUB-02-19XSV810/01.

Bradshaw, H.D. Jr, Ceulemans, R., Davis, J. and Stettler, R. (2000) Emerging model systems in plant biology: poplar (*Populus*) as a model forest tree. *Journal of Plant Growth Regulation* 19, 306–313.

Brown, G.R., Gill, G.P., Kuntz, R.L., Langley, C.H. and Neale, D.B. (2004) Nucelotide diversity and linkage disequilibrium in loblolly pine. *Proceeding of the National Academy of Sciences, United States of America* 101, 15255–15260.

Burton, G.W. (1956) Utilization of heterosis in pasture plant breeding. In: Neale, G.J. (ed.) *Proceedings of the International Grassland Congress, 7th*, Palmerston, New Zealand. 6–12 November 1956. Wright and Carman, Wellington, pp. 3–12.

Burton, G.W. (1972) Registration of 'Coastcross-1' bermudagrass. *Crop Science* 12, 125.

Burton, G.W. (1989) Registration of 'Merkeron' napiergrass. *Crop Science* 29, 1327.

Burton, G.W., Gates, R.N. and Hill, G.M. (1993) Registration of 'Tifton 85' bermudagrass. *Crop Science* 33, 644–645.

Byron, C.H. and Chu, H.L. (2007) Genetic improvement of *Saccharomyces cerevisiae* for xylose fermentation. *Biotechnology Advances* 25, 425–441.

Cai, Q., Aitken, K.S., Deng, H.H., Chen, X.W., Fu, C., Jackson, P.A. and McIntyre, C.L. (2005) Verification of the introgression of *Erianthus arundinaceus* germplasm into sugar-cane using molecular markers. *Plant Breeding* 124, 322–328.

Cameron, K.D., Phillips, I.S., Kopp, R.F., Volk, T.A., Maynard, C.A., Abrahamson, L.P. and Smart, L.B. (2008) Quantitative genetics of traits indicative of biomass production and heterosis in 34 full-sib F_1 *Salix eriocephala* families. *Bioenergy Research* 1, 80–90.

Caniato, F.F., Guimaraes, C.T., Schaffer, R.E., Alves, V.M.C., Kochian, L.V., Borem, A., Klein, P.E. and Magalhaes, J.V. (2007) Genetic diversity for aluminum tolerance in sorghum. *Theoretical and Applied Genetics* 114, 863–876.

Carroll, A. and Somerville, C. (2009) Cellulosic biofuels. *Annual Reviews in Plant Biology* 60, 165–182.

Casler, M.D., Vogel, K.P., Taliaferrro, C.M. and Wynia, R.L. (2004) Latitudinal adaptation of switchgrass populations. *Crop Science* 44, 293–303.

Cassidy, P. and Ashton, S. (2007) Bio-based products. In: Hubbard, W., Biles, L., Mayfield, C. and Ashton, S. (eds) *Sustainable Forestry for Bioenergy and Bio-based Products: Trainers Curriculum Notebook.* Southern Forest Research Partnership, Inc, Athens, Georgia, pp. 197–199.

Castro, F.B. and Machado, P.F. (1989). Digestive process evaluation of steam and pressure treated sugar-cane bagasse. *Boletim de Indústria Animal* 46, 213–217.

Cherney, J.H., Cherney, D.J.R., Akin, D.E. and Axtell, J.D. (1991) Potential of brown-midrib, low lignin mutants for improving forage quality. *Advances in Agronomy* 46, 157–198.

Cho, C.H., Aruga, S., Hatsu, M., Suzuki, T., Kawai, K. and Takamizawa, K. (2000) Development of medium by the combined treatment of steam explosion and chemical decomposition of peanut shells for production of xylitol by *Candida tropicalis. Journal of the Japan Society of Waste Management Experts* 11, 11–20.

Cho, C.H., Hatsu, M. and Takamizawa, K. (2002) The production of D-xylose by enzymatic hydrolysis of agricultural wastes. *Water Science and Technology* 45(12), 97–102.

Clifton-Brown, J., Chiang, Y.C. and Hodkinson, T.R. (2008) *Miscanthus*: genetic resources and breeding potential to enhance bioenergy production. In: Vermeris, W. (ed.) *Genetic Improvement of Bioenergy Crops*. Springer, New York, pp. 273–294.

Colberg, P.J. (1988) Anaerobic microbial degradation of cellulose, lignin, oligolignols, and monoaromatic lignin derivates. In: Zehnder, A.J.B. (ed.) *Biology of Anaerobic Microorganisms*. John Wiley and Sons, New York, pp. 333–372.

Cornelius, J. (1994) Heritabilites and additive genetic coefficients of variation in forest trees. *Canadian Journal of Forest Research* 24, 372–379.

Cote, W.A. (1981) Ultrastructure – critical domain for wood behavior. *Wood Science and Technology* 15, 1–29.

Cotter, J.L., Chinn, M.S. and Grunden, A.M. (2009) Ethanol and acetate production by *Clostridium ljungdahlii* and *Clostridium autoethanogenum* using resting cells. *Bioprocess Biosystems Engineering* 32, 369–380.

Cowling, E.B. and Kirk, T.K. (1976) Properties of cellulose and lignocellulosic materials as substrates for enzymatic conversion processes. *Bioprocess Biosystems Engineering* 6, 95–123.

Davis, J.M. (2008) Genetic improvement of *Populus* for bioenergy. In: Vermeris, W. (ed.) *Genetic Improvement of Bioenergy Crops*. Springer, New York, pp. 377–396.

Davis, L., Jeon, Y.J., Svenson, C., Rogers, P., Pearce, J. and Peiris, P. (2005) Evaluation of wheat stillage for ethanol production by recombinant *Zymomonas mobilis. Biomass and Bioenergy* 29, 49–59.

Delgenes, J.P., Moletta, R. and Navarro, J.M. (1996) Effects of lignocellulose degradiation products on ethanol fermentations of glucose and xylose by *Saccharomyces cerevisiae, Zymomonas mobilis, Pichia stipitis,* and *Candida shehatae. Enzyme and Microbial Technology* 19, 220–225.

Dien, B.S., Nichols, N.N., O'Bryan, P.J. and Bothast, R.J. (2000) Development of new ethanologenic *Escherichia coli* strains for fermentation of lignocellulosic biomass. *Applied Biochemistry and Biotechnology* 84–86, 181–196.

Dien, B.S., Jung, H.G., Vogel, K.P., Casler, M.D., Lamb, J.F.S., Iten, L., Mitchell, R.B. and Sarath, G. (2006) Chemical composition and response to dilute-acid pretreatment and enzymatic saccharification of alfalfa, reed canarygrass, and switchgrass. *Biomass Bioenergy* 30, 880–891.

Fengel, D. and Wegener, G. (1989) Wood – chemistry, ultrastructure, reactions. Walter de Gruyter, Berlin.

Ferrara, M.A. and Kling, S.H. (1987) Estudos comparativos sobre o pré-tratmento de bagaço de cana, eucaliptoe capimelefante por 'steam explosion'. In: Seminário de hidrólise enzimática de biomassa. Maringá, Brazil, pp. 12.

Foster, C.D. and Mayfield, C. (2007) Bioenergy production in planted pine forests. In: Hubbard, W., Biles, L., Mayfield, C. and Ashton, S. (eds) *Sustainable Forestry for Bioenergy and Bio-based Products: Trainers Curriculum Notebook.* Southern Forest Research Partnership, Inc, Athens, Georgia, pp. 103–122.

Fox, T.R., Jokela, E.J. and Allen, H.L. (2004) The evolution of pine plantation silviculture in the Southern United States. In: Rausher, H.M. and Johnson, K. (eds) *General Technical Report SRS 75.* US Department of Agriculture, Forest Service, Southern Research Station, Asheville, North Carolina, pp. 63–82.

Frederick, W.J. Jr, Lien, S.J., Courchene, C.E., DeMartini, N.A., Ragauskas, A.J. and Lisa, K. (2008) Production of ethanol from carbohydrates from loblolly pine: a technical and economic assessment. *Bioresource Technology* 99, 5051–5057.

Graham, R.L., Nelson, R., Sheehan, J., Perlack, R.D. and Wright, L.L. (2007) Current and potential US corn stover supplies. *Agronomy Journal* 99, 1–11.

Greene, N., Celik, F.E., Dale, B., Jackson, M., Jayawardhana, K., Jin, H., Larson, E.D., Laser, M., Lynd, L., MacKenzie, D., Mark, J., McBride, J., McLaughlin, S. and Saccardi, D. (2004) *Growing Energy: How Biofuels Can Help End America's Oil Dependence.* National Resources Defense Council, New York, pp. 78.

Hanna, W.W., Gaines, T.P., Gonzalez, B. and Monson, W.G. (1984) Effect of ploidy on yield and quality of pearl millet × napiergrass hybrids. *Agronomy Journal* 76, 969–971.

Hanna, W.W., Chaparro, C.J., Mathews, B.W., Burns, J.C. and Sollenberger, L.E. (2004) Perennial *Pennisetums*. In: Moser, L.E., Burson, B.L. and Sollenberger, L.E. (eds) *Warm-season (C_4) grasses.* Agronomy Monogram 45. ASA, CSSA, and SSSA, Madison, Wisconsin, pp. 503–545.

Hart, G.E., Schertz, K.F., Peng, Y. and Syed, N.H. (2001) Genetic mapping of *Sorghum biocolor* (L.) Moench QTLs that control variation in tillering and other morphological characters. *Theoretical and Applied Genetics* 103, 1232–1242.

Haussmann, B., Mahalakshmi, V., Reddy, B., Seetharama, N., Hash, C. and Geiger, H. (2002) QTL mapping of stay-green in two sorghum recombinant inbred populations. *Theoretical and Applied Genetics* 106, 133–142.

Heaton, E.A., Dohleman, F.G. and Long, S.P. (2008) Meeting US biofuel goals with less land: the potential of *Miscanthus. Global Change Biology* 14, 1–15.

Hondroyianni, D.K., Papakosta, A.A., Gagianas, K.A. and Tsatsarelis, E. (2000) Corn stalk traits related to lodging resistance in two soils of differing salinity. *Maydica* 45, 125–134.

Hoq, M.M., Alam, M.M., Mohiuddin, G. and Gomes, I. (1992) Enzymes biotechnology in developing countries: production and application of thermostable xylanase for jute fiber upgrading. In: Bose, M. (ed.) *Harnessing Biotechnology for the 21st Century.* Proceedings of the Ninth International Biotechnology Symposium and Exposition, Crystal City, Virginia. ACS, Washington, DC, pp. 568–573.

Horitsu, H., Yahashi, Y., Takamizawa, K., Kawai, K., Suzuki, T. and Watanabe, N. (1992). Production of xylitol from D-xylose by *Candida tropicalis*: optimization of production rate. *Biotechnology and Bioengineering* 40, 1085–1091.

Horwath, R.O. and Colonna, W.J. (1984) Process for oxidizing L-sorbitol to L-fructose. United States Patent Number 4467033.

Hubbard, W., Biles, L., Mayfield, C. and Ashton, S. (eds) (2007) *Sustainable Forestry for Bioenergy and Bio-based Products: Trainers Curriculum Notebook.* Southern Forest Research Partnership, Inc, Athens, Georgia.

Inderwildi, O.R., Jenkins, S.J. and King, D.A. (2008) Biofuels: mechanistic studies of hydrocarbon combustion and synthesis on noble metals. *Angewandte Chemie International Edition* 47, 5253–5255.

Jackson, B., Schroeder, R. and Ashton, S. (2007) Pre-processing and drying woody biomass. In: Hubbard, W., Biles, L., Mayfield, C. and Ashton, S. (eds) (2007) *Sustainable Forestry for Bioenergy and Bio-based Products: Trainers Curriculum Notebook.* Southern Forest Research Partnership, Inc, Athens, Georgia, pp. 139–142.

Jackson, M.G. (1977) Review article: the alkali treatment of straws. *Animal Feed Science and Technology* 2, 105–130.

Jackson, P. (1994) Genetic relationships between attributes in sugarcane clones closely related to *Saccharum spontaneum. Euphytica* 79, 101–108.

Jenkins, B.M., Baxter, L.L., Miles, T.R. Jr and Miles, T.R. (1998) Combustion properties of biomass. *Fuel Processing Technology* 54, 17–46.

Johnson, J.M.-F., Coleman, M.D., Gesch, R., Jaradat, A., Mitchell, R., Reicosky, D. and Wilhelm, W.W. (2007) Biomass-bioenergy crops in the United States: a changing paradigm. *The Americas Journal of Plant Science and Biotechnology* 1, 1–28.
Jones, M.B. and Walsh, M. (2001) *Miscanthus for Energy and Fibre*. James and James, London.
Jung, H.G. and Vogel, K.P. (1992) Lignification of switchgrass (*Panicum virgatum*) and big bluestem (*Andropogon gerardii*) plant parts during maturation and its effect on fiber degradability. *Journal of Science and Food Agriculture* 59, 169–176.
Khizzah, B.W., Miller, F.R. and Newton, R.J. (1993) Inheritance and heritability of heat tolerance in several sorghum cultivars during the reproductive phase. *African Crop Science Journal* 1, 81–85.
Kiniry, J.R., Schmer, M.R., Vogel, K.P. and Mitchell, R.B. (2008) Switchgrass biomass simulation at diverse sites in the Northern great plains of the US. *Bioenergy Research* 1, 259–264.
Klasson, K.T., Ackerson, M.D., Clausen, E.C. and Gaddy, J.L. (1992) Bioconversion of synthesis gas into liquid or gaseous fuels. *Enzyme and Microbial Technology* 14, 602–608.
Kondo, A. and Ueda, M. (2004) Yeast cell-surface display – applications of molecular display. *Applied Microbiology and Biotechnology* 64, 28–40.
Kormelink, F.J.M., Gruppen, H. and Voragen, A.G.J. (1993) Mode of action of (1,4)-a-D-arabinoxylan arabinofuranohydrolase (AXH) and a-L-arabinofuranosidases on alkali-extractable wheatflour arabinoxylan. *Carbohydrate Research* 249, 345–353.
Kszos, LA., Downing, M.E., Wright, L.L., Cushman, J.H., McLaughlin, S.B., Tolbert, V.R., Tuskan, G.A. and Walsh, M.E. (2001) Bioenegy feedstock development progam status report, ORNL/TM-2000/292 (http://www.ornl.gov/~webworks/cppr/y2001/rpt/108677_.pdf, accessed 20 January 2009).
Kubata, K.B., Horitsu, H., Kawai, K., Takamizawa, K. and Suzuki, T. (1992) Xylanase I of *Aeromonas caviae* ME-1 isolated from the intestine of a herbivorous insect (*Samia cynthia pryeri*). *Bioscience, Biotechnology and Biochemistry* 56, 1463–1464.
Kubata, K.B., Suzuki, T., Horitsu, H., Kawai, K. and Takamizawa, K. (1994) Purification and characterization of *Aeromonas cavie* ME-1 xylanase n which produces exclusively xylobiose from xylan. *Applied and Environmental Microbiology* 60, 531–535.
Kubata, K.B., Takamizawa, K., Kawai, K., Suzuki, T. and Horitsu, H. (1995) Xylanase IV, an exoxylanase of *Aeromonas caviae* ME-1 which produces xylotetraose as the only low-molecular-weight oligosaccharide from xylan. *Applied and Environmental Microbiology* 61, 1666–1668.
Kubata, K.B., Suzuki, T., Ito, Y., Naito, H., Kawai, K., Takamizawa, K. and Horitsu, H. (1997) Cloning and expression of xylanase I gene (xynA) of *Aeromonas caviae* ME-1 in *Escherichia coli*. *Journal of Fermentation and Bioengineering* 83, 292–295.
Kuhad, R.C., Singh, A. and Eriksson, K.-E.L. (1997) Microorganisms and enzymes involved in the degradation of plant fiber cell walls. In: Eriksson K.-E.L. (ed.) *Advances in Biochemical Engineering/Biotechnology*. Springer-Verlag, Germany, pp. 46–12.
Kundiyana, D., Bellmer, D., Huhnke, R. and Wilkins, M. (2006) 'Sorganol': production of ethanol from sweet sorghum. In: *Proceeding of the American Society of Agricultural and Biological Engineering Annual International Meeting, Portland, Oregon*. ASABE Paper No. 066070. ASABE, St Joseph, Michigan.
Lachke, A. (2002) Biofuel from D-xylose – the second most abundant sugar. General article. *Resonance* 7, 50–58.
Laere, K.M.J., Beldman, G. van and Voragen, A.G.J. (1997) A new arabinofuranohydrolase from *Bifidobacterium adolescentis* able to remove arabinosyl residues from double-substituted xylose units in arabinoxylan. *Applied Microbiology and Biotechnology* 47, 231–235.
Lamb, J.F.S., Jung, H.G., Sheaffer, C.C. and Samac, D.A. (2007) Alfalfa leaf protein and stem cell wall polysaccharide yields under hay and biomass management systems. *Crop Science* 47, 1407–1415.
Larson, S.R., Wu, X., Jones, T.A., Jensen, K.B., Chatterton, N.J., Waldron, B.L., Robins, J.G., Bushman, B.S. and Palazzo, A.J. (2006) Comparative mapping of growth habit, plant height, and flowering QTLs in two interspecific families of *Leymus*. *Crop Science* 46, 2526–2539.
Lauer, J.G., Coors, J.G. and Flannery, P.J. (2001) Forage yield and quality of corn cultivars developed in different eras. *Crop Science* 41, 1449–1455.
Leon, N. de and Coors, J.G. (2008) Genetic improvement of corn for lignocellulosic production. In: Vermeris, W. (ed.) *Genetic Improvement of Bioenergy Crops*. Springer, New York, pp. 185–210.
Lercher, J.A. (2009) Highly selective catalytic conversion of phenolic bio-oil to alkanes. *Angewandte Chemie International* 48(22), 4047–4050.
Li, G., Gu, M. and Chapman, K. (2004) *Sweet Sorghum*. China Agricultural Science and Technology Press, Beijing.

Lo, C.C., Chen, Y.H., Huang, Y.J. and Shih, S.C. (1986) Recent progress in *Miscanthus* mobilization program. *Proceedings of the International Society of Sugar Cane Technology* 19, 514–521.

McDougall, G.J., Morrison, I.M., Stewart, D., Weyers, J.D.B. and Hillman, J.R. (1993) Plant fibres: botany, chemistry and processing for industrial use. *Journal of the Science of Food and Agriculture* 62, 1–20.

McKeever, D.B. (2002) *Inventories of Woody Residues and Solid Wood Waste in the United States, 2002*. USDA Forest Service, Forest Products Laboratory, Madison, Wisconsin.

McKendry, P. (2002) Energy production form biomass (part 1): overview of biomass. *Bioresource Technology* 83, 37–46.

McLaughlin, S.B., Bouton, J., Bransby, D., Conger, R., Ocumpaugh, W., Parrish, D., Taliaferro, C., Vogel, K. and Wullschleger, S. (1999) Developing switchgrass as a bioenergy crop. In: Janick, J. (ed.) *Perspectives on New Crops and New Uses. Proceedings of the 4th National New Crops Symposium, Phoenix, Arizona.* American Society of Horticultural Science Press, Alexandria, Virginia, pp. 282–299.

McLaughlin, S.B. and Ksvos, L.A. (2005) Development of switchgrass (*Panicum virgatum*) as a bioenergy feedstock in the United States. *Biomass and Bioengineering* 28, 515–535.

Madhukar, G.R., Elmore, B.B. and Huckabay, H.K. (1996) Microbial conversion of synthesis gas components to useful fuels and chemicals. *Applied Biochemistry and Biotechnology* 57/58, 243–251.

Mandebvu, P., West, J.W., Froetschel, M.A., Hatfield, R.D., Gates, R.N. and Hill, G.M. (1999) Effect of enzyme or microbial treatment of bermudagrass forages before ensiling on cell wall composition, end products of silage fermentation and *in situ* digestion kinetics. *Animal Feed Science and Technology* 77, 317–329.

Mart'in, C., Galbe, M., Wahlbom, C.F., Hahn-Hägerdal, B. and Jönsson, L.J. (2002) Ethanol production from enzymatic hydrolysates of sugarcane bagasse using recombinant xylose-utilising *Saccharomyces cerevisiae. Enzyme and Microbial Technology* 31, 274–282.

Mart'in, C., Marcet, M., Almaza'n, O. and Jönsson, L.J. (2007) Adaptation of a recombinant xylose-utilizing *Saccharomyces cerevisiae* strain to a sugarcane bagasse hydrolysate with high content of fermentation inhibitors. *Bioresource Technology* 98, 1767–1773.

Matsuo, K., Chuenpreecha, T., Matsumoto, N. and Ponragdee, W. (2006) Eco-physical characteristics of *Erianthus* spp. and yielding abilities of three forages under conditions of cattle feces application. *JIRCAS Working Report* No. 30.

Ming, R., Moore, P.H., Wu, K.K., D'Hont, A., Tew, T.L., Mirkov, T.E., da Silva, J., Schnell, R.J., Brumbley, S.M., Lakshmanan, P., Jifon, J., Rai, M., Comstock, J.C., Glaszmann, J.C. and Paterson, A.H. (2006) Sugarcane improvement through breeding and biotechnology. *Plant Breeding Reviews* 27, 17–118.

Missaoui, A.M., Patterson, A.H. and Bouton, J.H. (2005) Investigation of genomic organization in switchgrass (*Panicum virgatum* L.) using DNA markers. *Theoretical and Applied Genetics* 110, 1372–1383.

Mohan, D., Pittman, C.U. and Steele, P.H. (2006) Pyrolysis of wood/biomass for bio-oil: a critical review. *Energy and Fuels* 20, 848–889.

Monson, W.G., Lowry, R.S. and Forbes, I. Jr (1969) *In vitro* nylon bag vs. two-stage *in vitro* digestion: comparison of two techniques for estimating dry-matter digestibility of forages. *Agronomy Journal* 61, 587–589.

Morgan, P.W., Finlayson, S.A., Childs, K.L., Mullet, J.E. and Rooney, W.L. (2002) Opportunities to improve adaptability and yield in grasses. *Crop Science* 42, 1791–1799.

Moser, L.E. and Vogel, K.G. (1995) Switchgrass, big bluestem, and indiangrass. In: Barnes, R.F., Miller, D.A. and Nelson, C.J. (eds) *Forages, Volume 1, An Introduction to Grassland Agriculture*. Iowa State University Press, Ames, Iowa, pp. 409–420.

Moser, L.E., Burson, B.L. and Sollenberger, L.E. (2004) Warm-season (C_4) grass overview. In: Mosher, L.E., Burton, B.L. and Sollenberger, L.E. (eds) *Warm-Season (C4) Grasses*. ASA/CSSA/SSSA, Madison, Wisconsin, pp. 1–15.

Mulkey, V.R., Owens, V.N. and Lee, D.K. (2008) Management of warm-season grass mixtures for biomass production in South Dakota, USA. *Bioresource Technology* 99, 609–617.

Najafpour, G. and Younesi, H. (2006) Ethanol and acetate synthesis from waste gas using batch culture of *Clostridium ljungdahlii. Enzyme and Microbial Technology* 38, 223–228.

Neale, D.B. (2007) Genomics to tree breeding and forest health. *Current Opinions of Genetics and Development* 17, 1–6.

Nehra, N.S., Becwar, M.R., Rottmann, W.H., Pearson, L., Chowdhury, K., Chang, S., Wilde H.D., Kodrzycki, R.J., Zhang, C., Gause, K.C., Parks, D.W. and Hinchee, M.A. (2005) Forest biotechnology: innovative methods emerging opportunities. *In Vitro Cellular and Developmental Biology – Plant* 41, 701–717.

Nigam, J.N. (2000) Continuous ethanol production from pineapple cannery waste using immobilized yeast cells. *Journal of Biotechnology* 80, 189–193.

Okamoto, T., Taguchi, H., Nakamura, K., Ikenaga, H., Kuraishi, H. and Yamasato, K. (1993) *Zymobacter palmae* gen. nov., sp. nov., a new ethanol-fermenting peritrichous bacterium isolated from palm sap. *Archives of Microbiology* 160, 333–337.

Okuda, N., Ninomiya, K., Katakura, Y. and Shioya, S. (2008) Strategies for reducing supplemental medium cost in bioethanol production from waste house wood hydrolysate by ethanologenic *Escherichia coli*: inoculum size increase and coculture with *Saccharomyces cerevisiae*. *Journal of Bioscience and Bioengineering* 105, 90–96.

Paterson, A.H. (2008) Genomics of sorghum. *International Journal of Plant Genomics* 2008, 362–451.

Patterson, A.H., Schertz, K.F., Lin, Y., Liu, S. and Chang, Y. (1995) The weediness of wild plants: molecular analysis of genes influencing dispersal and persistence of Johnsongrass, *Sorghum halepense* (L.) *Proceedings of National Academy of Science, United States of America* 92, 6127–6131.

Payne, J.H. (1991) *Cogeneration in the Cane Sugar Industry*. Elsevier, New York, pp. 338.

Pederson, J.F. and Rooney, W.L. (2004) Sorghums. In: Mosher, L.E., Burton, B.L. and Sollenberger, L.E. (eds) *Warm-Season (C4) Grasses*. ASA/CSSA/SSSA, Madison, Wisconsin, pp. 1057–1079.

Perlack, R.D., Wright, L.L., Turnhollow, A.F., Graham, R.L., Stokes, B. and Erbach, D.C. (2005) *Biomass as Feedstock for a Bioenergy and Bioproducts Industry: The Technical Feasibility of a Billion-ton Annual Supply*. DOE/GO-102005-2135, Oak Ridge National Laboratory, Oak Ridge, Tennessee, pp. 73.

Perrin, R., Vogel, K., Schmer, M. and Mitchell, R. (2008) Farm-scale production cost of switchgrass for biomass. *Bioenergy Research* 1, 91–97.

Peter, G.F. (2008) Southern pines: a resource for bioenergy. In: Vermeris, W. (ed.) *Genetic Improvement of Bioenergy Crops*. Springer, New York, pp. 397–419.

Quinby, J. (1967) The maturity genes of sorghum. *Advances in Agronomy* 19, 267–305.

Quinby, J. and Karper, R.E. (1954) Inheritance of height in sorghum. *Agronomy Journal* 46, 211–216.

Rae, A.M., Pinel, M.P.C., Bastien, C., Sabatti, M., Street, N.R., Tucker, J., Dixon, C., Marron, N., Dillen, S.Y. and Taylor, G. (2008) QTL for yield in bioenergy *Populus*: identifying G×E interactions from growth at three contrasting sites. *Tree Genetics and Genomes* 4, 97–112.

Rahman, A.K.M.S., Sugitani, N., Hatsu, M. and Takamizawa, K. (2003) A role of xylanase, α-L-arabinofuranosidase, and xylosidase in xylan degradation. *Canadian Journal of Microbiology* 49, 58–64.

Rockwood, D.L. and Dippon, D.R. (1989) Biological and economic potentials of *Eucalyptus grandis* and slash pine as biomass energy crops. *Biomass* 20, 155–165.

Rooney, W. and Aydin, S. (1999) The genetic control of a photoperiod sensitive response in *Sorghum biocolor* (L.) Moench. *Crop Science* 39, 397–400.

Rooney, W., Blumenthal, J., Bean, B. and Mullet, J. (2007) Designing sorghum as a dedicated bioenergy feedstock. *Biofuels, Bioproducts and Biorefining* 1, 147–157.

Roth, B.E., Jokela, E.J., Martin, T.A., Huber, D.A. and White, T.L. (2007) Genotype × environment interactions in selected loblolly and slash pine plantations in the Southeastern United States. *Forest Ecology Management* 238, 175–188.

Rouhollah, H., Iraj, N., Gitil, E. and Sorah, A. (2007) Mixed sugar fermentation by *Pichia stipitis*, *Sacharomyces cerevisiae*, and an isolated xylose fermenting *Kluyveromyces marxianus* and their cocultures. *African Journal of Biotechnology* 6, 1110–1114.

Saballos, A. (2008) Development and utilization of sorghum as a bioenergy crop. In: Vermeris, W. (ed.) *Genetic Improvement of Bioenergy Crops*. Springer, New York, pp. 211–248.

Saha, B.C., Iten, L.B., Cotta, M.A. and Wu, Y.V. (2005) Dilute acid pretreatment, enzymatic saccharification and fermentation of wheat straw to ethanol. *Process Biochemistry* 40, 3693–3700.

Samac, D.A., Jung, H.G. and Lamb, J.F.S. (2006) Development of alfalfa (*Medicago sativa* L.) as a feedstock for production of ethanol and other bioproducts. In: Minteer, S. (ed.) *Alcoholic Fuels*. CRC Press, Boca Raton, Florida, pp. 79–98.

Sarath, G., Akin, D.E., Mitchell, R.B. and Vogel, K.P. (2008) Cell-wall composition and accessibility to hydrolytic enzymes is differentially altered in divergently bred switchgrass (*Panicum virgatum* L.) genotypes. *Applied Biochemistry and Biotechnology* 150, 1–14.

Schank, S.C., Smith, R.L. and Russo, S. (1989) Characterization of genetic variability among accessions and crosses of napiergrass, *Pennisetum purpurem*. In: Desroches, R. (ed.) *Proceedings of XVI International Grassland Conference, Nice, France*. The French Grassland Society, Versailles, France, *pp.* 341–342.

Sheehan, J., Thomas, S., McMillan, J., Hames, B. and Templeton, D. (2001) Corn stover – what do we know about what makes a good feedstock? (http://bioenergy.ornl.gov/02workshop/sheehan.ppt, accessed 26 January 2009).

Shinners, K.J. and Binversie, B.N. (2007) Fractional yield and moisture of corn stover biomass produced in the Northern US Corn Belt. *Biomass and Bioenergy* 31, 575–584.

Siry, J., Greene, D., Harris, T., Izlar, B., Hamsley, A., Eason, T., Tye, T., Baldwin, S. and Hyldahl, C. (2006) Wood supply chain efficiency and fiber cost – what can we do better. *Forest Products Journal* 56(10), 4–10.

Sjöström, E. (1993) *Wood Chemistry, Fundamentals and Applications*, 2nd edn. Academic Press, New York/London.

Smart, L.B. and Cameron, K.D. (2008) Genetic improvement of Willow (*Salix* spp.) as a dedicated bioenergy crop. In: Vermeris, W. (ed.) *Genetic Improvement of Bioenergy Crops*. Springer, New York, pp. 347–376.

Smith, C.W. and Frederiksen, R.A. (2000) *Sorghum: Origin, History, Technology and Production*. John Wiley, New York, pp. 824.

Subudi, P.K., Rosenow, D.T. and Nguyen, H.T. (2000) Quantitative trait loci for the stay green trait in sorghum [*Sorghum biocolor* (L.) Moench]: consistency across genetic backgrounds and environments. *Theoretical and Applied Genetics* 101, 733–741.

Taliaferro, C.M., Roquette, F.M. Jr and Mislevy, P. (2004) Bermudagrass and stargrass. In: Moser, L.E., Burson, B.L. and Sollenberger, L.E. (eds) *Warm-season (C_4) Grasses*. Agronomy Monogram 45. ASA, CSSA and SSSA, Madison, Wisconsin, pp. 417–475.

Taniguchi, M., Itaya, T., Tohma, T. and Fujii, M. (1997) Ethanol production from a mixture of glucose and xylose by co-culture of *Pichia stipitis* and a respiratory deficient mutant of *Saccharomyces cerevisiae*. *Journal of Fermentation and Bioengineering* 83, 364–370.

Taylor, R.W. (1988) *Grain Sorghum: A Manual for Production and Marketing*. University of Delaware, Cooperative Extension Service, Newark, Delaware.

Tew, T.L. and Cobill, R.M. (2008) Genetic improvement of sugarcane (*Saccharum* spp.) as an energy crop. In: Vermeris, W. (ed.) *Genetic Improvement of Bioenergy Crops*. Springer, New York, pp. 249–272.

Theander, O. and Åman, P. (1978) Chemical compostion of some Swedish cereal straws. *Swedish Journal of Agricultural Research* 8, 189–194.

Tilley, J.M.A. and Terry, R.A. (1963) A two-stage technique for *in vitro* digestion of forage crops. *Journal of the British Grassland Society* 18, 104–111.

Timell, T.E. (1967) Recent advances in the chemistry of wood hemicelluloses. *Wood Science and Technology* 1, 45–70.

Toivola, A., Yarrow, D., Van Den Bosch, E., Van Dijken, J.P. and Scheffersi, W.A. (1984) Alcoholic fermentation of D-xylose by yeasts. *Applied and Environmental Microbiology* 47, 1221–1223.

Troeh, F.R. and Thompson, L.M. (2005) *Soils and Soil Fertility*. Blackwell Publishing, Oxford, UK.

Trybush, S., Jahodová, S., Macalpine, W. and Karp, A. (2008) A genetic study of a *Salix* germplasm resource reveals new insights into relationships among subgenera, sections and species. *Bioenergy Research* 1, 67–79.

Tuskan, G.A., DiFazio, S., Jansson, S., Bohlmann, J. and Grigoriev, I. (2006) The genome of black cottonwood, *Populus trichocarpa* (Torr. & Gray). *Science* 313, 1596–1604.

Varvel, G.E. and Wilhelm, W.W. (2008a) Cob biomass production in the Western corn belt. *Bioenergy Research* 1, 223–228.

Varvel, G.E. and Wilhelm, W.W. (2008b) Soil carbon levels in irrigated western corn belt rotations. *Agronomy Journal* 100, 1180–1184

Vergara, R., White, T.L., Huber, D.A. and Schmidt, R.A. (2007) Realized genetic gains of rust resistant selections of slash pine (*Pinus elliottii* var. *elliottii*) planted in high rust hazard zones. *Silvae Genetics* 56, 231–242.

Vogel, K.P. (2004) Switchgrass. In: Mosher, L.E., Burton, B.L. and Sollenberger, L.E. (eds) *Warm-Season (C4) Grasses*. ASA/CSSA/SSSA, Madison, Wisconsin, pp. 561–588.

Wasser, C.H. (1982) *Ecology and Culture of Selected Species Useful in Revegetating Disturbed Lands in the West*. USDA FWS/OBS-8256, Washington, DC.

Wayman, M. and Parekh, S.R. (1990) *Biotechnology of Biomass Conversion. Fuels and Chemicals from Renewable Resources*. Open University Press, Milton Keynes, UK, pp. 278.

Wear, D.N. and Greis, J.G. (2002) *Southern Forest Resource Assessment. General Technical Report SRS-53*. US Department of Agriculture, Forest Service, Southern Research Station, Asheville, North Carolina.

Wedin, D.A. (2004) Resource use, ecology and global change. In: Mosher, L.E., Burton, B.L. and Sollenberger, L.E. (eds) *Warm-Season (C4) Grasses*. ASA/CSSA/SSSA, Madison, Wisconsin, pp. 15–50.

Weimer, P.J. and Springer, T.L. (2007) Fermentability of eastern gamagrass, big bluestem and sand bluestem grown across a wide variety of environments. *Bioresource Technology* 98, 1615–1621.

Wilkie, K.C.B. (1979) The hemicelluloses of grasses and cereals. *Advances in Carbohydrate Chemistry and Biochemistry* 36, 215–262.

Woodard, K.R. and Prine, G.M. (1991) Forage yield and nutritive value of elephantgrass as affected by harvest frequency and genotype. *Agronomy Journal* 83, 541–546.

Wright, N.A. (1990) Screening of herbaceous species for energy crops on wet soils in Ohio. In: Janick, J. and Simon, J.E. (eds) *Advances in New Crops*. Proceedings of the 1st National Symposium on New Crops, 23–26 October 1988, Indianapolis, Indiana. Timber Press, Portland, Oregon, pp. 263–267.

Wullschleger, S.D., Yin, T.M., DiFazio, S.P., Tschaplinski, T.J., Gunter, L.E., Davis, M.F. and Tuskan, G.A. (2005) Phenotypic variation in growth and biomass distribution for two advanced-generation pedigrees of hybrid poplar. *Canadian Journal of Forest Resources* 35, 1779–1789.

Xin, Z., Wang, M.L., Barkley, N.A., Burow, G., Franks, C., Pederson, G. and Burke, J. (2008) Applying genotyping (TILLING) and phenotyping analyses to elucidate gene function in a chemically induced sorghum mutant population. *BMC Plant Biology* 8, 103.

Yanase, H., Yamamoto, K., Sato, D. and Okamoto, K. (2005) Ethanol production from cellobiose by *Zymobacter palmae* carrying the *Ruminocuccus albus* b-glucosidase gene. *Journal of Biotechnology* 118, 35–43.

Yanase, H., Sato, D., Yamamoto, K., Matsuda, S., Yamamoto, S. and Okamoto, K. (2007) Genetic engineering of *Zymobacter palmae* for production of ethanol from xylose. *Applied and Environmental Microbiology* 70, 2592–2599.

Zerbe, J.I. (2006) Thermal energy, electricity and transportation fuels from wood. *Forest Products Journal* 56, 1.

Zhang, M., Eddy, C., Deanda, K., Finkelstein, M. and Picataggio, S. (1995) Metabolic engineering of a pentose metabolism pathway in ethanologenic *Zymomonas mobilis*. *Science* 267, 240–243.

Zhu, X.G., Long, S.P. and Ort, D.R. (2008) What is the maximum efficiency with which photosynthesis can convert solar energy into biomass? *Current Opinions in Biotechnology* 19, 153–159.

7

Biodiesel from Oilseed Crops

DEV SHRESTHA AND JON VAN GERPEN

Introduction

Diesel engines play a central role in moving freight and other heavy-duty applications. In fact, more than 90% of the engines used in freight vehicles run on diesel. When an engine is designed, the designer optimizes the engine parameters to run effectively on a specific fuel. In the case of diesel engines, they are designed to run on diesel fuel. Diesel engines will not run satisfactorily on fuels that have significantly different physical, chemical or thermal properties from diesel fuel. If we would like to run a diesel engine on a different fuel, either the engine needs to be modified so that it will perform satisfactorily on that new fuel or the new fuel needs to be modified to make its properties similar to diesel fuel.

The first option is viable only if there is a large enough supply of fuel to justify a dedicated engine. Any unauthorized modification of an existing engine may void its warranty and it is not convenient for the operator to have to switch back and forth between diesel fuel and the new fuel. So, for an alternative fuel to be useful, it is better to match the new fuel's physical and chemical properties to diesel fuel so that it can be used without any engine modification. Table 7.1 shows some key property differences between diesel fuel, straight vegetable oil and biodiesel.

If the fuel properties are not matched carefully, the consequences may include increased emissions. For example, Tat *et al.* (2007) found that increases in NO_x from biodiesel could be explained by an advance in injection timing due to biodiesel use. They found that about half of the advance in injection timing was due to changes in the fuel injection pump as it responded to the increase in fuel delivery needed to overcome the power loss that resulted from biodiesel's lower energy content. The other half of the advance was due to the effects of the higher speed of sound, isentropic bulk modulus, viscosity and density of biodiesel fuel.

If straight vegetable oil is used in a diesel engine, as some people are promoting, the engine eventually will deteriorate. When the fuel is sprayed inside the engine through the nozzles, the heavier fuel droplets of vegetable oil will have greater momentum. As the flashpoint of vegetable oil is considerably higher, it takes longer for the oil droplets to ignite. Consequently, some of the droplets will reach the cylinder wall before they have time to burn completely. This is one of the reasons why, if a diesel engine is run on vegetable oil for a long time, a thick layer of partially burned material is deposited on the piston, injectors and in the oil sump, potentially ruining the engine. Therefore, engine manufacturers do not recommend running a

Table 7.1. Some key properties of diesel, biodiesel and soybean (*Glycine max*) oil (Song *et al*., 2000; Mittelbach and Remschmidt, 2005).

	Diesel	Soybean biodiesel	Soybean oil
Kinematic viscosity at 40°C (mm^2/s)	1.9–4.1	3.5–4.08	~ 30
Density (g/ml)	0.85	0.88	0.92
Cetane number	45	45–54.8	38
Flashpoint (°C)	52	141–171	254–282

diesel engine on straight vegetable oil. Even if vegetable oil is diluted with diesel fuel, the properties of the vegetable oil still change the diesel fuel and eventually may damage the engine.

Biodiesel Transesterification

Biodiesel consists of mono alkyl esters of long-chain fatty acids, which have similar properties to diesel fuel. Biodiesel has been used successfully in diesel engines without any engine modifications. Biodiesel can be used neat or in any blend ratio with diesel fuel. Chemically, the biodiesel reaction occurs when one molecule of triglyceride (vegetable oil or animal fat) breaks down and combines with three molecules of alcohol to produce three molecules of biodiesel and one molecule of glycerol (Fig. 7.1). Each molecule of triglyceride contains three fatty acid chains of varying length and saturation where R_1, R_2 and R_3 represent the various fatty acid chains. Vegetable oil and animal fat contain fatty acid chains of varying carbon length (usually from 12 to 22) and saturation. Saturation relates to the number of carbon–carbon double bonds in the oil and influences cold flow properties, oxidative stability and cetane number. Oils and fats are usually differentiated by their fatty acid profile.

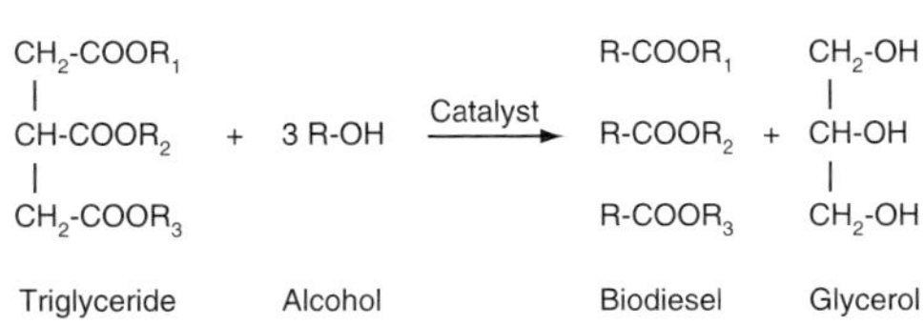

Fig. 7.1. Transesterification reaction.

The molecular weight of a particular fat or oil depends on its fatty acid profile. Fatty acid profile is the relative proportion of different fatty acids in an oil or fat. In order to determine the fatty acid profile, the oil is usually converted into alcohol esters of fatty acid and then their proportionate weight is measured using gas chromatography. The relative proportions of the esters are then converted back to corresponding amounts of fatty acids. The fatty acids of animal and plant origin are all straight carbon chains differing mainly in chain length and the number of double bonds (saturation). A saturated fatty acid chain with '*n*' number of carbons will have a chemical structure of CH_3-$(CH_2)_{n-2}$-COOH. Once the fatty acid profile of the oil is known, it is easy to calculate the average molecular weight of the oil. The molecular weight of a single fatty acid '*i*' can be calculated as:

$$MW_i = 14.027C - 2.016d + 31.9988 \qquad (7.1)$$

where '*C*' is the number of carbons and '*d*' is the number of double bonds. The average molecular weight of fatty acid mixture can be calculated by dividing the sum of all reported fatty acid weights by total moles in the mixture. Numerically,

$$\text{Average molecular weight of fatty acids} = \frac{\Sigma f_i}{\Sigma \frac{f_i}{MW_i}} \qquad (7.2)$$

where f_i is the weight fraction of a reported fatty acid. The molecular weight of the triglyceride (oil molecule) containing three fatty acids can be calculated as:

$$MW_i = 3 \times \text{Average molecular weight of fatty acids} + 38.049 \qquad (7.3)$$

The average molecular weight of the oil can be used to calculate the theoretical and actual amount of alcohol required to complete the reaction. The theoretical amount of methanol needed to complete the transesterification reaction with 1 kg of oil can be calculated as:

$$\text{Methanol} = \frac{96.12}{MW} kg \qquad (7.4)$$

where MW is the average molecular weight of the vegetable oil from Equation 7.3. The completed reaction of oil and methanol produces glycerol and biodiesel, with the weights of each given by:

$$\text{Glycerol} = \frac{92.1}{MW} kg \qquad (7.5)$$

$$\text{Biodiesel} = 1 + \frac{4.02}{MW} kg \qquad (7.6)$$

Example problem

For the fatty acid profile of the soybean oil shown in Table 7.2, calculate the average molecular weight of soybean oil. Calculate the theoretical amount of methanol needed to complete the biodiesel reaction with 100 kg of oil. Also estimate the amount of biodiesel produced.

Solution

The fatty acids are usually expressed in a form such as C18:1, meaning the fatty acid having a carbon chain length of 18 (C = 18 for Equation 7.1) and having 1 double bond (d = 1 for Equation 7.1). The molecular weight for each fatty acid can be calculated using Equation 7.1 and plugging in the value of 'C' and 'd'. The results are shown in the last column of Table 7.2. From this table and using Equation 7.2, the average molecular weight of the fatty acids is calculated as:

Table 7.2. Example fatty acid profile for soybean oil.

Fatty acid	Wt (%)	MW_i (from Equation 7.1)
Palmitic (C16:0)	8	256.43
Stearic (C18:0)	4	284.48
Oleic (C18:1)	25	282.47
Linoleic (C18:2)	55	280.45
Linolenic (C18:3)	8	278.44

Average molecular weight of fatty acids =

$$\frac{8+4+25+55+8}{\frac{8}{256.43}+\frac{4}{284.48}+\frac{25}{282.47}+\frac{55}{280.45}+\frac{8}{278.44}} = \frac{100}{0.3586}$$

or = 278.86 g/mol.

Therefore, the average molecular weight of soybean oil would be:

$$MW = 3 \times 278.86 + 38.05 = 874.63 \text{ g/mol}$$

In this particular example, the sum of the mass fraction of fatty acids in the numerator is 100, as the total fatty acid profile theoretically should be 100%. However, if some fatty acids are a very small amount, they are frequently dropped from the analysis table. In such cases, the numerator will be less than 100. By normalizing with the sum of all reported fatty acids in the numerator, the average molecular weight is always based on 100% of the fatty acids. However, it should be noted that the equation assumes the molecular weights of the unreported fatty acids to be the same as the average molecular weight of the reported fatty acids.

If the above soybean oil is transesterified using methanol, which has a molecular weight of 32.04, it will produce three molecules of methyl ester and one molecule of glycerol. Equation 7.4 gives the theoretical weight of methanol needed to react with 1 kg of soybean oil as:

$$\text{Methanol} = \frac{96.12}{874.63} = 0.110 \text{ kg}$$

In reality, as much as 100% excess of this theoretical amount of methanol is used to drive the transesterification reaction to completion. Therefore, the actual methanol added could be as high as 0.22 kg/kg of soybean oil. The excess methanol is not reacted and some of it is recovered after the reaction from the crude glycerol and biodiesel. The recovered methanol is used again in the biodiesel reaction. Equations 7.4 and 7.5 can be used to estimate the weight of biodiesel and glycerol produced as:

$$\text{Glycerol} = \frac{92.1}{874.63} = 0.105\ \text{kg}$$

$$\text{Biodiesel} = 1 + \frac{4.02}{874.63} = 1.005\ \text{kg}$$

Scaling to 100 kg of oil, the theoretical amount of methanol needed would be 11 kg and the reaction will produce 10.5 kg of glycerol and 100.5 kg of biodiesel. It is worth noting that the weight of biodiesel is approximately equal or, to be exact, biodiesel weight is only 0.5% more than the weight of the oil.

If the calculation needs to be carried out on a volume basis, the density of oil and biodiesel should be measured, as they vary from one feedstock to another. Assuming the approximate density of soybean oil to be 0.92 g/cc, the density of methanol as 0.7918 g/cc, the density of biodiesel as 0.88 g/cc and the density of glycerol as 1.261 g/cc, the volume of each component can be approximated as:

100 kg of soybean oil = 108.69 l of oil
11 kg of methanol = 13.89 l of methanol
10.5 kg of glycerol = 8.33 l of glycerol
100.5 kg of biodiesel = 114.21 of biodiesel

In this example, the volume of biodiesel is about 5.05% more than the volume of oil. This is because the density of biodiesel is lower than the density of the oil. The theoretical methanol used is about 13% of the soybean oil volume. The exact percentage depends on the fatty acid profile of the soybean oil, the percentage of triglycerides in the vegetable oil and the degree of completeness of the reaction. As a rule of thumb, the volumetric yield of biodiesel is usually considered to be 1 l of biodiesel for 1 l of vegetable oil.

Processing Biodiesel

Figure 7.2 shows a schematic diagram of the process to produce biodiesel from vegetable oils and animal fats. The oil, methanol and catalyst are combined in a reactor where the

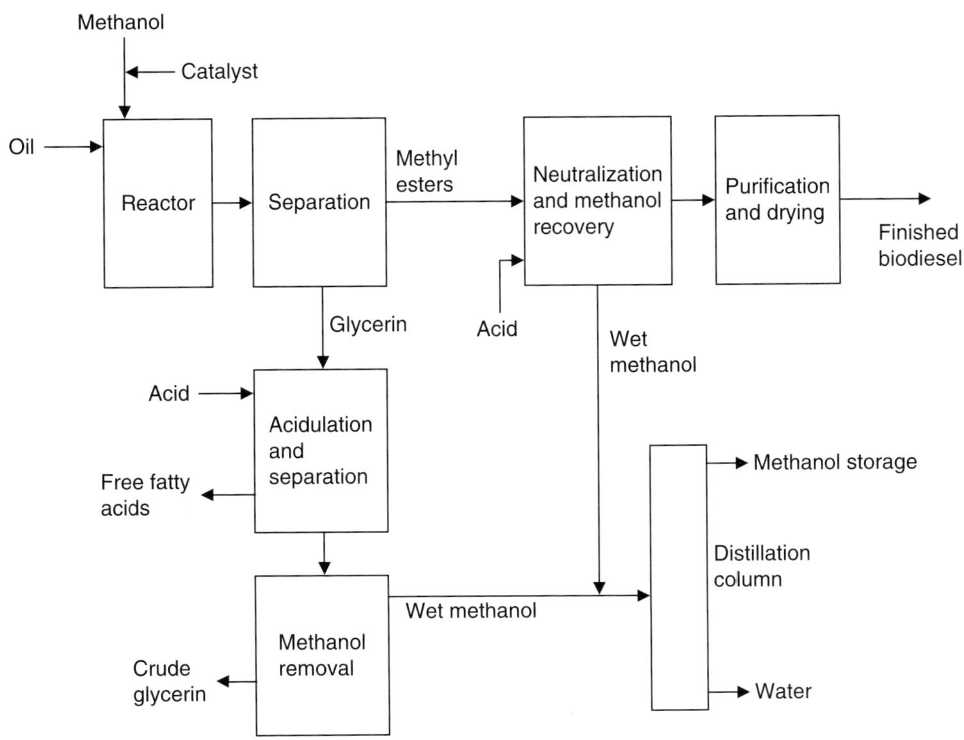

Fig. 7.2. Schematic of a biodiesel production process.

conversion to methyl esters and glycerine occurs. This reactor can take many forms, including a stirred tank reactor or a plug flow reactor (Van Gerpen, 2005). Some researchers have explored reactors that take advantage of ultrasonic excitation (Stavarache *et al.*, 2007) or cavitation (Ji *et al.*, 2006; Kelkar *et al.*, 2008) to enhance the mixing of the oil and alcohol. After the reaction is complete and the glycerin has been separated, the process is divided between the two streams. The biodiesel stream must be neutralized and have its methanol removed. Then, any soap or other contaminants such as catalyst, residual methanol or free glycerin must be removed. This is done most commonly with water washing, but the use of adsorbents and ion exchange resins, which do not require water, are becoming more popular. If water is used, then the purification process must be followed by a drying process to remove the water. This is usually performed with a vacuum flash process (Van Gerpen, 2005).

The glycerin stream must also be upgraded before it has commercial value. The glycerin is usually mixed with an acid to neutralize the catalyst and split the soaps into free fatty acids and salt. The free fatty acids are insoluble and thus form a separate phase that can be removed by settling or by centrifugation. After the free fatty acids have been removed, the methanol in the glycerin is vaporized, leaving glycerin that is 80–95% pure and which can be sold for further refining or for direct use as fuel or animal feed. The methanol removed from the biodiesel and glycerin streams will often accumulate water and this should be removed with a distillation column before the methanol is reused.

This process, as described, is typical of the technique used to produce most biodiesel. However, some oils contain high levels of contaminants that make it impossible to produce biodiesel using this approach. The techniques necessary to use these feedstocks are discussed in the next section.

Processing High Free Fatty Acid Feedstocks

The most important quality parameter that affects the processing of vegetable oils and animal fats into biodiesel is the free fatty acid (FFA) content. These FFAs are products of the breakdown of triglycerides through hydrolysis reactions with water. The FFAs interfere with the usual alkali-catalysed conversion of oil to biodiesel because they react with the catalyst to form soap, which deactivates the catalyst. As the biodiesel industry strives to compete with petroleum-based fuels, there is greater interest in using lower cost, high FFA feedstocks. These feedstocks require additional processing steps if used to produce biodiesel.

There are a large number of different processing options for high FFA feedstocks. The most common procedure when the FFA level is 5% or less is to add extra alkali catalyst and allow the FFAs to react with the catalyst and produce soap. Then, these soaps can be removed either before the transesterification process or after. When the soaps are removed before transesterification, the process is known as caustic stripping. This option has the advantage that other contaminants such as phospholipids can be removed at the same time. Some sulfur compounds found in animal fats may also be removed. In addition, if performed as a pretreatment, it is possible to use a less expensive alkali, such as sodium hydroxide, than that which will be used for transesterification. If the soap is left in the reaction mixture, then enough catalyst needs to remain to catalyse the transesterification reaction. The soaps that are present will mostly separate with the glycerin fraction and those that partition into the methyl esters must be removed with water washing, with a solid adsorbent, or with an ion exchange resin.

An undesirable feature of converting FFAs to soap and then removing them from the oil or fuel is that the yield loss generally will be at least twice the percentage of FFAs. The separation is never ideal and considerable usable oil is lost with the soaps. This oil, as well as the FFAs in the soap, can be recovered, but this is usually not economical. A more selective method for removing FFAs is steam stripping and this has become popular with feedstocks containing 5–10% FFA. This process involves heating the oil to 290°C under high vacuum and distilling the FFAs from the triglycerides. This high temperature degrades the oil somewhat, resulting in a darker colour. The process produces a stream of FFAs that could be recycled but is mostly sold as an animal feed ingredient.

Both caustic stripping and steam stripping remove the FFAs so they are not converted to biodiesel. Glycerolysis can be used to lower the FFA level by reacting the FFAs with glycerin to form mono- and diglycerides (Sonntag, 1982). The basic reaction is shown in Fig. 7.3.

Note that water is a product of the reaction and will inhibit the completion of the reaction if not removed. The reaction is usually conducted at 200–220°C with an inert gas sparge or vacuum to remove the water. The monoglycerides that are formed will be transesterified along with the triglycerides during the alkali-catalysed process.

The final technique, and the most commonly used for feedstocks containing over 10% FFAs, is acid-base processing. In this technique, originally described by Keim (1945), the oil is reacted initially with methanol using an inorganic acid, such as sulfuric acid, as a catalyst. While this catalyst is slow to convert triglycerides to biodiesel, it is much more effective for converting FFAs to methyl esters. When this process has proceeded to the point where the FFA level is low, then the acid and the water that is formed in the reaction are removed and a base catalyst is added so the transesterification process can proceed in the normal manner. As with glycerolysis, this process converts the FFAs to useful fuel so it has a much better yield than caustic or steam stripping. The reverse of the acid-base process has also been proposed (Sprules and Price, 1950). This process involves adding enough alkali catalyst to convert the FFAs to soap, as with caustic stripping, along with the amount needed to catalyse the conversion of the triglycerides to methyl esters. Then, sulfuric acid is added to the entire mixture to split the soap back to FFAs and then esterify the FFAs to methyl esters.

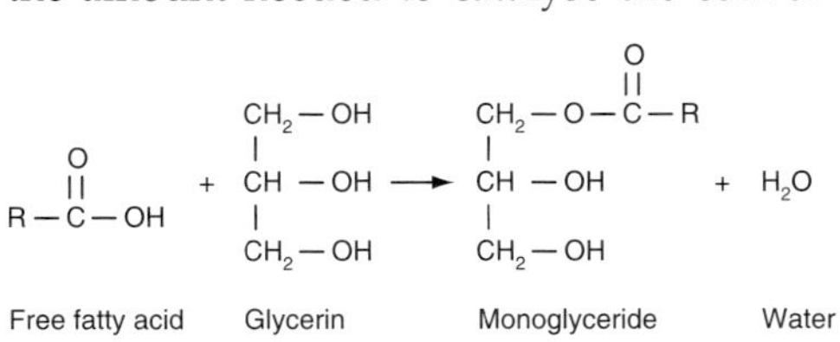

Fig. 7.3. Free fatty acid and glycerin may react to produce monoglyceride and water.

Cold Flow Issues with Biodiesel

Although biodiesel can be made from any oil feedstock that allows ASTM (American Society for Testing and Material) D6751 in the USA or EN (European Committee for Standardization) 14214 in Europe to be met, all properties of biodiesel are not the same. One of the most notably different properties of biodiesel is its cloud point. ASTM D6751 does not specify the required cloud point for biodiesel, but requires that the cloud point be reported to the customer. If the producer is not careful to select an appropriate biodiesel feedstock, it can lead to gelling of biodiesel in cold weather.

Biodiesel must meet the ASTM D6751 standard in the USA and EN 14214 in Europe. These standards are followed by several other countries outside of these regions. Even after biodiesel meets the specifications, the cold flow behaviour for different biodiesel fuels will vary depending on their feedstock fatty acid profile.

In the northern parts of the USA and other cold regions of the world, one of the major concerns among biodiesel users is its unfavourable cold flow properties. Handling and blending 100% biodiesel (B100) in cold weather can be difficult. This limits the use of biodiesel during the winter season. Petroleum-based No. 2 diesel fuel generally has a lower cloud point (CP) and pour point (PP) than biodiesel (Peterson *et al.*, 1997). ASTM (2003a) defines the CP for petroleum products and biodiesel fuels as the temperature of a liquid specimen when the smallest observable cluster of wax crystals first appears upon cooling under prescribed conditions. ASTM (2003b) defines the PP as the lowest temperature at which movement of the test specimen is observed under prescribed test conditions. Cold flow properties of biodiesel depend on many factors, including oil feedstock and the type of alcohol used. Peterson *et al.* (1997) compared ethyl and methyl esters of four biodiesel feedstocks on the basis of fuel characteristics and short-term engine performance tests. They reported a 16°C CP for tallow biodiesel compared to −12°C for US No. 2 diesel fuel. Even though biodiesels meet the standard, they may have different cloud and pour points (Table 7.3).

Low temperature engine operability is usually measured with a low-temperature flow test (LTFT) or cold filter plugging point (CFPP). ASTM D6751 for biodiesel quality specification does not specify the required CP for sale in a particular region, but it requires the producers to disclose the CP of B100 biodiesel. CP and PP have been used routinely to characterize the cold flow operability of diesel fuels. Chiu *et al.* (2004) showed the LTFT was a non-linear function of CP and PP. The non-linear coefficient showed that, for the same CP and PP, the LTFT was lower for fuel with a lower percentage of biodiesel. Dunn and Bagby (1995) showed that both LTFT and CFPP of formulations containing at least 10% by volume of methyl esters were linear functions of CP.

When a heterogeneous mixture of liquid is cooled from a liquid state to near its cloud point, the fraction that has the highest freezing point starts to crystallize and form a cloud nuclei. An individual crystal is too small to be seen by the naked eye, but as the temperature decreases, crystalline growth and agglomeration continue until the crystals become large enough to be visible in the form of a cloud which identifies the cloud point (Chandler *et al.*, 1992). In pure biodiesel, it is the saturated fraction that crystallizes out first and forms the cloud seed. Once the cloud seed is present, it is easy for other molecules to agglomerate because the molecules go to a lower state of free energy by doing so (Brice, 1973). The higher the fraction of saturates, the higher will be the cloud point. Therefore, any fraction in the biodiesel matrix, including impurities such as monoglycerides, that crystallize out at high temperature can play a role as the cloud seed, making the overall cloud point higher.

The most common impurities in biodiesel are free glycerin and monoglycerides. The melting point of monoglycerides is the highest, followed by saturated methyl esters (methyl palmitate, methyl stearate and methyl arachidate), followed by free glycerol and finally unsaturates (Table 7.4). Biodiesel contains methyl esters with different melting points. Even though biodiesel fractions such as methyl stearate have a high melting point, they do not necessarily crystallize at their melting point temperature. This is because the higher melting point fraction behaves as a solute that is dissolved in lower melting methyl esters (DeMan, 2000). Crystals are formed only when the solution becomes saturated with the solute.

Table 7.3. Cloud point and pour point for biodiesels made from some common feedstocks.

Biodiesel feedstock	Cloud point	Pour point
Soybean oil[a]	1	0
Canola oil[a]	0	−9
Palm oil[a]	17	15
Jatropha oil[b]	8	6
Tallow[c]	12–17	9

[a]Moser, 2008.
[b]Vyas *et al.*, 2009.
[c]Mittelbach and Remschmidt, 2005.

Methods to lower cloud and pour points

Different techniques have been used to lower the cloud and pour points of biodiesel for cold weather operation. Winterization is the

Table 7.4. Melting point of common biodiesel constituents and impurities (O'Connor, 1960; Mittelbach and Remschmidt, 2005).

Chemical name	Melting point °C
Methyl palmitate (methyl hexadecanoate)[a]	30.5
Methyl stearate (methyl octadecanoate)[a]	39.1
Methyl oleate (methyl *cis*-9-octadecanoate)[a]	−20
Methyl linoleate (methyl *cis,cis*-9,12-octadecadienoate)[a]	−35
Methyl arachidate (methyl eicosanoate)[a]	54.5
Glycerol[b]	18
1-Mono-palmitin[b]	74
1-Mono-stearin[b]	79
1-Mono-olein[b]	32

[a]Constituents of biodiesel.
[b]Common biodiesel impurities.

process of removing saturated methyl esters by causing crystallization by cooling and then separating the high melting components by filtration. Lee *et al.* (1996) found that the CP of a common soybean biodiesel could be reduced to −7.1°C through winterization with a yield loss of 26%. Davis *et al.* (2007) used soybean methyl ester fractionation by urea and methanol for producing modified biodiesel with a CP as low as −45°C. The process takes advantage of clathrates, which form between urea and long-chain saturated methyl esters. In either case, a significant amount of high CP biodiesel fraction is removed. Winterization is generally not an efficient way of improving cold flow properties because of the high yield loss. It is also not practical in many cases to store the high CP fraction for summer use or to transport it to a warmer climate region, so it must be used as a lower value fuel.

The use of a branched chain alcohol is an alternative way to reduce the CP. Isopropyl and 2-butyl esters of normal soybean oil crystallized at 7–11°C and 12–14°C lower, respectively, than the corresponding methyl esters (Lee *et al.*, 1995). However, use of isopropyl alcohol is more expensive and the reaction is harder to complete than for methanol.

Different fuel additives for diesel and biodiesel are commercially available to improve the cold flow properties. Dunn *et al.* (1996) studied the effect of 12 cold flow additives for petroleum diesel on the cold flow behaviour of biodiesel. They concluded that the additives improved the PP of diesel/biodiesel blends significantly but did not affect the CP greatly. Many additives contain some proprietary components, such as copolymers of ethylene, vinyl acetate or other olefin-ester copolymers. Because of these proprietary compounds, the impact of cold flow additives on biodiesel from different types of feedstock such as canola, mustard and used vegetable oil needs to be determined experimentally.

Shrestha *et al.* (2008) studied the effect of different fuel additives to improve cold flow issues with biodiesel and found that the additives in general worked better for ethyl esters than for methyl esters. The average reduction in CP and PP for neat mustard methyl esters was 0.3°C and 7.2°C, respectively, compared to 3°C and 19.4°C for mustard ethyl ester. However, the additives reduced the PP of petroleum diesel to below −36°C in all cases studied.

Another way to improve the CP of biodiesel is to blend with another biodiesel with low CP. This has been shown to be an effective technique for reducing the cloud point of palm oil. Moser (2008) was able to obtain CFPP values for palm oil at or less than 0°C through blending with other methyl esters.

Uses of Glycerol

As noted earlier, making biodiesel produces glycerol approximately 10% by weight of oil. This is equivalent to 0.35 kg of glycerol produced per 3.78 l of biodiesel. The US National Biodiesel Board (NBB, 2009) estimated the US production of biodiesel in 2008 to be 2.65 billion l and in the process 245 million kg of glycerol would be produced. Pure glycerin was historically valued at US$1.32–1.98/kg, used primarily in the manufacture of various other personal care products. However, since the biodiesel industry picked up, the price of glycerin has declined. The price of 99.5% pure glycerin was reposted as US$0.60/kg in 2006 (Miller-Klein, 2006). However, the price of refined glycerin climbed back up to about US$1.10–1.32/kg in 2008. The glycerol produced immediately after the reaction is called crude glycerol as it consists of almost all the catalyst used and about half of the excess methanol. Thompson and He (2006) found that crude glycerol was about 15% by weight of soybean oil and it had about 68% of glycerin. Kotrba (2007) reported the price of crude glycerol reached as low as US$0.04/kg in 2006. Refining of crude glycerol to commercial grade glycerol costs about US$0.44/kg.

Because of the low glycerol prices and inelastic market, researchers have been looking for new uses of glycerol. Thompson and He (2006) reported that crude glycerol might be used as animal feed. The crude glycerol from soybean had about 75% carbohydrate, 8% fat and 0.05% protein. Since crude glycerol also contains methanol and catalyst, it could be toxic to some animals and further research is needed to confirm the level of crude glycerol that would be safe to add to animal feed.

Johnson and Taconi (2007) have listed over 20 chemical compounds that potentially could be made through oxidation or reduction of the glycerol. Among those chemicals, propylene glycol, propionic acid, acrylic acid, propanol, isopropanol, allyl alcohol and acrolein have significant price differentials and sufficient market capacity. Compounds such as lactic acid and malonic acid also garner higher prices but lack large market capacities.

Johnson and Taconi (2007) have also listed several other compounds that could be made from glycerol through biological conversion. A wide variety of microorganisms can utilize glycerol, as it is often formed as an intermediate in both the aerobic and anaerobic metabolism of lipids and glucose. Combinations of microorganism and type of metabolism could be used to produce various chemical species. The bacteria *Enterobacter aerogenes* has been demonstrated to convert glycerol anaerobically to H_2 and ethanol (Ito *et al.*, 2005). Citric acid, acetic acid, succinic acid and lactic acid are some of the other products that can be produced through microbial activities.

Despite these possibilities of converting crude glycerol into several other chemicals, the current industry practice has not exploited the full potential of its benefits. The reasons are high cost of installation or technology not fully developed. Because of the high price of glycerol in the past, the technologies for converting glycerol into other chemicals had a limited use. Now that the price of glycerol is decreasing and more glycerol is available, we should, in future, see more plants having an onboard facility or integrated bioreactor system to produce multiple output streams.

Biodiesel Economics

As we have seen from the above discussion, the production of biodiesel requires an almost equal amount of vegetable oil and about 11% by weight of methanol. The co-products are biodiesel and glycerol. In practice, to drive the reaction towards completion, almost 100% excess of methanol or about 20% by weight of methanol is used. Sodium or potassium methoxide are used as the catalyst.

A feasibility study conducted by the University of Idaho for southern Idaho, USA, estimates the following cost proportions for a 15.1 million l/year plant based on a used oil and tallow biodiesel plant model (Crockett *et al.*, 2006).

Cost to produce 1 l biodiesel:

Oil – 1 l or 0.91 kg. The weight depends on feedstock.
Methanol – 0.175 l. After methanol recovery.

Other costs per litre of biodiesel production:
Cost of catalyst – US¢1.9
Plant construction, operation and management cost – US¢5.7
Free fatty acid treatment if used – US¢1.6
Freight charges – US¢1.3
Total cost other than oil and methanol – US¢11

It should be noted that the costs provided above are only a general guideline and may vary depending on the plant size, location and technology used. Treatment of FFAs is necessary for high FFA feedstocks such as used vegetable oil. For virgin oils with FFA levels less than 0.5%, removal of FFAs is usually not necessary.

Looking at the vegetable oil market, it can be seen that soybean oil prices have fluctuated quite a bit in recent years. The soybean oil price climbed from about US$0.44/kg in early 2005 to almost US$1.54/kg in early 2008 and back down to US$0.66/kg in early 2009 (CBOT, 2009). Similarly, methanol prices have also fluctuated from about US$0.25/l in early 2005 to US$0.66/l in early 2008 and back to US$0.19/l in January 2009 (Methanex, 2009). This excessive fluctuation in the prices of both oil and methanol has caused many biodiesel production facilities to go out of business.

The base price for biodiesel production from oil feedstock can be calculated as:

Estimated production cost in US$/l of biodiesel = 0.9 × Cost in US$/kg of oil + 0.175 × Cost in US$/l of methanol + 0.106

Note that the biodiesel production process also produces crude glycerol, which may have some economic value. The revenue from glycerol can be used to offset some of the cost in biodiesel production.

In October 2004, the US Congress passed a biodiesel tax incentive, structured as a federal excise tax credit, as part of the American Jobs Creation Act (JOBS Act) of 2004. The credit amounts to one cent for each percentage point of 'first-use' biodiesel blended with petroleum diesel (and one-halfpenny per cent for recycled oils). Thus, blending B20 made from soy, canola and other vegetable oils would receive a US¢5.28/l excise tax credit, while blenders of B5 would receive a US¢1.32 tax credit. For more information refer to NBB (2005) publications. If such credits are available, they can be taken into account when making business decisions for biodiesel. However, these credits are renewed on an annual basis so they add considerable risk to the profitability of the business.

Biodiesel Production Potential

Biodiesel can be made from any oil or fat of vegetable or animal origin. Oil and fat are chemically similar compounds, with the only major difference in the amount of saturation, which makes oils liquid and fats solid at room temperature. One of the major hurdles to biodiesel becoming a significant source to replace petroleum fuel is its limited supply. Only excess oil and fat from the food industry is used in making biodiesel, because of the cost.

World biodiesel production potential

The Food and Agriculture Organization (FAO) of the United Nations estimated that the average consumption of vegetable oil and animal fat in 2003 was 30.14 and 8.22 g/capita/day, respectively, for food (FAO, 2009). The animal fat total includes butter and cheese. While the fat in butter could be converted to methyl esters, the large amount of short-chain esters would keep the fuel from satisfying the flashpoint requirement of the ASTM specification. Out of 8.22 g of animal fat consumed, butter was 2.74 g and cheese was 5.48 g. Comparing the vegetable oil and animal fat consumed in 1993, vegetable oil consumption has increased by about 22% (from 24.66 g/capita/day in 1993), but consumption of animal fats has not increased. It should be noted that total fat consumed by humans is reportedly much higher than the sum of vegetable oil and animal fat reported above. As pointed out by Grigg (1999), much of the fat is consumed as invisible fat in meat, milk and cereals. In fact, FAO estimates showed that the total fat consumption from all products was 79 g/capita/day in 2003 compared to 68 g/capita/day reported for 1993.

As noted above, the consumption of fat as vegetable oil and animal fat is about 38.36 g/capita/day (30.14 g of vegetable oil and 8.22 g of animal fat), which is equivalent to an annual consumption of oil and fat of about 14 kg/capita/year. The world population in 2008 was estimated to be 6.7 billion (CIA, 2009). Therefore, worldwide consumption of oil and fat is estimated to be about 93.8 million t for food. At the same time, the world production of oil and fat is estimated to be about 165.65 million t (Gunstone, 2002). This leaves a total of 71.85 million t of oil and fat for industrial uses. Theoretically, if all of this excess oil is used for making biodiesel, it will produce 82.8 billion l of biodiesel, which is about 10% of the worldwide on-highway diesel use. It should be noted that there are several other industrial uses of oil and fat, that will compete for the available oil and fat. The assumption has also been made that no changes are made in consumption patterns or that new feedstock sources do not become available that would increase oil and fat supplies.

Although soybean oil is the major feedstock for biodiesel in the USA, it is not the major feedstock in other parts of the world. Rapeseed oil is the major source of oil in Europe. The most common variety of rapeseed grown is similar to canola grown in North America. In fact, rapeseed oil contributes about 85% of the oil for world biodiesel production, followed by sunflower seed oil, soybean oil and palm oil (Mittelbach and Remschmidt, 2005). According to the European Biodiesel Board (EBB, 2008), European (EU-27) production of biodiesel reached 5.7 million t (1.7 billion gallons) compared to US production of 1.7 billion l (450 million gallons) in 2007. Germany is the largest producer of biodiesel among EU countries, accounting for about half of the total European biodiesel production.

In the East Asian countries, palm oil is the major feedstock for biodiesel. The annual average production of palm oil is expected to be about 31.4 million t/year over the period 2006–2010 and is expected to be about 43.4 million t/year for 2016–2020. Palm oil is forecast to be the world's most produced and internationally traded edible oil by 2012 (Brown and Michael, 2005). Malaysia is currently the world's largest producer and exporter of palm oil. Malaysia produces 44% and Indonesia produces 42% of world palm oil. In 2005, the total amount of palm oil produced was 33.7 million t (Basiron, 2007). Palm oil yield is significantly higher than soybean oil. Palm yields about 4000 kg of oil/ha (about 4675 l/ha) compared to 375 kg/ha (about 440 l/ha) for soybeans (Hansen, 2007).

Palm oil is edible oil used around the world in such foods as margarine, shortening, cooking oil, soups, sauces, crackers and other baked goods, and confectionary products. After soybean oil, it is the most widely used oil. It is highly versatile and can be substituted for hard animal fats (butter and lard); for soy, olive or canola liquid vegetable oils; and for partially hydrogenated vegetable oil, which is a staple of the baking, fast food and other industries (Brown and Michael, 2005).

In some countries, such as India and Mexico, the use of edible oil for biodiesel is prohibited if it is its first use. A viable alternative in such countries is jatropha, which produces inedible oil and thus does not compete with food. There is a rapidly growing interest in using jatropha as a feedstock for the production of biodiesel, particularly because jatropha is a wild plant that can grow in dry and marginal lands without irrigation (Prueksakorn and Gheewala, 2008). The oil from jatropha can be converted as easily to biodiesel that meets the American and European standards as any other oil feedstock (Achten *et al.*, 2008).

Prueksakorn and Gheewala (2008) reported the net energy value for jatropha to be 236 GJ/ha/year. Net energy value is the difference in energy output from the crop and all energies that go into producing the crop, transportation, crushing and transesterification. This reported gain of 236 GJ is equivalent to a gain of 7189 l of biodiesel/ha/year after subtracting the energy used for biodiesel production. The energy gain is based on the assumption of jatropha seed production of 12.5 t/ha/year, which is the highest yield at the end of the 5th year on irrigated land reported by Jatropha World (2007). However, critics (Openshaw, 2000; Achten *et al.*, 2007, 2008)) are sceptical about data extrapolated from a single plant. Openshaw (2000) reported that in Mali, jatropha seed yield varied from 2.5–3.5 t/ha/year. This is equivalent to about 935 l of biodiesel/ha/year (100 gallon/acre/year). According to Achten *et al.* (2008), jatropha seed yield is still a difficult issue to characterize and the actual mature seed yield per hectare per year is not known, since systematic yield monitoring has started only recently. Earlier reported figures exhibit a very wide range (0.4–12 t/ha/year) of production potential for jatropha oil. It was indicated in their paper that the yield from a single tree could not be extrapolated to estimate yield, as planting distances affect seed yield greatly. The authors pointed out that the yield from jatropha depended on the amount of water available, soil characteristics, plant spacing and several other climatic factors.

Another drawback of jatropha oil is that it can be high in FFA. Jatropha oil contains about 14% FFA (Tiwari *et al.*, 2007), which is far beyond the limit of the 1% FFA level that can be converted into biodiesel easily by transesterification using an alkaline catalyst. Acid pretreatment is usually necessary to convert high FFA oils into biodiesel (Canakci and Van Gerpen, 2001). Therefore, the transesterification conversion cost of jatropha biodiesel may be higher than for oils such as soybean or rapeseed oil. In summary, biodiesel from jatropha has both opportunities and challenges. Jatropha is currently being researched to realize its potential and overcome the challenges. One day, jatropha may prove to be a significant source of oil for biodiesel.

US biodiesel production potential

In the USA, soybeans are the major source of oil for making biodiesel. In 2007, the total distillate fuel sale in the USA was 238 billion l. Of that, on-highway diesel fuel accounted for about 151 billion l (EIA, 2008). The annual average increase in demand has been about 3.78 billion l/year.

In contrast to the enormous demand for diesel fuel, biodiesel production in 2008 was only about 2.65 billion l (NBB, 2009). The total US biodiesel production capacity was about 9.8 billion l, so obviously plants were operating at around 20% of their full capacity. The major hurdle is the availability of low-cost feedstocks to compete with fluctuating petroleum diesel prices.

Just how much feedstock will be available for producing biodiesel at a certain price depends on market dynamics and supply and demand. In essence, vegetable oil will always have more preference as food than fuel. People are and will be willing to pay more for their food than for fuel. NASS (2009) data show that in 2005, consumption of oil and fat as food was about 11.5 million t, amounting to 38.8 kg/capita in the USA. Similarly, non-food consumption as soap, pet food, paint and other industrial products has averaged about 3.12 million t. Therefore, the total US market for fat and oil was about 14.7 million t in 2005. Total oil production from all plant and animal sources totalled about 18 million t (Table 7.5).

The price of oil products varies depending on their source. One of the lowest cost fats is inedible tallow, with an average price in 2005 of US¢42.18/kg compared with soybean oil at US¢50.67 /kg and corn oil at US¢62.64 /kg (NASS, 2009). Another source of fat for biodiesel is rendered yellow grease from restaurants and other food processing facilities. According to the National Renderers Association (NRA, 2008), 606,000 t of yellow grease were collected in 2005. Crockett *et al.* (2006), in their feasibility study, reported an average price of US¢24/kg for yellow grease. The lowest cost feedstocks that do not require extra processing are the most likely source of feedstock for biodiesel production. The price for yellow grease may vary significantly, depending on geographical location and quality. Before 2008, the tax credit incentive available for biodiesel from virgin oil was not the same as biodiesel made from yellow grease, but this has been changed.

Table 7.5. US oil and fat production for 2005.

	Thousand tonne	Million lb
Soybean[a]	9,250	20,387
Milk fat[b]	2,940	6,480
Inedible tallow[c]	2,208	4,866
Maize[a]	1,126	2,482
Edible tallow[c]	790	1,741
Cottonseed[a]	431	950
Rapeseed /canola[a]	412	909
Lard[b]	353	779
Sunflower[a]	247	544
Linseed[a]	145	320
Groundnut[a]	82	181
Safflower[a]	30	66
Olive[a]	2	4
Total oil and fat production	18,017	39,709

[a]FAO, 2009.
[b]NASS, 2009.
[c]NRA, 2008.

From Table 7.5, it is clear that the major source for oil in the USA is from soybean oil and since soybean oil is the lowest cost vegetable oil, it is the No. 1 feedstock for biodiesel production. The USA annually exports 2.72 million t of vegetable oil and animal fat, of which tallow accounts for about 0.91 million t and soybean oil accounts for 0.45 million t. The USA also imports oils and fats totalling about 2.27 million t, of which the major categories include rapeseed/canola oil, palm oil and olive oil. Therefore, after accounting for imports and exports, surplus oil and fat production for 2005 was 2.9 million t. This amount of oil and fat will produce about 3.21 billion l of biodiesel. This can be seen as the theoretical maximum amount of biodiesel that can be made from first use vegetable oil and animal fat without restricting the supply of food oils. Used vegetable oils such as restaurant grease can also be used to produce biodiesel. In fact, claims have been made that as much as 11.3 billion l of used cooking oil is drained from deep-fat fryers in the USA every year (Pahl, 2005). Therefore, in the future, it would not be surprising if most of the biodiesel were made from used cooking oil. As long as soybean oil is the primary feedstock for biodiesel, as it is now, it is hard to envision the biodiesel industry exceeding 3.78 billion l of production. Compared to the 238 billion l of distillate fuel consumed each year, 3.78 billion l of biodiesel is only 1.6% of the total distillate fuel use. In the best-case scenario, if all of the waste cooking oil and surplus oil and fat could be used for making biodiesel, about 6% of the distillate fuel could be replaced by biodiesel.

Future Directions

Currently, biodiesel is considered a renewable substitute for fossil diesel fuel. According to the US Energy Independence and Security Act (EISA) of 2007, the term 'renewable biofuel' means fuel that is produced from renewable biomass and that is used to replace or reduce the quantity of fossil fuel used for transportation fuel. This legislation created a Renewable Fuel Standard (RFS) that mandates the production and use of renewable fuels to displace petroleum-based fuels. It requires the US Environmental Protection Agency (EPA) to evaluate the extent to which a fuel can be considered renewable in terms of its impact on global climate change. Renewable biomass includes planted crops and crop residues, planted trees and tree residues, animal waste material and animal by-products, forest products and algae. The renewable fuel programme also amended the US Clean Air Act of 2005 to ensure that transportation fuel sold or introduced into commerce in the USA (except in non-contiguous states or territories), on an annual average basis, contained at least the applicable volume of renewable fuel. The applicable volume of renewable fuel for the calendar years 2006 through 2022 is shown in Table 7.6.

The term 'advanced biofuel' means a renewable fuel, other than ethanol derived from maize starch, that has life cycle greenhouse gas emissions that are at least 50% less than baseline life cycle greenhouse gas emissions. This includes ethanol from other than maize starch, biomass-based diesel, biogas and butanol. The Renewable Fuel Standard requires the use of 1.89 billion l of biomass-based diesel fuel in 2009 and 3.78 billion l by 2012 (Table 7.6). Biomass-based diesel includes biodiesel that has life cycle greenhouse gas emissions of at least 50% less than the baseline life cycle greenhouse gas emissions. The term 'life cycle greenhouse gas emissions' means the aggregate quantity of greenhouse gas emissions (including direct emissions and significant indirect emissions such as significant emissions from land use changes) related to the full fuel life cycle. This includes all stages of fuel and feedstock production and distribution, from feedstock generation or extraction through the distribution, delivery and use of

Table 7.6. Applicable volume of renewable fuel in billion litres (equivalent billion gallons shown in brackets), (based on the US Energy Independence and Security Act (EISA) of 2007).

Year	Renewable fuel	Advanced biofuel	Cellulosic biofuel	Biomass-based diesel
2006	15.12 (4.0)			
2007	17.77 (4.7)			
2008	34.02 (9.0)			
2009	41.96 (11.1)	2.27 (0.6)		1.89 (0.5)
2010	48.95 (12.95)	3.59 (0.95)	0.38 (0.1)	2.46 (0.65)
2011	53.73 (13.95)	5.10 (1.35)	0.95 (0.25)	3.02 (0.8)
2012	57.46 (15.2)	7.56 (2.0)	1.89 (0.5)	3.78 (1.0)
2013	62.56 (16.55)	10.40 (2.75)	3.78 (1.0)	
2014	68.61 (18.15)	14.18 (3.75)	6.62 (1.75)	
2015	77.49 (20.5)	20.79 (5.5)	11.34 (3.0)	
2016	84.11 (22.25)	27.41 (7.25)	16.07 (4.25)	
2017	90.72 (24.0)	34.02 (9.0)	20.79 (5.5)	
2018	98.28 (26.0)	41.58 (11.0)	26.46 (7.0)	
2019	105.84 (28.0)	49.14 (13.0)	32.13 (8.5)	
2020	113.40 (30.0)	56.70 (15.0)	39.69 (10.5)	
2021	124.74 (33.0)	68.04 (18.0)	51.03 (13.5)	
2022	136.08 (36.0)	79.38 (21.0)	60.48 (16.0)	

the finished fuel to the ultimate consumer, where the mass values for all greenhouse gases are adjusted to account for their relative global warming potential. Sheehan *et al.* (1998) found that the life cycle CO_2 emission reduction from soybean oil-based biodiesel was 78.45% less than petroleum diesel. This issue gained great attention in 2008 when Timothy Searchinger and eight co-authors (Searchinger, 2008; Searchinger *et al.*, 2008) presented an argument that increased use of biofuels in the USA would result in greater demand and higher prices for crops such as maize and soybean. This places increased pressure for the conversion of forest and grassland to crop production not only in the USA, but in developing countries such as Brazil, China and India.

When the acres are converted, Searchinger assumes that 25% of the carbon in the soil and 100% of the carbon in plants will be released to the atmosphere. This sudden injection of carbon dioxide would offset the benefits derived from the use of biofuels. Searchinger claims that the greenhouse gas savings that result from the use of soybean-based biodiesel will take 320 years to offset the carbon emissions resulting from the release of carbon caused by the land use change. Similar calculations for maize-based ethanol give a payback time of 167 years. Searchinger concludes that use of cropland to expand biofuels will probably exacerbate global warming rather than reduce it.

The greatest difficulty with including indirect land use changes is the change in perspective. Life cycle analysis (LCA), which is a well-developed science, attempts to characterize actual practice in fuel production and utilization. Incorporation of indirect land use change into LCA shifts the focus from actual practice to prediction of future outcomes. These outcomes are frequently predetermined by the assumptions involved in establishing the scenarios to be investigated. In Searchinger's case, the assumption that world demand for food and feed is inelastic so that conversion of land to biofuels in the USA must lead to conversion of non-crop land to crop production inevitably leads to a negative impact of biofuel production. Since biofuel is estimated to result in an increase in carbon dioxide release, Searchinger concludes that government policy should discourage biofuels. A more productive use of this type of analysis is to investigate a wide range of scenarios and then use the results to identify potential negative impacts and develop strategies to mitigate these impacts. For example, Kim *et al.* (2009) have presented a study of biofuel production in the USA that incorporates sustainable cropping management practices which shows that payback times for Searchinger's case can be reduced to as little as 3 years.

At the time of the preparation of this chapter, the EPA is scheduled to release their methodology for incorporating the effects of indirect land use changes into calculations of the renewability of biofuels. To participate in the RFS, biodiesel must reduce greenhouse gases by 50% or more. It is currently uncertain whether biodiesel will be included in the RFS.

The US government has been supportive of biodiesel production through tax incentives. The biodiesel tax incentive is a federal excise tax credit that brings lower-cost biodiesel to biodiesel consumers. The credit equates to US¢1/% of biodiesel in a fuel blend made from agricultural products like vegetable oils. This has helped biodiesel to be comparable in price with petroleum. Originally, the tax credit was scheduled to expire in December 2006; however, it has been extended and is now in effect until December 2009. The tax incentive may be extended again after this date.

In Europe as well, tax incentives for biofuels have triggered a dramatic increase in the demand for biodiesel. Total EU biodiesel production for 2007 was over 5.7 million t, an increase of 16.8% from the 2006 figures (EBB, 2008). In several European countries, particularly Germany, France and Italy, the production of biodiesel has also been boosted through tax reductions and exemptions. Currently, these three countries dominate the European biodiesel market, with Germany contributing more than half of all production. Guaranteeing tax reductions and exemptions, the European governments intend to increase the share of biofuels in total EU fuel consumption to 5.75% by 2010 (Frondel and Peters, 2007). With government mandates and regulations for energy independence and environmental protection, the use of biofuel is likely to increase. The EU has a target to provide 20% of its total energy use from bioenergy and to provide 10% of its transportation liquid fuels from biofuels such as ethanol and biodiesel. Among all energy needed, transportation liquid fuels are the hardest to replace. Therefore, in the foreseeable future, there will be a high demand for substitutes for liquid petroleum fuels. So far, ethanol and biodiesel are the only practical replacements for gasoline and diesel fuel. Therefore, there is no doubt that there will be a continued growth in biofuel production.

References

Achten, W.M.J., Mathijs, E., Verchot, L., Singh, V.P., Aerts, R. and Muys, B. (2007) Jatropha biodiesel fueling sustainability? *Biofuels Bioproducts and Biorefining* 1, 283–291.

Achten, W.M.J., Verchot, L., Franken, Y.J., Mathijs, E., Singh, V.P., Aerts, R. and Muys, B. (2008) Jatropha bio-diesel production and use. *Biomass and Bioenergy* 32, 1063–1084.

ASTM (2003a) Standard test method for cloud point of petroleum products D 2500-02. In: *Annual Book of ASTM Standards*. American Society for Testing and Materials, West Conshohocken, Pennsylvania, pp. 886–889.

ASTM (2003b) Standard test method for pour point of petroleum products D97-02. In *Annual Book of ASTM Standards*. American Society for Testing and Materials, West Conshohocken, Pennsylvania, pp. 87–94.

Basiron, Y. (2007) Palm oil production through sustainable plantations. *European Journal of Lipid Science and Technology* 109, 289–295.

Brice, J.C. (1973) Growth kinetics. In: Wohlfarth, E.P. (ed.) *The Growth of Crystals from Liquid*. American Elsevier, New York, pp. 78–127.

Brown, E. and Michael, F.J. (2005) *Cruel Oil: How Palm Oil Harms Health, Rainforest and Wildlife*. Center for the Science in the Public Interest, Washington, DC.

Canakci, M. and Van Gerpen, J. (2001) Biodiesel production from oils and fats with high free fatty acids. *Transactions of the ASAE* 44, 1429–1436.

CBOT (2009) Soybeans futures settlement. Chicago Board of Trade (http://www.cbot.com/cbot/pub/page/0,3181,1272,00.html, accessed 20 February 2009).

Chandler, J.E., Horneck, F.G. and Brown, G.I. (1992) The effect of cold flow additives on low temperature operability of diesel fuels. In: *Proceedings of SAE International Fuels and Lubricants Meeting and Exposition*. SAE International, Warrendale, Pennsylvania.

Chiu, C.W., Schumacher, L.G. and Suppes, G.J. (2004) Impact of cold flow improvers on soybean biodiesel blend. *Biomass and Bioenergy* 27, 485–491.

CIA (2009) The World Fact Book. Central Intelligence Agency (https://www.cia.gov/library/publications/the-world-factbook/print/xx.html, accessed 20 February 2009).

Crockett, J., Peterson, C. and Mann, P. (2006) Feasibility study for the commercial production of biodiesel in the treasure valley of Idaho. University of Idaho, Boise, Idaho.

Davis, R.A., Mohtar, S. and Tao, B.Y. (2007) Production of low-temp biodiesel through urea clathration. In: *Proceedings of ASABE*. American Society of Agricultural and Biological Engineers (ASABE), St Joseph, Missouri.

DeMan, J.M. (2000) Relationship among chemical, physical and textural properties of fats. In: Widlak, N. (ed.) *Physical Properties of Fats, Oils and Emulsifiers*. AOCS, Champaign, Illinois, pp. 79–95.

Dunn, R.O. and Bagby, M.O. (1995) Low-temperature properties of triglyceride-based diesel fuels: transesterified methyl-esters and petroleum middle distillate/ester blends. *Journal of the American Oil Chemists Society* 72, 895–904.

Dunn, R.O., Shockley, M.W. and Bagby, M.O. (1996) Improving the low-temperature properties of alternative diesel fuels: vegetable oil-derived methyl esters. *Journal of the American Oil Chemists Society* 73, 1719–1728.

EBB (2008) The EU biodiesel industry statistics. European Biodiesel Board (http://www.ebb-eu.org/stats.php, accessed 16 February 2009).

EIA (2008) Fuel oil and kerosene sales 2007. DOE/EIA-0535(07). Energy Information Administration, DOE, Washington DC.

FAO (2009) FAO statistical databases. Food and Agriculture Organization (FAO) (http://faostat.fao.org/, accessed 1 January 2009).
Frondel, M. and Peters, J. (2007) Biodiesel: a new Oildorado? *Energy Policy* 35, 1675–1684.
Grigg, D. (1999) The fat of the land: a geography of oil and fat consumption. *GeoJournal* 48, 259–268.
Gunstone, F.D. (2002) Production and trade of vegetable oils. In: Gunstone, F.D. (ed.) *Vegetable Oil in Food Industry Composition, Properties and Uses*. CRC Press, Florida, pp. 1–17.
Hansen, S. (2007) Feasibility study of performing a life cycle assessment on crude palm oil production in Malaysia. *The International Journal of Life Cycle Assessment* 12, 50–58.
Ito, T., Nakashimada, Y., Senba, K., Matsui, T. and Nishio, N. (2005) Hydrogen and ethanol production from glycerol-containing wastes discharged after biodiesel manufacturing process. *Journal of Bioscience and Bioengineering* 100, 260–265.
Jatropha World (2007) Jatropha plantation. Jatropha World (http://www.jatrophabiodiesel.org/jatrophaPlantation.php?_divid=menu2, accessed 10 February 2009).
Ji, J.B., Wang, J.L., Li, Y.C., Yu, Y.L. and Xu, Z.C. (2006) Preparation of biodiesel with the help of ultrasonic and hydrodynamic cavitation1. *Ultrasonics* 44, E411–E414.
Johnson, D.T. and Taconi, K.A. (2007) The glycerin glut: options for the value-added conversion of crude glycerol resulting from biodiesel production. *Environmental Progress* 26, 338–348.
Keim, G.I. (1945) Treating Fats and Fatty Oils. US Patent 2,383,601. Awarded 28 August 1945.
Kelkar, M.A., Gogate, P.R. and Pandit, A.B. (2008) Intensification of esterification of acids for synthesis of biodiesel using acoustic and hydrodynamic cavitation 1. *Ultrasonics Sonochemistry* 15, 188–194.
Kim, H., Kim, S. and Dale, B.E. (2009) Biofuels, land use change, and greenhouse gas emissions: some unexplored variables. *Environmental Science and Technology* 43, 961–967.
Kotrba, R. (2007) The glycerin spread. *Biodiesel Magazine* (http://www.biodieselmagazine.com/article.jsp?article_id=1797, accessed 10 February 2009).
Lee, I., Johnson, L.A. and Hammond, E.G. (1995) Use of branched-chain esters to reduce the crystallization temperature of biodiesel. *Journal of the American Oil Chemists Society* 72, 1155–1160.
Lee, I., Johnson, L.A. and Hammond, E.G. (1996) Reducing the crystallization temperature of biodiesel by winterizing methyl soyate. *Journal of the American Oil Chemists Society* 73, 631–636.
Methanex (2009) Historical methanol posted prices. Methanex Corporation (http://www.methanex.com/products/documents/MxAvgPrice_Dec232008.pdf, accessed 2 February 2009).
Miller-Klein (2006) Impact of biodiesel production on the glycerol market. HGCA (http://www.hgca.com/document.aspx?fn=load&media_id=3605&publicationId=2363, accessed 19 March 2009).
Mittelbach, M. and Remschmidt, C. (2005) *Biodiesel The Comprehensive Handbook*, 2nd edn. Boersedruck Ges.m.b.H, Vienna.
Moser, B.R. (2008) Influence of blending canola, palm, soybean, and sunflower oil methyl esters on fuel properties of biodiesel. *Energy and Fuels* 22, 4301–4306.
NASS (2009) Statistics of oilseeds, fats, and oils. In: *Agricultural Statistics 2008*. National Agricultural Statistics Service, USDA, Washington, DC, pp. III-1–III-30.
NBB (2005) Biodiesel Tax Credit Implementation. (http://www.biodiesel.org/news/taxincentive/Biodiesel%20Tax%20Credit%20NBB%20Issue%20Breif.pdf, accessed 10 January 2009).
NBB (2009) Estimated US biodiesel production by fiscal year. National Biodiesel Board (http://www.biodiesel.org/pdf_files/fuelfactsheets/Production_Graph_Slide.pdf, accessed 10 January 2009).
NRA (2008) North Americal Rendering, the source of essential high quality products. National Renderer's Assocoation (http://nationalrenderers.org/, accessed 20 March 2009).
O'Connor, R.T. (1960) X-ray defraction and polymorphism. In: *Fatty Acids: Their Chemistry, Properties and Uses (Part I)*. Interscience Publication, New York, pp. 285–378.
Openshaw, K. (2000) A review of Jatropha curcas: an oil plant of unfulfilled promise. *Biomass and Bioenergy* 19, 1–15.
Pahl, G. (2005) *Biodiesel, Growing a New Energy Economy*. Chelsea Green Publishing, White River Junction, Vermont.
Peterson, C.L., Reece, D.L., Hammond, B.L., Thompson, J.C. and Beck, S.M. (1997) Processing, characterization, and performance of eight fuels from lipids. *Applied Engineering in Agriculture* 13, 71–79.
Prueksakorn, K. and Gheewala, S.H. (2008) Full chain energy analysis of biodiesel from Jatropha curcas L. in Thailand. *Environmental Science and Technology* 42, 3388–3393.
Searchinger, T. (2008) The impacts of biofuels on greenhouse gases: how land use change alters the equation. The German Marshall Fund of the United States (http://www.gmfus.org/publications/article.cfm?id=385, accessed 23 February 2009).

Searchinger, T., Heimlich, R., Houghton, R.A., Dong, F.X., Elobeid, A., Fabiosa, J., Tokgoz, S., Hayes, D. and Yu, T.H. (2008) Use of US croplands for biofuels increases greenhouse gases through emissions from land-use change. *Science* 319(5867), 1238–1240.
Sheehan, J., Camobreco, V., Duffield, J., Graboski, M. and Shapouri, H. (1998) Life cycle inventory of biodiesel and petroleum diesel for use in an urban bus. NREL/SR-580-24089. National Renewable Energy Laboratory, US Department of Energy, Golden, Colorado.
Shrestha, D.S., Van Gerpen, J., and Thompson, J. (2008) Effectiveness of cold flow additives on various biodiesels, diesel, and their blends. *Transactions of the ASABE* 51, 1365–1370.
Song, C., Hsu, C.S. and Mochida, I. (2000) Introduction to chemistry of diesel fuel. In: *Chemistry of Diesel Fuels*. Taylor and Francis, New York, pp. 1–60.
Sonntag, N.O.V. (1982) Glycerolysis of fats and methyl-esters – status, review and critique 2. *Journal of the American Oil Chemists Society* 59, A795–A802.
Sprules, F.J. and Price, D. (1950) Production of fatty esters. United States Patent 2,494,366. Awarded 10 January 1950.
Stavarache, C., Vinatoru, M. and Maeda, Y. (2007) Aspects of ultrasonically assisted transesterification of various vegetable oils with methanol 1. *Ultrasonics Sonochemistry* 14, 380–386.
Tat, M.E., Van Gerpen, J. and Wang, P.S. (2007) Fuel property effects on injection timing, ignition timing, and oxides of nitrogen emissions from biodiesel-fueled engines. *Transactions of the ASABE* 50, 1123–1128.
Thompson, J.C. and He, B.B. (2006) Characterization of crude glycerol from biodiesel production from multiple feedstocks. *Applied Engineering in Agriculture* 22, 261–265.
Tiwari, A.K., Kumar, A. and Raheman, H. (2007) Biodiesel production from jatropha (*Jatropha curcas*) with high free fatty acids: an optimized process. *Biomass and Bioenergy* 31, 569–575.
Van Gerpen, J. (2005) Biodiesel processing and production. *Fuel Processing Technology* 86, 1097–1107.
Vyas, A.P., Subrahmanyam, N. and Patel, P.A. (2009) Production of biodiesel through transesterification of Jatropha oil using KNO3/Al2O3 solid catalyst. *Fuel* 88, 625–628.

8

Industrial Oil Types and Uses

Lou Honary

Introduction

The National Petrochemical Refiners Association (NPRA), after an annual survey of oil refiners, has published detailed information on the amount of various industrial and automotive lubricants produced in the USA. Based on a review of the data from NPRA, about 9.5 billion litres of lubricants are produced in the USA annually, with 4.2 billion litres being industrial lubricants and the rest considered automotive lubricants. While this is a significant volume, it is only about 1% of the total petroleum used in the USA.

Vegetable oils have a long history of use in the fuel, lubricating fluid and energy transfer media. But, their usage has been tied to the availability or lack thereof of petroleum. For example, prior to the discovery of petroleum, vegetable oils and animal fats were primary sources of lubricating products. During both World Wars, for example, there was an increased usage of vegetable oils for fuel, lubrication and energy transfer. Other examples include the Arab oil embargo of 1973 and the speculative increase in petroleum prices to the near US$150/barrel level in 2008, all of which resulted in an increased interest in the use of vegetable oils for lubricants and, more recently, for fuels.

Primarily a phenomenon starting in the 1960s, concern for the environment has been a consistent force for the use of vegetable oils in industrial applications. The environment factor promises to be a more steady and persistent force in promoting the use of vegetable oils than the fluctuating availability of petroleum. At the same time, performance and economic viability of bio-based lubricants and grease products have found equal importance. Another important factor introduced has been the food versus fuel controversy, which has resulted in an interest in non-edible oils and industrial crops. Either because of imposed environmental regulations or society's desire for alternatives to petroleum, the interest and increased use of vegetable-based lubricants and machinery greases will continue.

Because this chapter deals with *bio-based* lubricants and products, the emphasis will be placed on the use of vegetable oils and their properties for such applications. Furthermore, the chapter provides information in brief on the chemistry of vegetable oils, the general performance requirement of selected lubricant products, standard test procedures required for their approval when applicable and the performance reports on some of the vegetable oil-based lubricants.

Petroleum Base Oils – Paraffinic and Naphthenic

Base oils, in general, are derived from light and heavy crude based on the type and sources of crude petroleum. The presence of bitumen and heavy distillates in the oil, which are heavier than other components, make the oil heavy (crude). Some heavy crude could contain over 70% bitumen. These oils are suitable for heavier petroleum derivatives like base oils for lubricants. Light crude, on the other hand, is refined for gasoline and other fuels. Crude oils from the Middle East are usually light crude; while crude oils from parts of the world including continental Europe and Asia, North and South America and the North Sea are heavy crude.

In addition to this heavy/light classification, which is based on composition and density, the oils are also divided into paraffinic and naphthenic. Paraffinic oils come from light crude and contain a higher percentage of light gases, including gasoline. Naphthenic oils, on the other hand, come from heavy crude. Oil refineries may be focused on making speciality oils like base oil for hydraulic oils from heavy crude or naphthenic oils, or focused on fuel production from light crude or paraffinic oils.

Basic chemistry of crude oils

Like the vegetable oils, mineral oils are made primarily of hydrocarbon or atoms of carbon and hydrogen bound together to form molecules of different structures. In addition to the paraffinic and naphthenic structures, these molecules could be aromatic or polycyclic aromatic. Figure 8.1 shows the basic structures of hydrocarbon.

The paraffinic oils (alkanes) have either straight or branched structures. These oils in general contain waxes, resulting in reduced flowability at low temperatures. During the process of refining, the normal alkanes or waxes are removed. This process is similar to winterization of vegetable oils, where the oil is cooled to the point where waxes are solidified and then filtered out. As some waxes may require much colder temperatures to solidify, some refineries are equipped to solvent de-wax the oil by mixing it with a solvent and then cooling it and allowing it to crystallize. The wax is then removed through filtration. These 'deep de-waxing' processes could produce paraffinic oils with pour points down to –30°C.

The naphthenic oils (alkenes) have cyclic structures and are sometimes referred to as cycloalkanes. They often have 6 carbon rings (sometimes 5 or 7) and offer great solvency properties, as well as cold temperature flowability.

The aromatic oils have at least one ring of six carbon atoms and alternating double and

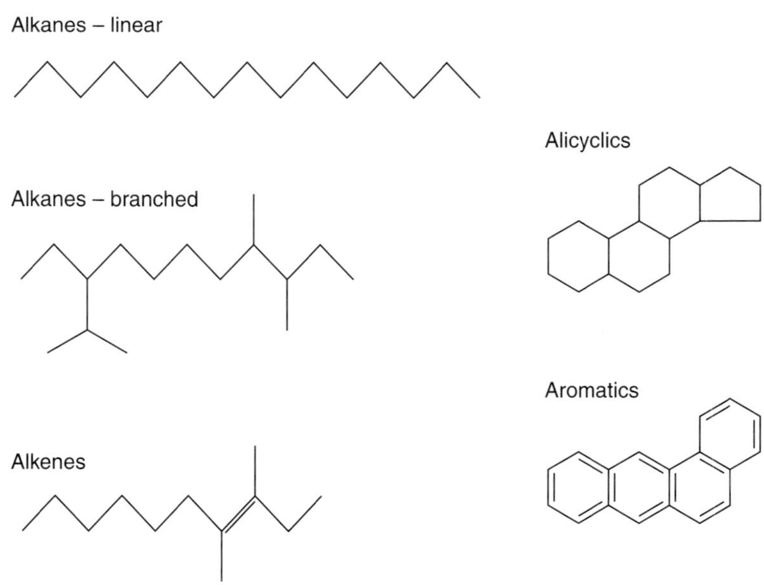

Fig. 8.1. Structure of hydrocarbons.

single bonds. Most of the sulfur and nitrogen in the oil are bound to aromatic structures; giving the aromatic oils properties that are different from the straight structured paraffinic and the cyclic structured naphthenic oils. When the oil has several cyclic aromatic rings adjacent to each other, it is referred to as polycyclic, as opposed to monocyclic, containing one cyclic ring. Cyclic aromatics and normal alkanes occur in oil as single molecules, while others appear in various molecular structures. Most other elements such as nitrogen, sulfur and oxygen molecules in the oils are bound to the aromatic structures. When the aromatic structures are separated from the oil, these elements too are removed. Each of these elements in small quantities in the oil could have some impact, both positive and negative. For example, nitrogen acts as an oxidation inhibitor and as a passivator for copper. Sulfur compounds provide high extreme pressure and anti-wear properties, but their presence in the oil could cause corrosion in copper materials. Sulfur too could inhibit oxidation by destroying peroxides. Oxygen in oils could accelerate oxidation, although the percentage of oxygen in fresh oil is minute.

Vegetable Oils

A major portion of oils and fats for human consumption come from plants and animals. The world production of edible oils is made up of 71% vegetable oils, 26% animal fat and 2% fats from marine species. It is estimated that soybean oil, for example, contributes to nearly one-third of the world's oilseeds, with almost one-half produced in the USA.

In general, according to the US Department of Agriculture (USDA), vegetable oils may be classified into three categories based on their production, use and volume: (i) major oils, (ii) minor oils and (iii) non-edible oils.

Major oils are those known for human or animal feed consumption and often play important economic roles in the regions producing them. They include: soybean, palm, rapeseed, sunflower, cottonseed, coconut, groundnut, olive, palm kernel, linseed and sesame.

Minor oils are those that are known for their uses but do not match the large production magnitude of major oils. These oils have fatty acid profiles that could make them effective for industrial uses. They include: niger, mango kernel, poppy, cocoa bean, shea, hempseed, grape seed, perilla, Chinese vegetable tallow, Ethiopian mahogany, German sesame, watermelon seed, avocado and apricot seed.

Non-edible oils comprise the third category where the oil is extracted exclusively for industrial use. While the majority of oil plants are cultivated for food applications, a number of non-edible oils such as linseed, castor and tung are grown commercially for their unique chemical make-up that is important to the industry. They are used in industrial applications such as soaps, paints, varnishes, resins, plastics, and in agrochemicals. But, their use is also being considered for industrial lubricant applications. Examples of these oils include: linseed, castor, neem, mahua, karanja, undi, kusum, khakan, pisa, kokum, nahor, sal and dhupa (Pryde, 1982; Bhatia, 1983; Weiss, 1983; Bringi, 1987).

The popularity of biofuels over the past few years has resulted in significant investment of public and private capital for the development of non-edible alternative crop oils. Although different in end-use, many industrial crops and special processes developed for biofuels have applications in bio-based lubricants as well. Additionally, the attention given to the negative health effect of trans-fats has reinvigorated the development of special varieties of oilseeds like the low linolenic and high oleic soybeans by major US agrobusinesses.

The Association for Advancements of Industrial Crops (AAIC) has a list of several alternative industrial crops that its members are working on. Most noteworthy are crops like cuphea, camelina, canola, castor, lesquerella, groundnut and pennycress. Some of these crops, like camelina, have reached commercial production stage and reasonably large acreages are being produced in the western USA. A brief description of some of these industrial crops will appear later in this chapter.

The future technologies will encompass the old fatty acids, the newer genetically enhanced high oleic varieties and more sophisticated and economical chemically modified, high functioning esters derived from a large variety of raw materials. These developments will not replace completely the use of petroleum for industrial and automotive lubricants, but they will capture a significant portion of those markets.

Chemistry of Vegetable Oils Relating to Lubricants

Triglycerides form the basic structure of fats and oils, which consist of one molecule of glycerol and three molecules of fatty acids, such that they form either liquid (oil) or solid (fat). Because of the three fatty acids, the resulting molecule is called a triglyceride. A triglyceride is illustrated in Fig. 8.2.

Fatty acids contain a chain of carbon atoms combined with hydrogen (forming hydrocarbon). They terminate in a carboxyl group. If the three fatty acids are alike, the molecule is a simple triglyceride; if they are different, it is a mixed triglyceride.

Each carbon atom along the chain has the ability to hold two hydrogen atoms. The fatty acid is saturated if all hydrogen atoms are in place; in other words, when all available carbon valences for hydrogen are satisfied. If two adjacent carbons are missing hydrogen atoms, the carbons bond doubly to one another, creating a point of unsaturation. If there is more than one double bond, the fatty acid is polyunsaturated, as compared to monounsaturated when there is only one double bond. The relative amounts of these fatty acids vary for the different vegetable oils. When one mole of the triglyceride molecule is hydrolysed, it would give three moles of fatty acids and one mole of glycerol.

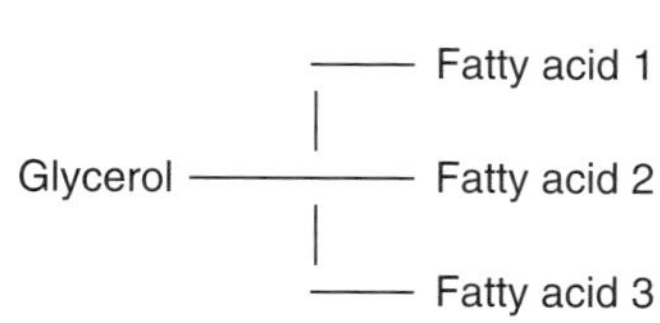

Fig. 8.2. Structure of a triglyceride.

Unsaturation is related inversely to the liquidity of the oil or its melting point; and related directly to its solubility and chemical reactivity. With an increase in unsaturation, the melting point goes down (higher liquidity) while solubility and chemical reactivity increase, resulting in oxidation and thermal polymerization. Saturated oils, in general, show more oxidative stability but have high melting points (lower liquidity), like palm oil, which has high oxidation stability but is solid at room temperature, thus limiting its use for liquid lubricants application unless modified.

Unsaturation may be distributed in different ways along the chain, which impacts the chemical properties of both the fatty acid and the glyceride. The distribution in a chain with a single bond may be such that the two hydrogen atoms from each side of the double bond lie on the same side of the chain. This is a *cis* configuration (Fig. 8.3) where the chain is shaped into a rigid arc and the hydrogen atoms are toward the outside of the arc. Alternately, the hydrogen atoms may be positioned on opposite sides of the chain. In this case, the configuration is a *trans*.

For a normal single-bond atom, there is freedom of rotation, but there is rigidity at a double bond and only two fixed positions of *cis* and *trans* are possible. Because the saturated fatty acids have no double bonds to distort the chain, they pack more easily into crystal forms and, therefore, have higher melting points than unsaturated fatty acids of the same length. They are also less vulnerable to oxidation. This property also allows for winterization of the vegetable oils, described later.

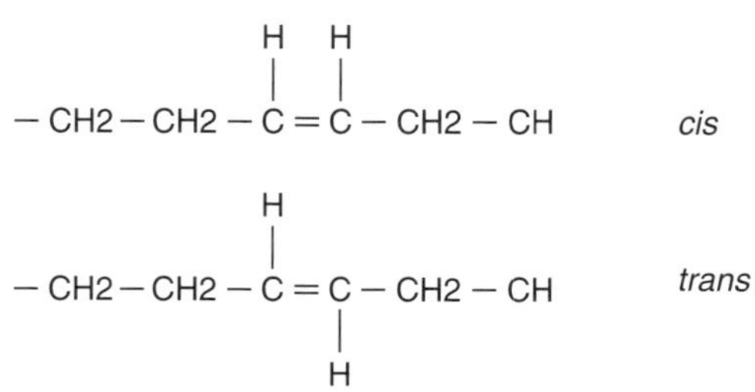

Fig. 8.3. Hydrogen atoms on one side (*cis*) or on both sides of the chain (*trans*).

Properties of Vegetable Oils

Vegetable oils have many advantages and some shortcomings when considered for use in industrial lubricants and hydraulic fluids. Most importantly, unless modified, they lack oxidation stability. Oxidation stability of vegetable oils is dependent on the position and degree of unsaturation of the fatty acids that are attached to the glycerol molecule. For example, the majority of soybean oil fatty acid composition is comprised of conjugated double bonds, which make it more susceptible to oxidation. Conventional soybean oil contains approximately 52% linoleic acid, which has two conjugated double bonds, and 7% linolenic acid, which contains three conjugated double bonds. If left untreated, the use of these oils could lead to increased oxidation and, consequently, increased viscosity. In extreme cases, if the oil continues to oxidize in use, it could lead to polymerization and formation of polymer films in the oil. To avoid oxidation in use, the vegetable oil is either chemically modified and/or antioxidants are used to increase oxidation stability. Hydrogenation, chemically adding hydrogen to the double bonds, is one method used to increase oxidative stability. Unfortunately, the melting point is also increased and can result in a product that is solid or semi-solid at room temperature. Oilseeds that are genetically enhanced and have higher oxidation stability are more conducive for use in industrial lubricants and hydraulic oils.

The longer the fatty acid carbon chain, the higher is the melting point. Double bonds within the carbon chain lower the melting point significantly. Vegetable oils, due to their fatty acid structure, tend to freeze at relatively higher temperatures than their mineral oil counterparts. A pour point comparison of hydraulic fluid using both mineral oil and soybean oil as base fluid is shown in Table 8.1. For applications where hydraulic oil or industrial lubricants are exposed to sub-zero temperatures, a mixture of vegetable oils and mineral or synthetic oils could be used. Mixing, however, impacts other properties of vegetable oils, including viscosity index and flash/fire points, as well as compatibility with elastomers and other components.

Vegetable oils, due to their polarity, adhere to metal surfaces for better metal–metal separation. Also, due to a higher viscosity index relative to petroleum oils, they are more stable as the temperature changes. For example, soybean oil has a viscosity index of about 220, with a viscosity of 30.69 at 40°C and a viscosity of 7.589 at 100°C. Comparable naphthenic base oil with a viscosity of 37.95 at 40°C and a viscosity of 5.295 at 100°C would have a viscosity index of 53. Since the high viscosity index results in a more stable viscosity

Table 8.1. Viscosity, viscosity index and pour points of selected oils and identical hydraulic fluids utilizing soybean oil and mineral oil-based fluids.

	Pour point (°C)	Viscosity at 40°C	Viscosity at 100°C	Viscosity index
Description	ASTM D6749	ASTM D445	ASTM D445	ASTM D2270
Refined high oleic soy oil	−16	31.19	8.424	200
Crude conventional soy	−6	31.69	7.589	222
Mineral oil – ISO VG 100	−50	20.58	3.684	28
Mineral oil – ISO VG 500	−32	96.21	9.040	53
Mineral oil blend of 57–43% (of ISO VG 100 and 500)	−49	37.95	5.295	53
Hydraulic fluid with crude conventional soy	−4	32.26	7.592	217
Hydraulic fluid with high oleic soy	−4	39.14	8.412	199
Hydraulic fluid with mineral oil blend	−11	25.24	4.248	46

when temperatures change, lower viscosity vegetable oil-based hydraulic fluid could be used in applications where higher viscosity petroleum oil is required. As an example, ISO Viscosity Grade (VG) 46 hydraulic fluid made from vegetable oil may be suitable for applications where an ISO VG 68 from petroleum oil is specified. Honary (1999) reported on the viscosity of base soybean oils and formulated soybean oil-based and petroleum oil-based tractor hydraulic fluids. Table 8.1 shows the viscosity, viscosity index and the pour points of soybean oils, mineral oils and identical hydraulic fluid packages utilizing both soybean oil and mineral oil as base fluids. The mixture of ISO VG 100 and ISO VG 500 was prepared to create a viscosity range closer to soybean oils. The difference in viscosity index is significant, with the soybean oil showing almost four times higher viscosity index than the petroleum mineral oils.

Tribological characteristics of the selected base oils and finished hydraulic fluids are shown in Table 8.2. Soybean oil shows better lubricating properties, as indicated by the 4-ball wear (ASTM D4172), 4-ball extreme pressure (ASTM D2783), pin and vee, (ASTM D3233A) and tapping torque (ASTM D5619) results.

The flash and fire points of vegetable oils are consistently and considerably higher than equivalent viscosity mineral oils. Typically, fire points of vegetable oils are greater than 300°C. This property is suitable for creation of hydraulic fluids and industrial lubricants that could meet some fire retardancy standards, including those of the factory mutual standards in the USA. Metalworking fluids made from vegetable oils show less tendency to burn, and hydraulic applications like building elevators could benefit from the fire safety aspect of this property of vegetable oils. Table 8.3 presents viscosity and viscosity index of groups I, II and III for comparison.

There are many factors that affect the fatty acid make-up of vegetable oils. In addition to their natural structure, changes in the growing conditions and geographic location and factors such as exposure to daylight, light intensity and quality impact the properties of vegetable oils. Because the fatty acid composition of oils and fats is unique, their characteristics are different. One important process that can be used to affect the types of fatty acid present is partial hydrogenation, which results in the formation of small amounts of geometrical and positional isomers of the present unsaturated acids.

Table 8.2. Tribological characteristics of selected oils and finished hydraulic fluids.

	Load wear index (weld point kg)	4-Ball wear scar (mm)	Pin and vee force (lb) – torque (lb-in)	Tapping torque (N-m)
Description	ASTM D2783	ASTM D4172	ASTM D3233 A	ASTM D5619
High oleic soy oil	21.87 (160)	0.626	Broke @ 1755.64 lb-f Torque = 31.8	8.198
Crude conventional soy	26.74 (160)	0.589	Broke @ 1656.94 lb-f Torque = 53.1	8.027
Mineral oil – ISO VG 100	12.63 (126)	0.663	Broke @ 567.64 lb-f Torque = 52.8	11.193
Mineral oil – ISO VG 500	17.06 (126)	1.238	Broke @ 520.3 lb-f Torque = 93	11.073
Mineral oil blend of 57%:43% (of ISO VG 100 and 500)	13.96 (126)	0.810	Broke @ 215.41 lb-f Torque = 51.1	10.99
Hydraulic fluid with crude conventional soy	33.98 (200)	0.510	N/A	N/A
Hydraulic fluid with high oleic soy	26.82 (160)	0.529	N/A	N/A
Hydraulic fluid with mineral oil blend	17.45 (126)	0.572	N/A	N/A

Table 8.3. Physio-chemical characteristics of mineral oils.

Properties	ISO VG32 group	ISO VG32 Group II	ISO VG32 Group III
Appearance	Clear yellowish liquid	Colourless liquid	Colourless liquid
Kinematic viscosity at 40°C (cSt)	29.15	29.65	37.55
Kinematic viscosity at 100°C (cSt)	5.14	5.37	6.43
Viscosity index	105	116	123

Hydrogenation

Since most triglycerides contain both saturated and unsaturated fatty acids, hydrogenation can be used to affect their fatty acid contents. Simply stated, hydrogenation is a way of saturating the double bonds. In the geometrical isomers, the *cis* structures are partially converted to the *trans* form. In the positional isomers, however, the original *cis*-9 double bonds, such as the oleic acid, are converted partially to a mainly double bond at other positions in the chain (Erickson *et al.*, 1985). Some oils like soybean oil have higher levels of linolenic acid, which is a polyunsaturated fatty acid. Its presence in the oil can lead to a higher degree of auto-oxidation, which in turn results in off-flavour and odour (rancidity). When considering a vegetable oil for industrial lubricants, the oils with the lowest amount of linoleic would be more suitable. Through partial hydrogenation, the linolenic acid can be lowered by conversion. Direct addition of hydrogen to the double bond of an unsaturated fatty acid involves overcoming a considerable energy barrier. However, both hydrogen and unsaturated bonds are absorbed readily at the surface of a catalyst such as nickel. In this case, the energy barrier is smaller and the reaction can be much faster; and in the process results in release of energy. Also, the removal of the reaction products from the surface of nickel requires overcoming a modest energy barrier before more energy is released. When hydrogenation takes place, the net energy release for a drop of one unit in iodine value is sufficient to raise the temperature of the oil by approximately 1.7°C. This, of course, depends on the specific heat of the oil, which varies with temperature. The exothermic heat of reaction has been computed as 1.7 BTU/lb or 0.942 kcal/unit drop in iodine value (Patterson, 1994).

The iodine value (IV) is used to express the degree of hydrogenation. The IV is defined as the number of grams of iodine absorbed under standard conditions by 100 g of fat. It represents the degree of unsaturation in the fatty acid chain. The saponification number, defined as the number of milligrams of potassium hydroxide required to saponify 1 g of fat, is a measure of the average molecular weight of fatty materials. With partial hydrogenation, the amounts of linolenic and linoleic acid contents are reduced, while increasing the amount of the monoene content. This not only saturates naturally occurring *cis* double bonds but also isomerizes them to a higher melting *trans* form. The fully hydrogenated oil is solid at room temperature. For industrial applications, the oil normally is partially hydrogenated to maintain the minimum desired liquidity while improving oxidation stability.

Winterization

Erickson *et al.* (1985) provide a simple description of winterization which involves chilling the fat at a prescribed rate and allowing the solid portions to crystallize. Then, through some sort of filtration, the solids can be separated. If the oil is allowed to cool too rapidly, small crystals are formed that are more difficult to filter. Filtering cold oil, which presents a higher resistance to flow, requires energy and is time-consuming, thus increasing the cost of the process. This process parallels dewaxing of petroleum oils, explained earlier.

Esters

As most natural products containing petroleum or vegetable oils have impurities and waxes, their properties are not uniform. Synthetic products, on the other hand, are made by synthesizing distinct and purer components of the lubricants so that their properties are clearly predictable. In general, esters may be defined as natural and synthetic, with natural esters being derived from vegetable oils or animal fats. The fully synthetic ester base oils may be categorized in two groups: (i) diacid esters and (ii) polyol esters, both of which are biodegradable with lower toxicity (Bergstra, 2004). According to Randles *et al.* (1989), many compounds have been investigated as base stocks for synthetic lubricants, including: polyalphaolefins (PAO), alkylated aromatics, polybutenes, aliphatic diesters, polyesters, polyalkyleneglycols and phosphate esters. Silicone, borate esters, perfluroethers and polyphenolene ethers are in limited use due to their higher costs and performance limitations.

Synthetic esters are made using acid and alcohols and, due to their purity, offer thermal stability and cold temperature performance far superior to their base materials. The chemistry of vegetable oils is generally similar in that they are triglycerides; or esters derived from glycerin and a range of fatty acids from C12 to C18 (Bergstra, 2004). Figure 8.4 shows a triglyceride, a trimethlolpropane and an adipate.

While they are more expensive to produce, they provide the needed oxidation stability and cold temperature flowability to be considered for use in the more demanding lubricant applications.

Esters made from petroleum base stock are made using a fundamental process involving esterification, filtration and distillation. The fundamental reaction for manufacturing esters is that of acid + alcohol → ester + water and the reaction being reversible and driven to conclusion by the use of excess alcohol and removal of water as it forms. Accordingly, heat and catalysts are used to react alcohol and acid; with possible catalysts being sulfuric acid, *p*-toluene sulfonic acid, tetra alkyl titanate, anhydrous sodium hydrogen sulfate, phosphorous oxides and stannous octonate. Heat for typical reaction would be 230°C with 50–60 mm Hg pressure. The alcohol and water vapours would need to be condensed and removed. At the conclusion of the reaction process, the untreated acid can be

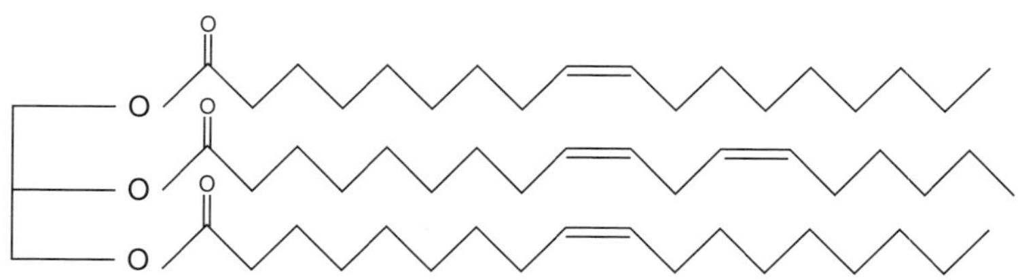

A triglyceride ester

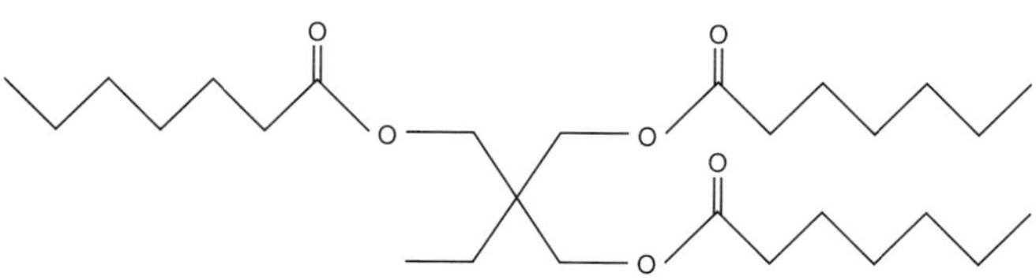

A trimethylolpropane ester

An adipate ester (diacid) ester

Fig. 8.4. A triglyceride, a trimethylolpropane and an adipate.

neutralized by sodium carbonate or calcium hydroxide and removed by filtration.

Similarly, polyol esters are produced by reacting a polyhydric alcohol with a monobasic acid. Examples of alcohols used include neopentyl glycol (NPG), trimethylol propane (TMP) or pentaerythritol (PE). Figure 8.5 shows some of the esters and their feedstock and Table 8.4 lists the physical properties of these esters.

Esters for biofuels

Another means of changing properties of a vegetable oil to become more comparable with those of low viscosity fuels and oils is by converting them chemically to monoesters. Diesel fuels have a viscosity of 4 cSt at 40°C and esters made for biofuels have similar level viscosities. Here too, the process used to make this conversion involves reacting an alcohol with the vegetable oil in the presence of a catalyst. In this process from each triglyceride molecule, three monoester molecules and a glycerol molecule are obtained. The glycerol, a by-product, is removed by water extraction. The final ester product is referred to as methyl ester if methyl alcohol is used or called ethyl ester if ethyl alcohol is used. These processes are well known and are readily available from numerous sources online and in print.

Complex esters

Complex esters have higher viscosity and higher molecular weight than the common esters, offering advantages for some applications. Meng and Dresel (2001) explain the process for complex esters as first esterifying the diol with dicarboxylic acid and then, depending on the desired product, this ester is reacted with either carboxylic acid or a monoalcohol (Meng and Dresel, 2001).

Some of these complex esters, for example, may be made via the reaction of a polyol, dicarboxylic acid and monoalcohol as an end-capping agent. Key structural features of these esters and basic structure of the alcoxy group from the end-capping monoalcohols are presented in Fig. 8.6.

Estolides

The chemical modification and esterification of oil, especially vegetable oils, are carried out to change the properties of the base material for special uses. High oxidation stability and lower cold temperature flowability are most desired for vegetable oils in lubricants uses. In an endeavour to synthesize base oils that are suitable for use in formulating bio-based lubricants, attempts are made to create flexibility in the use of the base raw materials. One example of these new chemical modification techniques is estolides. Estolides are designed with the idea of being able to use various vegetable oils or animal fats as input. Cermak and Isbell (2002) describe the creation of estolides for industrial and automotive lubricants (Fig. 8.7).

Simply described, the estolides were created by 'chemically connecting different unsaturated fatty acids. These are the building blocks of high-oleic oils, such as sunflowers, canola and lesquerella'. Using only the fatty acid components, Cermak and Isbell (2004) produced branched chains of either saturated or unsaturated oleic estolides whose performance in various tests rivalled that of mineral oil-based base oils. These estolides showed up to –30°C for the unsaturated oleic estolides and –40°C for the saturated ones; and oxidative breakdown of 200 and 400 min in the Rotary Bomb Oxidation Test (RBOT) and 200 min for comparable mineral engine oil (Cermak and Isbell, 2004). Table 8.5 gives the oxidation stability and viscosity-related data for three distilled estolides: oleic estolide 2-ethylhexyl ester, coco-oleic estolide 2-ethylhexyl ester and a monomer.

Oxidative Stability of Vegetable Oils

When dealing with the performance of vegetable oils in industrial applications, the percentage of some of the fatty acids, particularly

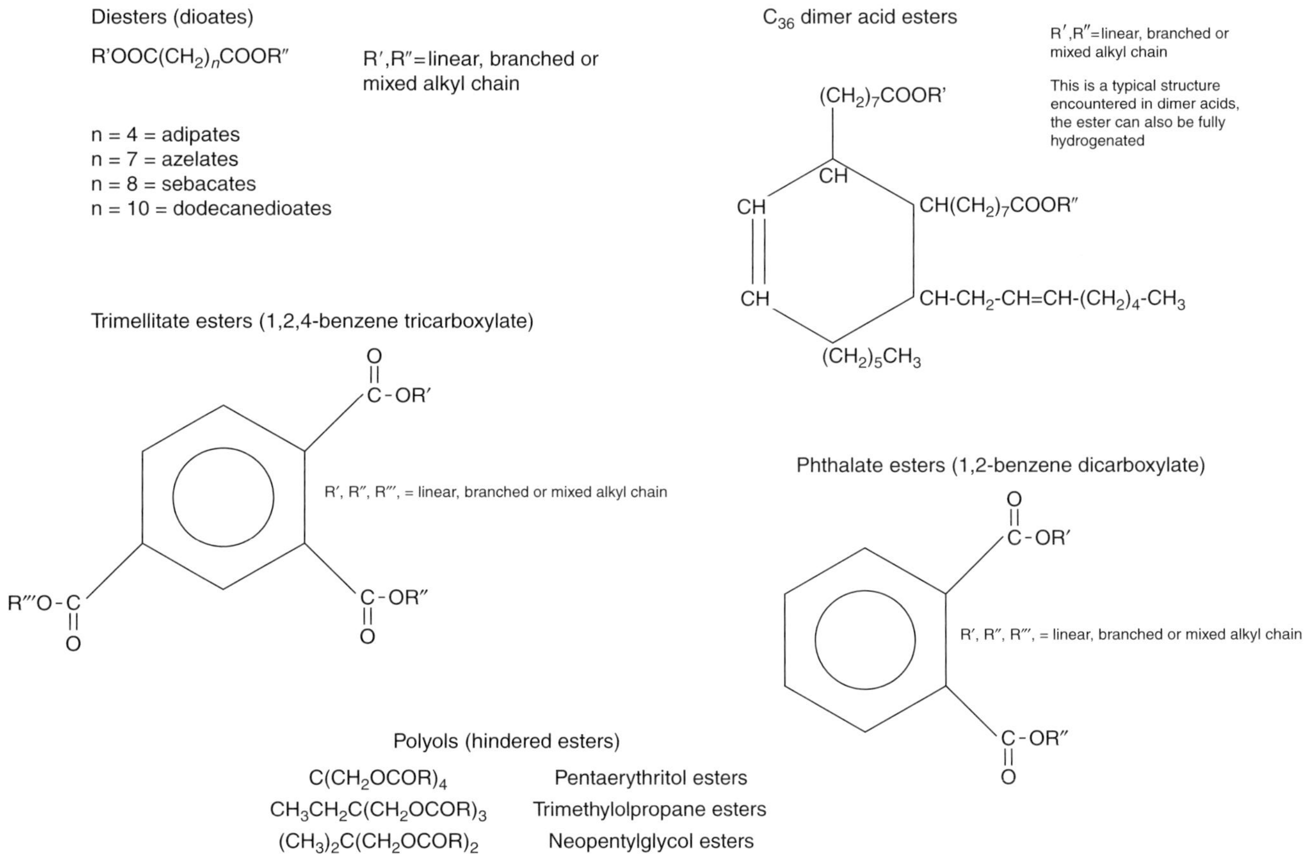

Fig. 8.5. Examples of esters and their feedstock (Mortier and Orszulik, 1997 with kind permission of Springer Science and Business Media).

Table 8.4. Physical properties of esters (Mortier and Orszulik, 1997 with kind permission of Springer Science and Business Media).

	Diesters	Phthalates	Trimelliates	C36 dimer esters	Polyols	Polyoleates
Viscosity at 40°C	6–46	29–94	47–366	13–20	14–35	8–95
Viscosity at 100°C	2–8	4–9	7–22	90–185	3–6	10–15
Viscosity index	90–170	40–90	60–120	120–150	120–130	130–180
Pour point (°C)	−70 to −40	−50 to −30	−55 to −25	−50 to −15	−60 to −9	−40 to −5
Flashpoints	200–260	200–270	270–300	240–310	250–310	220–280
Thermal stability	Good	Very good	Very good	Very good	Excellent	Fair
Conradson carbon	0.01–0.06	0.01–0.03	0.01–0.40	0.20–0.70	0.01–0.10	?
Biodegradability (%)	75–100	46–88	0–69	18–78	90–100	80–100
Costs (PAO = 1)	0.9–2.5	0.5–1.0	1.5–2.0	1.2–2.8	2.0–2.5	0.6–1.5

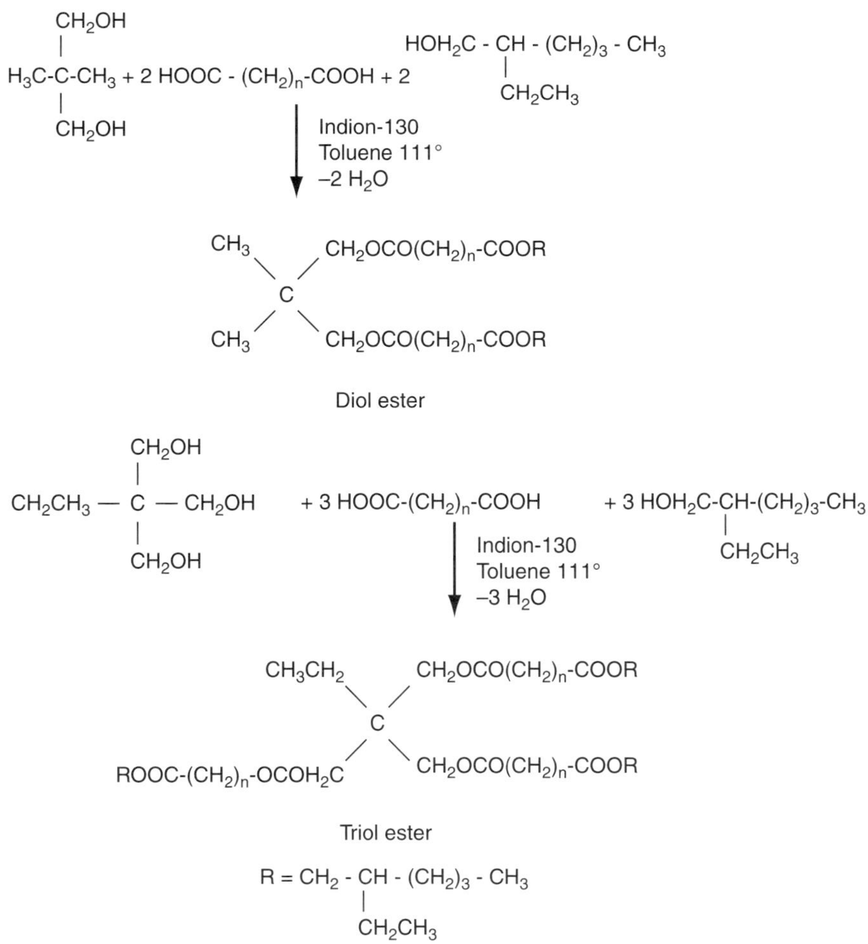

Fig. 8.6. Key structural features of certain complex esters and basic structure of the alcoxy group from the end-capping monoalcohols.

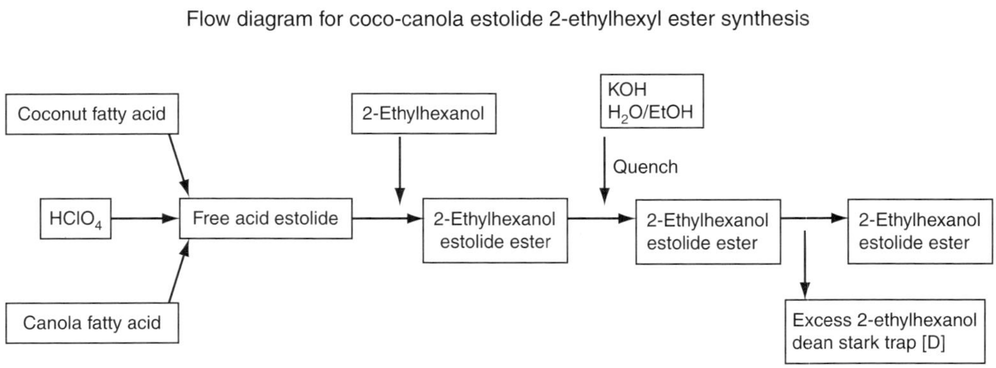

Fig. 8.7. Example of a process for synthesizing estolides. (Reprinted from Cermak and Isbell, 2002 with permission from Elsevier).

Table 8.5. Oxidation stability and viscosity data for three estolides.

		OSI time	Viscosity at 40°C	Viscosity at 100°C	Viscosity index
08-241	Oleic estolide	18.18	98.17 cSt	15.40 cSt	167
08-242	Coco-oleic estolide	22.25	76.58 cSt	12.59 cSt	165
08-243	Oleic estolide with monomer	15.45	40.30 cSt	8.20 cSt	185

oleic, linolenic and euricic, are taken into consideration as they impact the oil's resistance to oxidation. Canola oil, for example, is very low to trace on euricic acid (low-euricic version of rapeseed). Most speciality oils used for lubricant preparation are high in oleic and often low in linolenic acids. A vegetable oil that has poor oxidative stability, once exposed to the high pressure, high temperature, air and metallic surfaces in an industrial application like a hydraulic system, will thicken quickly, with a noticeable change in its viscosity. Highly oxidized oil could begin to polymerize with layers of plastic forming on its exposed surface. Once polymerization is initiated, it will continue to propagate until terminated. Eventually, the entire body of the oil could polymerize. The process could be used deliberately by 'blowing' air or oxygen into the oil while heating. Blown versions of vegetable oils are available on the market and are used among other applications as viscosity modifiers. Figure 8.8 shows a picture of soybean oil exposed to heat above 150°C and air forming plasticized layers on the surface. This is perhaps the most important factor in determining the use of vegetable oils in any industrial application.

Determination of oxidation stability

From a practical point of reference, oxidation stability refers to the ability of the oil to maintain its properties, mainly its viscosity, when exposed to the specific operating conditions. Since vegetable oils have been a main ingredient in the food industry, the majority of methods dealing with their oxidative stability have been created through the efforts of the food industry and, by association, the chemical industry. Oxidation stability methods used in the lubricants and grease industry are based on petroleum and its derivative oils and are not often able to determine the stability of vegetable oils. Bio-based lubricants researchers and developers have been relying on the use of standards created by the American Oil Chemists' Society (AOCS) as a well-modified version of standards created by the American Society for Testing and Materials (ASTM). Others at the University of Northern Iowa National Ag-Based Lubricants (NABL) Center have used hydraulic pump tests and field evolution to create reference materials for possible use with vegetable-based lubricants. Below are examples of some of the test methods used to determine oxidation stability of vegetable oils.

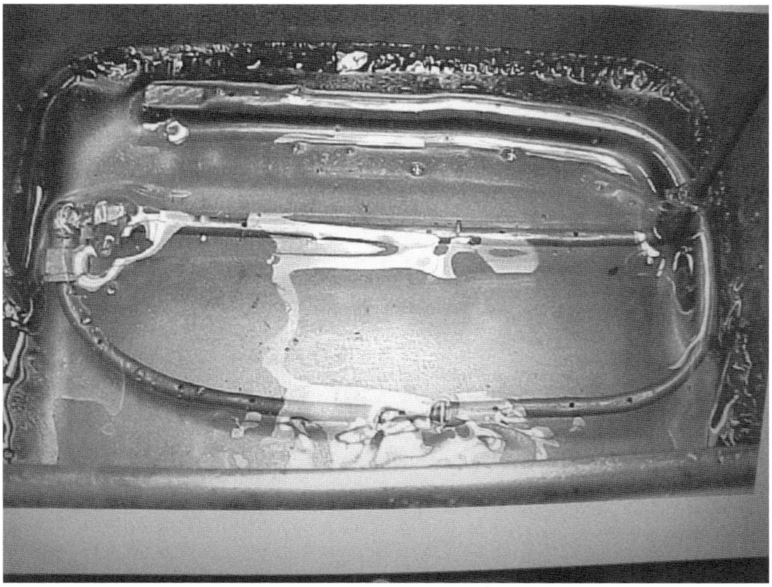

Fig. 8.8. Soybean oil oxidized using heat and air.

Active oxygen method (AOCS Cd 12-57)

In the active oxygen method (AOM), oxygen is bubbled through an oil or fat which is held at 36.6°C (97.8°F). Oil samples are withdrawn at regular intervals and the peroxide value (PV) is determined. The AOM is expressed in hours and is the length of time needed for the PV to reach a certain level. AOM is used as a specification for fats and oils. AOM hours tend to increase with the degree of saturation or hardness. This method largely has been replaced with the oxidative stability index (OSI).

Peroxide value (PV) (AOCS 8b-90)

This is a test for measuring oxidation in fresh oils and is highly sensitive to temperature. Peroxides are unstable radicals formed from triglycerides. A PV value over 2 is an indicator that the product has a high rancidity potential and could fail on the shelf.

Oxidative stability instrument (AOCS Cd 1 2b-92)

Like AOM, in the oxidative stability instrument (OSI), instead of pure oxygen, regular shop air is used and is simpler to operate than AOM. A conductivity probe monitors conductivity of deionized water as evaporatives from test oil are emitted into the deionized water (Fig. 8.9).

Oxidative stability index values are expressed in hours as the lower the number of hours, the lower the stability of the oil. For lubricants like hydraulic fluids that tend to reside in the system for hundreds or thousands of hours, a high OSI value for the base oil and a yet higher value for the formulated products will be needed. OSI values can be correlated with other oxidation tests. But, for vegetable oil-based lubricants, it is best to establish a relationship between field test results and the OSI values. Table 8.6 shows the OSI values for selected vegetable oils.

Rancimat

Rancimat works on the basis of the same principle of operation as the OSI. This procedure accelerates the oxidation process by exposing the sample to elevated temperatures while pumping air into it. It has been approved recently by the ASTM as a part of the biodiesel standard test ATM D6751 (Fig. 8.10).

Honary (1995) reported that using a hydraulic pump test would expose the vegetable oil to industrial conditions that were the duplicate of many standard test methods. For example, he reported that most untreated

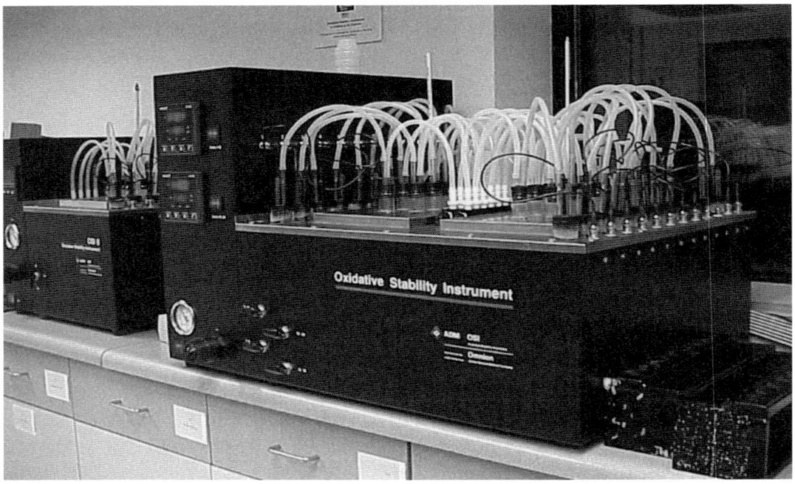

Fig. 8.9. Oxidation stability instrument.

Table 8.6. OSI values for selected vegetable oils.

Oil	OSI (hours)
Apricot kernel	23.42
Avocado	18.53
Babassu	57.8
Castor	105.13
Coconut	75.38
Maize	3.73
Cottonseed	4.35
Flaxseed	1.17
Grapeseed	2.83
Hempseed	0.10
Jojoba – refined	42.15
Jojoba – golden	38.3
Macadamia	6.87
Olive	5.08
Palm	21.52
Poppyseed	17.86
Ricebran	20.82
Ricinoleic acid	117.1
Safflower	17.98
Sesame	5.8
Soy	17.67
Sunflower	10.23
Walnut	16.48

vegetable oils meet ASTM wear protection requirements of hydraulic pumps because they naturally adhere better to metal surfaces, preventing boundary lubrication, but the oil also breaks down and oxidizes. In ASTM D2882, currently designated as ASTM D7043, the untreated crude soybean oil showed 40 mg wear, which was under the 50 mg passing level. The lack of oxidation stability, however, resulted in increased viscosity of the oil and, in extreme cases, led to polymerization.

As a result, in studying the performance of vegetable oil, vis-à-vis resistance to oxidation, some hydraulic pump tests could be used with the primary purpose of monitoring the changes in viscosity. Empirical research combined with field test observations indicate that an increase of less than 10% in viscosity for an oil tested in an ASTM D2271 (modified ASTM D7043) or equivalent is desirable. It is also observed that although the addition of additive packages to vegetable oils improves many of the oils' characteristics, it could impact its natural lubricity negatively.

For vegetable-based hydraulic fluid, tests that are longer in duration, such as ASTM D2271 (1000 h), are more desirable than shorter-term tests such as ASTM D2882 (100 h). Increased oleic acid combined with a reduced percentage of linolenic acid, as developed through new genes, could present improved oxidative stability, as well as a reduced cost of the base oil, due to elimination of the need for chemical modification such as hydrogenation.

Tables 8.7 and 8.8 show the change in viscosity for a number of vegetable oils when exposed to the hydraulic pump test conditions requiring 1000 h of exposure to 1000 psi at 79°C. In order to show the relationships between fatty acid make-up and stability, as

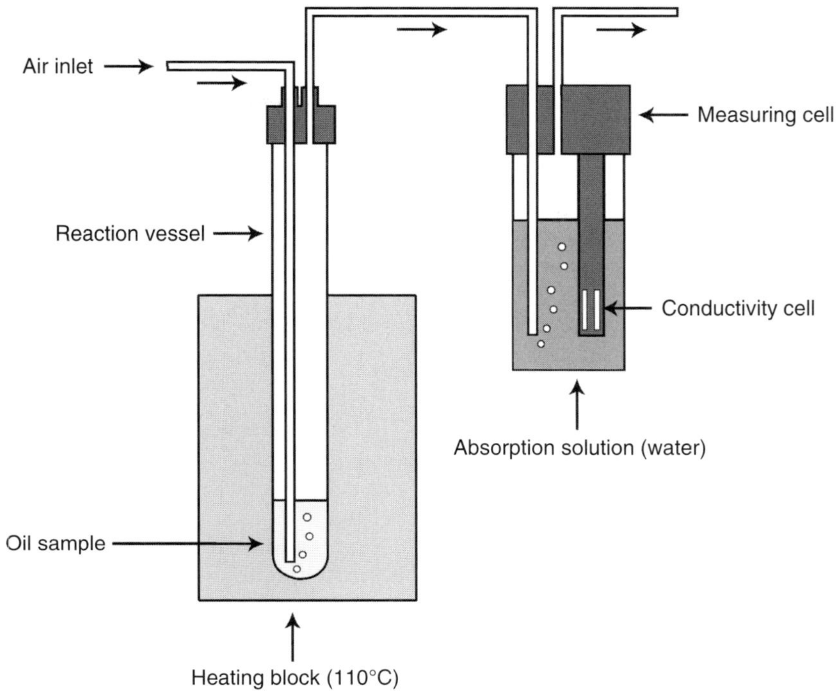

Fig. 8.10. Components of Rancimat with oil sample and conductivity probe in deionized water.

Table 8.7. Changes in the viscosity of oils as tested in ASTM D2271, 1000 h pump test.

Oil type	Oleic acid (%)	Δ Viscosity (cSt)
Crude soybean oil (hexane extracted)	23.4	43.86
Low linolenic soybean oil	32.9	20.57
Partially hydrogenated soybean oil	37.5	24.18
Crude soybean oil (mechanically extracted)	23.4	19.15
High oleic sunflower oil	78.2	19.24
Ultrahigh oleic sunflower oil	86.8	16.23
High oleic canola oil (Supplier 1)	76	14.08
High oleic canola oil (Supplier 2)	76.5	19.53
Palm oil	38.8	12.97
Meadowfoam	62.5	32.55

Table 8.8. Changes in the viscosity of formulated oils as tested in ASTM D2271, 1000 h pump test.

Oil type	Δ Viscosity (cSt)
Soybean oil-based tractor hydraulic fluid	4.60
Canola oil-based tractor hydraulic fluid	2.93
Rapeseed oil-based hydraulic fluid	37.2
Petroleum-based hydraulic fluid	1.5
50–50 Blend of soybean oil-based tractor hydraulic fluid and petroleum-based hydraulic fluid	1.6

shown by change in the viscosity, the per cent of oleic acid (18:1) for each oil is listed. From this table, it is shown that the higher oleic acid content improves oxidation stability, resulting in less of an increase in the viscosity of the oil. In cases where the oleic acid contents are the same, then the per cent of other fatty acids, especially linoleic (18:2) and linolenic acid (18:3), was considered the reason for the higher change in viscosity.

An example of the viscosity increase in the ASTM D2271 is shown in Fig. 8.11. Untreated high oleic soybean oil was tested and compared to the viscosity increase for

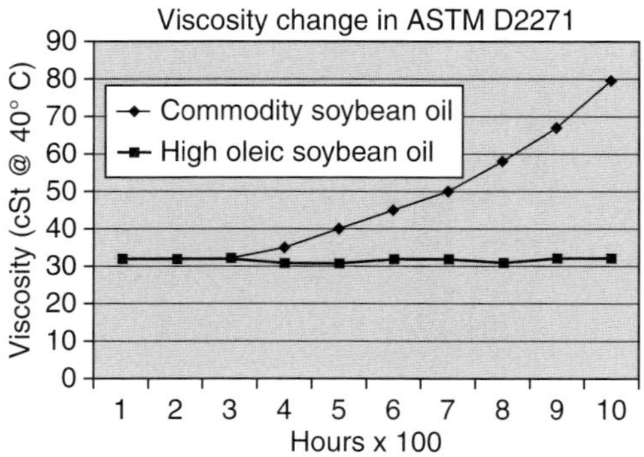

Fig. 8.11. Changes in viscosity for commodity and high oleic soybean oils in ASTM D2271.

conventional crude soybean oil. The viscosity increase for the commodity soybean oil was about 43.86 cSt or 146%, with the high oleic soybean oil showing negligible change in viscosity at about 1.2%. The high oleic soybean oil had an oleic acid content of 83% and a linolenic acid content of 1.75%.

Vegetable Oil Processing

Since different vegetable oils can be processed in nearly the same way, the processing of soybean oil, which is the largest seed oil produced, is described here.

According to Mounts (1985), extracting the oil from the oilseed requires three basic steps: (i) bean preparation, (ii) oil extraction and (iii) solvent stripping and reclamation. After processing, the oil may be degummed for use in food-related applications. Small on-the-farm extruder/expeller units are finding popularity among farmers wishing to extract the oil out of the seed; with the residual mean used for feed purposes. These units do not use the conventional oil-extraction techniques. Instead, the bean is forced through an extruder unit, which creates high pressure-induced temperatures. The shear and grinding processes result in the rupture of the cell walls. When the cell walls, particularly the oil cell walls, rupture, they release some of the natural tocopherols, which have antioxidant property, as well as some of the lecithin. Honary (1995) reported that crude soybean oil obtained through this process showed more oxidation stability using ASTM D2271 (hydraulic pump test) than the crude oil obtained through conventional hexane-extracted processing. This could be attributed to the retention of the natural antioxidants (including tocopherols) in the oil by the extrusion process. The oil is then extracted from the meal by the use of an expeller, which is essentially a mechanical press.

Refining

The crude oil contains a considerable amount of soluble and insoluble matters including gums. Refining is a purifying treatment designed to remove free fatty acids, phosphatides and gums, colouring matter, insoluble matter, settlings and other unsaponifiable materials from the triglycerides (Mounts, 1985). Accordingly, oils such as soybean oil are refined through chemical refining, caustic refining and physical or steam refining. Refining removes particles, such as free fatty acids, colouring and insoluble matters, phosphatides and so on, and is used as a purifying process. This prevents foaming, smoking and cloudiness when the oil is heated.

In the process of chemical refining, an alkali solution is used. The alkali combines with free fatty acids to form soaps that can be separated (process of saponification).

Phosphatides and gums absorb the alkali and are coagulated through hydration or degradation. Colouring is degraded, absorbed by gums, or made water-soluble by the alkali. The caustic process of refining involves analysing the amount of free fatty acids and neutral oil in order to determine the amount of caustic soda to be added. The formation of different density layers occurs as one layer containing most of the oil and the other layer most of the unwanted particles. These are separated using a centrifuge. Physical refining does not use caustic soda and does not have soapstock.

Treatment with phosphoric acids removes phosphatides. This oil can then be used in steam refining. Steam refining uses a stripping steam to remove odour from the oil, and it also removes free fatty acids.

Another related refining process is known as the Zenith process. This process removes non-fatty-acid substances with concentrated phosphoric acid. The acid treatment removes calcium and magnesium from gums.

Degumming

Gums are made up of phospholipids and non-triglycerides. Degumming involves the removal of phospholipids and other non-triglyceride materials. A by-product of degumming is lecithin, which can be used as an emulsifying agent. Treating crude soybean oil with caustic soda neutralizes free fatty acids, hydrolyses phosphatides and removes some coloured pigments and unsaponifiable matters.

Bleaching

Bleaching uses activated earth to absorb pigments, oxidation products, phosphatides, soaps and trace metals. This is an important step where natural clays or earths (Fuller's earth), which are basically hydrated aluminum silicates, are used for bleaching; or changing the colour of the oil to a neutral colour. The process is simple. The neutralized oil is mixed with the appropriate amount of clay, heated to the bleaching temperature and then filtered.

Deodorizing

Deodorization is a process of removing volatile substances and converting the oil into a bland-tasting, clear liquid (Brekke, 1985). Noticeable flavour and odour essentially will have disappeared when the free fatty acid content is lowered to 0.01–0.03% and the protein content is miniscule. The process of deodorizing involves heating, steam stripping and cooling the oil before exposing it to the atmosphere. The high temperature is needed to remove triglycerides by allowing the less volatile flavour and odour components present in the oil to evaporate. These substances must be volatilized to condense and subsequently be removed from the oil. Stripping steam is added to increase the rate of this process. The process removes the free fatty acids but the content cannot be reduced below 0.005% because of interaction with the stripping steam. Steam distillation also may result in the conversion of up to 25% of the linolenic acid present from the *cis* form to the *trans* form due to elevated temperature and time periods.

Reducing oxidation

The food industry has several methods to prevent oxidation. These methods can be applied to industrial lubricants uses as well and include: (i) exclusion of air during processing; (ii) cooling the deodorized oil before exposing it to the atmosphere; (iii) preventing exposure of oil to air by a cover of nitrogen; and (iv) addition of chemical antioxidants and metal excavengers (Brekke, 1985). Propyl gallate and tertiary butyl hydroquinone (TBHQ) may be added to increase oxidative stability and *t*-butylhydroxy toluene (BHT) or *t*-butyl hydroxyanisole (BHA) may be added to increase shelf life. These additives may be used in amounts of up to 0.01% singly or 0.02% in combination. The oil is then filtered after the deodorization process has been completed to remove any solids that have formed or been introduced. After the process is completed, the free fatty acid content should be less than 0.03% (gram weight) and peroxide values should be zero.

Brief Description of Selected Vegetable Oils

Soybean

Soybean is one of the largest crops grown in the world; and the USA is the leading producer of this crop. It is the world's most important oilseed. In addition to the USA, other countries producing soybeans include Brazil, Canada, Argentina, Paraguay, China, India, Indonesia, Korea, Thailand, Italy and

Romania. Important properties of soybean oil include:

Density	0.9075
Viscosity	30.52 at 40°C and 7.42 at 100°C
Viscosity index	224
Total acid number	0.10
Pour point	−9°C
Flashpoint	314°C
Gross heat of combustion (BTU/lb)	16,770

Salunkhe *et al.* (1992) documented that constituents impacting oxidative stability include more than 50% polyunsaturated fatty acids and about 15% saturated fatty acids, most of which is palmitic acid. The oil is usually hydrogenated to improve its oxidative stability. Optimized, hydrogenated soybean oils have been developed mainly for increased shelf life in the kitchen and stability during frying. The same technology can be applied for improving the oil's stability for industry uses such as in hydraulic fluids.

Palm oil

In addition to being a source of cooking oil, palm oil can be modified and used as fuel, lubricants and other non-food new uses such as palm ink. Oil palm is an important edible oil obtained from either palm fruit or from the palm kernel. Palm kernel is obtained as a minor product during the processing of oil palm fruit. Palm oil is solid at room temperature and has a neutral taste and can be extracted without the use of solvents, which meets standards for organic food processing. Honary (1995) reported that untreated palm olein tested in an ASTM D2271 hydraulic pump test showed high oxidative stability, as measured by changes in the oil viscosity. Palm oil contains some triglyceride species that are completely saturated. It consists of mostly: monoglycerides (48–55%) and diunsaturated glycerides (30–43%), with small quantities of saturated (6–8%) and unsaturated glycerides (6–8%).

Maiti *et al.* (1998) reported the fatty acid profile of palm oil to be myristic (0.516–6%), palmitic (35–40%), stearic (40–50%), oleic (40–50%) and linolenic (5–11%) acid contents. These values change for palm oil produced in different parts of the world and from various sources. Palm olein is the liquid fraction of palm oil and is used worldwide as cooking oil. Honary (1995) reported on the performance of an untreated palm olein as a hydraulic fluid with a supplier-reported fatty acid composition of lauric (0.2%), myristic (1.0%), palmitic (39.6%), stearic (4.6%), oleic (43.3%), linolenic (11%) and arachidic (0.3%). Accordingly, the oil contained 43.3% monounsaturates, 11.0% polyunsaturates and a high level of 45.7% saturates with 0% in trans acids. This type of oil showed excellent oxidative stability in a hydraulic pump test, but had the drawback of a very high melting point of about 23.9°C, being solid at room temperature.

Rapeseed

Typically grown in northern climates, rapeseed thrives well in a cool, moist climate and is grown extensively in northern Europe and Canada (Pryde, 1985). The zero-erucic variety contains 53% oleic acid and 11% linolenic acid. After soybean and palm, rapeseed is the largest produced vegetable oil in the world. Other countries growing rapeseed include India, China, Pakistan and Australia. Like soybean, the meal obtained after extraction of oil is used as animal feed.

Rapeseed refers to more than one plant species and is often used to denote the seeds derived from oil-yielding members of the Brassica family, including some mustard seeds grown for edible or industrial oil (Bengtsson *et al.*, 1972). *Brassica napus* and *B. campestris* are the two most important and widely grown species, with summer types grown in North America and a mixture of summer and winter types grown in Europe. *B. juncea* and *B. campestris* are grown in India and the Far East. Wild populations of *B. campestris* have been reported from different regions of Europe and Asia. *B. napus*, which is derived genetically from a natural hybridization of *B. campestris* and *B. oleracea*, occurs naturally in more restricted areas, mainly Europe and North Africa.

Rapeseeds, in general, contain about twice as much oil as soybeans, and the oil-free meal has only slightly less protein. Rapeseed oils exhibit a saponification value of 168–192 and an iodine value of 81–112. Vaisey and Klimaki (1985) found BHA/BHT with monoglyceride citrate to be ineffective against oxidation of canola oil. A polymeric antioxidant anoxomer effectively inhibited oxidative change in canola oil stored for 12 days at 65°C when added at levels of 2000–4000 ppm.

Sunflower oil

Sunflower oil has differing fatty acid composition if it is grown in northern as compared to southern climates. According to Erickson *et al.* (1985), oil from northern-grown seeds has a high linoleic acid content (64%) and low linolenic acid content (1%), whereas oil from southern-grown seeds contains a low linoleic acid content (49%) and a high oleic acid content (34 versus 21%).

Sunflower is known in two types: oilseed and non-oil or confectionary. The confectionary type is consumed as whole roasted seed and represents less than 10% of total sunflower production (Salunkhe *et al.*, 1992). The fatty acid composition of sunflower oil is primarily palmitic (7.2%), stearic (4.1%), oleic (16.2%) and a large portion of linoleic (72.5%) (Salunkhe *et al.*, 1992; Maiti *et al.*, 1998).

To extract the oil from sunflower seed, similar equipment and conditions as for soybean oil are used. After cleaning, drying and de-hulling, the oil is extracted by either mechanical extraction, prepress solvent extraction or by direct solvent extraction methods. Other processes such as bleaching, deodorizing and winterization are also used to prepare the oil for food uses. Sunflower oil is considered premium oil, as well as being one of the most palatable vegetable oils. Recent developments in genetic modification of the seeds have resulted in new high oleic and 'ultrahigh oleic' sunflower oils with high oxidative stability.

Nagley (1992) and Honary (1995) reported on the performance of sunflower oil for industrial applications. The physico-chemical characteristics of sunflower oil were as follows (Weiss, 1983):

Density at 60°C	0.894–0.899
Melting point (°C)	0
Smoke point (°C)	250*
Refractive index (at 25°C)	1.4597
Iodine value	128
Saponification value	191
Free fatty acid (%)	0.01–0.03*
Unsaponifiables (%)	0.3–0.5
AOM time (hours)	10–15*

*For refined, bleached, deodorized oils.

Maize

The oil content of the maize kernel, on a moisture-free basis, is about 5% as compared to soybean, which is about 20%. New genetic research has resulted in the development of the so-called high-oil maize, which contains about 8% oil. The high-oil maize, however, can be utilized directly for commercial oil extraction. The high-oil types are generally low yielders and need to be genetically improved for better agronomical performance. The oil is concentrated in the germ and is recovered both by wet milling, in the production of starch, and by dry milling, in the production of grits, meal and flour. More than 90% of the maize grain produced is processed and fed to animals in the Western world, whereas in Asia and Africa, almost all grain produced is used for human consumption by traditional processing without separating the germ (Salunkhe *et al.*, 1992).

Maize is a premium oil because of its high polyunsaturated fatty acid content and its low content (< 1 %) of linolenic acid. Leibovitz and Ruckenstein (1983) list the fatty acid composition of maize oil as follows:

Lauric acid	0.1%
Myristic acid	0.2%
Palmitic acid	11.8%
Palmiloleic	trace
Stearic acid	2.0%
Oleic acid	24.1%
Linoleic acid	61.9%
Linolenic acid	0.7%

The physico-chemical characteristics of maize oil are as follows:

Specific gravity	0.918–0.925
Density at 60°C	0.892–0.897
Titre (0°C)	18–20
Melting point (0°C)	0
Refractive index (at 25°C)	1.4596
Iodine value	103–133
Saponification value	187–195
Free fatty acids, as oleic (%)	0.03–4
Unsaponifiables (%)	1.2–2.8

Many other vegetable oils have been explored for industrial and hydraulic uses. Recently, due to environmental concerns, most grower groups in the USA have explored new uses of their vegetable oils. Dubois *et al.* (2007) have reported the fatty acid profile of 80 vegetable oils.

Industrial crops

The need for more crop-based oils for fuel and for lubricants has created an opportunity to explore the potential use of crops that are naturally grown in various parts of the world and are not necessarily edible. Research into the use of non-edible secondary crops has further intensified due to the food versus fuel controversy. As petroleum prices increased to over US$140/barrel in 2008, the increase in food prices was attributed to the growth of biofuels during the first half of the decade. As a result, considerable efforts have been made in identifying and qualifying natural oil crops with oil suitable for industrial and fuel uses. Especially when such crops are grown in arid lands, they offer an attractive alternative to both edible oils and petroleum.

In the USA, the Association for the Advancement of Industrial Crops has been leading exploration in the use of non-edible native crops that have oil and potential for use in biofuel and bio-based products.

Performance Properties of Industrial Lubricants

To develop bio-based lubricants and greases, in addition to the fatty acid profile of the oils, information on a number of physio-chemical properties is also essential in determining the needed property enhancements. Once such properties are known, by using chemical additives, the properties are enhanced so as to meet the requirements of the fully formulated final product. The physio-chemical properties to consider include:

1. Viscosity and viscosity index.
2. Flash and fire points.
3. Cold temperature performance:
  – pour point
  – Brookfield viscosity performance
  – storage at low temperatures.
4. Total acid number.
5. Iodine value.
6. Water content.
7. Rust and corrosion resistance:
  – humidity cabinet tests
  – copper corrosion.
8. Foaming characteristics.
9. Fluid compatibility:
  – compatibility with other fluids
  – elastomeric compatibility
  – compatibility with metallic components
  – homogeneity and stability of mixture
  – cleanliness.
10. Demulsibility.

Additionally, the oxidation stability of the oil would need to be evaluated using either:

11. Oxidation stability index; and/or
12. Viscosity stability in long-run pump tests.

After determining these properties, additives are used to enhance the performance using friction, wear and extreme pressure-testing instruments including:

– extreme pressure performance
  Timken OK
  4-ball extreme pressure
  pin and vee
  FZG rating (Forschungstelle fur Zahnrader und Getriebebau)
– anti-wear performance
  4-ball wear test
  pump tests.

Finally, once the products are formulated, an evaluation of the environmental performance of the final product will be conducted using any or all of the following test methods:

13. Biodegradability.
14. Plant toxicity.
15. Aquatic toxicity.
16. Per cent bio-based content.
17. Life cycle analysis according to methods developed by the National Institute of Standards and Technology (NIST).

Common performance requirements

Oxidation stability

For vegetable oil-based lubricants, it is most appropriate to use oxidation stability tests that are developed for vegetable oils. The common misconception that vegetable oils should also equal performance in tests that are developed for petroleum products could result in unduly eliminating some high-performance vegetable oils from evaluation. For example, thin film oxygen uptake (TFOUT) is commonly used for evaluation of base petroleum oils for engine oil formulation. TFOUT cannot differentiate between vegetable oils with different oxidation stability. Unpublished reports at the UNI-NABL Center has shown that vegetable oils of varied stability ranging in OSI from 7 to 500h, when tested in TFOUT, show low values of 17 and 18min. Similarly, petroleum oils will not perform the same way in the OSI as vegetable oils do, thus rendering the OSI useless for evaluating oxidation stability of petroleum oils. As recommended earlier, the test of viscosity over 1000h in the ASTM D2271, along with OSI or AOM values, could provide a better assessment of the stability of the oil for use in industrial lubricants.

Mineral oils oxidize when exposed to air and heat. As a rule of thumb, the rate of reaction of oil with oxygen increases by a factor of 2 for every 10°C. Nynas (2009) described the oxidation mechanism in three stages: (i) creation of free radicals by heat, ultraviolet light and/or mechanical shear; (ii) creation of peroxides by the reaction of the free radicals with oxygen; and (iii) the peroxide may further react and produce new radicals, alcohols, ketones, aldehydes and acids.

Viscosity

Defined as the (moving) fluid's resistance to flow, viscosity is normally tested using a kinematic viscosity method based on the ASTM D445, which uses velocity or the time it takes for a certain volume of oil, at a specified temperature, to pass through a standard capillary tube. Its unit of measurement is often expressed in centiStokes (cSt) or mm^2/s. Generally, the kinematic viscosity is tested at 40°C and at 100°C. The viscosity index (VI), which is *an indicator of the oil's resistance to change as temperature changes*, is then calculated using the values obtained at 40°C and 100°C, based on ASTM D2270. Vegetable oils have a higher VI than equivalent viscosity petroleum oils with the same viscosity. For example, soybean oil having a kinematic viscosity of 32 cSt at 40°C has a VI of 220, as compared to a mineral oil with the same viscosity having a VI of 95. This makes the viscosity of the vegetable oil more stable as the temperatures changes. Similarly, paraffinic oils have higher VI than naphthenic oils.

Dynamic viscosity may be defined as a measure of a (static) liquid's resistance to movement. Dynamic viscosity is Pose, often expressed in centiPose (cP). There is a relationship between kinematic and dynamic viscosity, using the same temperature (T) as follows:

Kinematic viscosity (cSt) T= dynamic (cP) at T/density in kg/dm^3 at T

Density

Density, measured in g/cm^3, is a measure of the mass of a material in a unit volume at 15°C. Measured according to ASTM D4502, density is affected by temperature. The higher the temperature, the lower the density, and the lower the temperature, the higher the density of the same material.

Flash and fire points

Using ASTM D56, the flash and fire points of vegetable oils are tested in the same manner as petroleum products. Vegetable oils in general have shown to have flash and fire points about 93.3°C higher than equivalent viscosity petroleum oils.

Boiling range

Since oils are not uniform in make-up and are made of different molecules with their own boiling points, a boiling range is established encompassing the boiling points of the various light or heavy constituents of the oil. There is a relationship between boiling range, molecular weight and viscosity. A higher molecular weight results in a higher viscosity and a higher boiling range.

Volatility

Volatility is associated with the flashpoint and is determined using the ASTM D972, which is designed to measure the loss in mass of a substance after 22h exposure to a standard temperature of 107°C. For lubricant applications, like metalworking fluids, lower volatility is desired to avoid the generation of mist and to reduce fire hazards.

Cold temperature properties

These properties are expressed in the form of cloud point and pour point. Pour point is measured using the ASTM D97 method and is the temperature at which the oil stops flowing. Cloud point is measured by determining the temperature at which the fats in vegetable oils or waxes in mineral oils crystallize. At the cloud point, the oil no longer follows the principles of Newtonian fluids; and becomes a multi-phase system of liquid and various crystallized or semi-crystallized matters. Naphthenic oils have lower pour points than paraffinic oils. Vegetable oils containing unsaturated fatty acids have lower pour points than those with a higher content of saturated fats. Table 8.9 (Nynas, 2009) shows the pour points of naphthenic and paraffinic oils.

Table 8.9. Pour points of naphthenic and paraffinic oils.

Viscosity in cSt at 40°C	Naphthenic pour point °C	Paraffinic pour point °C
30	–39	–39
110	–24	–3
400–500	–18	–5

Foaming properties

These are tested using ASTM D892. Typically, air is bubbled through sample oil which is at constant temperature in a bath, allowing the foam column to rise. After removing the air flow, the time it takes for the foam column to collapse is recorded. A fluid with good antifoam additive would not sustain the foam and the foam would collapse quickly.

Copper strip corrosion

The fluid is tested in an ASTM D130 copper corrosion test, which is rated by the changes in the colour of the test specimen, using reference colour charts.

Rust prevention

This is determined using ASTM D665, where humidity cabinets are employed to determine the rust protection of the sample oil in high humidity.

Neutralization number

Neutralization of vegetable oils is determined using ASTM D664.

Solubility

Solubility is an important property of base oils because often additives are used to enhance the natural properties of the base oil and to formulate products. A combination of the aniline point and viscosity gravity constant is used to determine the solubility of oils. The viscosity gravity constant (VGC) is determined using the ASTM D2501, which is a calculated dimensionless value. This is an indication of the solvency property of the fluid.

Aniline point

This is another property related to the solubility of the oil. Using the ASTM D611 method, it is determined as the lowest temperature at which an oil is fully miscible with an equal amount of aniline. The general description of aniline is C6H5NH2, which is a derivative of benzene and a colourless, oily liquid. The lower the aniline point, the better is solubility for that oil. In the ASTM D611, the temperature at which the aniline dissolves in the oil is considered the aniline point. Using a U tube, 10ml of the sample oil and 10ml of the aniline are heated until the aniline dissolves in the oil, and it is then cooled and temperature monitored until the aniline separates from the oil. This temperature is reported as the aniline point. The lower the aniline point, the higher the solubility of the oil.

Fluid quality

For some products like speciality hydraulic oils, the fluid is tested in ASTM D1500 and the numbers should match those of the manufacturer's recommendation.

Fluid compatibility

Some original equipment manufacturers (OEMs) have established fluid compatibility requirements for vegetable oils in anticipation of their mixing into petroleum-based oils. For example, Caterpillar has defined a BF-1 hydraulic fluid compatibility method to ensure that bio-based hydraulic oils could be mixed into petroleum-based hydraulic oils without negative interaction. Similarly, Deere and Company has a test of compatibility called JDQ 23, compatibility of lubricating oils.

Compatibility also refers to the compatibility of fluids with seals and rubber hoses, as well as propensity to staining metal such as copper and brass in metalworking applications.

Hydrolytic stability

Vegetable oils are prone to stability breakdown when water is introduced into the product during use. The ASTM D2619 hydrolytic stability of hydraulic fluids is often used to determine the stability of the oil in use where the presence of water from the atmosphere or the rest of the machine is possible.

Demulsibility

This refers to the ability of the oil to prevent water from mixing. Using ASTM D1401 (Standard Test Method for Water Separability of Petroleum Oils and Synthetic Fluids), the test sample should show separation of sufficient water having a volume of at least 37ml, observed before 20min has elapsed.

Toxicity

Bio-based fluids can be tested to determine if they present a health hazard when used as intended. Toxicity is measured according to the following:

1. *Aquatic toxicity.* A product's acute toxicity level of LC-50 is tested using ASTM D6081 'Toxicity studies – *Daphnia magna* for aquatic toxicity'.
2. *Toxicity, water hazard.* Acceptable product shall have a maximum rating of WGK 0.
3. *Toxicity, fish.* Acceptable bio-based products have acute fish toxicity level of LC-50 of greater than 1000PPM for amount measured in 48h.

Biodegradability

Products that have 60% biodegradability in 28 days, as per the OECD 301 *F*-test for biodegradability.

Biodegradable and Bio-based

Vegetable oils are in general considered biodegradable and meet the biodegradability standards of the ASTM and the Organization for Economic Cooperation and Development (OECD). Typical biodegradability results for base oils are shown in Table 8.10.

Table 8.10. Typical biodegradability results for base oils, Lubrizol Corporation Ohio, USA (Harold, 1993).

Mineral oil type	Amount biodegradable (%)
White mineral-based oil	25–45
Natural and vegetable oil	70–100
Polyalphaolefins (PAO)	5–30
Polyether	0–25
Polyisobutylene (PIB)	0–25
Phthalate and trimellitate esters	5–80
Polyols and diesters	55–100

In the USA, vegetable oil-based lubricants fall under the definition of bio-based, as defined by USDA. According to Honary (2007), discussions of bio-based materials must first draw the distinction between 'biodegradable' and 'bio-based'. Two widely used designations are 'readily' and 'inherently' biodegradable:

- According to the OECD pass levels for a readily biodegradable product, there must be 70% removal of dissolved organic carbon (DOC) and 60% removal of theoretical oxygen demand and theoretical carbon dioxide within 28 days (OECD 301 Guideline for Testing of Chemicals).
- Inherently biodegradable defines no timing or degree of biodegradability.

To avoid controversies of the term biodegradable, the USDA introduced the term 'bio-based', initially intended as a label for products comprised of 51% or more 'bio' materials. Later, USDA revised the approach to identify products based on their percentage of renewable materials. A broader definition of both 'bio-based' and 'biodegradable' has been provided by Honary (2007).

- Biobased: products containing natural renewable bio content like those made with agricultural materials (these products could be bio-based and biodegradable).
- Biodegradable: products that meet US (ASTM) or European (OECD) biodegradability requirements and could be made of biodegradable and or bio-based materials.

These definitions make a more clear distinction between the terms bio-based and biodegradable. It is essential to understand these terms, as they are not necessarily interchangeable. For example, there are some synthetic esters that are petroleum derived and are biodegradable (meets standards of biodegradability) but are not bio-based. In the case of lubricants and greases, they are tested for their carbon content to determine what percentage of their content comes from renewable carbon (bio) or what percentage of carbon comes from fossilized materials; this determines whether they are bio-based. USDA now relies on ASTM D6866-04 standards using radiocarbon and isotope ratio mass spectrometry analysis in order to determine the bio-based content of these items. These test methods look at content alone and do not address environmental impact, functionality or product performance.

Elemental analysis

In bio-based lubricants, often the list of the elements included in the product is as per the ASTM D5185 elemental analysis by induction plasma coupling (ICP) method. Elements identified include: Ag, Al, Ba, Ca, Cd, Cr, Fe, Mg, Mn, Mo, Na, Ni, P, Pb, Sn, Ti, V and Zn. Some specifications may not allow the use of certain metals like sulfur and zinc, which are considered undesirable, while they are commonly used for anti-wear and as extreme pressure property enhancers.

Cleanliness

Typically, the finished product, especially hydraulic fluids, is tested using particle counters to determine the amount of particles in sizes of 2, 5 and 15 μm. Using the International Standard Organization's method, an ISO cleanliness level using three numbers is specified. These numbers are indicative of the fluid's contamination level based on the size of the particles in microns. The tests are performed based on ASTM D1401.

Storage and shipping temperatures

Bio-based lubricant products are shipped with ambient temperatures in mind. When temperatures are within 10°C above its pour point, then heated truck and heated storage facilities will be needed.

Products intended to be stored more than 6 months may require a nitrogen barrier for quantities larger than 18.9 litres.

Future Prospects for Bio-based Lubricants

As the worldwide demand for petroleum grows, the global petroleum supply has become more constrained. Once more, renewable oil stocks are gaining favour as a potential solution throughout the increasingly industrialized world. Conventional oil crops, domesticated over many generations and produced in well-established and heavily mechanized operations, will no doubt play an important role. Widely known conventional crops will be the mainstay of this move to renewable oil stocks, especially in the earlier stages of transition.

As seen throughout the past decade, however, the increased demand for renewable oil stocks for biofuels production has also piqued the interest of both the research and finance communities in developing suitable alternative oil crops. As the growing human population places additional demands on food supply, non-edible oil crops will find more significance. Many indigenous, non-edible industrial oil crops can be grown efficiently in a variety of regions with local production technology. By producing these crops on marginal land, competition with the food supply is avoided.

Eventually, genetic modifications and transgenic technologies will provide non-edible oil crops with improved yields and enhanced performance for specific use as fuels and lubricants. Industrial crops hold the potential for decentralized lube and fuel production capabilities in an environment where the 'grow local, buy local' philosophy begins to apply, not only to food but also to lubes and fuels. In the future, industrial oil crops, with genetic improvements targeting varieties for specific oil properties and growing regions, will serve as a viable petroleum replacement, providing oil well suited to crop-based lubricants, chemicals, feedstocks for biopower and biofuels.

References

Bengtsson, L., Hofsten, A.V. and Loof, B. (1972) Botany of rapeseed. In: Appleqvist, L.A. and Ohlson, R. (eds) *Rapeseed: Cultivation, Composition, Processing and Utilization*. Elsevier, Amsterdam, pp. 36–44.

Bergstra, R. (2004) Green means go. *Lubes–n-Greases* 10, 36–42.

Bhatia, S.C. (1983) *Vegetable and Animal Oils*. Shree Publishing House, New Delhi.

Brekke, O.L. (1985) Deodorization. In: Erickson, D.R., Pryde, E.H., Brekke, O.L., Mounts, T.L. and Falb, R.A. (eds) *Handbook of Soybean Oil Processing and Utilization*. American Soybean Association, St Louis, Missouri, pp. 155–191.

Bringi, N.V. (1987) *Non-traditional Oilseeds and Oils of India*. Oxford and IBH, New Delhi.

Cermak, S.C. and Isbell, T.A. (2002) Physical properties of saturated estolides and their 2-ethylhexyl esters. *Industrial Crops and Products* 16, 119–127.

Cermak, S.C. and Isbell, T.A. (2004) Estolides – the next bio-based functional fluid. *Inform* 15, 515–517.

Dubois, V., Breton, S., Linder, M., Fanni, J. and Parmentier, M. (2007) Fatty acid profiles of 80 vegetable oils with regard to their nutritional potential. *European Journal of Lipid Science and Technology* 109, 710–732.

Erickson, D.R., Pryde, E.H., Brekke, O.L., Mounts, T.L. and Falb, R.A. (1985) *Handbook of Soybean Oil Processing and Utilization*. AOCS Press, Champaign, Illinois.

Harold, S. (1993) *Biodegradability: Review of the Current Situation*. The Lubrizol Corporation, Cleveland, Ohio.

Honary, L.A.T. (1995) Performance of Selected Vegetable Oils in ASTM Hydraulic Tests. SAE Technical Papers, Paper 952075.

Honary, L.A.T. (1999) Soybean-based hydraulic fluid, United States, 5972855 (http://www.freepatentsonline.com/5972855.html, accessed 12 July 2009).

Honary, L.A.T. (2007) Development and use of biobased and biodegradeable lubricants for the automotive industry. SAE 2007 Commercial vehicle engineering congress and exhibition (http://www.sae.org, 12 July 2009).

Leibovitz, Z. and Ruckenstein, C. (1983) Our experiences in processing corn (maize) germ oil. *Journal of American Oil Chemist Society* 60, 395–399.

Maiti, S., Hedge, M.R. and Chattopadhyay, S.B. (1998) *Handbook of Annual Oilseed Crops*. Oxford and IBH, New Delhi, pp. 279–289.
Meng, T. and Dresel, W. (2001) *Lubricants and Lubrication*. Wiley, Chichester, UK.
Mortier, R.M. and Orszulik, S.T. (1997) *Chemistry and Technology of Lubricants*. Blackie Academic and Professional Publishers, New York.
Mounts, T.L. (1985) Hydrogenation practices. In: Erickson, D.R., Pryde, E.H., Brekke, O.L., Mounts, T.L. and Falb, R.A. (eds) *Handbook of Soybean Oil Processing and Utilization*. AOCS Press, Champaign, Illinois, pp. 131–144.
Nagley, P.C. (1992) Environmentally acceptable lubricants. In: Mackenzie, S.L. and Taylor, D.C. (eds) *Seed Oils for Future*. AOCS Press, Champaign, Illinois, pp. 14–25.
Nynas, A.B. (2009) *Handbook of Naphthenic Specialty Oils for Greases*. Nynas Corporation, Sweden.
Patterson, H.B.W. (1994) *Hydrogenation of Fats and Oils: Theory and Practice*. AOCS Press, Champaign, Illinois.
Pryde, E.H. (1982) Non-food uses of vegetables oils. In: Wolff. I.A. (ed.) *Handbook of Processing and Utilization in Agriculture, Volume II Part 2. Plant Products*. CRC Press, Boca Raton, Florida, pp. 109–142.
Pryde, E.H. (1985) Physical properties of soybean oil. In: Erickson, D.R., Pryde, E.H., Brekke, O.L., Mounts, T.L. and Falb, R.A. (eds) *Handbook of Soybean Oil Processing and Utilization*. AOCS Press, Champaign, Illinois, pp. 33–47.
Randles, S.J., Robertson, A.J. and Cain, R.B. (1989) Environmentally friendly lubricants for the automotive and engineering industries. Presented at a Royal Society of Chemistry Seminar, York, UK.
Salunkhe, J.K., Chavan, J.K., Adsule, R.N. and Kadem, S.S. (1992) *World Oilseeds: Chemistry, Technology, and Utilization*. Van Nostrand Reinhold, New York.
Vaisey, M. and Klimaki, G. (1985) Effects of non-absorbable antioxidants on canola oil stability to accelerated storage and frying temperatures. *Canadian Institute of Food Science and Technology Journal* 18, 67–71.
Weiss, E.A. (1983) *Oilseed Crops*. Longman, London.

9

Improvement of Industrial Oil Crops

Denis J. Murphy

Introduction

Since the dawn of humanity, oil-bearing plants have been useful resources for both edible and non-edible applications. Well before the invention of agriculture more than ten millennia ago, oils from wild plant seeds and fruits were already being exploited by our hunter–gatherer ancestors for a wide range of non-food uses including lighting, fuels, lubricants and even some early forms of cosmetic applications – often in a ritual context. Some of our oldest crops that have been cultivated for over 7000 years include annual oilseeds like sesame, as well as perennial oil-rich fruits such as olive or oil palm (Davidson, 1999; Murphy, 2007b).

In 2008, the annual traded oil production from crop plants was almost 130 million t, the vast majority of which was used as edible vegetable oil (USDA, 2008). The proportion of plant oils used for non-edible or industrial purposes has fluctuated over the years, especially during the 20th century when petroleum and coal feedstocks competed increasingly with plant oils as sources of hydrocarbon-based products such as polymers, lubricants, fine chemicals and fuels. Industrial raw materials derived from fossil feedstocks such as oil or coal are generally termed petrochemicals, whereas those derived from plant sources are termed oleochemicals. Internationally traded vegetable oils are dominated by the 'big four' crops, which collectively make up over 87% of global production. The big four crops in order of annual oil production are oil palm, 35.6% (of which 31.8% is mesocarp oil and 3.8% kernel oil), soybean, 29.9%, rapeseed, 14.0%, and sunflower, 7.9%. Currently, only about 20% of global vegetable oil (25 million t) is used for non-fuel industrial applications as oleochemicals. Moreover, the recent increasing interest in supposedly carbon-neutral crop-derived biofuels has started to divert significant amounts of vegetable oil feedstocks away from food or industrial use and towards the large-scale production of biodiesel (Durrett *et al.*, 2008; Murphy, 2008).

Global biodiesel production reached 11 million t in 2008, with predicted continuing rises in the future. However, the rate of this rise started to slow down in 2008 as global economies entered a period of recession and the future of crop-based biodiesel as a growing part of oil crop utilization is now less certain. In the meantime, however, the biofuel phenomenon has created considerable uncertainty in vegetable oil markets, which had already been at near capacity in meeting the ever-increasing demand for edible oils from the growing developing country economies, especially the mega-economies of India and China (Murphy, 2007a). Government policies to subsidize biofuels have resulted in the diversion of edible grade oil from crops, such as rapeseed in Europe (often called canola in the Americas and Australia), where 90% of global biodiesel was consumed in 2008. A growing proportion of the Asian oil palm production, which is the major global source of both edible oil and many lauric-, oleic- and palmitic-based oleochemicals, has also been diverted to biodiesel. The consequence is that industrial oil users are now competing increasingly with edible and biodiesel users for plant-based feedstocks.

The segmentation of global vegetable oil markets between industrial, edible and biodiesel users will continue into the future, but the manner in which plant oil feedstocks are divided up between these three product categories is likely to be both volatile and unpredictable. Indeed, the severe impact of biofuel crops on food prices has led recently to a backlash that will probably see a reduced impetus for crop-based biofuels in the coming years (Gallagher, 2008). Already, in 2008, the EU reduced its ambitious targets for biofuels to replace 10% of fossil fuel use by 2020 (EurActive, 2008). A second factor that should reduce demand for crop-based biodiesel is the increasing research and development work on so-called second-generation biofuels that are not derived from existing food or industrial crops. For example, there is great interest in developing high oil-producing microalgae, grown on non-cropland such as shallow ponds in desert regions, as sources of biodiesel. These microalgal production systems would not compete directly with industrial and edible uses of existing oil crops, possibly relieving pressure on these markets and helping to stabilize prices (Hu *et al.*, 2008; Murphy, 2008).

Although the very long-term future of industrial oil crops looks secure, the many short-term uncertainties discussed above make it difficult to predict the immediate prospects for the industrial use of oil crop feedstocks. Continuing price volatility and economic uncertainty have the potential to act as brakes on the development of this sector, although this may be mitigated to some extent by government policies such as carbon taxes or subsidies for the industrial use of these renewable industrial feedstocks. Ironically, initial attempts by governments in North America and Europe to offset the use of fossil fuels by encouraging biofuel production from crops have distorted markets, driven up prices and potentially could threaten supplies of oleochemical feedstocks. One way forward is to use advanced breeding technologies to customize oil-rich plants so that they produce more valuable oleochemical intermediates for industry. This is in contrast to the present situation, where industrial oils tend to be derived from rather basic generic commodity feedstocks that normally require considerable downstream processing. For example, crops could be engineered to accumulate oils containing 90%+ of a specific fatty acid derivative, such as an epoxy or hydroxy compound, that is tailored for use in a particular set of markets. Similar breeding technologies are also being developed to increase crop oil yields, hence driving down costs and competing more favourably with petroleum-based products. In the remainder of this chapter, a brief overview of recent progress in the improvement of industrial oil crops will be presented.

Diversity of Fatty Acid Composition

It is perhaps ironic that while commercially grown oil crops have relatively restricted fatty acid compositions, non-crop oilseeds exhibit a wide range of acyl species, many of which have great industrial potential. Plant storage lipids, normally in the form of triacylglycerol oils, typically accumulate in the cotyledon or endosperm tissues of seeds. In major oilseed crops, such as rapeseed/canola or sunflower, storage lipids can constitute 40–50% of total seed weight. In plants with large oil-rich seeds, such as nuts, storage lipids can account for as much as 75% of total seed weight. Storage lipids are also found in abundance in certain oil-rich fruits, such as in the mesocarp tissue of olives, avocado, coconut and oil palm. The majority of commercially grown oil crops accumulate predominantly C-16 or C-18 saturated and unsaturated fatty acids in their storage lipids, and hence have relatively restricted acyl compositions. The major fatty acids comprising the world vegetable oil supply are therefore palmitic, oleic and linoleic acids (Table 9.1).

However, surveys of wild plant species have revealed an astonishing natural diversity of seed oils, many of which have potential value, either for nutritional or industrial use (Harwood *et al.*, 2007). Some measure of this diversity can be seen in Table 9.2, which shows that different plant species can accumulate high levels of a vast

Table 9.1. Percentage fatty acid composition of the 'big four' commodity oil crops. Note the predominance of palmitic, oleic and linoleic acids.

Fatty acid[a]	Oil palm[b]	Soybean	Rapeseed/canola	Sunflower
Palmitic 16:0	45	11	5	6
Stearic 18:0	5	4	1	5
Oleic 18:1	38	22	61	20
Linoleic 18:2	11	53	22	69
Linolenic 18:3	0.2	8	10	0.1

[a]Fatty acids are denoted by their carbon chain length followed by the number of double bonds.
[b]Mesocarp oil.

Table 9.2. Diversity of fatty acyl composition in selected plant storage oils.

Fatty acid				
Chain length/functionality	Common name	FA in oil (%)	Plant species	Uses
8:0[a]	Caprylic	94	*Cuphea avigera*	Fuel, foods
10:0	Capric	95	*Cuphea koehneana*	Detergents, foods
12:0	Lauric	94	Betel nut laurel	Detergents, foods
14:0	Myristic	92	*Knema globularia*	Soaps, cosmetics
16:0	Palmitic	75	Chinese tallow	Foods, soaps
18:0	Stearic	65	Kokum	Foods, confectionery
16:1 9c	Palmitoleic	40	Sea buckthorn	Cosmetics
18:1 9c	Oleic	78	Olive[b]	Foods, lubricants, inks
18:1 6c	Petroselinic	76	Coriander[b]	Nylons, detergents
18:1 9,10me, 9c Cyclopropene	Sterculic	50	*Sterculia foetida*	Insecticides, herbicides
18:2 9c, 12c	Linoleic	75	Sunflower	Foods, coatings
18:3 9c, 12c, 15c	α-Linolenic	60	Linseed[b]	Paints, varnishes
18:3 6c, 9c, 12c	γ-Linolenic	25	Borage[b]	Therapeutic products
18:1 9c, 12OH	Ricinoleic	90	Castor[b]	Plasticizers, lubricants
18:1 9c, 12epx	Vernolic	70	Ironweed	Resins, coatings
18:2 9c, 12trp	Crepenynic	70	*Crepis alpina*[b]	Coatings, lubricants
18:3 4-oxo, 9c, 11t, 13t	Licanic	78	Oiticica	Paints, inks
18:3 8t, 10t, 12c	Calendic	60	Calendula	Paints, coatings
18:3 9c, 11t, 13t	α-Eleostearic	70	Tung	Enamels, varnishes, resins, coatings
18:3 9c, 11t, 13c	Punicic	70	Pomegranate	Varnishes, resins, coatings
20:0	Arachidic	35	Rambutan[b]	Lubricants
20:1 11c	Eicosenoic	67	Meadowfoam[b]	Polymers, cosmetics
20:1 11c, 14OH	Lesquerolic	70	Lesquerella	Lubricating greases, resins, waxes, nylons, coatings, cosmetics
22:0	Behenic	48	Asian mustard	Lubricants
22:1 13c	Erucic	56	Crambe[b]	Polymers, inks
Wax 20:1/22:1	Jojoba wax	95	Jojoba[b]	Cosmetics, lubricants
24:1 15c	Nervonic	24	Honesty[b]	Pharmaceuticals

[a]c, *cis* double bond; t, *trans* double bond; epx, epoxy group; trp, triple bond; me, methylene.
[b]Indicates that genes have been isolated for synthesis of these novel fatty acids.

range of fatty acids with chain lengths from C-8 to C-24. In addition, numerous functionalities can occur in these fatty acids, including conjugated and non-conjugated double bonds, triple bonds, hydroxy, epoxy and oxo groups. Cyclopropenoids, cyclopentenoids and even fluoro fatty acids add even more diversity to the list of naturally occurring fatty acid species (Murphy, 1994). A more detailed list of over 100 oil-bearing plants with novel and potentially useful fatty acid profiles can be found in Gunstone and Harwood (2007).

The extreme diversity of naturally occurring storage lipid profiles in seed and fruit storage tissues, sometimes with dramatic variations in fatty acid composition even between genetic variants of the same species, indicates that the acyl compositions of these oils are not crucial to their major role, which is to act as energy and/or carbon stores. In contrast, it is well known that lipids involved in membrane structure or cell signalling processes have much more limited fatty acid compositions. This is because their acyl groups play vital roles in membrane functions such as the maintenance of fluidity and the activity of signalling and transporter proteins. Therefore, mutations that significantly alter the acyl composition of the major membrane (e.g. phospho- or galacto-lipids) or signalling (e.g. oxylipins) lipids are normally strongly selected against. But, in contrast, there is often little or no reduction in plant fitness when storage lipid profiles are altered quite drastically.

The properties, and hence the potential end-uses of a given plant oil, depend largely on its fatty acid constituents. Therefore, the wide range of naturally occurring fatty acids gives rise to an equally broad spectrum of potential end-uses. From the 1960s, the regional research division of the United States Department of Agriculture (USDA) at Peoria, Illinois, has been undertaking a survey of some of the amazing diversity in acyl lipid composition of oils from plants collected from around the world. From this and other studies, it has been found that there are many hundreds, and probably thousands, of plants, which are currently not grown as crops, but which have oil-rich seeds that accumulate novel and potentially industrially useful fatty acids and other types of non-acyl lipids. The USDA Peoria work can be accessed online at http://www.ncaur.usda.gov/nc/ncdb/search.html-ssi. As shown here in Table 9.2, naturally occurring plant oils have already been found with acyl chain lengths from C-8 to C-24. However, it is noteworthy that less than 0.1% of known plant species are grown currently in organized agriculture and fewer than 10% of known plant species have ever been surveyed for their oil content. This means that wild oil-bearing plants represent a scarcely tapped reservoir of useful genetic variation for lipid composition. There are at least 500,000 species of higher plant and many of these accumulate storage oils either in their seeds and/or fruits. More information on the full range of naturally occurring novel fatty acids is available in Lühs and Friedt (1994) or Gunstone and Harwood (2007). Given that we have surveyed only a fraction of oil-bearing plant species to date, it is also virtually certain that there is an even greater natural range of oils in plants than is reflected in Table 9.2. In addition to triacylglycerol-based oils, many plants accumulate other lipidic or non-aqueous compounds, including waxes and complex polymers such as rubber, gums and resins.

Designer Oil Crops

Many of the enzymes that generate 'unusual' fatty acids are mutated versions of enzymes such as desaturases or elongases that give rise to 'normal' membrane-located fatty acids. For example, the hydroxylases, conjugases, acetylenases and epoxidases (responsible, respectively, for the formation of hydroxy, conjugated, acetylenic and epoxy fatty acids) found in plants such as castor bean, tung, calendula and *Crepis* spp. are all variants of the familiar oleate desaturases that normally give rise to linoleic acid (Murphy and Piffanelli, 1998; Jaworski and Cahoon, 2003). The lack of any non-storage function of many seed and fruit lipids means that breeders can, in principle, manipulate their composition virtually at will. This has led to the concept of 'designer oil crops', namely transgenic varieties able to produce high yields of any one of the hundreds of fatty acid derivatives found in nature (Murphy, 1991, 1994). As we will see below, researchers have found that, in practice, the situation is more complex and a major challenge now facing breeders is to enable plants to accumulate exotic fatty acids in their storage oils without jeopardizing the overall patterns of crop growth and development.

In recent years, much of the focus on oil crop manipulation has involved the isolation of genes that regulate the formation of sometimes exotic, industrially useful fatty acids and transferring such genes to mainstream oil crops like soybean or rapeseed/canola. However, as we shall see below, this genetic engineering approach has not been without its problems. Quite apart from the technical problems of persuading an existing crop to accumulate novel fatty acids in the right place and in the appropriate quantity, there is the problem of managing the cultivation and processing of the new crop variety alongside conventional varieties of the same crop. For example, a soybean seed that contains regular edible soybean oil will look exactly the same as one that has been bred or genetically engineered to contain a non-edible and possibly toxic (to animals or people) industrial oil. This means that these seemingly identical (to the naked eye) varieties of mainstream oil crop would require strict segregation, both on-farm and throughout their downstream processing to finished products, to avoid the wasteful and potentially dangerous cross-contamination of one variety with another.

The requirement for this sort of segregation inevitably adds to the cost and complexity of oil production. However, this requirement for segregation need not be a 'deal breaker' for growers or processors, and nor is it necessarily fraught with risks for the food chain, as long as it is well managed. Already in several regions of Europe, high-oleic, edible-grade rapeseed/canola has been grown alongside identical-looking, high-erucic, industrial-grade versions of the same crop for over 40 years without any significant incidents of contamination. Here, segregation is achieved by imposing minimum separation distances between the two crop varieties, by good standards of farmer and processor management and by rigorous testing of seed batches at regular intervals after harvest.

Domestication of New Crops

An alternative strategy to transgenic 'designer oil crops' that is now beginning to receive more attention is the domestication of the original plants that made the exotic fatty acids in the first place, so that they can be grown as commercially viable crops in their own right. This involves domesticating entirely new oil crops that may have different cultivation requirements to existing crops. Previously, the prospect of a lengthy domestication process and the potentially costly development of new cultivation, harvesting and processing techniques was sufficiently daunting that only a few new industrial crops were ever developed. Thanks to the pioneering work of several US and European breeders over several decades (Jones and Wolff, 1960; Princen and Rothfus, 1984), several new industrial oil crops such as meadowfoam and *Cuphea* spp. have been developed, although production has always been on a relatively small scale compared to the major commodity oil crops (Knapp, 2003).

Meadowfoam seeds contain only 20–30% oil, which is about half the amount found in many established oilseed crops such as rapeseed or sunflower. However, this oil is uniquely enriched in long-chain C-20 and C-22 monounsaturated acids that make it the most oxidatively stable vegetable oil ever described (Eharan and Adhvaryu, 2005). Meadowfoam oil has many industrial uses, such as in lubricants, plasticizers, liquid or solid waxes and as estolides or silicone esters in coatings and adhesives. This exceptionally stable oil will also enhance the performance and stability of other cheaper oils when mixed with them, without loss of qualities of either oil. Most members of the genus *Cuphea* contain 30–36% oil and many species accumulate large amounts of short- or medium-chain acids, from C-8 to C-14 (see Table 9.2). These oils can be used to produce detergents, cosmetics, emulsifiers and solvents, and can be used in motor oils. Future prospects for newly domesticated oil crops may be considerably better, thanks to the use of modern advanced breeding technologies such as DNA-based marker-assisted selection and genomic profiling.

One of the main advantages of domesticating non-crop plants rather than developing transgenic versions of existing crops is that these non-crop plants are already adapted to accumulate their exotic fatty acids in the appropriate cellular compartment, namely in their storage oil bodies, without leakage into other lipid pools. In the native plants, these exotic fatty acids are hardly ever found in cell membranes or any other lipids where their presence could be damaging. In contrast, and as discussed in more detail below, it has been found that in transgenic oilseeds engineered to produce novel fatty acids, the latter can sometimes find their way into membrane lipid pools, with potentially deleterious consequences for the growth and development of the plant (Murphy, 1999; Thelen and Ohlrogge, 2002). A further advantage of novel oilseed crops is that the seed oils will already contain accessory stabilizing agents, such as antioxidants, which prevent the breakdown of some of the more highly reactive fatty acids such as conjugated polyunsaturates and those containing acetylenic bonds.

Although many potential new oil crops are excellent sources of useful products, such as novel fatty acids, they are frequently unsuitable for large-scale agriculture. This is because, in contrast with some of our more familiar crops, the new plants have not been optimized for agronomic performance. Hence, they tend to suffer from the usual characteristics of wild plants. For example, they tend to flower asynchronously throughout the summer and therefore do not produce their seeds at a single time, which makes harvesting very difficult. They often produce seedpods that are prone to shatter before or during harvest, resulting in a loss of many of the seeds. Often, the canopy architecture of the plant is not suitable for existing harvesting machinery. They may be susceptible to a variety of diseases or pests, including fungi and insects. Finally, in the case of oilseeds, although they may contain as much as 90% of a novel fatty acid in their seed oil, the overall oil yield may be relatively low. The improvement of these important agronomic characters requires the manipulation of numerous complex traits. In the past, researchers and companies may have been dismayed by the prospect of domesticating new species, citing the example of major crops such as wheat and rice, which are still being improved after more than 10,000 years of domestication. Nevertheless, we can now be more optimistic about the prospects for crop domestication (Murphy, 2007c).

Many of our newer commercial crops have been improved at a much more rapid rate than wheat over the past 50 years using modern breeding techniques. Examples of such crops include soybean, hybrid maize, rapeseed/canola and sunflower, which have only been grown as mainstream commercial crops for a century or less. Recent progress in identifying genes that may regulate important agronomic characters such as oil yield, flowering, height, disease resistance and canopy architecture now makes it feasible to consider the domestication of completely new species for agricultural use within a few decades (Murphy, 2007c). Therefore, in the future, it is likely that we will rely increasingly on a mixture of conventional, transgenic and newly domesticated oil crops for both industrial and edible products.

Biosynthesis of Potential Industrial Products

In order to manipulate fatty acid profiles in oil crops, it is crucial to understand the often-complex metabolic and regulatory mechanisms that underlie triacylglycerol accumulation in storage tissues. As biochemists and molecular geneticists have learned more about triacylglycerol accumulation over recent years, many of our previous ideas about the process have required revision and, in virtually all cases, triacylglycerol metabolism has turned out to be considerably more complex than originally thought. As shown in Fig. 9.1, the first committed step of fatty acid biosynthesis is the carboxylation of acetyl-CoA (Harwood, 2007). One of the aims of those seeking to produce commercially viable industrial oil crops is to increase overall storage oil yields, and this initial enzymatic step in fatty acid formation was an early target for biotechnologists. Although acetyl-CoA carboxylase is clearly an important rate-limiting step in fatty acid biosynthesis, and hence storage oil accumulation, efforts to increase oil yields in crops simply by overexpressing this enzyme have so far proved unsuccessful. In part, this failure may be due to the complex regulation of this key branch-point enzyme by mechanisms such as post-translational modifications and substrate/product activation/inhibition (Harwood, 2005).

Fatty acids in plants are synthesized *de novo* in the plastids. Acetyl-CoA is carboxylated to malonyl-CoA, which serves as the building block for the assembly of fatty acids on the multi-enzyme fatty acid synthetase complex. This enzyme complex can produce fatty acids ranging in length from C-8 to C-18 in plastids. Plastids are also the site of the initial desaturation reactions, which convert saturated fatty acids to monounsaturates. The C-8 to C-18 saturated and monounsaturated fatty acids are then exported from the plastids as acyl-CoAs for further modification on the endoplasmic reticulum. The endoplasmic reticulum is the site for the various fatty acid modification reactions found in different plant species that are responsible for the enormous diversity in oil composition shown in Table 9.2. These include acyl-chain elongation from C-18 up to C-24 and the numerous additional desaturase and desaturase-like reactions that can result in the formation of conjugated and non-conjugated polyunsaturated fatty acids, as well as their various hydroxy, epoxy and cyclopropene, etc., derivatives.

Storage lipids in plants typically accumulate as fluid cytosolic triacylglycerol or wax ester droplets. In desiccation-tolerant seeds, storage oil droplets are usually in the form of small oil bodies of about 1 μm diameter that are surrounded by a phospholipid monolayer and an additional annulus of specific proteins including oleosins and caleosins. Storage oil bodies in oil-rich fruits, such as avocado, olives or oil palm, tend to be larger than those in seeds and the annular proteins are less abundant (Leprince *et al.*, 1998). The final assembly of these complex macromolecular structures occurs on the endoplasmic reticulum. Acyl-CoAs produced by the various fatty acid modification enzymes described above are transferred to a glycerol 3-phosphate backbone, resulting in the progressive formation of mono-, di- and triacylglycerol derivatives. These three acylation reactions together comprise the Kennedy pathway and are catalysed by different acyltransferases. The differing specificities of the acyltransferases for different acyl-CoAs mean that these enzymes play an important role in regulating the final acyl composition of the storage lipids. Because the synthesis of storage triacylglycerols and membrane phospho- and galactolipids both occur on the endoplasmic reticulum, it has been proposed that these two processes must be segregated into discrete domains in order to avoid 'leakage' of acyl groups from one pool to another (Murphy, 2001). Recent ultrastructural evidence from plants and other organisms is consistent with this view (Man *et al.*, 2006; Robenek *et al.*, 2006; Shockey *et al.*, 2006).

In addition to the straightforward Kennedy pathway, there are several other routes for acyl exchange and acyl modification during triacylglycerol assembly (Fig. 9.2). The eventual acyl composition of storage triacylglycerols appears to be determined by a complex interaction of the various acyltransferases, acyl exchange and acyl modification enzymes. This situation has been described correctly as a 'quantitative trait', i.e. a genetic character determined by several (sometimes many) genes (Voelker *et al.*, 1996). Unfortunately, for efforts to modify oil composition by genetic engineering, quantitative traits are much more difficult, time-consuming and expensive to alter than simple monogenic traits. This is a major factor behind the slow pace of progress in producing GM oil crops compared to the sometimes optimistic projections from researchers (Murphy, 1991, 1992).

The fundamental mechanism of storage lipid assembly on the endoplasmic reticulum in plants is similar in many respects to that in animals. This is useful for researchers, but recent results have also highlighted the ever-increasing complexities surrounding the roles of storage lipids,

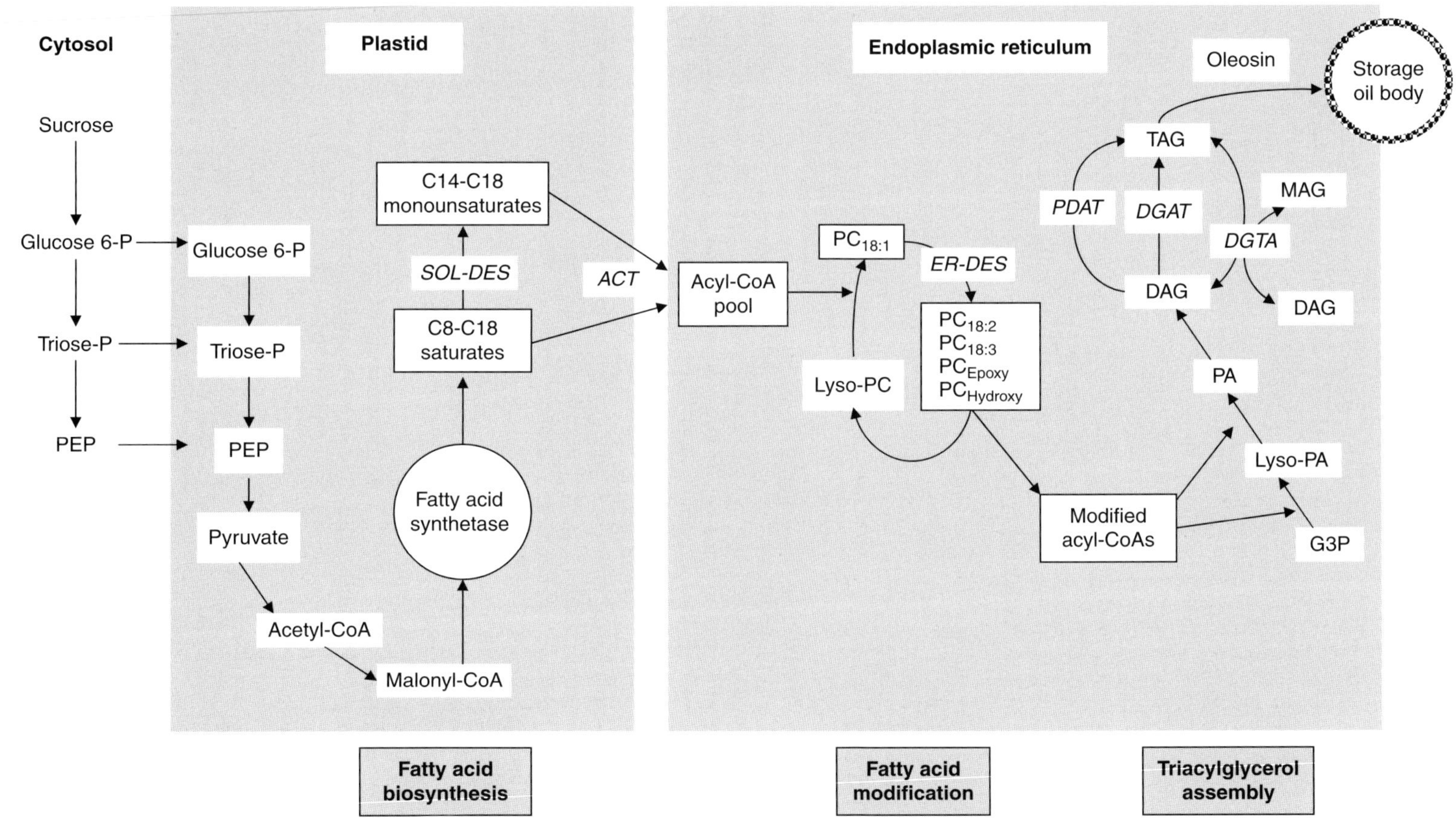

Fig. 9.1. Biosynthesis of storage lipids in plants (names of key enzymes are shown in italics). *Fatty acid biosynthesis de novo*. Sucrose is transported from

photosynthetic tissues into developing seeds, where it is converted in the cytosol of embryo and/or endosperm cells into precursors, such as glucose 6-phosphate and phosphoenolpyruvate, for onward transport into plastids for production of fatty acids. Acetyl-CoA and malonyl-CoA are the precursors for assembly of C-8 to C-18 saturated fatty acyl-ACPs on a plastidial multi-enzyme fatty acid synthetase complex. Plastids are also the site of the insertion of the first double bond by a soluble desaturase (*SOL-DES*) to produce fatty acid monounsaturates. Both unsaturates and monounsaturates are exported via an acyl-CoA transporter (*ACT*) from plastids to the endoplasmic reticulum (ER) for further processing. *Fatty acid modification.* Plastid-derived acyl-CoAs can be modified in the endoplasmic reticulum by a huge variety of enzymes to produce some of the hundreds of different fatty acids found in naturally occurring seed oils. However, since not all of these enzymes are present in any given plant species, non-transgenic oilseeds normally accumulate a relatively restricted range of fatty acids. Most fatty acid modification reactions occur via membrane-bound phosphatidylcholine (PC)-specific ER desaturases or desaturase-like enzymes (*ER-DES*) such as hydroxylases or epoxidases. Acyl-CoAs are then assembled into complex lipids on the ER. Similar ER-located pathways produce the various membrane lipids, storage lipids and also some signalling lipids, although recent evidence suggests that these pathways are spatially separated in discrete ER domains. Whereas storage oil bodies can accumulate virtually any type of fatty acid, the biological functions of membrane and signalling lipids require that they contain only a small range of C-16 and C-18 fatty acids. One of the challenges to producing oilseeds with novel acyl compositions is therefore to maintain the segregation of exotic fatty acids away from pools of membrane or signalling lipids. *Assembly of triacylglycerols.* Triacylglycerols are assembled via a complex process involving sequential acylation of a glycerol moiety (the traditional Kennedy pathway) plus extensive acyl editing via phosphatidylcholine-dependent desaturases or desaturase-like enzymes. The final conversion of diacylglycerol (DAG) into triacylglycerol (TAG) can occur via at least three enzymes: *DGAT*, acyl-CoA-dependent diacylglycerol acyltransferase; *PDAT*, phosphatidylcholine-dependent acyltransferase, or *DGTA*, diacylglyceroltransacylase. Nascent TAG droplets are coated with a phospholipid monolayer into which is embedded an annulus of specific proteins, such as oleosins and caleosins, hence forming the mature storage oil bodies that are released finally into the cytosol. G3P, glycerol 3-phosphate; MAG, monoacylglycerol; PA, phosphatidic acid.

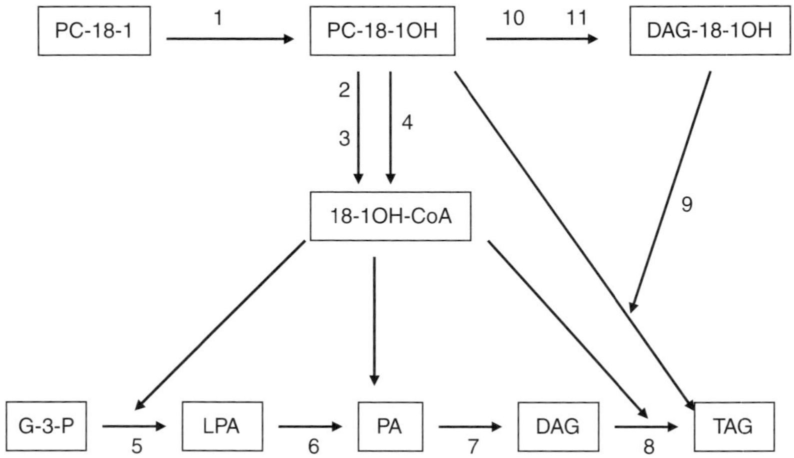

Fig. 9.2. Schematic diagram of storage triacylglycerol assembly. In this example, the formation of triacylglycerols containing ricinoleate groups is depicted. Ricinoleic acid is synthesized initially in the endoplasmic reticulum from oleoyl moieties esterified to phospholipids, primarily phosphatidylcholine (PC). Ricinoleate is then removed from the phospholipid pool and channelled into triacylglycerols through any of several coenzyme A (CoA)-dependent and/or CoA-independent acyltransferase reactions (adapted from Burgal *et al.*, 2008).

1. PC-dependent oleoyl hydroxylase.
2. Phospholipase A2.
3. Long-chain acyl-CoA synthetase.
4. Lysophosphatidylcholine acyltransferase.
5. Glycerol 3-phosphate acyltransferase.
6. Lysophosphatidic acid acyltransferase.
7. Phosphatidic acid phosphatase.
8. Diacylglycerol transacylase.
9. Phosphatidylcholine-dependent acyltransferase.
10. Phospholipase C.
11. CDP-choline:diacylglycerol cholinephosphotransferase.

and especially triacylglycerols, in all organisms. Following a great deal of research over the past few years, it has become apparent that storage lipids in all multicellular organisms, including plants, are far from being the inert end-products of metabolism that was once thought (Murphy, 2001). For example, leaves and meristematic tissues contain triacylglycerol-rich oil bodies that have roles in both long-term storage and in more short-term dynamic metabolic processes (Lersten *et al.*, 2006). These unexpected findings might complicate attempts to modify storage oils; for example, if some of the modified oils turn out to have unpredicted pleiotropic effects on leaf or meristem development. Once again, more research is required to elucidate the many roles of triacylglycerols in plants and the consequences of their manipulation by humans.

Breeding for Industrial Oils

Conventional breeding

During the past 50 years, plant breeders have achieved an impressive record of manipulating the storage lipid content of many of the major crop plants using conventional, non-transgenic approaches. Ironically, some of the most dramatic successes have involved improving the edible quality of crops that were hitherto used mainly for industrial applications. Two well-known examples are rapeseed and linseed. The modification of these two oilseed crops will be discussed here to illustrate the power of non-transgenic breeding for the modification of fatty acid profiles. Rapeseed/canola is now the third most important global oilseed crop. Like other members of the Brassicaceae, rapeseed normally accumulates a seed oil rich in the C-22 monounsaturate, erucic acid. This fatty acid has numerous industrial applications but is not regarded generally as having a high nutritional value. Following a breeding programme in Canada in the 1960s, new rapeseed varieties were developed based on naturally occurring mutants that accumulated less than 2% of erucic acid in the seed oil. This programme, which involved the screening of thousands of seeds, was only made possible by the recent invention of gas-liquid chromatography, which improved the speed and sensitivity of fatty acid analysis considerably (Murphy, 2007c). These new 'canola' varieties produced instead a high-oleic oil with a nutritional value similar to that of olive oil, which enabled rapeseed/canola to become a major oil crop that is now ranked third in the world after oil palm and soybean.

In the 1980s, breeders in Australia used chemically induced mutagenesis to produce new linseed varieties that made high quality edible oil, rich in linoleic acid (Green and Marshall, 1984). Conventional linseed produces an industrial-grade oil, high in α-linolenic acid, that is used for the manufacture of coatings, drying agents and putty. Although α-linolenic acid is a nutritionally desirable fatty acid, its presence in domestic vegetable oils would reduce shelf life due to rapid oxidation. Levels of α-linolenic acid were reduced by creating a double-mutant variety in which the seed-specific linoleate desaturase genes had been severely down-regulated. This meant that the new form of linseed, called Linola™, accumulated over 60% linoleic and less than 2% α-linolenic acid, which made it suitable to be marketed as high polyunsaturate food-grade oil with good oxidative stability and a long shelf life.

Conventional breeding approaches have been less successful in manipulating the fatty acid profiles of edible oil crops so that they produce industrial oils. This is due principally to the unusual and exotic nature of many industrially useful fatty acids. Most crop plants do not already contain the genes allowing them to accumulate such exotic fatty acids and therefore a transgenic approach is normally required. One important exception is oleic acid, which can be used both as premium-grade edible oil as well as high-grade industrial feedstock for the manufacture of a huge range of industrial products. Existing uses of high-oleic soybean oil include lubricating oils, greases, printing inks, plasticizers, electrical insulation, detergents, soaps, shampoos and disinfectants (Eharan and Adhvaryu, 2005; Sharma *et al.*, 2005). Oleic acid is a major component of all plants and is often abundant in seed and fruit oils, which means that many plant oils have the potential to act as feedstocks for some of the uses for soybean oil listed above. However, the value of oleate-rich oils as industrial feedstocks is often severely limited by the additional presence of oxidation-prone polyunsaturates, especially linoleic and α-linolenic acids. These fatty acids reduce the thermal performance and oxidative stability of many plant oils and therefore restrict their industrial uses. A major challenge for breeders has therefore been to reduce polyunsaturate levels in seed oils.

This challenge has been addressed with considerable success by breeders in the case of several major oil crops. For example, breeders in the former Soviet Union developed two different sunflower varieties with either high oleic (75%) or high linoleic oil (65%). Sunflower and safflower lines are now available with 75% oleate and < 1% α-linolenic (Murphy, 2007a). More recently, breeders in the USA and Europe have developed very high oleate and very low polyunsaturate lines of soybean and rapeseed/canola, which may have potential industrial applications. High-oleic soybean varieties with as much as 83% oleate and less than 3% α-linolenate have been developed (Rahmana *et al.*, 2001) and are now being marketed by major seed companies. Breeders have also developed other lines of soybean that have high levels of stearic acid (Rahmana *et al.*, 2003) and other nutritionally relevant fatty acids. Several high-oleic canola/rapeseed varieties have been developed that typically contain about 70–80% oleate, 15% linoleate and only 3% α-linolenate. Major seed companies such as Cargill, Dow Agrosciences and Bayer are now developing high-oleic varieties for various end-use markets, both edible and non-edible. By 2004, high-oleic rapeseed/canola was already being planted on about 250,000ha in Canada, which is 5% of the total area of canola cultivation (AgCanada, 2004). Some of the new high-oleate soybean oils have already been used as biodegradable lubricating fluids with relatively long working lives and low susceptibility to oxidation at high temperatures (Cahoon, 2003).

Efforts are also under way to produce very high-oleate varieties of the major global oil crop, oil palm. This is an especially challenging objective due to the slow growth and long generation time of oil palm trees, which take 5–7 years to flower and produce oil and 10–15 years before they reach their full commercial productivity. Over the past decade, several promising lines of germplasm have been identified from the African and South American centres of origin of the two domesticated oil palm species, *Elaeis guineensis* and *E. oleifera* (Maizura, 1999; Hayati *et al.*, 2004). For example, *E. oleifera* mesocarp oil can contain as much as 65% oleate. Currently, the world's largest oil palm germplasm collection is held by the Malaysian Palm Oil Board, and this and other collections increasingly are being examined and assessed by highly effective molecular genetic methods, as well as conventional biochemical methods (Moretzsohn *et al.*, 2002; Maizura *et al.*, 2006). By using DNA-based molecular markers, breeders hope to cross unimproved high oleate lines into existing commercial lines in order to produce a combination of altered fatty acid profile and high oil yield with good agronomic performance.

Transgenic oil crops

Despite the impressive achievements of conventional breeding listed above, the range of oils that can be produced by a given crop species is obviously limited by its genotype. This applies particularly to exotic fatty acids, which often result from unique mutations that have only ever occurred in a single group of plants. An example would be the Δ6 palmitate desaturase found in coriander and its relatives. This enzyme is a mutated version of the normal Δ12 palmitate desaturase found in other plants. The Δ6 palmitate desaturase enables coriander seed oil to accumulate over 85% of the industrially useful fatty acid, petroselinic acid. The only way to produce petroselinic acid in other oilseeds is, therefore, to transfer the Δ6 palmitate desaturase gene from coriander (and possible additional genes as well) into the crop of interest (Jaworski and Cahoon, 2003). Over the past 20 years, the desire to produce a greater range of oils, particularly for industrial purposes, has led to the repeated attempts to engineer transgenic oil crops, often with mixed success.

The manipulation of seed oil content via transgene insertion was one of the first successful applications of modern biotechnology in agriculture. Indeed, the first transgenic crop with a modified seed composition to be approved for unrestricted commercial cultivation in the USA was a lauric acid-rich (C-12) rapeseed/canola released by Calgene in 1995. This variety was produced by transferring a C-12-specific thioesterase from the California Bay plant into rapeseed. Although yields of lauric acid were relatively low in the early versions of this transgenic variety, the progressive addition of more transgenes eventually enabled Calgene breeders to produce a 60% lauric oil (Voelker *et al.*, 1996). However, even this relatively lauric-rich oil crop was unable to compete commercially with the much cheaper and more plentiful supplies of palm kernel oil, which contained only about 40% lauric acid. Therefore, the 'laurical' brand of transgenic rapeseed/canola was a commercial failure and was grown for a few seasons only in the mid-1990s in southern USA.

Many of the genes encoding key enzymes of fatty acid biosynthesis, modification and storage lipid assembly have now been isolated from a wide range of plant species, as shown in Table 9.2. Several transgenic rapeseed/canola and soybean varieties, with modified seed oils, are now available for commercial cultivation. In principle, biotechnologists have shown that it is possible to transfer genes from donor species into major crops, such as soybean and rapeseed/canola, in order to produce any fatty acid from C-8 to C-24 and with any given functionality, such as double or triple bonds, or hydroxy or epoxy groups. Some examples of the transfer of potentially oil-modifying genes into oilseeds and the resulting levels of novel fatty acids are shown in Table 9.3. Apart from rapeseed, which accumulates up to 60% lauric acid, the levels of the novel fatty acids in transgenic plants are relatively modest and remain far from achieving commercial viability.

As discussed below, the as-yet unrealized challenge for biotechnologists has been to produce novel fatty acids in transgenic plants at high enough levels to ensure their commercial viability (Thelen and Ohlrogge, 2002; Napier, 2007). At present, most novel fatty acids only accumulate at relatively low levels in transgenic species. The main reason for this situation is that simply transferring the relevant hydroxylase or desaturase gene into a plant does not necessarily mean that the corresponding hydroxy or polyunsaturated fatty acid will accumulate at high levels in the storage oil of the recipient plant. Indeed, despite over 20 years of sometimes ingenious efforts by molecular biologists, yields of novel fatty acids in most transgenic plants remain stubbornly low (Murphy, 1991, 1994, 1999; van de Loo *et al.*, 1995; Broun and Somerville, 1997; Cahoon *et al.*, 1999; Jaworski and Cahoon, 2003; Smith *et al.*, 2003).

Sometimes, as with the transgenic laurical rapeseed discussed above, novel fatty acid levels can be increased by transferring appropriate acyltransferase genes to ensure that the new fatty acids are efficiently assembled on to triacylglycerols. But it is not always predictable what additional enzymes/genes will be required to accumulate a given novel fatty acid. In coriander seeds, it is now clear that at least three specialized enzymes have evolved to ensure high levels of petroselinic acid accumulation in coriander seeds. In addition to the Δ6 palmitate desaturase that actually produces the petroselinic acid, a novel 3-ketoacyl ACP synthase and an additional acyl–ACP thioesterase are required (Cahoon and Ohlrogge, 1994; Dormann *et al.*, 1994; Mekhedov *et al.*, 2001).

The role of diacylglycerol acyltransferases in improving novel fatty acid yields has been demonstrated recently in transgenic *Arabidopsis* plants. When these plants expressed a castor bean hydroxylase on

Table 9.3. Examples of transgenic plants modified to produce potential industrial oils. In the majority of cases, which are not shown here, only very low levels of novel fatty acids were produced – as in the example from coriander shown here.

Fatty acid[a]	Donor species	FA in donor species (%)	Recipient species	FA in recipient species (%)
Lauric 12:0	California Bay	65	Rapeseed	60[b]
Petroselinic 18:1 6c	Coriander	80	*Arabidopsis*	< 1[b]
Ricinoleic 18:1-OH	Castor bean	90	*Arabidopsis*	26[c]
Vernolic 18:1 9c,12OH	*Crepis palaestina*	60	*Arabidopsis*	15[b]
Crepylinic 18:2 9c,12trp	*Crepis alpina*	70	*Arabidopsis*	25[b]
α-Eleostearic 18:3 9c,11t,13t	*Mormordica charantia*	65	Soybean	17[b]
Calendic 18:3 8t,10t,12c	*Calendula officinalis*	60	*Arabidopsis*	20[d]

[a]c, *cis* double bond; t, *trans* double bond; trp, triple bond.
[b]Jaworski and Cahoon, 2003.
[c]Burgal *et al.*, 2008.
[d]Cahoon *et al.*, 2006.

its own, they were able to accumulate 17% ricinoleic acid. However, when they were transformed additionally with a castor bean type-2 acyl-coenzyme A:diacylglycerol acyltransferase (DGAT2), the plants were able to accumulate as much as 26% ricinoleic acid plus almost 5% of a hydroxylinoleate species, possibly densipolic acid (12-OH 18:2 Δ9, 15), making a total of over 31% hydroxy fatty acids in their seed oil (Burgal *et al.*, 2008). While this is still far below the 90% levels of ricinoleic acid found in castor bean oil, and has been achieved in a model plant only, it nevertheless represents a useful incremental step towards the goal of engineering high-purity industrial oils in transgenic oil crops. In another case, the transformation of soybean plants with a DGAT2 gene from the fungus *Umbelopsis ramanniana* (formerly *Mortierella*) had quite a different result. In this case, the presence of the exogenous DGAT2 did not alter the fatty acid composition, but instead increased the overall flux of carbon towards triacylglycerol, resulting in significant increase in seed oil yields (Lardizabal *et al.*, 2008). Evidence from several plant species is now beginning to suggest that DGAT2 can sometimes play an important role in the accumulation of exotic fatty acids in storage triacylglycerols (Kroon *et al.*, 2006; Shockey *et al.*, 2006).

A further problem in obtaining high levels of novel fatty acids is that in some crop species, such as rapeseed/canola, not all of the novel acyl groups are necessarily channelled to storage lipids. Instead, some of the exotic fatty acids sometimes accumulate on membrane lipids. It appears that the accumulation of novel fatty acids in membrane lipids often triggers a regulatory mechanism, which results in the removal of the novel fatty acids and their breakdown via the β-oxidation and glyoxylate cycle pathways. This is one reason why some transgenic plants are unable to accumulate high levels of novel fatty acids. It is also an important reminder of the complexity of metabolic regulation in plants and the difficulties of manipulating this process via the insertion of one, or a few, transgenes.

Although transgenic approaches to oil modification have focused mainly on the introduction of exotic fatty acids, they have also been used to downregulate existing genes in order to decrease the levels of unwanted fatty acids. Most frequently, this has involved the use of technologies such as antisense or RNA interference to suppress genes encoding linoleate desaturases in order to reduce levels of oxidation-prone α-linolenate in seed oils. These approaches involve the insertion of various versions of the gene in question, such as a backwards (or antisense) copy or a partial version of its corresponding RNA. In the latter case, the strategy is called RNA interference, or RNAi. These inserted DNA or RNA segments interfere with expression of the target gene, suppressing its activity either partially or completely. In one recent study, a combined antisense-hairpin RNAi approach was used with considerable effect to modify the seed oil of the model species *Arabidopsis thaliana*, which is a relative of the oil crop, rapeseed/canola (Nguyen and Shanklin, 2009).

Popular targets of antisense or RNAi gene suppression programmes have been the genes encoding oleate and linoleate desaturases, with the aim of reducing levels of oxidation-prone polyunsaturates in seed oils. Downregulation of oleate and/or linoleate desaturases will also normally result in an increase in levels of oleate in seed oils. As discussed previously, high-oleate oils are desirable for both edible and many industrial applications and have been developed successfully in several major oil crops by non-transgenic methods. However, several companies have also developed high-oleate, low-polyunsaturate oils using various transgenic routes. The following high-oleate transgenic lines have been developed but not necessarily commercialized to date: rapeseed/canola with 89% oleate; *B. juncea* (Indian mustard) with 73% oleate; soybean with 90% oleate; and cottonseed with 78% oleate (Murphy, 2007a).

Segregation of Novel Fatty Acids from Membrane Lipids

The cellular membranes of all organisms are crucial to their metabolism and survival. Biological membranes are made up of a lipid bilayer into which is embedded the various proteins that mediate such processes as transport, respiration, photosynthesis and signal transduction. The fatty acid composition of a given membrane is regulated tightly and the presence of inappropriate acyl groups can lead to serious disruption of membrane, and hence cellular, function that can impair development or even be fatal. In general, plant membrane lipids contain only a very restricted range of C-16 and C-18 fatty acids. However, as noted above, some oilseeds can accumulate well over 90% of membrane-incompatible fatty acids in their seed oil (Badami and Patil, 1981; van de Loo, 1993; Millar *et al.*, 2000). These novel fatty acids normally are excluded either largely or completely from membrane lipids in the seeds and other parts of the plant. This is probably because the special physical and chemical properties of many novel fatty acids can perturb the integrity of the membrane structure, such that they are deleterious to the cell (Millar *et al.*, 2000). Hence, fatty acids such as very long and medium chains and oxygenated species are found only in very small amounts in phospholipids in seed tissues and are completely absent from non-seed tissues (Bafor *et al.*, 1990, 1991; Stahl *et al.*, 1995; Wiberg *et al.*, 1997). In plants that accumulate fatty acids that differ significantly from those of membrane lipids, specific mechanisms have evolved that prevent the 'leakage' of unwanted acyl groups into membrane lipid pools. Biochemically speaking, this is not a trivial task, because storage lipids and many membrane lipids are assembled on the same organelle – the endoplasmic reticulum.

As yet, we do not understand completely the mechanisms by which some plants are able to segregate unusual fatty acids away from membranes (Millar *et al.*, 2000). We know that these mechanisms are very important from some of the earlier attempts to modify oils in transgenic plants. In one case, the leakage of a novel fatty acid, the C-18 saturate:stearic acid, resulted in very poor seed germination rates in a transgenic variety of rapeseed/canola that had been engineered to have an elevated stearate oil for use in the manufacture of edible spreads (Thompson and Li, 1997). It was found that a small amount of stearate had leaked into cell membranes, resulting in a reduction in membrane fluidity and impairment of function that affected the development of the entire plant. In this case, the transgenic rapeseed/canola accumulated only about 40% stearate in its storage oil and just 3–5% leaked into cell membranes. However, other oilseeds like mangosteen can accumulate over 65% stearate in their seed oils without any detectable leakage into cell membranes or damage to their growth potential. In another case, we found that transgenic rapeseed plants expressing a coriander Δ6 desaturase gene initially accumulated moderate levels of petroselinic acid in their developing seeds, but the novel fatty acid subsequently disappeared as the seeds matured (Murphy *et al.*, 1999). The explanation was that some of the petroselinic acid in the transgenic rapeseed plants had 'leaked' into membrane lipids and triggered a protective response whereby their β-oxidation pathway had been upregulated in order to remove the potentially destabilizing fatty acid. Similar observations were made in the case of transgenic rapeseed accumulating lauric acid (Eccleston and Ohlrogge, 1998).

Two main hypotheses have been proposed to explain the sort of lipid segregation that plants like mangosteen seem capable of, but seem to be lacking in major oil crops like rapeseed/canola. The first hypothesis is that there is a spatial compartmentation of membrane and storage lipid synthesis in specific membrane domains of the endoplasmic reticulum. The other hypothesis is that there is a selective accumulation of the novel fatty acids in triacylglycerols that is controlled by desaturases and acyltransferases of differing specificities (Millar *et al.*, 2000). For example, petroselinic acid is synthesized by a divergent acyl-ACP desaturase in the plastid, but ricinoleic acid is synthesized by a diverged oleate desaturase in the endoplasmic reticulum. It is conceivable that this mechanism may involve different channelling pathways for segregation of storage lipids from membrane lipids. Also, petroselinic acid has only one double bond position difference from common acyl groups, while ricinoleic acid has a polar hydroxy group so that petroselinic acid may be more compatible with membrane lipids and not segregated away from the membrane lipids as strictly as is ricinoleic acid (Millar *et al.*, 2000). Research is currently under way to address these issues but until we understand more about fatty acid segregation, the production of most exotic fatty acids in transgenic crops will remain more of an aspiration than a reality (Thelen and Ohlrogge, 2002).

Biopolymers

As an alternative to modifying triacylglycerol-based storage lipids in oil crops, their biosynthetic pathways can be diverted to the production of other useful carbon-based compounds such as biopolymers. For example, there is great interest in using genetically engineered crops for the large-scale production of biodegradable polymers such as polyhydroxyalkanoates (PHAs). Such biopolymers potentially could substitute for conventional plastics currently derived from non-renewable petroleum feedstocks (Braunegg *et al.*, 1998; Stevens, 2002; Philip *et al.*, 2007; van Beilen and Poirier, 2007; Mooney, 2009). Virtually all conventional plastics currently are made from non-renewable petroleum-derived intermediates such as adipic acid and vinyl chloride. Moreover, such chemical processes are relatively energy-intensive and frequently produce undesirable by-products that can be costly and difficult to dispose of. Finally, many petroleum-based plastic products are difficult to recycle and can take decades or longer before they break down in landfill sites. An attractive alternative to conventional plastics is to use naturally occurring biopolymers that are renewable, biodegradable, non-polluting and less energy-intensive to manufacture.

Some soil bacteria, such as *Ralstonia eutrophus* or *Alcaligenes latus*, are able to accumulate up to 80% of their mass in the form of non-toxic biodegradable PHA polymers. The PHAs accumulate as dozens of tiny solid granules within the cytoplasm of bacterial cells. The PHAs are made up of β-hydroxyalkanoate subunits that are synthesized from acetyl-CoA via a relatively short pathway involving as few as three enzymes for the most common PHA, polyhydroxybutyrate (Steinbüchel and Füchtenbusch, 1998). Currently, PHAs are made industrially via a bioreactor process involving the incubation of bacterial cultures in contained vessels using carbon sources such as sugars, plant oils or crop by-products. Once the bacterial cultures reach stationary phase, the cells are broken up and the granules extracted and purified. An early version of the resulting polymeric material developed by ICI in the 1980s was called Biopol® and was able to act as a technical substitute for several petroleum-derived thermoplastics, but had the serious drawback of costing as much as tenfold more to produce.

Native PHA-synthesizing bacteria have relatively specific carbon source requirements in bioreactor production, which limits their utility, but this drawback has been overcome by transferring the PHA operon into established bioreactor species such as *Escherichia coli* (Schubert *et al.*, 1988; Kidwell *et al.*, 1995). Following the development of plant genetic engineering technologies in the 1980s, it was realized that the three bacterial genes could also be transferred into crops, which hopefully would then accumulate PHAs on an agricultural scale. Crop-based PHA production would eliminate the need for a costly, energy-requiring indoor bioreactor process. Moreover, the carbon feedstocks would be derived directly from photosynthesis rather than existing organic feedstocks such as crop-based sugars. Hence, plant-derived PHAs had the prospect of being much cheaper and capable of production on a much larger scale that might make them competitive with petroleum-derived plastics. This prospect has led several companies, including successively Zeneca, Monsanto and Metabolix, to attempt to develop transgenic rapeseed or other crop species that express the bacterial genes responsible for PHA biosynthesis. It was soon realized that PHA biosynthesis in the cytoplasm of plants was not viable due to the diversion of acetyl-CoA that was normally required for the formation of flavonoids and phytosterols (Nawrath *et al.*, 1994). However, providing the PHAs accumulated in the plastids and not in the cytosol, it was possible to obtain modest yields of the polymer from either leaves or seeds (Houmiel *et al.*, 1999; Snell and Peoples, 2002). Although oil crops such as rapeseed or oil palm (Murphy, 2007a) have been the most common production platforms for biopolymers, there has been a recent report of PHA accumulation in the lignocellulosic biomass crop, switchgrass (Somleva *et al.*, 2008).

A major and as yet unresolved technical hurdle is how to extract biopolymers from plant tissues in an efficient and cost-effective manner. Unlike bacteria, plant tissues often contain complex cross-linked polysaccharides, polyphenols and highly lignified structures that can severely disrupt the normal extraction procedures used to obtain biopolymers such as PHAs. It has been reported that there are several patent filings for PHA extraction from plants based on non-halogenated solvent extraction at high temperature, plus procedures that allow simultaneous extraction of PHAs and oils from oilseed crops using differential solvent extraction (Mooney, 2009). However, none of these methods seem particularly environmentally friendly, which may be a major issue for a product, such as a biopolymer, whose whole raison d'être is based on its environmental credentials. To date, none of these extraction methods have been published for commercial-scale extraction from plants and further progress on PHAs may be stymied until this challenge can be overcome.

Another complexity is that polyhydroxybutyrate, which is the most widespread PHA, is a rather brittle plastic and is not suitable for most applications. The best-performing plastics are co-polymers of polyhydroxybutyrate with other PHAs, such as polyhydroxyvalerate (PHBV). Although the production of such co-polymers as PHBV in transgenic plants is considerably more difficult than that of single-subunit polymers, progress has been made recently in this area (Slater *et al.*, 1999). Therefore, while there are several companies and academic laboratories still attempting to make commercially extractable PHAs in plants, it seems unlikely that these environmentally friendly products will be commercially available for quite a few years to come. Meanwhile, in the absence of a viable plant source, bacterial-based bioreactor production of PHAs is continuing to progress (Kim and Dale, 2005; Dias *et al.*, 2006). The latest version is a PHA called Mirel™ which, starting in 2009, is going to be made in bacterial cultures using a maize feedstock for commercial-scale production of 50,000 t/year at a plant in Clinton, Iowa, USA (Mooney, 2009). While this production method should reduce costs from the average US$14/kg for pure carbon substrates such as glucose, the costs of bioreactor-based biopolymers will still be considerably in excess of petroleum-based plastics such as polyethylene. Therefore, interest definitely will remain in the future development of plant-based production platforms for these renewable, carbon-neutral, environmentally friendly biopolymers.

Chemical and Biotechnological Transformations of Industrial Oils

So far, this chapter has focused on the production of oil-based feedstocks or more advanced oleochemical intermediates in crops. However, even relatively refined oils, such as a 95% pure triricinolein, will still require some downstream chemical and/or biotechnological conversion to generate specific oleochemical products at the required degree of purity. Less refined oils will require even more processing and therefore the oleochemicals industry will still be doing a great deal of research and development into appropriate technologies, whatever the success or otherwise of breeders in producing new crop-based oils. Downstream conversion and diversification transformations sometimes require biotechnological procedures such as lipase-catalysed interesterification or transesterifications. Alternatively, they may involve strictly chemical processes such as epoxidation or hydroxylation. A comprehensive list of chemical transformations of plant oils is given in Chapters 5 and 8 of *The Lipid Handbook* (Knothe *et al.*, 2007; Lie *et al.*, 2007).

Some of the most commonly used oleochemical feedstocks for transformations include the following: oleic acid from very high-oleate sunflower; linoleic acid from soybean; linolenic acid from linseed; erucic acid from high-erucate rapeseed; and ricinoleic acid from castor bean. These feedstocks are useful for chemical transformations because they contain one or more C-C double bonds in addition to the carboxy group. After a relatively slow start, novel oils from several new crops are also becoming industrially available, although not as yet in large quantities. Examples include petroselinic acid from coriander, calendic acid from *Calendula officinalis*, α-eleostearic acid from tung oil, santalbic acid from *Santalum album* and vernolic acid from *Vernonia galamensis* (Metzger and Bornsheuer, 2006). Useful new products include environmentally friendly industrial fluids and lubricants, insulating fluids for electric utilities and additives to asphalt. In addition to modern methods of synthetic organic chemistry, enzymatic and microbial transformations can be used with fatty compounds for the selective functionalization of alkyl chains. Some of the products of such syntheses include long-chain di-acids, ω-hydroxy fatty acids and ω-unsaturated fatty acids. It is also possible to open up C-C double bonds via chemical epoxidation to produce such advanced intermediates as polyetherpolyols. Finally, the purification of biologically produced and therefore enantiomerically pure fatty acids from oil crops provides the basis for the synthesis of high-value non-racemic building blocks in the manufacture of fine chemicals.

Key Targets for Future Industrial Oil Crops

One of the prerequisites for the commercial viability of future industrial oil crops is that they should accumulate single fatty acid species or very specific fatty acid mixtures or other lipid-derived feedstocks (such as PHAs) at the highest possible yields and purity (Cahoon *et al.*, 2007; Dyer *et al.*, 2008). This needs to be achieved by breeders without compromising other important agronomic characteristics of the crop. It will also be necessary to ensure that such industrial crops can be grown, harvested and processed on a large scale without affecting adjacent crops that might be destined for animal feed or for the human food chain. A list of some key targets for researchers, breeders and other developers of industrial oil crops is given below.

Tailored oil composition

This target means that the chosen fatty acid for a particular end-use should make up the vast bulk of the triacylglycerol oil of a crop, ideally at least 80–90%, in order to reduce downstream purification costs. As described previously, this goal has been difficult to achieve due to the genetic and biochemical complexities that underlie storage lipid composition in plants. Further progress here will depend on more information from biochemical and molecular genetics researchers about the identity of target genes to be selected by breeders. We already have an extensive 'shopping list' of such genes that includes those encoding acyltransferases, phospholipases and thioesterases, as well as the actual fatty acid modification enzymes, most of which are modified desaturases. However, more precise knowledge will be required about which particular genes to select in each case of fatty acid modification. Hence, the target gene list for producing ricinoleate in rapeseed may be different from producing ricinoleate in soybean, or producing petroselinate in rapeseed. Given the current rate of progress, it may be some time before the goal of specifically tailored homogeneous oil compositions can be achieved.

Alternatively, for some applications, it might be desirable to produce a specific mixture of fatty acids in seed oil rather than a single fatty acid species. In some cases where the target fatty acids are commonly occurring C-18 species, this has already been achieved by breeders. The best example is very high-oleic oils that also contain some linoleate, but with virtually no saturates or α-linolenate. As we saw in the section on breeding for industrial oils, high-oleic oils with these specifications have been developed via several breeding routes and are now commercially available for most of the major oilseed crops.

High oil yield

An important target is that a given industrial oil should be accumulated at the highest possible yields by the crop in question; for example, simply increasing the proportion of oil in seeds of a crop like meadowfoam from its current low levels of 20% to a value of 30% would result in an increased crop oil yield of 50%, without affecting any other production costs. This could increase the market uptake of such minor crops significantly and stimulate further efforts to improve them to supply renewable oleochemicals to industry. Some minor oil crops such as meadowfoam yield only 160–240 kg/ha, whereas rapeseed yields > 1 t/ha and oil palm yields 3–4 t/ha. These yield values could be increased by a breeding research programme focused primarily on oil yield. Recent discoveries about genes, such as diacylglycerol acyltransferases, that can affect oil yield significantly will give breeders new targets and increase the chances of success. Given the immense yield and logistical advantages of tropical perennial oil crops such as oil palm, which can be harvested continually year-round and does not require annual re-sowing, it may also be appropriate to invest more into the modification of such crops to serve as efficient, large-scale sources of industrial oils. The manipulation of tropical crops is still in its infancy and much of their immense potential as cost-effective sources of oleochemical feedstock remains untapped.

Use of co-products

The commercial viability of any industrial oil can be increased considerably if other co-products of the crop are also exploited for profit, rather than representing an additional expense for their separation and disposal. Examples include the protein 'cake' found in many oilseeds, which can often be processed to make animal feed. In other crops, the vegetative parts of the plant such as fibres, stems and trunks can be used. For example, the remainder of oil palm fruits and trees already can be processed into products such as fibreboard, fertilizer, paper and various composite materials including polymers and building materials (Rozman *et al.*, 2003; Ratnasingam *et al.*, 2008). Research is also under way to use other co-products for the production of high-value health care products.

Management and processing of industrial oil crops

To date, there has been relatively little research into the kinds of management and processing systems that will be required for many new industrial oil crops. Most of these crops have been developed separately on a piecemeal and rather empirical basis, with relatively little dissemination of useful methodologies or best practice. Among the novel challenges that might face the grower and user of completely new crops are differing requirements for the sowing, cultivation, harvesting, oil extraction and downstream processing of such crops. This might entail the purchase or hire of new kinds of equipment on-farm. Cultivation of the new crops might also affect the management of existing crops being grown on the same farm. The 'designer crop' concept brings a special set of management challenges that have not been faced previously by the majority of commodity crop growers. Moreover, since, by definition, these crops will be transgenic and designed solely for industrial use, there might be additional sensitivities about their cultivation close to edible and/or non-transgenic varieties of the same crop.

There already have been several unfortunate cases of transgenic crops that were not approved or intended for food use that inadvertently either entered or almost entered the human food chain (Murphy, 2007c). For example, in 2000, an unapproved (for food use) transgenic variety of maize called StarLink was discovered in the human food chain throughout the USA. As a result, millions of tonnes of maize seed, flour and food products such as tortillas were recalled for disposal. The disaster cost Aventis CropScience an estimated US$500 million, as it was required to buy back the contaminated maize and to compensate various injured parties (Harl *et al.*, 2003). In 2002, a small number of transgenic maize plants expressing a pig vaccine were found growing in a soybean crop destined for use in human food and/or animal feed (Hileman, 2003).

Discovery of this contamination led to incineration of the fields and destruction of more than 12,000 t of soybeans, at a cost to the biotech company concerned of about US$3.5 million. In an unprecedented action, this company was also fined US$250,000 by the US government (Ellstrand, 2003). In both of the above cases, contamination was minor and there was little or no health risk. But this fact is not as important as the perception created about the future potential for such crops to affect the human food chain and for this reason, growers and processors of designer industrial crops will need to pay special attention to management and quality control at all stages of production from sowing to end-product manufacture.

Future Prospects

For the foreseeable future, oil crops will continue to serve primarily as sources of edible vegetable oils for a global population that is projected to rise by as much as three billion to over nine billion people by 2050. The recent substantial diversion of plant oils towards the biodiesel sector is likely to be transient as second-generation biofuels are developed. In the longer term, as fossil-derived hydrocarbons inevitably become depleted and therefore more expensive, plant oils gradually will begin to supplant them in more and more applications. However, and unlike the picture a few decades ago, it now appears that this process will take many decades and the speed of the transition from petrochemicals to oleochemicals will depend crucially on factors such as the health of the global economy, progress in research and development into new plant oils and wider political developments. Examples of the latter include government policies such as carbon taxes or renewables obligations that might encourage the use of plant oils and stimulate a more rapid rate of research and development into industrial plant oils. While it is unlikely that there will be a sudden resurgence in industrial demand for plant oils during the next few years, it is virtually certain that the coming decades will see a gradual but steady increase in the use of oleochemical rather than petrochemical feedstocks across all sectors.

This developmental timescale – measured in decades rather than in months or years – is ideal for a considered series of programmes to bring onstream a new generation of improved industrial oil crops using the many new biotechnologies now at our disposal. It may take another decade or so to resolve the problem of producing transgenic crops with 70–90% levels of a given fatty acid. Recent progress here has been encouraging, but we still have a long way to go from typical current best levels of 25–30%. Increasing oil yields will be equally challenging. The best temperate oilseed crops currently yield about 1 t/ha, but that figure is based on a seed oil content of 40%, which could be increased towards values of 60–70% already found in some nuts. Oil yields of the major global crop, oil palm, are currently about 3–4 t/ha, but trees are available that can yield over 10 t/ha. If we can increase global oil yields by 50% and improve fatty acid purity in plant oils, there will be sufficient production to meet demand for edible oils and also to sustain a significant growth in price-competitive feedstocks for a widening range of industrial uses.

In order to meet these aspirations, it will be desirable to make use of all available plant breeding methods, as well as to continue with efforts to understand the biochemical and cellular processes that underlie oil accumulation, and related aspects of non-storage lipid metabolism, in plant tissues. The three key breeding approaches should be: (i) advanced non-transgenic breeding of existing oil crops including use of mutagenesis and wide crosses for gene manipulation; (ii) transgenic methods to bring new genes into existing crops, but with the realization that most oil modifications will be complex traits involving the addition of several key genes, not all of which may be identified at present; and (iii) domestication of new oil crops using advanced molecular breeding methods such as marker-assisted selection. The choice of which approach to adopt will vary from case to case, but all of them have the potential to deliver useful results. As we have learned over the past two decades, none of these methods is particularly rapid but, equally, neither is the required timescale for commercializing the new generation of industrial oil crops. By investing relatively modest sums now in applied breeding research and development into new oil crops and in basic research into lipid metabolism, we could lay the groundwork for development of the sustainable and environmentally friendly plant-based hydrocarbon feedstocks that will surely become increasingly needed as fossil carbon resources become depleted during the rest of the 21st century.

References

AgCanada (2004) The United States Canola Industry: Situation and Outlook. *Agriculture and Agri-Food Canada Bi-weekly Bulletin* 17(4) (http://www.agr.gc.ca/mad-dam/e/bulletine/v17e/v17n04_e.htm, accessed March 2009).

Badami, R.C. and Patil, K.B. (1981) Structure and occurrence of unusual fatty acids in minor seed oils. *Progress in Lipid Research* 19, 119–153.

Bafor, M., Jonsson, L., Stobart, A.K. and Stymne, S. (1990) Regulation of triacylglycerol biosynthesis in embryos and microsomal preparations from the developing seeds of *Cuphea lanceolata. Biochemical Journal* 272, 31–38.

Bafor, M., Smith, M.A., Jonsson, L., Stobart, A.K. and Stymne, S. (1991) Ricinoleic acid biosynthesis and triacylglycerol assembly in microsomal preparations from developing castor-bean (*Ricinus communis*). *Biochemical Journal* 280, 507–514.

Beilen, J.B. van and Poirier, Y. (2007) Prospects for biopolymer production in plants. *Advances in Biochemical Engineering and Biotechnology* 107,133–151.

Braunegg, G., Lefebvre, G. and Genser, K.F. (1998) Polyhydroxyalkanoates, biopolyesters from renewable resources: physiological and engineering aspects. *Journal of Biotechnology* 65, 127–161.

Broun, P. and Somerville, C. (1997) Accumulation of ricinoleic, lesquerolic, and densipolic acids in seeds of transgenic Arabidopsis plants that express a fatty acyl hydroxylase cDNA from castor bean. *Plant Physiology* 113, 933–942.

Burgal, J., Shockey, J., Lu, C., Dyer, J., Larson, T., Graham, I. and Browse, J. (2008) Metabolic engineering of hydroxy fatty acid production in plants: RcDGAT2 drives dramatic increases in ricinoleate levels in seed oil. *Plant Biotechnology Journal* 6, 819–831.

Cahoon, E.B. (2003) Genetic enhancement of soybean oil for industrial uses: prospects and challenges. *AgBioForum* 6, 11–13 (http://www.agbioforum.org, accessed March 2009).

Cahoon, E.B. and Ohlrogge, J.B. (1994) Metabolic evidence for the involvement of a D4-palmitoyl-acyl carrier protein desaturase in petroselinic acid synthesis in coriander endosperm and transgenic tobacco cells. *Plant Physiology* 104, 827–837.

Cahoon, E.B., Carlson, T.J., Ripp, K.G., Schweiger, B.J., Cook, G.A., Hall, S.E. and Kinney, A.J. (1999) Biosynthetic origin of conjugated double bonds: production of fatty acid components of high-value drying oils in transgenic soybean embryos. *Proceedings of the National Academy of Sciences, United States of America* 96, 12935–12940.

Cahoon, E.B., Dietrich, C.R., Meyer, K., Damude, H.G., Dyer, J.M. and Kinney, A.J. (2006) Conjugated fatty acids accumulate to high levels in phospholipids of metabolically engineered soybean and Arabidopsis seeds. *Phytochemistry* 67, 1166–1176.

Cahoon, E.B., Shockey, J.M., Dietrich, C.R., Gidda, S.K., Mullen, R.T. and Dyer, J.M. (2007) Engineering oilseeds for sustainable production of industrial and nutritional feedstocks: solving bottlenecks in fatty acid flux. *Current Biology* 10, 236–244.

Davidson, A. (1999) *The Oxford Companion to Food*. Oxford University Press, Oxford, UK.

Dias, J.M.L., Lemos, P.C., Serafim, L.S., Oliveira, C., Eiroa, M., Albuquerque, M.G.E., Ramos, A.M., Oliveira, R. and Reis, M.A.M. (2006) Recent advances in polyhydroxyalkanoate production by mixed aerobic cultures: from the substrate to the final product. *Macromolecular Bioscience* 6, 885–906.

Dormann, P., Frentzen, M. and Ohlrogge, J.B. (1994) Specificities of the acyl–acyl carrier protein (ACP) thioesterase and glycerol-3- phosphate acyltransferase for octadecenoyl-ACP isomers – identification of a petroselinoyl-ACP thioesterase in Umbelliferae. *Plant Physiology* 104, 839–844.

Durrett, T.P., Benning, C. and Ohlrogge, J. (2008) Plant triacylglycerols as feedstocks for the production of biofuels. *Plant Journal* 54, 593–607.

Dyer, J.M., Stymne, S., Green, A.G. and Carlsson, A.S. (2008) High-value oils from plants. *The Plant Journal* 54, 640–655.

Eccleston, V.S. and Ohlrogge, J.B. (1998) Expression of lauroyl-acyl carrier protein thioesterase in *Brassica napus* seeds induces pathways for both fatty acid oxidation and biosynthesis and implies a set point for triacylglycerol accumulation. *Plant Cell* 10, 613–621.

Eharan, S.Z. and Adhvaryu, A. (2005) Non-food lipids. In: Murphy, D.J. (ed.) *Plant Lipids: Biology, Utilisation and Manipulation*. Blackwell, Oxford, UK.

Ellstrand, N.C. (2003) Going to 'great lengths' to prevent the escape of genes that produce speciality chemicals. *Plant Physiology* 132, 1770–1774.

EurActive (2008) EU agrees 10% 'green fuel' target in renewables deal (http://www.euractiv.com/en/transport/eu-agrees-10-green-fuel-target-renewables-deal/article-177812, accessed March 2009).

Gallagher, E. (2008) The Gallagher review of the indirect effects of biofuels production. Renewable Fuels Agency, UK (http://www.renewablefuelsagency.org/_db/_documents/Report_of_the_Gallagher_review.pdf, accessed March 2009).

Green, A.G. and Marshall, D.R. (1984) Isolation of induced mutants in linseed (*Linum usitatissimum*) having reduced linolenic acid content. *Euphytica* 33, 321–328.

Gunstone, F.D. and Harwood, J.L. (2007) Occurrence and characterization of oils and fats. In: Gunstone, F.D., Harwood, J.L. and Dijkstra, A.J. (eds) *The Lipid Handbook*, 3rd edn. Taylor and Francis, UK, pp. 37–141.

Harl, N.E., Ginder, R.G., Hurgurgh, C.R. and Moline, S. (2003) The StarLink™ Situation. Iowa State University Extension, Ames, Iowa (http://www.extension.iastate.edu/grain/resources/publications/buspub/0010star.PDF, accessed March 2009).

Harwood, J.L. (2005) Fatty acid biosynthesis. In: Murphy, D.J. (ed.) *Plant Lipids: Biology, Utilisation and Manipulation*. Blackwell, Oxford, UK, pp. 27–66.

Harwood J.L. (2007) Lipid metabolism. In: Gunstone, F.D., Harwood, J.L. and Dijkstra, A.J. (eds) *The Lipid Handbook*, 3rd edn. Taylor and Francis, UK, pp. 637–702.

Harwood, J.L., Evans, M., Ramji, D.P., Murphy, D.J. and Dodds, P.F. (2007) Medical and agricultural aspects of lipids. In: Gunstone, F.D., Harwood, J.L. and Dijkstra, A.J. (eds) *The Lipid Handbook*, 3rd edn. Taylor and Francis, UK, pp. 703–782.

Hayati, A., Wickneswari, R., Maizura, I. and Rajanaidu, N. (2004) Genetic diversity of oil palm (*Elaeis guineensis* Jacq.) germplasm collections from Africa: implications for improvement and conservation of genetic resources. *Theoretical and Applied Genetics* 108, 1274–1284.

Hileman, B. (2003) ProdiGene and StarLink incidents provide ammunition to critics. *Chemical and Engineering News* 81, 25–33.

Houmiel, K.L., Slater, S., Broyles, D., Casagrande, L., Colburn, S., Gonzalez, K., Mitsky, T.A., Reiser, S.E., Shah, D., Taylor, N.B., Tran, M., Valentin, H.E. and Gruys, K.J. (1999) Poly(β-hydroxybutyrate) production in oilseed leukoplasts of *Brassica napus*. *Planta* 209, 547–550.

Hu, Q., Sommerfeld, M., Jarvis, E., Ghirardi, M., Posewitz, M., Seibert, M. and Darzins, A. (2008) Microalgal triacylglycerols as feedstocks for biofuel production: perspectives and advances. *Plant Journal* 54, 621–639.

Jaworski, J. and Cahoon, E.B. (2003) Industrial oils from transgenic plants. *Current Opinion in Plant Biology* 6, 178–184.

Jones, Q. and Wolff, I.A. (1960) The search for new industrial crops. *Economic Botany* 14, 56–68.

Kidwell, J., Valentin, H.E. and Dennis, D. (1995) Regulated expression of the *Alcaligenes eutrophus* pha biosynthesis genes in *Escherichia coli*. *Applied and Environmental Microbiology* 61, 1391–1398.

Kim, S. and Dale, B. (2005) Life cycle assessment study of biopolymers (polyhydroxyalkanoates) – derived from no-tilled corn. *International Journal of Life Cycle Analysis* 10, 200–210.

Knapp, S.J. (2003) Breeding and genetic engineering of Cuphea and meadowfoam. *International Atomic Energy Agency Technical Documents* (IAEA-TECDOCs) 1369, 123–124.

Knothe, G., Kenar, J.A. and Gunstone, F.D. (2007) Chemical properties. In: Gunstone, F.D., Harwood, J.L. and Dijkstra, A.J. (eds) *The Lipid Handbook*, 3rd edn. Taylor and Francis, UK, pp. 37–141.

Kroon, J.T., Wei, W., Simon, W.J. and Slabas, A.R. (2006) Identification and functional expression of a type 2 acyl-CoA:diacylglycerol acyltransferase (DGAT2) in developing castor bean seeds which has high homology to the major triglyceride biosynthetic enzyme of fungi and animals. *Phytochemistry* 67, 2541–2549.

Lardizabal, K., Effertz, R., Levering, C., Mai, J., Pedroso, M.C., Jury, T., Aasen, E., Gruys, K. and Bennett, K. (2008) Expression of *Umbelopsis ramanniana* DGAT2A in seed increases oil in soybean. *Plant Physiology* 148, 89–96.

Leprince, O., van Aelst, A.C., Pritchard, H.W. and Murphy, D.J. (1998) Oleosins prevent oil-body coalescence during seed imbibition as suggested by a low-temperature scanning electron microscope study of desiccation-tolerant and -sensitive oilseeds. *Planta* 204, 109–119.

Lersten, N.R., Czlapinski, A.R., Curtis, J.D., Freckmann, R. and Horner, H.T. (2006) Oil bodies in leaf mesophyll cells of angiosperms: overview and a selected survey. *American Journal of Botany* 93, 1731–1739.
Lie Ken Jie, M.S.F., Harwood, J.L., Gunstone, F.D., Cheung, W.H. and Lam, C.N.W. (2007) Synthesis. In: Gunstone, F.D., Harwood, J.L. and Dijkstra, A.J. (eds) *The Lipid Handbook*, 3rd edn. Taylor and Francis, UK, pp. 37–141.
Loo, F.J. van de, Fox, B.G. and Somerville, C. (1993) Unusual fatty acids. In: Moore, J.T.S. (ed.) *Lipid Metabolism in Plants*. CRC Press, Boca Raton, Florida, pp. 91–126.
Loo, F.J. van de, Broun, P., Turner, S. and Somerville, C. (1995) An oleate 12-hydroxylase from *Ricinus communis* L. is a fatty acyl desaturase homolog. *Proceedings of the National Academy of Sciences, United States of America* 92, 6743–6747.
Lühs, W. and Friedt, W. (1994) Non-food uses of vegetable oils and fatty acids. In: Murphy, D.J. (ed.) *Designer Oil Crops*. VCH Press, Weinheim, Germany, pp. 73–130.
Maizura, I. (1999) Genetic variability of oil palm (*Elaeis guineensis* Jacq.) germplasm collections using RFLP markers. PhD thesis, Universiti Kebangsaan, Malaysia.
Maizura, I., Rajanaidu, N., Zakri, A. and Cheah, S. (2006) Assessment of genetic diversity in oil palm (*Elaeis guineensis* Jacq.) using restriction fragment length polymorphism (RFLP). *Genetic Resources and Crop Evolution* 53, 187–195.
Man, W.C., Miyazaki, M., Chu, K. and Ntambi, J. (2006) Colocalization of SCD1 and DGAT2: implying preference for endogenous monounsaturated fatty acids in triglyceride synthesis. *Journal of Lipid Research* 47, 1928–1939.
Mekhedov, S., Cahoon, E.B. and Ohlrogge, J. (2001) An unusual seed-specific 3-ketoacyl-ACP synthase associated with the biosynthesis of petroselinic acid in coriander. *Plant Molecular Biology* 47, 507–518.
Metzger, J.O. and Bornsheuer, U. (2006) Lipids as renewable resources: current state of chemical and biotechnological conversion and diversification. *Applied Microbiology and Biotechnology* 71, 13–22.
Millar, A., Smith, M.A. and Kunst, L. (2000) All fatty acids are not equal: discrimination in plant membrane lipids. *Trends in Plant Science* 5, 95–101.
Mooney, B.P. (2009) The second green revolution? Production of plant-based biodegradable plastics. *Biochemical Journal* 418, 219–232.
Moretzsohn, M.C., Ferreira, M.A., Amaral, Z.P.S., Coelho, P.J.A., Grattapaglia, D. and Ferreira, M.E. (2002) Genetic diversity of Brazilian oil palm (*Elaeis oleifera* H.B.K.) germplasm collected in the Amazon Forest. *Euphytica* 124, 35–45.
Murphy, D.J. (1991) Designer oilseed crops. Genetic engineering of new oilseed crops for edible and non-edible applications. *Agro-Industry Hi-Tech* 5, 5–10.
Murphy, D.J. (1992) Modifying oilseed crops for non-edible products. *Trends in Biotechnology* 10, 84–87.
Murphy, D.J. (1994) *Designer Oil Crops*. VCH Press, Weinheim, Germany.
Murphy, D.J. (1999) Production of novel oils in plants. *Current Opinion in Biotechnology* 10, 175–180.
Murphy, D.J. (2001) The biogenesis and functions of lipid bodies in animals, plants and microorganisms. *Progress in Lipid Research* 40, 325–438.
Murphy, D.J. (2007a) Future prospects for oil palm in the 21st century: biological and related challenges. *European Journal of Lipid Science and Technology* 109, 296–306.
Murphy, D.J. (2007b) *People, Plants, and Genes: The Story of Crops and Humanity*. Oxford University Press, Oxford, UK.
Murphy, D.J. (2007c) *Plant Breeding and Biotechnology: Societal Context and the Future of Agriculture*. Cambridge University Press, Cambridge, UK.
Murphy, D.J. (2008) Future prospects for biofuels. *Chemistry Today* 26, 44–48.
Murphy, D.J. and Piffanelli, P. (1998) Fatty acid desaturases: structure mechanism and regulation. In: Harwood, J.L. (ed.) *Plant Lipid Biosynthesis: Recent Advances of Agricultural Importance*. Cambridge University Press, Cambridge, UK, pp. 95–1300.
Murphy, D.J., Fairbairn, D.J. and Bowra, D. (1999) Expression of unusual fatty acids in transgenic rapeseed causes induction of glyoxylate cycle genes. *John Innes and Sainsbury Laboratory Annual Report* 1998/99, John Innes Centre, Norwich, UK, pp. 44–46.
Napier, J.A. (2007) The production of unusual fatty acids in transgenic plants. *Annual Review of Plant Biology* 58, 295–319.
Nawrath, C., Poirier, Y. and Somerville, C. (1994) Targeting of the polyhydroxybutyrate biosynthetic pathway to the plastids of *Arabidopsis thaliana* results in high levels of polymer accumulation. *Proceedings of the National Academy of Sciences, United States of America* 91, 12760–12764.

Nguyen, T. and Shanklin, J. (2009) Altering Arabidopsis oilseed composition by a combined antisense-hairpin RNAi gene suppression approach. *Journal of the American Oil Chemists' Society* 86, 41–49.

Philip, S., Keshavarz, T. and Roy, I. (2007) Polyhydroxyalkanoates: biodegradable polymers with a range of applications. *Journal of Chemical Technology and Biotechnology* 82, 233–247.

Princen, L.H. and Rothfus, J.A. (1984) Development of new crops for industrial raw materials. *Journal of the American Oil Chemists' Society* 61, 281–289.

Rahmana, S.M., Kinoshita, T., Anai, T. and Takagi, Y. (2001) Combining ability in loci for high oleic and low linolenic acids in soybean. *Crop Science* 41, 26–29.

Rahmana, S.M., Anai, T., Kinoshita, T. and Takagi, Y. (2003) A novel soybean germplasm with elevated saturated fatty acids. *Crop Science* 43, 527–531.

Ratnasingam, J., Nyugen, V. and Ioras, F. (2008) Evaluation of some finishing properties of oil palm particleboard for furniture application. *Journal of Applied Sciences* 8, 1786–1789.

Robenek, H., Hofnagel, O., Buers, I., Robenek, M.J., Troyer, D. and Severs, N.J. (2006) Adipophilin-enriched domains in the ER membrane are sites of lipid droplet biogenesis. *Journal of Cell Science* 119, 4215–4224.

Rozman, H.D., Ahmadhilmi, K.R. and Abubakar, A. (2003) Polyurethane (PU)-oil palm empty fruit bunch (EFB) composites: the effect of EFBG reinforcement in mat form and isocyanate treatment on the mechanical properties. *Polymer Testing* 23, 559–565.

Schubert, P., Steinbüchel, A. and Schlegel, H.G. (1988) Cloning of the *Alcaligenes eutrophus* genes for synthesis of poly-β-hydroxybutyric acid (PHB) and synthesis of PHB in *Escherichia coli*. *Journal of Bacteriology* 170, 5837–5847.

Sharma, B.K., Adhvaryu, A., Perez, J.M. and Erhan, S.Z. (2005) Soybean oil-based greases: influence of composition on thermo-oxidative and tribochemical behavior. *Journal of Agriculture, Food and Chemistry* 53, 2961–2968.

Shockey, J.M., Gidda, S.K., Chapital, D.C., Kuan, J.C., Dhanoa, P.K., Bland, J.M., Rothstein, S.J., Mullen, R.T. and Dyer, J.M. (2006) Tung tree DGAT1 and DGAT2 have non-redundant functions in triacylglycerol biosynthesis and are localized to different subdomains of the endoplasmic reticulum. *Plant Cell* 18, 2294–2313.

Slater, S., Mitsky, T.A., Houmiel, K.L., Hao, M., Reiser, S.E., Taylor, N.B., Tran, M., Valentin, H.E., Rodriguez, D.J., Stone, D.A., Padgette, S.R., Kishore, G. and Gruys, K.J. (1999) Metabolic engineering of Arabidopsis and Brassica for poly(3-hydroxybutyrate-co-3-hydroxyvalerate) copolymer production. *Nature Biotechnology* 17, 1011–1016.

Smith, M.A., Moon, H., Chowrira, G. and Kunst, L. (2003) Heterologous expression of a fatty acid hydroxylase gene in developing seeds of *Arabidopsis thaliana*. *Planta* 217, 507–516.

Snell, K.D. and Peoples, O.P. (2002) Polyhydroxyalkanoate polymers and their production in transgenic plants. *Metabolic Engineering* 4, 29–40.

Somleva, M.N., Snell, K.D., Beaulieu, J.J., Peoples, O.P., Garrison, B.R. and Patterson, N.A. (2008) Production of polyhydroxybutyrate in switchgrass, a value-added co-product in an important lignocellulosic biomass crop. *Plant Biotechnology Journal* 6, 663–678.

Stahl, U., Banas, A. and Stymne, S. (1995) Plant microsomal phospholipid acyl hydrolases have selectivities for uncommon fatty acids. *Plant Physiology* 107, 953–962.

Steinbüchel, A. and Füchtenbusch, B. (1998) Bacterial and other biological systems for polyester production. *Trends in Biotechnology* 16, 419–427.

Stevens, E.S. (2002) *Green Plastics: An Introduction to the New Science of Biodegradable Plastics*. Princeton University Press, Princeton, New Jersey.

Thelen, J.J. and Ohlrogge, J.B. (2002) Metabolic engineering of fatty acid biosynthesis in plants. *Metabolic Engineering* 4, 12–21.

Thompson, G.A. and Li, C. (1997) Altered fatty acid composition of membrane lipids in seeds and seedling tissues of high-saturate canolas. In: Williams, J.P., Khan, M.U. and Lem, N.W. (eds) *Physiology, Biochemistry and Molecular Biology of Plant Lipids*. Kluwer, Dordrecht, The Netherlands, pp. 313–315.

USDA (2008) Oil crops year in review. Electronic Outlook Report from the Economic Research Service, June 2008 (http://usda.mannlib.cornell.edu/usda/current/OCS-yearbook/OCS-yearbook-06-18-2008_Special_Report.pdf, accessed March 2009).

Voelker, T.A., Hayes, T.R., Cranmer, A.M., Turner, J.C. and Davies, H.M. (1996) Genetic engineering of a quantitative trait: metabolic and genetic parameters influencing the accumulation of laurate in rapeseed. *Plant Journal* 9, 229–241.

Wiberg, E., Banas, A. and Stymne, S. (1997) Fatty acid distribution and lipid metabolism in developing seeds of laurate-producing rape (*Brassica napus* L.). *Planta* 203, 341–348.

10

Starch Characterization, Variety and Application

Jay-lin Jane, Clodualdo C. Maningat and Rungtiwa Wongsagonsup

Introduction

Starch is a polysaccharide produced as an energy reserve and is synthesized in amyloplasts of higher plants. It is the second largest biomass produced on earth, next to cellulose. Unlike cellulose (a β 1–4 linked glucan), which provides functions mainly for structure and support of plants, starch (an α 1–4 glucan with branches) is synthesized in granular forms and stored in seeds, tubers and roots. Starch is later used as the energy source for germination and growth of plants. Starch is also the major energy source for humans and animals. Grains, tubers and roots containing starch are used in food and feed by humans and animals. Starch in most plants consists of about 70–85% amylopectin (highly branched molecules) and 15–30% amylose (primarily linear molecules), which is known as normal starch. Many mutants of crops are found in nature, for example, waxy (*wx*) maize mutants produce starch consisting exclusively of amylopectin, known as waxy maize starch, and amylose-extender (*ae*) mutants of maize produce starch consisting of more than 50% amylose, known as high-amylose maize starch.

Starch granule biosynthesis originates from the hilum, the organic centre, and the synthesis occurs by apposition. Thus, amylopectin and amylose molecules are arranged radially in the granule, which displays a Maltese cross shape when viewed under a polarized-light microscope. The outer chains of amylopectin molecules are in double helical structures, which are organized to give a semi-crystalline structure for starch granules. Because of the semi-crystalline structure, starch granules do not disperse in water at an ambient temperature. Starch granules need to be heated to above the gelatinization temperature with the presence of water to dissociate the crystalline structure and disperse starch molecules. Native starch granules have a specific density of about 1.5 g/cm^3 (Imberty *et al.*, 1991). The granular structure gives efficient packing of starch molecules for energy storage in seeds, roots and tubers until it is ready to be used during germination. The granular structure of starch provides great advantages for its isolation by gravity during wet milling to produce low-cost commodity products. The granular structure also facilitates chemical modifications of starch with subsequent washing to produce various chemically modified starches. Modified starches provide superior functional properties for different applications in the food and other industries.

Starches isolated from different botanical sources or from different organs of the same plants have different chemical and physical structures and, in turn, different functional properties. For example, potato starch produces a paste of very high viscosity with a crystal clear appearance, whereas wheat starch paste displays a low viscosity and an opaque appearance. The different properties between the starches can be attributed to their amylose contents, branch-chain lengths of the amylopectin, phosphate-monoester derivatives, phospholipid contents and granule sizes.

In this chapter, general starch biosynthesis and granule development will be reviewed. Current understandings on molecular and granular structures of starch will be summarized and structures and properties of starches isolated from selected major crops, including maize, wheat, potato, rice and tapioca, and their applications will be discussed.

Starch Biosynthesis and Granule Development

Starch is synthesized in leaves using carbon fixed through photosynthesis in the daytime and mobilized at night and transferred into storage organs, including seeds, tubers, roots and fruits. Starch is also synthesized in many other organs in the plant. Starch biosynthesis occurs in amyloplasts in the storage organ using ADP-glucose as the substrate. ADP-glucose is produced from glucose-1-phosphate and ATP, catalysed by ADP-glucose pyrophosphorylase (AGPase). The activity of AGPase is regulated by several metabolites. Thus, the biosynthesis of ADP-glucose is considered the 'rate-limiting step' of starch biosynthesis (Smith *et al.*, 1997).

Starch granules are synthesized by starch synthase (SS), starch branching enzyme (BE) and starch debranching enzyme (DBE). There are many isozymes for each of the enzymes involved in the biosynthesis of starch. Activities of those isozymes depend on species and organs. Consequently, starches of different botanical sources possess different structures and display different properties. Functions of those enzymes are described briefly below. Readers are referred to several review articles on the subject (Smith *et al.*, 1997; Myers *et al.*, 2000; Nakamura, 2002).

Starch synthases (SSs)

The function of SS is to elongate starch chains by transferring a glucose unit from ADP-glucose to the non-reducing end of a growing chain of amylose or amylopectin, and the glucose is connected by an α 1–4 linkage. There are four SS isoforms commonly found in starch-storing organs and they are classified into SS1, SS2, SS3 and granular-bound starch synthase (GBSS1) on the basis of their primary amino acid sequences. The enzymes of SS1, SS2 and SS3 are present in soluble form and are primarily for amylopectin biosynthesis, whereas GBSS1 is bound with starch molecules and is required for amylose biosynthesis. GBSS1 is also involved in elongating chains of amylopectin, particularly for super-long branches (Yoo and Jane, 2002a). Each of the isoforms has its specific chain elongation pattern (Fujita *et al.*, 2006).

Activities of the SS isoforms vary in different plants. SS1 is the primary isoform in maize endosperm. The maize SS1 preferentially synthesizes short chains of DP 6–15 (Guan and Keeling, 1998) and is responsible for initiating the synthesis of the shorter A and B_1 chains of amylopectin (Commuri and Keeling, 2001). Mutation in SS1 in maize has a severe impact on the formation of the crystalline amylopectin. Fujita *et al.* (2006) report that SS1 in rice is responsible for elongating short chains of DP 6–7 (produced by branching enzyme) to DP 8–12. Further elongation of the chains in rice starch of the wild type is catalysed by SS2a and SS3a.

SS2 plays an essential role in the synthesis of intermediate-length branch-chains of the amylopectin in cereal endosperm. Japonica rice is deficient in SS2a, which results in having more short chains (A and B_1 chains) and fewer long chains in its amylopectin than the Indica rice counterpart. Consequently, the two rice varieties display different thermal properties and textures (Umemoto *et al.*, 1999, 2002). Maize sugary-2 (*su2*) mutant is also deficient in SS2a (Zhang *et al.*, 2004); the mutant suffers a low grain yield. The *su2* maize starch consists of fissured and lobe granules, which have very low gelatinization temperature and less crystalline structures (Perera *et al.*, 2001). The deficiency of starch granule protein-1 (SGP-1) (or SS2a) in wheat affects the starch granule morphology and amylopectin structures (Yamamori *et al.*, 2000). The mutant is enriched in chains of DP 6–10 and depleted in chains of DP 11–25. Barley *sex6* mutant is deficient in SS2a, which leads to novel properties of the starch (Morell *et al.*, 2003). The suppression in amylopectin synthesis of barley *sex6* mutant results in a phenotype having low amylopectin content and a relatively high amylose content (up to 71%). The amylopectin has a short branch-chain length and the starch has a low gelatinization temperature.

SS3 is the major isoform in tubers; a transgenic reduction of SS3 results in 80% loss in soluble enzyme activity in potato tubers and produces fissured starch granules (Edwards *et al.*, 1999). In maize, SS3 is coded by *Dull1* gene (Gao *et al.*, 1998). For the *du1* maize mutant, in which the SS3 is missing, the number of short branch-chains of amylopectin increases, the number of the long B chains substantially decreases, but the chain length of the long branch-chains increases (Wang *et al.*, 1993a). The peak chain length of the long B chains of *du waxy* amylopectin increases to DP 51 from DP 45 of the normal maize starch counterpart (Jane *et al.*, 1999).

GBSS1 is required for amylose biosynthesis. Waxy mutants of plants, missing the waxy gene that encodes the GBSS1 enzyme, produce starch without amylose, known as waxy starch. Different from other starch synthases, GBSS1 is almost all granule-bound and is found throughout the granule interior. GBSS1 is known to use malto-oligosaccharides of DP 2–7 as primers to synthesize amylose in isolated starch granules (Denyer *et al.*, 1996). Without added oligosaccharides, GBSS1 elongates the branch-chains of amylopectin, which can grow into super-long chains (Yoo and Jane, 2002a). The SS1, SS2 and SS3, which remain soluble and surrounding the outside of the granule, elongate starch chains at the non-reducing ends extending from the surface of the growing granule. The newly synthesized chains have freedom to move around and form double helices with one another. The double helical arrangement facilitates the branching enzyme to transfer one chain to the other in the double helix and produces a branch linkage (Borovsky *et al.*, 1979). GBSS1, however, elongates amylose chains that are physically surrounded by semi-crystalline amylopectin molecules in the granule. The elongated chains are confined in the semi-crystalline granule structure and have no opportunity to entangle with other chains for branching reactions to take place. Consequently, the chains remain essentially linear, known as amylose or super-long branch-chains of amylopectin.

Branching enzymes (BEs)

Branching enzymes catalyse cleavage of α 1–4 linkages and transfer the free C-1 to react with the C-6 hydroxyl group of glucose-unit in another chain and form an α 1–6 branch linkage. There are multiple isoforms of branching enzymes and they can be divided into two classes, BEI and BEII, on the basis of their primary sequence. The two classes are also known in pea as branching enzymes B and A, respectively (Smith *et al.*, 1997). *In vitro* reactions have shown that BEI prefers amylose as the substrate and transfers longer chains, whereas BEII prefers using amylopectin as the substrate and transfers shorter chains. BEIIb and BEI are the major branching enzymes in maize endosperm. When the BEIIb is missing in the endosperm of maize, the amylose-extender (*ae*) mutant produces starch consisting of high amylose content (> 50%) and amylopectin with more long branch-chains and fewer short branch-chains (Jane *et al.*, 1999). BEIIa, however, is the dominant isoform in barley (Sun *et al.*, 1998).

Debranching enzymes (DBEs)

DBEs, including isoamylase and pullulanase, catalyse hydrolysis of α 1–6 linkages, which have been proposed for playing a role in amylopectin biosynthesis (Myers *et al.*, 2000; Nakamura, 2002). Both maize and rice sugary-1 mutants, which do not have isoamylase, produce phytoglycogen in the endosperm instead of amylopectin. Two different mechanisms have been proposed for the function of DBEs during starch biosynthesis. One is the 'trimming model', in which SSs and BEs synthesize pre-amylopectin having extra branch-chains. DBEs remove the extra branch-chains and trim the molecule to become normal amylopectin (Myers *et al.*, 2000). Another model is the 'scavenging model'. In this model, DBEs are among scavenging enzymes, which hydrolyse soluble oligosaccharides and allow the glucose to be reused for starch biosynthesis to form granules (Smith, 2001). Once starch molecules are synthesized in the semi-crystalline granule structure, the molecules are no longer susceptible to DBEs hydrolysis. Without DBEs in the endosperm, the soluble oligosaccharides can grow continuously into phytoglycogen catalysed by SSs and BEs.

The biosynthesis of starch granules requires concerted actions of combinations of all the SS isoforms together with the branching and debranching enzymes. Attempts to grow starch granules *in vitro*, however, were not successful. These results suggest that a confined environment in a plastid is essential for synthesized starch chains to crystallize and develop a nucleus, which may be possible through a double helix or single helical complex with lipids. The chain elongation reactions then take place by adding glucose to the non-reducing end of the chain on the surface of the nucleus, gradually growing into a semi-crystalline starch granule. Starch molecules are arranged perpendicular to the surface of the granule, or radially in a spherical granule. In the granule, amylose and amylopectin molecules are synthesized by apposition and grow side by side, and amylose molecules are intertwined with amylopectin.

Granule development

During the growth of the maize kernel, small starch granules (1–4 μm diameter) first appear in the endosperm 5 days after pollination (DAP). The number of starch granules in the endosperm increases between 5 and 12 DAP, but the size of the granules remains similar (Li *et al.*, 2007). The size of starch granules increases to diameters of ~ 7 μm on 14 DAP and further increases to 23 μm on 30 DAP (Li *et al.*, 2007). Starch granule developments in rice and barley endosperm follow the same pattern. The starch content of maize endosperm increases from 1% on 8 DAP to 88.9% on 30 DAP, and the amylose content of the starch increases from 9.2% on 12 DAP to ~ 24.2% on 30 DAP and 45 DAP (mature and dried) (Table 10.1) (Li *et al.*, 2007). A substantial increase in the starch content of maize kernels between 12 DAP and 14 DAP coincides with the increase in activities of starch synthetic enzymes, including ADP-glucose pyrophosphorylase, hexokinase, granule-bound starch synthases and soluble starch synthases. The amylose content of the starch granules increases with kernel maturation, which coincides with the increase in the activity of granule-bound starch synthase (GBSS1) toward later dates of kernel development (Vrinten and Nakamura, 2000). Because starch granules are known to grow by apposition (Baba *et al.*, 1987), an

Table 10.1 Starch and amylose contents of the endosperm at different developmental stages of maize kernels (Li *et al.*, 2007, published with permission).

Days after pollination	Starch content (% dry weight)	Amylose content (%)
0	ND[a]	ND
8	1.0 ± 0.1	ND
10	1.5 ± 0.8	ND
12	2.0 ± 0.5	9.2 ± 0.8
14	10.7 ± 1.7	11.1 ± 0.6
20	68.3 ± 4.9	21.4 ± 0.9
30	88.9 ± 5.1	24.2 ± 0.8

[a]Not determined.

increase in the concentration of amylose with the maturation of the granule would result in more amylose located at the periphery of the starch granule. These results are confirmed by using the surface-gelatinization method; amylose is more concentrated at the periphery than at the core of the granule (Pan and Jane, 2000).

Branch-chain length distribution of amylopectin molecules changes at different maturation stages of kernel development. The average branch-chain lengths increase from DP 23.6 on 10 DAP to DP 26.7 on 14 DAP and then gradually decrease to DP 24.9 on 45 DAP (Li *et al.*, 2007). Amylopectin molecules isolated from maize starch of 10 DAP show a large proportion of very short chains (DP 5–8), which reduce with the maturation of the maize kernels. The unusually large proportion of DP 5–8 may reflect the lack of SS1 activity (Fujita *et al.*, 2006). The amylopectin of 10 DAP also lacks branch-chains longer than DP 55, resulting from the deficiency of soluble starch synthase activities. As the maize kernels develop to 12 DAP, the short branch-chains of DP 4–8 decrease, whereas the long branch-chains (DP > 55) increase and eventually reach the maximum length on 14 DAP (Li *et al.*, 2007). The branch-chain length distribution of amylopectin harvested on 20 DAP and later dates shows a typical bimodal distribution pattern, with the peak chain lengths at DP 14 and DP 48 for the short and long branch-chains, respectively. The difference in the branch-chain lengths are consistent with the gelatinization temperatures of the maize starch isolated at different development stages; the onset gelatinization temperature increases from 61.3°C (8 DAP) to the maximum of 69.0°C (14 DAP), and then decreases to 67.4°C (30 DAP) and 62.8°C (45 DAP) (Li *et al.*, 2007).

Starch Structures

Composition and chemical structures

Starch is composed of two major types of molecules, primarily linear amylose and highly branched amylopectin. Normal starch consists of about 75% amylopectin and 25% amylose; waxy starches consist mainly of amylopectin and 0–8% amylose; high-amylose starches consist of 40–85% amylose (Li *et al.*, 2008). Genetically modified starch has shown more than 90% amylose contents (Case *et al.*, 1998). In addition to amylose and amylopectin, most normal cereal starches also contain lipids and phospholipids (Morrison, 1984; Lim *et al.*, 1994), which have profound impacts on the pasting property of the starch (Yoo and Jane, 2002a). Most tuber and root starches and some cereal starches consist of phosphate monoester derivatives that are found exclusively on amylopectin molecules (Takeda and Hizukuri, 1982). Sugary-1 starch consists of phytoglycogen that is a water-soluble glucan with a highly branched structure and chain length substantially shorter than that of amylopectin. The presence of phytoglycogen is a result of the lack of starch debranching enzymes, as discussed earlier. Many starches, such as high-amylose maize starches (Wang *et al.*, 1993b; Kasemsuwan *et al.*, 1995) and sugary-2 starches (Perera *et al.*, 2001), also contain intermediate components that are branched molecules with smaller molecular weights and longer branch-chain lengths than amylopectin. The structures and properties of these components are discussed in the following sections.

Amylose

Amylose is primarily a linear glucan, consisting of α 1–4 linked D-glucopyranose with a few branches (Takeda *et al.*, 1984). The molecular size of amylose varies from *c.*500 anhydroglucose units (AGU) or DP 500 of high-amylose maize starch (Jane and Chen, 1992) to more than 6000 AGU or DP 6000 of potato starch (Hizukuri *et al.*, 1981). The number of branches increases with the molecular size of the amylose. Some amylose contains clusters of branches that resemble the clusters of amylopectin molecules (Takeda *et al.*, 1990), which result from transferring a cluster from an amylopectin molecule to amylose that is catalysed by the branching enzyme.

Amylose molecules have a random coil conformation in an aqueous solution. The hydrocarbon and the hydroxyl moieties of the anhydroglucose units prompt the formation of double helices by folding two linear starch chains with each other and having the hydrocarbon moiety of the chains folded inside of the helix to be away from water and reach a thermodynamically stable state. The formation of a double helical crystalline structure of starch during ageing is known as retrogradation, which causes bread staling and syneresis of gravy.

Many chemicals that consist of a hydrophobic moiety, such as alcohols and fatty acids (known as complexing agents), can form helical complexes with amylose. The amylose–helical complex has the complexing agent located inside of the hydrophobic cavity of the amylose helix as an inclusion. Depending on the size of the cross-section of the complexing agent, amylose can form single helices of different sizes to accommodate the diameters of the complexing agents. Amylose single helices of six, seven and eight glucose units per turn have been reported when amylose is complexed, respectively, with *n*-butyl alcohol (a linear molecule), iso-butyl alcohol (a branched molecule) and α-naphthol (a bulky molecule). The amylose–helical complex crystallites have been studied using electron microscopy (Yamashita *et al.*, 1973), X-ray diffractometry (Zobel, 1988) and enzymatic method (Jane and Robyt, 1984). X-ray diffraction analysis of the amylose single-helical complexes gives V-type patterns (Zobel, 1988).

Amylopectin

Amylopectin is a highly branched molecule, consisting of α 1–4 linked D-glucopyranose chains that are connected by α 1–6 branch linkages. Amylopectin has a very large molecular weight. Molecular weights of selected amylopectin, determined using high-performance size-exclusion chromatography and multi-angle laser-light scattering (HPSEC-MALLS) and refractive index detectors, are in the range of 7×10^7 to 5.7×10^9 g/mole (Yoo and Jane, 2002b). The results show that molecular weights of waxy starch amylopectins are larger than those of the normal starch counterparts.

Branch-chains of amylopectin molecules can be categorized into an A chain, which has the reducing end attached to a B or a C chain but does not carry any other chains. The B chain has its reducing end attached to another B or a C chain and also carries other A or B chains. The C chain is the only chain of the molecule carrying a free reducing end. The A chains are generally short and extend within one cluster. The B chains have different chain lengths, B_1 chains have lengths that extend within one cluster, B_2 chains extend through two clusters, B_3 chains through three clusters and so on (Hizukuri, 1986). The bimodal branch-chain length distribution of amylopectin differs from the single modal distribution of glycogen (Matsui *et al.*, 1996) and phytoglycogen (Yoo *et al.*, 2002). Branch-chain lengths of amylopectin vary with the botanical sources of the starch, which control the polymorphic forms of crystalline structure (Hizukuri *et al.*, 1983), gelatinization, pasting and retrogradation properties of the starch (Jane *et al.*, 1999).

Branch-chains of amylopectin are arranged in clusters (French, 1984). After native granular waxy maize starch was hydrolysed by acid at room temperature, Yamaguchi *et al.* (1979), using transmission electron microscopy, observed worm-like cluster structures that resisted acid hydrolysis. The worm-like cluster is attributed to the crystalline structure of outer chains of amylopectin molecules. The amorphous region of the waxy maize starch, consisting mainly of branch linkages, is hydrolysed during the acid hydrolysis. Amylose of normal starch is present in an amorphous form in the starch granule.

Starches that consist of amylopectin with more long branch-chains, such as potato, high-amylose maize and *ae* waxy maize starch, display the B-type X-ray pattern. Starches that consist of amylopectin with more short branch-chains, such as maize, wheat, barley, rice, taro and tapioca starch, display the A-type X-ray pattern. X-ray diffraction studies have shown that the A-type polymorph has a monoclinic unit packing, which has closely packed double helices and contains two water molecules in each unit

(Imberty *et al.*, 1991). The B-type polymorph has a hexagonal unit packing and contains *c.* 36 water molecules in each unit cell.

After exhaustive acid hydrolysis to remove the branch structures in the amorphous lamella of amylopectin in the starch granule, Naegeli dextrins remain. The Naegeli dextrins of the B-type polymorphic starches consist of a small number of branched molecules, whereas those of the A-type starches consist of large proportions of branched molecules (Jane *et al.*, 1997). These distinct structures reflect that many branch linkages, carrying short chains, are present within the crystalline region of amylopectin clusters in the A-type starch. On the contrary, there are few branch linkages carrying short branch-chains present in the B-type starches (Jane *et al.*, 1997). These results suggest that the A-type amylopectin carries many short branch-chains located in the middle of the crystalline region, which can be rearranged easily into the closely packed monoclinic unit cell. The rearrangement of the short branch-chain double helices also generates voids in the starch granules of the A-type polymorph (Fig. 10.1).

Extra-long branch-chains are found in amylopectin of many normal cereal starches, such as normal wheat, rice, barley and maize starches, but are not found in waxy cereal starch and potato starch (Hizukuri *et al.*, 1989; Yoo and Jane, 2002a). The extra-long branch-chains have linear structures similar to amylose and are responsible for giving a higher blue value and iodine affinity of normal cereal starch amylopectin than the waxy starch counterparts. The molecular size of the extra-long branch-chains in normal wheat starch is *c.*DP 720 (Yoo and Jane, 2002a). The concentration of the extra-long branch-chains of normal wheat starch amylopectin correlates to the dosage of waxy gene encoding GBSS, primarily for amylose biosynthesis. The results suggest that the extra-long branch-chains are synthesized by the GBSS.

Minor components of starch

Starch granules also contain minor components, including lipids, phospholipids and phosphate monoester derivatives. These

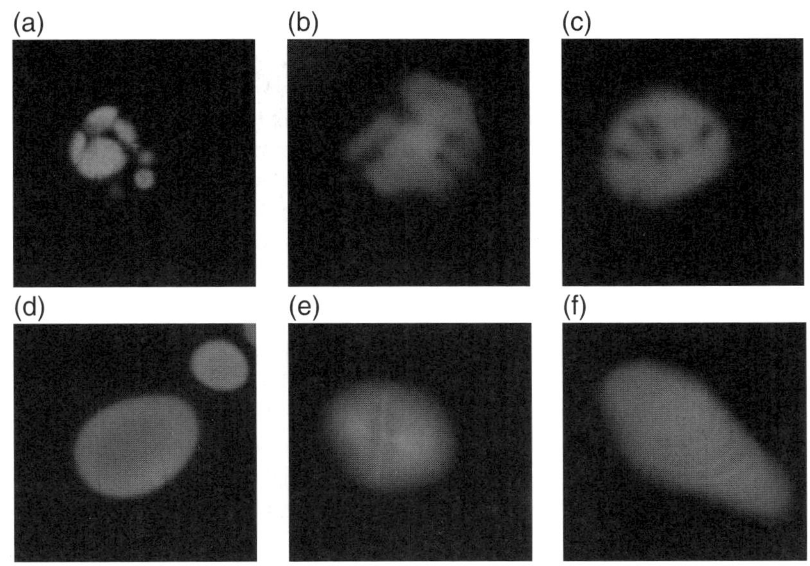

Fig. 10.1. Confocal laser-light scanning micrographs (CLSM) of starch granules. Starch was stained with rhodamine B and unbound dye was removed by rinsing with water and centrifuged immediately. (a) Sugary-2 maize starch; (b) waxy maize starch; (c) normal maize starch; (d) potato starch; (e) high-amylose maize VII starch; and (f) banana starch (from Jane, 2006, reprinted with permission).

minor components have significant impacts on the properties of starch. Lipids and phospholipids are commonly found in cereal starches. Normal cereal starches contain up to 1% lipids, and the level and structures of lipids vary with the botanical origin of the starch. For example, normal maize starch consists mainly of free fatty acids, triglycerides and little phospholipids; normal rice starch consists of more phospholipids and some free fatty acids; and wheat, barley, rye and triticale starches consist exclusively of phospholipids (Lim *et al.*, 1994; Morrison, 1995). Waxy cereal starches contain little lipids, and root and tuber starches contain very little or undetectable lipids (McPherson and Jane, 1999).

A substantial proportion of amylose in normal cereal starch is present in a helical complex with lipids, for example, 43% in normal rice starch, 33% in oat starch and 22% in normal maize starch (Morrison *et al.*, 1993), whereas other amylose is present in a random coil in the granule. Differential scanning calorimetry thermograms show melting peaks of amylose–lipid complex at higher temperatures (above 90°C) than the starch gelatinization temperature. The thermal transition peak of the amylose–lipid complex reappears on immediate rescan, indicating that the amylose–lipid complex formation is an instant reaction. The amylose–lipid complex found in starch granule is amorphous (type 1), which does not give a detectable V-type X-ray pattern and has a lower melting temperature (94–100°C) (Tufvesson *et al.*, 2003). After cooking or annealing, the amylose–lipid complex is reorganized to an orderly lamellar structure (type 2), displays the V-type X-ray pattern and has a higher melting temperature (100–125°C).

When helical complexes develop between starch chains and lipids, the helices hold amylose and amylopectin molecules together, which restrict granule swelling, increase the pasting temperatures and decrease the paste viscosities. Phospholipids in normal wheat starch are known to have greater impact on starch pasting properties than do free fatty acids in maize starch (Yoo and Jane, 2002a). The pasting temperature of normal wheat starch, which forms a helical complex with phospholipids, is substantially higher by 28.1°C than that of the waxy wheat starch, whereas the pasting temperature of normal maize starch, which forms the helical complex with free fatty acids, is 11.7°C higher than that of waxy maize starch (Yoo and Jane, 2002a). The viscosity difference between the waxy and normal wheat starch is also significantly greater than that between waxy and normal maize starch. Amylose–lipid complexes also render starch pastes opaque in colour.

In contrast to lipids and phospholipids, phosphate monoester derivatives found in potato and other tuber and root starches enhance the swelling of starch granules through charge repulsion between the negatively charged phosphate derivatives. Phosphate monoester derivatives are attached covalently to long branch-chains (average DP 42) of amylopectin molecules (Takeda and Hizukuri, 1982). Thus, potato starch displays low gelatinization and pasting temperatures, a very high peak viscosity (Jane *et al.*, 1999) and a crystal clear paste. In the presence of NaCl in the solution, the viscosity of potato starch reduces significantly. This is attributed to the masking of the negative charges of the phosphate monoesters.

Intermediate material is found in many mutants of starches, for example *ae*-mutants of maize starch (Wang *et al.*, 1993b; Kasemsuwan *et al.*, 1995) and sugary-2 maize starch (Perera *et al.*, 2001). The intermediate material has branched structures but smaller molecular weights than amylopectin. During the fractionation of amylose and amylopectin, the intermediate material does not form helical complex crystallites with *n*-butanol and remains in the supernatant with amylopectin.

Phytoglycogen is another water-soluble glucan and is found in sugary-1 mutant of maize and rice, which are deficient in the starch debranching enzyme (James *et al.*, 1995; Wong *et al.*, 2003). Phytoglycogen has shorter average branch-chains (DP 10.3) than amylopectin (DP 18.5 for waxy maize amylopectin) and also smaller molecular weights. The branch-chain lengths of phytoglycogen display a single-modal chain length distribution, which is different from the bimodal distribution of amylopectin.

Structure of Starch Granules

Starch granules of different botanical sources and different organs display different shapes and sizes (Jane *et al.*, 1994). Examples of representative starch granules are shown in Fig. 10.2.

The diameters of starch granules vary from submicron to more than 100 μm. Examples of the diameters of starch granules are potato starch 15–75 μm, wheat A-granules 18–33 μm and B-granules 2–5 μm, maize starch 5–20 μm, rice starch 3–8 μm and amaranth starch 0.5–2 μm (Jane *et al.*, 1994). Starch granules display spherical, oval, disk, polygonal, elongated, kidney and lobe shapes. Leaf starch has flat-shaped small granules with submicron diameter (Zeeman *et al.*, 1998). Wheat, barley, rye and triticale starches display bimodal granule size distributions consisting of disk-shaped, large A-granules and spherical, small B-granules. Rice and oat starches exist as compound granules, which are defined as multiple granules being synthesized within a single amyloplast. Thus, starch granules are packed together tightly and develop into polygonal irregular shapes.

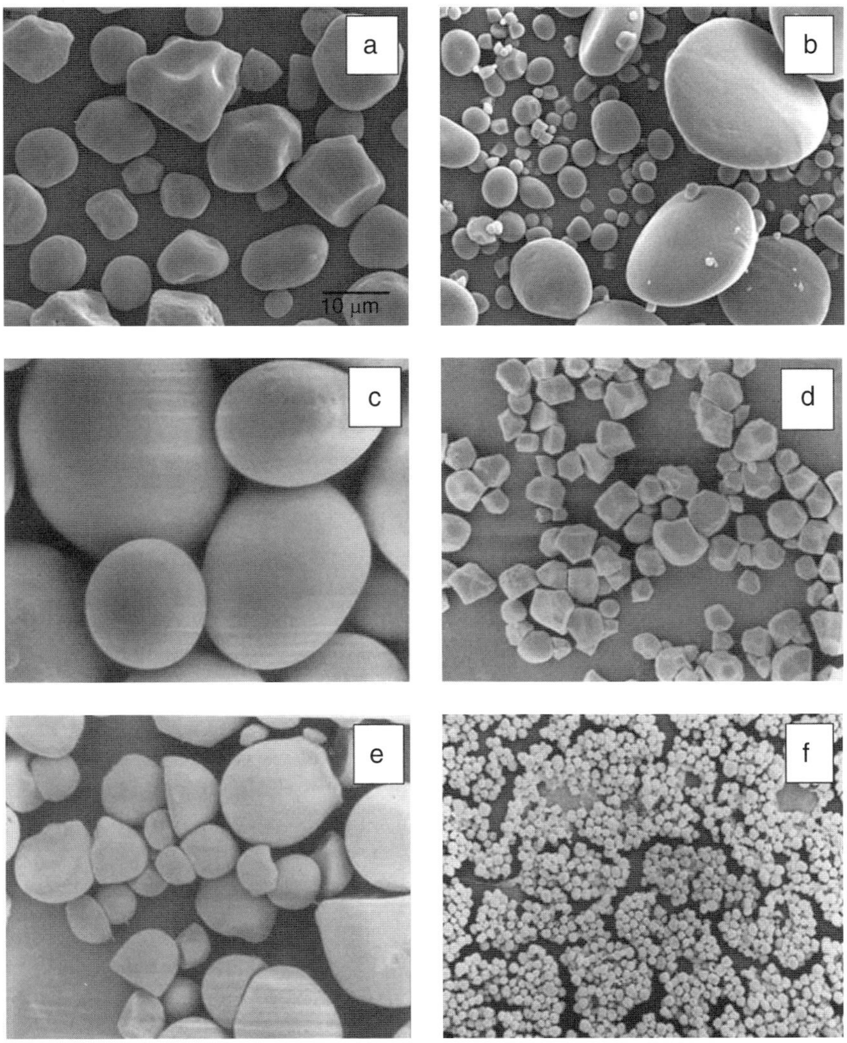

Fig. 10.2. Starch granules of different botanical sources. (a) Maize starch; (b) wheat starch; (c) potato starch; (d) rice starch; (e) tapioca starch; and (f) amaranth starch.

When viewed under a polarized-light microscope, starch granules display birefringence, known as the Maltese cross (Fig. 10.3a).

The Maltese cross birefringence reflects the radial arrangement of starch molecules in the starch granule. The centre of the Maltese cross, the hilum, is the organic centre of the granule where the biosynthesis of the starch granule is initiated. The hilum is not necessarily at the geometric centre of the granule; some can be located close to one end of the granule, reflecting the initiating site of starch biosynthesis.

Surface gelatinization of starch granules using saturated neutral salt solutions, such as LiCl and $CaCl_2$, has been used to investigate structures of starch molecules at different radial locations of a granule (Jane and Shen, 1993; Pan and Jane, 2000). Results show that the amylose content is greater at the periphery of the granule than at the hilum, which agrees with the fact that the amylose content of starch increases with the size and maturity of the starch granules. The same studies also reveal that amylopectin molecules at the inner part of the granule have longer branch-chains than do those at the periphery (Jane and Shen, 1993; Pan and Jane, 2000).

It has been demonstrated that amylose molecules are interspersed among amylopectin molecules. When intact native starch granules are subjected to chemical cross-linking reactions, amylose molecules are cross-linked on to amylopectin molecules but are not cross-linked to other amylose molecules (Jane *et al.*, 1992a; Kasemsuwan and Jane, 1994). Amylose in normal starch granule also provides a function of interacting with amylopectin and preserves the integrity of the granule during transition from the A-type to the V-type polymorph (Chen and Jane, 1994). A cross section of starch granule is shown in Fig. 10.3b. Starch chains of isolated starch granules can be elongated on the surface of the granule using a radioactive-labelled substrate, which is evidence of the absence of membrane on the surface of the granule (Baba *et al.*, 1987). The apparent shell or ghost on the surface of starch granules (Fannon and BeMiller, 1992) can be attributed to highly crystalline starch molecules, likely arising from association between amylose and amylopectin to give a tough layer resistant to cooking.

Properties of Starch

Starch granules go through transformation and provide diverse physical structures and properties for different applications (Fig. 10.4).

Granular starch absorbs ~ 30% water, by weight, after being soaked in water. The water is present in the amorphous region of

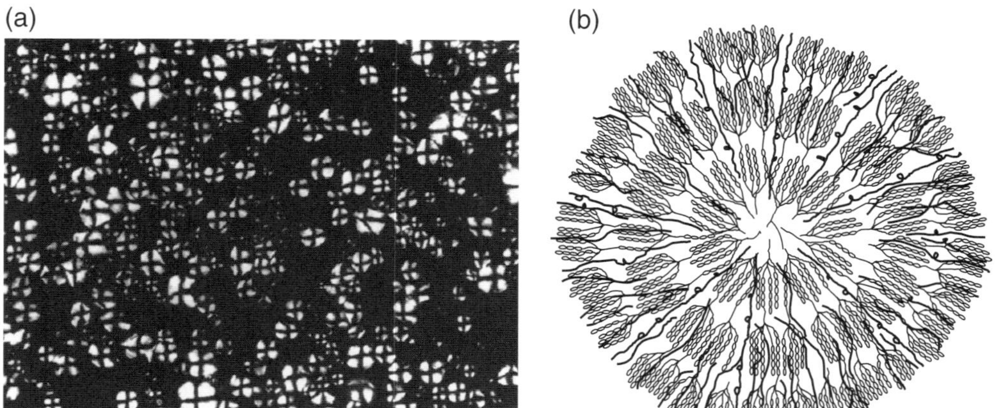

Fig. 10.3. Maltese cross (a) of starch granules viewed under polarized-light microscope and (b) cross section of starch granule. (Reprinted with permission of Taylor & Francis Group LLC-Books, from Jane, 2003; permission conveyed through Copyright Clearance Center, Inc.)

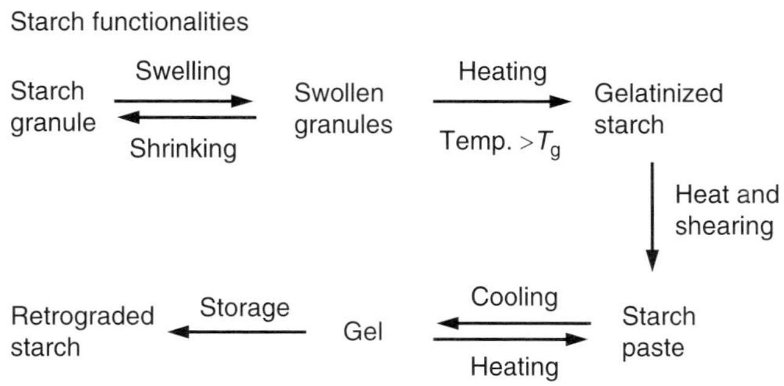

Fig. 10.4. Starch granule transformation.

the starch granule and can be evaporated on drying at a temperature below the gelatinization temperature of the starch. When starch granules are heated in the presence of excess water, the granules eventually lose the native crystalline structure and the Maltese cross shape. This process is known as gelatinization, and each starch has its own characteristic gelatinization temperature. When the gelatinized starch is heated continuously in excess water with shear, the starch granules swell, develop viscosity and become a paste. This process is collectively called pasting. On cooling, the viscosity increases and the starch molecules in the paste develop a network and form a gel. After an extended storage period or repeated freeze and thaw cycles, starch molecules crystallize and water squeezes out. The crystallization process is known as retrogradation and the water separation is called syneresis.

Starch gelatinization

Starches of different botanical sources and genetic backgrounds display different gelatinization properties, including gelatinization temperature, enthalpy change and melting of amylose–lipid complex. Gelatinization temperature of starch can be determined by using a light microscope equipped with a hot stage or by a differential scanning calorimeter (DSC). The temperature at which starch granules lose the Maltese cross in the presence of excess water is determined as the gelatinization temperature. Starch gelatinization shows an endothermic peak on the DSC thermogram. Without water or other proper plasticizers, starch will not gelatinize but will decompose when subjected to very high temperatures.

The gelatinization properties of starches vary substantially. For example, the onset gelatinization temperatures (T_o) vary from 47.8°C (sugary-2 maize starch) to 71.5°C (*ae* waxy maize starch); ranges of gelatinization temperatures vary from 6.6°C (barley starch) to 58.8°C (high-amylose maize VII starch); and enthalpy changes of starch gelatinization vary from 10 J/g (barley starch) to 22 J/g (*ae*-waxy maize starch). Percentage retrogradation of gelatinized starch after being stored at 5°C for 7 days varies from 4.3% (sweet rice starch) to 80.8% (high-amylose maize V starch) (Jane *et al.*, 1999). The gelatinization temperature of starch is highly correlated to the branch-chain length of the amylopectin (Wang *et al.*, 1993b; Yuan *et al.*, 1993; Shi and Seib, 1995). Starches that consist of amylopectin with more long branch-chains, such as high-amylose maize starches and *ae*-waxy maize starch, display higher gelatinization temperatures and greater enthalpy changes, gelatinization temperature ranges and percentage retrogradation. Potato starch amylopectin also has more long branch-chains but displays a very low gelatinization temperature (T_o = 58.2°C). This is attributed to the fact that amylopectin of potato starch contains a high level of phosphate monoester derivatives in addition to its B-type polymorph. The negative charges of the phosphate derivatives repel one another, destabilize the granular structure and reduce the gelatinization temperature.

Pasting of starch

Pasting properties are important for many applications of starch, such as thickening agents and sizing agents. Starches of different botanical sources, which have different amylose and lipid contents and branch structures of the amylopectin, display different pasting properties (Jane *et al.*, 1999). Normal cereal starches containing amylose and lipids display higher pasting temperatures, lower peak viscosities and shear-thinning but higher setback viscosities than do their waxy starch counterparts. The structure and content of lipids of the starch significantly affect the pasting properties. For example, normal wheat starch contains a high concentration of phospholipids (0.06%), whereas normal maize starch consists of mainly free fatty acids, triglycerides and a very low level of phospholipids (~ 0.01%) (Morrison, 1984; Lim *et al.*, 1994; Kasemsuwan and Jane, 1996). The two starches display significant differences in pasting properties, as described earlier. The extremely high pasting temperature and low peak viscosity of normal wheat starch are attributed to its amylose–phospholipid complex (Yoo and Jane, 2002a).

Waxy starches consist of a high proportion (almost 100%) of amylopectin, and amylopectin molecules in starch granules are primarily responsible for the swelling power of starch (Tester and Morrison, 1990). The high amylopectin contents of waxy starches contribute to the greater viscosity of waxy starch pastes. High-amylose starches, such as high-amylose maize (Jane *et al.*, 1999) and high-amylose barley (Song and Jane, 2000), display low viscosity, which can be attributed to their lower amylopectin contents and higher amylose and lipid contents.

Root and tuber starches, such as potato and tapioca, contain low levels of lipids and thus display lower pasting temperatures and higher peak viscosities (McPherson and Jane, 1999). Potato starch has an exceptionally low pasting temperature and high peak viscosity, which are attributed to its elevated content (~0.08%) of phosphate monoester derivatives and also its large granule size (diameter up to 75 μm). Pastes of normal cereal starches, such as normal wheat and normal maize, are more opaque than those of waxy and tuber starches. This is because of light being reflected by the dense, limited swollen granules of normal cereal starches.

Starch retrogradation

Dispersed amorphous starch in pastes, gel or solutions gradually develops double helical crystalline structures and loses its water-binding capacity. This process is known as starch retrogradation. Retrograded starch that consists of large crystalline size, such as retrograded amylose that has 31 AGU in the crystalline region, is highly resistant to enzyme hydrolysis (Jane and Robyt, 1984). Resistant starch is produced commercially from high-amylose maize starch (Sievert *et al.*, 1991) for bulking agent and is used in low-caloric diet food products.

Amylose molecules having linear structures are more prone to develop double helical crystallites. Amylopectin molecules with branched structures, in general, crystallize more slowly. The rate of retrogradation or crystallization of amylopectin depends on the branch-chain length. Amylopectin molecules that have long branch-chains, such as *ae*-waxy maize starch, crystallize faster than those possessing short branch-chains, such as waxy rice starch and sugary-2 maize starch (Jane *et al.*, 1999; Perera *et al.*, 2001).

Maize Starch

Compositions of maize (*Zea mays* L.) starch vary depending on genotypes. Normal maize starch contains about 25–30% amylose and 70–75% amylopectin. It consists of spherical and polygonal-shaped granules (Fig. 10.2a) and the granule size ranges from 5 to 25 μm in diameter (Mauro *et al.*, 2003). It exhibits peak viscosity of 152 rapid visco-units (RVU) on cooking, low breakdown and high setback, resulting from the influence of amylose and lipids (Jane *et al.*, 1999). It develops a strong gel on cooling. Representative structures and physico-chemical properties of normal maize starch, along with those of normal wheat, potato, rice and tapioca starches, are shown in Table 10.2.

Many recessive mutant genes of maize are present naturally and have their primary effect on the synthesis of starch or on a particular protein in maize. Identified recessive mutant genes in maize include amylose-extender (*ae*), brittle (*bt1* and 2), dull (*du*), floury (*fl*), horny (*h*), opaque (*o1* and 2), shrunken (*sh1* and 2), sugary (*su1* and 2) and waxy (*wx*), causing variations in amylose content or the total amount of starch accumulation (Wang *et al.*, 1992). The structural differences in the mutant maize starches result in different physical properties.

The *wx* and *ae* mutant maize starches are widely used commercially. The *wx* mutant lacks the waxy gene, resulting in essentially 100% amylopectin in the *wx* maize starch. The branching structure of *wx* maize starch is similar to that of normal maize amylopectin (Jane *et al.*, 1999). The gelatinization temperature and gelatinization range of *wx* maize starch are not different from those of normal maize starch, whereas *wx* maize starch displays larger enthalpy changes of gelatinization than normal maize starch, reflecting a higher percentage crystallinity of amylopectin (Jane *et al.*, 1999). Because amylose is absent at the periphery of the granules, the *wx* maize starch is more susceptible to amylase hydrolysis and thus has been used to improve feed digestibility (Boyer and Shannon, 2003). Enzyme digestibility of starch granules is negatively related to amylose content (Jane, 2007). The *wx* maize starch develops a higher

Table 10.2. Structures and physico-chemical properties of normal maize, wheat, potato, rice and tapioca starches.

	Maize	Wheat	Potato	Rice	Tapioca
X-ray pattern[a]	A	A	B	A	A/C_A
Amylose content[a] (%)					
Apparent amylose content	29.4	28.8	36.0	25.0	23.5
Absolute amylose content	22.5	25.8	16.9	20.5	17.8
DP_n of amylose[b]	990	830–1570	2110–4920	920–1100	2660
Amylopectin					
M_w ($\times 10^8$)[c]	4.9	3.1	1.7	26.8	0.7
Average branch-chain lengths[a]	24.4	22.7	29.4	22.7	27.6
Phosphorus content[d]					
Phosphate monoesters	Trace	0.001	0.089	0.013	0.008
Phospholipids	0.016	0.053	ND	0.048	ND
Gelatinization properties[a]					
T_o (°C)	64.1	57.1	58.2	70.3	64.3
T_p (°C)	69.4	61.6	62.6	76.2	68.3
ΔH (J/g)	12.3	10.7	15.8	13.2	14.7
Pasting properties[a,e]					
Pasting temperature (°C)	82.0	88.6	63.5	79.9	67.6
Peak viscosity (RVU)	152	104	702	113	173
Hot paste viscosity (RVU)	95	75	165	96	61
Final viscosity (RVU)	169	154	231	160	107
Setback viscosity (RVU)	74	79	66	64	46

[a]Data from Jane *et al.*, 1999.
[b]Data from Hizukuri, 2006.
[c]Data from Yoo and Jane, 2002b.
[d]Data from Lim *et al.*, 1994.
[e]Pastes consist of 8% (w/w, db) starch in water. ND, not detectable; DP_n, number-average degree of polymerization; M_w, weight-average molecular weight.

peak viscosity (205 RVU) than normal maize starch, followed by a rapid breakdown, with little setback (16 RVU) on cooling (Jane *et al.*, 1999). The *wx* maize starch produces a characteristic soft and translucent starch paste (Mauro *et al.*, 2003).

The *ae* mutant starch has a high apparent amylose content ranging from 50 to 80%, resulting from the loss of SBEIIb activity. The starch has large proportions of long B chains with average chain length of DP ~ 30, resulting in a high gelatinization temperature of *ae* maize starch compared with normal maize starch (Jane *et al.*, 1999). The *ae* maize starch displays very low enzyme digestibility, and a recently developed GEMS0067 *ae* starch contains a high-resistant starch content of up to 43.2%. The resistant starch content correlates to the amylose content ($r^2 = 0.99$) (Li *et al.*, 2008). The *ae* maize starch can be used for functional foods. It also has a high tendency to retrograde after being fully cooked under pressure and forms opaque gels and makes strong films. A dominant mutant *ae* (*Ae1-5180*) maize starch can be used to accelerate the development of high-amylose inbred or hybrid lines (Kasemsuwan *et al.*, 1995). Apparent amylose of the *Ae1-5180* dominant mutant maize starch is similar to that of the *ae* mutant (~ 56%).

The *su1* mutant starch is produced resulting from the loss of activity of debranching enzymes, mainly isoamylase (Rahman *et al.*, 1998). In addition to starch, the *su1* mutants accumulate a highly branched polysaccharide, phytoglycogen, to 25% or more of the kernel dry weight. Resulting from missing the SS2a activity, *su2* maize starch has a high proportion of short branch-chains of amylopectin (DP 6–12) and a high content of apparent amylose (33.5%) (Perera *et al.*, 2001). The starch has low crystallinity, as revealed by a weak X-ray diffraction pattern. The *su2* maize starch displays a very low gelatinization temperature and enthalpy changes compared with normal maize starch. Compared with normal and *wx* maize starches (Fig. 10.1), the *su2* maize starch has the largest internal surface, as revealed using confocal laser-light scanning microscope (CLSM). From these characteristics, the *su2* maize starch displays fast enzyme digestibility (Perera *et al.*, 2001) because of the large internal surface area of starch and short branch-chain lengths of amylopectin. The high amylose and lipid contents of *su2* maize starch restrict granule swelling, which results in high pasting temperature, very low peak viscosity and minimal breakdown.

The *du1* mutant starch results from the loss of SS3 activity and the changes in SBEIIa activity (Grimaud *et al.*, 2008). The *du1* maize starch has a relatively high amylose content (31.4%) (Wang *et al.*, 1993a). The *du1* maize amylopectin has a large number of short branch-chains and a reduced number of long B chains, but the chain lengths of the long B chains increases. The gelatinization temperature of *du1* maize starch is similar to that of normal maize starch, but the enthalpy change of gelatinization is lower (Wang *et al.*, 1992).

The *o2* maize contains an increased amount of lysine and less α-, β-, γ- and δ-zein. The lysine content of *o2* maize can be up to twice that of normal maize. It improves the amino acid composition and nutrition value (Mertz *et al.*, 1964). The *o2* maize starch contains low amylose and branch-chains of DP 13–24 in the amylopectin and, thus, is more susceptible to enzyme hydrolysis (Hasjim *et al.*, 2009). It has a large number of pinholes on the surface (resulting from endogenous amylase activity), reflecting greater starch digestibility, and lower swelling power than normal maize starch.

Many double-mutant maize starches have been produced by crossing different genotypes. Double-mutant starches provide additional modifications in the structure and physical properties of starch. For example, *ae wx* maize starch has long branch-chain lengths of amylopectin similar to *ae* maize starch, but does not have any amylose. The *ae wx* maize starch contains 34.5% apparent amylose content despite the starch containing no amylose. It exhibits higher gelatinization temperature and enthalpy change of gelatinization than *wx* maize starch. The long branch-chains of amylopectin of *ae wx* mutant starch complex with lipids and intertwine with other branch-chains to hold the integrity of starch granules during heating and shearing, resulting in a higher pasting temperature and smaller breakdown than waxy maize starch (Jane *et al.*, 1999).

Wheat Starch

Wheat (*Triticum aestivum* L.) starch and vital wheat gluten are economically important co-products of the wet processing of wheat flour. Commercial production methods have experienced significant progress over the years and the past few decades have shown the emergence of hydrocyclone and high-pressure disintegration processes as the favoured techniques (Maningat and Bassi, 1999; Maningat *et al.*, 2009). Approximately 6% of world wheat production is utilized for wet processing into starch and gluten (Oleson, 1994). Laboratory isolation of wheat starch normally follows a simple dough-washing procedure of wheat flour. Isolated starch granules are a binary mixture of large or A-granules with lenticular shape and small or B-granules with spherical shape (Fig. 10.2b). Diameters for A-granules vary from 17.0 to 20.2 μm by laser diffraction measurement or from 28.5 to 34.0 μm by image analysis technique (Wilson *et al.*, 2006). By comparison, the B-granules have dimensions of 3.9–4.2 μm and 8.2–10.0 μm, respectively. The A-granules represent about 75% by weight of wheat starch.

When viewed using a scanning electron microscope, the surface of isolated A-granules appears smooth and shows an equatorial groove (Lineback and Rasper, 1988). It contains surface pores, which are openings to serpentine-like channels that apparently penetrate radially into the hilum.

While considered a pure carbohydrate, wheat starch contains minor components in the interior and exterior of the granules. Very low levels of residual proteins, lipids, ash and dietary fibre are present (Maningat and Seib, 1997). Using a combination of HPLC, gel electrophoresis and tandem mass spectrometry, more than 150 surface-associated proteins were identified from commercially available wheat starches (Kasarda *et al.*, 2008). Many of these proteins, consisting of histones, purothionins and glutenins, have not been recognized previously as starch associated.

In hexaploid wheats, there are three copies of the Wx gene, and normal wheat has three active Wx alleles, termed *Wx-A1a*, *Wx-B1a* and *Wx-D1a*. These alleles code for the Wx protein isoforms, Wx-A1, Wx-B1 and Wx-D1, respectively. Variation in amylose content is related to deficiencies in one or

Table 10.3. Properties of waxy, partial waxy, non-waxy and SPG-1 null high-amylose wheat starch (Abdel-Aal *et al.*, 2002; Sasaki *et al.*, 2002; Yoo and Jane, 2002a; Bertolini *et al.*, 2003; Kim *et al.*, 2003; Tanaka *et al.*, 2006; Hung *et al.*, 2007, 2008a,b; Ral *et al.*, 2008.)

	Waxy	Partial waxy	Non-waxy	High-amylose
Granule size, μm	17.1	N/A	12.9–19.4	16.3
λ_{max}, nm	530–535	575	585–603	572–631
Blue value	0.05	0.28	0.34	0.38–0.42
Iodine affinity, %	0.04	4.3	5.2	N/A
Amylose content, %	1.0–3.2	17.1–28.6	20.2–37.7	28.0–42.2
X-ray diffraction pattern	A	A	A	A, C
Degree of crystallinity, %	30.0–34.8	11.1–28.3	21.6–27.7	9.4
Peak viscosity, RVU	302	197–216	201	59
Differential scanning calorimetry				
T_o, °C	55.6–61.0	51.1–58.3	53.9–62.5	47.2–57.5
T_p, °C	63.3–68.4	59.7–62.5	60.0–66.5	51.8–62.6
T_c, °C	68.2–88.5	67.0–72.5	66.7–73.4	56.1–69.0
ΔH, J/g	5.2–14.5	9.6–12.9	4.8–11.5	0.9–6.5
Amylose–lipid complex				
T_o, °C	N/A	94.4	92.4	86.8–96.5
T_p, °C	N/A	101.1	98.3	95.9–101.3
T_c, °C	N/A	104.4	102.5	100.9–106.9
ΔH, J/g	N/A	0.7–1.9	1.0–1.1	0.3–2.5

[a]N/A, not available.

more of the three Wx proteins. Waxy wheat lines with null alleles at all the three Wx loci do not express any of the Wx proteins, resulting in almost zero amylose content or ~ 100% amylopectin content (Kiribuchi-Otobe *et al.*, 2004). If null alleles exist at one or two Wx loci, the resulting product is partial waxy wheat with amylose content slightly lower than normal wheat. Waxy tetraploid wheat (*T. turgidum* L. var. *durum*), derived from a cross of the partial waxy hard red winter wheat 'Ike' and the durum 'Ben', has been reported (Grant *et al.*, 2001). Methods used to produce waxy wheat consist of traditional cross-breeding, chemical mutation or haploid breeding (Nakamura *et al.*, 1995; Hung *et al.*, 2008a). In contrast, a high-amylose wheat has been produced by back-crossing and generates starch granule protein-1 (SGP-1) null mutant, which enhances amylose content and generates altered amylopectin (Yamamori *et al.*, 2000). The SPG-1 null mutant starches have amylose content in the range of 28–36.9%, which is significantly higher than that of normal wheat starch (25.6%). Moreover, using the RNA interference technique for suppression of SBEIIa and SBEIIb yields high-amylose wheat starch with amylose content of > 70% (Regina *et al.*, 2006). Table 10.3 summarizes the properties of waxy, partial waxy, non-waxy and SGP-1 null high-amylose wheat starch.

Waxy wheat starch exhibits higher gelatinization enthalpy changes and crystallinity than normal and high-amylose wheat starches (Hung *et al.*, 2007). It displays higher swelling power and peak viscosity but lower pasting temperature and final viscosity than normal wheat starch. On the other hand, the SGP-1 null high-amylose wheat starch demonstrates higher amylose content, blue value and λ_{max}, but lower gelatinization temperature and crystallinity than normal wheat starch. Pastes of high-amylose wheat starch had lower peak, breakdown and final viscosity than that of normal wheat starch (Hung *et al.*, 2007). High-amylose wheat starch with > 70% amylose content has sickle-shaped A-granules (Regina *et al.*, 2006). Most of the granules (> 90%) are non-birefringent and have a decreased proportion of short branch-chain lengths of amylopectin (DP 4–12). It improves indices of gastrointestinal health in rats and has the potential to improve human health through its resistant starch content.

Potato Starch

Potato (*Solanum tuberosum* L.) typically has 18–20% solids with 13.5–15.0% starch corresponding to around 75% of the dry weight (Mitch, 1984). A range of 7–20% starch of the fresh weight was reported for 84 accessions from 57 potato species (Schittenhelm and Mengehartmann, 1992). Genotype and environmental factors altered total starch content, phosphorus content and granule size distribution (Haase and Plate, 1996).

The primary raw materials for potato starch production consist of cull potatoes or reclaimed potato starch from potato processing to make potato chips, French fries, dehydrated potatoes and other speciality products. Desirable properties of potato starch include its low gelatinization temperature, low tendency to retrograde, non-cereal flavour, bland taste, high viscosity, high water-binding capacity, translucent paste and relatively good stability.

Genotype, growing conditions, location, tuber size, tuber maturity and processing technologies determine the basic properties of starch (Bergthaller, 2004; Yoo *et al.*, 2009). Variation in properties of potato starches grown in a specific location even within the same cultivar exists due to the influence of harvest date (Zaidul *et al.*, 2007a). A late harvest date for six potato cultivars significantly increases granule size, phosphorus content, peak viscosity and breakdown viscosity and also leads to a significant but slight decrease in amylose content, pasting temperatures and DSC onset and peak temperatures. Representative structures and properties of normal potato starch are shown in Table 10.2.

Normal potato starch contains 20.1–31.0% amylose (Singh *et al.*, 2003). High-amylose potato starch with > 60% amylose has been produced by downregulation of the SBE gene and by additional inhibition of SBEI and SBEII activities (Schwall *et al.*, 2000). On the other hand, amylose-free potato starch can be obtained by induced mutation of the gene encoding GBSS or by antisense inhibition of the expression of GBSS (Visser *et al.*, 1997). The waxy potato starch has amylopectin with shorter branch-chains (average DP 25.8) than the normal potato starch counterpart (average DP 28.6). During pasting, the waxy potato starch displays a lower peak viscosity (410 RVU) than the normal counterpart (590 RVU) because there is no amylose to hold the integrity of swollen granules during heating and shearing (McPherson and Jane, 1999).

Simultaneous downregulation of GBSS, SS2 and SS3 genes produced a waxy potato starch with short-chain amylopectin and with improved freeze–thaw stability (Jobling *et al.*, 2002). In 2002, Jeffcoat and co-workers disclosed a genetically engineered modification of potato starch to suppress the formation of amylose (Jeffcoat *et al.*, 2002). This waxy potato starch when cross-linked and substituted to yield hydroxypropyl distarch phosphate, exhibited a much more viscous paste than similarly modified waxy maize and waxy rice starches.

The size of the round- to oval-shaped potato starch granules (Fig. 10.2c) ranges from 1 to 20 μm for small and from 20 to 110 μm for large granules (Singh *et al.*, 2003). Large granules demonstrated higher peak viscosity, breakdown viscosity and setback viscosity (Zaidul *et al.*, 2007b). Surface structures of potato starch granules previously have been reported to possess protrusions ranging in size from 10–50 to 50–300 nm in diameter (Baldwin *et al.*, 1998). These structures correspond to the 'blocklet' structures and are believed to be the ends of amylopectin clusters ranging in size from 1 to 5 or 5 to 30 clusters.

Starch granules from four potato cultivars were separated into large (40–85 μm), medium (25–40 μm) and small (5–20 μm) fractions (Singh and Kaur, 2004). The shape varied from ellipsoidal, cuboidal or irregular for large and medium granules to spherical or ellipsoidal for the small granules (Fig. 10.2c). The large granules possessed higher amylose content but lower swelling power than the small granules. With decreasing granule size, the DSC thermal transition temperatures increased, while the enthalpy change of gelatinization decreased.

Mealy potato cultivars were observed to have higher amylose content, swelling power, solubility, gelatinization range and enthalpy change of gelatinization, but exhibited lower gelatinization temperatures compared with the less mealy cultivars (Kaur *et al.*, 2002). Significant variation in starch content, phosphorus level and granule size distribution, but not amylose content, was observed due to genotype and environmental conditions (Haase and Plate, 1996). In 32 potato samples, a higher level of amylose in starch resulted in a lower peak viscosity and breakdown viscosity but higher setback viscosity (Zaidul *et al.*, 2007b). Correlation analysis using 105 potato samples showed that amylose content was correlated negatively with peak viscosity and breakdown viscosity but correlated positively with setback viscosity (Zaidul *et al.*, 2007a).

Naturally occurring covalently-bound phosphates are negatively charged and confer polyelectrolyte behaviour to potato starch (Bergthaller, 2004). The resulting Coulombic repulsion may explain its characteristic rapid swelling when heated in water, its high peak viscosity, good paste clarity and low rate of retrogradation of the paste. Phosphate esters also explain its sensitivity to cations. Phosphorus contents of starches isolated from 475 potato cultivars range from 308 to 1244 ppm (Noda *et al.*, 2007). Higher phosphorus content tends to increase swelling power, peak viscosity, breakdown viscosity and to decrease DSC onset and peak temperatures of gelatinization. Amylose content, granule size and enthalpy changes of gelatinization were not affected significantly by phosphorus levels. Phosphorus had a positive correlation with peak viscosity and breakdown viscosity but a negative correlation with setback viscosity (Zaidul *et al.*, 2007b). The unique combination of low concentrations of lipid and protein and the presence of covalently-bound phosphate groups in the amylopectin molecule makes potato starch more preferred in certain food and non-food applications.

Rice Starch

Isolation of rice (*Oryza sativa* L.) starch granules from milled rice can be accomplished by using an alkali, detergent or enzymatic method (Maningat and Juliano, 1980;

Zhong *et al.*, 2009). The granule of rice starch is one of the smallest granules produced by cereal grains. Rice starch has a single granule size distribution (Fig. 10.2d). It measures 3–8 μm and exists as compound granules, having a diameter of up to 150 μm, and forms clusters containing between 20 and 60 individual granules (Juliano, 1985; Fitzgerald, 2004). The surface of the granule looks rather smooth when viewed by a scanning electron microscope. However, the entire surface was covered with globular structures or protrusions when examined with an atomic force microscope (Ayoub *et al.*, 2006).

Rice sold in commerce is classified by amylose content (Suwannaporn *et al.*, 2007) as low (< 20%), medium (21–25%) and high (26–33%). Gelatinization temperature of rice starch can also vary from low, intermediate or high (Maningat and Juliano, 1979; Juliano, 1984). Higher amylose content corresponded with increased DSC onset and peak gelatinization temperatures and decreased peak and trough viscosities (Park *et al.*, 2007a). Swelling power (Hagenimana and Ding, 2005) correlated negatively with amylose content ($r = -0.925$, $P < 0.01$).

Among five rice samples, peak viscosity, swelling power, crystallinity and enthalpy changes of gelatinization of starches decreased, but setback viscosity increased with amylose content (Noosuk *et al.*, 2003). Waxy rice starch had higher swelling volume, resulting in a higher viscosity than other rice starches. Critical concentrations for close packing for waxy, medium-amylose and high-amylose starch were 1.1, 1.5 and 2.5%, respectively. Representative structures and properties of normal rice starch are shown in Table 10.2.

Rice starch with high amylose content contains amylopectin with super-long chains (Hizukuri *et al.*, 1989). A highly positive relationship exists between super-long chain content and setback viscosity (Horibata *et al.*, 2004), indicating that these chains in amylopectin contribute to the setback of starch.

Depending on genetic background, some waxy rice starches exhibiting large proportions of chains with DP 11–16 relative to chains with DP 18–21 displayed low gelatinization temperature (Jane *et al.*, 1999). Other waxy rice starch varieties have similar branch structures to the normal rice starch and display slightly lower and higher gelatinization temperature (Villareal *et al.*, 1997). Waxy rice starch with high gelatinization temperature had a smaller DP 6–9 fraction but a higher DP 14–24 fraction than low gelatinization waxy rice starch (Shi and Seib, 1992).

An increase in the proportion of chains with DP 6–9 reduced the retrogradation rate of waxy rice starch (Shi and Seib, 1992). Lower gelatinization temperatures are attributed to starch with short branch-chain lengths of amylopectin, while large enthalpy changes are associated with longer branch-chain lengths (Jane *et al.*, 1999).

The fine structure of amylopectin affects the physico-chemical properties of starch in several ways. For example, debranched amylopectins from ten non-waxy rice starches showed chains with DP 12–22 increased gelatinization temperature and enthalpy changes of staled amylopectin, but chains with DP 6–9 had the opposite effect (Vandeputte *et al.*, 2003a,b).

For amylopectins derived from long, short/medium and waxy rice varieties, the weight-average molar mass (M_w) determined by HPSEC-MALLS analysis were 1.10×10^8, 1.81×10^8 and 2.47×10^8 g/mole, respectively (Park *et al.*, 2007b). Debranching of amylopectin by isoamylase yielded two fractions with the first fraction (mole% = 16–18) having weight-average DP_w of 76, 68 and 60, respectively, for the above rice varieties. This fraction correlated with increased DSC onset and peak transition temperatures and decreased peak, trough, final and breakdown viscosities (Park *et al.*, 2007a). The second fraction (mole% = 82–84) has average DP_w in the range of 20–22. Isoamylase-debranched amylose from the above rice varieties has M_w ranging from 2.78 to 5.88×10^5 g/mole.

A mutant rice, RS111, with high content of resistant starch is produced by gamma-ray irradiation (Yang *et al.*, 2006). Using the Englyst method, hot cooked rice from RS111 has resistant starch level of 8.2%.

Tapioca Starch

Cassava (*Manihot esculenta* Crantz) is a perennial tropical shrub grown for its starchy storage roots and is considered the most efficient starch producer under suboptimal growing conditions. It is one of the most important food crops in tropical Africa, Asia and Latin America, with production in 2006 estimated at 226 million t (FAO, 2008). The majority of the supply is grown in Thailand, which exports 76% of world trade (Anon., 2008). The root provides as much as one-third of the daily calories of nearly one billion people in 105 countries. A typical composition of cassava roots is 70% moisture, 24% starch, 2% fibre, 1% protein and 3% fat, minerals and sugar (Corbishley and Miller, 1984). The chemical composition is dependent on variety, soil type, climate and age of root. While cassava is the term usually applied in Europe and in the USA to the roots of the plant, tapioca is the name given to the products processed from cassava. Other names used are manioc or mandioca.

Tapioca starch granules are mostly round in shape with a flat surface on one side (Fig. 10.2e) containing a conical pit and a diameter ranging from 4–43 μm (Moorthy, 2002). Other shapes used to describe the granule are oval, truncated or cylindrical, and some compound granules have been observed. Depending on the method used, the crystallinity of tapioca starch could vary from 8–17% to as high as 38%, as reviewed by Moorthy (2002). It possesses A, C or mixed X-ray diffraction patterns with three major peaks at 2θ angles of 15.3, 17.1 and 23.5°. Representative structures and properties of tapioca starch are shown in Table 10.2.

The viscosity of tapioca starch is influenced by plant variety, growing location, time of harvest, age of roots, soil fertility, rainfall amount and starch manufacturing practices (Corbishley and Miller, 1984). Three cassava varieties from Sri Lanka yielded starch with apparent amylose content of 25.4–25.9%, DSC peak temperature of 59.4–64.4°C and enthalpy change of gelatinization of 13.0–14.1 J/g (Wickramasinghe *et al.*, 2009). Several workers reported the DP for tapioca amylose to range from 2660–3390 and an average chain length of 20 for amylopectin with a trimodal molecular weight distribution with peaks of DP 38, DP 18 and DP 11 (Hizukuri, 1986). Jeffcoat and co-workers (2003) reported a low-amylose tapioca starch possessing 2.7% amylose that could be cross-linked to increase process tolerance in food, pharmaceutical and industrial applications.

Food and Non-Food Applications of Starch

Applications of starch in foods

In general, cereal, root and tuber starches and their modified forms are used in foods for a variety of purposes: as thickeners, texturizers, gelling agents, stabilizers, adhesion/coating agents, dusting agents, flowing aid agents, crisping agents, fat replacers, or emulsifiers (Mason, 2009). Maize starch, in particular, is used as the primary substrate for the production of syrups, maltodextrins and sweeteners.

Native starches have limitations and do not perform satisfactorily in processed food products (Langan, 1986). Some starches exhibit low paste viscosity and poor texture and clarity. During processing, the viscosity decreases, resulting from a combination of shear, high acidity and high temperature. During storage of food products, these native starches have a tendency to retrograde and lose water (syneresis) at ambient or refrigerated temperature or during freeze–thaw cycles. Consequently, the starch displays poor adhesion properties in coated products and loses emulsifying abilities. To remedy these shortcomings, starch is modified by reacting with chemicals that are allowed by the government food agencies of many countries. Such modified starches exhibit enhanced paste viscosity, clarity and texture. They acquire improved stability during processing and storage and improved adhesion, emulsifying and water repellency properties.

Conventional modification techniques include cross-linking, substitution, oxidation, acid thinning and combinations of those modifications (Rutenberg and Solarek, 1984; Chiu and Solarek, 2009). Cross-linking reactions introduce intermolecular bridges by using bifunctional chemicals like phosphorus oxychloride, sodium trimetaphosphate or adipic anhydride. The resulting cross-linked starch has restricted granular swelling when heated in water and exhibits resistance to shear, high temperature and high acidity. Cross-linked starches are used in foods that are processed using continuous cooker, sterilizer and canning operations. Substitution reactions by treating native starch with propylene oxide, acetic anhydride or octenyl succinic anhydride introduce functional groups (esters or ethers), which increase viscosity, clarity and stability. Canned, frozen and baked foods are formulated with substituted starches. Oxidation reactions are conducted by treating starch with sodium hypochlorite, calcium hypochlorite or hydrogen peroxide. This reaction creates functional groups (carbonyl and carboxyl) and depolymerizes the starch molecule. Suitable applications for oxidized starches are in cake batters and breadings. Acid thinning is accomplished by treating starch with hydrochloric acid or sulfuric acid that results in glycosidic bond cleavage. The viscosity of the resulting starch product is diminished, but its gelling tendency is enhanced. Confectioneries are the predominant use of acid-thinned starches. In order to have a versatile starch with general applications in a wide variety of food products, combinations of modifications such as cross-linking and substitutions are performed.

Recently, the use of starch in foods for health and well-being reasons has surged, resulting from the preponderance of perceived physiological benefits attributed to resistant starch (Sharma *et al.*, 2008). While 'resistant starch' was first reported in 1982 through the early works of Englyst and co-workers (1982), its commercial value was not appreciated until its definition was developed officially and its health benefits verified. As defined by Asp (1992), resistant starch is the 'sum of starch and starch degradation products not absorbed in the small intestines of healthy individuals'. It is included in the definition of dietary fibre as established by the American Association of Cereal Chemists International, Institute of Medicine, Codex Alimentarius Commission and government agencies of many countries (Anon., 2001; Institute of Medicine, 2002). Four classes of resistant starch have been proposed (Sharma *et al.*, 2008) on the basis of mechanisms of enzyme indigestibility, including inaccessibility of starch to amylases due to physical entrapment (RS1), inherent granular structure of raw starch (RS2), molecular association of amylose or retrogradation (RS3) and chemical modification (RS4). Commercial sources of resistant starch are high-amylose barley and meals and flours from high-amylose maize (RS1), heat-moisture treated high-amylose maize starch (RS2), retrograded high-amylose maize starch, tapioca starch and regular maize starch (RS3), and chemically modified wheat, potato and tapioca starch (RS4). These prior-mentioned starches analysed primarily as insoluble dietary fibre when assayed by standard AOAC procedure, as in Method 991.43. Some RS4-type starches analysed primarily as soluble fibre, such as those from wheat dextrins, maize dextrins and soluble maize fibres. Numerous applications of resistant starch have been formulated in bakery, pasta, noodle and other food products (Maningat *et al.*, 2005a,b, 2006; Woo *et al.*, 2006).

Resistant starch, which is included in the definition of dietary fibre, demonstrates similar physiological benefits as dietary fibre. At present, resistant starch from different sources has attracted huge interest, mainly because of its health benefits and functional properties. As discussed in a number of reviews (Nugent, 2005; Sajilata *et al.*, 2006; Sharma *et al.*, 2008), resistant starch demonstrates the following physiological benefits in humans: modulation of blood glucose and insulin levels, positive effects on gastrointestinal health, increased absorption of minerals, prebiotic or bifidogenic effect, increased fat oxidation and fermentation into butyrate, which is touted to be protective against colorectal cancer. Applications of resistant starch in foods have become a trend among food manufacturers to improve the health and well-being of consumers.

Non-food applications of starch

The traditional major usage of starch is in paper and paperboard, textiles and biofermented products. Starch acts as an adhesive, thickener and film-forming, coating and sizing agent in paper and textile products, and as a source of fermentable carbohydrates for ethanol production (Roper, 2002; Maningat and Bassi, 2004; Peters, 2006; Ochoa *et al.*, 2007).

US production of paper in 2005 was 37.6 million t and paperboard was 45.7 million t (Maurer, 2009). Starch is a major component of many grades of paper (bleached or unbleached; coated or uncoated; wood-containing or wood-free; printing paper; packaging paper) and ranks third, after cellulose and mineral pigments, in consumption for papermaking and conversion processes. It functions as a flocculant and retention aid, as a bonding agent, as a surface sizing agent, as a binder for coatings and as an adhesive for corrugated boards. Major starch sources are derived from maize, potato, waxy maize, wheat and tapioca. In the USA, more than 40% of the 1.1 million t of industrial maize starch consumed for paper and paperboard production is chemically modified. Acid-thinned starches have low viscosity, which restricts use in paper mills but functions primarily for surface sizing at the size press or calender stack of paper machines. Oxidized starches are widely used as surface sizing agents or coating binders. Etherified starches also function as coating and surface sizing agents. Cationic starches act as retention or drainage aids in paper-forming processes and as a surface-sizing agent. Speciality-grades of paper use anionic starches as thickeners in coating colours or as binders in coatings. For the production of tissue and other sanitary grades of paper, aldehyde starches are used as a wet strength agent. Grafted starches are used as surface-sizing agents for a metered-size press.

New developments in paper production take advantage of unique technologies and compositions that address the special requirements of paper products. For example, surface treatment of paper by a mixture comprising 75–95% pigments and 5–25% bonding agent (starch) minimized moisture gradient between printed and non-printed surfaces and reduced waving (Oksanen *et al.*, 2008). A composition containing thinned starch with 2–4% octenyl succinate group imparts oil or grease resistance when coated on paper products (Tippit, 2008). A similar technology to impart resistance to oil and grease stains in paper products is described by Egan and Sharp (2004), who developed a starch-based coating composition consisting of chemically modified starch, a flexibility enhancing agent and a rheological agent. Yoon and Deng (2006) describe a clay–starch composite preparation that imparts higher paper strength than untreated clay by virtue of its increased bonding strength. Coating compositions of starch in admixture with protein in paperboard applications have been described (Bassi *et al.*, 2003a,b).

The principal use of starch in textile manufacture is as a sizing agent (Kirby, 1986). Individual fibres of yarn are shaped or formed into a warp, which is passed through a sizing solution to deposit a coating over the surface of the twisted warp. Penetration by the sizing agent adds stiffness, strength and abrasion resistance. The sized or coated yarn is then heated to dry and be ready for weaving.

Starch derived from feedstocks like cereal grains (maize, grain sorghum, wheat, rye, barley, triticale and rice) and tubers (potato, sweet potato and cassava) is used as a biofermentation substrate for the production of ethyl alcohol (Maningat and Bassi, 2004). Fuel ethanol serves as an oxygenate to control carbon monoxide emissions and as an additive in reformulated gasoline to reduce smog-forming emissions.

Building products like joint compounds and texturized ceiling coatings and oil well drilling fluids utilize the binding and thickening properties of starch (Maningat *et al.*, 2009). Biotechnological applications take advantage of starch and its partially hydrolysed products as a fermentation base for bulk production of a wide variety of organic acids, gums, antibiotics, vitamins and hormones (Tharanathan, 2005). In recent years, the uses of starch have been dominated by events that impact health and well-being in the food industry and, in the non-food sector, by the price volatility of energy and grain commodities and the dwindling oil supply.

Resulting from the burden of accumulating plastic waste and the resulting demand for 'green' or sustainable products, starch has been on centre stage as a source of biodegradable, bio-based or renewable materials (Sun, 2005). Bio-based polymer products derived from renewable agricultural and biomass feedstocks can form the basis for a portfolio of sustainable, environmentally responsible products. Two basic schemes have been adapted. The first is direct extraction from biomass to yield a series of natural polymer materials (cellulose, starch and proteins), fibres and vegetable oils that can form the platform on which polymer materials and products can be developed. Alternatively, the renewable resources/biomass feedstock can be converted to monomers by fermentation or hydrolysis and then further converted by chemical reaction to produce biodegradable polyesters like polylactic acid (Sun, 2005). Starch can also be hydrolysed, for example, to glucose, fermented to succinic acid and converted to 1,4-butanediol by catalytic hydrogenation (Kim *et al.*, 2005). A family of biodegradable aliphatic polyesters can be produced through polycondensation of 1,4-butanediol with aliphatic dicarboxylic acids like succinic acid to generate polybutylene succinate (Kim *et al.*, 2005). Blending of expensive aliphatic biodegradable polyester with low-cost starch is necessary to reduce production costs. Polyesters are modified to increase miscibility and compatibility with starch. Alternatively, compatibilizing or reactive blending is undertaken.

Because of poor interaction between starch granules and petroleum-derived plastics like polyethylene and polypropylene, the function of starch is relegated to a filler status (Narayan, 1991) where its small particle size appears to confer beneficial properties to starch-filled polyethylene film (Jane *et al.*, 1992b; Lim *et al.*, 1992). The development of composite or compounded resins has taken advantage of the use of compatibilizers, for example, maleated derivatives, to improve interaction and impart mechanical and thermal properties to starch-polyolefin blends. The patent literature describes the unique applications of bioresins in moulded pet chew articles (Nie *et al.*, 2006a,b; McCollum *et al.*, 2007), single-use disposable utensils (Nie *et al.*, 2007) and wood substitutes (Nie *et al.*, 2006c).

References

Abdel-Aal, E.-S.M., Hucl, P., Chibbar, R.N., Han, H.L. and Demeke, T. (2002) Physico-chemical and structural characteristics of flours and starches from waxy and non-waxy wheats. *Cereal Chemistry* 79, 458–464.

Anon. (2001) The definition of dietary fiber. *Cereal Foods World* 46, 112–126.

Anon. (2008) Tapioca starch: the outlook for 2008. In: Bentley, S., Nutt, T., Fry, J., Giradello, S. and Marchant, A. (eds) *Starch and Fermentation*. LMC International Ltd, Oxford, UK, pp. 1–12.

Asp, N.G. (1992) Resistant starch. Proceedings from the Second Plenary Meeting of EURESTA: European FLAIR Concerted Action No. 11 on Physiological Implications of the Consumption of Resistant Starch in Man. Preface. *European Journal of Clinical Nutrition* (Suppl. 2), S1.

Ayoub, A., Ohtani, T. and Sugiyama, S. (2006) Atomic force microscopy investigation of disorder process on rice starch granule surface. *Starch/Starke* 58, 475–479.

Baba, T., Yoshii, M. and Kainuma, K. (1987) Incorporation of glucose to the outer chain of starch granule. *Starch/Staerke* 39, 53–56.

Baldwin, P.M., Adler, J., Davies, M.C. and Melia, C.D. (1998) High resolution imaging of starch granule surfaces by atomic force microscopy. *Journal of Cereal Science* 27, 255–265.

Bassi, S., Cain, M.D., Hill, A.M., Maningat, C.C. and Nie, L. (2003a) Protein/starch paper coating compositions and method of use thereof. US Patent 6,517,625.

Bassi, S., Cain, M.D., Hill, A.M., Maningat, C.C. and Nie, L. (2003b) Protein/starch paper coating compositions and method of use thereof. US Patent 6,605,367.

Bergthaller, W. (2004) Developments in potato starches. In: Eliasson, A.-C. (ed.) *Starch in Food: Structure, Function and Applications*. CRC Press, Washington, DC, pp. 241–257.

Bertolini, A.C., Souza, E., Nelson, J.E. and Huber, K.C. (2003) Composition and reactivity of A- and B-type starch granules of normal, partial waxy and waxy wheat. *Cereal Chemistry* 80, 544–549.

Borovsky, D., Smith, E.E., Whelan, W.J., French, D. and Kikumoto, S. (1979) The mechanism of Q-enzyme action and its influence on the structure of amylopectin. *Archives of Biochemistry and Biophysics* 198(2), 627–631.

Boyer, C.D. and Shannon, J.C. (2003) Carbohydrates of the kernel. In: White, P.J. and Johnson, L.A. (eds) *Corn: Chemistry and Technology*, 2nd edn. American Association of Cereal Chemists, Inc, St Paul, Minnesota, pp. 289–311.

Case, S.E., Capitani, T., Whaley, J.K., Shi, Y.C., Trzasko, P., Jeffcoat, R. and Goldfarb, H.B. (1998) Physical properties and gelation behavior of a low-amylopectin maize starch and other high-amylose maize starches. *Journal of Cereal Science* 27(3), 301–314.

Chen, J. and Jane, J. (1994) Preparation of granular cold-water-soluble starches by alcoholic–alkaline treatment. *Cereal Chemistry* 71(6), 618–622.

Chiu, C.-W. and Solarek, D. (2009) Modification of starches. In: BeMiller, J. and Whistler, R. (eds) *Starch Chemistry and Technology*, 3rd edn. Academic Press, New York, pp. 629–655.

Commuri, P.D. and Keeling, P.L. (2001) Chain-length specificities of maize starch synthase I enzyme: studies of glucan affinity and catalytic properties. *The Plant Journal* 25(5), 475–486.

Corbishley, D.A. and Miller, W. (1984) Tapioca, arrowroot and sago starches: Production. In: Whistler, R.L., BeMiller, J.N. and Paschall, E.F. (eds) *Starch: Chemistry and Technology*, 2nd edn. Academic Press, New York, pp. 469–478.

Denyer, K., Clarke, B., Hylton, C., Tatge, H. and Smith, A.M. (1996) The elongation of amylose and amylopectin chains in isolated starch granules. *The Plant Journal* 10(6), 1135–1143.

Edwards, A., Fulton, D.C., Hylton, C.M., Jobling, S.A., Gidley, M., Rossner, U., Martin, C. and Smith, A.M. (1999) A combined reduction in activity of starch synthases II and III of potato has novel effects on the starch of tubers. *The Plant Journal* 17(3), 251–261.

Egan, P.A. and Sharp, S.R. (2004) Non-fluorocarbon paper having flexible starch-based film and methods of producing same. Canadian Patent Appl. 2467601 A1.

Englyst, H., Wiggins, H.S. and Cummings, J.H. (1982) Determination of the non-starch polysaccharides in plant foods by gas-liquid chromatography of constituent sugars as alditol acetates. *Analyst* 107, 307–318.

Fannon, J.E. and BeMiller, J.N. (1992) Structure of corn starch paste and granule remnants revealed by low-temperature scanning electron microscopy after cryopreparation. *Cereal Chemistry* 69, 456–460.

FAO (2008) Cassava for food and energy security. FAO Newsroom (http://www.fao.org/newsroom/en/news/2008/1000899/index.html, accessed 25 July 2008).

Fitzgerald, M. (2004) Starch in rice. In: Champagne, E.T. (ed.) *Rice: Chemistry and Technology*, 3rd edn. American Association of Cereal Chemists, Inc, St Paul, Minnesota, pp. 109–141.

French, D. (1984) Organization of starch granules. In: Whistler, R.L., Bemiller, J.N. and Paschall, E.F. (eds) *Starch Chemistry and Technology*, 2nd edn. Academic Press, New York, pp. 183–247.

Fujita, N., Yoshida, M., Asakura, N., Ohdan, T., Miyao, A., Hirochika, H. and Nakamura, Y. (2006) Function and characterization of starch synthase I using mutants in rice. *Plant Physiology* 140, 1070–1084.

Gao, M., Wanat, J., Stinard, P.S., James, M.G. and Myers, A.M. (1998) Characterization of *dull1*, a maize gene coding for a novel starch synthase. *The Plant Cell* 10, 399–412.

Grant, L.A., Vignaux, N., Doehlert, D.C., McMullen, M.S., Elias, E.M. and Kianian, S. (2001) Starch characteristics of waxy and non-waxy tetraploid (*Triticum turgidum* L. var. *durum*) wheats. *Cereal Chemistry* 78, 590–595.

Grimaud, F., Rogniaux, H., James, M.G., Myers, A.M. and Planchot, V. (2008) Proteome and phosphoproteome analysis of starch granule-associated proteins from normal maize and mutants affected in starch biosynthesis. *Journal of Experimental Botany* 59(12), 3395–3406.

Guan, H.P. and Keeling, P.L. (1998) Starch biosynthesis: understanding the functions and interactions of multiple isozymes of starch synthase and branching enzyme. *Trends in Glycoscience and Glycotechnology* 10(54), 307–319.

Haase, N.U. and Plate, J. (1996) Properties of potato starch in relation to varieties and environmental factors. *Starch/Starke* 48, 167–171.

Hagenimana, A. and Ding, X. (2005) A comparative study on pasting and hydration properties of native rice starches and their mixtures. *Cereal Chemistry* 82, 70–76.

Hasjim, J., Srichuwong, S., Scott, M.P. and Jane, J. (2009) Kernel composition, starch structure, and enzyme digestibility of *opaque-2* maize and quality protein maize. *Journal of Agricultural and Food Chemistry* 57(5), 2049–2055.

Hizukuri, S. (1986) Polymodal distribution of the chain lengths of amylopectins and its significance. *Carbohydrate Research* 147, 342–347.

Hizukuri, S. (2006) Starch: analytical aspects. In: Eliasson, A.C. (ed.) *Carbohydrates in Food*. CRC Press, Boca Raton, Florida, pp. 305–390.

Hizukuri, S., Takeda, Y. and Yasuda, M. (1981) Multi-branched nature of amylose and the action of debranching enzymes. *Carbohydrate Research* 94, 205–213.

Hizukuri, S., Kaneko, T. and Takeda, Y. (1983) Measurement of the chain length of amylopectin and its relevance to the origin of crystalline polymorphism of starch granules. *Biochimica et Biophysica Acta* 760(1), 188–191.

Hizukuri, S., Takeda, Y., Maruta, N. and Juliano, B.O. (1989) Molecular structures of rice starch. *Carbohydrate Research* 189, 227–235.

Horibata, T., Nakamoto, M., Fuwa, H. and Inouchi, N. (2004) Structural and physicochemical characteristics of endosperm starches of rice cultivars recently bred in Japan. *Journal of Applied Glycoscience* 51(4), 303–313.

Hung, P.V., Maeda, T. and Morita, N. (2007) Study on physicochemical characteristics of waxy and high-amylose wheat starches in comparison with normal wheat starch. *Starch/Starke* 59, 125–131.

Hung, P.V., Yasui, T., Maeda, T. and Morita, N. (2008a) Physicochemical characteristics of starches of two sets of near isogenic wheat lines with different amylose content. *Starch/Starke* 60, 34–40.

Hung, P.V., Maeda, T., Miskelly, D., Tsumori, R. and Morita, N. (2008b) Physicochemical characteristics and fine structure of high-amylose wheat starches isolated from Australian wheat cultivars. *Carbohydrate Polymers* 71, 656–663.

Imberty, A., Buleon, A., Vinh, T. and Perez, S. (1991) Recent advances in knowledge of starch structure. *Starch/Staerke* 43(10), 375–384.

Institute of Medicine (2002) *Dietary Reference Intakes: Energy, Carbohydrates, Fiber, Fat, Fatty Acids, Cholesterol, Protein and Amino Acids*. National Academies Press, Washington, DC.

James, M.G., Robertson, D.S. and Myers, A.M. (1995) Characterization of the maize gene *sugary1*, a determinant of starch composition in kernels. *The Plant Cell* 7, 417–429.

Jane, J. (2003) Starch: structure and properties. In: Tomasik, P. (ed.) *Chemical and Functional Properties of Food Saccharides*. CRC Press, Inc., Boca Raton, Florida, pp. 81–101.

Jane, J. (2006) Current understanding on starch granule structures. *Journal of Applied Glycoscience* 53, 205–213.

Jane, J. (2007) Structure of starch granules. *Journal of Applied Glycoscience* 54, 31–36.

Jane, J. and Chen, J.F. (1992) Effect of amylose molecular size and amylopectin branch chain length on paste properties of starch. *Cereal Chemistry* 69(1), 60–65.

Jane, J. and Robyt, J.F. (1984) Structure studies of amylose-V complexes and retrograded amylose by action of alpha amylases, and a new method for preparing amylodextrins. *Carbohydrate Research* 132(1), 105–118.

Jane, J. and Shen, J.J. (1993) Internal structure of the potato starch granule revealed by chemical gelatinization. *Carbohydrate Research* 247, 279–290.

Jane, J., Xu, A., Radosavljevic, M. and Seib, P.A. (1992a) Location of amylose in normal starch granules. I. Susceptibility of amylose and amylopectin to cross-linking reagents. *Cereal Chemistry* 69(4), 405–409.

Jane, J., Shen, L., Wang, L. and Maningat, C.C. (1992b) Preparation and properties of small-particle corn starch. *Cereal Chemistry* 69, 280–283.

Jane, J., Kasemsuwan, T., Leas, S., Zobel, H. and Robyt, J.F. (1994) Anthology of starch granule morphology by scanning electron microscopy. *Starch/Staerke* 46(4), 121–129.

Jane, J., Wong, K.-S. and McPherson, A.E. (1997) Branch-structure difference in starches of A- and B-type X-ray patterns revealed by their Naegeli dextrins. *Carbohydrate Research* 300(3), 219–227.

Jane, J., Chen, Y.Y., Lee, L.F., McPherson, A.E., Wong, K.S., Radosavljevic, M. and Kasemsuwan, T. (1999) Effects of amylopectin branch chain length and amylose content on the gelatinization and pasting properties of starch. *Cereal Chemistry* 76(5), 629–637.

Jeffcoat, R., Mason, W.R., Emling, J.L. and Chiu, C.-W. (2002) Stabilized or stabilized crosslinked waxy potato starch. US Patent 6,488,980 B1.

Jeffcoat, R., Hanchett, D.J. and Tayal, A. (2003) Process tolerant low amylose tapioca distarch adipates. US Patent Application Pub. No. US 2003/0094104 A1.
Jobling, S.A., Westcott, R.J., Tayal, A., Jeffcoat, R. and Schwall, G.P. (2002) Production of freeze–thaw stable potato starch by antisense inhibition of three starch synthase genes. *Nature Biotechnology* 20, 295–299.
Juliano, B.O. (1984) Rice starch: production, properties and uses. In: Whistler, R.L., BeMiller, J.N. and Paschall, E.F. (eds) *Starch: Chemistry and Technology*, 2nd edn. Academic Press, New York, pp. 507–528.
Juliano, B.O. (1985) Polysaccharides, proteins and lipids of rice. In: Juliano, B.O. (ed.) *Rice: Chemistry and Technology*, 2nd edn. American Association of Cereal Chemists, Inc, St Paul, Minnesota, pp. 59–174.
Kasarda, D.D., Dupont, F.M., Vensel, W.H., Altenbach, S.B., Lopez, R., Tanaka, C.K. and Hurkman, W.J. (2008) Surface-associated proteins of wheat starch granules: suitability of wheat starch for celiac patients. *Journal of Agricultural and Food Chemistry* 56, 10292–10302.
Kasemsuwan, T. and Jane, J. (1994) Location of amylose in normal starch granules. II. Locations of phosphodiester cross-linking revealed by phosphorus-31 nuclear magnetic resonance. *Cereal Chemistry* 71(3), 282–287.
Kasemsuwan, T. and Jane, J. (1996) Quantitative method for the survey of starch phosphate derivatives and starch phospholipids by ^{31}P nuclear magnetic resonance spectroscopy. *Cereal Chemistry* 73, 702–707.
Kasemsuwan, T., Jane, J., Schnable, P., Stinard, P. and Robertson, D. (1995) Characterization of the dominant mutant amylose-extender (*Ae1-5180*) maize starch. *Cereal Chemistry* 72(5), 457–464.
Kaur, L., Singh, N. and Sodhi, N.S. (2002) Some properties of potatoes and their starches II. Morphological, thermal and rheological properties of starches. *Food Chemistry* 79, 183–192.
Kim, H.-S., Yang, H.-S. and Kim, H.-J. (2005) Biodegradability and mechanical properties of agro-flour-filled polybutylene succinate biocomposites. *Journal of Applied Polymer Science* 97, 1513–1521.
Kim, W., Johnson, J.W., Graybosch, R.A. and Gaines, C.S. (2003) Physico-chemical properties and end-use quality of wheat starch as a function of waxy protein alleles. *Journal of Cereal Science* 37, 195204.
Kirby, K.W. (1986) Textile industry. In: Wurzburg, O.B. (ed.) *Modified Starches: Properties and Use*. CRC Press, Inc, Boca Raton, Florida, pp. 229–252.
Kiribuchi-Otobe, C., Kurita, K., Yanagisawa, T., Takeda, G. and Inazu, A. (2004) Interactions among *Wx* genes and effects of a mutated *Wx-D1e* allele on starch properties in wheat. *Cereal Chemistry* 81, 48–50.
Langan, R.E. (1986) Food industry. In: Wurzburg, O.B. (ed.) *Modified Starches: Properties and Uses*. CRC Press, Inc, Boca Raton, Florida, pp. 199–212.
Li, L., Blanco, M. and Jane, J. (2007) Physico-chemical properties of endosperm and pericarp starches during maize development. *Carbohydrate Polymers* 67(4), 630–639.
Li, L., Jiang, H., Campbell, M., Blanco, M. and Jane, J. (2008) Characterization of maize amylose-extender (*ae*) mutant starches. Part I: Relationship between resistant starch contents and molecular structures. *Carbohydrate Polymers* 74(3), 396–404.
Lim, S., Jane, J., Rajagopalan, S. and Seib, P.A. (1992) Effect of starch granule size on physical properties of starch-filled polyethylene film. *Biotechnology Progress* 8(1), 51–57.
Lim, S., Kasemsuwan, T. and Jane, J. (1994) Characterization of phosphorus in starch by ^{31}P-nuclear magnetic resonance spectroscopy. *Cereal Chemistry* 71(5), 488–493.
Lineback, D.R. and Rasper, V.F. (1988) Wheat carbohydrates. In: Pomeranz, Y. (ed.) *Wheat: Chemistry and Technology*, 3rd edn. American Association of Cereal Chemists, Inc, St Paul, Minnesota, pp. 277–372.
McCollum, J.E., Goettert, E.J., Nie, L., Bassi, S.D., Parker, M.D., Worthy, N. and Maningat, C.C. (2007) Gluten-starch pet chew article. US Patent Application Publ. 2007224131 A1.
McPherson, A.E. and Jane, J. (1999) Comparison of waxy potato with other root and tuber starches. *Carbohydrate Polymers* 40(1), 57–70.
Maningat, C.C. and Bassi, S.D. (1999) Wheat starch production. *Proceedings of International Starch Technology Conference*. University of Illinois, Urbana, Illinois, pp. 26–40.
Maningat, C.C. and Bassi, S.D. (2004) Fuel alcohol production. In: Wrigley, C., Corke, H. and Walker, C.E. (eds) *Encyclopedia of Grain Science*. Elsevier Academic Press, Oxford, UK, pp. 406–415.
Maningat, C.C. and Juliano, B.O. (1979) Properties of lintnerized starch granules from rices differing in amylose content and gelatinization temperature. *Starch/Starke* 31, 5–10.
Maningat, C.C. and Juliano, B.O. (1980) Starch lipids and their effect on rice starch properties. *Starch/Starke* 32, 76–82.

Maningat, C.C. and Seib, P.A. (1997) Update on wheat starch and its uses. In: *Proceedings of International Wheat Quality Conference*. American Association of Cereal Chemists, Inc, St Paul, Minnesota, pp. 261–284.

Maningat, C.C., Bassi, S., Woo, K., Dohl, C., Gaul, J. and Stempien, G. (2005a) Formulation of high-protein, high-fiber (low-carbohydrate), reduced calorie breads. *AIB Technical Bulletin* 27, 1–16.

Maningat, C.C., Dohl, C.T., Gaul, J.A., Stempien, G.J., Bassi, S.D., Ranjan, S. and Woo, K. (2005b) High protein, reduced-carbohydrate bakery and other food products. US Patent Application Publ. 20050037125.

Maningat, C.C., Bassi, S.D. and Woo, K. (2006) High-fiber, high-protein pasta and noodle products. US Patent Application Publ. 20060134295 A1.

Maningat, C.C., Seib, P.A., Bassi, S.D., Woo, K.S. and Lasater, G.D. (2009) Wheat starch: production, properties, modification and uses. In: BeMiller, J. and Whistler, R. (eds) *Starch: Chemistry and Technology*, 3rd edn. Academic Press, New York, pp. 441–510.

Mason, W.R. (2009) Starch use in foods. In: BeMiller, J. and Whistler, R. (eds) *Starch Chemistry and Technology*, 3rd edn. Academic Press, New York, pp. 745–795.

Matsui, M., Kakut, M. and Misaki, A. (1996) Fine structural features of oyster glycogen: mode of multiple branching. *Carbohydrate Polymers* 31(4), 227–235.

Maurer, H.W. (2009) Starch in the paper industry. In: BeMiller, J. and Whistler, R. (eds) *Starch Chemistry and Technology*, 3rd edn. Academic Press, New York, pp. 658–713.

Mauro, D.J., Abbas, I.R. and Orthoefer, F.T. (2003) Corn starch modification and uses. In: White, P.J. and Johnson, L.A. (eds) *Corn: Chemistry and Technology*, 2nd edn. American Association of Cereal Chemists, Inc, St Paul, Minnesota, pp. 605–634.

Mertz, E.T., Bates, L.S. and Nelson, O.E. (1964) Mutant gene that changes protein composition and increases lysine content of maize endosperm. *Science* 145(3629), 279–280.

Mitch, E.L. (1984) Potato starch: production and uses. In: Whistler, R.L., BeMiller, J.N. and Paschall, E.F. (eds) *Starch: Chemistry and Technology*, 2nd edn. Academic Press, New York, pp. 479–490.

Moorthy, S.N. (2002) Physico-chemical and functional properties of tropical tuber starches: a review. *Starch/Starke* 54, 559–592.

Morell, M.K., Kosar-Hashemi, B., Cmiel, M., Samuel, M.S., Chandler, P., Rahman, S., Buleon, A., Batey, I.L. and Li, Z. (2003) Barley *sex6* mutant lack starch synthase IIa activity and contain a starch with novel properties. *The Plant Journal* 34, 173–185.

Morrison, W.R. (1984) Plant lipids. *Research in Food Science and Nutrition* 5, 247–260.

Morrison, W.R. (1995) Starch lipids and how they relate to starch granule structure and functionality. *Cereal Foods World* 40(6), 437–438, 440–441, 443–446.

Morrison, W.R., Law, R.V. and Snape, C.E. (1993) Evidence for inclusion complexes of lipids with V-amylose in maize, rice and oat starches. *Journal of Cereal Science* 18, 107–109.

Myers, A.M., Morell, M.K., James, M.G. and Ball, S.G. (2000) Recent progress toward understanding biosynthesis of the amylopectin crystal. *Plant Physiology* 122, 989–997.

Nakamura, T., Yamamori, M., Hirano, H., Hidaka, S. and Nagamine, T. (1995) Production of waxy (amylose-free) wheats. *Molecular and General Genetics* 248, 253–259.

Nakamura, Y. (2002) Towards a better understanding of the metabolic system for amylopectin biosynthesis in plants: rice endosperm as a model tissue. *Plant Cell Physiology* 43, 718–725.

Narayan, R. (1991) Starch-based plastics. In: Chum, H.L. (ed.) *Biopolymers from Biobased Materials*. Noyes Data Corp, Park Ridge, New Jersey, pp. 90–114.

Nie, L., Bassi, S.D., Maningat, C.C. and Parker, M.D. (2006a) Pet treats with rough surface texture. US Patent Application Publ. 2006188632 A1.

Nie, L., Parker, M.D., Woo, K., Maningat, C.C. and Bassi, S.D. (2006b) Low calorie injection molded starch-based pet chew bodies. US Patent Application Publ. 2006193959 A1.

Nie, L., Bassi, S.D., Maningat, C.C. and Parker, M.D. (2006c) Starch-plastic composite resins and profiles made by extrusion and useful as a wood substitute. US Patent Application Publ. 2006194902 A1.

Nie, L., Parker, M.D., Bassi, S.D. and Maningat, C.C. (2007) Thermotolerant starch-polyester blends for use in making cutlery, cups, plates, bowls and packagings. US Patent Application Publ. 2007092745 A1.

Noda, T., Kottearachchi, N.S., Tsuda, S., Mori, M., Takigawa, S., Matsuura-Endo, C., Kim, S.-J., Hashimoto, N. and Yamauchi, H. (2007) Starch phosphorus content in potato (*Solanum tuberosum* L.) cultivars and its effect on other starch properties. *Carbohydrate Polymers* 68, 793–796.

Noosuk, P., Hill, S.E., Pradipasena, P. and Mitchell, J.R. (2003) Structure–viscosity relationships for Thai rice starches. *Starch/Starke* 55, 337–344.

Nugent, A.P. (2005) Health properties of resistant starch. *Nutrition Bulletin* 30, 27–54.

Ochoa, S., Yoo, A., Repke, J.-U., Wozny, G. and Yang, D.R. (2007) Modeling and parameter identification of the simultaneous saccharification-fermentation process for ethanol production. *Biotechnology Progress* 23, 1454–1462.

Oksanen, M., Sinkko, T., Varvemaa, J. and Haapoja, K. (2008) Paper and method for making paper. International Publication No. WO 2008/132283 A1.

Oleson, B.T. (1994) World wheat production, utilization and trade. In: Bushuk, W. and Rasper, V.F. (eds) *Wheat: Production, Properties and Quality*. Chapman and Hall, London, pp. 1–11.

Pan, D.D. and Jane, J. (2000) Internal structure of normal maize starch granules revealed by chemical surface gelatinization. *Biomacromolecules* 1(1), 126–132.

Park, I.-M., Ibanez, A.M., Zhong, F. and Shoemaker, C.F. (2007a) Gelatinization and pasting properties of waxy and non-waxy rice starches. *Starch/Starke* 59, 388–396.

Park, I.M., Ibanez, A.M. and Shoemaker, C.F. (2007b) Rice starch molecular size and its relationship with amylose content. *Starch/Starke* 59, 69–77.

Perera, C., Lu, Z., Sell, J. and Jane, J. (2001) Comparison of physicochemical properties and structures of *sugary-2* cornstarch with normal and waxy cultivars. *Cereal Chemistry* 78(3), 249–256.

Peters, D. (2006) Carbohydrates for fermentation. *Biotechnology Journal* 1, 806–814.

Rahman, A., Wong, K.-S., Jane, J., Myers, A.M. and James, M.G. (1998) Characterization of SU1 isoamylase, a determinant of storage starch structure in maize. *Plant Physiology* 117(2), 425–435.

Ral, J.-P., Cavanagh, C.R., Larroque, O., Regina, A. and Morell, M.K. (2008) Structural and molecular basis of starch viscosity in hexaploid wheat. *Journal of Agricultural and Food Chemistry* 56, 4188–4197.

Regina, A., Bird, A., Topping, D., Bowden, S., Freeman, J., Barsby, T., Kosar-Hashemi, B., Li, Z., Rahman, S. and Morell, M. (2006) High-amylose wheat generated by RNA interference improves indices of large-bowel health in rats. *Proceedings of the National Academy of Sciences of the United States of America* 103(10), 3546–3551.

Roper, H. (2002) Renewable raw materials in Europe – industrial utilization of starch and sugar. *NutraCos* 1, 37–44.

Rutenberg, M.W. and Solarek, D. (1984) Starch derivatives: production and uses. In: Whistler, R.L., BeMiller, J.N. and Paschall, E.F. (eds) *Starch Chemistry and Technology*, 2nd edn. Academic Press, Inc, New York, pp. 312–388.

Sajilata, M.G., Rekha, S.S. and Kulkarni, P.R. (2006) Resistant starch – a review. *Comprehensive Reviews in Food Science and Food Safety* 5, 1–17.

Sasaki, T., Yasui, T., Matsuki, J. and Satake, T. (2002) Comparison of physical properties of wheat starch gels with different amylose content. *Cereal Chemistry* 79, 861–866.

Schittenhelm, S. and Mengehartmann, U. (1992) Starch and amylose content as well as size and morphology of starch granules from tuber-bearing *Solanum* species. *Landbauforsch Volkenrode* 42, 117–126.

Schwall, G.P., Safford, R., Westcott, R.J., Jeffcoat, R., Tayal, A., Shi, Y.-C., Gidley, M.J. and Jobling, S.A. (2000) Production of very-high-amylose potato starch by inhibition of SBE A and B. *Nature Biotechnology* 18, 551–554.

Sharma, A., Yadav, B.S. and Yada, R.B. (2008) Resistant starch: physiological roles and food applications. *Food Reviews International* 24, 193–234.

Shi, Y.-C. and Seib, P.A. (1992) The structure of four waxy starches related to gelatinization and retrogradation. *Carbohydrate Research* 227, 131–145.

Shi, Y.-C. and Seib, P.A. (1995) Fine structure of maize starches from four *wx*-containing genotypes of the W64A inbred line in relation to gelatinization and retrogradation. *Carbohydrate Polymers* 26(2), 141–147.

Sievert, D., Czuchajowska, Z. and Pomeranz, Y. (1991) Enzyme-resistant starch. III. X-ray diffraction of autoclaved amylomaize VII starch and enzyme-resistant starch residues. *Cereal Chemistry* 68(1), 86–91.

Singh, N. and Kaur, L. (2004) Morphological, thermal, rheological and retrogradation properties of potato starch fractions varying in granule size. *Journal of the Science of Food and Agriculture* 84, 1241–1252.

Singh, N., Singh, J., Kaur, L., Sodhi, N.S. and Gill, B.S. (2003) Morphological, thermal and rheological properties of starches from different botanical sources. *Food Chemistry* 81, 219–231.

Smith, A.M. (2001) The biosynthesis of starch granules. *Biomacromolecules* 2, 335–341.

Smith, A.M., Denyer, K. and Martin, C. (1997) The synthesis of the starch granule. *Annual Review of Plant Physiology and Plant Molecular Biology* 48, 67–87.
Song, Y. and Jane, J. (2000) Characterization of barley starches of waxy, normal, and high amylose varieties. *Carbohydrate Polymers* 41(4), 365–377.
Sun, C., Sathish, P., Ahlandsberg, S. and Jansson, C. (1998) The two genes encoding starch-branching enzymes IIa and IIb are differentially expressed in barley. *Plant Physiology* 118, 37–49.
Sun, X.S. (2005) Plastics derived from starch and poly (lactic acids). In: Wool, R.P. and Sun, X.S. (eds) *Bio-based Polymers and Composites*. Academic Press, New York, pp. 369–410.
Suwannaporn, P., Pitiphumpong, S. and Champangern, S. (2007) Classification of rice amylose content by discriminant analysis of physicochemical properties. *Starch/Starke* 59, 171–177.
Takeda, Y. and Hizukuri, S. (1982) Location of phosphate groups in potato amylopectin. *Carbohydrate Research* 102(1), 321–327.
Takeda, Y., Shirasaka, K. and Hizukuri, S. (1984) Examination of the purity and structure of amylose by gel-permeation chromatography. *Carbohydrate Research* 132, 83–92.
Takeda, Y., Shitaozono, T. and Hizukuri, S. (1990) Structures of sub-fractions of corn amylose. *Carbohydrate Research* 199, 207–214.
Tanaka, Y., Miura, H., Fukushima, M., Ito, M., Nishio, Z., Kim, S.-J., Hashimoto, N., Noda, T., Takigawa, S., Matsuura-Endo, C. and Yamauchi, H. (2006) Physical properties of yellow alkaline noodles from near-isogenic wheat lines with different Wx protein deficiency. *Starch/Starke* 58, 186–195.
Tester, R.F. and Morrison, W.R. (1990) Swelling and gelatinization of cereal starches. I. Effects of amylopectin, amylose, and lipids. *Cereal Chemistry* 67(6), 551–557.
Tharanathan, R.N. (2005) Starch – value addition by modification. *Critical Reviews in Food Science and Nutrition* 45, 371–384.
Tippit, P.S. (2008) Starch-containing compositions for use in imparting oil or grease resistance to paper. US Patent Application Publ. 2008/0193784 A1.
Tufvesson, F., Wahlgren, M. and Eliasson, A.-C. (2003) Formation of amylose–lipid complexes and effects of temperature treatment. Part 1. Monoglycerides. *Starch/Staerke* 55(2), 61–71.
Umemoto, T., Terashima, K., Nakamura, Y. and Satoh, H. (1999) Differences in amylopectin structure between two rice varieties in relation to the effects of temperature during grain-filling. *Starch/Staerke* 51 (2–3), 58–62.
Umemoto, T., Yano, M., Satoh, H., Shomura, A. and Nakamura, Y. (2002) Mapping of a gene responsible for the difference in amylopectin structure between *japonica*-type and *indica*-type rice varieties. *Theoretical and Applied Genetics* 104, 1–8.
Vandeputte, G.E., Vermeylen, R., Geeroms, J. and Delcour, J.A. (2003a) Rice starches. I. Structural aspects provide insight into crystallinity characteristics and gelatinisation behaviour of granular starch. *Journal of Cereal Science* 38(1), 43–52.
Vandeputte, G.E., Vermeylen, R., Geeroms, J. and Delcour, J.A. (2003b) Rice starches. III. Structural aspects provide insight in amylopectin retrogradation properties and gel texture. *Journal of Cereal Science* 38(1), 61–68.
Villareal, C.P., Hizukuri, S. and Juliano, B.O. (1997) Amylopectin staling of cooked milled rices and properties of amylopectin and amylose. *Cereal Chemistry* 74, 163–167.
Visser, R.G.F., Suurs, L.C.J.M., Bruinenberg, P.M., Bleeker, I. and Jacobsen, E. (1997) Comparison between amylose-free and amylose containing potato starches. *Starch/Starke* 49, 438–443.
Vrinten, P.L. and Nakamura, T. (2000) Wheat granule-bound starch synthase I and II are encoded by separate genes that are expressed in different tissues. *Plant Physiology* 122, 255–264.
Wang, Y.J., White, P. and Pollak, L. (1992) Thermal and gelling properties of maize mutants from the Oh43 inbred line. *Cereal Chemistry* 69(3), 328–334.
Wang, Y.J., White, P., Pollak, L. and Jane, J. (1993a) Characterization of starch structures of 17 maize endosperm mutant genotypes with Oh43 inbred line background. *Cereal Chemistry* 70(2), 171–179.
Wang, Y.J., White, P., Pollak, L. and Jane, J. (1993b) Amylopectin and intermediate materials in starches from mutant genotypes of the Oh43 inbred line. *Cereal Chemistry* 70(5), 521–525.
Wickramasinghe, H.A.M., Takigawa, S., Matsuura-Endo, C., Yamauchi, H. and Noda, T. (2009) Comparative analysis of starch properties of different root and tuber crops of Sri Lanka. *Food Chemistry* 112, 98–103.
Wilson, J.D., Bechtel, D.B., Todd, T.C. and Seib, P.A. (2006) Measurement of wheat starch granule size distribution using image analysis and laser diffraction technology. *Cereal Chemistry* 83, 259–269.

Wong, K.-S., Kubo, A., Jane, J., Harada, K., Satoh, H. and Nakamura, Y. (2003) Structures and properties of amylopectin and phytoglycogen in the endosperm of *sugary-1* mutants of rice. *Journal of Cereal Science* 37(2), 139–149.

Woo, K., Bassi, S.D., Maningat, C.C., Zhao, L., Trompeter, E.E., Kelley, G.A.J., Ranjan, S., Gaul, J., Dohl, C.T., DeMeritt, G.K.J., Stempien, G.J. and Krehbiel, K.D. (2006) Pregelatinized chemically modified resistant starch products and uses thereof. US Patent Application Publ. 20060188631.

Yamaguchi, M., Kainuma, K. and French, D. (1979) Electron microscopic observations of waxy maize starch. *Journal of Ultrastructure Research* 69, 249–261.

Yamamori, M., Fujita, S., Hayakawa, K., Matsuki, J. and Yasui, T. (2000) Genetic elimination of starch granule protein, SGP-1, of wheat generates an altered starch with apparent high amylose. *Theoretical and Applied Genetics* 101, 21–29.

Yamashita, Y.-H., Ryugo, J. and Monobe, K. (1973) Electron microscopic study on crystals of amylose V complexes. *Journal of Electron Microscopy* 22, 19–26.

Yang, C.Z., Shu, X.L., Zhang, L.L., Wang, X.Y., Zhao, H.J., Ma, C.X. and Wu, D.X. (2006) Starch properties of mutant rice high in resistant starch. *Journal of Agricultural and Food Chemistry* 54, 523–528.

Yoo, S.-H. and Jane, J. (2002a) Structural and physical characteristics of waxy and other wheat starches. *Carbohydrate Polymers* 49(3), 297–305.

Yoo, S.-H. and Jane, J. (2002b) Molecular weights and gyration radii of amylopectins determined by high-performance size-exclusion chromatography equipped with multi-angle laser-light scattering and refractive index detectors. *Carbohydrate Polymers* 49(3), 307–314.

Yoo, S.-H., Spalding, M.H. and Jane, J. (2002) Characterization of cyanobacterial glycogen isolated from the wild type and from a mutant lacking of branching enzyme. *Carbohydrate Research* 337(21–23), 2195–2203.

Yoo, S.-H., Perera, C., Shen, J., Ye, L., Suh, D.-S. and Jane, J. (2009) Molecular structure of selected tuber and root starches and effect of amylopectin structure on their physical properties. *Journal of Agricultural and Food Chemistry* 57, 1556–1564.

Yoon, S.-Y. and Deng, Y. (2006) Clay-starch composites and their application in papermaking. *Journal of Applied Polymer Science* 100, 1032–1038.

Yuan, R.C., Thompson, D.B. and Boyer, C.D. (1993) Fine structure of amylopectin in relation to gelatinization and retrogradation behavior of maize starches from three *wx*-containing genotypes in two inbred lines. *Cereal Chemistry* 70(1), 81–89.

Zaidul, I.S.M., Yamauchi, H., Takigawa, S., Matsuura-Endo, C., Suzuki, T. and Noda, T. (2007a) Correlation between the compositional and pasting properties of various potato starches. *Food Chemistry* 105, 164–172.

Zaidul, I.S.M., Norulaini, N., Omar, A.K.M., Yamauchi, H. and Noda, T. (2007b) Correlations of the composition, minerals and RVA pasting properties of various potato starches. *Starch/Starke* 59, 269–276.

Zeeman, S.C., Northrop, F., Smith, A.M. and Ap Rees, T. (1998) A starch-accumulating mutant of *Arabidopsis thaliana* deficient in a chloroplastic starch-hydrolyzing enzyme. *The Plant Journal* 15(3), 357–365.

Zhang, X., Colleoni, C., Ratushna, V., Sirghie-Colleoni, M., James, M.G. and Myers, A.M. (2004) Molecular characterization demonstrates that the *Zea mays* gene sugary2 codes for the starch synthase isoform SSIIa. *Plant Molecular Biology* 54, 865–879.

Zhong, F., Li, Y., Ibanz, A.M., Oh, M.H., McKenzie, K.S. and Shoemaker, C. (2009) The effect of rice variety and starch isolation method on the pasting and rheological properties of rice starch pastes. *Food Hydrocolloids* 23, 406–414.

Zobel, H.F. (1988) Starch crystal transformations and their industrial importance. *Starch/Staerke* 40(1), 1–7.

11

In Planta Modification of Starch Quantity and Quality

Sarita Jaiswal, Seedhabadee Ganeshan, Monica Båga and Ravindra N. Chibbar

Introduction

Green plants have the unique capability to utilize light energy to fix carbon dioxide and water for producing primary carbohydrates. In leaves, most of the fixed carbon is converted to transitory starch, which is broken down into simple sugars at night. The simple sugars are transported to the storage organs, where storage starch is synthesized in the amyloplasts. Starch is the most abundant storage polysaccharide in seeds, fruits, tubers and roots of higher plants. Cereal and tuber starches contribute about 50% of the calories in the human diet. In addition, extracted starch is used as an ingredient in processed foods and feed products. Starch is a very abundant homoglucan polymer, which, when modified slightly, can cause significant changes in functional properties (Orthoefer, 1994; Jobling, 2004). Physical, chemical and enzymatic methods have been used to alter starch structure to suit different applications. The amenability of starch to modifications has made it a preferred plant-based raw material used in industries like textiles, cosmetics, paper and pulp, packaging, biodegradable plastic films, construction and mining. However, environmental concerns and costs associated with chemical modifications are major constraints to realizing the full potential of starch as an industrial raw material. Recent studies have improved our understanding of the biochemical and molecular mechanisms and genetic basis of starch biosynthesis in model and crop plants. Improved understanding of starch biosynthesis, combined with genomics technologies, can be used to change starch quantity, composition and structure in plants (*in planta*) genetically. Some of the naturally modified starches have changed starch structure and functionality for desired end-use in food, feed or as an industrial polymer. *In planta* modified starches are environmentally friendly and reduce costs associated with postharvest modifications.

Starch Biosynthesis

Storage starch is synthesized in amyloplasts and accumulated as crystalline or semi-crystalline, water-insoluble granules with varied shapes and sizes specific to botanical species. Starch granules are composed mainly of two glucan polymers: one-quarter is amylose, a predominantly linear molecule with very sparse branches, and three-quarters is amylopectin, a highly-branched molecule with complex structure. In addition, minute quantities of proteins, lipids and ash are also present in the starch granules. Amylose and amylopectin concentrations and structures influence properties and functionality of the starch granule and determine its end use (see Chapter 10, this volume). During the day, photosynthetic green chloroplasts synthesize starch and at night, degrade starch to provide a continuous supply of sugars to maintain leaf metabolism and to export to sink organs for synthesis of storage starch. The basic biochemical mechanism is the same for synthesis of transitory starch in the chloroplasts in green tissues and for accumulated starch in amyloplasts of storage organs. Initially, biochemical and genetic characterization of variant starch phenotypes in maize, pea, rice and barley were used to establish basic steps in starch biosynthesis. During the past decade, characterization of starch mutants in a model plant *Arabidopsis* (Zeeman *et al.*, 2007a) and a unicellular alga *Chlamydomonas* (Ball and Morell, 2003) has been used extensively to elucidate starch biosynthesis and principles of its control in plants. Starch biosynthesis is complex and requires the participation of several enzymes (see Chapter 10). There is a general consensus that ADP-glucose pyrophosphorylase (AGPase), soluble starch synthases (SSs), starch branching enzymes (SBEs) and starch debranching enzymes (DBEs; pullulanase, isoamylase) catalyse the final steps leading to amylopectin synthesis. Several of the enzymes exist in different isoforms, some of which vary in their subcellular distribution, enzyme specificity, temporal activity and interactions with other enzymes, making the starch synthesis pathway very complex. A minimal subset of 14 conserved starch biosynthetic enzymes (two AGPases, five SSs, three SBEs and four DBEs) is homologous in all plant species studied to date (Morell and Myers, 2005). Several lines of evidence suggest that the key starch biosynthetic enzymes are physically associated to form complexes within the amyloplasts. It is often observed that a mutation in one starch biosynthetic gene has pleiotroic effects on several other starch biosynthetic enzymes. Characterization of maize plants with double or triple SS and SBE mutations reveal non-additive effects at the biochemical level and a range of starch phenotypes (Yao *et al.*, 2004), inferring involvement of multi-enzyme complexes. The first direct evidence of the existence of protein complexes was demonstrated by co-immunoprecipitation of SBEI, SBEII and starch phosphorylase from isolated wheat endosperm amyloplasts (Tetlow *et al.*, 2004a,b). Subsequent studies in maize and wheat, utilizing combinations of yeast two-hybrid assays, affinity purification with immobilized ligands and immunoprecipitation, have shown the formation of protein complexes between starch synthases and starch branching enzymes (Tetlow *et al.*, 2008; Hennen-Bierwagen *et al.*, 2008, 2009). The function of these protein complexes *in vivo* is not well understood. Hannah and James (2008) speculate that positioning of SSs, SBEs and DBEs in these complexes might have direct effects on substrate binding or enzymatic activity, which subsequently might influence amylopectin organization and architecture.

Strategies for Genetic Modification of Starch

The discovery of natural variants in maize with changed starch properties was the first indication of *in planta* starch modification. Initially, maize and pea natural mutants were used to identify the reaction catalysed by the enzyme deficiency in the natural mutant. The diploid nature of the maize genome, combined with the availability of large natural variants with altered starch structure, facilitated the identification of major steps in starch biosynthesis (Smith *et al.*, 1997; Smith, 2001; Ball and Morell, 2003; Zeeman *et al.*, 2007b). Subsequently, induced mutations in a unicellular alga, *C. reinhardtii* (Ball and Morell, 2003), and in *Arabidopsis* were used to characterize precisely some of the finer aspects of starch biosynthesis (Zeeman *et al.*, 2007b). The main requirement for producing *in planta* modified starches is to induce genetic variation in starch biosynthetic enzymes. Some of the strategies used to induce genetic variation for plant improvement are described in the following sections.

Mutation-mediated improvements

The identification of naturally occurring variants and their utilization in plant breeding heralded a new era in *in planta* alteration for desirable traits. This strategy has been incorporated successfully into breeding programmes for developing plants with desirable traits. Extensive domestication of crop plants has resulted in limited genetic diversity, which makes it difficult to find plants with genetic variability and/or mutations in genes for desirable traits. Plant breeders have relied on land races and wild species, but poor agronomic performance in their progeny and limited sexual compatibility among parents has sometimes precluded widespread use of such approaches. Naturally occurring mutations, combined with induced mutations produced by the use of physical or chemical mutagenic agents, have provided crop improvement programmes with additional resources for the generation of variability to be incorporated into new improved varieties.

A number of naturally occurring starch biosynthetic gene mutants have been identified (Table 11.1) and some of these mutations in maize have been used to produce

Table 11.1. Naturally occurring starch mutants in selected crops.

Mutation	Species	Target gene/enzyme	Phenotype	Reference
Opaque-2	*Zea mays*	–	–	Tsai *et al.*, 1978
Shrunken-2	*Zea mays*	ADP-glucose pyrophosphorylase activity	70–75% decrease in starch	Tsai and Nelson, 1966
Waxy	*Zea mays*	Deficiency in granule-bound starch synthase 1 (GBSS I)	Reduced amylose/ amylose free	Nelson and Rines, 1962
Ae (amylase extendor)	*Zea mays*	Deficiency in starch branching enzyme II (SBE II)	Increased amounts of amylose and intermediate fractions compared with normal starch	Wang *et al.*, 1993; Kasemsuwan *et al.*, 1995; Tziotis *et al.*, 2004, 2005
Sugary-1	*Zea mays*	Deficiency in the debranching enzyme (DBE) of the isoamylase type	Highly branched amylopectin	Pan and Nelson, 1984; James *et al.*, 1995
Agps-m1	*Zea mays*	Small subunit of ADP-glucose pyrophosphorylase	Less transitory starch	Slewinski *et al.*, 2008
Tdy1-R Tie-dyed1	*Zea mays*	Defect in sucrose export from leaves	More transitory starch	Slewinski *et al.*, 2008
Riso 16	*Hordeum vulgare*	Lack the cytosolic small subunit of ADP-glucose pyrophosphorylase	Low starch, high beta glucan	Johnson *et al.*, 2003; Munck *et al.*, 2004
Shx	*Hordeum vulgare*	Major effect on primer-independent soluble starch synthase	Decreased starch content and containing smaller A-starch granules together with normal B-granules	Tyynela *et al.*, 1995
Wx	*Hordeum vulgare*	Granule-bound starch synthase I with a deletion in the 5′-non-coding region	Low-amylose	Patron *et al.*, 2002

Table 11.1.

Mutation	Species	Target gene/enzyme	Phenotype	Reference
Sex6	*Hordeum vulgare*	Mutation in the starch synthase IIa gene	High-amylose starch	Morell *et al.*, 2003; Clarke *et al.*, 2008
Amf	*Solanum tuberosum*	Lack GBSS	Amylose free	Vanderleij *et al.*, 1991
	Manihot esculenta	Lack SBE	Free sugar content	Carvalho *et al.*, 2004

starch with altered starch structure commercially (Pollak, 2003). Screening for naturally occurring mutations is time-consuming and resource demanding. To expedite variant generation, mutagens have been used. Using a transgenic approach in *Arabidopsis*, point mutations were shown to occur at a rate of about 10^{-7} to 10^{-8} events/base pair (Kovalchuk *et al.*, 2000) and could be accelerated using mutagens. By selecting the type and dosage of mutagens judiciously, it is possible to minimize wasteful changes and maximize identification of useful variants. Studies using different mutagens such as ethylmethane sulphonate (EMS), ultraviolet (UV) radiation, thermal neutrons and gamma-rays induced changes in waxy loci of rice and maize ranging from transcription injury to complete loss of the locus (Amano, 1981). The availability of mutants is useful in studying mechanisms involved in the anabolism and catabolism of starch, which will further enhance our ability to generate designer starches *in planta*. In this regard, *Arabidopsis*, although lacking storage starch organs like cereal grains or potato tubers, has the advantage of being amenable to mutation induction and genetic transformation. Thus, some level of understanding of starch synthesis can still be gained from model systems such as *Arabidopsis*. Table 11.2 summarizes some of the induced starch mutants.

Table 11.2. Artificially induced starch mutants in selected crops.

Source	Mutagen	Phenotype	Reference
Oryza sativa (Japanese)	Ethyl methane sulphonate	–	Juliano *et al.*, 1990
Clarkia xantiana	Ethyl methane sulphonate	More sucrose, less starch in leaves	Jones *et al.*, 1986
Triticum aestivum L. (Kanto 107 lacking Wx-A1 and Wx-B1 proteins)	Ethyl methane sulphonate	Altered flour-pasting profile	Yasui *et al.*, 2002
Oryza sativa (Isv-1)	Gamma irradiation	Altered starch viscosity parameters	Wu *et al.*, 2002
Manihot esculenta (PRC 60a)	Gamma irradiation	50% decrease in starch content and a significant reduction (30%) in amylose content	Joseph *et al.*, 2004
Oryza sativa (SA419)	Sodium azide	Position of grain on panicle had little effect on starch and amylose content	Jeng *et al.*, 2003
Solanum tuberosum	X-ray	Amylose free starch	Hovenkamp-Hermelink *et al.*, 1987
Oryza sativa (Norin 8)	^{32}P beta rays	–	Juliano *et al.*, 1990
Oryza sativa (Kinmaze)	*N*-methyl-*N*-nitrosourea	–	Juliano *et al.*, 1990

TILLING for starch biosynthetic genes

Reverse genetics studies have gained tremendous impetus with the advent of the TILLING (targeting induced local lesions in genomes) approach (McCallum *et al.*, 2000b), providing additional resources to plant breeders to utilize genetic variability in their breeding programmes. Besides allowing for a more targeted mutation strategy, TILLING reduces the time required and the number of mutant lines to be screened to identify desirable changes. Single base changes that are induced are comparable to naturally occurring aberrations. TILLING involves screening of DNA pools in M2 populations of EMS mutagenized seeds. EMS causes G/C (guanine/cytosine) to A/T (adenine/thymine) transitions at a rate of 99.5% in *Arabidopsis* (Greene *et al.*, 2003). The target gene from pooled DNA is then PCR amplified and screened enzymatically with *Cel1* nuclease extract obtained from celery (Till *et al.*, 2004, 2006). *Cel1* cleaves at the 3~-side of any mismatch in heteroduplexes between mutated and non-mutated variants (Oleykowski *et al.*, 1998). The cleaved products are resolved either by high performance liquid chromomatography (HPLC) separation or in sequence analysers (slab gel or capillary format). Pooling helps in identification of the particular mutant pools, instead of screening the mutant lines individually. The identified pool is then screened for the individuals constituting the pool to identify the actual mutant line. TILLING has the potential for generating an allelic series for a gene of interest with a range of phenotypes, and lethal and sterile alleles can also be assessed due to heterozygous states (Till *et al.*, 2004). EcoTILLING, a variant of the TILLING approach, was introduced as a tool to explore natural variation among *Arabidopsis* ecotypes (Henikoff and Comai, 2003; Comai *et al.*, 2004) by comparing a particular line with a sequenced reference genome, thereby enabling sequence diversity studies and establishing relatedness. TILLING also allows generation of high-resolution maps of mutations in a gene of interest, including silent, miss-sense and truncation mutations (Till *et al.*, 2004).

TILLING has been used successfully for both functional genomics, for example, in *Arabidopsis* (McCallum *et al.*, 2000a), as well as for crop improvement studies like those in wheat (Slade *et al.*, 2005; Dong *et al.*, 2009), maize (Henikoff *et al.*, 2004), rice (Wu *et al.*, 2005; Till *et al.*, 2007), barley (Caldwell *et al.*, 2004), sugarbeet (Hohmann *et al.*, 2005), *Lotus japonicus* (Perry *et al.*, 2003), soybean (Cooper *et al.*, 2008) and several grass species reviewed in Weil (2009).

Screening for variants among polyploid plant genotypes has always been an arduous task. Bread wheat has proven to be particularly difficult for such screening, due to its hexaploid nature and compensation effects of one genome for defectiveness in another. Thus, the TILLING approach offers a viable alternative for studying genome changes in wheat, especially with regards to starch biosynthetic pathways. A significant breakthrough in wheat TILLING for starch biosynthetic genes allowed the identification of a wide range of phenotypes for the *waxy* gene (Slade *et al.*, 2005). Recently, a modified wheat TILLING method was reported wherein more spikes per line were chosen for analysis, thereby decreasing the population size required for screening (Dong *et al.*, 2009).

Transposon-targeted mutagenesis

Transposable elements or mobile genetic elements can also cause changes in the genome. Two transposition mechanisms occur in nature. One is through DNA excision from a chromosome and reinsertion at a different location in the genome. The other mechanism involves an RNA intermediate prior to reinsertion into the genome, the so-called retransposons. There have been several reports on the identification of naturally occurring transposition events leading to mutant phenotype, especially starch mutants. For example, the wrinkled pea seed phenotype was attributed to a transposon insertion, similar to the *Activator (Ac)/ Dissociator* (*Ds*) family of transposons, in the *SBEI* gene (Bhattacharyya *et al.*, 1990). With the characterization of the sequences of transposable elements such as *Ac/Ds* (Doring *et al.*, 1984), *Enhancer (En)/Suppressor-mutator* (*Spm*) (Pereira *et al.*, 1986) and *Mutator* (*Mu*) (Barker *et al.*, 1984) from maize, it was possible to use a transposon tagging strategy to identify genes from transposon tagged populations. For example, the maize sugary

Su1 mutation was identified in Mutator background tagged lines (James *et al.*, 1995). However, this strategy was useful only in plants where the occurrence of endogenous transposons was common. Thus, to apply this approach to other plants, particularly crops, a transgenic approach was used wherein cloned transposable elements were introduced into other plants. Essentially, vectors containing the transposable elements are delivered using an appropriate gene transfer method (see the following section on genetic transformation) and the excision of the elements monitored in transgenic plants. For example, in the *Ac/Ds* system, the *Ac* element can transpose autonomously, unlike the *Ds* element, and can be used in a construct carrying a reporter gene (reviewed in Haring *et al.*, 1991), thus rendering the reporter gene non-functional. However, on excision of the *Ac* element, the reporter gene becomes functional but retains the footprint of the *Ac* element. In a two-element system, the elements are delivered separately with appropriate marker genes, and therefore the elements act in *cis* and *trans*, wherein the autonomous element will transpose and also allow transposition of the non-autonomous element in *trans*. This allows more transposition events to be studied. Furthermore, different populations of the autonomous and non-autonomous elements can be generated and by crossing the two lines, the non-autonomous element in the progeny can be activated. Such an approach was demonstrated with the *Ac/Ds* element in barley (Koprek *et al.*, 2000). The transposon tagging strategy is a valuable tool for studying modified starch phenotypes.

Genetic transformation

Gene transfer to plants has been one of the most significant breakthroughs in plant biotechnology. With the capability to transfer a gene of interest into a plant cell, tissue or organ, the expression of the gene could be assessed more precisely, whether in a down-regulated or overexpressed state. More importantly, the ability to regenerate whole new genetically altered plants was a major achievement in functional analysis of a gene and in studying its effects, if any, on other metabolic aspects of the plants. Thus, it is of interest to highlight some of the important developments leading to the production of transgenic plants. Fundamental to the generation of transgenic plants is the establishment of an efficient tissue culture regeneration system in order to obtain a high percentage of transformed plants. Since the concept of totiopotency was put forward by Haberlandt in 1902 (Haberlandt, 1902), much research has focused on the development of methods to culture plant cells and tissues *in vitro* and subsequently to regenerate complete plants from cell and tissue cultures. Generally, dicotyledonous plants have been more responsive to tissue culture regeneration than monocotyledonous plants, especially cereals. However, in the past two decades, tremendous improvements have been made and cereals can now be regenerated *in vitro* at high frequencies. This was facilitated by the development of specialized culture media formulations and optimizing the type of explants for culture and regeneration. For example, the use of mature embryos for regeneration from cereals has provided a highly expeditious regeneration system that requires fewer resources for the production of explants (Ganeshan *et al.*, 2003, 2006). With this system using the plant growth regulator, thidiazuron, mature embryos produced direct multiple shoots in a genotype-independent manner. A further advantage to using this approach is the circumvention of a callus phase, which tends to promote somaclonal variation (Larkin and Scowcroft, 1981), sometimes with negative consequences on the agronomic performance of the regenerated plants.

With the availability of tissue culture systems for a number of plants (for a review see Ganeshan *et al.*, 2002), interest in developing gene transfer methods was intensified in the seventies and eighties. The gene delivery systems investigated included both physical and biological methods. It was only by 1983 that the first reports on the production of transgenic plants using *Agrobacterium*-mediated DNA transfer were published (Barton and Chilton, 1983; Fraley and Horsch, 1983). Using the physical delivery method based on microprojectile bombardment invented by Sanford and co-workers, it was possible to transfer DNA to maize and tobacco cells (Weissinger *et al.*, 1987; Klein *et al.*, 1988a,b). To date, these two methods of DNA transfer are the most popular and versatile for plant transformation, although each one has advantages and disadvantages. For example, *Agrobacterium* was very successful in genetically transforming most dicotyledonous plant species but not monocotyledons plants such as cereals, as these species were considered to be outside its natural host range. However, successful *Agrobacterium*-mediated transformation of Japonica rice (Hiei *et al.*, 1994), Indica rice (Rashid *et al.*, 1996), barley (Tingay *et al.*, 1997) and wheat (Cheng *et al.*, 1997) heralded a new beginning for the transformation of cereals. Prior to rice being transformed using *Agrobacterium*, the particle bombardment method was used extensively for cereals. It has the advantage of circumventing the barrier imposed by the cell walls. The disadvantages are that it is considered expensive and integration of DNA into the genome is often complex, with multiple copies of the transgenes inserted. Particle bombardment generally remains the tool of choice for the transformation of cereals, because *Agrobacterium*-mediated transformation efficiency is often low among laboratories and the lack of consensus as to the integration complexity patterns.

Alternative physical DNA delivery methods with varying degrees of efficiency and ease of use include electroporation, liposomes, microinjection and silicon carbide whiskers. Similarly, improvised biological delivery methods include agro-infection and virus-induced gene silencing (VIGS). More recently, the direct uptake of antisense oligodeoxynucleotides (ODNs) has also been shown to be effective in transiently downregulating genes of interest (Sun *et al.*, 2005; Jansson *et al.*, 2007; Drinkwater *et al.*, 2010). Thus, a number of genes can be rapidly studied for their downregulation prior to designing constructs for more stable downregulation of such genes, in essence validating downregulation prior to investing time and resources into making constructs. Transformation methodologies using *Agrobacterium* without having recourse to tissue culture are another important development, leading to the '*in planta* transformation' system. Although still requiring refinements, its success has been demonstrated unequivocally in *Arabidopsis* (Bechtold *et al.*, 1993; Clough and Bent, 1998) and to some extent in other species such as *Brassica rapa* ssp. *chinensis* (Qing *et al.*, 2000), *Medicago truncatula* (Trieu *et al.*, 2000), *Hibiscus canabinus* (Kojima *et al.*, 2004), rice (Supartana *et al.*, 2005), wheat (Supartana *et al.*, 2006; Zale *et al.*, 2009) and tomato (Yasmeen *et al.*, 2009). Other developments likely to influence gene transfer to plants in the future are the use of nanoparticles, as demonstrated by Torney *et al.* (2007), and the use of bioactive beads (Sone *et al.*, 2002; Liu *et al.*, 2004; Murakawa *et al.*, 2008a,b).

The improvement in plant transformation methods and the identification and characterization of a number of genes involved in starch biosynthesis led to the possibility of tailoring for starches *in planta* in a very precise manner. The first reported *in planta* modification of starch was achieved in potato, wherein an antisense construct for the *GBSSI* gene was used to downregulate *GBSSI*, resulting in reduced amylose-free starch (Visser *et al.*, 1991). Subsequently, many reports have shown *in planta* alteration of starches with varying success rates. Antisense inhibition of ADP-glucose pyrophosphorylase (AGPase) in potato resulted in reduced starch and increased sucrose and glucose (Müller-Röber *et al.*, 1992), whereas overexpression of a bacterial unregulated AGPase increased starch concentration (Stark *et al.*, 1992). Transgenic cassava expressing an antisense *GBSS* gene led to amylose-free starch (Raemakers *et al.*, 2005). Table 11.3 lists some of the starch alterations achieved using a transgenic approach.

Manipulation of complex biosynthetic pathways by alteration of a single gene in the pathway generally leads to minimal change, which may not always translate into significant alterations in the entire biosynthetic mechanism. Starch biosynthesis is complex and single gene alterations are likely to be inconsequential for overall *in planta* starch modification, especially for polyploid species such as wheat. Recent advances in plant transformation technology using *Agrobacterium* and specialized vectors like binary bacterial

Table 11.3. Some examples of genetically engineered starch modifications in selected plants.

Target gene	Regulatory expression	Gene delivery method	Species	Phenotype (starch modification)
Starch quantity				
AGPase	Antisense-mediated downregulation	*Agrobacterium tumefaciens*	*Solanum tuberosum*	[a]Sugar-storing tubers, reduced starch concentration
AGPase	Overexpression	*Agrobacterium tumefaciens*	*Solanum tuberosum*	[b]2–30% increase in tuber starch concentration
AGPase large subunit	*Shrunken2* gene (*sh2r6hs*) incorporation enhances activity	Particle bombardment	*Triticum aestivum*	[c,d]Increased number of grains per spike, no change in starch concentration
Sh2r6hs	Deregulation	Particle bombardment	*Oryza sativa*	[e]Increased number of grains, no increase in starch concentration
AGPase	Modified sensitivity via *glgC16* introduction	*Agrobacterium tumefaciens*	*Zea mays*	[f]Increase in seed weight
Starch composition				
Reduced amylose				
GBSSI(Wx)	Antisense	Electroporation	*Oryza sativa* L.	[g,h]Reduced amylose
GBSSI	Antisense	*Agrobacterium tumefaciens*	*Solanum tuberosum*	[i,j]Reduced amylose
GBSSI	RNAi	*Agrobacterium tumefaciens*	*Manihot esculenta*	[k,l]Amylose-free plants
GBSSI	RNAi-mediated downregulation	*Agrobacterium tumefaciens*	*Ipomoea batatas* (sweet potato)	[m]Amylose free
Increased amylose				
SBEII	RNAi-mediated downregulation	*Agrobacterium tumefaciens*	*Triticum aestivum*	[n]Increased amylose up to 70%
Altered amylopectin architecture				
SS	Antisense	*Agrobacterium tumefaciens*	*Solanum tuberosum*	[o,p]Altered amylopectin architecture
SBE	Upregulation by heterologous expression of bacterial *glgB*	*Agrobacterium tumefaciens*	*Oryza sativa* L. cv. Nakdong	[q]Increased branching in amylopectin
Starch granule number				
LD inhibitor (LDI)	Antisense	Particle bombardment	*Hordeum vulgare*	[r]Reduced B-type starch granules, starch and amylose concentration, altered amylopectin structure

[a]Müller-Röber *et al.*, 1992; [b]Stark *et al.*, 1992; [c]Smidansky *et al.*, 2002; [d]Meyer *et al.*, 2007; [e]Smidansky *et al.*, 2003; [f]Wang *et al.*, 2007; [g]Shimamoto *et al.*, 1989; [h]Terada *et al.*, 2000; [i]Visser *et al.*, 1991; [j]Kuipers *et al.*, 1994; [k]Raemakers *et al.*, 2005; [l]Taylor *et al.*, 2004; [m]Otani *et al.*, 2007; [n]Regina *et al.*, 2006; [o]Marshall *et al.*, 1996; [p]Fulton *et al.*, 2002; [q]Kim *et al.*, 2005; [r]Stahl *et al.*, 2004.

artificial chromosome (BIBAC) vectors (Hamilton *et al.*, 1996, 1999; Hamilton, 1997) and transformation-competent artificial chromosome (TAC) vectors (Liu *et al.*, 1999) have opened up new possibilities for the insertion of large fragments of DNA into plant genomes. Although the long-term stability of such large inserts has not been assessed fully, several reports have addressed such stability issues and enhancement of transformation efficiencies with such large inserts in tomato (Frary and Hamilton, 2001), potato (Song *et al.*, 2003), rice (Nakano *et al.*, 2005) and maize (Vega *et al.*, 2008). The specialized vectors with the capacity for accepting large inserts have led to the possibility of using a multi-gene linking approach (Lin *et al.*, 2003) to place a number of cloned starch biosynthetic genes in tandem and study their effects on *in planta* changes in starch. Interaction studies among starch-specific biosynthetic enzymes can also be carried out with this approach, especially since it is known that many of the starch biosynthetic enzymes function as complexes (Tetlow *et al.*, 2008).

Among transgenic approaches to downregulate starch biosynthetic genes, RNA interference (RNAi) appears as a viable technique. RNAi is a post-transcriptionally occurring gene-silencing mechanism induced by double-stranded RNA. With carefully designed RNAi constructs, it is possible to downregulate gene(s) of interest. With regards to *in planta* starch studies, silencing of β-amylase8 (linked to temperature stress) by RNAi in transgenic *Arabidopsis* plants led to starch accumulation in the leaves, independent of cold treatment (Kaplan and Guy, 2005). High-amylose (~ 70%) wheat lines have been developed by RNAi silencing of two starch branching enzyme isoforms (SBEIIa and SBEIIb) (Regina *et al.*, 2006). Numerous studies relating to downregulation of starch biosynthetic genes using RNAi are currently being undertaken and it is foreseen that significant *in planta* alteration of starch will be possible. As mentioned above, use of antisense ODNs, as in the effective transient downregulation of the *GBSSI* gene in wheat spikes (Drinkwater *et al.*, 2010), augurs well for the design of RNAi constructs to explore the long-term stable downregulation of the *GBSSI* gene further.

Targets for Starch Modification

A better understanding of starch biosynthesis, isolation and characterization of starch biosynthetic genes, combined with molecular biological strategies to create and identify genetic variants, has resulted in changing starch quantity and quality in several crop plants. Targets for starch modification depend on the end use, but can be grouped as follows:

1. Increased storage starch concentration.
2. Modified storage starch properties.
3. Phytoglycogen accumulation.

Increased storage starch concentration

Starch accounts for two-thirds to three-quarters of the dry weight in cereal grains and potato tubers and is an important contributor to yield in these crops. However, the first step in starch biosynthesis is catalysed by AGPase, which is a highly-regulated enzyme responsible for the synthesis of monomer precursors of starch biosynthesis. AGPase enzymatic activity is subjected to several regulatory mechanisms which make it a major rate-controlling reaction affecting the amount of starch produced. The AGPase is an allosterically regulated enzyme whose activity is generally stimulated by 3-PGA and inhibited by inorganic orthophosphate (Pi). Plant species, tissue, plastid-type and subcellular location affect the relative sensitivity of AGPase to the allosteric effector and inhibitor molecules. In chloroplasts, AGPase activity is activated by micromolar concentrations of 3-PGA and inhibited by Pi (Ghosh and Preiss, 1966). The ratio of the allosteric effector and inhibitor molecules plays a key role in controlling starch synthesis in all photosynthetic tissues (Preiss, 1991). However, only dicot storage tissues AGPase show similar sensitivity to the two allosteric effector molecules. In monocots such as wheat and barley, the endosperm AGPase (majority is cytosolic) is less sensitive to the two allosteric effector molecules. Structurally, AGPase is a heterotetramer ($\alpha_2\beta_2$), made up of a large (α) subunit of

54–60 kD and a small subunit (β) of 51–55 kD, which are coded by multiple genes. The two AGPase subunit genes show tissue- and time-specific expression. Thus, transcriptional regulation can result in differential structural organization in different tissues over time. The two subunits have differential sensitivity to allosteric effector and inhibitor molecules. Therefore, differential expression of the two subunits can result in an AGPase with varying degrees of allosteric regulation suited for the specific metabolic demands of a specific tissue at a particular time. Redox modulation of AGPase was proposed recently to explain the dithiothreitol-mediated activation of the partially inactive heterotetrameric recombinant potato enzyme produced in *Escherichia coli* (Fu *et al.*, 1998). Dithiothreitol treatment disrupted the intermolecular bridge between the cysteine-82 residues of the two AGP small subunits to make the enzyme active. The redox activation of AGPase is influenced by sugar status and an increase triggers signalling pathways, which result in reduction, and hence greater activation, of AGPase (Geigenberger *et al.*, 2005; Kolbe *et al.*, 2005).

Approaches for increasing storage starch concentration include strategies to modulate the activity of highly regulated AGPase. In the first instance of *in planta* starch modification, an unregulated AGPase encoded by *E. coli* glgC16 expressed in potato (var. Russett Burbank) increased tuber starch concentration by ~30% (Stark *et al.*, 1992). In another approach to increase starch, site-specific mutagenesis was used in maize to generate a Pi insensitive variant AGPase large subunit. In this mutant, line starch concentration did not change, but an 11–18% increase in seed dry weight was observed (Giroux *et al.*, 1996). Transgenic wheat expressing the gene for the same variant large AGPase subunit (Giroux *et al.*, 1996) produced AGPase with reduced Pi sensitivity, but did not have an increased grain starch concentration. However, transgenic wheat had a 38% higher seed yield and a 31% increase in total plant biomass (Smidansky *et al.*, 2002). Similarly, Pi insensitive AGPase large subunit expression in transgenic rice did not increase grain starch concentration, but the seed weight was increased by 11–18% (Sakulsingharoj *et al.*, 2004). Another *E. coli* AGPase variant (G336D) had high enzymatic activity with or without the activator, fructose-1,6-biphosphate, higher substrate affinity and reduced affinity for an inhibitor, and when expressed in transgenic cassava, had twofold higher enzyme activity as compared to non-transgenic control plants. While there was no difference in tuber starch concentration, transgenic cassava plants had twofold more root and stem biomass (Ihemere *et al.*, 2006).

Potato starch concentration was increased by manipulating inorganic pyrophosphatase (PPi) activity. Transgenic potato expressing *E. coli* PPase showed a 20–30% increase in tuber starch concentration (Geigenberger *et al.*, 1998). Allosteric mutants with high affinity for 3-PGA and lesser sensitivity towards Pi inhibition have great potential for increasing starch content. A random mutagenesis approach can be used to generate upregulated AGPase variants (Greene *et al.*, 1998).

Modified storage starch properties

Starch properties are influenced by the concentration and fine structure of amylose and amylopectin in a starch granule. Starch synthases catalyse the transfer of a glucosyl moiety from ADP-glucose to the non-reducing end of α-(1–4) glucan primer. Plants produce several SS isoforms that carry similar C-terminal regions, whereas the N-termini show considerable variation. The five major classes of SS include GBSSI, SSI, SSII, SSIII and SSIV, of which all but SSIV have been shown to have distinct roles in determining starch composition and structure. The granule-bound starch synthase (GBSSI) is accumulated in granules and is required for amylose synthesis, whereas SSI and SSII are present in both granular and stromal fractions and participate in amylopectin synthesis. SSI synthesizes the shortest chains, while SSII elongates the chains and prepares them for crystallization. SSIII synthesizes the even longer chains that extend between the clusters (Ball and Morell, 2003). Recently, GBSSI has been shown to have similar function to SSIII, i.e. synthesis of the long B-chains that interconnect the amylopectin clusters (Ral *et al.*, 2006). Several wheat, cowpea and *Arabidopsis* expressed sequence tags (EST) ascribed to SSIV (sometimes called SSV) have been reported, but no corresponding protein has been characterized from any plant. Relative contributions of each SS differ between plant species and/or tissues (Smith *et al.*, 1997) and modulation of SS activities can result in alterations of amylopectin fine structure and/or starch granule morphology (Craig *et al.*, 1998; Edwards *et al.*, 1999; Morell *et al.*, 2003).

The branches on the amylose and amylopectin molecules are introduced by SBEs, which generate α-(1,6)-linkages by cleaving internal α-(1,4)-bonds and reattaching the released chain through the formation of an α-(1,6)-bond at a new site on the glucan molecule. The primary amino acid sequences of higher plant SBEs reveal two major classes: SBEI and SBEII. The two classes differ in glucan chain length transferred *in vitro* and their substrate specificities. SBEII transfer shorter chains and have a higher affinity for amylopectin, while SBEI transfers longer glucan chains and have a higher affinity for amylose (Guan and Preiss, 1993; Takeda *et al.*, 1993). Analysis of chimeric forms of SBEI and SBEII showed that N- and C-termini of these proteins played important roles in determining substrate preference, catalytic activity and chain length transfer (Kuriki *et al.*, 1997). A larger form of SBEI, SBEIc (Båga *et al.*, 2000; Peng *et al.*, 2000), has been identified in plants of the Triticeae tribe and is associated preferentially with the large A-type granules.

Starch with reduced amylose concentration

Granule-bound starch synthase 1 (GBSSI) is accumulated in starch granules and a deficiency in its activity results in waxy starch composed mainly of amylopectin. GBSSI is encoded by waxy (*wx*) locus in cereals and a mutation at this locus results in GBSSI deficiency. Further studies on waxy mutant indicated that *wx* dosage was not proportional to amylose content, and factors in addition to GBSS might have roles in amylose synthesis (Fujita *et al.*, 2001). These other factors may include small malto-oligosaccharides as primers or the availability of ADP-glucose as substrate (Denyer *et al.*, 2001). GBSS has two isoforms, GBSSI and II in wheat. In the case of barley, a second isoform, termed GBSSIb, which is 96.5% similar to wheat GBSSII and 65.3% similar to barley SSI, has been identified (Patron *et al.*, 2002). The GBSSII is expressed in the pericarp, which accumulates transient starch (James *et al.*, 2003).

Mutations at *wx* locus have been reported in almost all crop plants, including maize, rice, barley and wheat. In diploid plants, mutation at *wx* locus causes a deficiency in GBSSI and absence of amylose, which is easy to detect. However, in plants with multiple genomes such as allohexaploid wheat, a mutation in one genome is compensated for by the *wx* locus in the other two genomes, thus producing no detectable change in phenotype. During the past decade, it has been possible to identify mutations at *wx* locus in the three wheat genomes and, using plant breeding techniques, to produce triple null lines with less than 1% amylose (Nakamura *et al.*, 1995; Demeke *et al.*, 1999; Seib, 2000; Chibbar and Chakraborty, 2005). In hexaploid wheat, the GBSSI produced by the three A, B and D genomes show differential enzymatic activities in the order B > D > A (Miura and Sugawara, 1996). Therefore, combinations of null alleles at different wheat genomes results in grain starch with varying amylose concentrations between the normal (25%) and waxy with undetectable amylose concentration. A wheat line deficient in GBSSI from B and D genomes produced partially waxy starch (12% amylose) (Demeke *et al.*, 1999).

Starch with increased amylose concentration

Starch with increased amylose concentration is highly desirable for both food and non-food applications. There is a demand in the food industry for wheat starch with increased amylose, as it can be converted to resistant starch (RS) on heating and subsequent cooling. RS is not digested in the small intestine, but is broken down by the bacteria in the colon. As a result, RS acts as dietary fibre (DF), reduces the calories from food, has a low glycaemic index and is considered beneficial for colon health. Natural high-amylose variants have been identified in maize, pea, rice and barley, in which amylose concentration is up to twofold higher than in wild-type lines (Chibbar and Båga, 2003). In maize and rice, the high-amylose phenotype is caused by a deficiency in SBEIIb, also known as *amylose extender* (*ae*). In potato, downregulation of the corresponding starch branching enzyme gene (SBE A) had limited effect in reducing amylose concentration (Jobling *et al.*, 1999). However, when both the starch branching enzyme genes (SBE A and SBE B) were downregulated, starch with amylose concentration as high as 70% was produced (Schwall *et al.*, 2000). Recently in wheat, RNAi-mediated inhibition of SBEIIb resulted in wheat grain starch with 70% amylose (Regina *et al.*, 2006). In high-amylose barley (Himalaya), the mutation has been attributed to SSIIa deficiency (Morell *et al.*, 2003). Similarly, elimination of SSII polypeptides in wheat increased apparent amylose (colorimetric assays) concentrations to 30–37% (Yamamori *et al.*, 2000). In a similar approach, wheat lines deficient in SSII A and B genome polypeptides resulted in lines with grain starch containing up to 32% amylose (Chibbar *et al.*, 2005; Lan *et al.*, 2008).

Starch with changed amylopectin architecture

Amylopectin architecture is determined by the frequency and pattern of branch points and is an important determinant of grain quality. For example, fine structure of amylopectin has been associated with slow digestible starch in maize (Zhang *et al.*, 2008). The potential to produce a wide range of new starches through the alteration of glucan chain lengths, branching pattern and granule crystallinity has been suggested (Johnson *et al.*, 2001a). However, amylopectin architecture is governed by starch synthases and starch branching and debranching enzymes, and each of these have several isoforms and exist in multi-enzyme complexes (Tetlow *et al.*, 2004b). Therefore, changing amylopectin architecture by manipulating the expression of single genes has not been very successful. Dramatic reductions in SBEI activity by genetic engineering in potato (Safford *et al.*, 1998) or in maize knockout mutations (Blauth *et al.*, 2001) did not affect starch composition and/or amylopectin architecture. However, a rice mutant with reduced SBEI activity had amylopectin with more short chains and fewer long chains (Satoh *et al.*, 2003). These subtle changes lowered the onset of gelatinization as compared to wild-type starches. Simultaneous reduction of SSII and SSIII in transgenic potato resulted in starch with an increase in amylopectin short chains, and lowered the starch gelatinization temperature (Edwards *et al.*, 1999). In a rice mutant deficient in SSIIa, the relative number of short A chains was increased, while B chains were reduced, thus changing amylopectin architecture (Umemoto *et al.*, 2004). Similarly, a wheat triple null line deficient in SSII also showed an increase in short (dp 9) chains and a reduced number of slightly longer (dp 10–22) chains (Kosar-Hashemi *et al.*, 2007). In another approach, transgenic expression of a bacterial glycogen branching enzyme gene in rice resulted in a marked increase in the degree of branching of amylopectin with increased short branches (Kim *et al.*, 2005).

Starch granule size

Starch granule size is an important factor influencing physico-chemical characteristics of starch and industrial applications (Lindeboom *et al.*, 2004). Generally, each amyloplast contains one starch granule, as seen in wheat, barley, maize and sorghum, while compound starch granules are composed of several granules in rice and oat. Granule size shows large variations, with cow cockle having granules as small as 0.3–1.5 μm, while tuber starch granules could be up to 100 μm in size (Table 11.4). Four members of the Triticeae tribe, wheat, barley, rye and triticale, produce two distinct types of starch granules. Large A-type starch granules are lenticular or oval, with sizes ranging from 15 to 35 μm. Small B-type starch granules are round and less than 15 μm in diameter (Peng *et al.*, 2000). The two types of starch granules are physiologically distinct, synthesized at different times during grain development and have different composition and functional properties. Very little is known about the processes

Table 11.4. Size and shape of starch granules derived from different botanical sources.

Botanical name	Common name	Granule size (μm)	Granule shape
Tuber and root starches			
[a]*Solanum tuberosum*	Potato	15–110	Oval, spherical
[a]*Ipomea batatas*	Sweet potato	2–42	Round, oval and polygonal
[a]*Dioscorea abyssinica*	True yam	29.2	Round
[a]*Amorphophallus paeonfolius*	Elephant yam	3–30	Round, polygonal
[a]*Colocassia esculenta*	Taro	3–3.5	Round, polygonal
[a]*Manihot esculenta*	Cassava	5.4	Round
[a]*Maranta arundaceae*	West Indian arrowroot	10–16	Round, oval and polygonal
[a]*Canna edulis*	Queensland arrowroot	13–57.6	Oval, elliptical
[a]*Pueraria tuberosa*	Kuzu	3–23	Polygonal
[a]*Cucurbita foetidissima*	Buffalo gourd	2–24	Oval, elliptical
Cereal starches			
[b]*Triticum aestivum*	Wheat	2–38	Round, lenticular
[b]*Hordeum vulgare*	Barley	1–40	Round, lenticular
[b]*Oryza sativa*	Rice	3–8	Angular and polygonal
[b]*Zea mays*	Maize	5–25	Round and polygonal
[b]*Avena sativa*	Oat	5–15	Round
[b]*Sorghum bicolor*	Sorghum	4–24	Round, polygonal
[b]*Urochloa ramosa*	Millet	4–12	Round, polygonal
[c]*Fagopyrum esculentum*	Buckwheat	2.9–9.3	Round and polygonal
Pulse starches			
[a]*Pisum sativum*	Smooth pea	5–10	Reniform, simple
[a]*Pisum sativum*	Wrinkled pea	30–40	Reniform
[d]*Cicer arietinum* L.	Chickpea	17–20.1	Oval, spherical
[e]*Lens culinaris*	Lentil	2.5–25	Oval, elliptical
[f]*Phaseolus vulgaris*	Great northern bean	12–58	Oval, elliptical
Other			
[g]*Musa paradisiaca*	Banana	7–70	–

[a]Hoover, 2001.
[b]Blanshard, 1987.
[c]Qian *et al.*, 1998.
[d]Singh *et al.*, 2004.
[e]Hoover and Manuel, 1995.
[f]Sathe and Salunkhe, 1981.
[g]Núñez-Santiago *et al.*, 2004.

determining starch granule size and number. A few studies have indicated that SBEI may be involved. Studies using antisense SBEI constructs have reported increased granule size in transgenic potato (Flipse *et al.*, 1996). Isoamylase-type debranching enzymes have also been shown to affect starch granule size morphology and number (Burton *et al.*, 2002; Dinges *et al.*, 2003). It has also been suggested that in wheat, an SBEI isoform, SBEIc, may be involved in determining bimodal starch granule size distribution in this crop. Analysis of wheat starch granule proteins has revealed a difference in the abundance of SBEIc, an isoform of starch branching enzyme in small and large granules (Båga *et al.*, 2000; Peng *et al.*, 2000). SBEIc is a large (150 kD) protein that, in contrast to 87 kD SBE, is incorporated preferentially into the large starch granules, and thus associated with plants showing bimodal starch granule size distribution in the endosperm. Downregulation or a mutation in either SBEI or SBEIc can alter the proportion of A- and B-type granules. Table 11.4 summarizes starch granule properties in some plant species.

Phytoglycogen accumulation

Starch debranching enzymes hydrolyse α-(1,6)-branch points in glucan polymers. Two distinct DBE families with different substrate specificities are recognized in plants: isoamylase (ISA) and limit dextrinase (LDA). The ISA hydrolyses amylopectin and glycogen but cannot act on pullulan, whereas LDA can act on pullulan, amylopectin and beta-limit dextrins, but not glycogen. The ISA can be subdivided into three subfamilies: ISA1, ISA2 and ISA3 (Hussain *et al.*, 2003; Rahman *et al.*, 2003). Mutations or downregulation of genes encoding ISA1 protein in developing endosperms result in reduction of starch granules and accumulation of phytoglycogen, a water-soluble glucan polymer. Phytoglycogen-accumulating ISA mutants have been reported in maize, rice and barley (James *et al.*, 1995; Nakamura *et al.*, 1997; Burton *et al.*, 2002), *Chlamydomonas* cells (Mouille *et al.*, 1996; Posewitz *et al.*, 2004), *Arabidopsis* leaves (Zeeman *et al.*, 1998; Delatte *et al.*, 2005; Wattebled *et al.*, 2005) and potato tubers (Bustos *et al.*, 2004). In *Arabidopsis*, there are four genes coding for each of the three ISA and one LDA. Quadruple mutants lacking all the four DBS proteins do not synthesize starch granules but accumulate highly branched glucans which are neither similar to amylopectin or phytoglycogen (Streb *et al.*, 2008). A fraction of these highly branched glucans are present as discrete nanometer-scale particles, whose structure and properties are very different from normal amylopectin present in starch granules (Streb *et al.*, 2008). Similar to glycogen, phytoglycogen is highly branched, with several short chains (dp > 10), having more branch points positioned close together as compared to amylopectin. The presence of up to 30% phytoglycogen in cereal grains affects starch characteristics by reducing viscosity, gel formation and retrogradation rate with increased water-holding capacity (Johnson *et al.*, 2001a). High phytoglycogen starches have increased digestibility, which is a preferred characteristic for livestock feed (Johnson *et al.*, 2001b). However, sugary mutant cereal grains have reduced seed weight and germination, which can be an impediment in their commercial utilization.

Future Perspectives

It has been discovered that natural variants deficient in a starch biosynthetic enzyme accumulate starch with modified structure. In maize, a large number of *in planta* modified starches have been commercialized. However, most of the *in planta* modified starches are limited to changed amylose and amylopectin concentrations. Advances in biochemical and molecular biological strategies have resulted in the identification and characterization of most of the genes in the starch biosynthetic pathway. Initially, characterized genes were used to genetically engineer starch concentration and structure, but in most instances, genetically engineered single genes did not produce the expected changes in starch structure. Recent discoveries of complexes of starch biosynthetic enzymes may partially explain the inability of genetically engineered single genes to change starch structure. The availability of complete genome sequences of rice and *Arabidopsis*, and other crops fairly soon, will provide DNA sequences for all the genes involved in starch biosynthesis. The nucleotide sequences of the starch biosynthetic genes can be used to screen natural germplasm and identify variants which can be used in a plant breeding programme to develop varieties with *in planta* modified starch structure. To accelerate *in planta* starch modification, TILLING will be a very useful strategy, as not only has it already been demonstrated in several diploid crops but also in plants with complex genomes, such as wheat. Genetically engineering well-characterized genes to confer expected changes in starch structure will be one of the most viable technologies. The challenge for realizing the full potential of all the above technologies is to identify critical genes conferring the desirable change in starch structure. Immature spike culture to study cereal grain development is a very pragmatic approach for studying grain development and storage compound deposition in controlled environments (Ganeshan *et al.*, 2007). Immature spike culture combined with antisense oligonucleotide (ODN) technology has been proposed as a systems biology strategy to study the role of starch biosynthetic genes in grain development (Sun *et al.*, 2005). Improved understanding of starch biosynthesis and identification of critical genes conferring desired phenotypes, combined with available or induced genetic diversity, will result in *in planta* production of starches with desirable characteristics.

Acknowledgements

Canada Research Chairs, Canada Foundation for Innovation, Natural Science and Engineering Research Council, Saskatchewan Agricultural Development Fund and Saskatchewan Department of Industry are gratefully acknowledged for supporting research in our laboratories.

References

Amano, E. (1981) Genetic and biochemical characterization of waxy mutants in cereals. *Environmental Health Perspectives* 37, 35–41.

Båga, M., Nair, R.B, Repellin, A., Scoles, G.J. and Chibbar, R.N. (2000) Isolation of a cDNA encoding a granule-bound 152-kilodalton starch-branching enzyme in wheat. *Plant Physiology* 124, 253–263.

Ball, S.G. and Morell, M.K. (2003) From bacterial glycogen to starch: understanding the biogenesis of the plant starch granule. *Annual Review of Plant Biology* 54, 207–233.

Barker, R.F., Thompson, D.V., Talbot, D.R., Swanson, J. and Bennetzen, J.L. (1984) Nucleotide sequence of the maize transposable element *Mul. Nucleic Acids Research* 12, 5955–5967.

Barton, K.A. and Chilton, M.D. (1983) *Agrobacterium* Ti plasmids as vectors for plant genetic engineering. *Methods in Enzymology* 101, 527–539.

Bechtold, N., Ellis, J. and Pelletier, G. (1993) *In planta Agrobacterium*-mediated gene transfer by infiltration of adult *Arabidopsis thaliana* plants. *Comptes Rendus de l Academie des Sciences Serie Iii-Sciences de la Vie-Life Sciences* 316, 1194–1199.

Bhattacharyya, M.K., Smith, A.M., Ellis, T.H.N., Hedley, C. and Martin, C. (1990) The wrinkled-seed character of pea described by Mendel is caused by a transposon-like insertion in a gene encoding starch branching enzyme. *Cell* 60, 115–122.

Blanshard, J.M.V. (1987) Starch granule structure and function: a physicochemical approach. In: Gallaird, T. (ed.) *Starch: Properties and Potential.* Wiley, New York, pp. 16–54.

Blauth, S.L., Yao, Y., Klucinec, J.D., Shannon, J.C., Thompson, D.B. and Guilitinan, M.J. (2001) Identification of Mutator insertional mutants of starch-branching enzyme 2a in corn. *Plant Physiology* 125, 1396–1405.

Boyer, C. (1996) *Biochemical Genetics of Carbohydrate Metabolism in Source and Sink Tissue in Photoassimilate Distribution in Plants and Crops: Source–Sink Relationships.* Zamski, E. and Schaffer, A.A. (eds) MarcelDekker, Inc., New York, pp. 341–367.

Burton, R.A., Jenner, H., Carrangis, L., Fahy, B., Fincher, G.B., Hylton, C., Laurie, D.A., Parker, M., Waite, D., van Wegen, S., Verhoeven, T. and Denyer, K. (2002) Starch granule initiation and growth are altered in barley mutants that lack isoamylase activity. *Plant Journal* 31, 97–112.

Bustos, R., Fahy, B., Hylton, C.M., Seale, R., Nebane, N.M., Edwards, A., Martin, C. and Smith, A.M. (2004) Starch granule initiation is controlled by a heteromultimeric isoamylase in potato tubers. *Proceedings of the National Academy of Sciences of the United States of America* 101, 2215–2220.

Caldwell, D.G., McCallum, N., Shaw, P., Muehlbauer, G.J., Marshall, D.F. and Waugh, R. (2004) A structured mutant population for forward and reverse genetics in barley (*Hordeum vulgare* L.). *The Plant Journal* 40, 143–150.

Carvalho, L.J.C.B., de Souza, C.R.B., Cascardo, J.C.D., Bloch, C. and Campos, L. (2004) Identification and characterization of a novel cassava (*Manihot esculenta* Crantz) clone with high free sugar content and novel starch. *Plant Molecular Biology* 56, 643–659.

Cheng, M., Fry, J.E., Pang, S.Z., Zhou, H.P., Hironaka, C.M., Duncan, D.R., Conner, T.W. and Wan, Y.C. (1997) Genetic transformation of wheat mediated by *Agrobacterium tumefaciens. Plant Physiology* 115, 971–980.

Chibbar, R.N. and Båga, M. (2003) Genetic modification of primary metabolism – carbohydrates. In: Thomas, B., Murphy, D.J. and Murray, B.Bg. (eds) *Encyclopedia of Applied Plant Sciences*. Elsevier, Oxford, UK, pp. 449–459.

Chibbar, R.N. and Chakraborty, M. (2005) Waxy wheat. In: Abdel-Aal, E. and Wood, P. (eds) *Specialty Grains for Food and Feed*. American Association of Cereal Chemists, St Paul, Minnesota, pp. 143–165.

Chibbar, R.N., Båga, M., Ganeshan, S., Hucl, P.J., Limin, A., Perron, C., Ratnayaka, I., Kapoor, A., Verma, V. and Fowler, D.B. (2005) Carbohydrate modification to add value to wheat. In: Chung, O.K. and Lookhart, G.L. (eds) *Proceedings Third International Wheat Quality Conference*. Grain Industry Alliance, Manhattan, Kansas, pp. 75–84.

Clarke, B., Liang, R., Morell, M.K., Bird, A.R., Jenkins, C.L.D. and Li, Z. (2008) Gene expression in a starch synthase IIa mutant of barley: changes in the level of gene transcription and grain composition. *Functional and Integrative Genomics* 8, 211–221.

Clough, S.J. and Bent, A.F. 1998) Floral dip: a simplified method for *Agrobacterium*-mediated transformation of *Arabidopsis thaliana. Plant Journal* 16, 735–743.

Comai, L., Young, K., Till, B.J., Reynolds, S.H., Greene, E.A., Codomo, C.A., Enns, L.C., Johnson, J.E., Burtner, C., Odden, A.R. and Henikoff, S. (2004) Efficient discovery of DNA polymorphisms in natural populations by Ecotilling. *Plant Journal* 37, 778–786.

Cooper, J.L., Till, B.J., Laport, R.G., Darlow, M.C., Kleffner, J.M., Jamai, A., El-Mellouki, T., Liu, S., Ritchie, R., Nielsen, N., Bilyeu, K.D., Meksem, K., Comai, L. and Henikoff, S. (2008) TILLING to detect induced mutations in soybean. *BMC Plant Biology* 8: 9, doi: 10.1186/1471-2229-8-9.

Craig, J., Lloyd, J.R., Tomlinson, K., Barber, L., Edwards, A., Wang, T.L., Martin, C., Hedley, C.L. and Smith, A.M. (1998) Mutations in the gene encoding starch synthase II profoundly alter amylopectin structure in pea embryos. *Plant Cell* 10, 413–426.

Delatte, T., Trevisan, M., Parker, M.L. and Zeeman, S.C. (2005) Arabidopsis mutants Atisa1 and Atisa2 have identical phenotypes and lack the same multimeric isoamylase, which influences the branch point distribution of amylopectin during starch synthesis. *Plant Journal* 41, 815–830.

Demeke, T., Huel, P., Abdel-Aal, E.S.M., Baga, M. and Chibbar, R.N. (1999) Biochemical characterization of the wheat waxy A protein and its effect on starch properties. *Cereal Chemistry* 76, 694–698.

Denyer, K., Johnson, P., Zeeman, S. and Smith, A.M. (2001) The control of amylose synthesis. *Journal of Plant Physiology* 158, 479–487.

Dinges, J.R., Colleoni, C., James, M.G. and Myers, A.M. (2003) Mutational analysis of the pullulanase-type debranching enzyme of maize indicates multiple functions in starch metabolism. *Plant Cell* 15, 666–680.

Dong, C., Dalton-Morgan, J., Vincent, K. and Sharp, P. (2009) A modified TILLING method for wheat breeding. *The Plant Genome* 2, 39–47.

Doring, H.P., Tillmann, E. and Starlinger, P. (1984) DNA sequence of the maize transposable element dissociation. *Nature* 307, 127–130.

Drinkwater, J., Ganesgan, S., Jansson, C. and Chibbar, R.N. (2010) Anti-sense GBSSI oligonucleotides effectively down-regulate GBSSI gene expression in developing wheat (*Triticum aestivum* L.) endosperms. In *Wheat Science Dynamics: Challenges and Opportunities,* Dexter, J.E. and Chibbar, R.N. (eds) Agebios International, Jodhpur, India (in press).

Edwards, A., Fulton, D.C., Hylton, C.M., Jobling, S.A., Gidley, M., Rossner, U., Martin, C. and Smith, A.M. (1999) A combined reduction in activity of starch synthases II and III of potato has novel effects on the starch of tubers. *Plant Journal* 17, 251–261.

Flipse, E., Suurs, L., Keetels, C.J.A.M., Kossmann, J., Jacobsen, E. and Visser, R.G. F. (1996) Introduction of sense and antisense cDNA for branching enzyme in the amylose-free potato mutant leads to physico-chemical changes in the starch. *Planta* 198, 340–347.

Fraley, R.T. and Horsch, R.B. (1983) *In vitro* transformation of *Petunia* protoplasts by *Agrobacterium tumefaciens. Journal of Cellular Biochemistry Supplement*, Issue 7, Part B, 250.

Frary, A. and Hamilton, C.M. (2001) Efficiency and stability of high molecular weight DNA transformation: an analysis in tomato. *Transgenic Research* 10, 121–132.

Fu, Y., Ballicora, M.A., Leykam, J.F. and Preiss, J. (1998) Mechanism of reductive activation of potato tuber ADP-glucose pyrophosphorylase. *Journal of Biological Chemistry* 273, 25045–25052.

Fujita, N., Hasegawa, H. and Taira, T. (2001) The isolation and characterization of a waxy mutant of diploid wheat (*Triticum monococcum* L.). *Plant Science* 160, 595–602.
Fulton, D.C., Edwards, A., Pilling, E., Robinson, H.L., Fahy, B., Seale, R., Kato, L., Donald, A.M., Geigenberger, P., Martin, C. and Smith, A.M. (2002) Role of granule-bound starch synthase in determination of amylopectin structure and starch granule morphology in potato. *Journal of Biological Chemistry* 277, 10834–10841.
Ganeshan, S., Caswell, K.L., Kartha, K.K. and Chibbar, R.N. (2002) *Shoot Regeneration and Proliferation in Transgenic Plants and Crops*. Marcel Dekker, Inc, New York, pp. 69–84.
Ganeshan, S., Baga, M., Harvey, B.L., Rossnagel, B.G., Scoles, G.J. and Chibbar, R.N. (2003) Production of multiple shoots from thidiazuron-treated mature embryos and leaf-base/apical meristems of barley (*Hordeum vulgare*). *Plant Cell Tissue and Organ Culture* 73, 57–64.
Ganeshan, S., Chodaparambil, S.V., Baga, M., Fowler, D.B., Hucl, P., Rossnagel, B.G. and Chibbar, R.N. (2006) *In vitro* regeneration of cereals based on multiple shoot induction from mature embryos in response to thidiazuron. *Plant Cell Tissue and Organ Culture* 85, 63–73.
Ganeshan, S., Baga, M., Limin, A.E., Ratnayaka, I., Chodaparambil, S.V., Fowler, D.B. and Chibbar, R.N. (2007) Biotechnological approaches for improving the tolerance of wheat to abiotic stress. In: *International Conference on Sustainable Crop Production in Stress Environments: Management and Genetic Options*. Agrobios International, Jodhpur, India, pp. 297–309.
Geigenberger, P., Hajirezaei, M., Geiger, M., Deiting, U., Sonnewald, U. and Stitt, M. (1998) Overexpression of pyrophosphatase leads to increased sucrose degradation and starch synthesis, increased activities of enzymes for sucrose–starch interconversions, and increased levels of nucleotides in growing potato tubers. *Planta* 205, 428–437.
Geigenberger, P., Kolbe, A. and Tiessen, A. (2005) Redox regulation of carbon storage and partitioning in response to light and sugars. *Journal of Experimental Botany* 56, 1469–1479.
Ghosh, H.P. and Preiss, J. (1966) Adenosine diphosphate glucose pyrophosphorylase – a regulatory enzyme in biosynthesis of starch in spinach leaf chloroplasts. *Journal of Biological Chemistry* 241, 4491–4504.
Giroux, M.J., Shaw, J., Barry, G., Cobb, B.G., Greene, T., Okita, T. and Hannah, L.C. (1996) A single gene mutation that increases maize seed weight. *Proceedings of the National Academy of Sciences of the United States of America* 93, 5824–5829.
Greene, E.A., Codomo, C.A., Taylor, N.E., Henikoff, J.G., Till, B.J., Reynolds, S.H., Enns, L.C., Burtner, C., Johnson, J.E., Odden, A.R., Comai, L. and Henikoff, S. (2003) Spectrum of chemically induced mutations from a large-scale reverse-genetic screen in arabidopsis. *Genetics* 164, 731–740.
Greene, T.W., Kavakli, I.H., Kahn, M.L. and Okita, T.W. (1998) Generation of up-regulated allosteric variants of potato ADP-glucose pyrophosphorylase by reversion genetics. *Proceedings of the National Academy of Sciences of the United States of America* 95, 10322–10327.
Guan, H.P. and Preiss, J. (1993) Differentiation of the properties of the branching isozymes from maize (*Zea mays*). *Plant Physiology* 102, 1269–1273.
Haberlandt, G. (1902) Kulturversuche mit isolierten pflanzellen. *Sitzungsberichte der Koniglichen Preussischen Akademie der Wissenschaften Mathematische - Naturwissenschaftliche*, 111, 69–72.
Hamilton, C.M. (1997) A binary-BAC system for plant transformation with high-molecular-weight DNA. *Gene* 200, 107–116.
Hamilton, C.M., Frary, A., Lewis, C. and Tanksley, S.D. (1996) Stable transfer of intact high molecular weight DNA into plant chromosomes. *Proceedings of the National Academy of Sciences of the United States of America* 93, 9975–9979.
Hamilton, C.M., Frary, A., Xu, Y.M., Tanksley, S.D. and Zhang, H.B. (1999) Construction of tomato genomic DNA libraries in a binary-BAC (BIBAC) vector. *Plant Journal* 18, 223–229.
Hannah, L.C. and James, M. (2008) The complexities of starch biosynthesis in cereal endosperms. *Current Opinion in Biotechnology* 19, 160–165.
Haring, M.A., Rommens, C.M.T., Nijkamp, H.J.J. and Hille, J. (1991) The use of transgenic plants to understand transposition mechanisms and to develop transposon tagging strategies. *Plant Molecular Biology* 16, 449–461.
Henikoff, S. and Comai, L. (2003) Single-nucleotide mutations for plant functional genomics. *Annual Review of Plant Biology* 54, 375–401.
Henikoff, S., Till, B.J. and Comai, L. (2004) TILLING. Traditional mutagenesis meets functional genomics. *Plant Physiology* 135, 630–636.

Hennen-Bierwagen, T.A., Liu, F., Marsh, R.S., Kim, S., Gan, Q., Tetlow, I.J., Emes, M.J., James, M.G. and Myers, A.M. (2008) Starch biosynthetic enzymes from developing maize endosperm associate in multisubunit complexes. *Plant Physiology* 146, 1892–1908.

Hennen-Bierwagen, T.A., Lin, Q., Grimaud, F., Planchot, V., Keeling, P.L., James, M.G. and Myers, A.M. (2009) Proteins from multiple metabolic pathways associate with starch biosynthetic enzymes in high molecular weight complexes: a model for regulation of carbon allocation in maize amyloplasts. *Plant Physiology* 149, 1541–1559.

Hiei, Y., Ohta, S., Komari, T. and Kumashiro, T. (1994) Efficient transformation of rice (*Oryza sativa* L) mediated by *Agrobacterium* and sequence analysis of the boundaries of the T-DNA. *Plant Journal* 6, 271–282.

Hohmann, U., Jacobs, G. and Jung, C. (2005) An EMS mutagenesis protocol for sugar beet and isolation of non-bolting mutants. *Plant Breeding* 124, 317–321.

Hoover, R. (2001) Composition, molecular structure, and physicochemical properties of tuber and root starches: a review. *Carbohydrate Polymers* 45, 253–267.

Hoover, R. and Manuel, H. (1995) A comparative study of the physicochemical properties of starches from 2 lentil cultivars. *Food Chemistry* 53, 275–284.

Hovenkamp-Hermelink, J.H.M., Jacobsen, E., Ponstein, A.S., Visser, R.G.F., Vosscheperkeuter, G.H., Bijmolt, E.W., Devries, J.N., Witholt, B. and Feenstra, W.J. (1987) Isolation of an amylose-free starch mutant of the potato (*Solanum tuberosum* L.). *Theoretical and Applied Genetics* 75, 217–221.

Hussain, H., Mant, A., Seale, R., Zeeman, S., Hinchliffe, E., Edwards, A., Hylton, C., Bornemann, S., Smith, A.M., Martin, C. and Bustos, R. (2003) Three isoforms of isoamylase contribute different catalytic properties for the debranching of potato glucans. *Plant Cell* 15, 133–149.

Ihemere, U., Arias-Garzon, D., Lawrence, S. and Sayre, R. (2006) Genetic modification of cassava for enhanced starch production. *Plant Biotechnology Journal* 4, 453–465.

James, M.G., Robertson, D.S. and Myers, A.M. (1995) Characterization of the maize gene Sugary1, a determinant of starch composition in kernels. *Plant Cell* 7, 417–429.

James, M.G., Denyer, K. and Myers, A.M. (2003) Starch synthesis in the cereal endosperm. *Current Opinion in Plant Biology* 6, 215–222.

Jansson, C., Sun, C., Ganeshan, S. and Chibbar, R.N. (2007) Plant breeding: antisense ODN inhibition in *in vitro* spike cultures as a powerful diagnostic tool in studies on cereal grain development. In: *Progress in Botany*. Esser, K., Löttge, U., Beyschlag, W. and Murata, J. (eds) Springer Verlag, Berlin, pp. 179–190.

Jeng, T.L., Tseng, T.H., Wang, C.S., Chen, C.L. and Sung, J.M. (2003) Starch biosynthesizing enzymes in developing grains of rice cultivar Tainung 67 and its sodium azide-induced rice mutant. *Field Crops Research* 84, 261–269.

Jobling, S. (2004) Improving starch for food and industrial applications. *Current Opinion in Plant Biology* 7, 210–218.

Jobling, S.A., Schwall, G.P., Westcott, R.J., Sidebottom, C.M., Debet, M., Gidley, G.J., Jeffcoat, R. and Safford, R. (1999) A minor form of starch branching enzyme in potato (*Solanum tuberosum* L.) tubers has a major effect on starch structure: cloning and characterization of multiple forms of SBE A. *Plant Journal* 18, 163–171.

Johnson, L.A., Hardy, C.L., Baumel, C.P. and White, P.J. (2001a) Identifying valuable corn quality traits for starch production. *Cereal Foods World* 46, 417–423.

Johnson, L.A., Hardy C.L., Baumel, C.P., Yu, Y.-H. and Sell, J.L. (2001b) Identifying valuable corn quality traits for livestock feed. *Cereal Foods World* 46, 472–481.

Johnson, P.E., Patron, N.J., Bottrill, A.R., Dinges, J.R., Fahy, B.F., Parker, M.L., Waite, D.N. and Denyer, K. (2003) A low-starch barley mutant, Riso 16, lacking the cytosolic small subunit of ADP-glucose pyrophosphorylase, reveals the importance of the cytosolic isoform and the identity of the plastidial small subunit. *Plant Physiology* 131, 684–696.

Jones, T.W.A., Gottlieb, L.D. and Pichersky, E. (1986) Reduced enzyme activity and starch level in an induced mutant of chloroplast phosphoglucose isomerase. *Plant Physiology* 81, 367–371.

Joseph, R., Yeoh, H.H. and Loh, C.S. (2004) Induced mutations in cassava using somatic embryos and the identification of mutant plants with altered starch yield and composition. *Plant Cell Reports* 23, 91–98.

Juliano, B.O., Perez, C.M., Kaushik, R. and Khush, G.S. (1990) Some grain properties of Ir36-based starch mutants. *Starch-Starke* 42, 256–260.

Kaplan, F. and Guy, C.L. (2005) RNA interference of *Arabidopsis* beta-amylase8 prevents maltose accumulation upon cold shock and increases sensitivity of PSII photochemical efficiency to freezing stress. *Plant Journal* 44, 730–743.

Kasemsuwan, T., Jane, J., Schnable, P., Stinard, P. and Robertson, D. (1995) Characterization of the dominant mutant amylose-extender (Ael-5180) maize starch. *Cereal Chemistry* 72, 457–464.

Kim, W.S., Kim, J., Krishnan, H.B. and Nahm, B.H. (2005) Expression of *Escherichia coli* branching enzyme in caryopses of transgenic rice results in amylopectin with an increased degree of branching. *Planta* 220, 689–695.

Klein, T.M., Gradziel, T., Fromm, M.E. and Sanford, J.C. (1988a) Factors influencing gene delivery into *Zea mays* cells by high velocity microprojectiles. *Bio-Technology* 6, 559–563.

Klein, T.M., Harper, E.C., Svab, Z., Sanford, J.C., Fromm, M.E. and Maliga, P. (1988b) Stable genetic transformation of intact *Nicotiana* cells by the particle bombardment process. *Proceedings of the National Academy of Sciences of the United States of America* 85, 8502–8505.

Kojima, M., Shiojri, H., Nogawa, M., Nozue, M., Matsumoto, D., Wada, A., Saiki, Y. and Kiguch, K. (2004) *In planta* transformation of kenaf plants (*Hibiscus cannabinus* var. aokawa No. 3) by *Agrobacterium tumefaciens*. *Journal of Bioscience and Bioengineering* 98, 136–139.

Kolbe, A., Tiessen, A., Schluepmann, H., Paul, M., Ulrich, S. and Geigenberger, P. (2005) Trehalose 6-phosphate regulates starch synthesis via posttranslational redox activation of ADP-glucose pyrophosphorylase. *Proceedings of the National Academy of Sciences of the United States of America* 102, 11118–11123.

Koprek, T., McElroy, D., Louwerse, J., Williams-Carrier, R. and Lemaux, P.G. (2000) An efficient method for dispersing Ds elements in the barley genome as a tool for determining gene function. *Plant Journal* 24, 253–263.

Kosar-Hashemi, B., Li, Z.Y., Larroque, O., Regina, A., Yamamori, M., Morell, M.K. and Rahman, S. (2007) Multiple effects of the starch synthase II mutation in developing wheat endosperm. *Functional Plant Biology* 34, 431–438.

Kovalchuk, I., Kovalchuk, O. and Hohn, B. (2000) Genome-wide variation of the somatic mutation frequency in transgenic plants. *Embo Journal* 19, 4431–4438.

Kuipers, A.G.J., Jacobsen, E. and Visser, R.G.F. (1994) Formation and deposition of amylose in the potato-tuber starch granule are affected by the reduction of granule-bound starch synthase gene-expression. *Plant Cell* 6, 43–52.

Kuriki, T., Stewart, D.C. and Preiss, J. (1997) Construction of chimeric enzymes out of maize endosperm branching enzymes I and II: activity and properties. *Journal of Biological Chemistry* 272, 28999–29004.

Lan, H., Hoover, R., Liu, Q., Donner, E., Båga, M., Asare, E. and Chibbar, R.N. (2008) Impact of annealing on the molecular structure and physicochemical properties of normal, waxy and high amylose wheat starches. *Food Chemistry* 111, 663–675.

Larkin, P.J. and Scowcroft, W.R. (1981) Somaclonal variation – a novel source of variability from cell cultures for plant improvement. *Theoretical and Applied Genetics* 60, 197–214.

Lin, L., Liu, Y.G., Xu, X.P. and Li, B.J. (2003) Efficient linking and transfer of multiple genes by a multigene assembly and transformation vector system. *Proceedings of the National Academy of Sciences of the United States of America* 100, 5962–5967.

Lindeboom, N., Chang, P.R. and Tyler, R.T. (2004) Analytical, biochemical and physicochemical aspects of starch granule size, with emphasis on small granule starches: a review. *Starch-Starke* 56, 89–99.

Liu, H.B., Kawabe, A., Matsunaga, S., Murakawa, T., Mizukami, A., Yanagisawa, M., Nagamori, E., Harashima, S., Kobayashi, A. and Fukui, K. (2004) Obtaining transgenic plants using the bio-active beads method. *Journal of Plant Research* 117, 95–99.

Liu, Y.G., Shirano, Y., Fukaki, H., Yanai, Y., Tasaka, M., Tabata, S. and Shibata, D. (1999) Complementation of plant mutants with large genomic DNA fragments by a transformation-competent artificial chromosome vector accelerates positional cloning. *Proceedings of the National Academy of Sciences of the United States of America* 96, 6535–6540.

McCallum, C.M., Comai, L., Greene, E.A. and Henikoff, S. (2000a) Targeted screening for induced mutations. *Nature Biotechnology* 18, 455–457.

McCallum, C.M., Comai, L., Greene, E.A. and Henikoff, S. (2000b) Targeting induced local lesions in genomes (TILLING) for plant functional genomics. *Plant Physiology* 123, 439–442.

Marshall, J., Sidebottom, C., Debet, M., Martin, C., Smith, A.M. and Edwards, A. (1996) Identification of the major starch synthase in the soluble fraction of potato tubers. *Plant Cell* 8, 1121–1135.

Meyer, F.D., Talbert, L.E., Martin, J.M., Lanning, S.P., Greene, T.W. and Giroux, M.J. (2007) Field evaluation of transgenic wheat expressing a modified ADP-glucose pyrophosphorylase large subunit. *Crop Science* 47, 336–342.

Miura, H. and Sugawara, A. (1996) Dosage effects of the three Wx genes on amylose synthesis in wheat endosperm. *Theoretical and Applied Genetics* 93, 1066–1070.

Morell, M.K. and Myers, A.M. (2005) Towards the rational design of cereal starches. *Current Opinion in Plant Biology* 8, 204–210.
Morell, M.K., Kosar-Hashemi, B., Cmiel, M., Samuel, M.S., Chandler, P., Rahman, S., Buleon, A., Batey, I.L. and Li, Z.Y. (2003) Barley sex6 mutants lack starch synthase IIa activity and contain a starch with novel properties. *Plant Journal* 34, 172–184.
Mouille, G., Maddelein, M.L., Libessart, N., Talaga, P., Decq, A., Delrue, B. and Ball, S. (1996) Preamylopectin processing: a mandatory step for starch biosynthesis in plants. *Plant Cell* 8, 1353–1366.
Müller-Röber, B., Sonnewald, U. and Willmitzer, L. (1992) Inhibition of the Adp-glucose pyrophosphorylase in transgenic potatoes leads to sugar-storing tubers and influences tuber formation and expression of tuber storage protein genes 145. *Embo Journal* 11, 1229–1238.
Munck, L., Moller, B., Jacobsen, S. and Sondergaard, I. (2004) Near infrared spectra indicate specific mutant endosperm genes and reveal a new mechanism for substituting starch with (1 -> 3,1 -> 4)-beta-glucan in barley. *Journal of Cereal Science* 40, 213–222.
Murakawa, T., Kajiyama, S., Ikeuchi, T., Kawakami, S. and Fukui, K. (2008a) Improvement of transformation efficiency by bioactive-beads-mediated gene transfer using DNA-lipofectin complex as entrapped genetic material. *Journal of Bioscience and Bioengineering* 105, 77–80.
Murakawa, T., Kajiyama, S. and Fukui, K. (2008b) Improvement of bioactive bead-mediated transformation by concomitant application of electroporation. *Plant Biotechnology* 25, 387–390.
Nakamura, T., Yamamori, M., Hirano, H., Hidaka, S. and Nagamine, T. (1995) Production of waxy (amylose-free) wheats. *Molecular and General Genetics* 248, 253–259.
Nakamura, Y., Kubo, A., Shimamune, T., Matsuda, T., Harada, K. and Satoh, H. (1997) Correlation between activities of starch debranching enzyme and alpha-polyglucan structure in endosperms of sugary-1 mutants of rice. *Plant Journal* 12, 143–153.
Nakano, A., Suzuki, G., Yamamoto, M., Turnbull, K., Rahman, S. and Mukai, Y. (2005) Rearrangements of large-insert T-DNAs in transgenic rice. *Molecular Genetics and Genomics* 273, 123–129.
Nelson, O.E. and Rines, H.W. (1962) The enzymatic deficiency in the *waxy* mutant of maize. *Biochemical and Biophysical Research Communications* 9, 297–300.
Núñez-Santiago, M.C., Bello-Perez, L.A. and Tecante, A. (2004) Swelling-solubility characteristics, granule size distribution and rheological behavior of banana (*Musa paradisiaca*) starch. *Carbohydrate Polymers* 56, 65–75.
Oleykowski, C.A., Mullins, C.R.B., Godwin, A.K. and Yeung, A.T. (1998) Mutation detection using a novel plant endonuclease. *Nucleic Acids Research* 26, 4597–4602.
Orthoefer, F.T. (1994) *Corn Starch Modification and Uses in Corn: Chemistry and Technology*. American Association of Cereal Chemists, St Paul, Minnesota, pp. 479–499.
Otani, M., Hamada, T., Katayama, K., Kitahara, K., Kim, S.H., Takahata, Y., Suganuma, T. and Shimada, T. (2007) Inhibition of the gene expression for granule-bound starch synthase I by RNA interference in sweet potato plants. *Plant Cell Reports* 26, 1801–1807.
Pan, D. and Nelson, O.E. (1984) A debranching enzyme deficiency in endosperms of the *sugary*-1 mutants of maize. *Plant Physiology* 74, 324–328.
Patron, N.J., Smith, A.M., Fahy, B.F., Hylton, C.M., Naldrett, M.J., Rossnagel, B.G. and Denyer, K. (2002) The altered pattern of amylose accumulation in the endosperm of low-amylose barley cultivars is attributable to a single mutant allele of granule-bound starch synthase I with a deletion in the 5′-non-coding region. *Plant Physiology* 130, 190–198.
Peng, M.S., Gao, M., Baga, M., Hucl, P. and Chibbar, R.N. (2000) Starch-branching enzymes preferentially associated with A-type starch granules in wheat endosperm. *Plant Physiology* 124, 265–272.
Pereira, A., Cuypers, H., Gierl, A., Schwarzsommer, Z. and Saedler, H. (1986) Molecular analysis of the En/Spm transposable element system of *Zea mays*. *Embo Journal* 5, 835–841.
Perry, J.A., Wang, T.L., Welham, T.J., Gardner, S., Pike, J.M., Yoshida, S. and Parniske, M. (2003) A TILLING reverse genetics tool and a web-accessible collection of mutants of the legume *Lotus japonicus*. *Plant Physiology* 131, 866–871.
Pollak, L.M. (2003) The history and success of the public–private project on germplasm enhancement in maize (GEM). *Advances in Agronomy* 78, 45–87.
Posewitz, M.C., Smolinski, S.L., Kanakagiri, S., Melis, A., Seibert, M. and Ghirardi, M.L. (2004) Hydrogen photoproduction is attenuated by disruption of an isoamylase gene in *Chlamydomonas reinhardtii*. *Plant Cell* 16, 2151–2163.
Preiss, J. (1991) Biology and molecular biology of starch synthesis and its regulation. *Oxford Surveys of Plant Molecular Biology* 7, 59–114.

Qian, J., Rayas-Duarte, P. and Grant, L. (1998) Partial characterization of buckwheat (*Fagopyrum esculentum*) starch. *Cereal Chemistry* 75, 365–373.

Qing, C.M., Fan, L., Lei, Y., Bouchez, D., Tourneur, C., Yan, L. and Robaglia, C. (2000) Transformation of Pakchoi (*Brassica rapa* L. ssp. *chinensis*) by Agrobacterium infiltration. *Molecular Breeding* 6, 67–72.

Raemakers, K., Schreuder, M., Suurs, L., Furrer-Verhorst, H., Vincken, J-P., de Vetten, N., Jacobsen, E. and Visser, R.G.F. (2005) Improved cassava starch by antisense inhibition of granule-bound starch synthase I. *Molecular Breeding* 16, 163–172.

Rahman, S., Nakamura, Y., Li, Z., Clarke, B., Fujita, N., Mukai, Y., Yamamoto, M., Regina, A., Tan, Z., Kawasaki, S. and Morell, M. (2003) The sugary-type isoamylase gene from rice and *Aegilops tauschii*: characterization and comparison with maize and *Arabidopsis*. *Genome* 46, 496–506.

Ral, J.P., Colleoni, C., Wattebled, F., Dauvillee, D., Nempont, C., Deschamps, P., Li, Z.Y., Morell, M.K., Chibbar, R., Purton, S., D'Hulst, C. and Ball, S.G. (2006) Circadian clock regulation of starch metabolism establishes GBSSI as a major contributor to amylopectin synthesis in *Chlamydomonas reinhardtii*. *Plant Physiology* 142, 305–317.

Rashid, H., Yokoi, S., Toriyama, K. and Hinata, K. (1996) Transgenic plant production mediated by Agrobacterium in indica rice. *Plant Cell Reports* 15, 727–730.

Regina, A., Bird, A., Topping, D., Bowden, S., Freeman, J., Barsby, T., Kosar-Hashemi, B., Li, Z.Y., Rahman, S. and Morell, M. (2006) High-amylose wheat generated by RNA interference improves indices of large-bowel health in rats. *Proceedings of the National Academy of Sciences of the United States of America* 103, 3546–3551.

Safford, R., Jobling, S.A., Sidebottom, C.M., Westcott, R.J., Cooke, D., Tober, K.J., Strongitharm, B.H., Russell, A.L. and Gidley, M.J. (1998) Consequences of antisense RNA inhibition of starch branching enzyme activity on properties of potato starch. *Carbohydrate Polymers* 35, 155–168.

Sakulsingharoj, C., Choi, S.B., Hwang, S.K., Edwards, G.E., Bork, J., Meyer, C.R., Preiss, J. and Okita, T.W. (2004) Engineering starch biosynthesis for increasing rice seed weight: the role of the cytoplasmic ADP-glucose pyrophosphorylase. *Plant Science* 167, 1323–1333.

Sathe, S.K. and Salunkhe, D.K. (1981) Isolation, partial characterization and modification of the Great Northern Bean (*Phaseolus-Vulgaris* L) starch. *Journal of Food Science* 46, 617–621.

Satoh, H., Nishi, A., Yamashita, K., Takemoto, Y., Tanaka, Y., Hosaka, Y., Sakurai, A., Fujita, N. and Nakamura, Y. (2003) Starch-branching enzyme I-deficient mutation specifically affects the structure and properties of starch in rice endosperm. *Plant Physiology* 133, 1111–1121.

Schwall, G.P., Safford, R., Westcott, R.J., Jeffcoat, R., Tayal, A., Shi, Y.C., Gidley, M.J. and Jobling, S.A. (2000) Production of very-high-amylose potato starch by inhibition of SBE A and B. *Nature Biotechnology* 18, 551–554.

Seib, P. (2000) Reduced-amylose wheats and Asian noodles. *Cereal Foods World* 45, 504–512.

Shimamoto, K., Terada, R., Izawa, T. and Fujimoto, H. (1989) Fertile transgenic rice plants regenerated from transformed protoplasts. *Nature* 338, 274–276.

Singh, N., Sandhu, K.S. and Kaur, M. (2004) Characterization of starches separated from Indian chickpea (*Cicer arietinum* L.) cultivars. *Journal of Food Engineering* 63, 441–449.

Slade, A.J., Fuerstenberg, S.I., Loeffler, D., Steine, M.N. and Facciotti, D. (2005) A reverse genetic, nontransgenic approach to wheat crop improvement by TILLING. *Nat Biotech* 23, 75–81.

Slewinski, T.L., Ma, Y., Baker, R.F., Huang, M.S., Meeley, R. and Braun, D.M.(2008) Determining the role of tie-dyed1 in starch metabolism: epistasis analysis with a maize ADP-glucose pyrophosphorylase mutant lacking leaf starch. *Journal of Heredity* 99, 661–666.

Smidansky, E.D., Clancy, M., Meyer, F.D., Lanning, S.P., Blake, N.K., Talbert, L.E. and Giroux, M.J. (2002) Enhanced ADP-glucose pyrophosphorylase activity in wheat endosperm increases seed yield. *Proceedings of the National Academy of Sciences of the United States of America* 99, 1724–1729.

Smidansky, E.D., Martin, J.M., Hannah, L.C., Fischer, A.M. and Giroux, M.J. (2003) Seed yield and plant biomass increases in rice are conferred by deregulation of endosperm ADP-glucose pyrophosphorylase. *Planta* 216, 656–664.

Smith, A.M. (2001) The biosynthesis of starch granules. *Biomacromolecules* 2, 335–341.

Smith, A.M., Denyer, K. and Martin, C. (1997) The synthesis of the starch granule. *Annual Review of Plant Physiology and Plant Molecular Biology* 48, 65–87.

Sone, T., Nagamori, E., Ikeuchi, T., Mizukami, A., Takakura, Y., Kajiyama, S.I., Fukusaki, E.I., Harashima, S., Kobayashi, A. and Fukui, K. (2002) A novel gene delivery system in plants with calcium alginate micro-beads. *Journal of Bioscience and Bioengineering* 94, 87–91.

Song, J., Bradeen, J.M., Naess, S.K., Helgeson, J.P. and Jiang, J. (2003) BIBAC and TAC clones containing potato genomic DNA fragments larger than 100 kb are not stable in Agrobacterium. *Theoretical and Applied Genetics* 107, 958–964.

Stahl, Y., Coates, S., Bryce, J.H. and Morris, P.C. (2004) Antisense downregulation of the barley limit dextrinase inhibitor modulates starch granule size distribution, starch composition and amylopectin structure. *Plant Journal* 39, 599–611.

Stark, D.M., Timmerman, K.P., Barry, G.F., Preiss, J. and Kishore, GM. (1992) Regulation of the amount of starch in plant tissues by ADP glucose pyrophosphorylase. *Science* 258, 287–292.

Streb, S., Delatte, T., Umhang, M., Eicke, S., Schorderet, M., Reinhardt, D. and Zeeman, S.C. (2008) Starch granule biosynthesis in *Arabidopsis* is abolished by removal of all debranching enzymes but restored by the subsequent removal of an endoamylase. *Plant Cell* 20, 3448–3466.

Sun, C.X., Hoglund, A.S., Olsson, H., Mangelsen, E. and Jansson, C. (2005) Antisense oligodeoxynucleotide inhibition as a potent strategy in plant biology: identification of SUSIBA2 as a transcriptional activator in plant sugar signalling. *Plant Journal* 44, 128–138.

Supartana, P., Shimizu, T., Shioiri, H., Nogawa, M., Nozue, M. and Kojima, M. (2005) Development of simple and efficient *in planta* transformation method for rice (*Oryza sativa* L.) using *Agrobacterium tumefaciens. Journal of Bioscience and Bioengineering* 100, 391–397.

Supartana, P., Shimizu, T., Nogawa, M., Shioiri, H., Nakajima, T., Haramoto, N., Nozue, M. and Kojima, M. (2006) Development of simple and efficient in planta transformation method for wheat (*Triticum aestivum* L.) using *Agrobacterium tumefaciens. Journal of Bioscience and Bioengineering* 102, 162–170.

Takeda, Y., Guan, H.P. and Preiss, J. (1993) Branching of amylose by the branching isoenzymes of maize endosperm. *Carbohydrate Research* 240, 253–263.

Taylor, N., Chavarriaga, P., Raemakers, K., Siritunga, D. and Zhang, P. (2004) Development and application of transgenic technologies in cassava. *Plant Molecular Biology* 56, 671–688.

Terada, R., Nakajima, M., Isshiki, M., Okagaki, R., Wessler, S.R. and Shimamoto, K. (2000) Antisense waxy genes with highly active promoters effectively suppress waxy gene expression in transgenic rice. *Plant and Cell Physiology* 41, 881–888.

Tetlow, I.J., Morell, M.K. and Emes, M.J. (2004a) Recent developments in understanding the regulation of starch metabolism in higher plants. *Journal of Experimental Botany* 55, 2131–2145.

Tetlow, I.J., Wait, R., Lu, Z.X., Akkasaeng, R., Bowsher, C.G., Esposito, S., Kosar-Hashemi, B., Morell, M.K. and Emes, M.J. (2004b) Protein phosphorylation in amyloplasts regulates starch branching enzyme activity and protein–protein interactions. *Plant Cell* 16, 694–708.

Tetlow, I.J., Beisel, K.G., Cameron, S., Makhmoudova, A., Liu, F., Bresolin, N.S., Wait, R., Morell, M.K. and Emes, M.J. (2008) Analysis of protein complexes in wheat amyloplasts reveals functional interactions among starch biosynthetic enzymes. *Plant Physiology* 146, 1878–1891.

Till, B., Cooper, J., Tai, T., Colowit, P., Greene, E., Henikoff, S. and Comai, L. (2007) Discovery of chemically induced mutations in rice by TILLING. *BMC Plant Biology* 7, 11, doi:10.1186/1471-2229-7-19.

Till, B.J., Burtner, C., Comai, L. and Henikoff, S. (2004) Mismatch cleavage by single-strand specific nucleases. *Nucleic Acids Research* 32, 2632–2641.

Till, B.J., Zerr, T., Comai, L. and Henikoff, S. (2006) A protocol for TILLING and Ecotilling in plants and animals. *Nature Protocols* 1, 2465–2477.

Tingay, S., McElroy, D., Kalla, R., Fieg, S., Wang, M.B., Thornton, S. and Brettell, R. (1997) *Agrobacterium tumefaciens*-mediated barley transformation. *Plant Journal* 11, 1369–1376.

Torney, F., Trewyn, B.G., Lin, V.S.Y. and Wang, K. (2007) Mesoporous silica nanoparticles deliver DNA and chemicals into plants. *Nature Nanotechnology* 2, 295–300.

Trieu, A.T., Burleigh, S.H., Kardailsky, I.V., Maldonado-Mendoza, I.E., Versaw, W.K., Blaylock, L.A., Shin, H.S., Chiou, T.J., Katagi, H., Dewbre, G.R., Weigel, D. and Harrison, M.J. (2000) Transformation of *Medicago truncatula* via infiltration of seedlings or flowering plants with Agrobacterium. *Plant Journal* 22, 531–541.

Tsai, C.Y. and Nelson, O.E. (1966) Starch-deficient maize mutant lacking adenosine diphosphate glucose pyrophosphorylase activity. *Science* 151, 341–343.

Tsai, C.Y., Larkins, B.A. and Glover, D.V. (1978) Interaction of Opaque-2 gene with starch forming mutant genes on synthesis of zein in maize endosperm. *Biochemical Genetics* 16, 883–896.

Tyynela, J., Stitt, M., Lonneborg, A., Smeekens, S. and Schulman, A.H. (1995) Metabolism of starch synthesis in developing grains of the Shx shrunken mutant of barley (*Hordeum vulgare*). *Physiologia Plantarum* 93, 77–84.

Tziotis, A., Seetharaman, K., Wong, K.S., Klucinec, J.D., Jane, J.L. and White, P.J. (2004) Structural properties of starch fractions isolated from normal and mutant corn genotypes using different methods. *Cereal Chemistry* 81, 611–620.

Tziotis, A., Seetharaman, K., Klucinec, J.D., Keeling, P. and White, P.J. (2005) Functional properties of starch from normal and mutant corn genotypes. *Carbohydrate Polymers* 61, 238–247.

Umemoto, T., Aoki, N., Lin, H.X., Nakamura, Y., Inouchi, N., Sato, Y., Yano, M., Hirabayashi, H. and Maruyama, S. (2004) Natural variation in rice starch synthase IIa affects enzyme and starch properties. *Functional Plant Biology* 31, 671–684.

Vanderleij, F.R., Visser, R.G.F., Oosterhaven, K., Vanderkop, D.A.M., Jacobsen, E. and Feenstra, W.J. (1991) Complementation of the amylose-free starch mutant of potato (*Solanum-Tuberosum*) by the gene encoding granule-bound starch synthase. *Theoretical and Applied Genetics* 82, 289–295.

Vega, J.M., Yu, W.C., Han, F.P., Kato, A., Peters, E.M., Zhang, Z.J. and Birchler, J.A. (2008) Agrobacterium-mediated transformation of maize (*Zea mays)* with Cre-lox site specific recombination cassettes in BIBAC vectors. *Plant Molecular Biology* 66, 587–598.

Visser, R.G.F., Somhorst, I., Kuipers, G.J., Ruys, N.J., Feenstra, W.J. and Jacobsen, E. (1991) Inhibition of the expression of the gene for granule-bound starch synthase in potato by antisense constructs. *Molecular Genetics and Genomics* 225, 289–296.

Wang, F., Chen, S., Cheng, F., Liu, Y. and Zhang, G. (2007) The differences in grain weight and quality within a rice (*Oryza sativa* L.) panicle as affected by panicle type and source-sink relation. *Journal of Agronomy and Crop Science* 193, 63–73.

Wang, Y.J., White, P., Pollak, L. and Jane, J. (1993) Characterization of starch structures of 17 maize endosperm mutant genotypes with Oh43 inbred line background. *Cereal Chemistry* 70, 171–179.

Wattebled, F., Dong, Y., Dumez, S., Delvalle, D., Planchot, R., Berbezy, P., Vyas, D., Colonna, P., Chatterjee, M., Ball, S. and D'Hulst, C. (2005) Mutants of Arabidopsis lacking a chloroplastic isoamylase accumulate phytoglycogen and an abnormal form of amylopectin. *Plant Physiology* 138, 184–195.

Weil, C.F. (2009) TILLING in grass species. *Plant Physiology* 149, 158–164.

Weissinger, A., Tomes, D., Sanford, J., Kline, T. and Fromm, M. (1987) Microprojectile bombardment for maize transformation. *In Vitro Cellular and Developmental Biology* 23, A75.

Wu, D.X., Shu, Q.Y., Wang, Z.H. and Xia, Y.W. (2002) Effect of gamma irradiation on starch viscosity and physicochemical properties of different rice. *Radiation Physics and Chemistry* 65, 79-86.

Wu, J.L., Wu, C.J., Lei, C.L., Baraoidan, M., Bordeos, A., Madamba, M.R.S., Ramos-Pamplona, M., Mauleon, R., Portugal, A., Ulat, V.J., Bruskiewich, R., Wang, G.L., Leach, J., Khush, G. and Leung, H. (2005) Chemical- and irradiation-induced mutants of indica rice IR64 for forward and reverse genetics. *Plant Molecular Biology* 59, 85–97.

Yamamori, M., Fujita, S., Hayakawa, K., Matsuki, J. and Yasui, T. (2000) Genetic elimination of a starch granule protein, SGP-1, of wheat generates an altered starch with apparent high amylose. *Theoretical and Applied Genetics* 101, 21–29.

Yao, Y., Thompson, D.B. and Guiltinan, M.J. (2004) Maize starch-branching enzyme isoforms and amylopectin structure. In the absence of starch-branching enzyme IIb, the further absence of starch-branching enzyme Ia leads to increased branching. *Plant Physiology* 136, 3515–3523.

Yasmeen, A., Mirza, B., Inayatullah, S., Safdar, N., Jamil, M., Ali, S. and Choudhry, M.F. (2009) *In planta* transformation of tomato. *Plant Molecular Biology Reporter* 27, 20–28.

Yasui, T., Sasaki, T. and Matsuki, J. (2002) Starch properties of a bread wheat (*Triticum aestivum* L.) mutant with an altered flour-pasting profile. *Journal of Cereal Science* 35, 11–16.

Zale, J.M., Agarwal, S., Loar, S. and Steber, C.M. (2009) Evidence for stable transformation of wheat by floral dip in *Agrobacterium tumefaciens. Plant Cell Reports* 28, 903–913.

Zeeman, S.C., Umemoto, T., Lue, W.L., Au-Yeung, P., Martin, C., Smith, A.M. and Chen, J. (1998) A mutant of Arabidopsis lacking a chloroplastic isoamylase accumulates both starch and phytoglycogen. *Plant Cell* 10, 1699–1711.

Zeeman, S.C., Delatte, T., Messerli, G., Umhang, M., Stettler, M., Mettler, T., Streb, S., Reinhold, H. and Kotting, O. (2007a) Starch breakdown: recent discoveries suggest distinct pathways and novel mechanisms 15. *Functional Plant Biology* 34, 465–473.

Zeeman, S.C., Smith, S.M. and Smith, A.M. (2007b) The diurnal metabolism of leaf starch. *Biochemical Journal* 401, 13–28.

Zhang, G., Ao, Z. and Hamaker, B.R. (2008) Nutritional prpoerties of endosperm starches from maize mutants: a parabolic relationship between slowly disgestible starch and amylopectin fine structure. *Journal of Agricultural and Food Chemistry* 56, 4686–4694.

12

Cotton Production, Processing and Uses of Cotton Raw Material

B. Todd Campbell and Lori Hinze

Introduction

Cotton is one of the most significant agricultural commodities in the global market. From 2005 to 2007, the initial economic value of harvested 'raw' cotton averaged US$28 billion. Not taking into account additional economic value captured through cotton processing and associated by-products, this 3-year average world value of 'raw' cotton was equivalent to 20% of the 2007 World Bank (http://web.worldbank.org) estimated gross domestic product (GDP) of the Philippines (US$144 billion). According to the National Cotton Council of America, US cotton production and associated industries in 2002 accounted for US$27 billion in revenue and generated over 234,000 jobs.

While there are four species within the *Gossypium* genus cultivated across the world, the primary production area is limited to two of the New World tetraploid species, *G. hirsutum* and *G. barbadense*. The Asiatic diploid species, *G. arboreum* and *G. herbaceum*, are cultivated on a very limited land area in India, Eastern Asia and parts of Africa. Worldwide, *G. hirsutum* represents about 90% of cotton production and is known as Upland cotton. The widespread cultivation and popularity of Upland cotton is attributed primarily to its high yield potential and wide environmental adaptation. Upland cotton generally produces fibres with quality amenable to textile manufacturing machinery, but does not represent the highest quality cotton fibre available. Acala cotton, also known as extra long staple Upland, represents a special category of Upland cotton that produces fibres of higher quality compared to traditional Upland. Productivity, yield potential and adaptation of the Acala cottons are limited compared to traditional Upland. *G. barbadense*, also known as Pima, Egyptian or Sea Island cotton, is known to produce the finest quality fibres of any cultivated cottons, but also represents a lower degree of environmental adaptation and yield potential compared to Upland and Acala types. Acala and Pima cottons typically favour long growing seasons, dry and low humidity weather and locations with ample irrigation resources (May and Lege, 1999).

In this chapter, we will discuss the cotton supply chain, beginning with cotton production and ending with product manufacturing. Production statistics and practices, fibre processing and several uses of cotton products will be described. Several characteristics unique to cotton fibre relative to other types of textile fibres will be discussed. Since fibre represents the largest source of economic value gleaned from cotton production, focus will be placed on issues related to the production and manufacturing of cotton fibre. However, other products and by-products of cotton production will also be discussed. Within the scope of improving the productivity of cotton production systems, emphasis will be placed on past, current and future advances related to genetic improvement and biotechnology.

Cotton Production

Cotton production is the starting point of the cotton supply chain and generates harvested 'raw' cotton with a value of approximately US$28 billion. Cotton is produced in over 30 countries of the world and provides a major fibre source for textile manufacturers. In view of the entire cotton supply chain, the cotton production system plays a major role in determining both the quantity and quality of the cotton fibre produced for textile manufacturing processes. Broadly speaking, the cotton production system begins with cultivar selection and is determined by optimizing crop production practices related to temperature, soil, water, nutrients and plant protection. In the following sections, we will discuss cotton production systems, with special attention paid to production statistics and practices, while also discussing the role of genetic improvement and biotechnology.

Production statistics and practices

Consistently across the globe, raw cotton is handled and processed in the form of bales, with each bale representing about 218 kg (480 lbs) of raw cotton. Cotton production is amenable to the tropical and subtropical climates of the world between latitudes of ~42°N in the northern hemisphere to 30°S in the southern hemisphere. The northernmost production areas include the former Soviet Union and the People's Republic of China and southernmost areas include Australia. However, over 50% of the world's production occurs in latitudes above 30°N.

From 2005 to 2007, world cotton production averaged over 117 million bales spread across 31 countries in North America, Central America, South America, Africa, Europe, Asia and Australia. Table 12.1 shows the average lint yield and total production of the top 15 cotton producing countries from 2005 to 2007. Consistently over this 3 year period, the three largest cotton producing countries have been China, the USA and India. The average production of these three countries accounts for approximately 66% of the total world production. Other significant producers of cotton include Pakistan, Brazil, Uzbekistan, Turkey

Table 12.1. Average lint yield and production of the top 15 cotton producing countries from 2005 to 2007.

	Lint yield (kg/ha)				Production (× 1000 bales)			
Country	2005	2006	2007	3-year average	2005	2006	2007	3-year average
People's Republic of China	1,168	1,288	1,249	1,235	28,400	35,500	37,000	33,633
India	467	518	573	519	19,050	21,800	24,600	21,817
USA	931	912	995	946	23,890	21,588	19,207	21,562
Pakistan	714	663	603	660	10,165	9,900	8,900	9,655
Brazil	1,204	1,393	1,374	1,324	4,700	7,000	7,360	6,353
Republic of Uzbekistan	844	815	826	828	5,550	5,350	5,500	5,467
Turkey	1,288	1,327	1,222	1,279	3,550	3,800	3,100	3,483
Australia	1,814	2,027	1,742	1,861	2,800	1,350	620	1,590
Greece	1,211	924	983	1,039	1,975	1,400	1,325	1,567
Syria	1,422	1,038	1,252	1,237	1,600	1,020	1,150	1,257
Turkmenistan	354	435	472	420	975	1,200	1,300	1,158
Burkina	472	404	384	420	1,367	1,300	675	1,114
Egypt	743	863	914	840	938	975	1,000	971
Mali	419	375	345	380	1,003	792	450	748
Argentina	446	435	445	442	625	800	703	709

and Greece. Prior to 2007, Australia was also a major producer of cotton, but a recent drought resulted in decreased production during 2007 and 2008.

Many of the top cotton producing countries depend on cotton production to provide fibre for domestic textile industries and for a major export product. The import, export and domestic mill use data for the top 15 cotton producing countries from 2005 to 2007 are provided in Table 12.2. Recently, the primary manufacturers of textile products include mills inside China, India and Pakistan. Also among the top 15 cotton producing countries, China, Pakistan and Turkey represent major importers of cotton fibre. The USA, India and the Republic of Uzbekistan represent the primary exporters of cotton fibre.

Temperature, sunlight, soil, rainfall and/or supplemental irrigation, and supplemental nutrients are the primary resources necessary for cotton production worldwide (Waddle, 1984). Generally speaking, successful cotton production requires 160 days above 15°C and a minimum of 50 cm of water during the growing season. The three primary nutrients necessary for high-yielding cotton production systems are nitrogen, phosphorus and potassium. In addition to these required resources for cotton production, the baseline productivity of a cotton production system is determined by cultivar selection. Selecting the appropriate cultivar best suited for a specific production environment allows one to optimize cultural practices to maximize production potential.

The cultural practices used vary across the world and often are dependent on the availability of labour and mechanization. In countries such as the USA and Australia, cotton production is highly mechanized from planting through harvest. In countries with mechanized production practices, tractors and mechanical planters are used to plant seeds at a determined soil depth and row spacing. Supplemental fertilizers are applied to the soil according to soil test results using mechanical applicators. Weed and insect control measures are also applied using mechanical equipment and are usually based on the results of in-season field scouting to determine weed and/or insect problems. The cropping system used in cotton production varies, depending primarily on water availability. Examples of various cropping systems include continuous cotton, cotton double-cropped with winter crops and cotton in rotation with other summer crops. Over the past 10 years, conservation tillage practices have been adopted readily in the USA. These practices conserve soil water and reduce soil erosion in rainfed production systems (Bauer and Busscher, 1996). In comparison, in countries such as India and Pakistan, cotton production is highly dependent on human and animal power. These factors ultimately determine the cultural practices used in planting, fertility, irrigation, weed/insect/disease management and harvest.

Sustainable cotton production systems

With the widespread adoption of biotechnology-derived cotton cultivars, decreased pesticide use and adoption of conservation tillage practices have decreased the environmental impact of cotton production (Locke *et al.*, 2008). Since the primary energy source used to produce cotton is derived from free and abundant sunlight through photosynthesis, cotton production produces a net gain in energy when considering both fibre and seed (Matlock *et al.*, 2008; Keystone Center, 2009; Nelson *et al.*, 2009). In terms of greenhouse gas emissions, cotton production results in a neutral carbon footprint when considering only fibre production. The carbon footprint of cotton becomes positive when accounting for the additional carbon sequestered in the seed.

Genetic improvement

Over the past 100 years, genetics has played a large role in increasing the productivity of cotton production systems. Since *G. hirsutum* and *G. barbadense* represent the majority of cotton cultivation worldwide, genetic improvement research has focused on these two species. Together with improvements in production practices and mechanization, genetic improvement through cotton breeding has resulted in yield gains and increased production

Table 12.2. Total imports, mill use and exports of the top 15 cotton producing countries from 2005 to 2007.

	Imports (× 1000 bales)				Mill use (× 1000 bales)				Exports (× 1000 bales)			
Country	2005	2006	2007	3-year average	2005	2006	2007	3-year average	2005	2006	2007	3-year average
People's Republic of China	19,284	10,588	11,530	13,801	45,000	50,000	51,500	48,833	36	88	62	62
India	400	465	450	438	16,700	18,100	18,300	17,700	3,450	4,565	7,030	5,015
USA	28	19	12	20	5,871	4,935	4,609	5,138	17,549	13,010	13,653	14,737
Pakistan	1,615	2,305	3,900	2,607	11,500	12,500	12,400	12,133	288	217	270	258
Brazil	308	515	164	329	4,452	4,573	4,600	4,542	1,972	1,300	2,231	1,834
Republic of Uzbekistan	0	0	0	0	800	900	1,000	900	4,800	4,500	4,400	4,567
Turkey	3,501	4,029	3,268	3,599	6,900	7,300	6,000	6,733	216	303	370	296
Australia	0	0	0	0	60	55	50	55	2,884	2,129	1,219	2,077
Greece	20	20	20	20	400	350	300	350	1,350	1,250	1,050	1,217
Syria	0	0	0	0	725	725	725	725	825	375	450	550
Turkmenistan	0	0	0	0	425	430	440	432	550	700	800	683
Burkina	0	0	0	0	4	4	4	4	1,400	1,350	775	1,175
Egypt	525	500	450	492	1,000	1,000	950	983	435	370	629	478
Mali	0	0	0	0	20	23	23	22	1,025	850	500	792
Argentina	157	176	181	171	700	785	825	770	52	33	14	33

throughout the world. Whereas improved production practices maximize genetic potential, the genetic make-up of a specific strain or cultivar essentially determines the baseline production potential. Universally, the primary goal of cotton breeding programmes is to increase the genetic potential of cotton germplasm (or genetic resources) for important production-related traits (e.g. lint yield, disease resistance, pest resistance, fibre quality). Overall, there are many traits that must be heritable and considered concomitantly by cotton breeders. The importance of specific traits varies with the growing region or environment conditions and the requirements of the manufacturers and end-users of raw cotton.

The cotton flower is classified morphologically as a perfect flower. Although some growing environments promote cross-fertilization through the activity of external pollinators such as bumblebees and honeybees, the flower morphology generally favours self-fertilization. The self-fertilizing nature of cotton has resulted in the use of breeding methods developed for self-pollinated crops. In the early days of organized cotton breeding, most breeders developed new cultivars using mass selection and reselection schemes within an existing strain or cultivar (Campbell *et al.*, 2008). The strains or cultivars used were commonly heterogeneous and not homozygous or pure bred, which allowed for early cotton breeders to select promising individual plants and subsequently propagate larger amounts of the seed of selected plants. As new strains or cultivars were introduced into specific growing areas, the early cotton breeders practised these techniques and made significant progress.

However, as the frequency of new strain introductions began to decrease, cotton breeders began applying hybridization and selection techniques based on the principles of Mendelian genetics. Generally speaking, this procedure involves: (i) the selection and hybridization of two or more parental lines; (ii) a system to evaluate, inbreed and advance promising progeny; and (iii) a system to test selected lines after inbreeding. Over time, because it was believed to maximize genetic variation, this breeding system became the predominant method used to develop pure-line cultivars. Recurrent selection breeding methods have also been used with some success in cotton breeding in the USA at the USDA-ARS (United States Department of Agriculture – Agricultural Research Service) Pee Dee breeding programme in South Carolina and the New Mexico State University Acala breeding programme in New Mexico (Campbell *et al.*, 2009). The progress attributed to organized cotton breeding has been demonstrated in several studies in the literature, some of which have been summarized by Calhoun and Bowman (1999). Overall, lint yield gains attributable to breeding have been estimated at 9 kg/ha/year on average.

In addition to pure-line cultivar development programmes, a limited number of breeding programmes worldwide concentrate efforts to develop F_1 and F_2 hybrids instead of pure-line cultivars. Although significant yield advantages have been demonstrated from producing cotton hybrids commercially, their widespread adoption has been limited (Meredith and Brown, 1998; Campbell *et al.*, 2008). The lack of efficient and cost-effective hybrid seed production systems has limited hybrid production primarily to China and India. Commercially grown *G. hirsutum* × *G. hirsutum* (intraspecific) and *G. hirsutum* × *G. barbadense* (interspecific) hybrids have increased yield potential in areas of China and India.

Biotechnology

Biotechnology is a broad term often used to describe the manipulation of nucleic acids and/or proteins at the molecular level. Unlike classical genetic techniques that rely primarily on sexual hybridization, biotechnology offers the opportunity for genetic manipulations beyond the classical species, genus and organism boundaries. Since the 1980s and the first report of a successfully genetically engineered plant cell (Fraley *et al.*, 1983), plant biotechnology research efforts and successes have increased and impacted science and society significantly. Biotechnology research efforts have led to the commercialization of many biotechnology-derived agricultural seed products in three of the world's most important food, feed and fibre crops, which include soybean, corn and cotton.

Although biotechnology research includes the manipulation of a plant's endogenous genetic make-up within a species, the primary advances in cotton biotechnology have come through the directed transfer of genetic material from other organisms using genetic engineering. Cotton plant genetic engineering involves two primary processes including cell culture regeneration and genetic transformation. Commonly used procedures for cell culture regeneration and genetic transformation are reviewed by Paterson and Smith (1999). A rapid cell culture regeneration system is required to prepare cotton cells for genetic transformation and to regenerate transgenic cells into whole plants. Similar to other crop plants, cotton genetic transformation is genotype-dependent, primarily because many cotton genotypes are recalcitrant to cell regeneration. Although biolostic genetic transformation has been used effectively in cotton (Finer and McMullen, 1990), *Agrobacterium tumefaciens*-mediated transformation is the preferred method of genetic transfer. Overall, cotton represents one of the most prolific agricultural biotechnology success stories to date.

Since the introduction of the first biotechnology-derived cotton cultivars beginning in 1996, insect- and herbicide-resistant cotton cultivars have revolutionized the industry. Host-plant resistance to certain worm species is available through the development of cultivars and hybrids containing one or more *Bacillus thuringiensis* (*Bt*) genes. Resistance to broad-spectrum herbicides such as glyphosate and glufosinate has also been developed using biotechnology. Cultivars containing stacked versions of insect and herbicide resistance, in addition to insect or herbicide resistance alone, are now widely available and grown in many parts of the world. Cotton containing a *Bt* gene allows the plant to produce a protein toxic to certain worm species, while herbicide-tolerant cotton allows farmers to control a broad spectrum of weeds without harming their cotton crop. Seed products that contain stacked versions of Bt and herbicide-tolerant genes are also available and provide farmers the benefit of both products in a single cultivar. These biotechnology-derived seed products add biotic stress protection in the seed and decrease farmer risk in relation to weed and worm control. These products also provide environmental benefits by decreasing pesticide applications and increasing the adoption of soil and water conservation tillage practices. Several reports have documented a decrease in the amount of pesticide active ingredient applied that corresponds with the widespread adoption of biotechnology-derived cultivars with insect tolerance, herbicide tolerance and stacked insect/herbicide tolerance (Fernandez-Cornejo and Caswell, 2006; Sankula, 2006). According to James (2009), the global area planted to biotechnology-derived cotton was 0.8M ha in 1996 and rapidly increased to 15.5M ha in 2008. The rapid growth of biotechnology-derived cotton cultivars worldwide since 1996 (19-fold) represents an unprecedented level of adoption of new technology in only 12 years.

Fibre Processing

Fibre processing is an important part of the cotton supply chain. Cotton fibre processing begins after crop cultivation prior to harvest and concludes with the production of yarn. The yarn produced is used subsequently to manufacture many different textile products. Each step in fibre processing has a direct effect on end-product quality. In the following sections, we will describe the fibre processing system used in the USA. Many of the steps in the processing system are also widely used across the world, with some minor modifications based on a country's own economic, labour and political situation. Fibre quality is determined by the physical characteristics of cotton fibre and ultimately dictates what textile end-product is manufactured. The quality of cotton fibre is affected by each step in the cotton supply chain, including fibre processing.

Harvesting and ginning

For centuries, cotton has been harvested by hand. In countries including India and China, where there is a ready supply of available workers, cotton is still harvested by hand today. This method, while time-consuming and labour-intensive, produces better quality seed cotton with fewer broken fibres and less trash. In the USA, only small research plots are still harvested by hand. All production fields are harvested with mechanical harvesters.

Mechanical harvesters are frequently cited as one of the most important inventions that have revolutionized the cotton industry in the USA. Prior to the 1950s, when these mechanical pickers came to be in wide use, cotton was picked by hand at a rate of approximately 70–100 kg/day. The early mechanical harvesters could harvest 1135 kg/day (Christidis and Harrison, 1955). Today's modern mechanical harvesters have the ability to harvest 45,000 kg/day or more (Ratliff, 2007).

Approximately three-quarters of the cotton grown in the USA today is harvested using spindle-type cotton pickers (Mayfield *et al.*, 1999). These cotton pickers use rotating barbed spindles to pick the seed cotton from the plant. Once on the spindles, doffers remove the seed cotton and it is blown into a basket on the back of the machine. The remaining quarter of cotton acreage in the USA is harvested using cotton strippers. These machines use rollers together with rotating bats and brushes to remove seed cotton from the plants. In the process, any loose material from the plant, including burs, green bolls, branches and leaves, are also harvested; therefore, this method of harvesting accumulates more trash. As with spindle-type cotton pickers, all harvested material is blown into a basket on the back of the machine, where it remains until dumped into a boll buggy, trailer or module builder.

A boll buggy is a type of wagon used to increase the efficiency of a harvesting operation. A picker or stripper dumps its basket of seed cotton into a boll buggy in the middle of the field and continues to harvest. Meanwhile, the boll buggy is pulled to the end of the row, where it is dumped into a trailer or module builder for storage until the seed cotton is ginned. Module builders are another of the most beneficial technological innovations in the cotton industry. More than 90% of US cotton is now harvested using the module builder (Force, 2002). This technology was so innovative that in 2002, the American Society of Agricultural Engineers dedicated a historical landmark to it. Module builders are large rectangular units that press and compact seed cotton into modules that hold approximately 10 finished bales (Force, 2002). New technology has just been commercialized that allows smaller versions of these modules to be built onboard cotton harvesters. This technology provides even more efficiency and reduced labour costs as less equipment is needed during harvest. These modules of seed cotton are then loaded into special tilt-bed trailers and transported to gins.

The ginning process is a sequence of steps where the seed cotton is separated into its component parts: lint, seed and trash. The initial step in ginning is to condition and clean the seed cotton that comes from producers' fields, whether in modules or trailers. Any excess moisture in the seed cotton will cause trash to stick to the cotton. A lack of moisture leads to increased fibre breakage during ginning. Therefore, seed cotton must have optimum conditioning to reduce the negative impacts on fibre quality.

The seed cotton then travels to gin stands, where fibres are removed from the seed. There are two types of gins – saw gins and roller gins. Saw gins, the type of ginning system patented by Eli Whitney in 1794, are the third technological innovation that has dramatically improved the cotton industry. A single commercial gin stand is equipped with over 100 saws that grab the seed cotton and pull it through a slot narrow enough to prevent the seed from coming through (Baker and Griffin, 1984). These gins are used primarily with Upland cotton, while roller gin stands are used with extra long staple cottons. Roller gins existed before saw gins and were initially the churka type which used a pair of rollers to pinch and then pull the fibres from the seeds. Modern roller gin stands use a moving knife pressed against a single roller to remove the fibres from the seeds.

After passing through the gin stand, the raw fibres (now called lint) are cleaned once again and then packaged for commerce. A bale press uses hydraulic force to compress the lint into a universal density bale weighing approximately 218 kg. These bales are banded with steel straps to hold them together, sampled for classing to measure its fibre characteristics, wrapped to keep the lint clean and sent to a warehouse, textile mill, or exported. The cottonseed separated from the fibres at the gin stand is stored temporarily in a pile at the gin and eventually transported to oilseed mills for further processing, or used directly as livestock feed.

Seed cotton delivered to a gin is composed of cottonseed, lint and trash (any foreign material). Cotton pickers yield seed cotton with the least amount of trash. One bale of lint can be obtained from 681 kg of seed cotton containing about 55 kg of trash (Mayfield *et al.*, 1999). Cotton strippers accumulate a great deal more trash in the seed cotton. One bale of lint can be obtained from 1000 kg of seed cotton containing 363 kg of trash.

Classing, marketing and spinning

Classing cotton is a standardized procedure for measuring the physical characteristics of the cotton fibre. These physical characteristics ultimately affect the quality of the manufactured product. This system of classification was initiated in 1907 following a meeting of an international group of cotton industry representatives to discuss serious problems that had developed in the marketing of cotton. Up to that time, there were several cotton markets in the USA, with each using a different set of standards to determine price. Hence, cotton producers found it very difficult to determine a fair market price. In response, standard measurements for cotton were established (USDA-AMS, 2005).

All US cotton is classed from a sample of lint taken after the bale is formed. All lint samples are sent to one of 12 regional classing offices for official measurements. Beginning in 1995, official classing measurements including fibre length, length uniformity index, fibre strength, micronaire, colour and trash were taken using high volume instruments (HVI). Leaf grade and extraneous matter are determined visually by official classers. The standards for these measurements are established by the US Department of Agriculture. Once the quality of a cotton bale is determined, it can be marketed. The HVI classification measurements are accepted internationally and provide a basis for negotiating cotton market values.

The modern cotton marketing system is highly integrated between the activities of ginning, storing in warehouses, handling, delivery to mills and, ultimately, the production of goods bought by consumers. A large amount of classing data is shared during each of these activities, and the use of computers to store, manage and transfer this information has made operations more effective and created the opportunity for a more efficient marketing system. In addition to classing information, the marketing system also tracks the physical handling and transfer of ownership that occurs while moving cotton from farms to domestic and international textile mills. The production of cotton is vital to the economic stability of many countries, and cotton is a major factor in trade agreements throughout the world.

The New York Cotton Exchange is the oldest commodity exchange in New York, established in 1870 to facilitate the buying and selling of cotton raw material. These cotton exchanges and marketing firms are part of an effective marketing infrastructure that identifies the correct amount of cotton with a desired cotton quality (classification) from the range of US production environments. This cotton is selected to meet domestic and international manufacturing requirements. Modern textile spinning mills require specific fibre quality properties to produce a given textile. The matching of fibre properties from a cotton bale at a gin to the requirements of a textile mill requires specific information exchange that is facilitated through the marketing system.

Once the desired number of bales meeting specific fibre quality properties has reached a textile mill, the process of spinning that fibre into yarn begins. Modernization efforts have brought major changes to the US textile industry. Many of the operations in textile mills have been automated and updated equipment has greatly increased the speed of production. Spinning is a series of steps to clean the cotton and twist the fibres into yarn. To begin spinning, anywhere from six to greater than 50 bales are arranged in a 'lay down', a process where bales are opened and the compacted cotton is 'plucked' into smaller tufts and mixed together to blend uniformly the fibre properties of all bales in the lay down.

The blended cotton is cleaned and then goes through a picking machine that separates and aligns the fibres into a thin, mist-like web. This web is fed into a carding machine that gently twists it into a loose rope known as a card sliver. Several card slivers are then combined and drawn out into a single strand known as a drawn sliver. During roving, a drawn sliver is drawn out again into a single, smaller strand of fibres. Roving is the first step in ring spinning to make yarn.

There are three primary types of spinning that include ring, rotor and air-jet spinning, with ring spinning being the most common method used. During ring spinning, roving fibres are further drawn out into a tiny strand of fibres and twisted into yarn. Rotor (open-end) spinning is faster than ring spinning and its use is becoming more widespread. The larger card or drawn sliver is reduced to individual fibres and twisted into yarn. Roving is not required during open-end spinning. Air-jet spinning is another system that does not require roving and uses forced air rather than mechanical twisting to form yarn. This method is more expensive but is significantly faster than ring or rotor spinning. After spinning, the yarns are wound tightly around bobbins and are ready to be woven into fabric.

Fibre quality

Each step in fibre processing affects cotton fibre quality and ultimately contributes to the function and use of textile products. Cotton fibre quality is defined as the combination of physical fibre properties that affect the efficiency of yarn-spinning, weaving and other textile manufacturing processes, and end-product quality. The primary physical fibre characteristics of note include those related to fibre length, strength and fineness. According to Smith *et al.* (2008), the international fibre length desired for Upland cotton is 27.8 mm. Acala or extra long staple Upland is classified as having a fibre length greater than or equal to 32 mm, and the highest quality Pima cottons are classified as having a fibre length of 34.8 mm. Each of these physical properties are important for textile manufacturers, because the type of yarn-spinning technology used (rotor, air-jet or ring spinning) ultimately determines the fibre's physical properties or quality required for efficient fibre processing (May, 1999). Unfortunately, each of the three spinning technologies requires a different combination of fibre quality properties to operate most efficiently. Overall, ring spinning is facilitated by acceptable fibre length and length uniformity, high fibre strength and fine fibre. Rotor spinning is facilitated by high fibre strength and fineness ahead of fibre length. Air-jet spinning is facilitated by uniform fibre length and low short fibre content. The importance of these inherent fibre properties is such that the textile industry has invested extensively in devising instruments to measure the physical properties of fibres and to evaluate their significance for processing and textile uses.

The most commonly used method to measure the physical properties of cotton fibre is the HVI instrument. It is used to class each bale of cotton produced in the USA and other countries of the world. The standard HVI currently consists of determinations of fibre length, length uniformity, strength, micronaire and colour; while subjective determinations of leaf grade, preparation and extraneous matter are recorded by trained classing personnel. In recent years, the advanced fibre information system (AFIS) instrument (Uster Technologies AG, Memphis, Tennessee, USA) has been developed to measure a series of individual fibre properties as an alternative to HVI-based fibre property mean measures. Research studies relating to fibre quality properties and spinning performance have used the HVI, AFIS and additional instruments to measure specific fibre properties. Overall, the main fibre properties highly correlated with spinning performance and end product quality include those related to the length, strength, elongation and fineness of the fibre. These fibre properties are being used to determine the price premiums and discounts associated with the market value of raw cotton.

Since the HVI is the most commonly used instrument to measure fibre properties, a brief description of each fibre property measured is warranted. The upper half mean length of fibre is the average length of the longer half of a 'beard' of cotton. A 'beard' is formed when a cotton sample is held in a clamp and combed to make the fibres parallel. Length is measured in hundredths of an inch in the HVI system and converted to thirty-seconds of an inch. Length uniformity index is the ratio of the mean length of all fibres in a beard and the upper half mean length (Table 12.3). Since the fibres vary within a bale, length uniformity provides a measure of the variability within that bale. Fibre strength is the force required to break a fibre bundle measured in grams per tex (Table 12.4). Micronaire is a value for fibre fineness and maturity measured by airflow across a bundle of fibres. The fibre's fineness is important for determining the type of yarns that can be made from the fibre. The market value in terms of micronaire is at a premium at the middle of the distribution of micronaire readings (Table 12.5). Cotton bales measuring both above and below a base range are discounted. The colour measurement is represented by reflectance (greyness of the sample) and yellowness. Cotton that is very white generally is of higher value than cottons whose colour may have yellowed with exposure to elements before harvesting. The HVI system uses a video trashmeter to measure the area and count of trash particles on the lint sample. Harvested seed cotton contains pieces of dried leaves that contaminate the ginned fibre and are, therefore, trash particles. The amount of trash also influences the cotton's value since the textile mill must remove trash before processing. Leaf grade describes the leaf or trash content in the cotton.

Genetic studies have shown the primary physical fibre properties are heritable and can be improved through plant breeding (May, 1999). This is important because, as mentioned above, the baseline productivity of the production system, including fibre quality, is determined by a cultivar's genetic make-up. In addition to practising wise cultivar selection for maximum baseline genetic potential, the cotton farmer must watch environmental conditions closely during the time of harvest to minimize losses in yield or quality. Yield losses can occur when the seed cotton is completely dislodged from the plant by wind, rain, hail or sand storms. Fibre quality can also be impacted negatively by blowing debris into the cotton field that may be picked up during harvest and result in increased trash content. Cotton fibre can also be discoloured by green leaves and wet conditions that promote the growth of bacteria.

Any contaminants or moisture in the harvested cotton requires more processing at the gin to clean and condition the cotton. As machine processing increases, so does the opportunity for fibres to break, especially if they are too dry. Removing trash improves

Table 12.3. Fibre length uniformity distribution (data from USDA-AMS, 2005).

Designation	Length uniformity (%)
Very low	76.4 and below
Low	76.5–79.4
Average	79.5–82.4
High	82.5–85.4
Very high	85.5 and above

Table 12.4. Fibre strength distribution (data from USDA-AMS, 2005).

Designation	Strength (g/tex)
Weak	23.0 and below
Intermediate	24.0–25.0
Average	26.0–28.0
Strong	29.0–30.0
Very strong	31.0 and above

Table 12.5. Micronaire measurement distribution (data from Ramey, 1999).

Designation	Micronaire units
Discount	3.4 and below
Base	3.5–3.6
Premium	3.7–4.2
Base	4.3–4.9
Discount	5.0 and above

the overall appearance of the cotton, but fibre length is reduced. These quality properties have a significant impact on market price and end use of the cotton.

Extra processing at the gin generates more short fibres that affect yarn strength, appearance and breakage negatively. In addition, any trash remaining in the cotton when it is spun can knot the fibres and cause neps or imperfections to form in the yarn. Neps have different properties from the rest of the yarn that affect how the final textiles absorb dyes. The woven textile will have tiny spots that are not the same colour as the rest of the fabric and therefore lower the quality of the fabric.

Uses of Cotton Raw Material

Fibre

Cotton fibre represents about 90% of cotton's total economic value. Cotton fibre has unique properties that distinguish it from comparable fibres used to produce textile products. A cotton fibre is a single-celled epidermal hair that elongates from the surface of a cottonseed. This cell elongates for about 20 days while depositing a primary cell wall consisting of pectin, callose, a thin layer of cellulose and a surface cuticle (DeLanghe, 1986). Following primary cell wall formation, the secondary cell wall, consisting mainly of cellulose, is deposited and continues until approximately 45 days after pollination. At this time, the boll splits open, allowing the fibres to dry. As the fibres dry, the central vacuole of the fibre also dries and collapses so it is no longer cylindrical. The flattened fibres then twist upon themselves.

The cellulose in the cotton fibre gives it many attributes that provide an advantage for end uses of cotton. The length of cellulose chains is highly correlated with the strength of cotton fibres (Bennedict *et al.*, 1994). Cotton is hygroscopic, i.e. it pulls water into pores present throughout the layers of cellulose that comprise the fibre structure (Masson and Richards, 1906). This unique absorbing capacity lends a feeling of dryness when wearing cotton fabrics.

Cotton, along with linen, hemp, wool and silk, are natural, renewable fibres. Cotton, linen and hemp use sunlight directly to produce natural plant fibres from seeds (cotton) and stems (linen and hemp). In comparison, both wool and silk require processing of plant-based energy through an animal before a fibre is produced. Linen and hemp fabrics are very similar in appearance, but both wrinkle easily and threads can break if creased frequently. Cotton fabric can also wrinkle, but the fabric is highly resilient and will not tear easily if creased frequently. Both linen and cotton fabrics can be sterilized by boiling in hot water without damaging the fibre. Wool fabrics include mohair and cashmere from goats and fleece from alpacas, among others. Wool fabrics can be rough and are frequently heavy, suiting them for cooler weather apparel. Silk fibres are found in the cocoons of silkworms and are the strongest of all natural fibres. However, fabrics made from silk can shrink and are weakened by sunlight and perspiration. In contrast, cotton fibres are exposed to sunlight from the time the boll opens in a farmer's field to when those fibres are woven into a T-shirt or blue jeans worn by the very same farmer.

A second group of fibres is synthetic or man-made fibres. These fibres start as a liquid that is refined from petrochemicals and then forced through holes to produce filaments (threads) of exact dimensions. Fabrics created from synthetic fibres include nylon, polyester and spandex. Many of these fabrics cannot resist high temperatures, so care must be taken when drying and ironing. Because these fabrics are refined from petrochemicals, their price will increase as the price of petrochemicals increases. Petrochemicals are a non-renewable resource, while a natural fibre like cotton is renewable.

Recently, the apparel industry has seen an increasing trend in consumer preference for natural over synthetic fibres. In surveys reported by the Cotton Incorporated Lifestyle Monitor, 56% of women and 71% of men said that they preferred cotton for athletic clothing (Fig. 12.1). For the women respondents, this response is 12 percentage points higher than when the same question was asked in 2006. As the value of total imported apparel

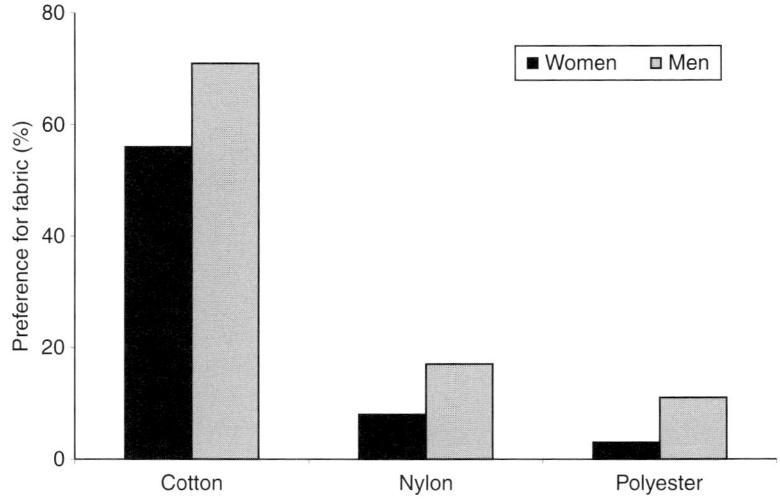

Fig. 12.1. Preference for cotton, nylon and polyester fabric types in athletic apparel for women and men during 2007–2008 (Cotton Inc, 2008b, 2009).

has climbed since the 1980s, the portion of cotton garments as part of that total has also increased. This implies that cotton apparel imports are increasing faster than total apparel imports, reflecting a growing consumer preference for cotton. Cotton's share of the apparel imports is now at a record high of 60.9% (Cotton Inc, 2008a).

The principal use of cotton fabric is to make textile products for clothing. The apparel manufacturing industry produces many well-known items like denim blue jeans; chambray blue work shirts; seersucker, corduroy and twill fabrics; along with trousers, dresses, suits, shirts, socks, sweaters, underwear and T-shirts made from cotton. At any time when the cotton fabric is made, finishing processes may be performed on the fabric. These processes include dyeing, bleaching and stonewashing, among others. They can add decorative patterns to fabric or make the fabric wrinkle- or stain-resistant. Fibre, yarn or fabric finishing treatments improve appearance, texture or performance of the end product. Cotton fabric is also made into home décor linens, including curtains, tablecloths, bed sheets and terrycloth used in towels. Other household items, such as carpets and rugs, cords and twine, furniture and automotive upholstery are made of woven cotton fabrics.

Another class of cotton products, non-wovens, utilizes waste products from the apparel industry. These products are bleached, composed of pure cellulose and do not have added chemicals that could interfere with medical conditions. Cotton personal care products are 'finish free', while synthetic products frequently are treated with chemicals to create a surface suitable for processing. In most applications for non-woven products, cotton's absorbency is its primary advantage. Surgical sponges, sanitary supplies and cosmetic applicators are made of non-woven cottons. In addition to the textile industry, cotton has a variety of applications in fishnets, coffee filters, tents, belting and hoses, bookbinding, tyres, babies' nappies and US paper currency (75% cotton and 25% linen), among others.

Cottonseed

For every 45 kg of fibre, cotton plants produce approximately 70 kg of cottonseed (NCPA, 2002). Historically, seed represents 12–15% of the crop's economic value. However, in recent years, the value of cottonseed has increased and is approaching 20–25% of the crop's economic value. Once the seed is separated from the fibres during ginning, it is sent to

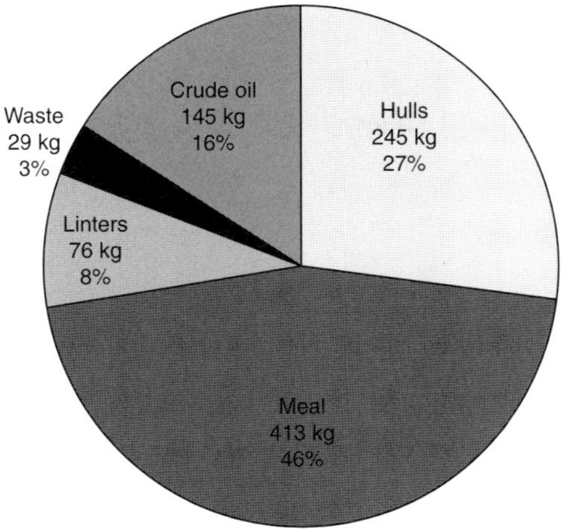

Fig. 12.2. Products obtained after crushing 1 t of ginned cottonseed (NCPA, 2002).

livestock producers and used directly as feed, or it is sent to oil mills and separated into oil, meal, hulls and linters (Fig. 12.2).

Whole cottonseed is a source of protein (20%), energy (87%) and fibre (22%) for livestock (Ely and Guthrie, 2008). Animal nutritionists recognize ginned cottonseed as a premium supplement for cattle and other ruminant animals (Blasi and Drouillard, 2002). Cottonseed is not fed to non-ruminant animals (poultry and pigs) because it contains gossypol, a toxic secondary metabolite.

Gossypol is found in pigment glands that are located throughout the cotton stem, leaves and seeds. Cotton breeders have attempted to eliminate gossypol for decades (Muramato, 1969; Dilday, 1986; Altman *et al.*, 1987; Vroh *et al.*, 1999). Breeders were able to remove gossypol from the cotton plant, but the plant became vulnerable to pests and suffered large economic losses. Recently, researchers have achieved a reduction of gossypol in the seed only, using RNAi to disrupt the gossypol biosynthesis pathway, specifically by interfering with expression of the δ-cadinene synthase gene during seed development (Sunilkumar *et al.*, 2006). Another approach for eliminating gossypol is removal of the glands that store gossypol in seeds. Research has been initiated to clone the Gl_2^e gene and link it in antisense with a seed-specific promoter so that no glands occur in seeds, while they remain unaltered in the rest of the cotton plant (Kohel *et al.*, 2001).

Current research activities are also investigating the pathways of lipid synthesis and developing transgenic plants to modify the fatty acid profile of cottonseed oil. On average, 145 kg of oil can be extracted from the cottonseed kernel after crushing 1 t of seed (Fig. 12.2). This represents 40–50% of the economic value of the cottonseed. Its fatty acid profile generally consists of about 73% unsaturated fatty acids (linoleic, oleic, palmitoleic and linolenic) and 27% saturated fatty acids (palmitic, stearic and myristic; Fig. 12.3).

Oleic acid in cottonseed has been increased by suppressing a Δ-12 fatty acid desaturase (FAD2) using a mutant allele and inserting this construct into cotton plants using *Agrobacterium*-mediated transformation (Chapman *et al.*, 2001). Increasing oleic acid adds to the stability of the oil and increases its health benefits. Other research has attempted to increase seed oleic acid by antisense suppression of FAD2 (Liu *et al.*, 2002; Sunilkumar *et al.*, 2005). Successful research efforts in producing these transgenic plants have not yet been adapted to commercial cotton cultivars.

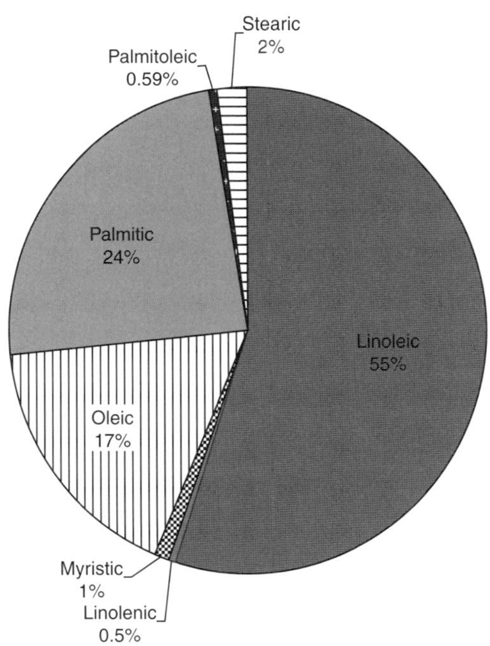

Fig. 12.3. Fatty acid composition of cottonseed cooking oil (NCPA, 2002).

Cottonseed oil has high levels of vitamin E and is essentially cholesterol free (NCPA, 2002). After refining, this oil can be used the same as any other vegetable oil (e.g. as a frying or salad oil) in margarine and shortening (Crisco™ – crystallized cottonseed oil), as a substitute for cocoa butter and in confectionary products.

Cottonseed meal is the embryo and cotyledon material that remains after the oil has been extracted from the cottonseed kernel. Meal accounts for over 33% of the value obtained from the seed (NCPA, 2002). Cottonseed meal is a very concentrated source of protein (41%) and is used primarily as a supplement in livestock feed.

Cottonseed hulls and linters have a wide range of uses (Fig. 12.4). Hulls are the outer coat of the seed and are also used in livestock feed rations for roughage due to their high fibre content. Hulls can be spread over flower and vegetable beds as a mulch to suppress weeds and enhance the soil. They are also used to cover the floor in poultry houses. Another unique use for hulls is in oil well drilling mud to thicken the mud so it stays in the drill hole rather than seeping into the surrounding ground. Hulls can be chemically treated to synthesize furfural, a compound used by the petrochemical industry for refining and making rubber and plastics.

Cotton linters are the short, fine fibres remaining on the seeds after they have been ginned. Linters are commonly used in products for their cellulose. They are used in the preparation of films, lacquers, explosives, plastics, paper and plasma television screens. These short fibres are used in absorbent cotton medical supplies, yarns for wicks, twine, rugs and mops, and felts for upholstery and mattresses.

Looking Toward the Future of Cotton: Continued Genetic and Genomic Advances

As cotton production and end-use processing continue to play a role in the world economy, the demand to increase cotton productivity and quality to meet world fibre needs will continue to grow. Plant breeding, genetic engineering, genetics and genomics research will play a large role in increasing cotton productivity and quality at the crop production level. In addition, synergistic efforts combining these research areas will be increasingly necessary to ensure improvement.

Public sector plant breeding programmes continue to broaden the genetic base of cotton and develop improved germplasm that can be used to develop commercial cultivars. Since 2000, 214 germplasm lines and/or cultivars have been released and registered in *Crop Science* or *Journal of Plant Registrations*. In addition, Jenkins *et al.* (2008) have developed and released a diverse random mating population that broadens the genetic base. In total, these germplasm and cultivar registrations represent advances in pest resistance, disease resistance, abiotic stress tolerance, increased fibre quality, increased lint yield and a broadened genetic base. Commercial cultivar development programmes are using these improved germplasms to develop the next generation of commercial cultivars.

Genetic engineering advances are continuing to revolutionize cotton production. Genetic engineering and transgenic cultivars have already made a huge impact on cotton production. Currently, there are transgenic commercial cultivars available that provide tolerance to the herbicides, glyphosate and glufosinate. Commercial cultivars with various forms of Bt genes are available, as well as cultivars with stacked herbicide tolerance. It has also been reported that transgenic cultivars with resistance to the broadleaf herbicide, Dicamba, will also be available in the near future (http://www.monsanto.com/products/pipeline/dicamba_tolerant_cotton.asp). As the discovery of novel genes continues to increase, the development of new transgenic cultivars containing these genes will continue to grow.

In dramatic fashion over the past decade, genetics and genomics research programmes have developed molecular resources and tools that are leading to the discovery of genes controlling traits important in cotton production. The discovery of cotton-specific putative genes or expressed sequence tags

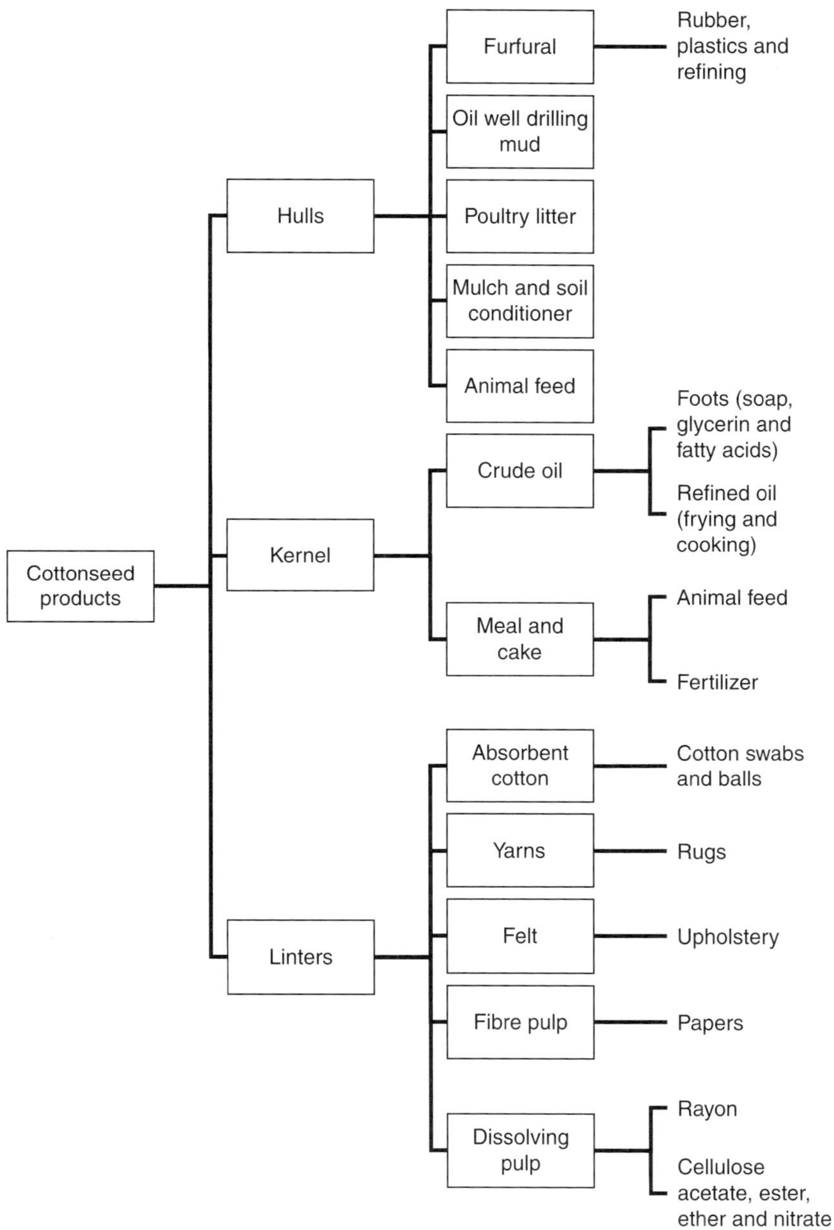

Fig. 12.4. Cottonseed products and their uses (adapted from Chaudhry and Guitchounts, 2003).

(EST) has increased dramatically over the past 10 years, and there are currently 374,292 total ESTs available at the National Center for Biotechnology Information (http://www.ncbi.nlm.nih.gov/). This resource provides the research community a database of genes that may control or associate with the function of an important cotton production trait. The discovery of cotton-specific DNA molecular markers has also increased dramatically over the past 10 years. The Cotton Marker Database (http://www.cottonmarker.org) currently contains information related to over 4500 simple sequence repeat DNA markers.

These DNA markers and others have been used to construct cotton chromosome maps and identify chromosomal regions of interest in cotton production. Specific chromosomal regions detected are commonly called quantitative trait loci (QTL) and represent putative genes. DNA markers and chromosome maps have led to the identification of specific chromosomes and QTL that influence important traits such as fibre quality. According to Chee and Campbell (2009), over 100 QTL affecting fibre length and over 80 QTL affecting fibre strength have been detected in a summary of fibre quality QTL experiments. Furthermore, Rong *et al.* (2007) have demonstrated that almost half of the QTL responsible for fibre quality are located in 21% of the total cotton genome. This information and that emerging from current experiments is providing cotton geneticists and breeders with a glimpse at the number, location and effect of the genes underlying important cotton production traits such as fibre quality.

Looking to the future, it is clear that biotechnology advances will continue and will impact cotton productivity. Commercial firms that have invested in cultivar development are improving the efficiency and effectiveness of plant breeding by using DNA markers. Not only are DNA markers being used increasingly to select for difficult-to-measure traits, but they are also being used to custom-design parental matings and predict the agronomic performance of their offspring (Eathington *et al.*, 2007; Campbell *et al.*, 2009). The emerging development of single nucleotide polymorphic (SNP) DNA markers also provides a virtually unlimited supply of DNA markers to develop new cultivars. In addition to DNA markers and their applications, emerging DNA sequencing technologies offer tremendous opportunity to determine the genome sequence of the species that make up the *Gossypium* genus. Genome sequencing projects are currently under way to sequence several of the diploid ancestor species of cultivated cotton. These sequencing projects, and ones yet to come, will provide a plethora of information concerning the structure and function of the genes in cotton. It is anticipated that genome sequence data will provide information to researchers that leads to future increased cotton productivity. It is likely that this information and other genomic information will increase lint yield and fibre quality potential, while indirectly improving the characteristics of fibre end-use products.

Summary

Cotton is the single most important natural fibre in the world and is used globally to manufacture a variety of textile products. It is cultivated and produced in over 30 countries across the world and is a major source of export income for several of these countries. Although the majority of cotton's value resides in the lint fibre, there are additional products and by-products captured through cotton production. Additional benefits are obtained from cottonseed products that include animal feeds and various oil-derived products. As an agricultural commodity, cotton production and its downstream products represent a significant portion of the global economy. Recent advances in production practices, genetics and biotechnology have revolutionized cotton production by increasing productivity and soil conservation practices, while also reducing pesticide use. In this chapter, we have discussed several key parts of the cotton supply chain, including cotton production, fibre processing, uses of cotton and prospects for future cotton productivity improvement. Cotton production and fibre processing are highly mechanized in many countries of the world and research efforts increasingly are focused on improving cotton fibre quality to meet the needs of textile manufacturers and the end-users of textile products. Future increases in cotton productivity will likely rest on the scientific community's ability to translate and apply the technological advances being made today through genomics and biotechnology. The enormous information generated from genomics and biotechnology advances offers the possibility to increase lint yield and fibre quality potential, while also improving the characteristics of fibre end-use products.

References

Altman, D.W., Stelly, D. M. and Kohel, R.J. (1987) Introgression of glanded-plant and glandless-seed trait from *Gossypium sturtianum* (Willis) into cultivated upland cotton using ovule culture. *Crop Science* 27, 880–884.

Baker, R.V. and Griffin, A.C. Jr (1984) Ginning. In: Kohel, R.J. and Lewis, C.F. (eds) *Cotton*. ASA-CSSA-SSSA, Madison, Wisconsin, pp. 397–436.

Bauer, P.J. and Busscher, W.J. (1996) Winter cover and tillage influences on coastal plain cotton production. *Journal of Production Agriculture* 9, 50–54.

Bennedict, C.R., Kohel, R.J. and Jividen, G.M. (1994) Crystalline cellulose and cotton fiber strength. *Crop Science* 34, 147–151.

Blasi, D.A. and Drouillard, J. (2002) Composition and feeding value of cottonseed feed products for beef cattle. *Bulletin MF-2538*. Kansas State University Agricultural Experiment Station and Cooperative Extension Service, Manhattan, Kansas.

Calhoun, D.S. and Bowman, D.T. (1999) Techniques for development of new cultivars. In: Smith, C.W. and Cothren, J.T. (eds) *Cotton: Origin, History, Technology, and Production*. John Wiley and Sons, Hoboken, New Jersey, pp. 361–414.

Campbell, B.T., Williams, V. and Park, W. (2009) Using molecular markers and field performance data to characterize the Pee Dee cotton germplasm resources. *Euphytica* 169, 285–301.

Campbell, B.T., Bowman, D.T. and Weaver, D.B. (2008) Heterotic effects in topcrosses of modern and obsolete cotton cultivars. *Crop Science* 48, 593–600.

Chapman, K.D., Austin-Brown, S., Sparace, S.A., Kinney, A.J., Ripp, K.G., Pirtle, I.L. and Pirtle, R.M. (2001) Transgenic cotton plants with increased seed oleic acid content. *Journal American Oil Chemical Society* 78, 941–947.

Chaudhry, M.R. and Guitchounts, A. (2003) *Cotton facts*. International Cotton Advisory Committee, Washington, DC.

Chee, P.W. and Campbell, B.T. (2009) Bridging classical and molecular genetics of cotton fiber quality and development. In: Paterson, A.H. (ed.) *Genetics and Genomics of Cotton*, Volume 3. Springer, New York, pp. 1–29.

Christidis, B.G. and Harrison, G.F. (1955) *Cotton Growing Problems*. McGraw-Hill, New York.

Cotton Incorporated (2008a) 2007 Cotton apparel import trends (http://www.cottoninc.com/SupplyChainInsights/2007-Cotton-Apparel-Import-Trends/2007-Cotton-Apparel-Import-Trends.pdf?CFID=13265265&CFTOKEN=99087361, accessed 26 March 2009).

Cotton Incorporated (2008b) Men prefer cotton's feel, features when working out (http://www.cottoninc.com/lsmarticles/?articleID=569, accessed 11 February 2009).

Cotton Incorporated (2009) Resolve to wear athletic apparel in 2009 (http://www.cottoninc.com/lsmarticles/?articleID=606&pub=Womenswear, accessed 11 February 2009).

DeLanghe, E.A.L. (1986) Lint development. In: Mauney, J.R. and Stewart, J.M. (eds) *Cotton Physiology*. The Cotton Foundation, National Cotton Council, Memphis, Tennessee, pp. 325–349.

Dilday, R.H. (1986) Development of a cotton plant with glandless seeds and glanded foliage and fruiting forms. *Crop Science* 26, 639–641.

Eathington, S.R., Crosbie, T.M., Edwards, M.D., Reiter, R.S. and Bull, J.K. (2007) Molecular markers in a commercial breeding program. *Crop Science* 47, S-154–163.

Ely, L.O. and Guthrie, L.D. (2008) Feeding whole cottonseed to dairy cows and replacements. *Special Bulletin 59*. University Georgia Cooperative Extension, Athens, Georgia.

Fernandez-Cornejo, J. and Caswell, M. (2006) US Department of Agriculture, Economic Research Service. Electronic Information Bulletin Number 11, pp. 1–30.

Finer, J.J. and McMullen, M.D. (1990) Transformation of cotton (*Gossypium hirsutum* L.) via particle bombardment. *Plant Cell Reports* 8, 586–589.

Force, B. (2002) Cotton module builder to be dedicated as historical landmark. *AgNews* 21 May, p. 30.

Fraley, R.T., Rogers, S.G., Horsch, R.B., Sanders, P.R., Flick, J.S., Adams, S.P., Bittner, M.L., Brand, L.A., Fink, C.L., Fry, J.S., Galluppi, G.R., Goldberg, S.B., Hoffmann, N.L. and Woo, S.C. (1983) Expression of bacterial genes in plant cells. *Proceedings National Academy of Sciences of the United States of America* 80, 4803–4807.

James, C. (2009) Global status of commercialized biotech/GM crops: 2008. ISAAA Briefs No. 39-2008.

Jenkins, J.N., McCarty, J.C. Jr, Gutierrez, O.A., Hayes, R.W., Bowman, D.T., Watson, C.E. and Jones, D.C. (2008) Registration of RMUP-C5, a random mated population of Upland cotton germplasm. *Journal of Plant Registrations* 2, 239–242.

Keystone Center (2009) Environmental resource indicators for measuring outcomes of on-farm agricultural production in the United States (http://keystone.org/spp/env-sustainag.html, accessed 17 May 2009).

Kohel, R.J., Yu, J.Z., Decanini, L.I., Dong, J. and Zhang, H.B. (2001) Cottonseed quality and creation of glanded plants with glandless seeds. *Proceedings Genetic Control of Cotton Fiber and Seed Quality Workshop*, 5–6 December 2000. Cotton Incorporated, Cary, North Carolina, pp. 271–277.

Liu, Q.L., Surinder, S. and Green, A. (2002) High-oleic and high-stearic cottonseed oils: nutritionally improved cooking oils developed using gene silencing. *Journal of American College of Nutrition* 21, 205S–211S.

Locke, M.A., Knight, S.S., Smith, S., Cullum, R.F., Yuan, Y. and Bingner, R.L. (2008) Environmental quality research in the Beasley Lake watershed, 1995–2007: succession from conventional to conservation practices. *Journal of Soil Water Conservation* 63, 430–442.

Masson, O. and Richards, E.S. (1906) On the hygroscopic action of cotton. *Proceedings. Royal Society London, A* 78, 412–429.

Matlock, M., Thoma, G., Nutter, D. and Costello, T. (2008) Energy life cycle assessment for global cotton production practices (http://www.uark.edu/ua/cars/documents/documents/Resources/Global%20Cotton%20LCA%20Report.pdf, accessed 17 May 2009).

May, L. (1999) Genetic variation in fiber quality. In: Basra, A.S. (ed.) *Cotton Fibers, Development Biology, Quality Improvement and Textile Processing*. Food Products Press, New York, pp. 183–230.

May, O.L. and Lege, K. (1999) Development of the world cotton industry. In: Smith, C.W. and Cothren, J.T. (eds) *Cotton: Origin, History, Technology, and Production*. John Wiley and Sons, Hoboken, New Jersey, pp. 65–98.

Mayfield, W.D., Anthony, W.S., Baker, R.V., Hughs, S.E. and Lalor, W.F. (1999) Ginning. In: Smith, C.W. and Cothren, J.T. (eds) *Cotton: Origin, History, Technology, and Production*. John Wiley and Sons, Hoboken, New Jersey, pp. 683–707.

Meredith, W.R. and Brown, J.S. (1998) Heterosis and combining ability of cottons originating from different regions of the United States. *Journal of Cotton Science*, 2, 77–84.

Muramato, H. (1969) Hexaploid cotton: some plant and fibre properties. *Crop Science* 9, 27–29.

NCPA (National Cottonseed Products Association) (2002) Cottonseed and its products. 10th edn (http://www.cottonseed.com/publications/cottonseedanditsproducts.asp, accessed 12 February 2009).

Nelson, R.G., Hellwinckel, C.M., Brandt, C.C., West, T.O., Ugarte, D.G.D.L.T. and Marland, G. (2009) Energy use and carbon dioxide emissions from cropland production in the United States. *Journal Environmental Quality* 38, 418–425.

Paterson, A.H. and Smith, R.H. (1999) Future horizons: biotechnology for cotton improvement. In: Smith, C.W. and Cothren, J.T. (eds) *Cotton: Origin, History, Technology, and Production*. John Wiley and Sons, Hoboken, New Jersey, pp. 415–432.

Ramey, H.H. (1999) Classing of fiber. In: Smith, C.W. and Cothren, J.T. (eds) *Cotton: Origin, History, Technology, and Production*. John Wiley and Sons, Hoboken, New Jersey, pp. 709–727.

Ratliff, B. (2007) Modern cotton production has deep Delta roots. *Mississippi Landmarks* 3, 4–6.

Rong, J., Feltus, F.A., Waghmare, V.N., Pierce, G.J., Chee, P.W., Draye, X., Saranga, Y., Wright, R.J., Wilkins, T.A., May, O.L., Smith, C.W., Gannaway, J.R., Wendel, J.F. and Paterson, A.H. (2007) Meta-analysis of polyploid cotton QTL shows unequal contributions of subgenomes to a complex network of genes and gene clusters implicated in lint fiber development. *Genetics* 176, 2577–2588.

Sankula, S. (2006) National Center for Food and Agriculture Policy (http://www.ncfap.org, accessed 11 February 2009).

Smith, C.W., Hague, S., Hequet, E., Thaxton, P.S. and Brown, I.N. (2008) Development of extra-long staple upland cotton. *Crop Science* 48, 1823–1831.

Sunilkumar, G., Campbell, L.M., Hossen, M., Connell, J.P., Hernandez, E., Reddy, A.S., Smith, C.W. and Rathore, K.S. (2005) A comprehensive study of the use of a homologous promoter in antisense cotton lines exhibiting a high seed oleic acid phenotype. *Plant Biotechnology Journal* 3, 319–330.

Sunilkumar, G., Campbell, L.M., Puckhaber, L., Stipanovic, R.D. and Rathore, K.S. (2006) Engineering cottonseed for use in human nutrition by tissue-specific reduction of toxic gossypol. *Proceedings National Academy of Sciences of the United States of America* 103, 18054–18059.

USDA-AMS (Agricultural Marketing Service – Cotton Division) (2005) Cotton classification: understanding the data (http://www.ams.usda.gov/AMSv1.0/getfile?dDocName=STELPRDC5074569, accessed 12 February 2009).

Vroh, B.I., Baudoin, J.P., Hau, B. and Mergeai, G. (1999) Development of high-gossypol cotton plants with low-gossypol seeds using trispecies bridge crosses and *in vitro* culture of seed embryos. *Euphytica* 106, 243–251.

Waddle, B.A. (1984) Crop growing practices. In: Kohel, R.J. and Lewis, C.F. (eds) *Cotton*. ASA-CSSA-SSSA, Madison, Wisconsin, pp. 233–263.

13

Research in Cotton Fibre Improvement

LORENZO ALEMAN AND RANDY D. ALLEN

Introduction

Plant trichomes are fine outgrowths that originate from cells of aerial epidermis. These surface hairs exhibit diverse morphologies and spatial distribution patterns. A range of biologically important roles has been ascribed to trichomes, depending on whether they are glandular, non-glandular, multicellular or unicellular. For example, glandular trichomes offer chemical protection against herbivores and pathogens or attract animals through secretory mechanisms. Non-glandular trichomes provide thermal regulatory functions to help protect against both excess heat and freezing, absorb water, protect the plant from excess irradiation and facilitate seed dispersal (Werker, 2000). While the majority of plant trichomes are multicellular, cotton is unique among crop plants in that it produces seed trichomes, or fibres, that consist of extremely elongated single cells. These single-celled fibres are considered to be the world's leading natural fibre and a mainstay of the global economy (Arpat *et al.*, 2004). Prized as an agricultural commodity for more than 8000 years, four separate cotton species have been domesticated independently for fibre (Wendel and Cronn, 2003). Two cotton species are from the Americas, *Gossypium hirsutum* and *G. barbadense*, and the other two from Africa-Asia, namely *G. arboretum* and *G. herbaceum*. The unique properties of cotton fibres allow them to be used for producing a wide range of products, the most important being textile fabrics. Thus, fibre characteristics are critical to the spinning and weaving industry and the quality of fibre in a particular bale determines its final use and influences the price paid for the crop (Arpat *et al.*, 2004).

Apart from their economic value, cotton fibres provide a unique and valuable model for research into cellular differentiation (Kim and Triplett, 2001; Haigler, 2007). The non-isotropic expansion of elongating cotton fibres makes it one of the most rapid cell expansion systems known, producing cells that are 1000–3000 times longer than they are wide, while cellulose biogenesis during fibre maturation produces a secondary cell wall that is approximately 95% cellulose, one of the purest forms of cellulose in nature (Ryser, 1999). A clearer understanding of the genetic and metabolic control of cellulose biogenesis can be used to increase yield and quality, as well as diversifying the product spectrum of fibre, which is pertinent to the applied objectives of renewable resource and bioenergy research. In this chapter, we provide background information as well as new research invested towards understanding this unique and valuable cell in terms of its structure, quality, use, genetic and environmental control of its development and modification.

Fibre Development and Cellulose Biogenesis

Reproductive growth in cotton begins about 4–5 weeks after planting, with the formation of floral buds in the apical part of the plant. This is followed in several weeks by flowering (anthesis) and the start of fruit (boll) development (Oosterhuis and Jernstedt, 1999). Cotton bolls are usually segmented into three to five compartments (locules) that hold the ovules via a connection called the funiculus. Each boll typically contains 20–30 ovules awaiting fertilization. As fibres arise from the seed integument of the ovule, they pass through four distinct yet overlapping stages of development: (i) initiation; (ii) elongation (primary cell wall synthesis); (iii) secondary cell wall synthesis (cellulose deposition); and (iv) maturation (Basra and Malik, 1984). Often, these developmental stages are temporally correlated to the time of anthesis as 'days post-anthesis', or DPA (Haigler *et al.*, 2005). Thus, DPA serves as a reference point for all subsequent events in seed and fibre development. However, it should be noted that environmental conditions could directly influence both the rate the fibres develop and the properties of the mature fibre.

Fibre initiation, elongation and primary cell wall development

By a process that is not yet fully understood, approximately one in four epidermal cells on a growing ovule are determined to become fibres (Stewart, 1975; Zhang *et al.*, 2007). The morphological development and ultrastructural characteristics of this event have been documented extensively by scanning and transmission electron microscopy (SEM and TEM, respectively) (Basra and Malik, 1984; Graves and Stewart, 1988; Ryser, 1999; Ruan *et al.*, 2001; Zhang *et al.*, 2007). It has been noted that fibre initials form within a few days to a few hours prior to anthesis, depending on the region of the ovule (Ryser, 1999). For example, in *G. hirsutum*, fibre initials are first visible at the chalazal end of the growing ovule (Oosterhuis and Jernstedt, 1999; Seagull, 2001). These initials appear as single-celled spherical protrusions that exhibit microtubule reorientation from longitudinal to a criss-cross arrangement above the immature epidermal (protoderm) surface (Stewart, 1975; Ryser, 1999; Seagull, 2001). This process, which also leads to enlarged nuclei and ring-shaped nucleoli in fibre initials, is rapid and quasi-synchronous as it progresses from the chalazal cap to the micropylar end of the ovule within 24–48 h (Stewart, 1975; Ryser, 1999). It has been estimated that on the day of anthesis (0 DPA), 25–30% of the epidermal cells are determined to become fibre cells, while by the end of 1 DPA approximately 80–90% of the seed surface is covered with cellular protrusions (Zhang *et al.*, 2007). While cells destined to become 'lint' fibres originate and end as single cells, non-fibre epidermal cells continue to divide and expand to increase the surface area of the developing ovule (Wu *et al.*, 2006).

At 2–3 DPA, spherical expansion ceases and the tips of the lint fibre initials become tapered and the direction of growth changes with longitudinal elongation exceeding the rate of diametric expansion (Ryser, 1999; Seagull, 2001). At 4 DPA, there is a second wave of fibre initiation that produce fuzz fibres (linters), which are distinguished from lint fibres by their cylindrical shape. Fuzz fibres do not reach the length of lint fibres, probably due to reduced turgor (Zhang *et al.*, 2007). The primary walls of elongating cells exhibit a delicate balance of controlled cell wall expansion and turgor pressure that transform them to 'hydraulic machines' (Peters *et al.*, 2000). It has been observed that the closure of plasmodesmata at the base of the fibre follows a significant rise of turgor and the rapid phase of fibre elongation (Ruan *et al.*, 2001). Moreover, Ruan *et al.* (2004) has shown that callose deposition and degradation correlate with the closure and opening of fibre plasmodesmata, respectively. Their analysis indicates that the deposition of callose may contribute directly or indirectly to the closure of plasmodesmata in cotton fibre. Plugging of plasmodesmata by callose could result in osmotic isolation and restriction of water movement out of the cells to maintain high turgor. Elongating fibres often have to cope with two conflicting forces. They must maintain a rigid cell wall to withstand the internal and external pressure, while allowing enough 'extensibility' for cell wall expansion during growth (Burgert and Fratzl, 2006). In contrast to the tip-growth characteristics of root hairs, pollen tubes and fungal hyphae, cotton fibre elongation occurs by extension of the wall over its entire surface or by diffuse growth (Tiwari and Wilkins, 1995). In the cytoplasm of these cells, smaller vacuoles coalesce into one large central vacuole, resulting in only a thin layer of cytoplasm between the vacuole membrane and the plasmalemma (Wakelyn *et al.*, 2007). During this process, cellulose microfibrils exhibit dynamic and random characteristics in their orientation as the cell is delimited by an interwoven network of carbohydrate and protein macromolecules made up of pectin, callose, a ~0.5 μm thin layer of cellulose (20–25%) and cuticle on the surface (Seagull, 1992; Haigler, 2007). However, as the cell continues to elongate, the sugar composition of the primary cell wall changes, leading to structural rearrangement. For example, cell walls of elongating cotton fibres contain large amounts of acidic polymers and non-cellulosic β-glucans including hemicellulose, and the extractable amounts of these carbohydrates increase as fibres grow and then decrease after elongation ceases (Tokumoto *et al.*, 2002). As in other plant cell types, Golgi bodies are involved in the biogenesis and modification of most of the primary cell wall polymers and in the formation of the plasma and vacuolar membranes (Ryser, 1999). Since cotton fibre elongation may be achieved initially by cell wall loosening (Ruan *et al.*, 2001), changes in extractable matrix polysaccharides would contribute to cell elongation. Tokumoto *et al.* (2002) noted that there was a rapid increase in the amount of xyloglucans in the early phase of fibre elongation and the average molecular mass of these molecules was reduced during the elongation stage. Hemicellulose xyloglucans serve as a structural polysaccharide of the plant primary cell wall and, together with cellulose, comprise around 70% of the cell wall dry mass. Hemicellulose is bound to cellulose with hydrogen bonds, covering and cross-linking proximal cellulose microfibrils (Vissenberg *et al.*, 2005; Michailidis *et al.*, 2009). Tokumoto *et al.* (2003) suggest that a massive turnover of xyloglucans contributes to cell wall loosening, leading to the rapid elongation of cotton fibre. In terms of biomechanics, when growing cells loosen their walls, they reduce turgor pressure and water potential simultaneously, and cause water uptake (Cosgrove, 1987). Because rapid cell expansion requires high hydraulic permeability of the tonoplast to support water entry into the vacuole, membrane channel proteins such as aquaporins may be involved in water transport across the tonoplast and plasma membrane during fibre elongation (Liu *et al.*, 2008). Depending on the genotype and, in many cases, the environmental conditions, lint fibres continue to elongate until reaching their final lengths at 20–25 DPA, resulting eventually in individual cells which range in length from 25 to 35 mm.

Recent observations of fibre development using cryo-field-emission scanning electron microscopy, reported by Singh *et al.* (2009), showed that elongating fibres adhered to one another through a cell wall layer, which they named the cotton fibre middle lamella. Thus, elongating fibres appear to be bound together into bundles that resemble a simple tissue. The cotton fibre middle lamella is rich in non-cellulosic polysaccharides including galacturonans that are typical of pectin. As fibre elongation terminates, the cotton fibre middle lamella degrades to release individual fibres. The authors suggest that the bundling of elongating fibres in a tissue-like structure could help coordinate fibre elongation for efficient packaging into the restricted space within the seed coat.

Fibre secondary cell wall development

The temporal boundary that separates the elongation phase from the subsequent synthesis of the secondary wall is not discrete. While it is estimated that the fibre elongation phase is from 5 to 20 DPA, the secondary cell wall development phase is thought to commence around 15 to 19 DPA. During this transition, callose, a (β-(1 ≥ 3)-D-glucan polymer is synthesized and deposited between the cellulose (β-(1 ≥ 4)-D-glucan) of the thickening secondary wall and the plasma membrane (Maltby *et al.*, 1979; Wakelyn *et al.*, 2007). Although immunolocalization studies have shown that callose and cellulose form within the same complex (Amor *et al.*, 1995: Salnikov *et al.*, 2003), it is yet to be determined what actual function callose plays in secondary cell wall formation (Cui *et al.*, 2001).

In the transitional phase to secondary cell wall development, the rate of cellulose biosynthesis increases by about 100-fold. Its deposition occurs in an ordered manner by membrane embedded terminal complexes (TC) called 'TC rosettes' (Roberts and Roberts, 2007). The term 'rosettes' refers to the hexagonal structures visualized by freeze fracture electron microscopy (Lai-Kee-Him *et al.*, 2002) that are assembled in the Golgi apparatus (Haigler and Brown, 1986). In general, TC rosettes are thought to function as 'nanospinnets' that move in the plane of the plasma membrane while extruding cellulose microfibrils across the cell surface (Brown and Montezinos, 1976). The hexagonal rosette structure of the terminal complex in vascular plants has been reported to contain 36 glucan chains (Herth, 1985), giving rise to the theory that each rosette subunit of the hexagon must synthesize six glucan chains. Therefore, if each rosette is composed of six particles, then it would follow that each particle is an assemblage of six cellulose synthase catalytic (CesA) proteins. Little is known about rosette assembly as biochemical approaches to isolate and identify the protein composition of the TC rosettes have been challenging. However, substantial evidence supports the notion that the CesA proteins are part of the complex responsible for glucan-chain elongation within rosettes, as seen in immunolabelling studies with an antibody to cotton CesA1 (Kimura *et al.*, 1999). In fact, cotton *CesA1* and *CesA2* were the first cellulose synthase genes to be characterized from a higher plant (Pear *et al.*, 1996), and much of our understanding of cellulose biosynthesis has been obtained from genetic analyses of these genes in other plant species. Nevertheless, accumulating evidence has led to the suggestion that certain CesA proteins are expressed during primary cell wall synthesis, while others are expressed during secondary cell wall synthesis (reviewed in Roberts and Roberts, 2007). This is consistent with observations indicating that cell wall polymers (including cellulose) from fibres at primary cell wall stages have lower molecular weights than the cellulose from fibres at the secondary wall stages (Timpa and Triplett, 1993). It is possible that the differences in cellulose molecular weight may be due to the temporal regulation of *CesA1* and *CesA2* gene expression (Wakelyn *et al.*, 2007) and/or the turnover of the CesA proteins during the two phases of fibre cell wall development, as reported by Jacob-Wilk *et al.* (2006).

Besides the shift in cellulose content that occurs with the developmental transition from primary (20–25%) to secondary wall (> 95%), there is also a shift in the orientation of microfibrils relative to the longitudinal axis of the fibre (Seagull, 1992). Microfibril orientation in plant cell walls often correlates with cytoskeletal elements including actin microfilaments and microtubules made up of α- and β-tubulins. Microfibril deposition during fibre primary wall development parallels the roughly transverse orientation of the cortical microtubule network, roughly 70–90° relative to the longitudinal axis of the fibre (Seagull, 1992). However, as fibre development transitions to secondary wall formation, the orientation of both the microtubule network and cellulose microfibril deposition undergoes a shift from transverse to a steeply helical winding layer, with microfibrils oriented at about 45–55° relative to the long axis of the fibre (Seagull, 1992). This shift in microfibril orientation to helical arrays has been proposed to be a major contributor to the cessation of the elongation process (Delmer, 1999), and differences between short and long staple in different cotton species may depend on the timing of reorientation of cellulose microfibrils in relation to elongation. Therefore, understanding events at this transition stage is key to understanding the limits of fibre elongation.

Fibre maturation

As the fibre secondary cell wall thickens between 22 and 32 DPA, the angle of microfibril orientation and of the co-aligned microtubules is reduced to about 20° (Seagull, 1992). However, interactions between microfibrils result in subtle changes in the order and degree of variability in orientation (Hseih, 1999). This phase of secondary wall synthesis is characterized by high rates of cellulose synthesis as fibre elongation and primary wall synthesis cease (Delmer, 1999). As secondary wall accumulation continues, there is gradual infilling of the cell lumen (illustrated in Seagull, 2001). Successive layers of cellulose microfibrils are deposited throughout the secondary cell wall. Each layer deposited exhibits a steeper helical gyre than the previous layer, resulting in a wall with a polylamellate construction. This structure causes the mature fibres to twist when dried and the formation of twist reversals along the length of the fibre (Seagull, 2001). The final cell wall can be up to 10 μm thick and 30 mm long, exhibiting about 100 reversals throughout the length of the fibre (Haigler, 2007).

Bolls dehisce when mature at about 40–45 DPA, leaving the fibres fully exposed to the elements. At this stage, the water content of the fibre decreases rapidly, ceasing most, if not all, metabolism during secondary wall synthesis. The cytoplasm dries against the inner surface of the wall and the enlarged lumen is left where the central vacuole was once located (Wakelyn *et al.*, 2007). As the fibres desiccate, the lumen collapses and their shape changes from cylindrical to a flattened twisted ribbon structure with a kidney-shaped cross section. The twisting and convolution of individual fibres allows them to twist together to form an entangled, three-dimensional network that permits the spinning of fibres into yarns (Seagull, 2001). Thus, yarn and fabric quality is correlated directly to the mechanical and physical properties of the fibre secondary cell wall.

Fibre Properties

Dried mature cotton fibres represent the culmination of the developmental phases described above. Combined, ovule fertilization, epidermal cell differentiation, elongation, secondary cell wall development and maturation are the biological parameters that determine the physical properties of the highly valuable cotton fibre (Seagull, 2001) and are correlated directly to the molecular structure of the cellulose (French and Johnson, 2007).

Physiochemical composition and structural properties of cellulose

Ultimately, mature cotton fibres are made up of ~95% cellulose. This cellulose, as with most polysaccharides, tends to crystallize into a variety of polyforms (French and Johnson, 2007). The crystallinity of cellulose, which is made up of long chains of glucose residues (5000–300,000), provides structural information when studied with neutron or X-ray diffraction. There are at least four recognized cellulose polymorphs, based on their diffraction patterns, named cellulose I–IV (see Wakelyn *et al.*, 2007; French and Johnson, 2007, for a complete review). Cellulose I is the predominant form found in nature. Two forms of cellulose I exist (Iα and Iβ) that have similar intra- and intermolecular hydrogen bonding, as well as crystal packing (Wakelyn *et al.*, 2007). While bacteria and algae make a high proportion of cellulose Iα, diffraction patterns of cellulose from higher plants identify it as cellulose Iβ (Atalla and Vanderhart, 1984). Non-crystalline cellulose exists naturally and cellulose crystallinity can be modified or destroyed by various chemical or physical treatments (French and Johnson, 2007). For example, cellulose II can result by subjecting the cellulose to cold, concentrated NaOH leading to disruption and decrystallization, while cellulose III is obtained from treatment with various amines. Two forms of cellulose IV (IV_I and IV_{II}) can result from high temperature treatments. Electron diffraction patterns taken from primary cell walls of cotton fibres displayed two strong equatorial reflections reminiscent of cellulose IV_I (Chanzy *et al.*, 1978). The authors propose that this polyform is explained best in terms of native cellulose I crystals having coherence along the length of the microfibrils, but with poor lateral organization of interchain hydrogen bonds within the network. This is supported by more recent data using Fourier transform infrared spectroscopy (FTIR) and solid-state ^{13}C NMR spectroscopy that indicates that the structure of cellulose IV_I is analogous to cellulose Iβ having various degrees of lateral disorder (Wada *et al.*, 2004). These authors propose that cellulose IV_I should be called cellulose Iβ. Cotton fibre secondary cell walls, on the other hand, are made up of the higher crystalline Iβ form, as are most secondary cell walls in other plant species (Haigler, 2007). However, there is ongoing debate whether cellulose I is a mixture of the two forms, with Iβ being predominant in plants (reviewed in Wakelyn *et al.*, 2007).

Despite the hydrophilic nature of cellulose, cotton fibres are often treated to produce forms that are more chemically and structurally compatible with downstream commercial applications. For example, mercerization (named after John Mercer) is the treatment of cotton fibres with a mild alkali such as aqueous NaOH to swell the cotton and improve dye affinity, chemical reactivity, dimensional stability, tensile strength, lustre and smoothness of cotton fabrics. Wakelyn *et al.* (2007) provide an in-depth review of such chemical treatments to cotton fibre as it pertains to textile manufacturing. The cellulosic fibre in its natural form exhibits a high degree of crystallinity due to its high degree of polymerization. It is estimated that a cotton fibre may contain as few as 20,000 monomeric D-glucopyranosyl units, giving it a large molecular weight with unusually strong and extensive hydrogen bonds. These properties, along with the fact that only a small fraction of the cellulose chain contains reducing ends, makes it difficult for chemical agents to penetrate the interior of the cellulose fibre. Therefore, fibres have to be 'swollen' prior to treatment. Subsequent treatments provide cotton products with useful, durable properties, including wrinkle resistance, water repellency, flame resistance and antimicrobial action (Wakelyn *et al.*, 2007).

Besides cellulose, raw-untreated cotton fibres also contain non-cellulosic compounds. These compounds are found in the cuticle of the fibre that serves to protect the fibre from water penetration and microbial degradation. The natural wax in the cuticle, which comprises an estimated 0.6% of a typical cotton fibre, is composed of long-chain saturated fatty acids and alcohols, resins, hydrocarbons and sterols (Wakelyn *et al.*, 2007). However, in the atypical cotton mutants displaying green lint or fuzz, this wax layer is present in higher amounts ranging from 14 to 17% relative to fibre weight (Ryser, 1999; Wakelyn *et al.*, 2007). The waxy polymer contains a high content of C_{22} fatty acids, which makes up a suberin-like compound located between the lamellae in the secondary wall that is not present in white or other naturally coloured fibres. The naturally occurring wax in lint fibres is beneficial during the early processing stages because the wax serves as a lubricant for proper spinning into yarn; however, this waxy layer must to be removed prior to chemical treatment to allow for penetration of aqueous solutions.

The primary cell wall is also a source for non-cellulosic compounds in cotton fibres (Buchala, 1999; Wakelyn *et al.*, 2007). These compounds accumulate during the elongation phase and the transitional phase from primary to secondary cell wall synthesis when cellulose biosynthesis is not the predominant process. Of the non-cellulosic polysaccharides, xyloglucans (primarily hemicellulose) play a dominant role in the structure of the primary cell wall, possibly by interconnecting adjacent cellulose microfibrils, thereby tethering or cross-linking microfibrils together via non-covalent hydrogen bonding (Vissenberg *et al.*, 2005). This creates an extensive, coherent xyloglucan–cellulose framework, which acts as a tension-bearing structure in the primary cells walls (Michailidis *et al.*, 2009). Found in the innermost layer of the primary wall, xyloglucan and cellulose together comprise about 70% of the cell wall dry mass. However, in the outermost layer of the primary wall, pectins make up 0.7–1.2% of the dry fibre weight.

Yet another source of non-cellulosic compounds is the lumen, which may contain residual metabolic compounds including sugars, namely glucose and fructose, organic acids and inorganic cations. These components typically compose only a small fraction of the total dry fibre mass (Wakelyn *et al.*, 2007).

Physical properties of fibre

In order to understand how genetic or chemical modifications can be used to improve the quality and utility of cotton fibres, it is necessary to understand their physical and structural properties. Because fibre quality is related directly to the market value of cotton fibres and its products, much ongoing research is directed toward its improvement. Thus, there has been great emphasis on measuring fibre quality and setting uniformity and constancy standards. Consequently, there is a wealth of information relating to the physical and structural development of cotton fibres as it relates to quality (Wakelyn *et al.*, 2007). Although there are different systems in place to estimate fibre quality, the high volume instrumentation (HVI) system is used currently by many industrialized countries as the standard method to appraise cotton quality. The most essential cotton fibre qualities are related to length, strength and micronaire (maturity and fineness), which typically are measured from ginned matured fibres in their dried state. However, it should be noted that there are other quality parameters that are used to class cotton fibres, such as colour grade, trash and neps (entangled clumps of fibre due to mechanical processing), which are not covered in this chapter.

Length determination of cotton fibres is carried out by instrumental measurements on a 'beard' of fibres that are prepared by a mechanical sampler that clamps fibres at random along their length (Ramey, 1999). After combing and brushing the fibres, the HVI system sensor reports the average or mean length of the longest one-half of the fibres to the nearest one-hundredth of an inch. This value is known as the upper half mean length. To calculate length uniformity, the ratio is taken between the mean length and the upper half mean length and expressed as a percentage. High length uniformity is considered to be above 85%, while anything below 77% is considered to be very low (Ramey, 1999). The advance fibre information system (AFIS) is the most reliable and rapid system used for the determination of short fibre content (SFC), which is defined as fibres less than 12.7 mm (McAlister *et al.*, 2003). It has been noted that variations in SFC are mostly the result of fibre breakage during the ginning process (May and Lege, 1999). However, cottons with low micronaire exhibited a higher level of short fibre content than their high micronaire counterparts (McAlister *et al.*, 2003), which would suggest that genotype and environmental conditions also contributed to SFC.

The strength of fibre can also be determined with HVI by applying force to the point of rupture on the same beard of fibre used for length determination. The measurement is recorded in grams force per tex (gf/tex), which measures the power to resist force per mass predetermined in the length scan (Wakelyn *et al.*, 2007). The relative strength of cotton fibres ranges from very strong (31 and above g/tex) to average (26–28 g/tex) to weak (23 and below g/tex). Some of the major factors that are believed to affect fibre strength are the degree of rigidity of the cellulosic chains within a fibre and its fibrillar and crystalline orientation, as well as the inter- and intramolecular hydrogen bonding (Hseih, 1999). In contrast to length, where it is determined primarily during the elongation phase, fibre strength may be due to both primary and secondary cell wall development, when cytoskeletal rearrangements (i.e. crystallinity and cellulose molecular weight) are very dynamic.

Micronaire is an indirect measurement of maturity and fineness. Maturity makes a reference to fibre secondary wall thickness, while fineness is the mass per unit length. Using the HVI system, micronaire is often determined by passing air through a predetermined mass of fibres that are compressed to a fixed volume. The air permeability of that sample gives the micronaire reading. Mature fibres that resist compression have high air permeability and therefore a high micronaire value. On the other hand, fine or immature fibres tend to compress easily, resulting in a lower air permeability and lower micronaire reading. Immature fibres are often a result of poor secondary wall development. For example, fibres without secondary walls lack rigidity and body and exist only in clumps (Goynes *et al.*, 1995). This is evident through SEM analysis of fibre cross sections, where immature fibres fail to display the thick, bean-shaped characteristics of mature fibres. However, it should be noted that within a given sample, there is a mix of both mature and immature fibres, and the quality stamp is given to samples with a higher ratio of mature fibres that is often dependent on genotype and the environmental history of the plant.

Regulation of Fibre Development

Fibre quality is correlated directly with the developmental events that occur during fibre formation. Fibre initiation, elongation, secondary cell wall development and maturation are genetically regulated, but are also strongly affected by the environmental conditions faced by the plant during its life cycle. Throughout cotton development, the plant perceives both internal and external cues that alter the physiological, metabolic and cellular programmes that ultimately determine the final characteristics of the fibre. Understanding fibre biology in terms of these cues has been slow in coming. However, with the advent of an *in vitro* ovule-culture system and progress made in unravelling the genetic mechanisms that regulate fibre growth and development, we are now poised to develop a deeper understanding of how the developmental complexity of the fibre cell is influenced by biotic and abiotic effectors.

Metabolic control of fibre development

Comparisons between the cotton fibre transcriptome and metabolome at different stages of development have shown that stage-specific events can be characterized by their transcript and metabolite profiles (Gou *et al.*, 2007). The upregulation and downregulation of genes is dependent on the stage of fibre development, as are the metabolic pathways that are used. For example, during fibre initiation and elongation, fibre cells must synthesize and assemble primary cell wall constituents while maintaining a balance between turgor pressure and extensibility. During the transitional phase from primary to secondary cell wall synthesis, the cell shifts its metabolism to meet the demand for cellulose synthesis by redirecting energy to carbohydrate metabolism and cell wall synthesis. This shift in cellular function corresponds with the unique metabolic demands of the two major events in the fibre cell, namely, cell elongation and cellulose deposition in secondary wall formation.

Using a gene expression and GC/MS-based metabolite profiling approach, Gou *et al.* (2007) identified seven metabolic pathways including secondary metabolites, fatty acid and carbohydrate metabolism that function during cotton fibre development. At 3 DPA, metabolite analysis revealed high levels of sucrose, which correlate with increased expression of eight aquaporin-like genes. This combination supports the build-up of turgor by increasing the osmotic potential and accelerating the rate of water uptake, respectively. Aquaporins are present in the plasma membrane (PIPs) and the tonoplast (TIPs) and are essential for cell expansion. Liu *et al.* (2008) characterized the expression of cotton aquaporin genes *GhPIP1–2* and *GhTIP1* and found these genes to be expressed highly and preferentially at 5 DPA, further supporting their important roles during cotton fibre cell expansion. Fibre cell elongation also requires that the cell wall be loosened for expansion. α-Expansins play a major role in cell wall weakening and disassembly in processes such as ripening, abscission and certain developmental pathways, including pollen tube growth and xylem formation (McQueen-Mason *et al.*, 2007). In cotton fibre, four genes that belong to the α-expansin family were highly expressed during the outgrowth and rapid elongation stages, but were downregulated when cells entered the secondary cell wall synthesis stage (Gou *et al.*, 2007). Similarly, genes encoding putative xyloglucan endotransglycosylases (XTHs), which are involved in cell wall remodelling, have been characterized in cotton recently, and some XTH members have been shown to be expressed preferentially during the early stages of fibre elongation (Lee, 2006; Michailidis *et al.*, 2009).

Thaker *et al.* (1999) demonstrated that the osmolyte, malate, might play an important role during rapid cell elongation. This was based on the activity measurements of malate-synthesizing enzymes such as phosphoenolpyruvate carboxylase (PEPC) and malate dehydrogenase (MDH). PEPC and

MDH activities were high during the elongation stage of fibre development, whereas NADPH-MDH activity (an antagonist of PEPC) was low. This is consistent with findings that *PEPC* and *MDH* expression levels are higher in fibres from long staple cultivars than in those from short staple cultivars (Basra and Malik, 1983). Other genes implicated in the elongating cell are the plasma membrane proton translocating-ATPase (PM-H^+-ATPase) and vacuolar proton translocating-ATPase (V-ATPase) (Benedict *et al.*, 1999). V-ATPase is known to be involved in driving solute movement into vacuoles for maintaining turgor, whereas PM-H^+-ATPase transports H^+ out of the cytosol, acidifying the apoplast and changing the extensibility of the cell wall.

Lipids are an integral part of membrane and cell wall synthesis. Gou *et al.* (2007) reported the upregulation of lipid biosynthetic genes and lipid metabolism at 6 DPA that was maintained throughout the elongation phase. In accordance with the amounts of fatty acids in fibre cells, genes that encode enzymes such as acyl-CoA-binding protein, fatty acid elongase, 3-keto-acyl-CoA synthase, β-ketoacyl-CoA synthase and ω-3 fatty acid desaturase and very long-chain fatty acids condensing enzyme were upregulated at this stage and greatly reduced at 21 DPA. This is consistent with findings that lipid metabolizing enzymes and lipid transfer proteins, which recently have been shown to induce cell wall extension in *in vitro* assays (McQueen-Mason *et al.*, 2007), are particularly highly expressed in fibre cells (Song and Allen, 1997; Orford and Timmis, 1998; Ji *et al.*, 2003). During fibre elongation, two predominant respiratory pathways, the oxidative pentose phosphate pathway (OPPP) and glycolysis, provide energy and the conversion of substrates to intermediates required for biosynthesis. The enzyme activity levels in these pathways vary with the demand for respiratory products (Thaker *et al.*, 1999). For example, measured activity of glucose 6-phosphate dehydrogenase (G6PDH) and 6-phosphogluconate dehydrogenase (6PGDH) was high during cell elongation up to 15 DPA, before falling to negligible levels at 30 DPA and 24 DPA, respectively. Thus, increased activity of OPPP enzymes could reflect the demand for NADPH and intermediates in the regulation of carbon channelling during the elongation phase, and this is further supported by increased hexose kinase activity (Thaker *et al.*, 1999).

At the onset of secondary cell wall formation, data gathered from transcript and metabolite profiles clearly demonstrate dynamic changes in metabolism that centre on cellulose synthesis (Gou *et al.*, 2007). Thus, metabolic pathways that are active during fibre elongation are downregulated with the onset of secondary wall formation. This is evident in the reduction of G6PDH and 6PGDH activity, indicating a transition in metabolic priorities (Thaker *et al.*, 1999). To illustrate this, pectin, a polysaccharide component of primary cell walls, is synthesized in part by UDP-glucose 6-dehydrogenase and UDP-D-glucuronate 4-epimerase. These enzymes, which convert UDP-glucose into UDP-D-glucuronate and then UDP-galacturonate, are downregulated during the secondary wall synthesis stage. In light of the view that UDP-glucose serves as an immediate substrate for cellulose polymerization in cotton fibre, downregulation of enzymes that compete for UDP-glucose makes metabolic sense (Guo *et al.*, 2003). Interestingly, the activity of the glycolytic enzymes, aldolase and pyruvate kinase increase on the shift to secondary cell wall deposition, indicating a role in cellulose synthesis (Thaker *et al.*, 1999). Metabolite profiling showed that glucose, and to some extent, fructose, accounted for about 50% of the total polar phase metabolites in rapidly elongating fibre cells, but decreased to 9% at 21 DPA, indicating an increase in carbohydrate utilization for cellulose synthesis (Guo *et al.*, 2003). The demand for carbon in secondary cell wall synthesis is further supported by an increase of both gene expression and activity of pectin degrading enzymes, such as β-galactosidase and β-arabinosidase.

In-depth reviews by Delmer (1999) and, more recently, Haigler (2007) discussed the carbon flux into cellulose. In the models presented by these authors, UDP-glucose, derived from a variety of enzymatic reactions, is the immediate substrate for cellulose synthesis. One source of UDP-glucose is through the hydrolysis of sucrose-by-sucrose synthase (SuSy) in the following reaction: Suc + UDP <-> UDP-glucose + fructose.

Although it is not determined conclusively whether the cytosolic (S-SuSy) or the membrane-associated (M-SuSy) enzyme supplies the substrate for cellulose synthesis, substantial evidence indicates that M-SuSy is likely to be the predominant enzyme that channels UDP-glucose to cellulose, while S-SuSy partitions carbon for general metabolic needs (Haigler, 2007). This evidence comes from the observation that more than 50% of total SuSy protein is tightly associated with the plasma membrane, paralleling the patterns of cellulose deposition during secondary wall synthesis (Amor *et al.*, 1995; Salnikov *et al.*, 2003). Furthermore, it was shown that sucrose and not UDP-glucose was the preferred substrate for cellulose synthesis, indicating that a direct, energy-saving mechanism for channelling UDP-glucose to cellulose synthase was in place. However, UDP-glucose for cellulose synthesis could also be supplied by UDP-glucose pyrophosphorylase in the following reaction: glucose-1-P + UTP <–> UDP-glucose + PPi (Carpita and Delmer, 1981; Waefler and Meier, 1994). It should be noted, however, that production of UDP-glucose through this reaction requires more energy input than from SuSy (Haigler, 2007).

Ultimately, all carbon comes from imported sucrose or resynthesized sucrose within the cell. Besides SuSy, cell wall and vacuolar invertases also catalyse the breakdown of sucrose into glucose and fructose. On the other hand, sucrose-phosphate synthase (SPS) can exert control over carbon allocation by resynthesizing sucrose-6-phosphate irreversibly in the following reaction: fructose 6-phosphate + UDP-glucose –> sucrose-phosphate + UDP, followed by the production of sucrose by sucrose-phosphate phosphatase (SPP). This sucrose cycling can be useful for controlling metabolic processes efficiently at the different stages of fibre development (Haigler *et al.*, 2001).

Environmental effects on fibre development

Growth of cotton plants under unsuitable environmental conditions such as temperature extremes, water deficit and salinity stress face reduced growth and productivity resulting from loss of fruit and altered fibre development. McMichael *et al.* (1973) found that water deficit stress before 14 DPA led to boll abscission, but beyond that 'window of susceptibility' abscission generally did not occur. However, water deficit stress during fibre elongation or secondary wall synthesis leads to decreased fibre length and maturity, respectively (reviewed in Cothren, 1999). Although it is not fully known how fibre quality is affected by stress, it could be due, at least in part, to the accumulation of signalling molecules such as abscisic acid (ABA) that are known to increase in response to stress (Finkelstein *et al.*, 2002). In cotton, ABA produced in response to water deficit and heat stress, for example, induces stomatal closure and lowers leaf water potential; these responses affect photosynthesis and accumulation of carbon assimilate negatively (Cothren, 1999). ABA is recruited as an internal signal to mediate some aspects of the physiological response to environmental stresses. ABA has been shown to regulate plant responses to drought, cold and high temperature (reviewed in Marion-Poll and Leung, 2006). Under these stressful conditions, ABA levels increase in vegetative tissues, triggering adaptive responses that are essential for their survival and productivity. Under drought conditions, ABA induces stomatal closure, minimizing water loss through transpiration (Finkelstein *et al.*, 2002). Water stress in the root zone also increases ABA levels in the root tissue (Zhang *et al.*, 1987), which is then transported to the leaves. Many of the biochemical and physiological changes result from ABA-induced changes in gene expression patterns. Moreover, Dasani and Thaker (2006) reported that there was an inverse correlation between final fibre length and ABA levels in three different cultivars. In a long staple cultivar, rapid ABA accumulation started after fibre had attained peak elongation growth, while in a short staple cultivar, ABA accumulation was observed even during elongation growth. Greater inhibition of fibre elongation was observed in cultured ovules of short and middle staple cultivars than in ovules of a long staple cultivar when the media were supplemented with ABA.

It is yet to be determined if and how these changes ultimately affect cellulose synthesis in the fibre.

There are substantial data that show cotton fibre cellulose deposition and the degree of polymerization are affected by cool temperatures and, furthermore, that this process might be more sensitive than respiration (Haiger, 2007). Fibres exposed to cool temperatures have a prolonged period of elongation and reduced rate of secondary wall thickening, giving rise to growth rings (Basra and Saha, 1999). Temperatures less than 27°C can affect cellulose deposition in the secondary wall negatively through the disruption of photoassimilate production, transport and uptake, the availability of respiration-derived energy, or direct and/or indirect effects on enzyme activity and kinetics (Roberts *et al.*, 1992). The decrease in cell wall synthesis during cool nights could relate to the metabolic pathways that partition the substrate for cellulose synthesis at different developmental stages (see below; Haigler, 2007). For example, Haigler *et al.* (2001) proposed a model to indicate that, under stress conditions, cells could shift from an M-SuSy (thought to channel UDP-glucose to cellulose synthase) to the soluble isoform (S-SuSy), reflecting a down-regulation of cellulose synthase.

Hormonal regulation of fibre development and the Bt ovule-culture system

A culture method for growing cotton fibres on the upper surface of immature ovules (–3 to 2 DPA) floating on liquid media was first developed by Beasley and Ting (1973, 1974). This system has provided a useful experimental model for plant-grown fibre. Although there are obvious limitations in the use of cultured ovules, since the ovules are detached from the maternal tissue, this system has been used extensively to study the physiology, biochemistry and genetic control of the developing fibre under controlled conditions. In this section we cover the most recent studies using this system. For a complete review, see Davidonis (1999), Triplett (2000) and Kim and Triplett (2001).

Experimental results obtained through systematic study of cotton ovules cultured in the presence/absence of different compounds (mostly hormones) that induce/inhibit fibre production have elucidated the function of several biosynthetic and signalling pathways involved in fibre development. For example, it has been known for decades that gibberellic acid (GA) and auxin promote fibre development, something that has been corroborated by further experimentation. However, until recently, ethylene was considered to be an inhibitor of fibre growth. Using cultured ovules, Shi *et al.* (2006) revealed that ethylene actually promoted fibre elongation. In their experiment, measured amounts of ethylene released from cultured ovules correlated with the expression of 1-aminocyclopropane-1-carboxylic acid oxidase1-3 (ACO) genes and the rate of fibre growth. Furthermore, exogenously applied ethylene promoted robust fibre expansion, whereas the presence of an inhibitor of ethylene synthesis specifically suppressed fibre growth. Moreover, ethylene induced the expression of SuSy, tubulin 1 and expansin genes, indicating that this hormone played a role in fibre development by activating genes that mediated cell wall synthesis, wall loosening, or cytoskeleton arrangement. Similarly, this cultured-ovule system was used to confirm a connection between brassinosteroid and fibre development (Sun *et al.*, 2005). Results from those experiments showed that suppression of brassinosteroid (BR) biosynthesis in cotton ovules by treatment with the BR biosynthesis inhibitor brassinazole (Brz) inhibited fibre formation, while addition of low concentrations of brassinolide in combination with GA and auxin were found to promote fibre elongation (Sun *et al.*, 2004, 2005). Additionally, the expression of XTH and expansin genes increased in ovules treated with brassinolide and was suppressed by Brz treatment, establishing a correlation between brassinosteroid and fibre development. However, Shi *et al.* (2006) noted that BR was less effective than ethylene in promoting fibre cell elongation in their experiments. Interestingly, brassinosteroid and Brz affected both fibre length and ovule size, whereas ethylene and the inhibitor affected only fibre length. Although the actions of BR and ethylene on fibre elongation have not been reported to be interdependent, there does appear to be cross-talk between these hormone signalling pathways in cotton, since BR induces the expression of an ethylene biosynthetic gene and ethylene induces BR biosynthetic genes, suggesting that each hormone modulates the synthesis of the other positively.

Besides testing compounds known to promote or inhibit growth, the ovule-culture system has also made it possible to discover and confirm other, non-classical signalling molecules. Qin *et al.* (2007) elegantly showed that saturated very long-chain fatty acids (VLCFAs; C20:0–C30:0) applied exogenously in ovule culture medium promoted fibre cell elongation significantly, whereas acetochlor, which inhibited VLCFA biosynthesis, abolished fibre growth. This effect could be overcome by addition of lignoceric acid (C24:0). These authors also showed that ethylene nullified inhibition by acetochlor, while lignoceric acid was inactive in the presence of an ethylene biosynthesis inhibitor, indicating that VLCFAs might act upstream of ethylene. Furthermore, lignoceric acid induced a rapid and significant increase in ACO transcript levels that resulted in substantial ethylene production. Treating cultured ovules with exogenous phytohormones has showed that individual genes can be induced by different agents. As an example, He *et al.* (2008) characterized 19 different β-tubulin genes (*GhTub1–19*), which encoded the major components of microtubules, and demonstrated that individual genes could be induced specifically by different hormones. In their experiment, *GhTUB1, 3, 9, 12* and *16* responded to BR treatment, whereas *GhTUB1, 3* and *12* were also upregulated by lignoceric acid. Ethylene-induced expression of *GhTUB5, 9* and *12*, and GA was able to induce expression of *GhTUB1* and *3* only, indicating that different hormones might promote fibre cell elongation via different mechanisms. The response of a single gene to different hormones would capacitate the plant cell to produce the same output via different signalling pathways. It is currently not known if or where these signals converge. However, using the cultured ovule system, more comprehensive experimentation using transcript profiling with microarray technology is being carried out to identify genes that are regulated differentially and/or synergistically by different hormones.

The development of a novel transient gene expression assay has also been made possible with the use of cultured ovules. Kim *et al.* (2002) reported an expression system to analyse promoter activities expressed in cotton fibre by using cultured ovules and biolistic transformation techniques. In their report, they showed that the expression of the β-glucuronidase (*GUS*) reporter gene under the control of the fibre and secondary cell wall stage-specific *GhCesA4* promoter was consistent with the expression of the *GhCesA4* gene *in planta*. Although there are limitations to this assay, it could provide a rapid approach to identify fibre stage-specific promoters.

Molecular Genetics of Fibre Development

Understanding the cotton genome will provide an important foundation to understand the functional and agronomic significance of genome variation in cotton species. The genus *Gossypium* includes between 40 and 45 diploid species ($2n = 26$) and 5 allotetraploids ($2n = 52$). These species have been categorized into eight genome groups, designated A through G and K, on the basis of chromosome pairing affinities (Endrizzi *et al.*, 1984). The genomes of the five tetraploid species are designated $(AD)_1$ through $(AD)_5$. Four of these *Gossypium* species are cultivated, including two allotetraploids (*G. hirsutum* and *G. barbadense*) and two diploids (*G. herbaceum* and *G. arboreum*). More than 90% of the world's cotton production is from *G. hirsutum*, typically called Upland cotton, while *G. barbadense*, also known as Sea Island, Pima or Egyptian cotton, accounts for around 8% of total production. The diploid species *G. herbaceum* and *G. arboreum* provide the final 2% of the world's cotton. A-genome cotton species are thought to have originated in Africa/Asia while the D-genome species are endemic to subtropical America. Presumed hybridization between an immigrant A-genome species and a local D-genome species and subsequent polyploidization has led to the AD-genome allopolyploids (Wendel *et al.*, 1995). The A-subgenome of these polyploid cottons is related most closely to the genome of *G. herbaceum* (Desai *et al.*, 2006), while the D-subgenome is related most closely to the genome of the extant diploid, *G. raimondii* or *G. gossypioides* (Wendel *et al.*, 1995). Variation in the haploid genome sizes of these species is apparent, with D-genome species such as *G. raimondii* estimated to be approximately 880 Mb, while the DNA content of the A-genome species *G. arboreum* is approximately 1.75 Gb and the tetraploid *G. hirsutum* is approximately 2.5 Gb (Hendrix and Stewart, 2005). The variation in DNA content between diploid species reflects the relative copy number of various repetitive DNA sequences including retrotransposon-like elements (Hawkins *et al.*, 2006), while the DNA content of the allotetraploid approximates the sum of the A- and D-genome progenitors (Liu *et al.*, 2001). Cytogenetic stocks have been developed in cotton, including artificially derived AD-genome polyploids (Beasley, 1942) and individual chromosome addition and substitution lines (Stelly, 1993). These cytogenetic stocks provide valuable resources for cotton genetics research.

Advances in cotton genomics

The first molecular linkage map of *Gossypium* was developed from an interspecific F_2 *G. hirsutum* × *G. barbadense* population (Reinish *et al.*, 1994). As initially reported, this map, which was based on restriction fragment length polymorphisms (RFLPs), contained 705 loci that were assigned to 41 linkage groups and spanned 4675 cM. The most complete genetic map reported to date comprises 2584 loci at 1.74 cM intervals and covers all 13 homeologous chromosomes of the allotetraploid cotton genome (Rong *et al.*, 2005). Subsequent maps based on PCR-based DNA markers have also been developed that include 1029 loci mapped to 26 linkage groups with an average distance between loci of 5.32 cM (He *et al.*, 2007). The linkage groups of the genetic maps have been assigned to their corresponding chromosomes by using the available cotton aneuploid stocks (Stelly, 1993; Reinish *et al.*, 1994) and fluorescent *in situ* hybridization using bacterial artificial chromosomes that contain genetic markers as probes (Wang *et al.*, 2006a).

Genome-wide integrated genetic and physical maps are required for large-scale genome sequencing and for many other aspects of genome research, including marker development, positional cloning and EST mapping (Zhang and Wu, 2001). Whole-genome physical maps have been constructed for *Arabidopsis thaliana* (Marra *et al.*, 1999), rice (Chen *et al.*, 2002), maize (Coe *et al.*, 2002) and soybean (Wu *et al.*, 2004). Although a number of large-insert bacterial artificial chromosome (BAC) and plant transformation-competent binary large-insert plasmid clone (BIBAC) libraries have been constructed for cotton (Dong *et al.*, 2001; Tomkins *et al.*, 2001; Yu *et al.*, 2002), no genome-wide physical map or chromosome contig map has yet been reported for any *Gossypium* species. The recently reported integrated genetic and physical map of the homeologous chromosomes 12 and 26 of *G. hirsutum* demonstrates the feasibility of linking a variety of genetic mapping data accurately to a physical map (Xu *et al.*, 2008).

The primary purpose for the development of genetic maps is to locate genes that affect qualitative and quantitative traits. The availability of DNA-based genetic markers associated with genes that confer important agronomic traits will provide a dependable and cost-effective method for selection of progeny with desirable genotypes in breeding programmes. Qualitative traits are those that are controlled typically by single genes and the phenotypic variation in segregating progeny falls into discrete classes. More than 200 qualitative traits have been identified in either diploid or tetraploid cotton species (Endrizzi *et al.*, 1984). These traits include variants for leaf shape, pollen colour, plant colour, lint colour, pubescence, bract shape, etc. Since few of these traits are of agronomic interest, relatively little effort has been made to map them or identify the underlying lesions. Mapped genes for qualitative traits that are related to agricultural productivity include those for leaf shape, fibre development, disease and insect resistance and fertility restoration (Ulloa *et al.*, 2006).

Quantitative traits tend to show continuous variation in a segregating population and are considered typically to be controlled interactions of multiple genetic loci. Quantitative trait loci (QTLs) that affect a wide range of traits, including yield and yield components, fibre quality, flowering date, plant architecture and resistance to diseases and pests have been identified in cotton. Interestingly, although the D-subgenome of tetraploid cotton is derived from an ancestor related to *G. raimondii*, which does not produce spinnable fibres, several of the QTLs that influence fibre quality traits are found in the D-subgenome (Rong *et al.*, 2007). This suggests that genetic complementation between the subgenomes plays a role in determining fibre characteristics and may help explain why genetic improvement of fibre quality in tetraploid cottons has been more productive than in the diploids (Jiang *et al.*, 1998). In addition, QTLs identified that affect fibre quality differ in location and effect between different studies (Draye *et al.*, 2005; Rong *et al.*, 2007). This suggests strong interaction between genotype and environmental conditions and indicates that the search for fibre quality QTLs is far from complete (Chee *et al.*, 2005a,b).

Use of QTL-linked molecular markers for molecular assisted selection (MAS) for genetic improvement of cotton fibre quality shows significant promise. For example, using a *G. anomalum* introgression population that segregated for fibre quality traits, Zhang *et al.* (2003) identified molecular markers linked to a major fibre strength QTL. This QTL, *QTLFS1*, explained greater than 30% of the phenotypic variation in this population. Use of a specific marker linked tightly to this locus in large-scale screening for the presence or absence of this QTL could be useful for development of commercial cultivars with improved fibre strength (Guo *et al.*, 2003; Shen *et al.*, 2005). A stable fibre length QTL, *qFLD2-1*, identified by Wang *et al.* (2006b), could also be valuable for use in MAS-based breeding programmes.

The most detailed genomic maps are, of course, full genomic DNA sequences. In addition to the physical positions of genes, they also provide information about all of the constituent components of the genome and any functional information that can be inferred from the sequences. Whole genome sequences of several plant and animal species are available, including model plant species *Arabidopsis* and rice, along with several other agronomically important species. Preliminary steps towards sequencing the *Gossypium* genome are now under way. Hawkins *et al.* (2006) constructed whole-genome shotgun libraries for comparative sequencing between *G. raimondii*, *G. herbaceum*, *G. exiguum* and *Gossypioides kirkii*. In addition, the first phase sequencing of the *G. raimondii* genome has been supported by the US Department of Energy Joint Genome Institute, and a recent collaboration between Texas Tech University and the National Center for Genome Research has undertaken high-throughput sequencing of the *G. kirkii* genome. Despite the rapid technological development in the field of high-throughput DNA sequencing and a steep decline in associated costs, sequencing the genomes of angiosperm plants such as cotton remains a daunting task. Although large amounts of sequence data can now be generated very quickly, the problems associated with sequencing and assembling large genomes such as that of cotton are a difficult challenge. For this reason, sequencing efforts will focus initially on gene-rich areas of the genome. These sequences contain the vast majority of the functional genetic elements within the genome and assembly of these sequences is facilitated greatly by comparison with known cDNA sequence from extensive expressed sequence tag (EST) databases.

Cotton functional genomics

Nearly 300,000 ESTs are available for the *Gossypium* species in GenBank. Of these ESTs, approximately 63% are from *G. hirsutum*, while fewer than 1% are from *G. barbadense*, 14% are from *G. arboreum*, 23% from *G. raimondii* and less than 1% are from *G. herbaceum*. Nearly 70% of these ESTs were generated from mRNAs isolated from fibre or ovules and most (~ 90%) were from early stages of fibre development (initiation and elongation). Cluster analyses of these ESTs has resulted in the identification of between 51,000 and 55,000 unique coding sequences (unigenes) (Udall *et al.*, 2006; Yang *et al.*, 2006), which provides an estimate of the number of transcribed genes in the cotton genome.

EST databases serve as the basis for the development of DNA microarrays for comparative analysis of global transcriptional patterns. As the database coverage increases and the technology to produce microarrays progresses, several generations of microarray resources have become available to the cotton research community. As the first step towards defining the cotton fibre transcriptome, Arpat *et al.* (2004) used oligonucleotide microarrays derived from non-redundant fibre ESTs to develop expression profiles for comparison of transcripts in fibres at 10 and 24 dpa. More than 2500 transcripts that were downregulated during the transition from fibre elongation to secondary wall development were identified, along with 81 transcripts that were expressed preferentially during secondary cell wall synthesis. Comparison of transcript profiles of wild-type cotton plants and mutant cotton plants with fibreless or reduced fibre phenotypes have been reported (Wilkins and Arpat, 2005; Wu *et al.*, 2006; Lee *et al.*, 2007). Reduced expression of only 13 transcripts was detected in the fibreless mutants using assays performed with cDNA microarrays that contain approximately 10,000 hybridization targets (Wu *et al.*, 2006). These included transcripts for proteins such as lipid transfer protein, α-expansin and sucrose synthase, which are known to be involved in cell elongation, along with transcripts for putative regulatory proteins including Myb and homeodomain transcription factors. The cotton *Myb* gene (*GhMYB25*) was shown to be expressed primarily in ovules and fibre initials and its potential role in fibre development was investigated in transgenic plants. Ectopic expression of this gene in tobacco plants resulted in an increased number of highly branched leaf trichomes, suggesting that the *GhMYB25* gene could play a role in fibre initiation or elongation. This suggestion was confirmed recently by characterization of transgenic cotton plants with altered expression of *GhMYB25* (Machado *et al.*, 2009). Suppression of *GhMYB25* expression using an RNAi-based gene construct resulted in a reduced number of trichomes on aerial plant organs and ovules. *GhMYB25*-silenced cotton plants showed delayed fibre elongation, resulting in shorter fibres. Ectopic overexpression of *GhMYB25* in transgenic cotton led to an increase in both leaf trichomes and cotton fibres, although plants with high levels of expression were sterile.

Genomic research has provided a wealth of data about the evolution and function of the cotton genome. Application of this knowledge to the improvement of cotton varieties is now beginning. Identification of molecular markers linked closely to important agronomic traits, including fibre quality characteristics, will provide important tools for the development of MAS breeding methods, and identification of fibre-expressed genes through high-throughput transcript profiling will produce numerous candidates with strong potential for crop improvement.

Genetic Engineering Efforts to Alter Fibre Characteristics

Since the first introduction of transgenic cotton in 1995 by erstwhile Calgene, Inc (of Davis, California, USA; later bought by Monsanto Company, St Louis, Missouri), there has been an increasing number of biotechnology traits commercialized. So far, these have all been input traits for herbicide tolerance and bollworm control. These commercial transgenic cotton crops have been tested across a range of environments and have been adopted rapidly by cotton growers worldwide. This has resulted in significant environmental and human health and welfare benefits. However, the challenges of population growth, competing crops, environmental change and economic downturns will certainly require the development and adoption of new biotechnology solutions. In this respect, development of input traits to increase the quality and quantity of fibre through transgenic technology has lagged behind. Although many genes that are expressed in developing cotton fibres have been identified, the functions of relatively few have been determined experimentally.

Table 13.1 shows a list of recently reported functionally characterized cotton fibre genes. Genetic transformation of cotton is labour-intensive and time-consuming; therefore, it is not surprising that the functions of many cotton genes have been confirmed only by expression in a model plant or microbial species and relatively few have been tested in cotton.

Manipulating metabolic pathways

Of the numerous metabolic pathways involved in cotton fibre development, carbon portioning and the subsequent events leading to cellulose synthesis for cell wall formation are by far the best studied when it comes to identifying targets for metabolic engineering. As discussed previously, cotton fibres are a strong sink for carbon. Thus, it is expected that fibre cell development is highly sensitive to changes in the carbon flux that feeds cellulose synthesis. Ruan *et al.* (2003) examined the role of SuSy in cotton fibre and seed development by suppressing the expression of SuSy using reverse genetics. By transforming cotton plants with a transgene that included a part of the *SuSy* cDNA in either sense or antisense orientation driven by the constitutive segmented-7 promoter (S7) from *subterranean clover stunt virus*, they were able to demonstrate that SuSy played a rate-limiting role in fibre initiation and elongation. In those experiments,

Table 13.1. Functional characterization of cotton genes in heterologous systems.

Cotton gene	Gene type	Method of characterization	Phenotype/function	Fibre expression	Reference
FAD2-3 *FAD2-4*	Phosphatidylcholine desaturases	Genetic complementation of *Arabidopsis fad2-1*[a]	Restored wild-type fatty acid composition and growth characteristics	Developing fibres	Zhang *et al.*, 2009
ECR1 *ECR2*	Trans-2-enoyl-CoA reductase	Genetic complementation of yeast *tsc13Δ* mutant[b]	Fatty acid elongation enzymes harbouring non-classical NADPH-binding sites at their C termini	3–10 DPA	Song *et al.*, 2009
CPK1	Ca^{+2} dependent protein kinase	Ectopic expression in *Arabidopsis*	Subcellular localization, promoter analysis and *in vitro* assay of kinase activity	9–15 DPA	Huang *et al.*, 2008
MYB2	Transcription factor	Genetic complementation of *Arabidopsis gl1*[c]	Rescued trichome formation and induced seed-trichome production. Trichome-specific expression	0–9 DPA	Wang *et al.*, 2004; Shangguan *et al.*, 2008
GID1	GA-receptor	Genetic complementation of rice *gid1*[d]	Rescued plant stature	Developing fibres	Aleman *et al.*, 2008
SLR1	DELLA, GA signalling repressor	Ectopic expression in *Arabidopsis*	Reduced plant stature	Developing fibres	Aleman *et al.*, 2008
CSD1 *CSD2* *CSD3*	Cu/Zn superoxide dismutases	Ectopic expression in *Arabidopsis*	*Gh*CSD1 localizes to the cytosol, *Gh*CSD2a localizes to plastids and *Gh*CSD3 is translocated to the cell wall	Differential expression	Kim *et al.*, 2008
KCS13/ *CER6*	3-Ketoacyl-CoA synthase	Genetic complementation of *Arabidopsis cut1*[e]	Restored stem length and leaf size	Elongation	Qin *et al.*, 2007

Table 13.1.

Cotton gene	Gene type	Method of characterization	Phenotype/function	Fibre expression	Reference
BIN2	Shaggy-like protein kinase	Ectopic expression in *Arabidopsis*	Reduced growth and similar phenotypes to the semi-dominant *bin2* mutant plants		Sun and Allen, 2005
TTG1 *TTG3*	WD-repeat protein	Genetic complementation of *Arabidopsis ttg1*[f]	Restored wild-type trichome development	0–24 DPA	Humphries *et al*., 2005
PFN1	Profiling	Ectopic expression in tobacco	Subcellular localization, promoter analysis and *in vitro* assay	6–18 DPA	Wang *et al*., 2005
KCR1 *KCR2*	3-Ketoacyl-CoA reductases	Genetic complementation of yeast haploid *ybr159wΔ* mutant[g]	Fatty acid elongation enzymes	5–10 DPA	Qin *et al*., 2005
GhRac1	Rac/Rop GT Pase	Ectopic expression in *E. coli*	Production of a functional GTPase as shown by *in vitro* enzyme activity assay	8–12 DPA	Kim and Triplett, 2004
BRI1	BR-receptor	Genetic complementation of *Arabidopsis bri1*[h]	Rescued growth to wild-type phenotype		Sun *et al*., 2004
WBC1	ATP-binding cassette transporter of the WBC	Ectopic expression in *Arabidopsis*	Short siliques	5–9 DPA	Zhu *et al*., 2003

[a]FATTY-ACID DESATURAS2 mutant; [b]yeast mutant accumulates ceramides that harbour fatty acids shorter than 26 carbons; [c,d]GIBBERELLIC-INSENSITIVE DWARF1 mutant; [e]CUT1 mutant, KCS1 and CER6 (CUT1) are responsible for the biosynthesis of VLCFAs with chain lengths longer than C22; [f]TRANSPARENT TESTA GLABRA1; [g]deficient in 3-ketoacyl-CoA reductase activity; [h]BRASSINOSTEROID INSENSITIVE1mutant.

SuSy suppression inhibited fibre initiation and elongation, as well as seed development, depending on the site of SuSy repression. For example, repression of SuSy in the seed coat of some transgenic lines affected only fibre development, while repression of SuSy in the endosperm and embryo of other transgenic lines blocked the development of the entire seed. Although this is the first report to show phenotypic changes in fibre development in transgenic cotton plants, it is yet to be demonstrated whether overexpression of SuSy leads to modified or enhanced fibre properties. However, in another report, Ruan *et al.* (2005) provided additional data to support the important role of SuSy in fibre development. This comes from examining the expression patterns of SuSy in a lintless cotton mutant (*fls* or *SL1-7-1*) that produces only fuzz-like short fibres with delayed initiation (Ruan *et al.*, 2005). Immunolocalization revealed delayed expression of SuSy in the mutant seed-coat epidermis that correlated temporally and spatially with the initiation of the fibre cells. The short fibres from the mutant were different from fibres from wild-type plants in that their basal ends enlarged immediately after initiation, while the majority of the wild-type fibres did not show this enlargement until the end of elongation. Because the suppression of SuSy in transgenic cotton plants reduced the length of both lint and fuzz-like short fibres, it was suggested that high expression of SuSy played a crucial role in the growth of both lint and fuzz fibres. Furthermore, Ruan *et al.* (2004, 2005) also showed that plasmodesmata gating might be as important as SuSy expression in an elongating fibre cell. In their hypothesis, transient closure of plasmodesmata maintains high turgor and drives rapid elongation in fibre cells (Ruan *et al.*, 2001). In this study, they provide evidence that callose deposition and degradation may be involved in the closure and reopening of plasmodesmata, respectively. They further suggest that the expression of a β-1,3-glucanase gene (*GhGluc1*) may play a role in this process by degrading callose selectively during the end of cell elongation. In this sense, reducing or delaying callose degradation in fibre cells may lead to prolonged closure of the plasmodesmata and extend the elongation period.

Another target of interest in altering the carbon flux to cellulose synthesis in cotton fibre has been sucrose-phosphate synthase (SPS). Previously, Babb and Haigler (2001) showed that increased SPS activity correlated with a higher rate of cellulose synthesis during secondary wall formation and hypothesized that enhancement of sucrose supply by overexpressing SPS in cotton could improve fibre quality. To test this hypothesis, Haigler *et al.* (2007) generated transgenic cotton plants harbouring the spinach SPS gene driven by the constitutive CaMV 35S promoter. Transgenic lines with high SPS activity were shown to produce fibre with higher micronaire and maturity ratio associated with greater thickness of the cellulosic secondary wall than wild-type plants. However, these differences were seen only when plants were grown in growth chambers with cool nights and low light below the canopy, an effect most likely due to post-transcriptional regulation of the heterologous SPS. Nevertheless, these data provide supporting evidence for the hypothesis that increasing carbon flux via SPS to cellulose synthase may be a viable approach to improve fibre quality. It would be interesting to see if overexpressing a different and less regulated SPS gene would lead to improved fibre quality under normal conditions. In fact, the most beneficial means of increasing sucrose synthesis has been by the use of the maize SPS gene in tomato, *Arabidopsis*, tobacco and rice (Worrell *et al.*, 1991; Signora *et al.*, 1998; Ono *et al.*, 1999). Overexpression in tomato resulted in elevated leaf SPS, and the maize enzyme was unregulated with respect to normal light/dark modulation (Worrell *et al.*, 1991; Galtier *et al.*, 1993). The enhanced SPS activity was associated with an increase in light- and CO_2-saturated rate of photosynthesis and, under ambient conditions, an increased sucrose:starch ratio in leaves (Galtier *et al.*, 1993) and increased partitioning of fixed carbon into sucrose (Micallef *et al.*, 1995). Similarly in rice, increased SPS activity promoted the export of carbon from leaves and photosynthesis was enhanced under elevated CO_2 (Ono *et al.*, 2003).

Modifying hormone levels/signalling and pathways

Analyses of hormone levels in developing fibres, as well as exogenous application of various hormones to cotton flowers, bolls or cultured ovules, have allowed for assessment of the influence of gibberellins (GA), auxins, cytokinins, abscisic acid, ethylene and brassinosteroids on fibre development. Hormones known to promote fibre development include GA, auxins, ethylene and brassinosteroids, whereas cytokinins and abscisic acid appear to inhibit fibre growth (Triplett, 2000). Although similar effects on gene expression result from the addition of different hormones (GA, auxin, BR, for example), it is still not clear if and how these signals converge. Thus far, there have been only a few attempts to modulate hormone levels or alter signalling mechanisms in transgenic cotton to improve fibre characteristics, and these experiments have focused on auxin, cytokinins, brassinosteroids and GA.

Evidence indicates that auxin (indole acetic acid, IAA) is important for epidermal cell differentiation into fibres (Beasley and Ting, 1973, 1974; Seagull and Giavalis, 2004). The synthesis of IAA involves two enzymatic reactions. The first step is the conversion of tryptophan to indole-3-acetamide (IAM) by tryptophan monooxygenase (IAAM). IAM is then hydrolysed to IAA by inoleacetamide hydrolase (IAAH). To determine if increasing auxin levels in cotton had any effect on fibres, the *A. tumefaciens iaaM* and *iaaH* genes were introduced independently into cotton driven by a fibre-specific promoter (John, 1999). Reports from those experiments indicated that there was no discernible effect of increased free

IAA content between transgenic and wild-type plants. Similarly, a gene construct containing the *A. tumefaciens* isopentyenyl transferase (*IPT*) gene driven by the CaMV 35S promoter was introduced into cotton (John, 1999). IPT catalyses the condensation of AMP and isopentenyl pyrophosphate to form isopentenyl AMP, a precursor of cytokinins. Although these plants exhibited significant changes in plant morphology, John (1999) reported that fibre traits were unchanged when compared to wild-type plants.

Thus far, the only transgenic cotton plants with altered phytohormone signalling that are reported to show an effect on fibre traits are those with alterations in BR signalling and BR biosynthesis. Preliminary experiments using the cultured-ovule system showed that suppression of BR biosynthesis in cotton ovules by treatment with the BR biosynthesis inhibitor brassinazole (Brz) inhibited fibre formation, whereas addition of low concentrations of BL, in combination with GA and auxin, was found to promote fibre elongation (Sun *et al.*, 2004, 2005). A cotton gene that encoded an ortholog of the BR receptor (*GhBRI1*) was cloned and characterized (Sun *et al.*, 2004). The biological function of *GhBRI1* was confirmed by transgenic complementation of dwarf BR-insensitive mutant *Arabidopsis* plants homozygous for the weak *BRI1-5* allele with the cotton gene. Ectopic expression of *GhBRI1* resulted in recovery of the normal plant growth phenotype, demonstrating that the *GhBRI1* gene encoded a function BR receptor protein. To evaluate further the role of the BR signalling pathway in cotton fibre development, transgenic cotton plants were created that expressed transgenes designed either to overexpress or to suppress the expression of the *GhBRI1* (Y. Sun, S. Veerabomma and R.D. Allen, unpublished data). Although antisense suppression of *BRI1* expression in many transgenic cotton plants resulted in stunted growth and sterility, some plants with moderate levels of *BRI1* suppression grew normally, were fertile and produced viable seed. Overexpression of *BRI1* in transgenic cotton plants had no noticeable effect on growth and development. As shown in Table 13.2, analysis of the characteristics of cotton fibre from these plants indicated that alteration of BR responsiveness had little effect on fibre length. However, suppression of *BRI1* expression by about 50% strongly inhibited secondary cell wall development, resulting in fibres with reduced maturity. Conversely, overexpression of *BRI1* led to increased fibre cellulose deposition, resulting in thicker secondary cell walls and higher micronaire. These alterations in fibre cell wall development corresponded with changes in cellulose synthase gene expression, indicating that secondary wall deposition was regulated, at least in part, by BR-regulated modulation of cellulose synthase gene expression. These results indicate that, while BR1-dependent signalling may not limit cotton fibre elongation *in planta*, it appears to be required for normal maturation of cotton fibres through the deposition of cellulose into the secondary cell wall.

Recent results from analyses of transgenic cotton plants with altered expression of the BR biosynthetic gene *GhDET2* indicate that

Table 13.2. Comparison of cotton fibre characteristics using HVI and AFIS analyses. Mature fibres from non-transgenic plants (C312) and transgenic plants that antisense suppress (*asBRI1*) or overexpress (*oeBRI1*) BRI1 were tested, along with fibres from *immature* mutant plants (*im/im*) and wild-type control plants (TM-1). Four independent lines for each transgenic genotype were tested. Data are means of a minimum of three assays per line.

	HVI			AFIS
Genotype	Micronaire	Length (mm)	Strength (g/tex)	Fineness (millitex)
C312	4.6	31.50	30.0	193
asBRI1	3.7[a]	32.00	32.2	171
oeBRI1	5.1[a]	28.96	33.4	204
im/im	2.8[a]	30.73	32.1	162

[a]Indicates results that differ significantly from those of non-transgenic plants ($P < 0.05$).

BR biosynthesis may, in fact, be necessary for fibre initiation and elongation (Luo *et al.*, 2007). *GhDET2* encodes a steroid 5d-reductase which is considered to catalyse a major rate-limiting step in BR biosynthesis. Expression of antisense *GhDET2* by the 35S promoter in cotton reduced regeneration rates, and viable antisense *GhDET2* transgenic plants exhibited stunted growth and fruit abortion within 3–5 DPA. Scanning electron microscopy showed that fibre initiation at 0–1 DPA was inhibited compared to wild-type plants, and fibre elongation could be restored by application of exogenous BR to ovules in culture. On the other hand, transgenic plants that expressed the sense-*GhDET2* transgene under control of the CaMV 35S promoter exhibited more normal growth. However, as in the antisense plants, *GhDET* overexpressing plants were sterile. However, transgenic cotton plants that expressed *GhDET2* under the control of a promoter from a seed coat-specific gene (*FBP7*) from petunia grew and developed normally (Luo *et al.*, 2007). Analysis of the fibres produced by these plants showed that seed coat-specific expression of *GhDET2* increased fibre number and length. Therefore, these authors assert that BR is critical for normal initiation and elongation of cotton fibre cells, which suggests that modulation of BR biosynthesis may improve fibre quality or yield. Additional research is needed to understand the regulation of BR biosynthesis and signalling to develop the most effective strategies for the modification of phytohormone signalling to optimize cotton fibre yield and quality.

The cotton GA-receptor (GhGID1a) and a DELLA protein that acts as a transcriptional repressor of GA-signalling (GhSLR1b) were first characterized functionally in rice and *Arabidopsis*, respectively (Aleman *et al.*, 2008). However, the development of GhGID1a overexpressing cotton plants is now under way and preliminary analyses of transgenic cotton plants that overexpress GhSLR1b confirm our previous results from *Arabidopsis*. Interestingly, while the transgenic plants exhibited a semi-dwarf phenotype, consistent with the suppression of GA responses, HVI analysis showed that fibre quality was not altered significantly in these plants compared to non-transformed plants (Aleman, Abdel-Mageed and Allen, unpublished). It should be noted that expression of the native GA receptor genes (*GhGID1a* and *GhGID1b*) and GA-biosynthetic genes (*GhGA20ox* and *GhGA3ox*) was elevated significantly in these plants, indicating that the repression of GA-responsive gene expression mediated by GhSLR1 overexpression could be partially compensated for by increased GA biosynthesis and GA sensitivity. This compensation may be strengthened in ovules due to a burst in GA levels during pollination and fertilization. Olimpieri *et al.* (2007) showed that *GA20ox1* transcript levels were low in tomato ovules before anthesis and that they increased substantially on pollination and fertilization, leading to increased GA biosynthesis.

Targeting cell wall genes

As discussed previously, an elongating fibre cell must withstand internal and external pressure but allow enough extensibility for cell wall expansion during growth. Turgor is the internal pressure that drives elongation, but growth is limited by the extensibility of the primary cell wall. Turgor is increased in elongating fibres by osmotic adjustment via the accumulation of intracellular solutes. Ruan *et al.* (2001) showed that the permeability of plasmodesmata of fibre cells was regulated developmentally and correlated with the expression of solute transport proteins. However, even with high turgor pressure, fibre cells will not elongate unless the primary cell wall becomes extensible. Regulation of cell wall extensibility is thought to be by various glycosylases, along with glycosyl transferases and expansins (see Cosgrove, 1997, for a review). Xyloglucan endotransglucosylase/hydrolase is able to transfer a high molecular weight portion from a donor xyloglucan to a suitable acceptor such as a xyloglucan-derived nonasaccharide. Thus, XTHs can increase plant cell expansion by cutting and rejoining intermicrofibrillar xyloglucan chains, causing wall loosening (Fry *et al.*, 1992; Nishitani, 1997). To investigate the role of cell wall loosening enzymes such as XTH in developing cotton fibres,

Allen *et al.* (2000) developed transgenic cotton plants that expressed constitutively a transgene that encoded cotton fibre XTH (*GhXTH*). Transgenic seedlings had XTH-specific activities that were nearly twofold higher than non-expressing plants. It should be noted that the constitutive overexpression of cotton XTH in transgenic plants did not appear to affect the growth of the vegetative plants as the transgenic lines were indistinguishable from untransformed control cotton plants with respect to plant height, time to first flower or other traits. Fibres from the XTH-expressing lines were significantly longer than fibre from non-transformed and transgenic control plants (expressing β-glucuronidase). Additionally, XTH-expressing segregants had average fibre lengths of 33 mm compared to 27 mm for non-expressing segregants. Fibres from T_1 XTH-expressing plants were also somewhat stronger than fibres from control plants, but these differences were only statistically significant in greenhouse-grown plants. No significant differences in fineness (micronaire) were detected between XTH-expressing and non-expressing plants. To determine whether these XTH-expressing plants also produced longer fibres under field conditions, segregating T_1 plants from several independent transgenic lines were planted in replicate plots in Lubbock, Texas, and the plants were maintained using typical irrigated production methods. As in the greenhouse, the mean length of fibre from XTH-expressing plants was 33 mm, while fibre from non-expressing segregant plants of the same transgenic line averaged 27 mm in length. In these studies, inheritance of the transgene cassette corresponded with fibre lengths of 31 mm or greater, while plants that did not carry the transgene had fibres less than 28 mm in length. These data indicate that levels of XTH activity limit cell wall extensibility in elongating cotton fibres and the XTH transgene acts as a fully dominant fibre quality allele that is stable through at least three sexual generations. However, since the XTH-expressing plants do not show significant morphological effects on vegetative plants, it is possible that XTH activity does not limit cell expansion during plant growth. This result is in agreement with the findings of McQueen-Mason *et al.* (1993), who reported that XTH was neither sufficient nor necessary for cell wall expansion in hypocotyls. Interestingly, while expression of expansin increases during the elongation phase of fibre development, then decreases as the secondary cell wall synthesis stage begins, XTH levels remain relatively constant (Shimizu *et al.*, 1997; Ruan *et al.*, 2001). Therefore, since XTH activity may be limiting, increased expression of this gene via transgene expression may have a more substantial effect on fibre elongation than expansin.

Introducing novel traits to cotton fibres

We are aware of two reports of experiments designed to evaluate the use of genetic engineering to produce cotton fibres that express novel traits. In the first experiment, which was reported more than a decade ago, cotton was transformed with gene constructs that expressed the *Alcaligenes eutrophus* genes for acetoacetyl-CoA reductase (*phaB*) and polyhydroxyalkanoate sythase (*phaC*). These enzymes catalyse the production of the bioplastic aliphatic polyester poly-D-(–)-3 hydroxybutyrate, or PHB (John and Keller, 1996). Examination of the fibres by electron microscopy and HPLC provided evidence that the new polymer produced in the transgenic plants was PHB. More importantly, PHB granules in the fibres of these transgenic plants led to altered fibre properties. The fibres containing PHB exhibited changes in thermal properties, with improved thermal insulating characteristics. A measurable decrease in the rate of heat uptake and cooling in transgenic fibres, when compared to wild-type fibres, indicated a higher heat capacity.

An additional recent report on the introduction of a heterologous gene to cotton aimed at introducing a novel trait is focused on fibroin, one of the two proteinaceous polymers that make up silk fibres produced by silkworm insects and spiders (Sabajo *et al.*, 2008). Fibroin is a natural keratin protein that exhibits no apparent physiological activity. Up to 90% of the fibroin molecule consists of repeat sequences composed of the hydrophobic amino acids, glycine and alanine, that form antiparallel β-fold layers and give the protein its high tensile and breaking strength. Li *et al.* (2008) introduced a *fibroin* gene from silkworm under the control fibre promoter, GAE6-3A, into cotton by genetic transformation. The *fibroin* gene was identified in six independent transgenic plants and *fibroin* transcripts accumulated in 15 DPA fibres from homozygous lines. Additionally, microscopic analysis showed that, compared to fibre from control plants, the surface structure of mature fibre in the six lines was distorted significantly, with an increased number of convolutions or twists. Although the accumulation of *fibroin* transcripts in 15 DPA fibres from the homozygous lines was reported, fibroin protein levels in these plants or evidence for the incorporation of fibroin into the fibre was not present. It would be interesting to know where the fibroin protein is targeted and if it serves as a supporting scaffold for cotton fibre as a means to improve its characteristics, such as strength and thermal qualities.

Use of model plant systems in cotton fibre research

Substantial amounts of gene information relating to cotton cell wall synthesis and plant development in general have come, and continue to come, from research in *A. thaliana*. For example, genetic approaches in *Arabidopsis* have revealed the complexity of genes involved in cellulose production through the analysis of mutants. Phenotypes of mutant plants affected in cellulose synthesis can be identified readily, depending on whether the lesion affects cellulose deposition in primary or secondary cell walls. Defects in primary cell walls result in reduced cell elongation in dark-grown hypocotyls associated with exaggerated radial expansion, whereas secondary wall defects lack the secondary cell wall thickening that line the lumen of the xylem elements and interfasicular fibres without any obvious growth defects (Hematy and Hofte, 2006). Several excellent reviews discuss the different *Arabidopsis* mutants used to reveal the many important genes involved in cell wall synthesis (Scheible and Pauly, 2004; Hematy and Hofte, 2006; Geisler *et al.*, 2008). It is reported that as many as ten *Ces*A genes exist in *Arabidopsis*, and a few of these genes have been characterized thoroughly by the use of *radial swelling* (*rws*) and *irregular xylem* (*irx*) mutants (reviewed in Hematy and Hofte, 2006). In addition to cellulose synthases, a number of other pertinent genes were identified through mutant analyses. These include: *KORRIGAN1*, a membrane-bound cellulase (Nicol *et al.*, 1998); *KOBITO*1, a type II membrane protein with predicted endoglucanse activity (Pagant *et al.*, 2002); *COBRA* and *COBRA-like* GPI-anchored extracellular protein (Hauser *et al.*, 1995; Schnidelman *et al.*, 2001); along with glucosidase I (*knf*), glucosidase II (*rsw*3) and mannose 1-phosphate guanylyltransferase (*Cyt*1), all of which are involved in *N*-linked glycan formation (Gilmor *et al.*, 2002; Burn *et al.*, 2002; and Lukowitz *et al.*, 2001, respectively); and sterol biosynthesis genes (*fk*, *hyd*1 and *smt1/chp*) (Hematy and Hofte, 2006 and references therein). The knowledge gained though isolation and biochemical characterization of cell wall mutants in *Arabidopsis* has led to significant gains in our understanding of cell wall biogenesis and serves as a point of departure for translational genomics research into cotton fibre. Identification of cotton orthologs through bioinformatics followed by functional characterization in cotton is a viable strategy for the genetic exploration of fibre development (Haigler *et al.*, 2005). A good example of progress made using this approach comes from studies reported by Peng *et al.* (2002), who have shown that Korrigan may be required for cellulose synthesis by cleaving sitosterol-cellodextrins and providing sitosterol-β-glucoside, a primer for glucan polymerization by CesA in cotton fibre membranes. These results provide the basis of a model for cellulose synthesis in fibres that can be investigated further.

With the coming advances in cotton genomics, many new research tools will become available to study the development of this valuable seed trichome. We anticipate that, at some time in the relatively near future, the full sequence of the cotton genome will become available to the cotton research community. However, our ability to identify the biological functions of these genes remains a limiting step. The complex genome of cotton, combined with a lack of genetic resources and an inefficient transformation system, will continue to make this task difficult.

Conclusion and Perspectives

Cotton is among the most attractive targets for biotechnology, yet progress towards fibre improvement has been relatively slow. Though significant progress has been made by cotton breeders to overcome the negative genetic linkages between fibre quality and yield, most of the progress in the biotechnology arena has been focused on input traits such as herbicide tolerance and insect resistance. We anticipate that the recent identification of closely linked molecular markers that can be used for marker-assisted selection of specific fibre quality QTLs will lead to much more rapid progress in this area. As genetic and physical maps improve and genomic sequences become available, opportunities to identify the genes that underlie these QTLs will arise. Functional confirmation of the role of these genes in fibre development will require testing in transgenic plants. Work directed at modifying cotton fibre properties through transgenic modifications has been initiated and significant progress is being made both with single effect genes such as XTH and SPS and with regulatory factors such as hormone receptors and transcription factors. Future synergy created by the combination of genome information, molecular breeding and transgenic modification will provide important new tools for the genetic improvement of cotton fibre quality.

The nearly synchronous development and accumulation of almost pure cellulose make cotton fibre development an excellent model system for research into the biosynthesis of cellulose in secondary cell walls of plants. Transgenic technology will also provide numerous opportunities to create speciality fibres with unusual properties not found in nature. Early attempts to produce bioplastic and to express silk proteins in cotton fibres provide an intriguing glimpse into the possible applications of this technology. Incorporation of additional cell wall components or alterations in the deposition patterns of cellulose or other cell wall materials could lead to fibres with unique properties. Thus, it may be possible to modify cotton fibre characteristics directly to make them more amenable for processing or for use in a greater variety of specific industrial uses. Therefore, the future of the cotton fibre looks bright. Improved breeding strategies that utilize genomic resources will lead to continued progress in cotton yield and quality, while new technologies may produce cotton with novel characteristics that can open new markets for this important biological fibre.

References

Aleman, L., Kitamura, J., Abdel-Mageed, H., Lee, J., Sun, Y., Nakajima, M., Ueguchi-Tanaka, M., Matsuoka, M. and Allen, R.D. (2008) Functional analysis of cotton orthologs of GA signal transduction factors GID1 and SLR1. *Plant Molecular Biology* 68, 1–16.

Allen, R.D., Burns, T.H., Light, G. and Fokar, M. (2000) Investigating the role of xyloglucan endotransglycosylases in cotton fiber quality. In: Benedict, C. (ed.) *Genetic Control of Cotton Fiber and Seed Quality*. Cotton Inc, Raleigh, North Carolina, pp. 166–174.

Amor, Y., Haigler, C.H., Johnson, S., Wainscott, M. and Delmer, D.P. (1995) A membrane-associated form of sucrose synthase and its potential role in synthesis of cellulose and callose in plants. *Proceedings of the National Academy of Sciences of the United States of America* 92, 9353–9357.

Arpat, A.B., Waugh, M., Sullivan, J.P., Gonzales, M., Frisch, D., Main, D., Wood, T., Leslie, A., Wing, R.A. and Wilkins, T.A. (2004) Functional genomics of cell elongation in developing cotton fibres. *Plant Molecular Biology* 54, 911–929.

Atalla, R.H. and Vanderhart, D.L. (1984) Native cellulose: a composite of two distinct crystalline forms. *Science* 223, 283–285.

Babb, V.M. and Haigler, C.H. (2001) Sucrose phosphate synthase activity rises in correlation with high-rate cellulose synthesis in three heterotrophic systems. *Plant Physiology* 127, 1234–1242.

Basra, S.A. and Malik, C.P. (1983) Dark metabolism of CO_2 during fibre elongation of two cotton differing in fibre lengths. *Journal of Experimental Botany* 34, 1–9.

Basra, S.A. and Malik, C.P. (1984) Development of the cotton fibre. *International Review of Cytology* 87, 65–113.

Basra, S.A. and Saha, S. (1999) Growth regulation of cotton fibres. In: Basra, A.S. (ed.) *Cotton Fibres: Developmental Biology, Quality Improvement, and Textile Processing*. The Haworth Press Inc, Binghamton, New York, pp. 47–64.

Beasley, C.A. and Ting, I.P. (1973) The effects of plant growth substances on *in vitro* development from unfertilized cotton ovules. *American Journal of Botany* 60, 130–189.

Beasley, C.A. and Ting, I.P. (1974) Effects of plant growth substances on *in vitro* fibre development from unfertilized cotton ovules. *American Journal of Botany* 61, 188–194.

Beasley, J.O. (1942) Meiotic chromosome behavior in species, species hybrids, haploids, and induced polyploids of *Gossypium*. *Genetics* 27, 25–54.

Benedict, C.R., Kohel, R.J. and Lewis, H.L. (1999) Cotton fibre quality. In: Smith, C.W. and Cothren, J.T. (eds) *Cotton: Origin, History, Technology, and Production*. John Wiley and Sons Inc, New York, pp. 269–288.

Brown, R.M. and Montezinos, D. (1976) Cellulose microfibrils – visualization of biosynthetic and orienting complexes in association with plasma-membrane. *Proceedings of the National Academy of Sciences of the United States of America* 73, 143–147.

Buchala, A.J. (1999) Non-cellulosic carbohydrates in cotton fibres. In: Basra, A.S. (ed.) *Cotton Fibres: Developmental Biology, Quality Improvement, and Textile Processing*. The Haworth Press, New York, pp. 113–130.

Burgert, I. and Fratzl, P. (2006) Mechanics of the expanding cell wall. In: Verbelen, J.P. and Vissenberg, K. (eds) *The Expanding Cell*. Springer, Berlin, pp. 191–215.

Burn, J.E., Hurley, U.A., Birch, R.J., Arioli, T., Cork, A. and Williamson, R.E. (2002) The cellulose-deficient Arabidopsis mutant rsw3 is defective in a gene encoding a putative glucosidase II, an enzyme processing N-glycans during ER quality control. *Plant Journal* 32, 949–960.

Carpita, N.C. and Delmer, D.P. (1981) Concentration and metabolic turnover of UDP-glucose in developing cotton fibres. *Journal of Biological Chemistry* 256, 308–315.

Chanzy, H., Imada, K. and Vuong, R. (1978) Electron diffraction from the primary wall of cotton fibres. *Protoplasma* 94, 299–306.

Chee, P., Draye, X., Jiang, C.X., Decanini, L., Delmonte, T.A., Bredhauer, R., Smith, C.W. and Paterson, A.H. (2005a) Molecular dissection of interspecific variation between *Gossypium hirsutum* and *Gossypium barbadense* (cotton) by a backcross-self approach: I. Fibre elongation. *Theoretical and Applied Genetics* 111, 757–763.

Chee, P., Draye, X., Jiang, C.X., Decanini, L., Delmonte, T.A., Bredhauer, R., Smith, C.W. and Paterson, A.H. (2005b) Molecular dissection of interspecific variation between *Gossypium hirsutum* and *Gossypium barbadense* (cotton) by a backcross-self approach: III. Fibre length. *Theoretical and Applied Genetics* 1111, 772–781.

Chen, M., Presting, G., Barbazuk, W.B., Goicoechea, J.L., Blackmon, B., Fang, G., Kim, H., Frisch, D., Yu, Y., Sun, S., Higingbottom, S., Phimphilai, J., Phimphilai, D., Thurmond, S., Gaudette, B., Li, P., Liu, J., Hatfield, J., Main, D., Farrar, K., Henderson, C., Barnett, L., Costa, R., Williams, B., Walser, S., Atkins, M., Hall, C., Budiman, M.A., Tomkins, J.P., Luo, M., Bancroft, I., Salse, J., Regad, F., Mohapatra, T., Singh, N.K., Tyagi, A.K., Soderlund, C., Dean, R.A. and Wing, R.A. (2002) An integrated physical and genetic map of the rice genome. *Plant Cell* 14, 537–545.

Coe, E., Cone, K., McMullen, M., Chen, S.S., Davis, G., Gardiner, J., Liscum, E., Polacco, M., Paterson, A., Sanchez-Villeda, H., Soderlund, C. and Wing, R. (2002) Access to the maize genome: an integrated physical and genetic map. *Plant Physiology* 128, 9–12.

Cosgrove, D.J. (1987) Wall relaxation and the driving forces for cell expansive growth. *Plant Physiology* 84, 561–564.

Cosgrove, D.J. (1997) Relaxation in a high-stress environment: the molecular bases of extensible cell walls and cell enlargement. *Plant Cell* 9, 1031–1041.

Cothren, J.T. (1999) Physiology of the cotton plant. In: Smith, C.W. and Cothren, J.T. (eds) *Cotton: Origin, History, Technology, and Production*. John Wiley and Sons, New York, pp. 207–268.

Cui, X.J., Shin, H.S., Song, C., Laosinchai, W., Amano, Y. and Brown, R.M. (2001) A putative plant homolog of the yeast beta-1,3-glucan synthase subunit FKS1 from cotton (*Gossypium hirsutum* L.) fibres. *Planta* 213, 223–230.

Dasani, S.H. and Thaker, V.S. (2006) Role of abscisic acid in cotton fibre development. *Russian Journal of Plant Physiology* 53, 62–67.
Davidonis, G.H. (1999) Cotton fibres *in vitro*. In: Basra, A.S. (ed.) *Cotton Fibres: Developmental Biology, Quality Improvement, and Textile Processing*. The Haworth Press Inc, Binghamton, New York, pp. 65–84.
Delmer, D. (1999) Cellulose biosynthesis in the developing cotton fibres. In: Basra, A.S. (ed.) *Cotton Fibres: Developmental Biology, Quality Improvement, and Textile Processing*. The Haworth Press Inc, Binghamton, New York, pp. 85–112.
Desai, A., Chee, P.W., Rong, J., May, O.L. and Paterson, A.H. (2006) Chromosome structural changes in diploid and tetraploid A genomes of *Gossypium*. *Genome* 49, 336–345.
Dong, J., Kohel, R., Zhang, H. and Yu, J. (2001) Bacterial artificial chromosome (BAC) libraries constructed from the genetic standard of upland cottons. *Proceedings of IX International Conference on the Status of Plant and Animal Genome Researches*, San Diego, California, San Diego, CA, January 13–17, 2001.
Draye, X., Chee, P., Jiang, C.X., Decanini, L., Delmonte, T.A., Bredhauer, R., Smith, C.W. and Paterson, A.H. (2005) Molecular dissection of interspecific variation between *Gossypium hirsutum* and *G. barbadense* (cotton) by a backcross-self approach. *Theoretical and Applied Genetics* 111, 764–771.
Endrizzi, J.E., Turcotte, E.L. and Kohel, R.J. (1984) Qualitative genetics, cytology, and cytogenetics. In: Kohel, R.J. and Lewis, C.F (eds) *Cotton*. American Society of Agronomy, Madison, Wisconsin, pp. 81–129.
Finkelstein, R.R., Gampala, S.S. and Rock, C.D. (2002) Abscisic acid signaling in seeds and seedlings. *Plant Cell* 14 Suppl, S15–45.
French, A.D. and Johnson, G.P. (2007) Cellulose shapes. In: Brown, R.M. and Saxena, I.M. (eds) *Cellulose: Molecular and Structural Biology*. Springer, Dordrecht, The Netherlands, pp. 257–282.
Fry, S.C., Smith, R.C., Renwick, K.F., Martin, D.J., Hodge, S.K. and Matthews, K.J. (1992) Xyloglucan endotransglycosylase, a new wall-loosening enzyme activity from plants. *Biochemical Journal* 282, 821–828.
Galtier, N., Foyer, C.H., Huber, J., Voelker, T.A. and Huber, S.C. (1993) Effects of elevated sucrose-phosphate synthase activity on photosynthesis, assimilate partitioning, and growth in tomato (*Lycopersicon esculentum* var UC82B). *Plant Physiology* 101, 535–543.
Geisler, D.A., Sampathkumar, A., Mutwil, M. and Persson, S. (2008) Laying down the bricks: logistic aspects of cell wall biosynthesis. *Current Opinion in Plant Biology* 11, 647–652.
Gilmor, C.S., Poindexter, P., Lorieau, J., Palcic, M.M. and Somerville, C. (2002) Alpha-glucosidase I is required for cellulose biosynthesis and morphogenesis in Arabidopsis. *Journal of Cell Biology* 156, 1003–1013.
Gou, J.Y., Wang, L.J., Chen, S.P., Hu, W.L. and Chen, X.Y. (2007) Gene expression and metabolite profiles of cotton fibre during cell elongation and secondary cell wall synthesis. *Cell Research* 17, 422–434.
Goynes, W., Ingber, B. and Triplett, B. (1995) Cotton fibre secondary wall development-time versus thickness. *Textile Research Journal* 65, 400–408.
Graves, D.A. and Stewart, J.M. (1988) Chronology of the differentiation of cotton (*Gossypium hirsutum* L.) fibre cells. *Planta* 175, 254–258.
Guo, W., Zhang, T., Shen, X., Yu, J.Z. and Kohel, R.J. (2003) Development of SCAR marker linked to a major QTL for high fibre strength and its usage in molecular-marker assisted selection in upland cotton. *Crop Science* 43, 2252–2256.
Haigler, C. (2007) Substrate supply for cellulose synthesis and its stress sensitivity in the cotton fibre. In: Brown, R.M. and Saxena, I.M. (eds) *Cellulose: Molecular and Structural Biology*. Springer, Dordecht, The Netherlands, pp. 147–168.
Haigler, C. and Brown, R.M. (1986) Transport of rosettes from the golgi apparatus to the plasma membrane in isolated mesophyll cells of *Zinnia elegans* during differentiation to tracheary elements in suspension culture. *Proplasma* 134, 111–120.
Haigler, C.H., Ivanova-Datcheva, M., Hogan, P.S., Salnikov, V.V., Hwang, S., Martin, K. and Delmer, D.P. (2001) Carbon partitioning to cellulose synthesis. *Plant Molecular Biology* 47, 29–51.
Haigler, C., Zhang, D. and Wilkerson, C.G. (2005) Biotechnological improvement of cotton fibre maturity. *Physiologia Plantarum* 124, 285–294.
Haigler, C.H., Singh, B., Zhang, D., Hwang, S., Wu, C., Cai, W.X., Hozain, M., Kang, W., Kieddaisch, B., Strauss, R.E., Hequet, E.F., Wyatt, B.G., Jividen, G.M. and Holaday, A.S. (2007) Transgenic cotton over-producing spinach sucrose phosphate synthase showed enhanced leaf sucrose synthesis and improved fiber quality under controlled environmental conditions. *Plant Molecular Biology* 63, 815–832.

Hauser, M.T., Morikami, A. and Benfey, P.N. (1995) Conditional root expansion mutants of Arabidopsis. *Development* 121, 1237–1252.

Hawkins, J.S., Kim, H., Nason, J.D., Wing, R.A. and Wendel, J.F. (2006) Differential lineage-specific amplification of transposable elements is responsible for genome size variation in *Gossypium. Genome Research* 16, 1252–1261.

He, D., Lin, Z.X., Zhang, X.L., Nie, Y.C., Guo, X.P., Zhang, Y.X. and Li, W. (2007) QTL mapping for economic traits based on a dense genetic map of cotton with PCR-based markers using the interspecific cross of *Gossypium hirsutum* × *Gossypium barbadense. Euphytica* 153, 181–197.

He, Y.J., Guo, W.Z., Shen, X.L. and Zhang, T.Z. (2008) Molecular cloning and characterization of a cytosolic glutamine synthetase gene, a fibre strength-associated gene in cotton. *Planta* 228, 473–483.

Hematy, K. and Hofte, H. (2006) Cellulose and cell expansion. In: Verbelen, J.P. and Vissenberg, K. (eds) *The Expanding Cell*. Springer, Berlin, pp. 33–56.

Hendrix, B. and Stewart, J.M. (2005) Estimation of the nuclear DNA content of gossypium species. *Annals of Botany (London)* 95, 789–797.

Herth, W. (1985) Plasma-membrane rosettes involved in localized wall thickening during xylem vessel formation of *Lepidium sativum* L. *Planta* 164, 12–21.

Hseih, Y.L. (1999) Sturctural development of cotton fibres and linkages in fibre quality. In: Basra, A.S. (ed.) *Cotton Fibres: Developmental Biology, Quality Improvement, and Textile Processing*. The Haworth Press Inc, Binghamton, New York, pp. 137–166.

Huang, Q.S., Wang, H.Y., Gao, P., Wang, G.Y. and Xia, G.X. (2008) Cloning and characterization of a calcium dependent protein kinase gene associated with cotton fibre development. *Plant Cell Reports* 27, 1869–1875.

Humphries, J.A., Walker, A.R., Timmis, J.N. and Orford, S.J. (2005) Two WD-repeat genes from cotton are functional homologues of the *Arabidopsis thaliana* TRANSPARENT TESTA GLABRA1 (TTG1) gene. *Plant Molecular Biology* 57, 67–81.

Jacob-Wilk, D., Kurek, I., Hogan, P. and Delmer, D.P. (2006) The cotton fibre zinc-binding domain of cellulose synthase A1 from *Gossypium hirsutum* displays rapid turnover *in vitro* and *in vivo. Proceedings of the National Academy of Sciences of the United States of America* 103, 12191–12196.

Ji, S.J., Lu, Y.C., Feng, J.X., Wei, G., Li, J., Shi, Y.H., Fu, Q., Liu, D., Luo, J.C. and Zhu, Y.X. (2003) Isolation and analyses of genes preferentially expressed during early cotton fibre development by subtractive PCR and cDNA array. *Nucleic Acids Research* 31, 2534–2543.

Jiang, C., Wright, R.J., El-Zik, K.M. and Paterson, A.H. (1998) Polyploid formation created unique avenues for response to selection in Gossypium (cotton). *Proceedings of the National Academy of Sciences of the United States of America* 95, 4419–4424.

John, M.E. (1999) Genetic engineering strategies for cotton fibre modification. In: Basra, A.S. (ed.) *Cotton Fibres: Developmental Biology, Quality Improvement, and Textile Processing*. The Haworth Press Inc, Binghamton, New York, pp. 271–292.

John, M.E. and Keller, G. (1996) Metabolic pathway engineering in cotton: biosynthesis of polyhydroxybutyrate in fibre cells. *Proceedings of the National Academy of Sciences of the United States of America* 93, 12768–12773.

Kim, H.J. and Triplett, B.A. (2001) Cotton fibre growth *in planta* and *in vitro*. Models for plant cell elongation and cell wall biogenesis. *Plant Physiology* 127, 1361–1366.

Kim, H.J. and Triplett, B.A. (2004) Characterization of GhRac1 GTPase expressed in developing cotton (*Gossypium hirsutum* L.) fibres. *Biochimica et Biophysica Acta–Gene Structure and Expression* 1679, 214–221.

Kim, H., William, M.Y. and Triplett, B. (2002) A novel expression assay system for fibre-specific promoters in developing cotton fibres. *Plant Molecular Biology Reporter* 20, 7–18.

Kim, H.J., Kato, N., Kim, S. and Triplett, B. (2008) Cu/Zn superoxide dismutases in developing cotton fibres: evidence for an extracellular form. *Planta* 228, 281–292.

Kimura, S., Laosinchai, W., Itoh, T., Cui, X.J., Linder, C.R. and Brown, R.M. (1999) Immunogold labeling of rosette terminal cellulose-synthesizing complexes in the vascular plant *Vigna angularis. Plant Cell* 11, 2075–2085.

Lai-Kee-Him, J., Chanzy, H., Muller, M., Putaux, J.L., Imai, T. and Bulone, V. (2002) *In vitro* versus *in vivo* cellulose microfibrils from plant primary wall synthases: structural differences. *Journal of Biological Chemistry* 277, 36931–36939.

Lee, J. (2006) Characterization of the XTH gene family in cotton. PhD dissertation. Texas Tech University, Lubbock, Texas.

Lee, J.J., Woodward, A.W. and Chen, Z.J. (2007) Gene expression changes and early events in cotton fibre development. *Annals of Botany (London)* 100, 1391–1401.

Li, F., Wu, S., Lu, F., Chen, T., Ju, M., Wang, H., Jiang, Y., Zhang, J., Guo, W.Z. and Zhang, T. (2008) Modified fibre qualities of the transgenic cotton expressing a silkworm fibroin gene. *Chinese Science Bulletin* 54, 1210–1216.

Liu, B., Brubaker, C.L., Mergeai, G., Cronn, R.C. and Wendel, J.F. (2001) Polyploid formation in cotton is not accompanied by rapid genomic changes. *Genome* 44, 321–330.

Liu, D.Q., Tu, L.L., Li, W., Li, Y.J., Zhu, L.F. and Zhang, X.L. (2008) Characterization and expression of plasma and tonoplast membrane aquaporins in elongating cotton fibres. *Plant Cell Reports* 27, 1385–1394.

Lukowitz, W., Nickle, T.C., Meinke, D.W., Last, R.L., Conklin, P.L. and Somerville, C.R. (2001) Arabidopsis cyt1 mutants are deficient in a mannose-1-phosphate guanylyltransferase and point to a requirement of N-linked glycosylation for cellulose biosynthesis. *Proceedings of the National Academy of Sciences of the United States of America* 98, 2262–2267.

Luo, M., Xiao, Y., Li, X., Lu, X., Deng, W., Li, D., Hou, L., Hu, M., Li, Y. and Pei, Y. (2007) GhDET2, a steroid 5alpha-reductase, plays an important role in cotton fibre cell initiation and elongation. *Plant Journal* 51, 419–430.

McAlister D., III, Foulk, J. and Harrison, R. (2003) Advanced fibre information system length measurement of cottons hand-sorted by length group. *Journal of Cotton Science* 7, 217–223.

McMichael, B.L., Jordan, W.R. and Powell, R.D. (1973) Abscission processes in cotton: induction by plant water deficit. *Agronomy Journal* 65, 202–204.

McQueen-Mason, S., Fry, S.C., Durachko, D.M. and Cosgrove, D.J. (1993) The relationship between xyloglucan endotransglycosylase and *in-vitro* cell wall extension in cucumber hypocotyls. *Planta* 190, 327–331.

McQueen-Mason, S., Le, N.T. and Brocklehurst, D. (2007) Expansins. In: Verbelen, J.P. and Vissenberg, K. (eds) *The Expanding Cell*. Springer, Berlin, pp. 117–138.

Machado, A., Wu, Y., Yang, Y., Llewellyn, D.J. and Dennis, E.S. (2009) The MYB transcription factor *GhMYB25* regulates early fibre and trichome development. *Plant Journal* 59, 52–62.

Maltby, D., Carpita, N.C., Montezinos, D., Kulow, C. and Delmer, D.P. (1979) Beta-1,3-Glucan in Developing Cotton Fibres - Structure, Localization, and Relationship of Synthesis to That of Secondary Wall Cellulose. *Plant Physiology* 63, 1158-1164.

Marion-Poll, A. and Leung, J., (2006) Abscisic acid synthesis, metabolism and signal transduction. In: Hedden, P. and Thomas, S.G. (eds) *Plant Hormone Signaling*. Blackwell Publishing Ltd, Oxford, UK, pp. 1–35.

Marra, M., Kucaba, T., Sekhon, M., Hillier, L., Martienssen, R., Chinwalla, A., Crockett, J., Fedele, J., Grover, H., Gund, C., McCombie, W.R., McDonald, K., McPherson, J., Mudd, N., Parnell, L., Schein, J., Seim, R., Shelby, P., Waterston, R. and Wilson, R. (1999) A map for sequence analysis of the *Arabidopsis thaliana* genome. *Nature Genetics* 22, 265–270.

May, O.L. and Lege, K.E. (1999) Development of the world cotton industry. In: Smith, C.W. and Cothren, J.T. *Cotton: Origin, History, Technology, and Production*. John Wiley and Sons Inc, New York, pp. 65–98.

Micallef, B.J., Haskins, K.A., Vanderveer, P.J., Roh, K., Shewmaker, C.K. and Sharkey, T.D. (1995) Altered photosynthesis, flowering, and fruiting in transgenic tomato plants that have an increased capacity for sucrose synthesis. *Planta* 196, 327–334.

Michailidis, G., Argiriou, A., Darzentas, N. and Tsaftaris, A. (2009) Analysis of xyloglucan endotransglycosylase/hydrolase (XTH) genes from allotetraploid (*Gossypium hirsutum*) cotton and its diploid progenitors expressed during fibre elongation. *Journal of Plant Physiology* 166, 403–416.

Nicol, F., His, I., Jauneau, A., Vernhettes, S., Canut, H. and Hofte, H. (1998) A plasma membrane-bound putative endo-1,4-beta-D-glucanase is required for normal wall assembly and cell elongation in Arabidopsis. *EMBO Journal* 17, 5563–5576.

Nishitani, K. (1997) The role of endoxyloglucan transferase in the organization of plant cell walls. *International Review of Cytology* 173, 157–206.

Olimpieri, I., Siligato, F., Caccia, R., Soressi, G., Mazzucato, A., Mariotti, L. and Ceccarelli, N. (2007) Tomato fruit set driven by pollination or by the parthenocarpic fruit allele are mediated by transcriptionally regulated gibberellin biosynthesis. *Planta* 226, 877–888.

Ono, K., Ishimaru, K., Aoki, N. and Ohsugi, R. (1999) Transgenic rice with low sucrose-phosphate synthase activities retain more soluble protein and chlorophyll during flag leaf senescence. *Plant Physiology and Biochemistry* 37, 949–953.

Ono, K., Sasaki, H., Hara, T., Kobayashi, K. and Ishimaru, K. (2003) Changes in photosynthetic activity and export of carbon by overexpressing a maize sucrose-phosphate synthase gene under elevated CO2 in transgenic rice. *Plant Production Science* 6, 281–286.

Oosterhuis, D.M. and Jernstedt, J. (1999) Anatomy and morphology of cotton. In: Smith, C.W. and Cothren, J.S. (eds) *Cotton: Origin, History, Technology, and Production*. John Wiley and Sons. New York, pp. 175–206.

Orford, S.J. and Timmis, J.N. (1998) Specific expression of an expansin gene during elongation of cotton fibres. *Biochimica et Biophysica Acta* 1398, 342–346.

Pagant, S., Bichet, A., Sugimoto, K., Lerouxel, O., Desprez, T., McCann, M., Lerouge, P., Vernhettes, S. and Hofte, H. (2002) KOBITO1 encodes a novel plasma membrane protein necessary for normal synthesis of cellulose during cell expansion in Arabidopsis. *Plant Cell* 14, 2001–2013.

Pear, J.R., Kawagoe, Y., Schreckengost, W.E., Delmer, D.P. and Stalker, D.M. (1996) Higher plants contain homologs of the bacterial celA genes encoding the catalytic subunit of cellulose synthase. *Proceedings of the National Academy of Sciences of the United States of America* 93, 12637–12642.

Peng, L.C., Kawagoe, Y., Hogan, P. and Delmer, D. (2002) Sitosterol-beta-glucoside as primer for cellulose synthesis in plants. *Science* 295, 147–150.

Peters, W.S., Hageman, W. and Tomos, D. (2000) What makes plants different? Principles of extracellular matrix function in 'soft' plant tissues. *Comparative Biochemistry and Physiology Part A: Molecular and Integrative Physiology* 125, 151–167.

Qin, Y.M., Pujol, F.M.A., Shi, Y.H., Feng, J.X., Liu, Y.M., Kastaniotis, A.J., Hiltunen, J.K. and Zhu, Y.X. (2005) Cloning and functional characterization of two cDNAs encoding NADPH-dependent 3-ketoacyl-CoA reductased from developing cotton fibres. *Cell Research* 15, 465–473.

Qin, Y.M., Hu, C.Y., Pang, Y., Kastaniotis, A.J., Hiltunen, J.K. and Zhu, Y.X. (2007) Saturated very-long-chain fatty acids promote cotton fibre and Arabidopsis cell elongation by activating ethylene biosynthesis. *Plant Cell* 19, 3692–3704.

Ramey, H.H. Jr (1999) Classing fibre. In: Smith, C.W. and Cothren, J.T (eds) *Cotton: Origin, History, Technology, and Production*. John Wiley and Sons Inc, New York, pp. 709–727.

Reinish, A.J., Dong, J.M., Brubaker, C.L., Stelly, D.M., Wendel, J.F. and Paterson, A.H. (1994) A detailed RFLPmap of cotton, *Gossypium hirsutum* × *Gossypium barbadense*: chromosome organization and evolution in a disomic polyploid genome *Genetics* 138, 829–847.

Roberts, A.W. and Roberts, E. (2007) Evolution of the cellulose synthase (*CesA*) gene family: insights from green algae and seedless plants. In: Brown, R.M. and Saxena, I.M. (eds) *Cellulose: Molecular and Structural Biology*. Springer, Dordrecht, The Netherlands, pp. 17–29.

Roberts, E.M., Rao, N.R., Huang, J.Y., Trolinder, N.L. and Haigler, C.H. (1992) Effects of cycling temperatures on fibre metabolism in cultured cotton ovules. *Plant Physiology* 100, 979–986.

Rong, J., Feltus, F.A., Waghmare, V.N., Pierce, G.J., Chee, P.W., Draye, X., Saranga, Y., Wright, R.J., Wilkins, T.A., May, O.L., Smith, C.W., Gannaway, J.R., Wendel, J.F. and Paterson, A.H. (2007) Meta-analysis of polyploid cotton QTL shows unequal contributions of subgenomes to a complex network of genes and gene clusters implicated in lint fibre development *Genetics* 176, 2577–2588.

Rong, J.K., Pierce, G.J., Waghmare, V.N., Rogers, C.A.J., Desai, A., Chee, P.W., May, O.L., Gannaway, J.R., Wendel, J.F., Wilkins, T.A. and Paterson, A.H. (2005) Genetic mapping and comparative analysis of seven mutants related to seed fibre development in cotton. *Theoretical and Applied Genetics* 111, 1137–1146.

Ruan, Y.L., Llewellyn, D.J. and Furbank, R.T. (2001) The control of single-celled cotton fibre elongation by developmentally reversible gating of plasmodesmata and coordinated expression of sucrose and K+ transporters and expansin. *Plant Cell* 13, 47–60.

Ruan, Y.L., Llewellyn, D.J. and Furbank, R.T. (2003) Suppression of sucrose synthase gene expression represses cotton fibre cell initiation, elongation, and seed development. *Plant Cell* 15, 952–964.

Ruan, Y.L., Xu, S.M., White, R. and Furbank, R.T. (2004) Genotypic and developmental evidence for the role of plasmodesmatal regulation in cotton fibre elongation mediated by callose turnover. *Plant Physiology* 136, 4104–4113.

Ruan, Y.L., Llewellyn, D.J., Furbank, R.T. and Chourey, P.S. (2005) The delayed initiation and slow elongation of fuzz-like short fibre cells in relation to altered patterns of sucrose synthase expression and plasmodesmata gating in a lintless mutant of cotton. *Journal of Experimental Botany* 56, 977–984.

Ryser, U. (1999) Cotton fibre initiation and histodifferentiation. In: Basra, A.S. (ed.) *Cotton Fibres: Developmental Biology, Quality Improvement, and Textile Processing*. The Haworth Press Inc, Binghamton, New York, pp. 1–45.

Sabajo, C., Behzad, F., Yuan, X.-F. and Bayat, A. (2008) Silk: a potential medium for tissue engineering. *Eplasty* 8, 438–446.

Salnikov, V.V., Grimson, M.J., Seagull, R.W. and Haigler, C.H. (2003) Localization of sucrose synthase and callose in freeze-substituted secondary-wall-stage cotton fibres. *Protoplasma* 221, 175–184.

Scheible, W.R. and Pauly, M. (2004) Glycosyltransferases and cell wall biosynthesis: novel players and insights. *Current Opinion in Plant Biology* 7, 285–295.

Schnidelman, G., Morikami, A., Jung, J., Baskin, T.I., Carpita, N.C., Derbyshire, P., McCann, M. and Benfey, P.N. (2001) COBRA encodes a putative GPI-anchored protein, which is polarly localized and necessary for oriented cell expansion in Arabidopsis. *Genes and Development* 15, 1115–1127.

Seagull, R. (1992) A quantitative electron microscopic study of changes in microtubule arrays and wall microfibril orientation during the *in vitro* fibre development. *Journal of Cell Science* 101, 561–577.

Seagull, R. (2001) Fibre development and maturation. In: Seagull, R. and Alspaugh, P. (eds) *Cotton Fibre Development and Processing: An Ilustrated Overview*. International Textile Center, Lubbock, Texas, pp. 32–55.

Seagull, R. and Giavalis, S. (2004) Pre- and post-anthesis application of exogenous hormones alters fibre production in *Gossypium hirsutum* L. cultivar Maxxa GTO. *Journal of Cotton Science* 8, 105–111.

Shangguan, X.X., Xu, B., Yu, Z.X., Wang, L.J. and Chen, X.Y. (2008) Promoter of a cotton fibre MYB gene functional in trichomes of Arabidopsis and glandular trichomes of tobacco. *Journal of Experimental Botany* 59, 3533–3542.

Shen, X., Guo, W., Zhu, X., Wu, Y., Huang, N. and Zhang, T. (2005) Molecular mapping of QTLs for fibre qualities in three diverse lines in Upland cotton using SSR markers. *Molecular Breeding* 15, 169–181.

Shi, Y.H., Zhu, S.W., Mao, X.Z., Feng, J.X., Qin, Y.M., Zhang, L., Cheng, J., Wei, L.P., Wang, Z.Y. and Zhu, Y.X. (2006) Transcriptome profiling, molecular biological, and physiological studies reveal a major role for ethylene in cotton fibre cell elongation. *Plant Cell* 18, 651–664.

Shimizu, Y., Aotsuka, S., Hasegawa, O., Kawada, T., Sakuno, T., Sakai, F. and Hayashi, T. (1997) Changes in levels of mRNAs for cell wall-related enzymes in growing cotton fibre cells. *Plant Cell Physiology* 38, 375–378.

Signora, L., Galtier, N., Skot, L., Lucas, H. and Foyer, C. (1998) Over-expression of sucrose phosphate synthase in *Ababidopsis thaliana* results in increased foliar sucrose/starch ratios and favours decreased foliar carbohydrate accumulation in plants after prolonged growth with CO2 enrichment. *Journal of Experimental Botany* 49, 669–678.

Singh, B., Avci, U., Eichler Inwood, S., Grimson, M.J., Landgraf, J., Mohnen, D., Sørensen, I., Wilkerson, C.G., Willats, W.G.T. and Haigler, C.H. (2009) A specialized outer layer of the primary cell wall joins elongating cotton fibers into tissue-like bundles. *Plant Physiology* 150, 684–699.

Song, P. and Allen, R.D. (1997) Identification of a cotton fibre-specific acyl carrier protein cDNA by differential display. *Biochimica et Biophysica Acta* 1351, 305–312.

Song, W.Q., Qin, Y.M., Saito, M., Shirai, T., Pujol, F.M., Kastaniotis, A.J., Hiltunen, J.K. and Zhu, Y.X. (2009) Characterization of two cotton cDNAs encoding trans-2-enoyl-CoA reductase reveals a putative novel NADPH-binding motif. *Journal of Experimental Botany* 60, 1839–1848.

Stelly, D.M. (1993) Interfacing cytogenetics with the cotton genome mapping effort. In: Herber, D.J. and Richter, D.A. (eds) *Proceedings of the Beltwide Cotton Production Conference*. New National Cotton Council of America, Memphis, Tennessee, pp.1545–1550.

Stewart, J.M. (1975) Fibre initiation on the cotton ovule (*Gossypium hirsutum*). *American Journal of Botany* 62, 723–730.

Sun, Y. and Allen, R.D. (2005) Functional analysis of the BIN2 genes of cotton. *Molecular Genetics and Genomics* 274, 51–59.

Sun, Y., Fokar, M., Asami, T., Yoshida, S. and Allen, R.D. (2004) Characterization of the brassinosteroid insensitive 1 genes of cotton. *Plant Molecular Biology* 54, 221–232.

Sun, Y., Veerabomma, S., Abdel-Mageed, H.A., Fokar, M., Asami, T., Yoshida, S. and Allen, R.D. (2005) Brassinosteroid regulates fibre development on cultured cotton ovules. *Plant Cell Physiology* 46, 1384–1391.

Thaker, V.S., Rabadia, V.S. and Singh, Y.D. (1999) Physiological and biochemical changes associated with cotton fibre development. VII Carbohydrate metabolism. *Acta Physiologiae Plantarum* 21, 57–61.

Timpa, J.D. and Triplett, B. (1993) Analysis of cell-wall polymers during cotton fibre development. *Planta* 189, 101–108.

Tiwari, S.C. and Wilkins, T.A. (1995) Cotton (*Gossypium hirsutum*) seed trichomes expand via diffuse growing mechanism. *Canadian Journal of Botany-Revue Canadienne De Botanique* 73, 746–757.

Tokumoto, H., Wakabayashi, K., Kamisaka, S. and Hoson, T. (2002) Changes in the sugar composition and molecular mass distribution of matrix polysaccharides during cotton fibre development. *Plant and Cell Physiology* 43, 411–418.

Tokumoto, H., Wakabayashi, K., Kamisaka, S. and Hoson, T. (2003) Xyloglucan breakdown during cotton fibre development. *Journal of Plant Physiology* 160, 1411–1414.

Tomkins, J.P., Peterson, D.G., Yang, T.J., Main, D., Wilkins, T.A., Paterson, A.H. and Wing, R.A. (2001) Development of genomic resources for cotton, *Gossypium hirsutum* L.: BAC library construction, preliminary STC analysis, and identification of clones associated with fibre development. *Molecular Breeding* 8, 255–261.

Triplett, B.A. (2000) Cotton ovule culture: a tool for basic biology, biotechnology and cotton improvement. *In Vitro Cellular and Developmental Biology – Plant* 36, 93–101.

Udall, J.A., Swanson, J.M., Haller, K., Rapp, R.A., Sparks, M.E., Hatfield, J., Yu, Y., Wu, Y., Dowd, C., Arpat, A.B., Sickler, B.A., Wilkins, T.A., Guo, J.Y., Chen, X.Y., Scheffler, J., Taliercio, E., Turley, R., McFadden, H., Payton, P., Klueva, N., Allen, R., Zhang, D., Haigler, C., Wilkerson, C., Suo, J., Schulze, S.R., Pierce, M.L., Essenberg, M., Kim, H., Llewellyn, D.J., Dennis, E.S., Kudrna, D., Wing, R., Paterson, A.H., Soderlund, C. and Wendel, J.F. (2006) A global assembly of cotton ESTs. *Genome Research* 16, 441–450.

Ulloa, M., Brubaker, C. and Chee, P. (2006) Cotton. In: Kole, C. (ed.) *Genome Mapping and Molecular Breeding*. Springer, New York, pp. 1–49.

Vissenberg, K., Fry, S.C., Pauly, M., Hofte, H. and Verbelen, J.P. (2005) XTH acts at the microfibril-matrix interface during cell elongation. *Journal of Experimental Botany* 56, 673–683.

Wada, M., Heux, L. and Sugiyama, J. (2004) Polymorphism of cellulose I family: reinvestigation of cellulose IV. *Biomacromolecules* 5, 1385–1391.

Waefler, U. and Meier, H. (1994) Enzyme activities in developing cotton fibres. *Plant Physiology and Biochemistry* 32, 697–702.

Wakelyn, P.J., Bertoniere, N.R., French, A.D., Thibodeaux, D.P., Triplett, B.A., Rousselle, M.A., Goynes, W.R., Edwards, J.V., Hunter, L., McAlister, D.D., and Gamble, G.R. (2007) *Cotton Fibre Chemistry and Technology*. CRC Press, Boca Raton, Florida.

Wang, B.H., Wu, Y.T., Huang, N.T., Zhu, X.F., Guo, W.Z. and Zhang, T.Z. (2006b) QTL mapping for plant architecture traits in upland cotton using RILs and SSR markers. *Acta Genetica Sinica* 33, 161–170.

Wang, H.Y., Yu, Y., Chen, Z.L. and Xia, G.X. (2005) Functional characterization of *Gossypium hirsutum* profilin 1 gene (GhPFN1) in tobacco suspension cells. *Planta* 222, 594–603.

Wang, K., Song, X., Han, Z., Guo, W., Yu, J.Z., Sun, J., Pan, J., Kohel, R.J. and Zhang, T. (2006a) Complete assignment of the chromosomes of *Gossypium hirsutum* L. by translocation and fluorescence in situ hybridization mapping. *Theoretical and Applied Genetics* 113, 73–80.

Wang, S., Wang, J.W., Yu, N., Li, C.H., Luo, B., Gou, J.Y., Wang, L.J. and Chen, X.Y. (2004) Control of plant trichome development by a cotton fibre MYB gene. *Plant Cell* 16, 2323–2334.

Wendel, J.F. and Cronn, R.C. (2003) Polyploidy and the evolutionary history of cotton. *Advances in Agronomy* 78, 139–186.

Wendel, J.F., Schnabel, A. and Seelanan, T. (1995) Bidirectional interlocus concerted evolution following allopolyploid speciation in cotton (Gossypium). *Proceedings of the National Academy of Sciences of the United States of America* 92, 280–284.

Werker, E. (2000) Trichome diversity and development. *Advances in Botanical Research Incorporating Advances in Plant Pathology* 31, 1–35.

Wilkins, T.A. and Arpat, A.B. (2005) The cotton fibre transcriptome. *Physiologia Plantarum* 124, 295–300.

Worrell, A.C., Bruneau, J.M., Summerfelt, K., Boersig, M. and Voelker, T.A. (1991) Expression of a maize sucrose phosphate synthase in tomato alters leaf carbohydrate partitioning. *Plant Cell* 3, 1121–1130.

Wu, C., Sun, S., Nimmakayala, P., Santos, F.A., Meksem, K., Springman, R., Ding, K., Lightfoot, D.A. and Zhang, H.B. (2004) A BAC- and BIBAC-based physical map of the soybean genome. *Genome Research* 14, 319–326.

Wu, Y., Machado, A.C., White, R.G., Llewellyn, D.J. and Dennis, E.S. (2006) Expression profiling identifies genes expressed early during lint fibre initiation in cotton. *Plant Cell Physiology* 47, 107–127.

Xu, Z., Kohel, R.J., Song, G., Cho, J., Alabady, M., Yu, J., Koo, P., Chu, J., Yu, S., Wilkins, T.A., Zhu, Y. and Yu, J.Z. (2008) Gene-rich islands for fibre development in the cotton genome. *Genomics* 92, 173–183.

Yang, S.S., Cheung, F., Lee, J.J., Ha, M., Wei, N.E., Sze, S.H., Stelly, D.M., Thaxton, P., Triplett, B., Town, C.D. and Jeffrey Chen, Z. (2006) Accumulation of genome-specific transcripts, transcription factors and phytohormonal regulators during early stages of fibre cell development in allotetraploid cotton. *Plant Journal* 47, 761–775.

Yu, J.Z., Kohel, R.J., Zhang, H.B., Dong, J.M., Sun, S.K. and Steele, N.L. (2002) Toward development of a whole-genome, BAC/BIBAC-based integrated physical/genetic map of the cotton genome using the Upland genetic standard TM-1: BAC and BIBAC library construction, SSR marker development, and physical/genetic map integration. *Proceedings of the 3rd Workshop of the International Cotton Genome Initiative ICGI*. Supplement Edition of *Chinese Cotton Science*, Nanjing, China.

Zhang, D., Pirtle, I., Park, S., Nampaisansuk, M., Neogi, P., Wanjie, S.W., Pirtle, R. and Chapman, K.D. (2009) Identification and expression of a new delta-12 fatty acid desaturase (FAD2-4) gene in upland cotton and its functional expression in yeast and *Arabidopsis thaliana* plants. *Plant Physiology and Biochemistry* 47, 462–471.

Zhang, D.Y., Zhang, T.Z., Sang, Z.Q. and Guo, W.Z. (2007) Comparative development of lint and fuzz using different cotton fibre-specific developmental mutants in *Gossypium hirsutum*. *Journal of Integrative Plant Biology* 49, 1038–1046.

Zhang, H.B. and Wu, C. (2001) BAC as tools for genome sequencing. *Plant Physiology and Biochemistry* 39, 195–209.

Zhang, J.H., Schurr, U. and Davies, W.J. (1987) Control of stomatal behavior by abscisic-acid which apparently originates in the roots. *Journal of Experimental Botany* 38, 1174–1181.

Zhang, T., Yuan, Y., Yu, J., Guo, W. and Kohel, R.J. (2003) Molecular tagging of a major QTL for fibre strength in Upland cotton and its marker-assisted selection. *Theoretical and Applied Genetics* 106, 262–268.

Zhu, Y.Q., Xu, K.X., Luo, B., Wang J.W. and Chen X.Y. (2003) An ATP-binding cassette transporter GhWBC1 from elongating cotton fibers. *Plant Physiology* 133, 580–588.

14

Bast Fibres: From Plants to Products

JONATHAN Y. CHEN AND FRANK LIU

Bast Fibre Overview

Bast fibre type and usage history

The UN General Assembly Resolution 61/189 declared the year 2009 as International Year of Natural Fibres. A general goal of the campaign for the Natural Fibre Year was to renew the natural fibre profile and highlight the value of natural fibres for lifting human living standards and for helping rural economic development. While the Natural Fibre Year praises the wonders of natural fibres for meeting human necessities, it also sets a moment for people to think about the future impact of natural fibres on an upcoming bio-based economy.

Bast fibres are a subgroup of plant fibres. Different from seed fibres or leaf fibres, bast fibres are obtained from the skin of a plant stem. Ramie, flax, hemp, kenaf, jute and nettle are major bast fibre plants widely used for producing consumer and industrial fibres. Sugarcane and bamboo also have stem skin from which fibre can be extracted. But these plants are not used primarily as fibre crops.

The flax plant, also called common flax or linseed, is one of the oldest fibre crops in the world. The use of flax fibre for textile fabrics (linen) dates back to 5000 years ago (Oplinger *et al.*, 2009). Found in the early Egyptian tombs, linen cloth was used as a mummy wrap. Commercial production of flax fibre started in the mid-1700s. After the invention of the cotton gin at the end of the 1700s, the production of flax fibre fell below that of cotton fibre. Today, flax plants can be found in different regions of the world, with over 150 plant species. They can be grown for both fibre and seed. EU countries, Russia, China and Canada are among the major fibre flax producers in the international bast fibre market. The USA and Canada produce seed flax to meet human nutritional need for flaxseed oil and food ingredients.

The ramie plant has over 5000 years of cultivation history. The literature states that China was the first country in history to produce ramie fibre for textile applications (Li, 1982). The nickname for ramie, Chinagrass, reflects the influence of the Chinese cultivating technique for ramie production on the global ramie market. Currently, the volume of China's ramie fibre production still accounts for 90% of the world total ramie fibre market (Kozlowski *et al.*, 2005). Because ramie is the only bast fibre that most resembles cotton fibre in both fineness and colour, it is used primarily for apparel applications in today's global consumer product market.

Hemp is a high-yield herbaceous annual plant grown around the world. Both hemp stalks and seeds are high-value agricultural raw products. Hemp stalks provide long and strong staple fibre that can be widely used for diverse end uses, such as textiles and apparels, paper products, housing materials and automotive interior parts. Hemp seeds can be used directly in the production of diverse food products. Hemp oil extracted from hemp seeds is the richest known source of essential polyunsaturated fatty acids (linoleic and linolenic acids) that are excellent ingredients for nutraceuticals and personal care products. Hemp is also a potential energy source. By bioprocessing, hemp can be converted into many bioenergy products, such as fuel pellets, liquid fuels and gas (Gibson, 2005; Thames, 2005).

Kenaf is a warm-season annual fibre crop. Its early cultivation goes back to 4000 BC in West Africa. In different parts of the world, kenaf has many other names such as mesta (India, Bengal), stockroot (South Africa), Java jute (Indonesia) and ambari (Taiwan), etc. Current major farming areas are located in South-eastern Asia such as China, India, Malaysia, Indonesia and Thailand (Cook, 1960). Apart from textile applications, kenaf applications are also extended to many industrial areas like automobile, agriculture, construction, chemical process and packaging. Major end-use products include apparel fabrics and plastic/fibre composites from the fibre; oil and chemical absorbents, animal bedding and horticulture potting mix from the core; and livestock feed from the leaf. Kenaf is a relatively new type of fibre crop in the USA. Early production of kenaf in the USA can be dated back to the 1940s, with an initiative to use it as a non-wood fibre alternative for pulp and paper manufacture. This promotion was highlighted in the 1980s, when a kenaf demonstration project was sponsored by USDA (United States Department of Agriculture) and an 82/18 kenaf/bleached-kraft newsprint successfully passed the test in the pressrooms of four American newspapers. However, the development of kenaf production for the paper and pulp market is experiencing a difficult time. The major problems that producers face are a fierce competition from papermakers dominating the commodity paper market, a lack of financial support and public awareness, and the length of time required for the establishment of sustainable kenaf production and marketing.

Jute is also a warm-season, rainfed fibre crop produced mainly in South-eastern Asia. India and Bangladesh are the two largest jute producers in the world, from which jute yields account for 91% of the world jute production (Wikipedia, 2009). Jute refers to the plant genus *Corchorus*, which has over 30 different species. Among these species, white jute (*Corchorus capsularis*) and tossa jute (*C. olitorius*) are widely used in today's consumer applications and other industrial products. The history of jute cultivation is related closely to the Bengali culture because the early cultivation of jute began in eastern Bengal (formerly East Pakistan and now Bangladesh) and West Bengal of India. After the 1960s, jute production reduced dramatically because of the use of synthetic fibres. Jute fibre is used primarily for making packaging fabrics (sacks), carpet backing and other padding materials. Its high-end applications need to be investigated.

Current production capacity

Overall, world production of natural fibres trends a continuous decline since the use of synthetic fibres in the 1950s. The current production capacity of the major bast fibres is listed in Table 14.1. Cotton (seed fibre) and bagasse (sugarcane fibre residue) are also listed in the table for comparison purposes (Kozlowski *et al.*, 2005; Leslie *et al.*, 2008; FAO, 2009). The data indicate that jute is ranked number one in bast fibre production. The availability of jute is almost twice that of the sum of flax, kenaf, ramie and hemp. According to recent statistical data from the UN Food and Agricultural Organization (FAO), India, Bangladesh, China, Côte d'Ivoire, Thailand, Myanmar, Brazil, Uzbekistan, Nepal and Vietnam are the top ten jute producers in the world (Wikipedia, 2009).

Flax ranks second and is also well above kenaf, ramie and hemp in terms of production capacity. Flax farming areas are scattered throughout Russia, China and the EU countries. To date, the major purpose of growing flax is still for making flax fibre, which can be sold for textile

Table 14.1. Bast fibre production capacity (million tonne).

Fibre	World	USA
Jute	3.23	No production
Flax	0.97	5.72 (million bushels)
Kenaf	0.33	0.13
Ramie	0.28	No production
Hemp	0.07	No production
Cotton	112.9 (million bales)	19.2 (million bales)
Bagasse	100	20

applications at a price almost 50% higher than that of cotton. With an increased demand for biofuels and bio-based products, attention is also drawn to flaxseed, from which flaxseed oil can be extracted for biofuel refinery and the production of nutraceutical products. Ramie is produced mainly in China, with a total growing area of 1.2 million ha distributed around the central and upper part of the Yangtze River (Ouyang, 2007). Although ramie fibre is an ideal blender with cotton for textile and apparel end uses, utilization of ramie seed, leaf and root will lead to producing a value-added product chain including linoleic acid, medical herb and animal feed. World kenaf producers are primarily from developing countries. Kenaf production in Africa accounts for 2.9% of the world total. Producers in Latin America share 6.3% of the global production volume. India, China and Thailand are the top three kenaf producers, yielding 85% of the world total volume of kenaf (FAO, 2003). The capacity of the US kenaf production was estimated to be 3237 ha in 1997 (USDA, 2000). These data have not been updated because kenaf is not included in the national crop production report.

As hemp varieties with less than 0.3% THC (delta-9-tetrahydrocannabinol) have been legalized in the UK, Germany, Austria and Switzerland since 1990, followed by Canada and Australia lifting the prohibition for hemp cultivation in 1998, hemp production has shown a significantly upward trend in the global market. For example, in Germany and Austria alone, the use of hemp for the automotive industry had a 90% increase from 1999 to 2000, reaching an annual consumption of 2100 t (Karus and Kaup, 2002). Of the current annual global production of hemp fibre and tow, 27% is produced in the EU countries and 60% is produced in the Asian countries. In the USA, however, there is no commercial production of the industrial hemp crop, because of its status of the Schedule I controlled substance classified by the US Drug Enforcement Administration (DEA) (USDA, 2000). Consumer needs for hemp-related products rely only on the import of hemp fibre and hemp products.

Bast Fibre Basic Structure and Properties

Molecular structure

All bast fibres are cellulosic fibres, i.e. their major chemical component is cellulose. Structural differences among these bast fibres result largely from chemical contents. Apart from cellulose, hemicellulose and lignin are also the basic chemical components in bast fibres. The molecular structure of these bast fibres is often called lignocellulose structure. This structure is illustrated in Fig. 14.1. A single fibre unit of the bast fibre is formed by chains of cellulose molecules and an amorphous matrix composed of hemicellulose, lignin, pectin and other substances. Cross-linking of the cellulose chains relies on H-bonds. More cellulose chains and H-bonding nodes produce more crystalline regions. As a result, the fibre is longer and stronger.

Table 14.2 provides the chemical content information for the major bast fibres discussed above (Batra and Turner, 1998; Franck, 2005). It can be seen that the cellulose contents of flax, hemp, ramie, jute and kenaf are at the same level. Their hemicelluloses contents are also similar. However, the lignin contents for these bast fibres show a fairly large variation. Flax and hemp can be considered as the same group. Ramie is close to cotton, which has 0% lignin. Kenaf is similar to waste bagasse, having a lignin content above 20%. The higher the lignin content, the lower the bast fibre quality.

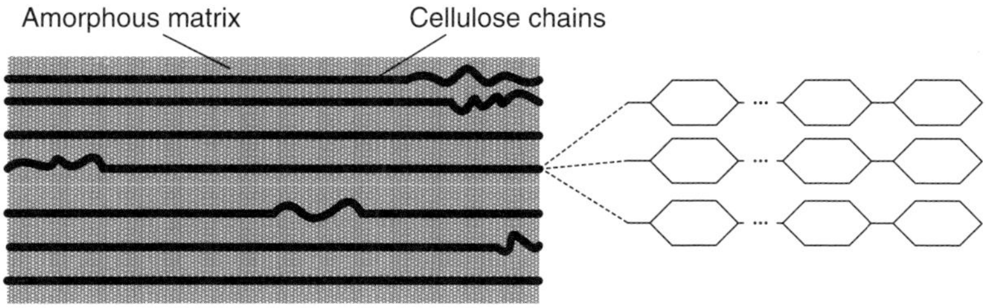

Fig. 14.1. Illustration of bast fibre molecular structure.

Table 14.2. Bast fibres chemical contents (%).

Fibre	Cellulose	Hemicellulose	Lignin	Pectin	Fat/wax
Flax	62–71	16–18	2.0–2.5	1.8–2.0	1.5
Hemp	67–75	16–18	2.9–3.3	0.8	0.7
Ramie	68–76	13–14	0.6–0.7	1.9–2.1	0.3
Jute	59–71	12–13	11.8–12.9	0.2–4.4	0.5
Kenaf	65	13	21.6	–	–
Bagasse	46	25	19.9	–	3.5
Cotton	92–95	5.7	0	1.2	0.6

Morphological structure

Bast fibres are obtained from fibrous plant stems. In general, a typical biological structure of the fibre plant stem consists of a bark layer, a bast layer and a stem core (Fig. 14.2). The bark layer is a thin skin (cuticle) to hold bast fibre bundles and protect the whole stem. The bast layer includes a primary fibre layer, where bast fibres are extracted, and a secondary fibre layer (phloem or real bast). The stem core has two parts: xylem (woody tissue) and pith.

Compared to the structure of a fibre plant stem, bast fibres produced from the plant stem are tiny in diameter and short in length. The techniques of advanced optical microscopy and scanning electron microscopy are often used to examine the bast fibre morphological structure, i.e. the fibre longitudinal (surface) and cross-sectional morphology. Figures 14.3–14.5 exhibit three SEM photomicrographs showing the surface and cross-sectional structures of kenaf, ramie and hemp fibres. Their surface images are fairly similar but the cross-sectional forms are different. As a summary, the morphological features of the major bast fibres are listed in Table 14.3.

Mechanical properties

Like any other industrial materials, bast fibres have mechanical behaviour in response to different physical actions during production and consumption. Commonly discussed mechanical deformations for bast fibres include extension, bending, lateral compression and abrasion. Among these mechanical deformations, tensile property is the most important property to determine. According to material mechanics, tensile strength (break force divided by the sam-

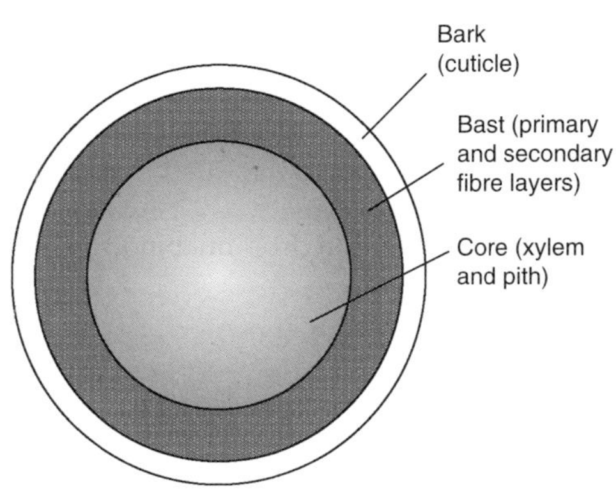

Fig. 14.2. Cross section of bast fibre plant stem.

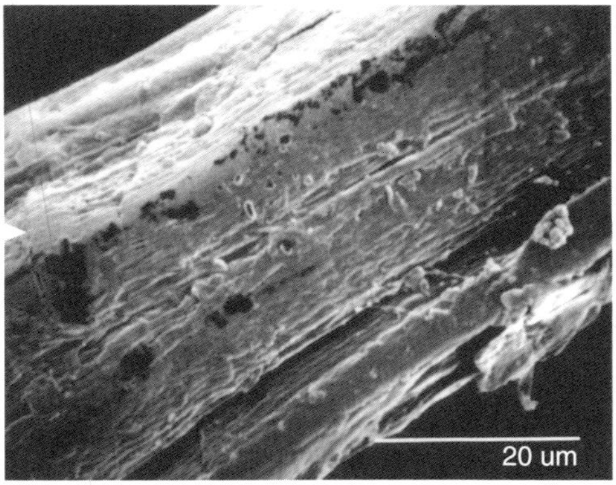

Fig. 14.3. Kenaf surface.

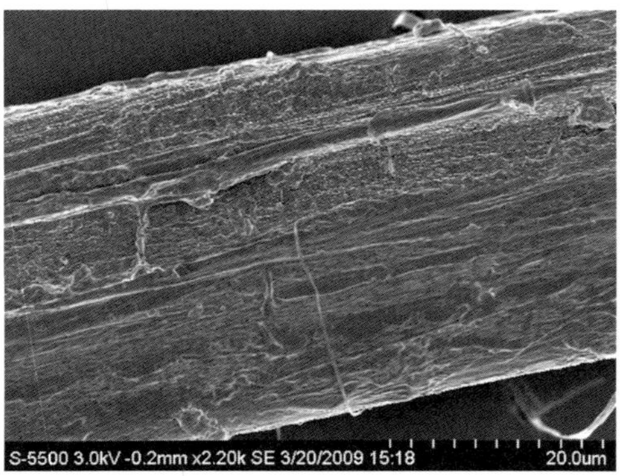

Fig. 14.4. Ramie surface.

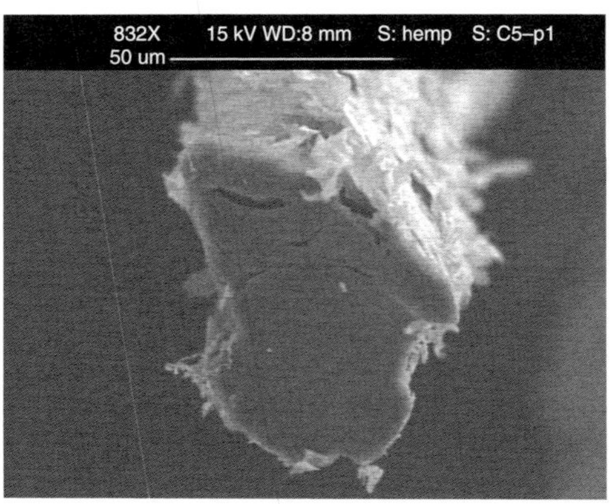

Fig. 14.5. Hemp cross section.

ple area subject to the break force), break elongation and tensile modulus are basic parameters to describe material tensile property. However, because the bast fibres are not completely elastic, measurement of the tensile modulus depends on different definitions. Also, because the cross-sectional area of the bast fibres is not uniform along the direction of the fibre axis, and is too tiny to measure in practice, a special term called tenacity is often used to express fibre tensile strength. The fibre break tenacity is determined by the break force divided by fibre density (1 tex = 1 g of fibre weight with 1000 m of fibre length). As an outlook, Table 14.4 lists the tensile strength and elongation of the major bast fibres (Lewin and Pearce, 1985; Franck, 2005). It can be noticed that the data in the table show a considerably large range of variation. This makes a direct data comparison difficult.

Physical and chemical properties

The major physical properties of bast fibres discussed here refer to fibre length, fineness, linear density, volume density and moisture regain. Table 14.5 summarizes some published data in terms of these physical properties. Because the bast fibres grow in the form of a bundle, fibre bundle length is often measured. The bundle lengths listed in the table actually indicate the plant stem lengths. Overall, ramie has the longest single fibre length, followed by flax and hemp. The single fibre length of jute and kenaf is the shortest. Fibre width and linear density both describe the bast fibre fineness. But fibre width is usually measured by microscopic image methods. It does not consider the irregularity of the fibre cross-sectional area and fibre volume density. In contrast, fibre linear density is determined by both fibre length and fibre weight. Using fibre linear density to express fibre fineness is more practical and reliable. The volume density of all bast fibres is identical. The moisture regain describes the bast fibres' ability to absorb water. It is defined as a ratio of the weight of absorbed water over the weight of the dry fibre sample. Referring to Table 14.5, it can be seen that the moisture regain of ramie and jute is higher than that of flax and hemp.

The reactions of bast fibres to chemical and biological substances determine the durability and life cycle of bast fibre products in many end uses. In general, bast fibres have good resistance to bases and oxidants, but have weak resistance

Table 14.3. Morphological characteristics of major bast fibres.

Fibre	Longitudinal	Cross-sectional
Ramie	Twisted wide ribbon-looking with longitudinal and transverse cracks	Compressed bean shape with radial cracks; line lumen slightly open in the middle
Flax	Smooth rod with some cross markings	Irregular polygonal shape with narrow, round or oval lumen
Hemp	Similar to flax	Similar to flax
Kenaf	Smooth rod with longitudinal grooves	Compressed oval shape with oval lumen
Jute	Smooth rod with longitudinal grooves	Oval or irregular polygonal shape with wide open lumen
Cotton	Twisted ribbon-looking	Kidney and bean shape with line or oval lumen in the middle
Bagasse	Rugged surface with longitudinal lines	Similar to kenaf but with smaller lumen

Table 14.4. Bast fibre tensile property.

Fibre	Break tenacity (gf/tex)	Break elongation (%)	Tensile strength (GPa)
Flax	23.4–72.0	1.5–5	0.90
Hemp	27–63	1.5–5	0.31–0.39
Ramie	40.5–79.2	1.5–5	0.29
Jute	18.0–56.7	1–2	0.22–0.53
Kenaf	70	3.5	0.18
Cotton	27–44	3.0–9.5	0.28–0.84

Table 14.5. Bast fibre physical properties (Anding, 1978; Batra, 1983).

	Length (mm)					
Fibre	Bundle	Single	Width (micron)	Linear density (tex, g/1000 m)	Volume density (g/cm³)	Moisture regain (%)
Flax	250–1200	13–60	12–30	0.2–1.8	1.4	7
Hemp	1000–3000	5–55	16–50	0.3–2.2	1.4	8
Ramie	1500	40–250	16–64	0.5–0.7	1.4	12
Jute	1500–3600	0.8–6	5–25	1.4–3.0	1.4	12
Kenaf	900–1800	1.5–11	14–33	1.9–2.2	1.4	–
Bagasse	25–200	2–4	88	6.5–14.0	–	–
Cotton	–	15–56	14–21	0.1–0.3	1.27	8

to acids. Bast fibres are biodegradable and apt to develop mildew and rot under wet conditions.

Bast Fibre Production

Agricultural production of bast fibre crops, jute, kenaf and hemp, is distributed in different geographical regions. Jute and kenaf are warm-season, annual and cultivated almost exclusively in South-east and East Asian countries like India, Bangladesh, China, Thailand, Indonesia, Vietnam and Nepal. Hemp grows mostly in mild temperate climates. Countries producing hemp include Russia, Australia, Austria, Canada, China, Great Britain, North Korea, Hungary, Romania, Poland, France and Italy. It is illegal to produce hemp commercially in the USA, although some states allow hemp cultivation for research purposes. Flax grows well in temperate climates. Countries producing flax include Canada, China, India, the USA, northern Europe and Russia.

Bast fibre production has exhibited a significant impact on regional economy. For example, jute and kenaf used to be, and still are, considered cash crops for millions of marginal farmers in Bangladesh and India. Historically, the production of bast fibres fluctuated, reflecting the changes in market demand and industrial development. The entrance of synthetic fibres in 1970, especially the price drop of polypropylene during 1988–1990 and 1998–2000 and the increased bulk handling of agricultural commodities, has resulted in significantly reduced demand for bast fibres. Between 1998 and 2000, worldwide production of jute and allied fibres declined to an average of 2.6 million t, as compared to the record production of 6.0 million t in 1985 (FAO, *Medium-term prospects for agricultural commodities*). From 2001 to 2007, the total production of jute and allied fibres fluctuated between 2.4 and 3.2 million t. In 2007, total world production of hemp and flax was 68,839 t and 1.9 million t, respectively (source: Food and Agricultural Organization of the United Nations/Economic and Social Department/The Statistical Division, http://faostat.fao.org).

In the past decade, bast fibres have drawn more attention from the industrial world due to their biodegradable and renewable natures, which makes them potential raw materials for many industrial applications. Commercial products made from bast fibres include composite, thermoplastic and environmental cleaners. More importantly, bast fibres could become the source of biofuel production because of their rich cellulosic content and high productivity. Diversified use of bast fibres could lead to market demand for these fibres in the future.

Plant cultivation

Jute

Jute belongs to the Tiliaceae family growing primarily in tropical areas. Bangladesh and India are primary jute producing countries due to their favourable hot and humid monsoon climates. Jute is also cultivated in China, Brazil, Thailand and Myanmar, but the acreage is shrinking gradually because the return from jute cultivation has become less competitive. White jute (*C. capsularis* L.) and tossa jute (*C. olitorius* L.) are two popular cultivars. They are distinguished easily by the seed shapes at maturity. White jute has round seed pods containing 30–50 seeds and tossa jute has oval-shaped seed pods containing 100–120 seeds.

Normally, soil temperature above 15°C and day length over 12.5 h are suitable for jute sowing. The seed is broadcast at about 7.5–10.0 kg/ha for white jute and 4.5–6.0 kg/ha for tossa jute (Wood, 1995). Sowing too early causes early flowering, resulting in low productivity and inferior fibre quality. Late sowing shortens the jute growing season and full physiological and morphological maturity cannot be achieved. During early establishment, young seedlings are subject to injury caused by low temperature, waterlogging, drought and diseases. After the seedlings reach 30–40 cm high, jute enters a rapid growing stage, when it can grow as fast as 4–5 cm/day. During this stage, the plant develops strong tolerance to waterlogging and can survive 5–12 days in submerged conditions.

Primary jute root systems can grow up to 1 m vertically into the soil. Tropical climates in the jute belt include a relative humidity of 70–80%, annual rainfall of 1600 mm or more, with at least 250 mm of monthly precipitation in March, April and May, and mean temperature of 18–30°C. To adapt to soil water saturation and sometimes long periods of waterlogging conditions, the plant develops secondary roots above the soil surface and adventitious root systems on the base of stems.

Jute is a short-day annual. Day length affects fibre productivity and quality, flowering and seed production. Early flowering has been an obstacle in achieving high fibre production and quality. In some regions, it is common practice to plant a low latitude jute variety in high latitude areas in order to prolong vegetative growth. The optimum temperature for growth is 25–38°C. Jute, especially at seedling stage, does not tolerate low temperature.

Jute adapts to a wide range of soil types but grows best in fertile soils rich in organic matter with pH 6.0–6.5. Because jute is grown mostly in plain alluvial soil in the main production areas, it needs little fertilizer. Applying nitrogen, phosphorus and potassium is still an agricultural practice used to improve fibre yield and quality.

On average, jute can yield 1.7t of fibres/ha. Highest fibre yield has been observed at a planting population around 300,000 plants/ha (Wu and Zhang, 1982). With improved breeding techniques, new varieties have been developed and released for higher yielding, disease and drought resistance (Fig. 14.6). Jute fits well in rotating cropping systems with rice, maize and cotton.

Kenaf

In the late 1950s, substantial research was conducted on kenaf because of a research programme initiated by the USDA to evaluate a number of crops for possible paper pulp production. Kenaf was identified as the most promising crop for pulp and papermaking. Therefore, abundant data are available for kenaf on agricultural production, cultivation techniques, field management, processing, harvesting, transporting, storage and cropping systems.

Kenaf grows in tropical and subtropical warm climates. The optimum temperature for growth is around 25°C. The effective accumulated temperature needs to be above 3000°C for fibre harvesting and above 3300°C for seed harvesting. Kenaf is a short-day annual which can grow to a height of 5m in 4–5 months. Its yields of 6–10t (new varieties may reach 12t) of dry weight per acre per year are generally 3–5 times greater than the yield for Southern pine trees, which can take from 7–40 years to reach harvestable size (source: http://www.hort.purdue.edu/newcrop/cropfactsheets/kenaf.html).

Kenaf sowing is conducted normally by broadcasting at a seeding rate of 22.5–30.0kg/ha at a soil depth of 2cm. Plant population is an important factor affecting fibre yield and fibre quality. A study showed that a yield of 10.2t/ha could be obtained at a plant population of 270,000–300,000/ha (IJO, 1999).

Kenaf's primary root system can grow as deep as 2m in the soil. Like jute, kenaf also develops adventitious roots on the base of stems during waterlogging, which allows it to survive prolonged waterlogging. After establishment, the plant can tolerate waterlogging for a sustainable period of time. Studies showed that after growing above a height of 50cm, kenaf could survive submerged for up to 9 days (Lu, 1993).

All soils are adapted for kenaf cultivation, including acid, peat, alluvial, silt loam, sandy loam, clay loam, alkaline and saline soils. Kenaf shows a similar tolerance to salinity as cotton and performs better than wheat and maize under salinity conditions. Its wide ecological adaptability allows it to grow in marginal lands which cannot grow food crops. In addition, cultivation of kenaf helps conserve the environment and soil fertility.

Fig. 14.6. New jute and kenaf varieties developed in the Fujian Agricultural University, China: (a) jute meifeng-4; (b) Fuhong variety 992.

A study showed that cultivation of 1 ha of jute or kenaf would return 3.84 t of biomass to the soil in the form of leaves, roots and retting waste (Khandakar, 1995). Kenaf cultivation does not require a lot of agricultural inputs like field management and use of fertilizers, herbicides and pesticides.

A good collection of kenaf gemplasm has made it possible to develop a number of varieties. Breeding efforts have been focused on high biomass and fibre production, improved tolerance to unfavourable environmental conditions like drought, flooding, salinity, acidity and root-knot nematode disease. Similar to jute, early flowering is a problem in kenaf production. Developing a photo-insensitive variety has been high on the breeding agenda. Fujian Agricultural University has obtained successfully a variety with delayed flowering through sending the seeds into space. Fuhong 952 is another variety with both high yield and fibre quality developed in the same institute (Fig. 14.6). Other commercially cultivated varieties include: Everglades 41, Tainung 2 and Cuba 2032.

Hemp

Hemp is an annual herbaceous plant. The use of hemp for clothes and paper goes back at least 10,000 years in China. Because of its THC (Δ^9-tetrahydrocannabinol) content, the crop is prohibited from commercial cultivation in some countries like the USA. Commercially, hemp is cultivated for its fibres and seeds and is generally called industrial hemp. The variety grown for harvesting fibres usually has a long stem with little branching.

Hemp seeds can germinate at 1–3°C, but sowing usually takes place when the soil temperature reaches 8–10°C. A short period of low temperature will not injure young seedlings. Tolerance to low temperature allows early sowing, which has been a practice to improve fibre production. Normally, for the harvesting of fibre, hemp is sown at a seeding rate of 60–75 kg/ha.

Industrial hemp is grown best in mild climates with an optimum temperature range of 19–23°C and an annual rainfall of at least 700 mm. Mild acid soil (pH 6.0–6.5) favours hemp growth. A late maturing variety of hemp can grow to a height of 4.0–5.0 m in 160–210 days.

Hemp is a short-day plant. Growing southern varieties in the north has proved to be an effective practice to increase fibre production.

Male and female flowers of hemp are borne on separate plants with a ratio of near 1:1 in the field condition. Normally, male plants produce about 18–20% fibres on dry weight base and female plants 15–16%. Furthermore, the fibres produced from male plants exhibit higher elasticity and strength than the fibres from female plants. Therefore, male plants are preferred for fibre production and female plants for seed production. Because male and female plants mature 30–40 days apart, hemp harvesting time can be managed with more flexibility.

The use of nitrogen is relatively high for hemp cultivation as compared with other kinds of fertilizers. In cropping systems, legume species rotate well with hemp. The use of herbicide is also limited because hemp has a high planting density and dense foliage, which prevents weeds from growing.

Flax

Of the more than 100 species within *Linum*, flax is the most cultivated. It is cultivated for both fibre and oilseeds. Several varieties of oilseed flax developed by plant breeders in North Dakota, Minnesota and Canada are available. Because of its nutrient value, flax seeds have been used increasingly for food additives and ingredients for health products. Flax varieties harvested for fibres are normally taller than those harvested for oilseed and have much lower seed yields.

Although the optimum temperature for flax seed germination is 20–25°C, the seed can germinate at low temperatures of 1–3°C. Sowing can commence as soon as the soil temperature reaches 7–8°C. A seeding rate of 80–110 kg/ha is necessary to obtain a good plant stand (Wood, 1995).

Under normal conditions, flax grows to 60–120 cm, with little branching at the top. Compared to other bast fibre crops, it has a relatively shorter growing season. The crop is usually grown for 70–80 days for harvesting fibre and 90–110 days for harvesting seeds. During its rapid growing period, which is about 30 days, the plant can grow as fast as 3–5 cm/day. Sufficient water and nutrients are necessary to sustain growth during this period. On an average, flax consumes 200–215 g of water for every 0.5 kg of dry matter accumulated. Water is an important factor limiting fibre yield and quality. Breeding flax for drought tolerance has remained one of the important breeding efforts.

Since flax is a long-day plant, the fibres develop well in day lengths of 13–16 h. As a temperate crop, it requires gradual temperature increase for healthy vegetative growth. A sudden change in temperature and drastic temperature fluctuations between day and night hamper growth.

Flax grows well in alluvial and loam soils with pH 6.5–7.0. Sandy and clay soils are not suitable for its cultivation because of the low organic content or poor drainage property. The plant has moderate fertility needs, similar to those of spring oats. Flax is not very competitive with weeds because it has small leaves that do not adequately shade the ground below. It is not recommended to grow flax in fields known to have a heavy population of cool-season weeds, such as wild mustards (source: http://www.jeffersoninstitute.org/pubs/flax.shtml).

Flax should not be grown continuously on the same field to avoid disease build-up and deterioration of soil fertility. The crop fits well in rotating cropping systems with maize, soybean and sugarcane. Double cropping is also a practice to improve profitability from planting flax.

Plant harvesting

Harvesting of bast fibres is a labour-intensive process. Manual harvesting is still a common practice in areas where production is not on an industrial scale. For small farmholders, plants are mostly pulled by hand and bundled. The bundles are either left in the fields for drying and defoliation or transported to retting areas. Sometimes bark portions (ribbon) are separated by using a ribboner or decorticator, and the ribbon alone is retted.

Jute can be harvested any time after 90 days old until the seeds are mature. However, the proper time for jute harvesting is when the fruits start to form. Harvesting too early will result in lower yield and weak fibres. On the other hand, delayed harvest will result in coarse and rough fibres. Hemp male and female plants mature at different times, which provides more flexibility in managing hemp harvesting. Flax is harvested for fibre production after approximately 100 days' growth. To ensure fibre quality, the usual practice is to harvest flax when the lower two-thirds of the stem has turned yellow and the leaves have shed (Wood, 1995).

Mechanical harvesting is common for hemp and flax. However, smallholders still harvest manually by cutting the plants at 2–3 cm above the soil. The advantage of manual harvesting is the maximization of fibre yield and fibre length. The mechanization level for jute and kenaf harvesting is particularly lower than that for hemp and flax. Because of the potential for using kenaf for pulp and papermaking, mechanization of kenaf harvesting has been studied extensively by researchers in the USA and Japan. In the USA, sugarcane harvesters have been modified to harvest, process and haul kenaf. In general, mechanical harvesting is still a bottleneck for large-scale production of bast fibres, especially for jute and kenaf.

Bast Fibre Extraction

Bast fibre extraction begins with harvesting. Different harvesting approaches determine the quality of the bast fibres to be extracted. In the past, harvesting was based mainly on a manual operation. This allowed bast stems to be processed carefully. As a result, extracted bast fibres could be of the best quality (i.e. suitable for textile applications). Today, this traditional labour-intensive approach for bast fibre harvesting and processing is only affordable in developing countries. In the USA and EU countries, bast fibre production relies on automatic processing equipment. Fibre production automation renders a high productivity but, in the meantime, requires a high capital investment for fibre producers. Thus, farmers in many developing countries cannot afford this modern approach.

The flow chart in Fig. 14.7 describes the major steps involved in bast harvesting and extracting. It reveals that the major operations of bast fibre production are within the bale-to-bale processing circle. Some integrated automatic fibre processing lines are commercially available. These processing lines have the following common mechanical units: bale opening, stalk breaking, fibre scutching, sorting, line fibre cutting and bale press. When a fibre stalk bale is put on the feeding station of the line, the bale is opened and fibre stalk is spread out to form a continuous layer using an opener and divider. The fibre stalk breaking unit crushes the woody core inside the stalk into short pieces without damaging the fibre bundle. The crushed fibre stalk is fed through retaining rollers to a scutcher to separate the fibre and broken woody core (shives). The scutcher is a rotating drum with spring steel blades mounted on its periphery. A basic requirement for scutching is to keep a full length of fibre bundles. To enhance the efficiency of fibre separation and cleaning, usually more than two drums are used in a processing line for performing rough and fine scutching. The purpose of sorting is to separate scutched short fibre (tow fibre) and shives from full-length fibre (long staple line fibre). Horizontal or angled shaking screens are used for removing short fibre and shives during fibre bundle transporting. After the sorting process, line fibre can be sent directly to a bale press for packing, or can be cut into short staple fibre and packed with a bale press. Short staple tow fibre can also be pressed for packing after further cleaning and carding.

Bast Fibre Characterization

After the bast fibres are extracted from plant stems, they are packed and shipped to purchasers for downstream processes. To determine the fibre market value, bast fibre quality needs to be graded before entering the market for trading. Characterization of the bast fibres involves many physical attributes and testing methods. Among the most important parameters are fibre dimensions, fibre strength, fibre colour, maturity and trash content. However, in the manufacture of natural fibres, industrial standards and advanced testing instruments are developed mainly for cotton and wool, because the cotton and wool industries are well established. For bast fibre producers and end-users, there is still a lack of industrial standards and measuring equipment commercially available for evaluating fibre quality. The following subsections discuss briefly some basic approaches for characterizing bast fibres.

Fibre physical dimensions

Fibre length and fineness are two primary measures for bast fibre dimensions. As described before, the fibre length of a single bast fibre or bast fibre staple is on a millimetre scale, mostly within 100 mm. In current industrial practice, the measurement of bast fibre length still relies on manual measurement because there are no automatic instruments designed especially for bast fibre application.

Fibre fineness refers to the cross-sectional size of bast fibres. This size is on a micron

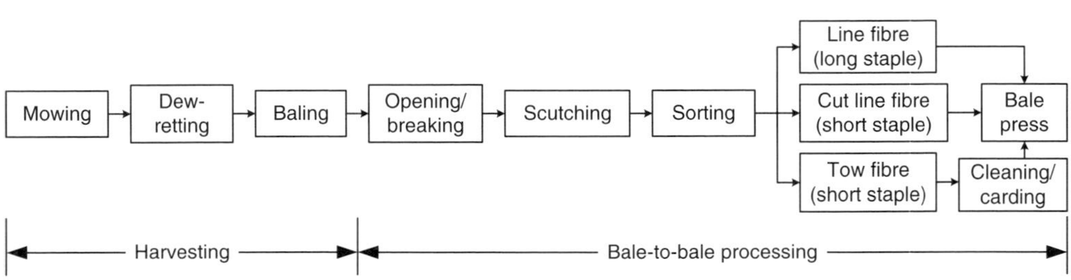

Fig. 14.7. Processing of bast fibre.

scale and is often measured by lateral width (thickness), diameter, or cross-sectional area. In most cases, the cross-sectional shape of the bast fibres is irregular and varies along the fibre longitudinal direction. Therefore, determination of the fibre diameter and cross-sectional area becomes difficult. A practical way of solving this is to measure fibre width using a microscope. Some free software packages for basic image analysis, such as Scicon, can be used for manual measurement. To improve testing efficiency, some products of advanced image analysis software are available for automatic and customer-defined measurements.

A major commercial instrument capable of measuring fibre cross-sectional diameter and area is the DIA-STRON's dimensional measurement system. This instrument was designed originally for wool and hair measurement and has now extended to plant fibres for evaluating the fibre cross-sectional area. It uses a scanning laser micrometer to measure the mean diameter of a single fibre by rotating the tested fibre sample at a measuring point. The laser beam can scan a maximum distance of 24 mm along the fibre axis.

Fibre linear density is an alternative for describing fibre fineness. This term is widely adopted by fibre producers, fibre traders and end-users because of a simple test method and statistically robust test result. Determination of the fibre linear density can refer to ASTM D1577-96, a direct weighing method to measure the fibre linear density in accordance with its definition. For bast fibres, however, the test should be conducted under the standard temperature–humidity condition. ASTM D7025-04 provides a guide to assess clean flax fibre fineness using an airflow instrument. This standard testing method can also be applied to ramie, kenaf, jute and hemp fibres.

Fibre tensile strength

Fibre tensile strength not only is the most important mechanical property, but also plays a critical role in fibre quality grading. Once again, unlike wool and cotton fibres for which some advanced automatic instruments for tensile strength test have already been developed, bast fibres rely largely on a universal tensile tester for evaluating fibre tensile property. The fibre tensile test can be performed by clamping a single fibre or a fibre bundle. By breaking a single fibre, tested results will have small deviation, but the testing is time-consuming. By breaking a fibre bundle, it helps to reduce testing time, but the tested data will involve a large variation.

ASTM D1294-05 defines a standard method for testing the tensile strength of wool fibre bundles using a universal tensile tester. This method can be a reference for evaluating the tensile strength of bast fibres. In particular, the procedure of preparing test fibre specimens is very useful. Meanwhile, there is also a standard test method (ASTM D1445-05) developed for measuring breaking strength and elongation of cotton fibre bundles using the Pressley instrument or Stelometer. In some cases, this standard method can also be used for bast fibres in the practice of fibre strength evaluation.

Fibre colour and cleanness

Fibre colour is a visual attribute of raw bast fibres. It varies depending on fibre species, growing regions, production technologies and processing stages. Overall, fibre colour reflects some important aspects of fibre product quality. Therefore, the fibre colour is also an essential factor included in the grading criteria of raw bast fibres. Evaluation of fibre colour can be subjective or objective. The subjective way is a traditional way that relies on a group of trained inspectors to make judgements by comparing raw fibres to reference fibre samples. This approach becomes more and more inconvenient and costly for the highly competitive and dramatically consolidated fibre industry and fibre-related business. With the latest technological development of colour measurement, fibre colour is no longer numerically immeasurable. Many colour measurement instruments, like those produced by Datacolor, HunterLab and Minolta, can be used to meet routine needs for fibre colour testing with affordable costs. As a result, fibre colour grading can be implemented on a digital basis and fibre colour information can be communicated with numerical values. Some industrial standards such as ASTM E308, ASTM E313, AATCC Procedure 6 and CIE15.2 provide further information about different colour scales and colour measurement procedures. In addition to fibre colour, the fibre surface may contain some spots that are not removed during fibre processing. These spots can affect fibre visual quality negatively.

Fibre cleanness refers to the impurity that remains in the fibre bundles. Impurity contents include plant leaves, stem particles, soil, trash, etc. To obtain the highest fibre grade, these impurity substances need to be removed thoroughly during fibre extraction. On the other hand, extra fibre cleaning tends to increase processing cost.

Bast Fibre End-use Applications

Apparel fabric production

Use of bast fibres for apparel fabric production is a high-end application. It can yield high-value fashion products. But, in the meantime, it requires best-quality bast fibres (finer, stronger and cleaner, with uniform length). Ramie and flax are two favourite bast fibres for apparel application because of their characteristics suitable for textile processes. They can be used either to produce pure ramie and flax fabrics or to produce multifibre blended fabrics, depending on the different end-uses. For example, pure flax fabrics (linen) are well known for tablecloths. Ramie-cotton or ramie-rayon blended fabrics are often used for casual apparels. The ramie-rich and flax-rich apparel fabrics provide excellent moisture absorption and a cool feel for summer apparel products.

The textile processes of apparel fabric production are illustrated in Fig. 14.8. This long processing chain usually involves yarn mills, fabric mills, the wet processing industry and garment manufacturers. When bast fibre bales (short staple from either cut line fibre or tow fibre) enter yarn mills, the first processing stage is fibre preparation. This includes bale opening, fibre cleaning and fibre blending (e.g. an addition of cotton or other synthetic fibres to blend with a bast fibre). The cleaned fibre blend is then fed into a carding machine for combing. Through the carding process single fibres become more parallel to each other in the bundles; impurities are further removed and the fibre blend is mixed more evenly. Out of the carding machine, the fibre bundles become a long fibre strand called a sliver. In the spinning stage, the fibre sliver is fed into a drawing machine for drafting. Fibre sliver size is reduced significantly by mechanical drafting. After drawing, the thin size sliver is fed into a roving frame to reduce the sliver size further to a suitable size for spinning. A spinning machine is used to convert the fibre sliver into a yarn with specific yarn counts. Fabric mills use yarns to make apparel fabrics through weaving or knitting processes. The fabrics from fabric mills are called grey cloth and need to go through dyeing and finishing before they can be sent to apparel producers for making up. Fabric dyeing and finishing is always a wet process that requires the consumption of a large volume of water and energy, and heavy use of dyestuff and chemicals. Because the textile wet processing industry is capital-intensive and environment-sensitive, the promotion of bast fibre for textile applications is a balance issue of profitability, ecological responsibility and economic development.

Non-woven fabric production

Bast fibres are not only a natural resource for making textile products, but also an important

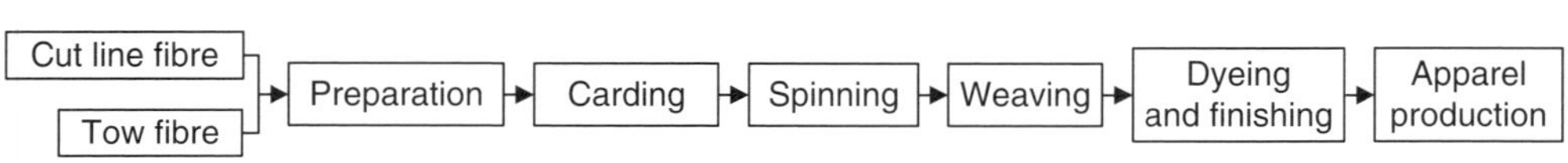

Fig. 14.8. Processing of bast fibre for textile and apparel applications.

type of feedstock for making industrial materials. They are ideal for non-woven fabric production. Non-woven fabric production using bast fibres features a shorter processing route and higher productivity. In most cases, it does not need dyeing and finishing, so that the production approach is eco-friendly. Figure 14.9 describes the major approach to non-woven fabric production. The first stage is web forming. There are three methods of making bast fibre webs. The carded web is produced using a carding machine similar to the equipment used in yarn production. The carding technique suits most bast fibres and enables the production of high-quality staple non-wovens. It is simple and cost-effective and therefore is the most popular way for staple fibre web formation. The airlaid web is formed by an aerodynamic system that can suck airborne fibre to deposit on a revolving condenser screen where a continuous random fibre web structure is formed. The airlaid method is suitable for processing low-quality bast fibres and, overall, the web strength is not high. The wetlaid web is formed by an aqua system in which the fibres are dispersed and swollen. The aquaborne fibre is pumped on to a moving screen with water filtering. A continuous fibre deposit on the screen forms a random fibre web. A squeeze machine and a dryer are needed to dehydrate the wet web. The wetlaid technique is similar to the paper-forming method and is suitable for processing short staple fibre that usually cannot be processed by the carding and airlaying machines.

After the web formation, fibre webs need to be bonded mechanically or chemically to meet end-use specifications in strength, thickness and density. Needlepunching and high pressure water-jetting are widely used mechanical bonding techniques. Calendering is a typical thermomechanical bonding method using polymer fibre or powder as a binder. Impregnating with liquid binders is a chemical bonding approach. Figure 14.9 illustrates the needlepunching process. On the needlepunching machine, a metal plate-mounted needle array moves up and down to allow needles to pass through the web in a direction perpendicular to the web surface. After the needlepunching, non-woven fabrics are produced and are wound as roll goods for distribution.

Applications of bast fibre non-woven fabrics cover a wide range of industrial sectors, such as agriculture, automotive and construction. For instance, thin bast fibre non-wovens can be used as a soil covering material (nourishing layer) in farm fields. Dense bast fibre non-wovens can be used as noise barriers or noise absorbers in the building industry. Research has demonstrated that the use of bast fibres for automotive composite applications has many advantages in both technical and economic aspects. These bio-based composites can enhance mechanical strength and acoustic performance, reduce material weight and processing time, lower production costs, improve passenger safety and shatterproof performance under extreme temperature changes, and improve biodegradability for the auto interior parts. The following example exhibits the conversion of a bast fibre non-woven into a 3D auto interior part. On a stamp forming press illustrated by Fig. 14.10, a 50/50 flax/polypropylene non-woven felt (1200 g/m^2) is mounted on

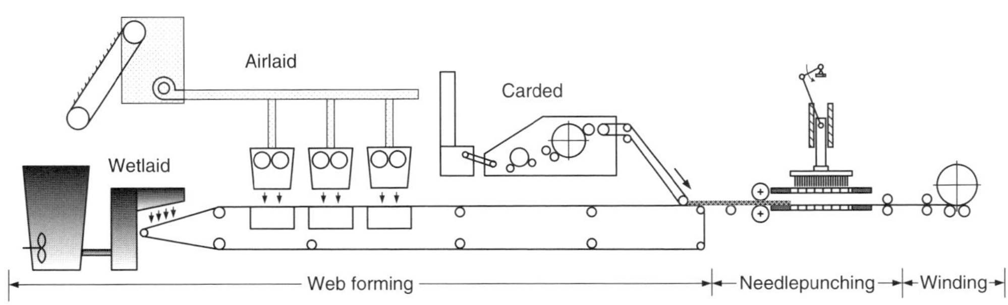

Fig. 14.9 Bast fibre non-woven fabrication.

a frame in an IR heating area for preheating 5 min at a temperature of 180°C. After preheating, the sample is put in a die for moulding. During this stamp press, the counter die keeps the heating temperature of 180°C and compresses the sample with a 3 MPa (30 bars) pressure for 15 min. Figure 14.11 shows the produced 3D composite part.

Bioenergy and bio-based product production

Current US energy supplies include 39% share from petroleum, 24% share from natural gas, 23% share from coal, 8% share from nuclear and 6% share from renewable resources (Lana *et al.*, 2004). According to the national energy policy described in the US 2002 Farm Bill, a vision goal for the increase of bioenergy and bio-based products was set as this: 10% biofuel share and 18% bio-based product share by the year 2020 (Biomass Technical Advisory Committee, 2002). In today's economic stimulation for the biomass industry, more and more attention is drawn to new technologies for converting lignocellulosic biomass into biofuels, biopolymers and consumer chemicals. Focuses have been on cellulose, hemicelluloses and lignin, the three major polymer components in lignocellulosic biomass. Bast fibres, being very rich in these chemical contents, are becoming an important type of energy feedstock for producing biofuels and bio-based non-textile products. Bast fibre feedstock most probably will be used for combustion because it has high cellulose content and high-grade feedstock quality.

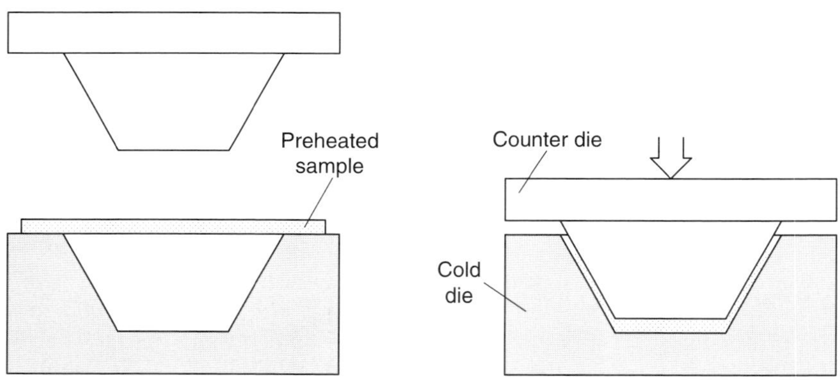

Fig. 14.10. Stamp forming press.

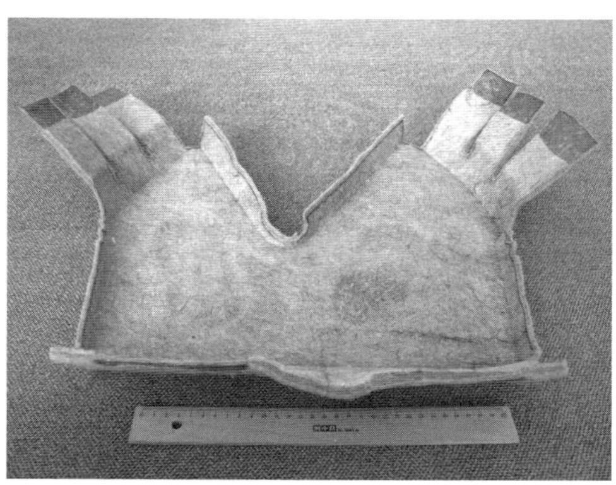

Fig. 14.11. Bast fibre composite part.

Bast fibre crop by-product utilization

The production of bast fibres produces two major by-products: bast fibre crop seed and woody stem core. These by-products are also a good biomass resource and can be utilized for making value-added products. Fibre seed nut and flour are favourite ingredients for healthy food because of rich nutritional contents in protein, fatty acid, vitamins and dietary fibre. For example, flax seed is sold on the Internet at an average price of US$14.50/kg. Fibre seed oil is widely used for food products (salad oil, margarine, etc.), nutraceutical products (such as flaxseed oil, Internet price US$18.53 for 250 gels), personal care products (shampoo, hand cream, etc.), technical products (lubricant, solvent, coating, varnish, printing ink, etc.) and biofuel products (bioethanol and biodiesel). The woody core separated from fibre stem can also be used for livestock feed, animal bedding (litter), mulch, oil/chemical absorbent, building insulator and packaging material. A commercial application example is a kenaf core-based absorbent, Kenaf Absorb, that has been produced for cleaning up oil/chemical spills in oil fields and refineries. This product is distributed by Fisher-Stevens, Inc.

Summary and Further Trends

The bast fibres are the oldest natural fibres in history for human cultivation and consumption. The production of bast fibre crops is still a major livelihood for many people living in undeveloped areas and an essential resource for rural economic development. However, the bast fibre industry is not well established compared to the cotton and wool industries that were developed in the industrial revolution era of the 1700s and to the man-made fibre industry invented in the 1950s. A major reason for this is that the production of bast fibres is an agriculture-based industry suitable for developing in countries with a large population of farmers. At present, the world's total production capacity of bast fibres is still largely dependent on small-scale farmers and processors scattered in many of the developing countries. Even within agriculture, the development of bast fibres is always behind food products, the first priority of agriculture. Another reason is that the manufacture of man-made fibres is a petroleum-based industry requiring advanced chemical processing technologies. The huge petrochemical industry in developed countries ensures the production of man-made fibres with high productivity, low cost, engineered quality and customized end uses. In contrast to this, the current state of the art for bast fibre production is far-fetched to be able to compete with the technologies developed for the manufacturing of man-made fibres. The impact of bast fibres on today's economy has been diminishing over the past half century.

Although the consumption of bast fibres in apparel textile applications is less than that of cotton and wool, there are many industrial niche areas where bast fibres have a stronghold. The use of bast fibres for non-woven composite fabrication is a typical example. Other industrial end uses cover agricultural mulch, horticulture potter, automotive interior composite, building insulator, noise absorber, marine rope, paper pulp and thermoset moulds. With the current economic focus on renewable energy and bio-based products, more effort is being made to find new and sustainable biomass feedstock for the ever-increasing need to produce more bioenergy and bio-based products in place of petro-based energy and consumer products. This 'going green' initiative opens a new opportunity for bast fibres to restore their role as an important renewable resource for a bio-based economy. Future technological development still seems multidirectional, but the following progress trends can be envisaged.

In the technical aspect of bast fibre cultivation, more focus will be on plant genetics and molecular biology for producing different grades of bast fibre crops. As for bast fibre crops for fibre production, research is needed for the modification of cellulose chain arrangement and amorphous matrix bonding, so that single bast fibre uniformity can be improved and fibre bundles can be separated more easily. In terms of energy-grade bast fibre crops, the emphasis will be on molecular modification that enables cellulose chains to break easily during hydrolysis and glucose fermentation to take place easily.

With respect to the technology innovation of bast fibre extraction, more efforts will be needed on the development of highly efficient, flexible and affordable automatic equipment that can be used in harvesting and bale-to-bale processing. Current products of automatic equipment for fibre harvesting and processing are limited and capital-intensive. In addition, for trading and quality control needs, standard testing methods and computerized instruments for bast fibre evaluation need to be developed urgently. Without industrial standards and commercial testing equipment, the bast fibre industry lacks competitiveness with other fibre industries.

Finally, in cellulose biomass research and development, the challenge is whether bast fibre crops can become competitive energy feedstock in the production of biofuel products. At present, research foci are mainly on converting agricultural residues (maize stover, bagasse, woodchips, etc.) into bioethanol, biodiesel and other energy products. Research efforts are needed to investigate the feasibility of bast fibre crops in bioenergy conversion. If bast fibre crops are proven to be advantageous cellulose feedstock for producing biofuels and bio-based products, it will lead to a renaissance of the bast fibre industry.

References

Anding, R.G. (1978) *The Microbial Generation of Methane from Alkali Treated Bagasse*. MSc thesis Louisiana State University, Baton Rouge, Louisiana.

Batra, S.K. (1983) Other long vegetable fibres: abaca, banana, sisal, henequen, flax, hemp, sunn, and coir. In: Lewin, M. and Pearce, E.M. (eds) *Handbook of Fibre Science and Technology*. Marcel Dekker, New York.

Batra, S. and Turner, A.J. (1998) The structure of textile fibres. In: Jarman, C. (ed.) *Plant Fibre and Processing: A Handbook*. Intermediate Technology Publications, Rugby, UK.

Biomass Technical Advisory Committee (2002) *Vision for Bioenergy and Biobased Products in the United States* US Biomass Technical Advisory Committee, Washington, DC.

Cook, J.G. (1960) *Handbook of Textile Fibres*. Merrow Publishing, Watford, UK.

FAO (2003) Medium-term prospects for agricultural commodities (http://www.fao.org/docrep/006/y5143e/y5143eg.htm).

FAO (2003) The production and consumption of kenaf in China (Rep. No. ESC-Fibres Consultation No 03/6). Food and Agriculture Organization of the United Nations, Salvador, Brazil.

FAO (2009) FAOSTAT – Crop production (http://faostat.fao.org/site/567/default.aspx#ancor, accessed 7 July 2009).

Franck, R.R. (2005) Overview. In: Franck, R.R. (ed.) *Bast and Other Plant Fibres*. Woodhead Publishing Ltd, Cambridge, UK, pp. 1–22.

Gibson, K. (2005) Hemp: a substance of hope. *Journal of Industrial Hemp* 10, 75–83.

IJO (1999) Final report – yield performance and cost of jute cultivation in Bangladesh, China, India, Nepal, and Thailand. International Jute Organization, Dhaka.

Karus, M. and Kaup, M. (2002) Natural fibres in the European automotive industry. *Journal of Industrial Hemp* 7, 119–131.

Khandakar, A.L. (1995) *Jute Cultivation and the Environment*. Bangladesh Jute Research Institute, Dhaka.

Kozlowski, R., Rawluk, M. and Barriga-Bedoya, J. (2005) Ramie. In: Franck. R.R. (ed.) *Bast and Other Plant Fibres*. Woodhead Publishing Ltd, Cambridge, UK, pp. 208–209.

Lana, P., Bull, S.R., Pacheco, M.A. and Benner, J.P. (2004) DOE/National Association of State Universities and Land Grant Colleges (NASULGC) Biomass and Solar Energy Workshops (http://www.nrel.gov/docs/gen/fy04/36831.pdf, accessed 23 July 2009).

Leslie, M., MacDonald, S. and Kiawu, J. (2008) Cotton and wool outlook (http://usda.mannlib.cornell.edu/usda/ers/CWS//2000s/2008/CWS-12-12-2008.pdf, accessed 23 July 2009).

Lewin, P.M. and Pearce, E.M. (1985) Fibre chemistry. In: Batra, S. (ed.) *Handbook of Fibre Science and Technology*. Marcel Dekker, Inc, New York.

Li, Z. (1982) *Cultivating Technique for High-yield Ramie (Chinese)*. Hunan Publishing House of Science and Technology, Changsha, China.
Lu, H.R. (1993) *Cultivation of Fibre Crops in China*. China Agriculture Publishing, Beijing.
Oplinger, E.S., Oelke, E.A., Doll, J.D., Bundy, L.G. and Schuler, R.T. (2009) Flax (http://www.hort.purdue.edu/newcrop/afcm/flax.html, accessed 22 July 2009).
Ouyang, X. (2007) Current production capacity of ramie in China. Personal communication.
Suzuki, M., Ito, M., Sasaki, Y., Takano, M. and Konishi, Y. (2003) Development of a mechanized cultivation system for the resource plant – kenaf. In: Liu, A. (ed.) *International Development of Kenaf and Allied Fibres*. CCG International, St Paul, Minnesota.
Thames, S.F. (2005) Hemp as a potentially important crop and area research (http://naihc.org/Thames0305.pdf, accessed 22 July 2009).
USDA (2000) Industrial hemp in the United States: status and market potential (Rep. No. ERS AGES No. 001). USDA Economic Research Service, Washington, DC.
Wikipedia, the free encyclopedia (2009) Jute (http://en.wikipedia.org/wiki/Jute, accessed 22 July 2009).
Wood, I.M. (1995) *Fibre Crops – New Opportunities for Australian Agriculture*. The Australian Department of Primary Industries, Queensland, Australia.
Wu, X. and Zhang, M. (1982) Technique of cultivating jute for high yield. *China Fibre Crops* 2, 1–3.

15

Bast Fibre Processing and Uses

Holger Fischer and Jörg Müssig

Introduction

Bast fibres as a renewable resource are good candidates for usage in high-performance composite materials, as well as in yarns for textile production. Bast in the form of long fibre bundles has a long tradition in the textile industry but, at present, has only limited use in this sector. Besides the economic crisis of the textile industry, the availability of bast fibre of a suitable quality and adequate price is a problem.

The variety of fibres in nature is tremendous. Depending on the function of natural fibres, these entities are located in different parts of the plant. The most important plant fibre is cotton. A cotton fibre as a seed hair from the point of view of botanical classification is not a fibre but a trichome. Natural fibres can be found in the stems of monocotyledonous and dicotyledonous plants and in various positions in dicotyledonous and gymnosperm trees (Eder and Burgert, 2010). Another category of plant fibres is called mesocarp fibres, like coir. Figure 15.1 gives an overview of the 'natural fibres' that can be used for technical applications.

The Nova-Institute (Hürth, Germany) carried out a series of surveys to collect reliable data on the market for natural fibres as composite materials in the Austrian and German automotive industry. The result is summarized in Fig. 15.2. It depicts clearly that the usage of natural fibres (not including cotton and wood fibres) is increasing permanently, although the growth rate dropped to less than 3% in 2004 and 2005. In 2005, the use of natural fibres (except cotton and wood) reached 19,000 t for the first time. The share of the different fibre types has varied over the past few years due to price fluctuations: European flax reached a high price level during 2002 and 2003, resulting in increased use of jute, kenaf, sisal (*Agave sisalana*), coir and abacá, and later stagnated due to decreasing flax prices.

Based on the data of the association of the German automotive industry (VDA), 5.2 million passenger cars were produced in 2004 (VDA, 2005) and 5.4 million in 2005 (VDA, 2006) in Germany alone. Combined with the data given in Fig. 15.2, it is possible to calculate the average use of natural fibres per vehicle: 3.6 kg/car in 2004 and 2005 (Karus *et al.*, 2006). The total statistics are listed in detail in Table 15.1. If this usage is enlarged to other countries and models, the potential for natural fibres is about 80,000–160,000 t/year, calculated from 16 million vehicles produced yearly in the EU (Müssig *et al.*, 2003).

Textile production from bast fibre usually involves mechanical decortication and subsequent chemical separation in NaOH. The resulting fibre bundles are further

  Industrial Crops and Uses (ed. Bharat P. Singh)

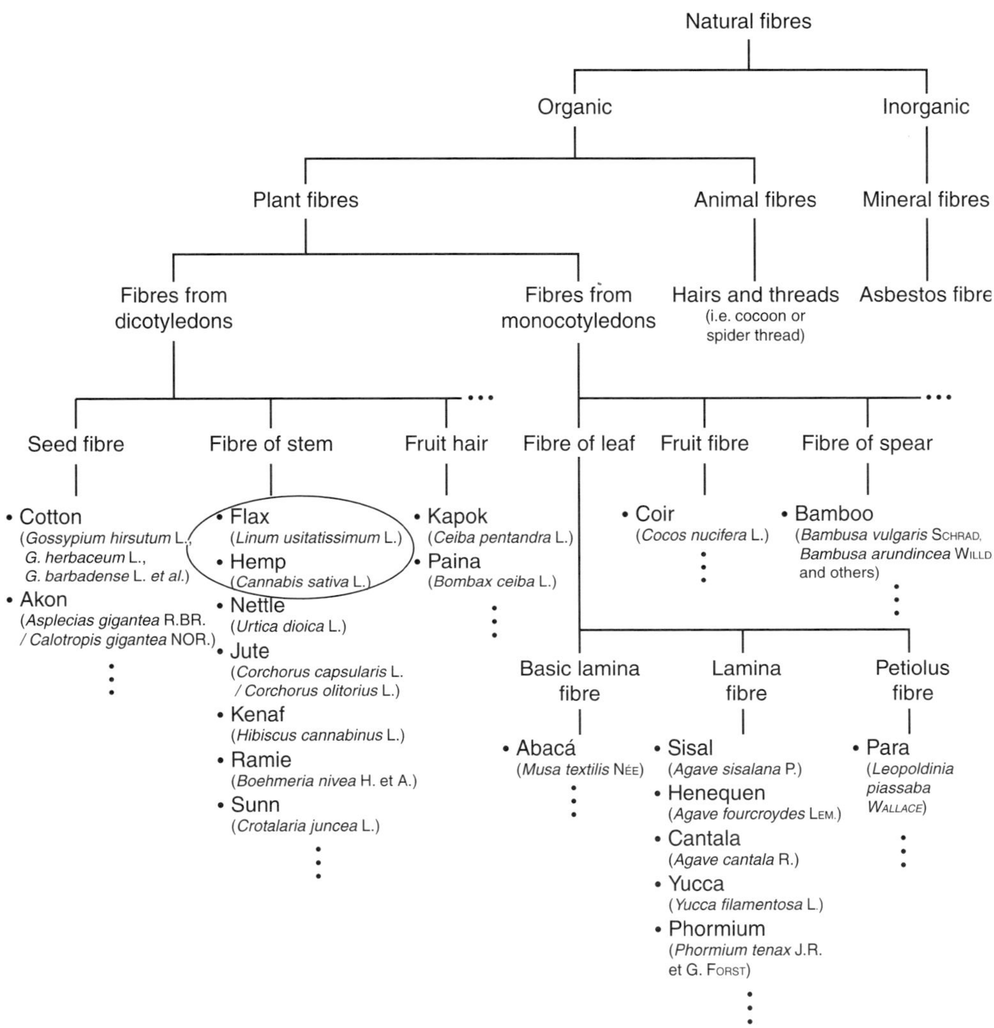

Fig. 15.1. Overview of natural fibres (adapted from Müssig, 2001b).

processed with conventional wool or cotton spinning systems. This process consumes large amounts of water, chemicals and energy. Microbial (Leupin and Leupin, 2002; Amaducci *et al.*, 2005) or enzymatic separation with potentially low energy consumption have been suggested alternatives to the current processing procedure.

The main task in processing is to dissolve the pectin species connecting the single fibres in order to reduce the size of the fibre bundles. Recent and ongoing projects have considered various aspects of the production chain of hemp for textile destination. In the EU project HEMP-SYS (2005), a production chain for hemp textile based on 'green decortication' (processing of hemp stems not retted in the field) and biological retting of the fibre bundles has been developed. The results of this project indicate that the production of hemp textile is possible with long fibre bundles in a traditional wet spinning process, while the possibility of using shorter fibre bundles with a staple length of wool or cotton fibres should also be explored. It has been observed that microbiological retting results in a good separation and refining of the bundles, but a subsequent treatment has to be carried out to

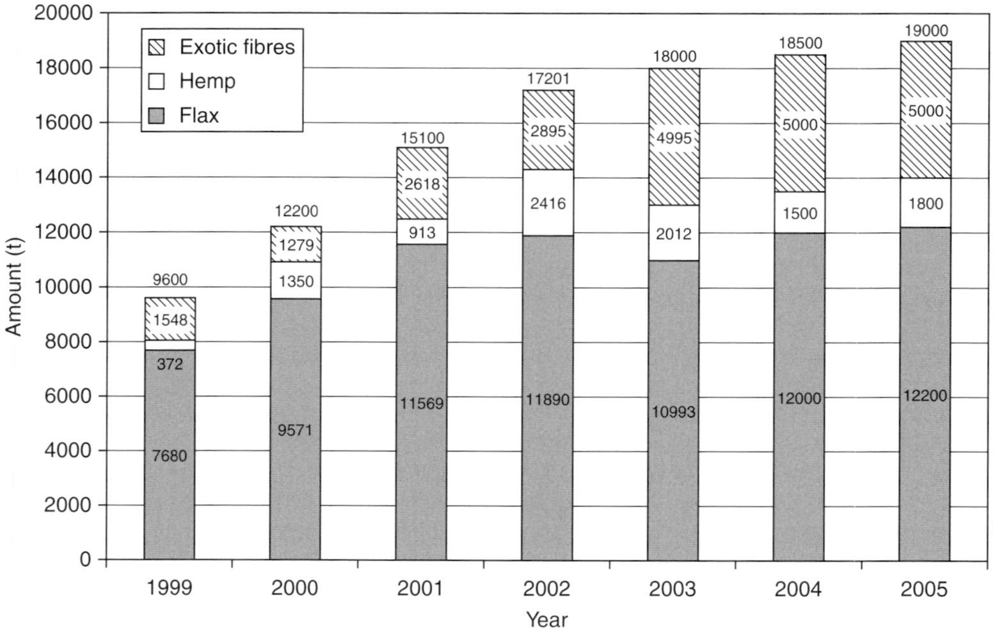

Fig. 15.2. Usage of natural fibres (except wood and cotton) for composites in the automotive industry (data from Karus *et al.*, 2006). Total amounts given above the bars.

Table 15.1. Use of natural fibres in the German automotive industry in 2005 (data from Carus *et al.*, 2008).

Fibre type	Main applications	Fibre use (t)	Average natural fibre fraction in composite	Natural fibre reinforced composites (t)	Average use of natural fibres per vehicle (kg)
Natural fibres (plant fibres except cotton and wood)	Interior space (middle- and upper-class passenger cars)	19,000	*c.*30–65%	*c.*30,000	3.3 (only passenger cars: 3.6)
Wood fibres	Interior space (passenger cars and trucks)	*c.*27,000	*c.*70%	*c.*40,000	*c.*4.5
Recycled cotton	Interior space (passenger cars) and driver's cab (trucks)	*c.*45,000	*c.*71%, resp. 57%[a]	*c.*79,000	*c.*8.2
Total	Trucks and passenger cars	*c.*90,000	–	*c.*150,000	*c.*16

[a]The fibre fraction in composites with recycled cotton is *c.*71%. This consists of est. 80% cotton, 5–10% wool and 10–15% other fibres. Thus, the resulting share of cotton is 57%.

remove residues of degraded material that encrust the fibres.

When natural fibres are used in composites, the fraction in the material depends on the type of processing. The share of natural fibres, except cotton, in composites using natural fibre thermoplastics, natural fibre thermoset and wood fibre thermoset processing technology in 2005 averaged 46.2, 55.2 and 84.7%, respectively (Karus *et al.*, 2006). Bearing in mind that the European automotive industry uses only coarse separated European bast fibre bundles in low-price applications (press moulding, thermoplastics), a better separation would become essential to fulfil the needs of advanced applications and injection moulding.

Bast Fibre Structure

A fibre in a plant structure is the smallest intact unit (Vincent, 2000). These units are rarely found as individual cells in the plant, but are mostly assembled into bundles. Müssig and Martens (2003) mention that testing a fibre or a bundle can lead to very different strength values. Based on this knowledge, it is essential to differentiate appropriately between fibre and fibre bundle. Figure 15.3 gives an overview about what is meant by a single fibre and a fibre bundle. For flax and hemp, the single fibre has a mean width of about 20 μm and a length of approximately 20 mm.

Cleaning raw bast simultaneously and achieving finer qualities is possible by enzymatic processes (see the section on hemp fibre separation later in this chapter). An intensive screening or exact knowledge of the pectin structure is necessary in order to choose an optimal enzyme mixture with a high pectinolytic effect.

Investigations on the structure of flax pectins were carried out, for example, by Mooney *et al.* (2001) and Morvan *et al.* (1989, 1990a,b). The cross-links between flax fibres are reported as a network of β-(1,4)-glucans, β-(1,4)-galactans and β-(3,6)-galactans (Morvan *et al.*, 2003), whereas the hemp pectins are identified as polysaccharides containing galacturonic acid, rhamnose and galactose units in variable ratios, with a disaccharide repeating unit consisting of α-rhamnose and α-galactose (Vignon and Garcia-Jaldon, 1996).

Another topic that has to be mentioned in this context is the cross-linking of pectins

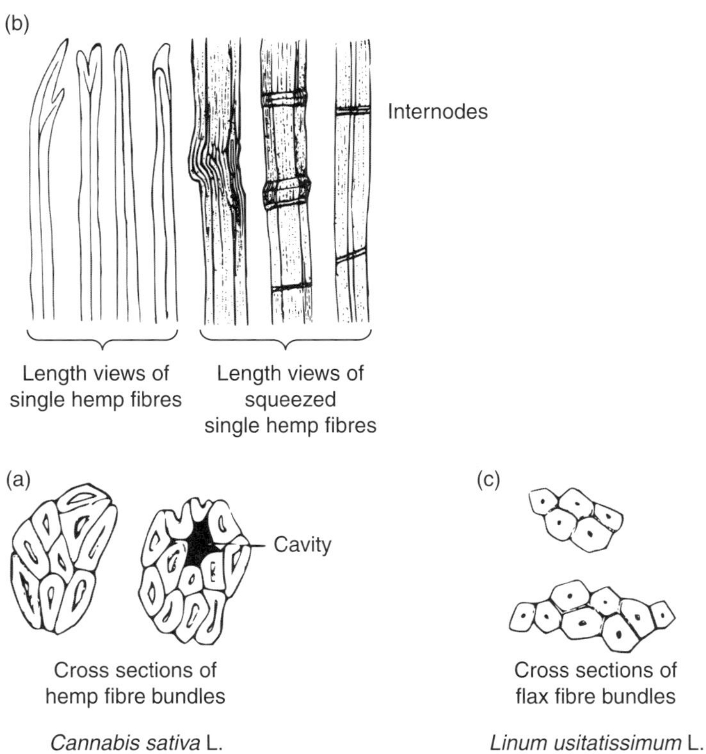

Fig. 15.3. (a) Cross section of hemp fibre bundles; (b) length views of single hemp fibres with and without internodes; and (c) cross section of flax fibre bundles.

by connecting carboxyl groups with Ca^{2+}. Consequently, research work has been carried out to examine the effect of chelating agents in the separation of flax (Adamsen *et al.*, 2002a,b). EDTA (ethylenediaminetetraacetic acid) as a chelating agent is so effective in the dissolution of flax pectins that this process is used as an analytical method to measure the fibre content (van den Oever *et al.*, 2003). On the other hand, treatment of hemp with EDTA under identical conditions has no significant influence on bundle refinement (Fischer *et al.*, 2006a).

This difference makes it clear that the treatment of hemp or flax by the same enzyme will not necessarily produce the same results. Even for the well-studied process of enzymatic flax retting, it is not clear which enzymes are required for an efficient retting process, since there is no good correlation between pectinase activity and retting efficiency (Zhang *et al.*, 2000). Hemp treatment by pectinolytic enzymes like BioPrep™ 3000 L (Fischer *et al.*, 2004a) or Texazym™ BFE (Fischer *et al.*, 2008a) has been shown to bring about a reduction of fibre odour. This interesting effect presumably is caused by the reduction of pectin content.

The effectiveness of the enzymatic hemp treatment is improved substantially by a preceding soda treatment. Soda as a first reaction step has a chemical cleaning effect at the fibre bundle surface. This enables the enzyme in the next step to act directly on the pectin layers between the fibre cells (Fischer *et al.*, 2006a). Without the soda pretreatment, the high specificity of the enzyme becomes a disadvantage.

Hemp fibre is reported to contain about 1.9% pectin, 16.1% hemicellulose and 3–4% lignin (Krässig *et al.*, 1989; Crônier *et al.*, 2005; Gutiérrez *et al.*, 2006). The chemical separation process (i.e. alkaline boiling) removes the pectins and reduces the lignin content of the fibre bundles substantially (Wang *et al.*, 2003; Qu *et al.*, 2005). Complete removal of pectin from hemp by enzymatic treatment using a pectate lyase (EC 4.2.2.2) has been reported (Ouajai and Shanks, 2005).

Separation Techniques

The separation process from the stalks to the final fibre bundles or single fibres starts with processing in the field. After harvest and a possible dew or water retting process, the stalks are collected and transported to the decortication and separation facilities. One should distinguish between decortication and separation. While in the decortication process, the wooden part of the stalk and the fibre bundles are separated from each other; the separation process is used to refine the fibre bundles. There are many different possible separation processes but only the mechanical and, to some extent, the chemical separation are commonly used in the industry (Fig. 15.4). The microbiological process takes place in the traditional retting processes like dew retting or water retting, while the enzymatical separation as an industrial implemented process is in a transition state from research to industry.

The biotechnological treatment of stalks as an industrial process at moderate temperatures and with a low environmental impact can be a good and sustainable alternative. Such processes have been proven to work at the laboratory and the pilot plant scale with some commercial and special enzymes (Schlüter *et al.*, 2003; Fischer *et al.*, 2004a), as well with selected strains of bacteria inoculated in controlled retting pools (TOSCANAPA, 2003).

Enzymatic separation of bast fibre bundles has been carried out at the laboratory scale by different researchers. There are some reports of successful enzymatic separation of ramie, numerous work has been carried out on flax by Akin and co-workers (Akin *et al.*, 2001, 2002; Adamsen *et al.*, 2002a), but only a small number of articles can be found reporting on enzymatic hemp separation (Buschle-Diller *et al.*, 1994; Dreyer *et al.*, 2002).

In order to help readers develop a comprehensive understanding of the separation techniques, it is essential to confine the description to one plant species, as plant fibre characteristics vary greatly between species and the approaches taken also vary accordingly. We have selected hemp for this deliberation since

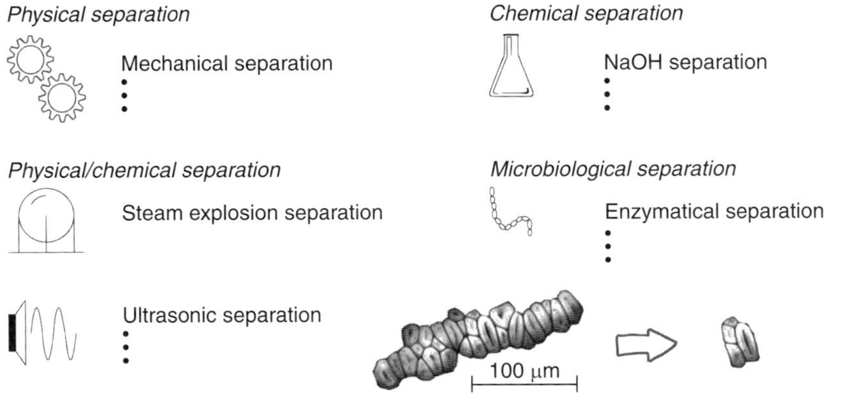

Fig. 15.4. Separation techniques to separate and refine fibre bundles (adapted from Müssig, 2001a).

our laboratory has been involved intensively in researching this crop.

Hemp Fibre Separation

In order to achieve a better comparability, in this chapter data will be presented based on experiments using identical raw material. Three lots of hemp (variety Fedora, grown near Neerstedt, Germany) were used: Hemp01 (grown 2000), GDE02 and KGE02 (both grown 2001). After field-retting the stems were decorticated and coarse separated on a DEMTEC™ line with four drums (Demaitre B.V., Belgium) in 2001 (Hemp01) and 2002 (GDE02, KGE02) at the company AGRO-Dienst (Neerstedt, Germany). A detailed description of the industrial equipment was published by Müssig (2001a). The general mechanical and morphological characteristics of these raw fibre lots are listed in Table 15.2. Descriptions for these analyses are published as follows: fineness as OFDA mean value (Baxter *et al.*, 1992; Drieling *et al.*, 1999); collective strength by Stelometer (Müssig *et al.*, 2003); A_{1000} value (Bluhm and Müssig, 1999) and pure fibre content according to Bredemann (1942).

In order to obtain data comparable to fibres in industrial processing, the fibre bundles were coarse separated preceding each fineness or strength analysis. This is necessary because fibre bundles after chemical or enzymatical separation have weak remaining internal bonds. This makes them appear as coarse as before processing, although these bonds are directly opened in the industrial carding, which is the initial process step in yarn and composite production. The device used (a self-developed laboratory coarse separator 'CS') consists of a serrated cylinder (Ø 261 mm), fed by a rotating roll (Ø 32 mm), with a distance of 20 mm between them. The results of separation are comparable with industrial carding techniques (Müssig, 2001b). Subsequently, fibres were treated with soda and enzyme. These treatments are reported in detail by Fischer *et al.* (2006a).

The influence of mechanical, chemical and enzymatical hemp separation on the lignin and hemicellulose content was determined by the following method:

1. The lignin content of the hemp samples was determined as micro kappa number according to TAPPI Standard UM 246 (1991).

Table 15.2. Characteristics of the hemp fibre lots used in this work.

Sample name	A_{1000} value	Fineness (OFDA mean value) (μm)	Collective strength (Stelometer) (cN/tex)	Pure fibre content (%)
Hemp01	1.7	43.4	42.8	70.4
GDE02	1.3	44.9	42.6	70.2
KGE02	1.1	42.5	47.3	71.4

The hemp sample has to be disintegrated in 80 ml water by using a mixer or disintegrator and thermostated to 25°C. Then 10 ml of 0.1N $KMnO_4$ solution and 10 ml 4N H_2SO_4 are added and the reaction is allowed to proceed for 10 min. Finally, 2 ml 1N KI solution is added and titrated using 0.01N $Na_2S_2O_3$ solution. The kappa number is the volume (in millilitres) of 0.lN potassium permanganate solution consumed by 1 g of moisture-free fibres under the conditions mentioned above. The results are corrected to 50% consumption of the permanganate added. For this reason, the sample amount has to be found iterative to consume 30–70% (ideally 50%) of the permanganate.

2. The hemicellulose content of the fibre bundles was determined by measuring the alkali solubility in NaOH according to DIN 54356 (1977). The fibres are treated in NaOH solution of different concentrations at room temperature. The soluble fraction is then oxidized by $K_2Cr_2O_7$ in acidic conditions (H_2SO_4) and the excess dichromate is titrated using $(NH_4)_2Fe(SO_4)_2$ solution. The soluble fraction in 18% NaOH (S_{18}) is reported to represent the hemicelluloses, the S_5 fraction represents the subfraction of so-called 'gum' and the fraction of low molecular cellulose is represented by subtraction of S_{18} from S_{10} (Frohberg, 2005).

Effect of separation treatments on lignin and hemicellulose content

Effect of chemical treatment on lignin content

All three lots of hemp were examined in different process steps: after soda washing and after chemical separation. In addition, the raw and chemical separated fibre bundles were coarse separated (CS) and again tested for kappa number. The results are presented in Fig. 15.5.

Obviously, all lots have very similar lignin contents. In general, it can be stated that the kappa number decreases from > 20 for raw fibres to < 10 for chemical separated fibre bundles, indicating a lignin reduction of 50% by the NaOH treatment. This is in accordance with the literature reporting a reduction of lignin content by alkali treatment (e.g. Wang *et al*., 2003), but is stronger than the reported reduction of about 20% in alkali concentrations of 0.1–2% (Qu *et al*., 2005).

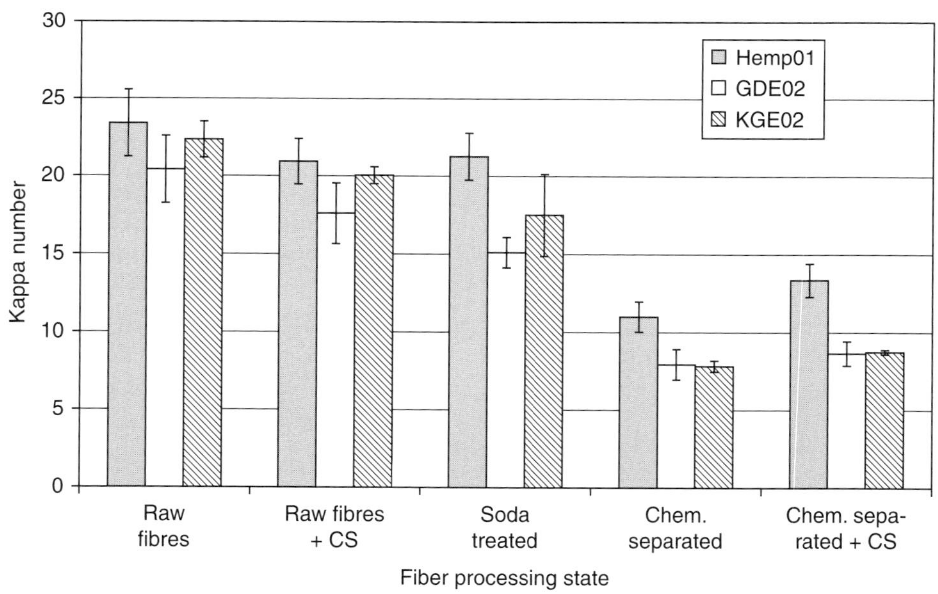

Fig. 15.5. Influence of raw material and processing step on the lignin content of hemp. CS, coarse separated (data from Fischer *et al.*, 2006b.)

Comparing the various lots of hemp tested here, there is only a small difference between them: in general, lot Hemp01 shows the highest kappa number of all tested specimens but, for the chemical separated variants only, the difference is significantly outside the experimental error. A possible reason for this effect may be the higher degree of retting (A_{1000}-value of 1.7 for Hemp01 compared to 1.1 and 1.3, respectively, for the other lots). The three lots compared here are of the same cultivar, grown in the same region and under nearly identical conditions. The retting process removes the easily soluble pectins first, whereas the fibre and lignin fraction remain nearly unchanged (if no over-retting takes place). So, the strongest retted of the nearly identical variants in this test should have the highest relative lignin content.

Effect of enzymatic treatment on lignin content

In the experiments reported by Fischer *et al.* (2004b), Texazym™ BFE was the commercially available enzyme with the biggest potential for fine separation of hemp. The samples from this series of experiments were tested for lignin content. The results are displayed in Fig. 15.6. All the enzymatic-treated variants were coarse separated so, for comparison, the data of the coarse separated raw material and the coarse separated chemical separated fibre bundles were included. All samples were from the same lot, GDE02.

The effect of enzymatic separation on lignin content is a little weaker than the effect of chemical separation. While the chemical NaOH treatment reduces the kappa number by ~50%, the enzymatic treatment reduces the lignin content by only ~35%. There is no significant difference between Texazym™ BFE and Texazym™ DLG, although the first is a pure pectinase and the latter is reported as hemicellulase with additional ligninase activity. The kappa value for combination of both Texazym variants (here applied in amounts of 10 ml each) appears to be slightly higher but is not significantly different.

Effect of chemical treatment on hemicellulose content

For different raw materials, the hemicellulose content was analysed as the soluble fraction in 18% NaOH solution (S_{18}) according to DIN 54356 (1977). In addition, the S_5 and

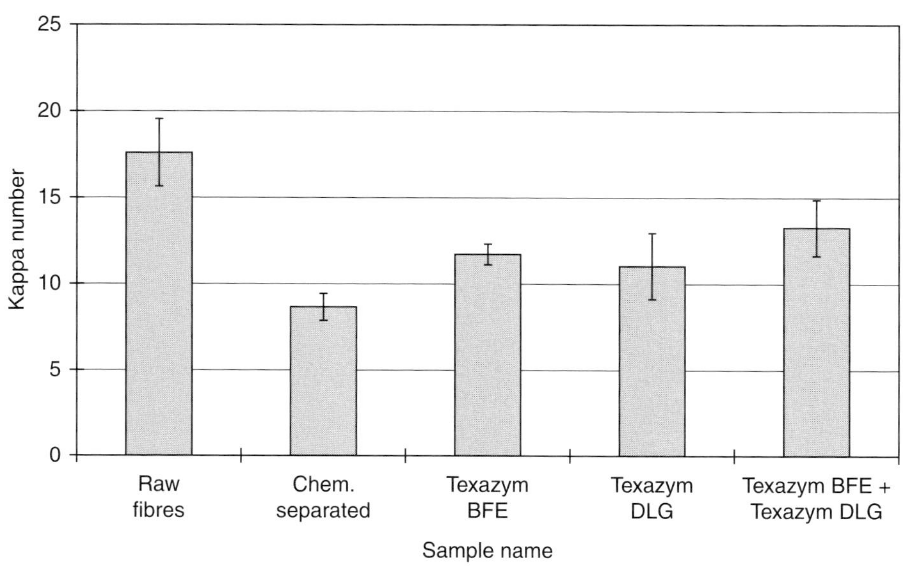

Fig. 15.6. Kappa number of enzymatic-treated hemp compared to chemically separated fibre bundles and raw material (data from Fischer *et al.*, 2006b).

S_{10} fractions were determined to identify the 'gum' fraction and the content of low molecular cellulose fragments. The results are displayed in Fig. 15.7. All samples described here have been coarse separated, as described by Fischer *et al.* (2004b).

It can be seen that there are considerable differences in hemicellulose content for the different raw hemp lots compared here, although all of them have similar pure fibre contents, tenacity and fineness. Again, the only difference between the lots is the degree of retting, with A_{1000} being 1.1 for KGE02, 1.3 for GDE02 and 1.7 for Hemp01, having the same tendency as the hemicellulose content. The 'gum' fraction of the hemicellulose follows the same tendency.

The fraction of low molecular cellulose fragments is very small compared to the hemicellulose. The data must be calculated by subtracting two similar values, causing a relatively high uncertainty, which is bigger than the value itself in the case of GDE02 and Hemp01. For this reason, there is not a reliable basis for detailed discussion of this value.

Summing up, it can be stated that it is possible to evaluate the hemicellulose content and the 'gum' fraction by this method, requiring only standard laboratory equipment. If there are only small fractions of low molecular cellulose, this method is not useful.

Effect of enzymatic treatment on hemicellulose content

In Fig. 15.8, the results of hemicellulose contents for Hemp01 as raw hemp, enzymatically treated and chemically separated variant are displayed. All these samples have been coarse separated, as described above.

It is obvious that the hemicellulose content is only about 20% reduced by the enzymatic treatment but is reduced dramatically (nearly 90%) by the chemical separation. The kappa numbers of the samples were included in the diagram. Comparing these to the hemicellulose content, the influence of enzymatic or chemical separation appears to be nearly the same for lignin content, but very different for hemicellulose content.

Conclusions on chemical fibre composition

Data from the literature on the measurement of natural samples are often not easy to

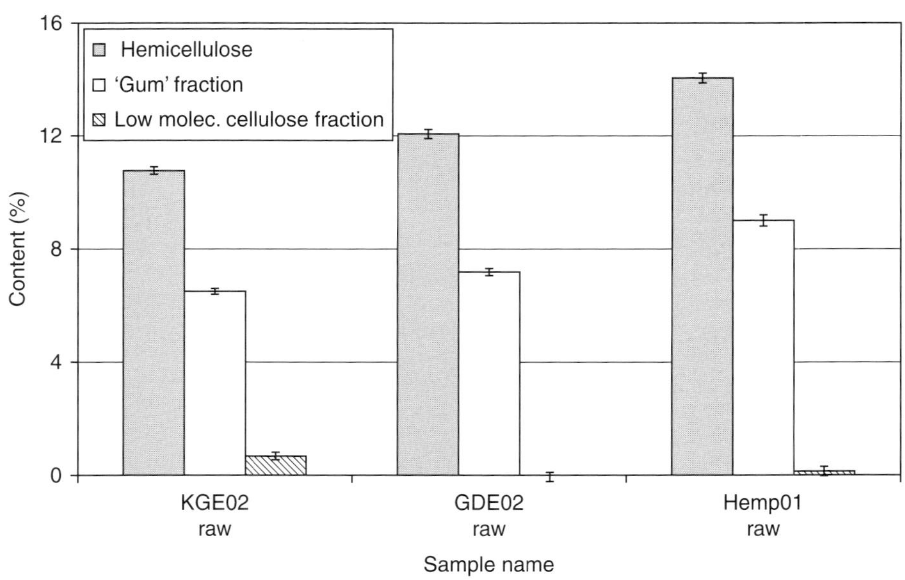

Fig. 15.7. Content of hemicellulose and related fractions of different raw materials (data from Fischer *et al.*, 2006b).

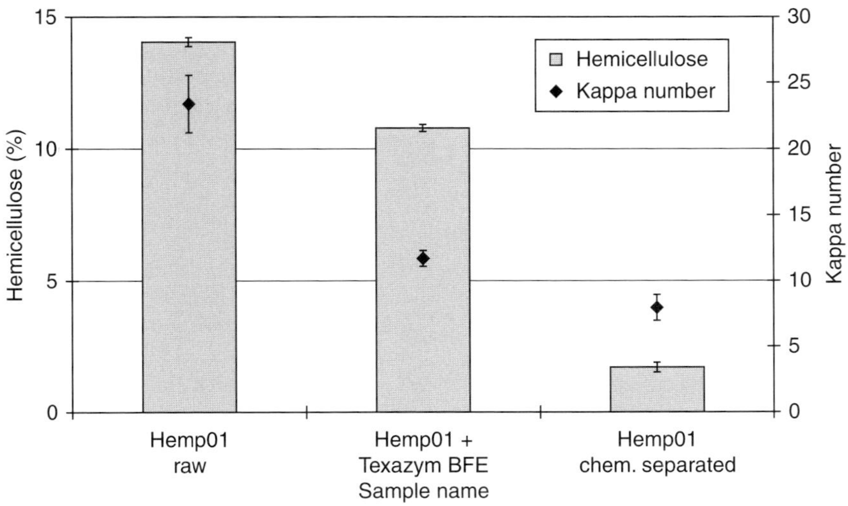

Fig. 15.8. Content of hemicellulose and kappa numbers for enzymatic and chemically separated variants of Hemp01 (data from Fischer *et al.*, 2006b).

compare due to the natural variations of the material. Another problem is often the lack of information about the individual properties of the tested materials. For this reason, the properties of the raw materials used here have been presented in detail. The results of the kappa measurements display clearly the necessity of this approach, depicting significant changes of lignin contents caused even by pure mechanical processing steps.

The degree of retting is the parameter found to have a remarkable influence not only on the constitution of the raw fibre bundles, but also causing remaining differences along the whole process chain. Comparing the raw hemp, there is a tendency to a higher lignin content for stronger retted variants, which is more significant for the enzymatic or chemical separated variants than for the raw materials. Retting also influences the hemicellulose content. The samples with the highest degree of retting displayed the highest hemicellulose content as well.

Comparing the influence of enzymatic and chemical treatment, it becomes obvious that they cause a similar reduction of lignin content but that reduction of hemicellulose content is much bigger by chemical separation than by enzymatic separation.

Effect of pretreatment on subsequent enzymatic treatment

In enzymatic hemp treatment, the essential standard parameters like type of enzyme, temperature and duration of enzyme action are usually reported in the literature; less intensively reported are the washing, drying steps, etc. But there are some more influencing factors like pretreatment or the sequence of process steps. The results briefly presented here have been published in detail by Fischer *et al.* (2006a).

Hemp (lot Hemp01) pretreatment in low concentrations of NaOH had only a minor effect on fineness, comparable to boiling in water. Better effects were achieved by treatment in low concentrations of soda (sodium carbonate, Na_2CO_3; Fig. 15.9). With increasing soda concentration, an increasing fibre bundle refinement of an additional 5–8 μm to the pure water effect is observed. The collective strength is kept at a slightly better level than with NaOH treatment. Good refinement is achieved even with 0.05 or 0.2% (w/v) soda concentration. Enlarging the concentration to 0.5% gives slightly better results for fineness and tenacity, but causes a 2.5-fold higher soda demand.

Examination of the influence of the processing steps in the enzyme treatment of hemp is displayed exemplarily in Fig. 15.10 with Lyvelin™, which is normally used as endopolygalcturonase in flax treatment.

Enzyme only treatment at 50°C/1h has no better effect than water treatment at 90°C and thus only has the advantage of a lower reaction temperature. A subsequent soda treatment (0.2% soda, 90°C/1h) gives only a minor improvement compared to the pure soda treatment at 90°C. A simple change in the sequence of these process steps improves the bundle refinement with an additional 3 µm without additional loss of tenacity. A subsequent third process step

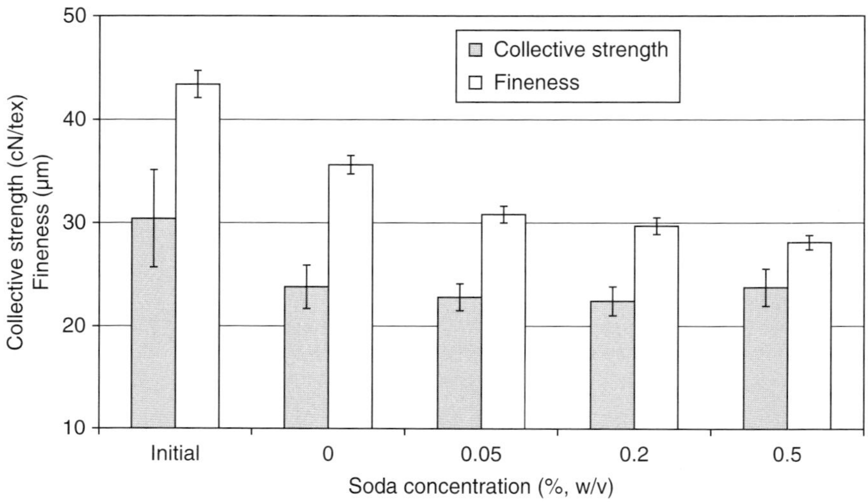

Fig. 15.9. Hemp pretreatment by different soda concentrations at 90°C. Error bars: standard deviation for collective strength, 95% level of accuracy for OFDA mean value (data from Fischer *et al.*, 2006a).

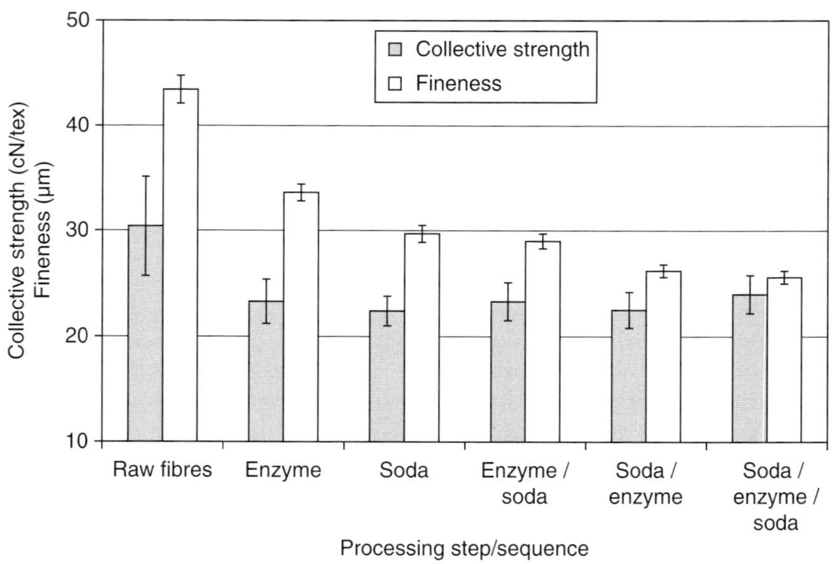

Fig. 15.10. Influence of the number and sequence of different process steps on hemp quality. Error bars: standard deviation for Stelometer/collective strength, 95% level of accuracy for OFDA mean value (data from Fischer *et al.*, 2006a).

with additional soda treatment displays only minor improvements.

We obtained best results by a pretreatment with small soda concentrations and subsequent enzyme treatment. An optimal effect can only be achieved in a sequence of soda first and enzyme second. Soda as a first reaction step has a chemical cleaning effect on the fibre bundle surface (for details and SEM pictures, see Fischer *et al.*, 2006a). This enables the enzyme in the next step to act directly on the pectin layers between the fibre cells. Without the prewashing step, the enzyme can only act at those places where the pectins are directly accessible at the surface and are not covered by other plant compounds.

Hemp separation using different commercial enzymes

These experiments were carried out at the laboratory scale (10 g of hemp fibres in 200 ml solution). Preceding the enzymatic treatment, hemps were washed in soda solution as described above (0.2% Na_2CO_3 at 90°C for 1 h). Enzyme treatments were carried out using BioPrep™ 3000 L (pectate lyase, EC 4.2.2.2; Adamsen *et al.*, 2002a), obtained from Novozymes A/S Denmark, Baylase™ EVO (diluted preparation of BioPrep™ 3000 L), obtained from Bayer AG, Germany, Lyvelin™ (endopolygalcturonase, EC 3.2.1.15; Adamsen *et al.*, 2002a), obtained from Lyven SA, France, Texazym™ BFE (pectinase) and Texazym™ DLG (hemicellulase with additional ligninase activity), both obtained from INOTEX sro, Czech Republic. The experiments were carried out using 1 ml of enzyme solution (except Baylase™ EVO: 10 ml). The experiments were carried out at 50°C (Lyvelin and Texazym variants) or 60°C (BioPrep, Baylase) for 1 h in a shaking bath.

The efficiency of the different enzymes mentioned above in hemp separation by means of fineness is displayed in Fig. 15.11. The values of the raw material and of the only pretreated sample are added as blind values for comparison. Starting with the pure pectate lyase activities of BioPrep™ and Baylase™, it is obvious that the activity of the diluted Baylase species is nearly zero. It was used in amounts of 10 ml to compensate the dilution, but there is no significant change in fibre bundle width compared to the only pretreated variant. The other pectate lyase, BioPrep™ and Lyvelin™ as endopolygalcturonase, both display similar significant activities with a bundle refinement by 3–4 μm. Most effective in these experiments is the activity of the pectinase Texazym™ BFE with 8 μm

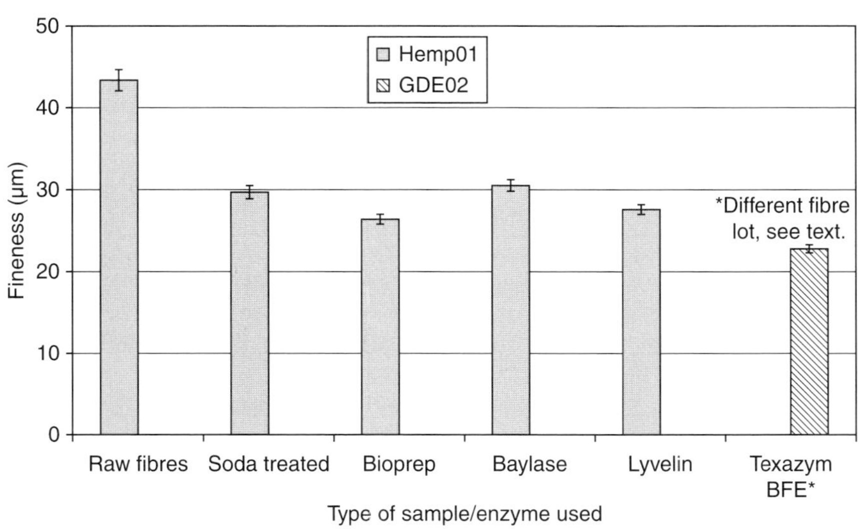

Fig. 15.11. Efficiency of different enzymes in hemp separation (data from Fischer *et al.*, 2006a).

compared to the only pretreated variant. It has to be mentioned in this context, that the Texazym™ experiment was performed on hemp lot GDE02, whereas the other experiments were conducted on lot Hemp01.

The influence of different raw materials on the result of hemp separation has been shown to be negligible. Hemp01 and GDE02 are nearly identical in fineness after soda pretreatment, as well as after enzymatic separation (Lyvelin™). The use of enzyme mixtures to combine the advantages of the single enzyme preparations in order to improve fineness better than by only one enzyme has been disadvantageous, for two reasons: first is the cost of two enzymes instead of one, which can be a knock-out criterion for the total process, and second is the enlarged risk of fibre damage. This effect was observed using both the Texazym™ variants, where the effect of the hemicellulase Texazym™ DLG was slightly enlarged by the pectinase Texazym™ BFE in combination. Unfortunately, there was a cellulytic side-activity of Texazym™ DLG, enlarged by the combination with the pectinase. The Stelometer tenacity broke down from originally > 30 cN/tex to 19.5 cN/tex by treatment with only Texazym™ DLG, and to only 7 cN/tex by combination of both enzymes (Fischer *et al.*, 2004b).

In general, a task for future research in hemp separation will be the finding of an ideal enzyme mixture which is able not only to dissolve the hemp pectins between the single fibre cells better than the actual enzymes used, but also to be able to clean the fibre bundle surface. This would enable the process to be carried out in one step only, instead of currently being performed as a two-step process with soda pretreatment and subsequent enzymatic treatment.

Influence of hemp separation on textile and product properties/side effects

Hemp GDE02 after mechanical decortications and separation was used without further treatment and after enzymatic treatment (GDE02-enzyme: [1st] 0.2% soda, 1 h, 90°C; [2nd] 0.2% BioPrep™ 3000 L [Novozymes] and tenside 1 h, 55°C, pH 9; [3rd] water rinsing, drying and carding). Subsequently, GDE02 and GDE02-enzyme were processed into a multilayer web by using a conventional roller card (Spinnbau, Bremen, Germany) at NAFGO GmbH, Neerstedt, Germany (Müssig *et al.*, 2005). The samples were analysed regarding fineness and length. The results are given in Table 15.3.

As can be seen from Table 15.3, the enzymatic process has a strong influence on the fineness and length of fibre bundles. After separation and carding, single fibres can be found in the sample, while some coarser and longer bundles are still present in the sample.

To evaluate the potential to use enzymatic separated hemp in technical applications, the two hemp samples (GDE02 and GDE02-enzyme) were processed into needlefelts and further to composites.

Table 15.3. Results from fineness test (fineness FMT Shirley; 3 samples; 4 g; 2 measurements) and length test (length ALMETER; 3 samples; *c.* 0.6 g; A-beard) of the mechanically and enzymatically separated GDE02 hemp before and after carding.

Sample	Shirley P_L value	Almeter mean length, ML(Q) in mm	Almeter median, L(Q)50% in mm	The longest fibre bundles, L(Q)1% in mm
GDE02 – before card	12.8(± 0.4)	45.2	35.9	161.4
GDE02 – after card	16.0(± 0.0)	44.3	39.2	131.0
GDE02-enzyme – before card	37.2(± 1.2)	17.9	15.9	47.0
GDE02-enzyme – after card	40.0(± 1.8)	18.5	16.4	51.0

Production of needle-felts

GDE02 and GDE02-enzyme were carded with a conventional roller card (Spinnbau, Bremen, Germany) at NAFGO GmbH, Neerstedt, Germany. The multilayer webs (approximately 1000 g/m^2) were processed to needle-felts on an industrial machine. The roller-carded webs were pre-needled with a Heuer and Sohn (Tespe, Germany) pre-needle machine type ROV 12S/250/11/900. The main needle-felt production was carried out by a Fehrer (Linz, Austria) needle-felt machine type NL1226 with two needle boards, a working width of 2600 mm.

Compression moulding of the composites

After carding and needling, the needle-felts were cut into parts of 300 × 250 mm^2. The felts were dried before pressing for 3 h at 105°C in a forced air oven. The pre-cut parts were centred on the bottom pressure plate with four aluminium slats, 2 mm thick, used as spacers. The aluminium slats were positioned around the pre-cut parts. A hydraulic press type KV 215.01 (Rucks Maschinenbau GmbH, Glauchau, Germany; nominal pressure force: 1000 kN; maximum temperature: 330°C; pressure table: 600 × 600 mm^2) was used for composite production. The resin used was an epoxy thermoset resin type Larit L-135 (100:35), company Lange+Ritter GmbH, Gerlingen, Germany with a hardener type Larit 135. The composites were produced at a pressure of 100 bar, a temperature of 80°C and a pressing time of 1.5 h. The fibre content of the composites was 30 and 40 mass%. The composites were cut into specimens with a moulding cutter for the different mechanical tests.

FOGGING TEST. The fogging test was done according to procedure B of DIN 75201-G (1992; gravimetric measurement). The sample mass was 5 g for the felts and 12 g for the flock. According to the standard, two samples were tested for felt and flock samples. The results of the fogging test of the felts and the epoxy composites are shown in Fig. 15.12.

It can be summarized that the freshly enzymatically separated hemp fibre bundles are much finer compared to the mechanically separated hemp fibre bundles (see Table 15.3). This results in much higher inner surface in the GDE02-enzyme sample, which increases the potential for higher fogging (substances which can be emitted from the fibre surface under fogging conditions). As shown in Fig. 15.12, the impregnation with the epoxy

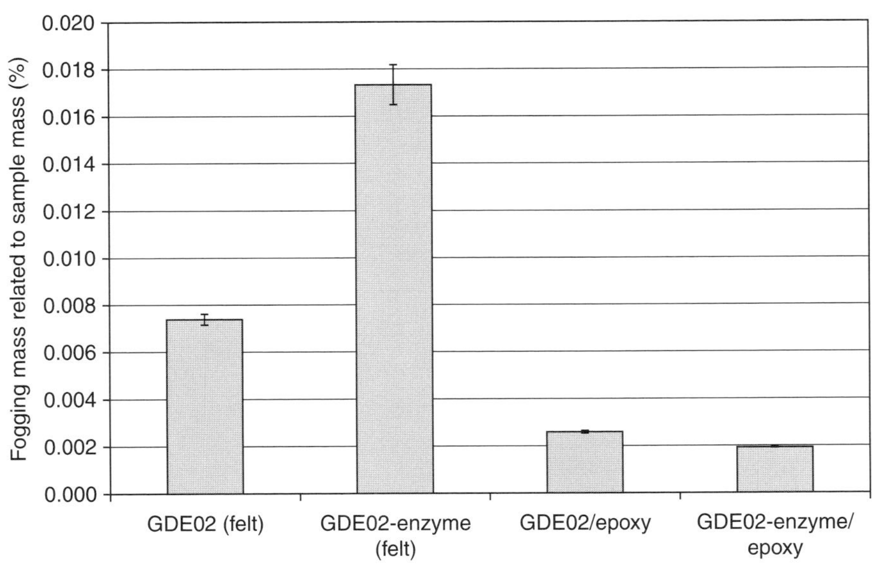

Fig. 15.12. Results from the fogging test of the felts and composites of mechanically and enzymatically separated hemp.

resin reduces significantly the possibility for an emission of fogging relevant substances from the fibre surface. Identical effects were observed for the odour emissions of GDE02 presented below.

CHARPY IMPACT TEST. For the Charpy impact strength test unnotched, rectangular specimens of 80 × 10 × 2 mm^3 were manufactured and tested according to DIN EN ISO 179 (2006). Impact strength was measured on a Thwing-Albert FRANK testing machine (type 53302) (Birkenau, Germany) operating with a pendulum length of 225 mm.

The results of the Charpy impact test are summarized in Fig. 15.13. Impact increased with increasing fibre mass content in the composites. The influence of the refinement of hemp fibre bundles by enzymatical separation on composite properties such as impact behaviour is obvious. Using finer fibre bundles will increase the impact and less fibre mass is necessary to reach the same impact value. The same trends were observed for hemp after steam explosion (Müssig, 2002).

Fibre odour

Raw hemp GDE02 was first washed in soda solution and subsequently treated by different Texazym variants. These treatments originally were conducted in order to improve fibre bundle separation and to improve mechanical fibre characteristics. The results of these experiments have been reported elsewhere by Fischer *et al.* (2004a,b), but in fact these treatments also influence odour emissions (Fig. 15.14). Even pure soda treatment causes an increase in odour intensity and the different enzymes also cause a more or less distinct increase of odour intensity. The reason for this effect is the increased fineness of the fibre bundle. The increase in specific surface is accompanied by an increase in odour intensity (Fischer and Lohmeyer, 2009). As displayed above, there is an identical effect on the fogging behaviour of hemp.

A different approach is the treatment of malodorant hemp displayed in Fig. 15.15. This hemp was mildew-contaminated due to over-retting, and the odour intensity of raw fibre bundles FH-S08 was reduced by more than 75% with this process, whereas the hedonic assessment was improved dramatically from –3 to slightly positive values. Furthermore, the scalability of this process to industrial scale was proven.

Subsequently, needle-felts made of odour-optimized HS08-2 fibre bundles have been used to produce automotive door trim panels (hemp/PUR) as demonstrative parts to give evidence that this process enables the use of hemp lots such as odour-contaminated raw hemp, which are normally unusable in industrial processing (Fischer *et al.*, 2008a). Again, the same tendency for fogging emissions (cf. the paragraph above in this section) was observed: the impregnation in PUR-matrix decreased

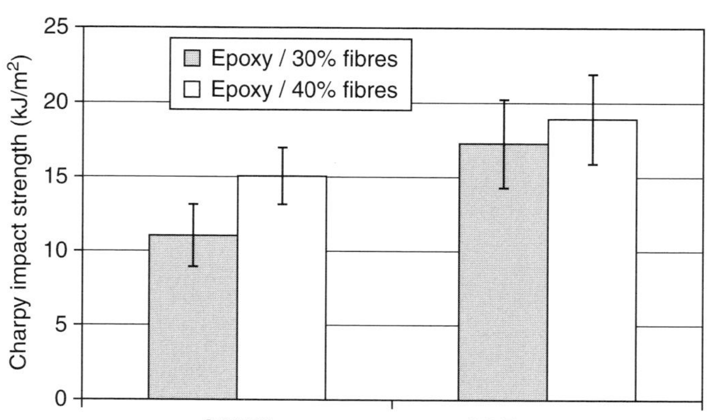

Fig. 15.13. Influence of fibre bundle fineness on composite characteristics.

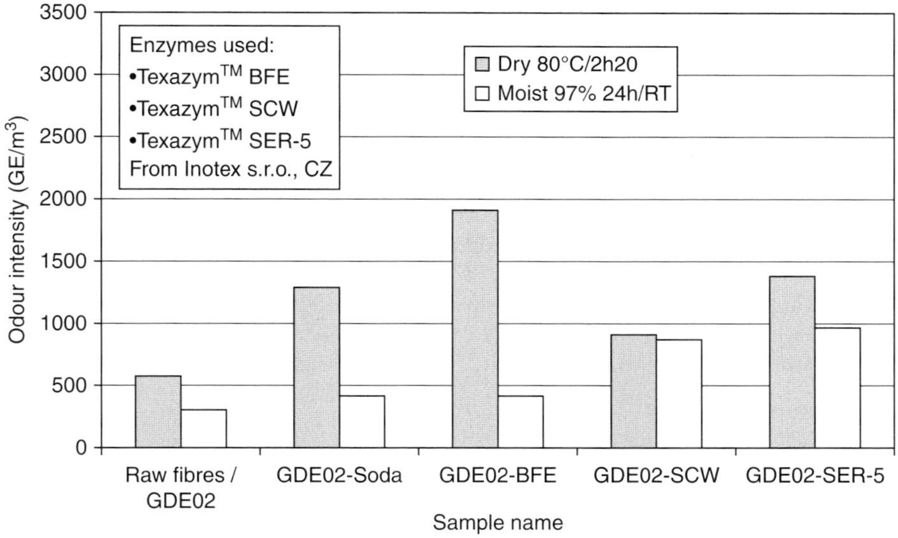

Fig. 15.14. Effect of enzymatic treatment on odour intensity of raw hemp lot GDE02.

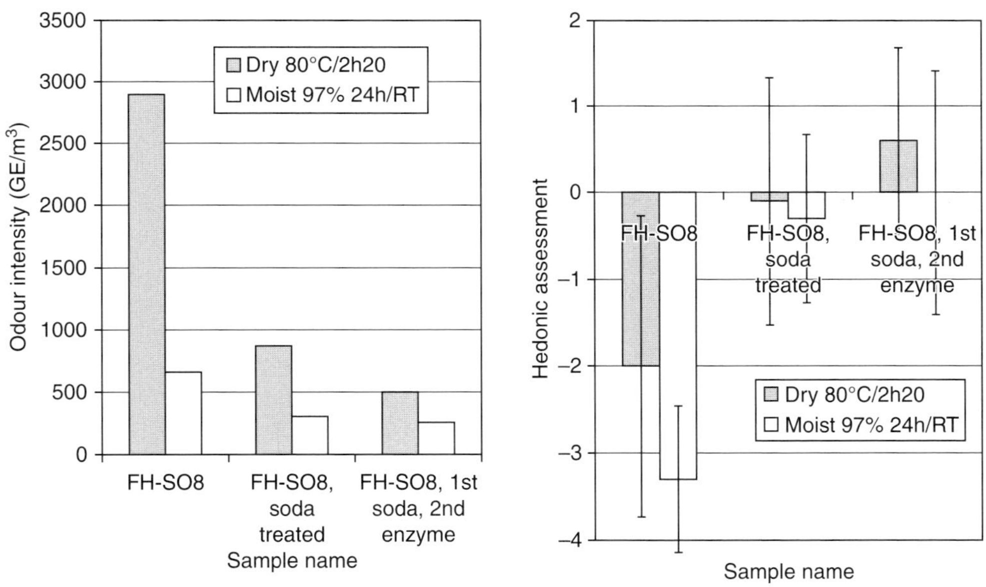

Fig. 15.15. Effect of enzymatic treatment on odour intensity (left) and hedonic assessment (right) of mildew-contaminated hemp lot FH-S08 (data from Fischer *et al.*, 2008a).

the level of odour emissions substantially and again lower emission was observed for the enzyme-treated variant.

In general, these investigations show that malodour of bast fibres can be caused by different sources. If these odours have to be reduced or removed, this cannot be done by one identical treatment, but the type of treatment has to be adequate. Different types of smell and possible treatments have been examined (Fischer *et al.*, 2008b), for example:

- Overheated fibre bundles smell like smouldering fire and could be treated chemically by reducing agents. An unsolved problem here is that overheating normally occurs during compound formation, and there is no pathway known to the authors to treat fibres inside compounds by such a process.

- Textile finishes with positive and negative effects – no recommendation was possible for or against the use of textile finishes in odour optimization.
- Odour caused by mildew contamination was extremely well removed by enzymatic treatment, as displayed above.

The last case in particular is not always a reason for customer complaint because bast fibre bundles are normally field retted. This desired fungal process enables a good separation of bark and stem and is essential for a good quality fibre bundle (Müssig, 2001a). But, if over-retting takes place, it causes not only the start of fibre damage but also mildew contamination, with its typical smell.

In general, there was a recommendation for processing companies to avoid unfavourable odour problems by: (i) avoiding the use of too much textile finishes; (ii) avoiding too high or too long thermal stress during processing; and (iii) either avoiding mildew-contaminated fibre lots or treating them by the enzymatic process described above in order to achieve a good-quality product (Fischer *et al.*, 2008b).

Quality Management

The aspects of quality management in industrial bast fibre processing have been discussed intensively over the past few years. A comprehensive presentation can be found, for example, in Müssig *et al.* (2006). For this reason, we will point out the main aspects only briefly here.

When bast fibres are to be used as a resource for industrial products, the establishment of appropriate quality management becomes essential to ensure constant product quality. The set-up of a quality control system has to start with the selection of an appropriate cultivar for the location, depending on climatic conditions, soil quality, etc.

During the period of plant growth, climate data like precipitation, temperature, etc., should be recorded as information supplied with the fibres. After mowing the plants, this must be continued during the period of field retting to enable retting to be stopped in optimal time. The degree of retting depends greatly on temperature and moisture during this time and, beside visual assessment of the stem colour, it is possible to analyse the degree of retting by NIR spectroscopy as A_{1000}-coefficient directly in the field (Bluhm and Müssig, 1999). Results of the development of retting degree are shown exemplarily in Fig. 15.16.

Fibre quality is strongly dependent on the type and time of harvesting and the intensity of retting. In addition, the type of mowing equipment influences the course of retting and, if inappropriate equipment is used, severe fibre damage can be the result (Müssig *et al.*, 1999).

Finally, the degree of retting influences the processability of bast fibre bundles greatly. Unretted plants display higher fibre loss in decortication and carding and a higher remaining shive content, combined with the disadvantage of coarser fibre bundles (Fig. 15.17). Thus, the final quality after mechanical coarse separation not only is a result of this processing step, but also is influenced by retting and the harvesting equipment used. Consequently, fibre properties should be examined before harvesting and several times during the retting process to ensure an optimal result of retting.

The measurements necessary for quality control can be done easily by using methods adapted and optimized for bast fibre and fibre bundle testing recommended by Müssig *et al.* (2006):

- Tensile strength by Stelometer, even if testing only a small number of fibre collectives, gives reliable data if the fibre fineness (i.e. the number of fibre bundles in the collective) is considered. To get physically exact data, for example, for a numerical simulation of the behaviour of a natural fibre-reinforced plastic, it is necessary to use the single-element test, e.g. by using Dia-Stron.

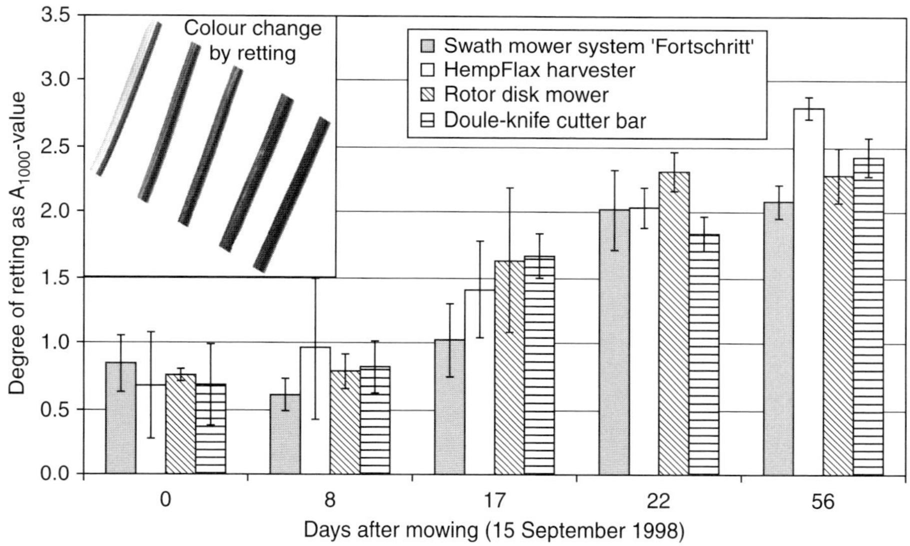

Fig. 15.16. Increase of retting degree versus type of harvesting equipment (based on data from Müssig and Martens, 2003).

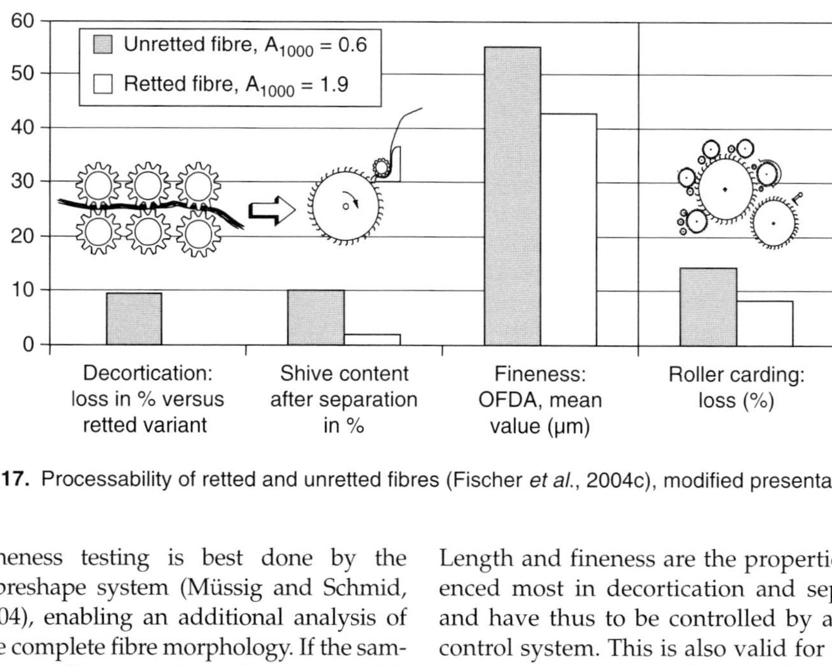

Fig. 15.17. Processability of retted and unretted fibres (Fischer *et al.*, 2004c), modified presentation.

- Fineness testing is best done by the Fibreshape system (Müssig and Schmid, 2004), enabling an additional analysis of the complete fibre morphology. If the samples are fine enough, analysis is possible by OFDA or Laserscan for fine fibres, but is accompanied by the risk of errors due to cut-off of coarser fibres because both these systems were developed for wool.
- Length, if in normal production with fibre bundles < 250 mm, can be controlled easily by an Almeter, combined with a modified preparation unit (Drieling *et al.*, 1999).

Length and fineness are the properties influenced most in decortication and separation and have thus to be controlled by a quality control system. This is also valid for the subsequent carding step, because fibre bundle length influences the needle-felt properties directly.

Finally, fineness is crucial for the selection of optimal hemp lots for special products. The tensile strength of needle-felts, for example, is dependent on the fineness of the lots used. Similar effects are observed in the production of natural fibre-reinforced thermosets. If the felt structure is penetrated completely by the polymer, the mechanical properties are enhanced significantly. On the other hand, too small intermediate spaces between the fibre bundles can prevent a complete penetration by the polymer. In such cases, usage of coarser fibre bundle lots can be of advantage.

Conclusions

Nowadays, bast fibres are established raw materials, especially in the automotive industry. For traditional reasons, these fibre bundles are mechanically decorticated and separated. Since the type of separation has direct influence on the properties, quality could be improved substantially by choosing another type of separation.

Enzymatic separation in particular offers great potential for designing tailored fibre properties. As shown in the previous sections, it has a positive influence on fibre bundle fineness, thermal stability, fogging and odour emission due to changes in pectin, lignin and hemicellulose content of the material. Finer fibre bundles lead to higher composite tenacity and impact strength. Fogging and odour emissions are first increased due to a higher specific surface, but this has been shown to be a temporary effect for odour and in a composite as a final product, the emissions are lower for enzymatic-treated variants. Furthermore, enzymatic treatment can reduce or remove malodour of mildew-contaminated lots and enables use of these fibres as in non-contaminated lots.

These additional process steps are combined with additional costs, but this is counterbalanced by a higher fibre and product quality. This will open the way for usage of bast fibres in more advanced applications in the future. For such applications, consistant raw material qualities are essential, making the existence of a reliable quality control system necessary.

If this results in increased use of domestic fibre plants, it would not only be of ecological but of economic advantage as well: it would combine short transport distances within the whole value-added chain with the possibility to valorize natural resources in constant qualities.

References

Adamsen, A.P.S., Akin, D.E. and Rigsby, L.L. (2002a) Chelating agents and enzyme retting of flax. *Textile Research Journal* 72(4), 296–302.

Adamsen, A.P.S., Akin, D.E. and Rigsby, L.L. (2002b) Chemical retting of flax straw under alkaline conditions. *Textile Research Journal* 72(9), 789–794.

Akin, D.E., Foulk, J.A., Dodd, R.B. and McAlister, D.D. III (2001) Enzyme-retting of flax and characterization of processed fibres. *Journal of Biotechnology* 89, 193–203.

Akin, D.E., Foulk, J.A. and Dodd, R.B. (2002) Influence of flax fibres of components in enzyme retting formulations. *Textile Research Journal* 72(6), 510–515.

Akin, D.E., Condon, B., Sohna, M., Foulk, J.A., Dodd, R.B. and Rigsby, L.L. (2007) Optimization for enzyme-retting of flax with pectate lyase. *Industrial Crops and Products* 25(2), 136–156.

Amaducci, S., Müssig, J., Zatta, A. and Pelatti, F. (2005) HEMP SYS: design, development and up-scaling of a sustainable production system for HEMP textiles: an integrated quality SYStems approach. How to affect hemp fibre quality? In: *Textiles for Sustainable Development. FAO/ESCORENA International Conference/The Conference of the FAO/ESCORENA European Cooperative Research Network on Flax and other Bast Plants, Session 9 – Economical Aspects of Textile Production*, 23–27 October 2005. CD-ROM Proceedings. CSIR, Port Elizabeth, South Africa and Institute of Natural Fibres, Poznan, Poland 9 pp.

Baxter, B.P., Brims, M.A. and Taylor, T.B. (1992) Description and performance of the optical fibre diameter analyser (OFDA). *Journal of the Textile Institute* 83(4), 507–526.

Bluhm, C. and Müssig, J. (1999) Ansätze zur reproduzierbaren Röstgradmessung mit dem Ziel der Bereitstellung einheitlicher Hanffasern. In: *BRAIN-Themenkonferenz 'Ermittlung von Qualifizierungsbedarf für den Hanfanbau' im Rahmen des EU-Projektes BRAIN, Bad Sassendorf, Germany, Haus Düsse, 16 April 1999*. Conference proceedings. nova-Institut GmbH und Institut für Internationale Sozialforschung - INFIS e.V. Hürth, 7 pp.

Bredemann, G. (1942) Die Bestimmung des Fasergehaltes bei Massenuntersuchungen von Hanf, Flachs, Fasernesseln und anderen Bastfaserpflanzen. *Faserforschung* 16, 15–39 [in German].

Buschle-Diller, G., Zeronian, S.H., Pan, N. and Yoon, M.Y. (1994) Enzymatic hydrolysis of cotton, linen, ramie, and viscose rayon fabrics. *Textile Research Journal* 64(5), 270–279.

Carus, M., Müssig, J. and Gahle, C. (2008) *Naturfaserverstärkte Kunststoffe*. Fachagentur Nachwachsende Rohstoffe e.V. (FNR), 2008, Gülzow, Germany, 40 pp (http://www.fnr.de, accessed 02/07/2009).

Crônier, D., Monties, B. and Chabbert, B. (2005) Structure and chemical composition of bast fibers isolated from developing hemp stem. *Journal of Agricultural and Food Chemistry* 53, 8279–8289.

DIN 54356 (November 1977) Bestimmung der Alkalilöslichkeit von Zellstoff. DIN Deutsches Institut für Normung e.V. [in German].

DIN 75201 (September 1992) Bestimmung des Foggingverhaltens von Werkstoffen der Kraftfahrzeug-Innenausstattung. DIN Deutsches Institut für Normung e.V. [in German].

DIN EN ISO 179-1 (May 2006) Kunststoffe – Bestimmung der Charpy-Schlageigenschaften – Teil 1: Nicht instrumentierte Schlagzähigkeitsprüfung, DIN Deutsches Institut für Normung e.V. [in German].

Dreyer, J., Müssig, J., Koschke, N., Ibenthal, W.-D. and Harig, H. (2002) Comparison of enzymatically separated hemp and nettle fibre to chemically separated and steam exploded hemp fibre. *Journal of Industrial Hemp* 7(1), 43–59.

Drieling, A., Bäumer, R., Müssig, J. and Harig, H. (1999) Möglichkeiten zur Charakterisierung von Festigkeit, Feinheit und Länge von Bastfasern. *Technische Textilien* 42(4), 261–262 [and in English: p. E66].

Eder, M. and Burgert, I. (2010) Natural fibres – function in nature. In: Müssig, J. (ed.) *Industrial Applications of Natural Fibres: Structure, Properties and Technical Applications*. John Wiley and Sons, Chichester, UK, pp. 23–39.

Fischer, H. and Lohmeyer, B. (2009) Geruchsverhalten von Bastfasern: Optimierung für den Einsatz in industrieller Produktion. In: Bledzki, A.K. et al. (eds.) *11th Workshop, Odour and Emissions of Plastic Materials, Kassel/Germany, 30–31 March 2009*. Institut für Werkstofftechnik, Kunststoff- und Recyclingtechnik, Universität Kassel, Germany, pp. 5-1–5-9.

Fischer, H., Müssig, J., Geppert, N. and Bluhm, C. (2004a) Beurteilung des Geruchspotenzials von Naturfasern für den Einsatz im Automobilbereich. In: AVK-TV (ed.) *Proceedings 7th International AVK-TV Conference for Reinforced Plastics and Thermoset Moulding Compounds*, 28–29 September 2004, Baden-Baden, Germany. Frankfurt, pp. B11-1–B11-8 [in German].

Fischer, H., Müssig, J., Bluhm, C., Marek, J. and Antonov, V. (2004b) Enzymatic modification of hemp fibres for sustainable production of high quality materials: comparing different commercial enzymes. In: Team of Authors (eds): *11th International Conference STRUTEX (Proceedings)*. Technical University of Liberec, Czech Republic, pp. 301–308.

Fischer, H., Müssig, J. and Bluhm, C. (2004c) Enzymatic separation of hemp fibres – aspects of quality control along the value-added chain. In: Mackiewicz-Talarczyk, M. and Csiszár, E. (eds) *COST Action 847, Textile Quality and Biotechnology, The Joint Meeting of WG/1 and WG/2. (Documentation of the Meetings WG/1 and WG/2, Budapest, Hungary 23–24 September 2004)*. Department of Plastics and Rubber Technology, Budapest University of Technology and Economics, Budapest, p. 15.

Fischer, H., Müssig, J. and Bluhm, C. (2006a) Enzymatic modification of hemp fibres for sustainable production of high quality materials: influence of processing parameters. *Journal of Natural Fibers* 3(2/3), 39–53.

Fischer, H., Abdulajeva, D. and Müssig, J. (2006b) Enzymatic modification of hemp fibres for sustainable production of high quality materials: effects on chemical fibre composition. In: Choe, E.K. and Park, E.Y. (eds) *Proceedings of the 4th International Conference on Textile Biotechnology, 4–7 June 2006, COEX, Seoul, Korea*. Hanrimwon Publishing Co Ltd, Seoul, p. 33.

Fischer, H., Gerardi, H., Knittel, D. and Antonov, V. (2008a) Removal of odour from bast fibres in industrial scale by chemical and enzymatic treatment. In: *15th International Conference STRUTEX (Proceedings)*. Technical University of Liberec, Liberec, Czech Republic, pp. 331–338.

Fischer, H., Knittel, D. and Opwis, K. (2008b) Verbesserung der Geruchseigenschaften von Naturfasern zur Öffnung neuer Märkte in den Bereichen Technische Textilien und Verbundwerkstoffe.

In: *Forschungsberichte aus dem Faserinstitut Bremen* (ISSN 1618–7016) 43, November 2008, 100 pp. ISBN 978-3-837-07213-6. [in German]. BoD, Norderstedt.
Frohberg, K. (2005) Chemical pulp evaluation methods and their applicability to recycled pulp. *PTS News* 5, 26–29.
Gutiérrez, A., Rodríguez, I.M. and del Río, J.C. (2006) Chemical characterization of lignin and lipid fractions in industrial hemp bast fibers used for manufacturing high-quality paper pulps. *Journal of Agricultural and Food Chemistry* 54, 2138–2154.
HEMP-SYS (2005) Design, development and up-scaling of a sustainable production system for HEMP textiles: an integrated quality SYStems approach – HEMP-SYS. Research project: EU – Fifth Framework Programme (Quality of Life Management of Living Resources)/Contract No: QLK5-CT-2002-01363 (http://www.hempsys.net/, accessed 17/05/2006).
Karus, M. and Kaup, M. (2001) Naturfasereinsatz in der europäischen Automobilindustrie. *Technische Textilien* 44, 258–262.
Karus, M. and Kaup, M. (2002) Natural fibres in the European automotive industry. *Journal of Industrial Hemp* 7(1), 119–131.
Karus M., Ortmann S., Gahle, C. and Pendarovski, C. (2006) Use of natural fibres in composites for the German automotive production from 1999 till 2005. nova-Institut, Hürth, Germany (http://www.nova-institut.de/pdf/06-12_nova_NF-CompositesAutomotive.pdf, accessed 20/11/2007).
Kaup, M., Karus, M. and Ortmann, S. (2003) Naturfasereinsatz in Verbundwerkstoffen in der deutschen und österreichischen Automobilindustrie – Status 2002, Analyse und Trends (http://www.nova-institut.de/pdf/Markt_NF-Verbund.pdf, accessed 15/10/2004). Short version in English (http://www.nova-institut.de/pdf/03-03_Market_NF_Composite.pdf, accessed 15/10/2004).
Krässig, H., Schurz, J., Steadman, R.G., Schliefer, K. and Albrecht, W. (1989) Cellulose. In: Elvers, B., Hawkins, S., Ravenscroft, M. and Schulz, D. (eds) *Ullmann's Encyclopedia of Industrial Chemistry.* Volume A28. Verlag Chemie VCH, Weinheim, Germany, pp. 375–419,
Leupin, M. and Leupin, R.E. (2002) Ein Rotorgarn aus reinem Hanf. In: *Neujahrsschrift der ETH Zürich.* Institut für Automatisierte Produktion der Eidgenössischen Technischen Hochschule, Zürich, pp. 12–25 [in German].
Mooney, C., Stolle-Smits, T., Schols, H. and de Jong, E. (2001) Analysis of retted and non-retted flax fibres by chemical and enzymatic means. *Journal of Biotechnology* 89, 205–216.
Morvan, C., Abdul-Hafez, A., Morvan, O., Jauneau, A. and Demarty, M. (1989) Étude physicochimique de polysaccharides extraits de lin sous-roui. *Plant Physiology and Biochemistry* 27(3), 451–459.
Morvan, C., Jauneau, A., Voreux, H., Morvan, O. and Demarty, M. (1990a) Contribution des ciments aux charactéristiques physicochimiques des fibres de lin: application à un rouissage enzymatique. *Comptes Rendus des Séances de la Société de Biologie et de ses Filiales* 184, 18–30.
Morvan, C., Jauneau, A., Flaman, A., Millet, J. and Demarty, M. (1990b) Degradation of flax polysaccharides with purified endo-polygalacturonase. *Carbohydrate Polymers* 13, 159–163.
Morvan, C., Andème-Onzighi, C., Girault, R., Himmelsbach, D.S., Driouich, A. and Akin, D.E. (2003) Building lax fibres: more than one brick in the walls. *Plant Physiology and Biochemistry* 41, 935–944.
Müssig, J. (2001a) *Neue technische Textilien aus heimischen Pflanzenfasern (Hanf und Nessel) (AiF 11918N).* FIBRE – 2001. Faserinstitut Bremen e.V., Bremen, 98 pp. [in German].
Müssig, J. (2001b) *Untersuchung der Eignung heimischer Pflanzenfasern für die Herstellung von naturfaserverstärkten Duroplasten – vom Anbau zum Verbundwerkstoff.* Fortschritt-Bericht VDI, Series 5, No. 630. VDI Verlag GmbH, Düsseldorf, Germany, 215 pp. [in German].
Müssig, J. (2002) Influence of fibre fineness on the properties of natural fibre composites. In: *Paper No. 5, 4th International Wood and Natural Fibre Composites Symposium*, Kassel 10–11 April 2002. Institut für Werkstofftechnik, Kunststoff- und Recyclingtechnik (Hrsg. und Veranst.), Universität Gh Kassel, Germany, pp. 5-1–5-12.
Müssig, J. (2003) EIHA's classification of hemp fibre by colour grades. In: *First International EIHA Conference.* Documentation of the First International Conference of the European Industrial Hemp Association (EIHA), 23–24 October 2003. European Industrial Hemp Association (EIHA), nova-Institut, Hürth, Germany, pp. 1–23.
Müssig, J. and Martens, R. (2003) Quality aspects in hemp fibre production – influence of cultivation harvesting and retting. *Journal of Industrial Hemp* 8(1, 11–32.

Müssig, J. and Schmid, H.G. (2004) Quality control of fibers along the value added chain by using scanning technique – from fibers to the final product. In: Anderson, I.M., Price, R., Clark, E. and McKernan, S. (eds) *Microscopy and Microanalysis 2004*. 1–5 August 2004, Savannah, Georgia, USA. Volume 10, Supplement 2. Press Syndicate of the University of Cambridge, pp. 1332CD–1333CD.

Müssig, J., Martens, R. and Harig, H. (1999) Der Einfluß von Ernte-, Aufbereitungs- und Verarbeitungstechnik auf die Eigenschaften hanfnadelfilzverstärkter Polymerwerkstoffe. In: Bledzki, A.K. (ed.) *2nd International Wood and Natural Fibre Composites Symposium, Kassel/Germany, 28–29 June 1999*. Institut für Werkstofftechnik, Kunststoff- und Recyclingtechnik, Universität GH Kassel, Kassel, Germany, pp. 21-1–21-13 [in German].

Müssig J., Schlüter M. and Dreyer, J. (2003) Neues nachhaltiges Produktionsverfahren zur Herstellung innovativer Materialien für technische Anwendungen und Textilien mittels biotechnologischer Modifikation von Naturfasern. Presentation in the Workshop *Nachhaltigkeit biotechnologischer Prozesse und Produkte des BMBF-Förderschwerpunkt – Nachhaltige BioProduktion des Bundesministerium für Bildung und Forschung BMBF*. Projektträger Jülich – PTJ, DECHEMA e.V. (Veranst.), Frankfurt/Main, 3–4 November 2003.

Müssig, J., Fischer, H. and Bluhm, C. (2005) Natural fibre reinforced plastics based on enzyme treated bast fibres – mechanical properties and odour behaviour. In: Messe Erfurt AG and Forschungsvereinigung Werkstoffe aus Nachwachsenden Rohstoffen e.V. Rudolstadt (WNR) (Organizer and Editor): *Materials Made from Renewable Resources*, 5th International Symposium, 1–2 September 2005, Erfurt, Germany. Messe Erfurt and WNR, p. 54.

Müssig, J., Cescutti, G. and Fischer, H. (2006) Le management de la qualité appliqué à l'emploi des fibres naturelles dans l'industrie. In: Bouloc, P. (ed.) *Le chanvre industriel — production et utilisations*. Editions France Agricole, Paris, pp. 235–269 [in French].

Ouajai, S. and Shanks, R.A. (2005) Morphology and structure of hemp fibre after bioscouring. *Macromolecular Bioscience* 5, 124–134.

Qu, L.J., Zhu, S.F., Liu, M.J. and Wang, S.Y. (2005) The mechanism and technology parameters optimization of alkali-H_2O_2 one-bath cooking and bleaching of hemp. *Journal of Applied Polymer Science* 97, 2279–2285.

Schlüter, M., Meyer, M., Risse, S., Räbiger, N., Fischer, H., Müssig, J. and Bluhm, C. (2003) Sustainable production of high quality hemp fibres by enzymatic modification. In: DECHEMA e.V. (ed.) *27th International Exhibition-Congress on Chemical Engineering, Environmental Protection and Biotechnology*, 19–23 May 2003, Frankfurt am Main. Abstracts of the lecture groups Bioprocess Engineering, Tissue Engineering and Organ Repair, Systems Biology, IT in Life Sciences, Targets and Drugs, Protein Production, Food Technology, Pharmaceutical Production Technology. DECHEMA, Frankfurt/Main, Germany, p. 103.

Schnegelsberg, G. (1999) *Handbook of Fibres – Theory and Systematic of Fibres*. Theories and Systems in Technique and Economy, Volume 1. Deutscher Fachverlag, Frankfurt am Main, Germany [in German].

TAPPI Standard UM 246 (1991) Tappi Method UM 246. Micro kappa number. In: *TAPPI Useful Methods 1991*. TAPPI Press, Atlanta.

Thomé, O.W. (1905) Flora von Deutschland, Österreich und der Schweiz. Band 3, 2. Auflage, Gera, Germany, Verlag Friedrich von Zezschwitz, Verlag Flora von Deutschland (this book is part of www.biolib.de, Kurt Stueber, 2007), Flachs, Tafel 387, 1905.

TOSCANAPA (2003) Project TOSCANAPA financed by the Tuscany Region, Italy 2003 (http://www.toscanapa.it, accessed 16/10/2004).

van den Oever, M.J.A., Bas, N., van Soest, L.J.M., Melis, C. and van Dam, J.E.G. (2003) Improved method for fibre content and quality analysis and their application to flax genetic diversity investigations. *Industrial Crops and Products* 18, 231–243.

VDA Verband der Automobilindustrie (Hrsg.) (2005) Auto Jahresbericht 2005. Frankfurt/Main (http://www.vda.de/de/publikationen/publikationen_downloads/index.html, accessed 01/10/2008) ISSN 0171-4317.

VDA Verband der Automobilindustrie (Hrsg.) (2006) Auto Jahresbericht 2006. Frankfurt/Main (http://www.vda.de/de/publikationen/publikationen_downloads/index.html, accessed 01/10/2008) ISSN 0171-4317.

Vignon, M.R. and Garcia-Jaldon, C. (1996) Structural features of the pectic polysaccharides isolated from retted hemp bast fibres. *Carbohydrate Research* 296(1–4), 249–260.

Vincent, J.F.V. (2000) A unified nomenclature for plant fibres for industrial use. *Applied Composite Materials* 7, 269–271.

Wang, H.M., Postle, R., Kessler, R.W. and Kessler, W. (2003) Removing pectin and lignin during chemical processing of hemp for textile applications. *Textile Research Journal* 73, 664–669.

Zhang, J., Henriksson, G. and Johansson, G. (2000) Polygalacturonase is the key component in enzymatic retting of flax. *Journal of Biotechnology* 81, 85–89.

16

Plant Dyes

R. Siva

Introduction

The colours exhibited by various organic and inorganic molecules are a natural phenomenon, the perception of which requires light. Visible 'white' light such as that produced by the sun is actually a mixture of diversely coloured light rays. When a beam of sunlight passes through a glass prism, it separates into bands of many different colours in the following order: violet, indigo, blue, green, yellow, orange and red; abbreviated VIBGYOR. The differently coloured light rays have different wavelengths in the visible region of 400–800 nm, with red wavelengths being the longest and violet wavelengths being the shortest. This separation and arrangement of individual colours is known as the colour spectrum. The absorption of light depends on the structure or constitution of the colourant molecules containing various chromophores, which absorb light at different wavelengths to display a plethora of colours.

When light strikes a fabric surface, it absorbs some wavelengths and reflects others. For example, fabric material perceived blue by the human eye is absorbing violet, indigo, green, orange and red wavelengths but reflecting the blue colour to be captured by the eye. But if a surface reflects all wavelengths and absorbs none, the eye perceives the material to be white, because white light encompasses the full spectrum. If the surface reflects none of the wavelengths but absorbs all of them, the fabric is perceived to be black, which is the absence of reflected wavelengths.

Natural colouring matters are dyes and pigments occurring in plants, animals, birds, insects, bacteria, fungi and minerals. A spectrum of beautiful colours ranging from yellow to black exists in the above sources. Many plants and some animals have been identified as a potential source of colouring content and some of them have been used as natural dyes for quite some time. Dyeing is the process of extracting colour-producing agents from plants and animals and their application to the surface of an object to change its colour. Dyes are coloured substances that have an affinity with the substrate to which they are applied. Dye is generally applied in an aqueous solution and may require a mordant to improve the fastness of the dye on to the fibre.

Plants possess a wide array of colours in different parts of their body like leaves, stem, flowers, root, etc., and they do have a reason for retaining those colours. For example, green pigment in leaves contains chlorophyll essential for capturing the sun's energy to convert it to chemical energy for plant use. Colours in flowers are adaptations that attract insects and other animals that, in turn, pollinate and help plants reproduce. Some plants have colourful fruits which attract animals that eat them, inadvertently spreading the plant's seeds in their faeces. Other pigments may help protect plants from diseases. There are thousands of plant pigments, but presently the role of only a few is known.

Although plants exhibit a wide range of colours, not all of these pigments can be used as dyes. There are many reasons for this: some will not dissolve in water, some cannot adsorb on to fibres, whereas others will fade when washed or exposed to air or sunlight. It is not obvious from looking at plant colours which one will produce vibrant dyes. Some plants may have more than one colour suitable for extraction as dye. The hue may vary according to the time of year the plant is picked, the growth condition and the nature of the soil, among other things.

History

The use and origin of natural dyes, dyestuffs and dyeing has a long past and many dyes go back to prehistory. There is no doubt that primitive man tried different plants for their various properties, including colouring. When they plucked fruits or flowers from the plants, it allowed them to observe the staining properties of various colours. Human bones in prehistoric Neolithic graves have been found powdered with coloured mineral pigments, suggesting that these people used the colours of the earth to add decoration to their clothing and bodies. Men used natural matter to stain hides, decorate shells and feathers, and paint their story on the walls of ancient caves. Colours from different berries were mixed with mud for use in cave paintings. The Bhimbhetka Cave Paintings in India are the site of ancient cave paintings dating from the Upper Paleolithic and Mesolithic period (8000–5000 BC) (Fig. 16.1).

Primitive dyeing techniques included sticking plants to fabric or rubbing crushed pigments into the cloth. The methods became more sophisticated with time and techniques were developed using natural dyes from crushed fruits, berries and other plants, which were boiled into the fabric and which gave light and water fastness

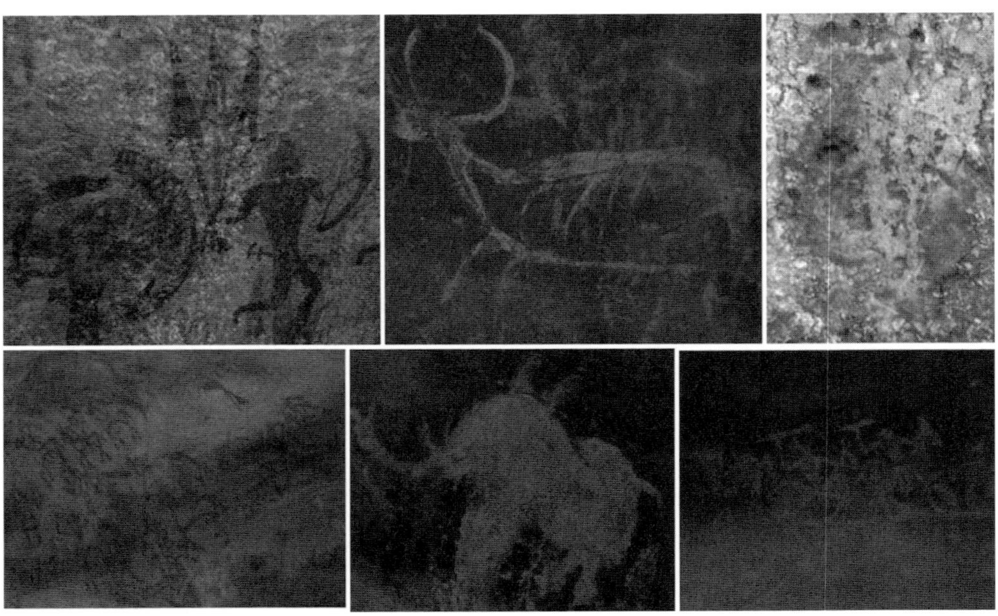

Fig. 16.1. Bhimhetka Cave Paintings with natural dye.

(resistance). Some well-known ancient dyes include madder, a red dye made from the roots of *Rubia tinctorum*, blue indigo from the leaves of *Indigofera tinctoria* and yellow from the stigmas of the saffron plant and from turmeric.

After the accidental synthesis of mauveine by William Henry Perkin in Germany in 1856 and its subsequent commercialization, the use of coal tar dyes began to compete with natural dyes. The advent of synthetic dyes caused rapid decline in the use of natural dyes, which were replaced completely by the former within a century. Today, dyeing is a complex and specialized science. Nearly all dyestuffs are now produced from synthetic compounds. This means that costs have been greatly reduced and certain application and wear characteristics have been greatly enhanced. But many practitioners of the craft of natural dyeing (i.e. using naturally occurring sources of dye) maintain that natural dyes have a far superior aesthetic quality which is much more pleasing to the eye. On the other hand, many commercial practitioners feel that natural dyes are non-viable on the grounds of both quality and economics. In the West, natural dyeing is now practised only as a handicraft, synthetic dyes being used in all commercial applications. Some craft spinners, weavers and knitters use natural dyes as a particular feature of their work. Some dyes for special purposes continue to be derived from natural sources; for example, dyes for lipstick are still obtained from *Bixa orellana* and *Lithospermum erythrorhizon* and eye shadow from indigo.

Recently, apprehension has grown that synthetic dyes release harmful chemicals that are allergenic, carcinogenic and detrimental to human health. Ironically, Germany, which invented the synthetic dye, became the first country to ban certain azo dyes (compounds bearing the functional group R-N=N-R′, in which R and R′ can be either aryl or alkyl) in 1996 due to adverse environmental concerns (Singh and Singh, 2002).

Dyes might have been discovered accidentally, but their use has become such a part of humanity's customs that it is very difficult to imagine a modern world without dyes. Without the multicoloured dresses of men, women and children, the world would certainly be drab.

Plant Dyes and Modrants

All plants contain various pigment compounds, but only a select few possess colours of choice with the right properties for dye use. Some important dye colours and their principal plant sources are: red from *Carthamus tinctorius*, *Caesalpinia sappan*, *R. tinctorum*, *Haematoxylon campechianum*, *Rumex dentatus*, *Morinda tinctoria*, *Mallotus philippinensis* and *Coccus lacca*; yellow from *Solidago grandis*, *Tectona grandis*, *Tagetes* species, *Crocus sativus* and *Butea monosperma*; blue colour from *Indigofera* species and *Isatis tinctoria*; black from *Alnus glutinosa*, *Loranthus pentapetalus* and *Terminalia chebula*; orange from *B. orellana* and *Dahlia* species.

A mordant is defined as a substance which aids the chemical reaction between the dye and the fibre in order for dye to be absorbed. Not all dyes need mordant to help them adhere to the fabric. Dyes needing no mordant, such as lichens and walnut hulls, are called substantive. However, if a mordant is required for adherence, such dyes are termed adjective. In general, natural dyes are substantive, requiring mordant to fix to the fabric and prevent the colour from either fading with exposure to light or washing out. While some fabrics such as silk and wool can be coloured simply by dipping in the dye, others such as cotton require a mordant.

There are three types of mordants: (i) metallic – metal salts of aluminum, chromium, iron, copper and tin; (ii) tannin – myrobalan and sumach, commonly used in the textile industry (for black colour); and (iii) oil – mainly used in dyeing turkey-red colour from madder. The main function of oil mordant is to form a complex with alum, used as the main mordent. The most common mordants are alum (usually used with cream of tartar, which helps evenness and brightens slightly); iron (or copper) (which saddens or darkens colours, bringing out green shades); tin (usually used with cream of tartar, which blooms or brightens colours, especially reds, oranges and yellows); and blue vitriol (which saddens colours and brings out green shades).

Moderating methods can be separated into three categories: (i) pre-mordanting – substrate is treated with the mordant and then dyed; (ii) meta-mordanting – mordant is added to the dye bath itself; and (iii) post-mordanting – dyed material treated with a mordant. The methods have different effects on the shade obtained after dyeing, and also on the fastness properties. Shade of colour also depends on the dye and the substrate. It is therefore necessary to choose a proper method to obtain the required shade and fastness by optimization of parameters. Containers used for dyeing must be non-reactive (enamel, stainless steel). Brass, copper or iron pots will do their own mordanting.

Important Dye-yielding Plants

The colours produced by plant dyes include red, yellow, blue, brown and combinations of these. Almost all parts of the plants such as the root, bark, leaf, fruit, wood, seed and flower produce dyes. Over 2000 pigments are synthesized by various plant parts, of which only about 160 have been exploited commercially. Nearly 450 taxa are known to yield dyes in India alone (Chandramouli, 1995), of which at least 50 plants are considered important commercially; 10 of these are from roots, 4 from bark, 5 from leaves, 7 from flowers, 7 from fruit, 3 from seeds, 8 from wood and 3 from gums and resins. Table 16.1 shows some of the important traditional dye-yielding plants and their exploited pigment. The content or amount of dye present in a plant varies greatly, depending on the season as well as the age of the plant.

Table 16.1. Characteristics of important dye-yielding plants.

Plant name	Dye colour	Dye active ingredient	Dye distribution in plant
Catechu, Black cutch	Brown, black	Catechin and catechutanic acid	Dye constituents distributed throughout the heartwood, 4–7% by weight
Malabar nut	Yellow	Adhatodic acid, carotein, lutolin, quercetin	
Annatto	Orange, red	Bixin, norbixin	Dye content 5–6% of seed weight
Flame of the forest	Yellow or orange	Butrin	
Safflower	Yellow, red	Carthamin	Carthamin 3–6% of the weight of the flower
Turmeric	Yellow	Curcumin	Curcumin 5.4–8.7 of the weight of the rhizome
Indigo plant	Blue	Indigotin, indicant	Best grade contains 70–90% indigotin in dried leaves; content varies according to season and age of the plant
Henna	Orange	Lawsone	Lawsone in 1.0–1.4% concentration present in the dried leaves
Kamala tree	Red	Rottlerin	Rottlerin 1.4–3.7% of fruit fresh weight
Great morinda, Indian mulberry	Yellow, red	Morindone	Roots used for dyeing, 3- to 4-year-old roots dug, dried and sorted for use
Indian madder	Red	Alizarin, rubicholric acid	
Red sandalwood	Red	Santalin	≈ 16% colouring matter santalin (santalic acid) in the wood
Punica granatum	Yellow	Petargonidon 3,5, diglucoside	

Several of the plants used for dye extraction are classified as medicinal, and some among them have been shown to possess remarkable antimicrobial activity. *Punica granatum* L. and many other common natural dyes are reported as potent antimicrobial agents owing to the presence of a large amount of tannins (Hussein *et al.*, 1997). Several other sources of plant dyes rich in naphthoquinones, such as lawsone from *Lawsonia inermis* L. (henna), juglone from walnut and lapachol from alkanet, are reported to exhibit antibacterial and antifungal activity (Gerson, 1975; Wagner *et al.*, 1989).

Singh *et al.* (2005) studied the antimicrobial activities of some natural dyes. The optimized natural dye powders of *Acacia catechu*, *Kerria lacca*, *R. cordifolia* and *R. maritimus* obtained from commercial industries show antimicrobial activities. Lycopene – a carotenoid pigment responsible for the red colour in tomato, watermelon, carrot and other fruit and used as a colour ingredient in many food formulations – has received considerable attention in recent years because of its possible role in the prevention of chronic diseases such as prostate cancer (Clinton, 1998; Rao and Agarwal, 1999).

Chemistry of Natural Dyes

A dye molecule has two principal chemical groups, viz. chromophores and auxochromes. The chromophore, usually an aromatic ring, is associated with the colouring property. It has unsaturated bonds such as -C=C, =C=O, -C-S, =C-NH, -CH=N-, -N=N- and -N=O, whose number decides the intensity of colour. The auxochrome helps the dye molecule to combine with the substrate, thus imparting colour to the latter (Krishnamurthy, 1999).

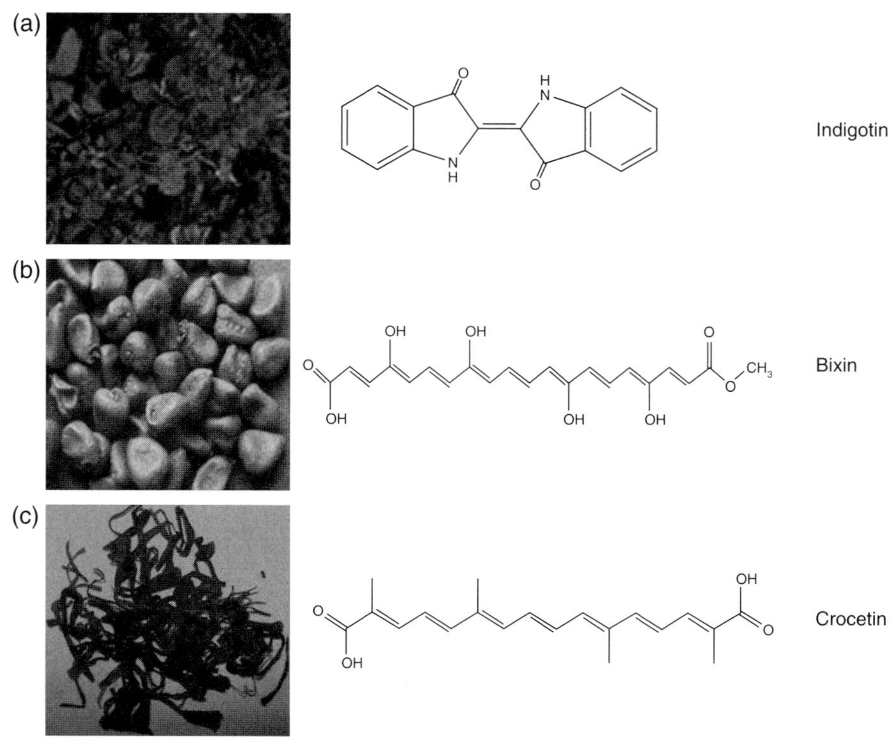

Fig. 16.2. Chemical structure of important pigments from plants: (a) *Indigofera indica*; (b) *Bixa orellana*; (c) *Crocus sativus*.

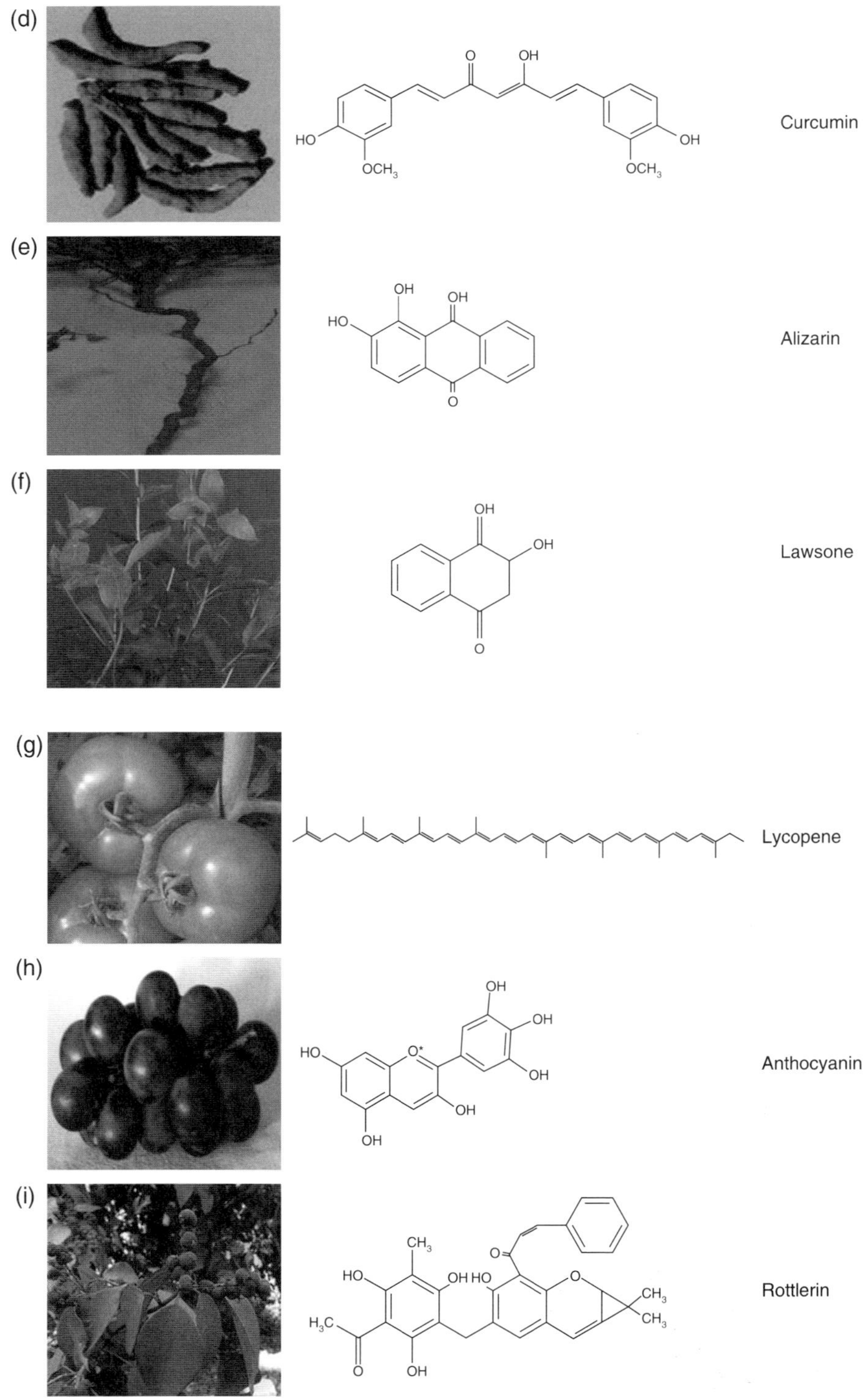

Fig. 16.2. Continued. (d) *Curcuma longa*; (e) *Oldenlandia umbellata*; (f) *Lawsonia inermis*; (g) *Lycopersicon esculentum*; (h) *Vitis vinifera*; (i) *Mallotus phillipinensis*.

Types of Dyes

Dyes can be classified based on their chemical structure, sources, method of application, colour, etc. A brief description of important dyes follows.

Indigo

This is one of the most ancient natural dyes used in textiles by humans. It is also considered to be the most important dye of plant origin. The indigo plant grows to about 0.6–0.9 m high and is cultivated in India and other countries. The whole plant is used for the extraction of dye and the extracted dye is supplied in powder form. The dye gives a deep blue colour on wool, silk, cotton, etc.

Anthroquinone

Some of the most important red dyes are based on the anthroquinone structure. These are obtained from both plants and insects. They form complexes with metal salts and the resultant metal–complex dyes have good fastness to light.

While madder and cochineal from insects are the most common sources of anthraquinone, it is also found in senna pods, cascara sagrada, alder buckthorn, rhubarb root, yellow dock and aloes.

Naphthoquinones

Lawsone (2,4-hydroxy, 1-4 naphthoquinone), otherwise known as hennaotannic acid, is the most prominent member of this class of dye obtained from *L. inermis*.

Flavones

Many pale-yellow flowers, such as carnations and *Cyclamen persicum*, contain 2′,4,4′,6′-tetrahydroxy chalcone (THC) 2′glucoside (isosalipurposide). *Carthamus tinctorius*, commonly known as safflower, produces a rare yellow chalcone glucoside, safflomin A, which is used as a colourant. Most of the Asteraceae family produces pale-yellow 6′-deoxy chalcones. Aurones, a class of rare flavonoids, give brighter yellow flowers than chalcones and are found in a limited number of species, such as snapdragon, cosmos and members of the genus *Limonium*. Flavonols and flavones are very pale-yellow and are mostly invisible to the human eye (Tanaka *et al*., 2008). Most of the natural yellow colours are hydroxyl and methoxy derivatives of flavones and isoflavones. As they absorb UV, which insects recognize, they give colour and patterns to flowers to attract insects.

Anthocyanins

A total of 19 types of anthocyanidins, aglycons or chromophores of anthocyanins are known, but there are only six major ones: pelargonidin, cyanidin, peonidin, delphinidin, petunidin and malvidin. Their colour depends greatly on the number of hydroxyl groups on the B-ring: the larger the number of groups, the bluer the colour. *O*-Methylation of anthocyanins has a slight reddening effect.

Carotenoids

Carotenoids are isoprenoid compounds (mostly C-40) with polyene chains that may contain up to 16 conjugated double bonds. More than 700 naturally occurring carotenoids are known. Carotenoids differ from anthocyanins and betalains in that they play essential roles in plant life. They are reported to have photoprotective functions during photosynthesis (Green and Durnford, 1996; Niyogi, 2000) and provide substrates for biosynthesis of the plant growth regulator, abscisic acid (ABA), and perhaps other hormones as well (Auldridge *et al*., 2006). Carotenoids such as zeaxanthin, violaxanthin, antherxanthin and lutein invariably are found in leaves and stems. Capsanthin and capsorbin, keto-carotenoids that contain one and two acyl-cyclo-pentanol rings, respectively, are the typical carotenoids of red pepper, while tomato is rich in lycopene. The typical examples of this group as dyestuffs are annatto and saffron.

Dye Extraction and Fabric Dyeing Process

The dye is generally prepared by boiling the crushed powder with water, but sometimes it is left to steep in cold water. The solution then obtained is used generally to dye coarse cotton fabrics. Alum is generally used as a mordant. The flowers of *B. monosperma* yield an orange-coloured dye, which is not fast and is washed away easily. For the purpose of colouring, the material is steeped in a hot or cold decoction of the flowers. A more permanent colour is produced either by first preparing the cloth with alum and wood ash, or by adding these substances to the dye-bath. The indigo dye is produced by steeping the plant in water and allowing it to ferment. This is followed by oxidation of the solution with air in a separate vessel. *M. philippinensis* yields an orange colour, used for dyeing silk and wool. To prepare the annatto dye, the fruits are collected when nearly ripe. The seeds and pulp are removed from the mature fruit and macerated with water. Thereafter, they are either ground into an 'annatto paste' or dried and marketed as annatto seeds. Sometimes, when the seeds and pulp are macerated with water, the product is strained through a sieve and the colouring matter which settles out is collected, partially evaporated by heat and finally dried in the sun.

The technology for the production of natural dyes could vary from simple aqueous to complicated solvent systems to sophisticated supercritical fluid extraction techniques, depending on the product and purity required. Purification may entail filtration, reverse osmosis or preparatory HPLC (high pressure liquid chromatography), and drying of the product may be by spray, under vacuum or using a freeze-drying technique.

Advantages and Limitations of Plant Dyes

Plant dyes are less toxic and polluting than synthetic dyes. While some of the synthetic dyes are believed to be carcinogenic, no such threat is associated with plant dyes. Further, plant dyes provide harmonizing, gentle, soft and subtle colours that create a restful effect.

Dyes from plants, however, have several limitations. Tedious extraction of the colouring component from the raw material, low colour value and longer dyeing time make the cost of dyeing with natural dyes considerably higher than with synthetic dyes. Some of the natural dyes are fugitive and need a mordant for enhancement of their fastness properties. A number of metallic mordants used with natural dyes are hazardous. There are a number of problems, like the availability of dye species and adequacy of harvest volume, difficulties in the collection of plants, lack of standardization, inadequate technical knowledge of extraction and dyeing, that limit the wide-scale commercial application of plant dyes.

Conclusion

Until the turn of the 20th century, all colours came from the natural world, as there was no other means by which they could be derived. Therefore, they were widely used and traded. Nowadays, most of the colours used in commercial textile dyeing are synthetic. They are synthesized, by various means, from by-products of fossil fuels, e.g. aniline and other aromatic derivatives. Natural dyes are rarely used in modern dyeing, except by specialist companies and craft dyers.

It is a common misconception that natural dyes produce only beiges and browns and washed-out shades. In reality, vibrant, fast, natural colours can be produced which are comparable with and often surpass the colours of synthetics.

Although a large number of plants possess the colouring property, only a few have been exploited so far. More detailed studies and scientific investigations need to be made in order to assess the real potential and availability of natural dye-yielding resources and for the propagation of species in great demand on a commercial scale.

Biotechnological and other modern techniques are required to improve the quality and quantity of dye production.

Lack of the availability of precise technical knowledge on extraction and dyeing techniques is a handicap for wide-scale commercial dye use. The extra time needed for dyeing with natural dyes pushes the cost considerably higher than with synthetic dyes. Finally, there is an urgent need for proper collection, documentation, assessment and characterization of dye-yielding plants and their dyes, as well as research on overcoming the limitation of natural dyes.

References

Auldridge, M.E., McCarty, D.R. and Klee, H.J. (2006) Plant carotenoid cleavage oxygenase and their apocarotenoid products. *Current Opinion in Plant Bio*logy 9, 316–321.

Chandramouli, K.V. (1995) *Sources of Natural Dyes in India – A Compendium with Regional Names.* Patriotic and People Oriented Science and Technology (PPST) Foundation, Adayar, Madras, India.

Clinton S.K. (1998) Lycopene: chemistry, biology, and implications for human health and disease. *Nutrition Reviews* 56, 35–51.

Gerson, H. (1975) Fungi toxicity of 1,4-napthoquinones to *Candida albicans* and *Trichophyton mentagrophytes. Canadian Journal of Microbio*logy 21,197–205.

Green, B.R. and Durnford, D.G. (1996) The chlorophyll–carotenoid proteins of oxygenic photosynthesis. *Annual Review of Plant Physiology and Plant Molecular Biology* 47, 685–714.

Hussein, S.A.M., Barakat, H.H., Merfort, I. and Nawwar, M.A.M. (1997) Tannins from the leaves of *Punica granatum. Photochemistry* 45, 819–823.

Krishnamurthy, K.V. (1999) *Methods in Cell Wall Cytochemistry*, CRC Press, Boca Raton.

Niyogi, K. (2000) Safety valves for photosynthesis. *Current Opinion in Plant Bio*logy 3, 455–460.

Rao, A.V. and Agarwal, S. (1999) Role of lycopene as antioxidant carotenoid in the prevention of chronic diseases. *Nutrition Research* 19, 305–323.

Singh, R., Jain, A., Panwan, S., Gupta, D. and Khare, S.K. (2005) Antimicrobial activity of natural dyes. *Dyes and Pigments* 66, 99–102.

Singh, V. and Singh, R.V. (2002) Healthy hues. *Down to Earth* 11, 25–31.

Tanaka, Y., Sasaki, N. and Ohmiya, A. (2008) Biosynthesis of plant pigments: anthocyanins, betalains and carotenoids. *The Plant Journal* 54, 733–749.

Wagner, H., Kreher, B., Lotter, H., Hamburger, M.O. and Cordell, G.A (1989) Structure determination of new isomeric naphthol[2,3,-b] furan-4,9-diones from *Tabebuia avellanedae* by the selective INEPT technique. *Helvetica Chimica Acta* 72, 659–667.

17

Natural Rubber

Hari P. Singh and Bharat P. Singh

Sources of Natural Rubber

Over 7500 plant species spread over seven families and 300 genera synthesize rubber (Cornish *et al.*, 1993). However, the amount and quality of rubber produced by the majority does not meet the standards required for commercial production. Some well-known examples of rubber sources are the commercial rubber tree, *Hevea brasiliensis*, *Euphobiaceae* family; the rubber plant, *Ficus elastica*, *Moraceae* family; and guayule, *Parthenium argentatum*, and Russian dandelion, *Taraxacum kok-saghyz*, Compositae family. Most of the rubber-synthesizing species grow in tropical and subtropical climates, but some species such as Russian dandelion and other species of the Compositae family have temperate origins. The *Hevea* rubber tree is the original source of natural rubber (NR) and still supplies most commercial needs. The severe reaction of people allergic to *Hevea* latex products has intensified the search for alternative latex sources. Guayule has emerged as a plant producing hypoallergenic latex with commercial potential.

History of *Hevea* Rubber

The original name of the rubber tree, 'caoutchouc', is a French version of the name given by the Amazonian Indians meaning 'weeping wood'. The name was changed to India rubber tree by the British around 1770 in recognition that its rubber could wipe out from paper the marks of black lead pencil.

Some uses of rubber have been known for centuries. The Amazonian Indians made balls of rubber by smoking the milky, white latex from the rubber tree. The natives also waterproofed their outer garments and shoes by brushing with latex and allowing the coating to dry. The commercial use of NR started in 1791 with the patent by the English manufacturer, Samuel Pearl, of a method for waterproofing clothes using a solution of rubber in turpentine. The use of rubber for waterproofing was further advanced when in 1823 Scottish chemist and inventor, Charles Macintosh (1766–1843), began the manufacture of double-textured garments known as 'mackintoshes'. He circumvented the 'sticky when hot and brittle when cold' problem of single-textured rubber-coated garments by concealing the coal tar naphtha solution of rubber between two pieces of fabric.

The next giant leap in rubber commercialization is credited to the discovery in 1839 by American inventor, Charles Goodyear (1800–1860), of a process that turned rubber hard and prevented it from softening or dissolving in various solvents. The process consisted of mixing rubber with sulfur and litharge (PbO) on a hot stove, which converted it into a heavily cross-linked insoluble and infusible compound. The cross-linking between linear chains of rubber molecules allowed the rubber material to retain elasticity but prevented slippage of linkages. This curing process was called 'vulcanization' after the ancient Roman god of fire and metalworking. The chemical reactions taking place between rubber and sulfur during the vulcanization process are still not well understood. Quickly after, factories to manufacture rubber shoes and mechanical goods were started in the USA and UK.

In the early days, rubber was gathered by natives in the Amazon region, Central America and Africa and brought by ship captains to be resold to rubber manufacturers. As the demand for rubber grew, regular buying expeditions into the interiors of both South America and Africa began. Rubber was an important export commodity for Brazil until 1940 (Dean, 1987). At present, Brazil and other adjoining countries do not figure prominently as a rubber source, due mainly to the destruction of the crop by South American leaf blight (SALB), caused by *Microcyclus ulei*.

Baulkwill (1989) has narrated the domestication of the rubber tree. Henry Wickham, a naturalist, was commissioned by Clement Markham of the British India Office to collect rubber plant seeds from South America. Wickham collected about 70,000 seeds from the Rio Tapajoz region and delivered them in June 1876 to the Royal Botanic Gardens, Kew, UK, for multiplication. The seeds were germinated at Kew and of the 2700 germinated seeds, 1911 were sent to the Botanical Gardens in Sri Lanka, where 90% of them survived. During September 1877, another 100 rubber tree seedlings, specified as 'cross material', were also sent from Kew to Sri Lanka. The Royal Botanic Gardens also sent 22 seedlings, without specifying whether they were from the Wickham collection or 'cross material', to the Singapore Botanic Gardens, where they were multiplied and distributed in Malaya. Seedlings of the Wickham collection of Sri Lanka were also distributed to different parts of the world.

The first plantation of bud-grafted rubber tree seedlings was established during 1918 in Sumatra using Ct3, Ct9 and Ct38 clones (Dijkman, 1951; Tan *et al.*, 1996). Bud-grafted clones became very popular because of higher yields. Breeding and selection during those 70 years produced well-known clones such as RRIM 501, RRIM 600, RRIM 712, PB 217, PB 235, PB 260, RRII 105, RRIC 100, IRCA 18, IRCA 230, IRCA 331 and BPM 24 (Priyadarshan, 2003).

Of the total 7.97 million t of rubber produced during 2003, more than 90% was produced in Asian countries, with Thailand leading with 2.3 million t, followed by Indonesia, India, Malaysia, China, Vietnam, Sri Lanka, Philippines, Cambodia and Myanmar in descending order of production (Sekhar, 2004).

Composition, Chemistry and Properties of *Hevea* Latex

Composition

Natural rubber is the coagulated sap (latex) obtained from the rubber tree. The latex is a polydisperse aqueous serum (C-serum) with suspension of various types of negatively charged particles. C-serum also contains water-soluble proteins and enzymes and can be considered as latex cytosol. Of the most important in the serum cytosol is the presence of rubber biosynthesis pathway enzymes (Koyama *et al.*, 1996; Tangpakdee, *et al.*, 1997) and rubber biochemical intermediaries (Wititsuwannakul *et al.*, 1990).

Rubber particles and lutoid particles are the two major particulate matters in the suspension constituting 30–45% and 10–12%, respectively, of the total mass. A third type of particle known as Frey–Wyssling complexes are also present.

Rubber particles range in size from 0.01 to 5.0 μm, with the majority between 0.1–2.0 μm and the most frequent particle size in the vicinity of 1.0 μm (Wycherley, 1992). Schoon and van der Bie (1955) suggested that the range in particulate size resulted from larger particles being formed by the association of smaller particles. The particle shape ranges from spherical to pear-shaped and, in certain cases, appears to be of a clonal character. A rubber particle of 1.0 μm contains hundreds of molecules of the hydrocarbon and is enveloped in proteins and lipids. The protein envelope is approximately 0.1 μm thick, carries a negative charge and confers colloidal stability to the rubber particles. Rubber particles contain upwards of 80 different proteins from 5 to over 200 kDa in size (Cornish, 2001).

Lutoids are membrane-bound bodies, mostly larger in size than rubber particles, suspended in B-serum. Lutoids flocculate aqueous suspension of rubber particles in latex, resulting in the formation of microfloccs (Southorn and Edwin, 1968). Flocculating apparently is moderated by the ambient C-serum.

Frey–Wyssling complexes are much less numerous than lutoids. They are spheroid in shape, 3–6 μm in size and are bound by a double membrane. Dickenson (1969) suggested Frey–Wyssling complexes were the possible sites of rubber biosynthesis. The presence of carotene and polyphenoloxidase and a double membrane layer has led to its tentative identification as a type of plastid (Tangpakdee, 1998).

Chemistry

Rubbers are polymeric compounds, i.e. the molecules are large sized, composed of repeating small molecules (known as monomers) connected together by covalent chemical bonds. The monomers are made of carbon and hydrogen. English chemist, Michael Faraday (1791–1867), in 1826 first analysed and reported the empirical (simplest) formula of NR to be C_5H_8 (Fig. 17.1). In 1862, another English chemist, Charles Hanson Greville Williams (1829–1910), distilled NR to obtain the pure monomer. He further showed that the monomer, named isoprene, polymerized to a rubbery product and suggested that NR was a linear polymer of isoprene.

Rubber is an elastomer, i.e. it possesses the ability to regain its original shape after being deformed by tension, compression, or shear. Rubbers are virtually incompressible because the molecules straighten out when pulled but regain the natural molecular arrangement instantaneously when the force is released. This is possible because of high molecular weights (more than 1×10^6 Da) and a flexible polymer chain. The repeating units have the *cis* configuration [*cis* 1,4-polyisoprene, (C_5H_8)], i.e. chain extensions are on the same side of the ethylene double bond, essential for elasticity. The *trans* configuration (with chain extensions on opposite sides of the ethylene double bond) would make the polymer hard.

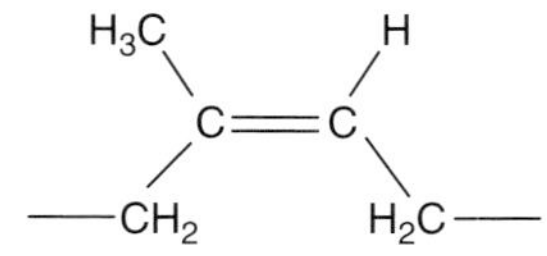

Fig. 17.1. Structure of isoprene unit.

Properties

Due to its unique molecular structure and high molecular weight, crude rubber possesses valuable properties (Table 17.1) of elasticity, plasticity, strength, durability, abrasion resistance, impact resistance, efficient heat dispersion, electrical non-conductivity and resistance to water. Artificially produced polymers have not been able to match NR for these properties.

Among the drawbacks of NR is the fact that it is thermoplastic, i.e. it hardens in winter, softens and becomes sticky in summer, can be attacked by solvents and smells bad. Severe allergy caused to some individuals by protein in the *Hevea* latex is also concerning. Complete allergy protein removal is not easy or cost-effective and impacts latex product performance negatively.

Botany, Origin and Ecology of the Rubber Tree

The genus *Hevea* comprises of ten species (Webster and Paardekooper, 1989; Wycherley, 1992). Intercrossing has been reported between species (Clement-Demange *et al.*, 2000). According to Priyadarshan and Gonçalves

Table 17.1. Properties of natural rubber (UNCTAD, 2009, with permission).

Name	Natural rubber	Natural polyisoprene
Molecular behaviour	Glass transition temperature (°C)	–70
	Melting temperature (°C)	25
	Hardness range (shore A)	30–100
	Maximum tensile strength (at 70 F, psi)	4000
	Maximum elongation (at 70 F, %)	750
Advantages	Physical resistance	Excellent resilience Excellent tear strength Excellent abrasion resistance Excellent impact strength Excellent cut growth resistance Good compression set
	Environmental resistance	Excellent water resistance Good–excellent low temperature flexibility Fair–good oxidation resistance
	Chemical resistance	Good resistance to alcohols and oxygenated solvents Fair–good resistance to acids
Limits	Environmental resistance	Poor ozone resistance Poor sunlight resistance Very little flame retardance
	Chemical resistance	Poor oil and gasoline resistance Poor resistance to (aliphatic and aromatic) hydrocarbon solvents

(2003), *H. brasiliensis* is an amphidiploid ($2n = 4x = 36$). However, for practical purposes, *Hevea* is considered a diploid genus ($2n = 2x = 36$). Further details on the taxonomy of *Hevea* can be obtained from Wycherley (1992) and Priyadarshan and Clement-Demange (2004).

H. brasiliensis is a tall, erect tree with fairly smooth and grey-coloured bark (Fig. 17.2). The trunk is cylindrical and relatively slim but frequently swollen towards the base, having upward extending branches. The rubber tree can attain a height over 40 m in the wild, but usually is 20–25 m tall under cultivation. Tree lifespan is over 100 years. The root system is made up of a taproot and multiple creeping roots. The leaf is lanceolate-elliptic to broadly lanceolate, long acuminate, basally acute or cuneate, varying between 3–11 cm long and 2–8 cm wide, mid-vein extending to the apex but not calloused. The arrangement of the leaves is storied; each storey consisting of a cluster of spirally arranged trifoliate glabrous leaves with nectaries in the leaflet insertion region. The rubber tree has a deciduous habit; wintering lasts for 6–10 weeks when the old leaves are shed, followed by formation of new leaves and flowering. The flower is incomplete, monoecious, produced in pyramid-shaped panicles in the axils of leaves. The male flowers are small and numerous and female flowers are larger. The ovary is tricarpellary syncarpous, which forms into a three-lobed dehiscent capsule holding the seed. Flowers are bright or creamy yellow having an extremely pungent aroma. The rubber tree flowers once a year and the flowers are cross-pollinated by insects. The fruits are large with hard outer coats and ripen in 5–6 months after fertilization. Seeds are variable in size and shape but usually ellipsoidal, ventrally somewhat compressed, 15–32 mm long and 15–24 mm wide, basally grey-brown with dark-brown mottling and contain an oily endosperm.

The natural habitat of *Hevea* species is the Amazon basin of South and Central America, covering Bolivia, Brazil, Colombia, Ecuador, French Guiana, Guyana, Peru, Surinam and Venezuela. However, very limited rubber tree cultivation is practised in South America due

Fig. 17.2. Rubber plantation in Vietnam's Dak Lak Province.

to severe damage to the trees by blight in the early 20th century.

The rubber tree prefers a tropical climate with the temperature in the range of 24–28°C and annual rainfall of 2000–4000 mm spread throughout the year. Temperature and rainfall define its prime growing areas between 10° latitudes on either side of the equator. New cultivars needing rainfall as low as 1500 mm/year and dry season of up to 5 months have now been developed. Rubber trees grow satisfactorily up to 600 m elevation and in most soils with adequate drainage. High winds in growing areas cause severe damage to the trees.

Rubber Biosynthesis

The latex vessel is the factory for rubber synthesis in the *Hevea* plant. The raw materials for the synthesis of rubber come from different sources. Water and minerals absorbed from the soil are carried in the xylem vessels located in the wood near the cambium. The source of sugars, amino acids, hormones and other molecules are leaves and are carried by sieve tubes, also situated close to the cambium. The latex vessels are not connected directly to these raw material sources, but a horizontal system composed of rays provides the necessary connection.

The rubber biosynthesis pathway is presented in Fig. 17.3. The building unit of rubber is the compound isopentenyl pyrophosphate (IPP). Tracer experiments led to the discovery of the mevalonic acid (MVA) pathway for IPP synthesis (Bandurski and Teas, 1957; Lynen, 1969; Chappell, 1995). Later, an alternate pathway, known as methylerythritol 4-phosphate (MEP), was discovered (Rohmer *et al.*, 1993; Schwarz, 1994). The two pathways are localized in different compartments; while the MVA pathway produces IPP in the cytosol, the site for the MEP pathway IPP is the plastids. The high volume of latex produced would suggest the presence of both pathways in the rubber tree, although experimental evidence is yet to be provided (Chow *et al.*, 2007).

The catalyst for polymerization into rubber compound is a rubber particle-bound enzyme or enzyme complex (Madhavan

(a)

Cytoplasm

Gycolysis

↓ *Pyruvate kinase*

Pyruvate

⇣ *Pyruvate dehydrogenase complex*

Acetyl-CoA

↓ *Acetyl-CoA acetyl transferase*

Acetoacetyl-CoA

↓ *HMG-CoA synthase*

HMG-CoA (3-hydroxy-3-methylglutaryl-CoA)

↓ *HMG-CoA reductase*

Mevalonate

↓ *Mevalonate kinase*

Mevalonate phosphate

↓ *Phosphomevalonate kinase*

Mevalonate pyrophosphate

↓ *Mevalonate-5-pyrophosphate decarboxylase*

IPP (Isopentenyl-5-pyrophosphate)

(b)

Plastid

Pyruvate + GA3P (glyceraldehyde-3-phosphate)

↓ *Deoxy-D-xylulose phosphate synthase*

DXP (1-deoxy-D-xylulose-5-phosphate)

↓ *Deoxy-D-xylulose phosphate reductase*

MEP (2-C-methyl-D-erythritol-4-phosphate)

⇣ *Enzyme complex*

MECPP (2-C-methyl-D-erythritol-2,4-cyclopyrophosphate)

↓ *Hydroxy-methyl-butenyl pyrophosphate synthase)*

HMBPP ((E)-4-hydroxy-3-methyl-2-butenyl pyrophosphate)

↓ *Hydroxy-methyl-butenyl pyrophosphate reductase*

IPP (Isopentenyl-5-pyrophosphate)

Fig. 17.3. Biosynthesis of natural rubber: (a) MVA pathway in cytoplasm; (b) MEP pathway in plastid.

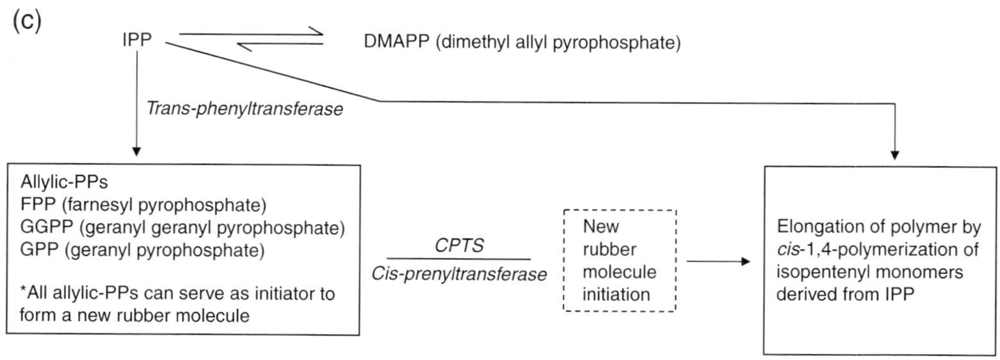

Fig. 17.3. Continued. (c) IPP to natural rubber.

et al., 1989; Cornish and Siler, 1996); the identity of rubber particle-bound enzyme transferase is not very clear. The catalytic activity has been attributed to two different proteins, viz. the 14.6 kD 'rubber elongation factor' (Dennis and Light, 1989) and the 24 kD 'small rubber particle protein' (Oh *et al.*, 1999). However, research data appear to favour the role for 24 kD protein more convincing than for 14.6 kD protein (Cornish, 1993).

Natural rubber is synthesized from allylic pyrophosphate (allylic-PP) and IPP. One allylic-PP molecule initiates the reaction and the rubber polymer is then made by the sequential condensation of IPP with magnesium cations as a cofactor releasing diphosphate at each condensation (Archer *et al.*, 1963) (Fig. 17.3). The rubber molecule is released from the enzyme at the end of elongation.

Many different allylic-PPs, including DMAPP (dimethyl allyl-PP, 5 carbons), GPP (geranyl-PP, *trans* 10 carbons), FPP (*trans*, *trans*-farnesyl-PP, all-*trans* 15 carbon) and GGPP (geranyl geranyl-PP, *trans* 20 carbon), can initiate rubber biosynthesis (Cornish and Siler, 1995). Evidence from nuclear magnetic resonance (NMR) suggests that FPP is the principal initiator of rubber biosynthesis in *H. brasiliensis*. Apparent K_m for FPP is also lower than for GPP or DMAPP (Cornish, 2001).

The rate of rubber molecule initiation increases with increased IPP and initiator concentration (Cornish and Siler, 1995). It is reasonable to expect the rate of initialization to increase with initiator concentration until all active rubber transferases have initiated a rubber molecule.

IPP incorporation rate is a function concentration when above IPP K_m (Cornish *et al.*, 2000). Very little is known about what stops molecule size elongation. Tanaka (1989) opined that molecule dephosphorylation and release might involve esterification, cyclization or hydrolysis.

The size of the rubber molecule is affected by the substrate concentration and identity. The higher the allylic-PP initiator concentration, the lower the molecular weight of the rubber biosynthesized. On the other hand, rubber molecular weight increased with the IPP concentration (Cornish and Siler, 1995; Cornish *et al.*, 2000). It was suggested that competition occurred between free allylic-PP and the allylic-PP end of the growing polymer for the allylic-PP binding site.

Cultivation of Rubber Trees

Propagation

Rubber trees are propagated vegetatively with grafted plants. There are two main grafting methods, viz. brown budding and green budding. Brown budding is the traditional method of grafting in which brown buds about 1 year old are grafted on to a rootstock of the same age. In green budding, both the rootstock and the buds are about 3–5 months old. Brown-budded stumps are hardier than green-budded stumps, but the budding success is much higher for green budding. Another advantage of green budding is that the rootstocks can be grown in polyethylene bags in the summer and transplanted directly in the field during the next planting season. Nowadays, green budding is the most accepted method of propagation. In another grafting variation, budding is performed at 7–8 weeks of age. The procedure of grafting is the same, irrespective of age.

Rootstocks are generated by germinating high quality seeds as soon as harvested because seeds deteriorate quickly. Polyethylene bags filled with potting soil are used for green budding and germinators filled with sand for brown budding. The grafting is performed when seedlings reach the desired age. Seedlings are prepared for budding by making two cuts 4–5 cm long and 2 cm apart in the stem a few centimetres above the ground and joining the two cuts at the lower end. Care is exercised to confine the cuts to the bark without injuring the wood. Usually, buds for grafting are collected from source nurseries specializing in high-yielding clones within 24 h of the operation. Woody materials associated with the buds are removed and the buds are placed under the rootstock slit and wrapped to keep the union intact. After 3 weeks, the tapes are removed and the strips of rootstock barks are cut vertically at the top. If the graft is successful, the scion should be well joined by this time. The rootstock is decapitated once the bud commences growth.

Crop management

Land is cleared manually or by machine before planting. Planting is done along the contour on sloping land. Planting density varies depending on clone, soil and climatic conditions, usually varying from 450 to 550 plants/ha.

Induction of branching, pruning and thinning are important operations for shaping plants properly. Immature tree shoots should be pruned regularly, making sure to maintain one to two leaf layers under the main shoot to create favourable conditions for photosynthesis and nutrition of the tree.

Rubber is usually raised as a monocrop. The objective of the management practices under a monocrop cropping system is to reduce the immaturity phase and maximize latex production during the mature phase. Weed control is the main priority during the early growth phase. Young trees are mulched to preserve moisture and keep weeds out from the plant vicinity. Traditionally, manual weed control and planting of a groundcover strip between the rows until the canopy closes have been practised. Leguminous creepers have been found to be the best cover crops as they provide nitrogen to the soil, grow fast, tolerate shade and drought, do not have much disease and pest problems, and are in minimum competition with the rubber plants for nutrients (Varghese and Abraham, 2005). Larger plantations also use chemical weed control for both broadleaved weeds and grasses.

Fertilization during the immature stage hastens growth to reach maturity faster. Fertilizer recommendations for different rubber growing regions vary depending on the soil nutrient content. Soil analysis as well as leaf tissue analysis is used in many areas as the basis of fertilizer recommendation to growers. George (1962) reported higher rubber yields with N, P and K fertilizer combination.

Small growers also produce intercrops for the first 2–3 years to generate income when the rubber tree is still growing. Intercrops should not be planted too close to the rubber plants to prevent competition and to avoid soil erosion on sloped lands. Irrigation is common in rubber nurseries. Traditional areas of rubber production receive high rainfall and trees are seldom irrigated. Irrigation of trees out of the traditional zone is often practised during the dry period. Jessy *et al.* (1994) reported that summer irrigation during the young phase could reduce the immaturity period by 6 months to 1 year, even in the traditional growing areas. Vijayakumar *et al.* (1998) found that in some drought prone, non-traditional rubber growing areas of India, irrigation amounting to 50% of the estimated crop water requirement shortened the immaturity period of the trees from 10 years to 6 years. Philip (1997) observed that irrigation was a good safeguard against cold winters in the north-eastern part of India.

In order to protect the plantation from summer fire, vegetation and debris from an approximately 6 m wide band around the plantation are cleared at the beginning of the dry season to create a ring of bare earth as a fire barrier.

Rubber tapping

Latex is the cytoplasmic content of laticifers from the vessel system just under the outer bark of the tree (Fig. 17.4). The laticifers are formed from chains of contiguous cells that are arranged in rings parallel to the vascular cambium. Between the laticifers in each ring, there are anastomoses that allow the withdrawal of latex from a large area of bark by means of a single tapping cut (Peng *et al.*, 2009).

Rubber trees are ready for tapping at 5–10 years of age, peak yields occur at 12–15 years of age and yields taper to an uneconomic level by the time trees are 25–30 years old. Usually, the criterion to start tapping is when the tree trunk attains a circumference of 50 cm at 1 m high. Rubber production fluctuates with the season and is normally low during the rainy season.

Tapping involves periodically cutting bark on the trunk to sever latex vessels (Fig. 17.5). It is started at the highest conveniently reachable point for the tapper. The cut is made slanting down at an angle of 25–30° from the horizontal with a knife for the purpose of exposing the maximum number of latex vessels per length of incision. The cut penetrates within 1 mm of the cambium, depending on the skill of the tapper. Care is exercised to leave the cambium layer uninjured to allow bark to regenerate for future tapping. The latex exudes from the fresh incision for 1–3 h,

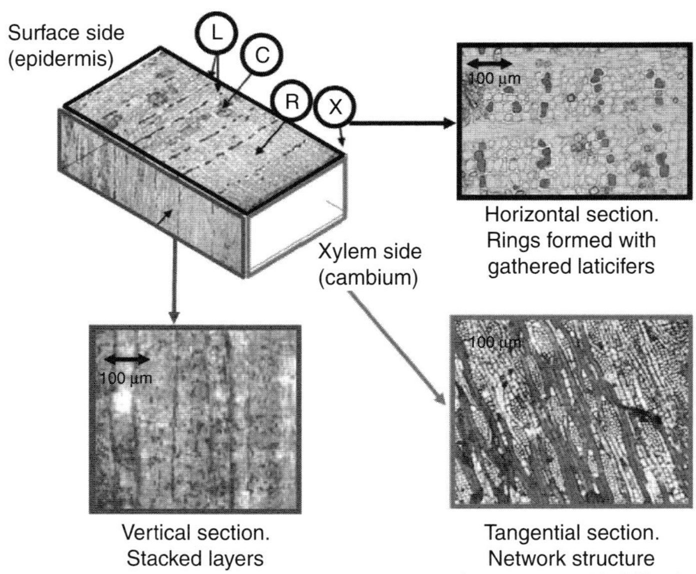

Fig. 17.4. Histological study of the laticifer of *Hevea brasiliensis*. Horizontal, tangential and vertical sections were made from vascular tissue of an 8-year-old *Hevea* tree and microscopic analysis was performed to clarify the structure of the laticifer. Oil-red-O was employed for staining the laticifer. L, laticifer; C, cork; R, ray parenchyma; X, xylem (Hayashi, 2009, with permission).

Fig. 17.5. Tapping panel on a rubber tree.

yielding about 50 g of latex. It is advisable to begin tapping early in the day when the turgor pressure of the latex is high, allowing easy flow. A small quantity of latex dries on the bark and is collected the next morning before tapping. Tapping is repeated progressively down the trunk on alternate days in half-circumference panels. One person can tap 200–300 trees in 3–4 h. Trees are usually allowed to rest after heavy tapping. Labour is the highest recurring cost of rubber cultivation.

Increase in tapping intensity, generally known as slaughter tapping, is usually followed for 2–3 years before replanting. Little consideration is made of bark consumption and wounding of the cambium; the only consideration is to extract as much latex as possible before the trees are taken out.

Latex coagulates within a few hours of tapping if preservative is not added in order to offset the actions of enzymes and bacteria. Ammonia is the most common preservative used because it can be removed easily by aeration for processes requiring low ammonia content.

Consistent gains in rubber yields have been achieved by replacement of old unselected plant material with new improved clones. Breeding has resulted in substantial yield gains, from 650 kg/ha in unselected seedlings during the 1920s to 1600 kg/ha in the best clones during the 1950s and further to 2500 kg/ha during the 1990s (Priyadarshan, 2007). Improved tapping techniques and use of stimulants have also contributed to the yield gains and reduced labour costs.

Stimulation of rubber latex production

Back in the 1950s, Rhines, working at the US Rubber Company Research Centre in Wayne, New Jersey, USA, reported that plant hormones stimulated latex flow (Rhines, 1958). A yield increase of 20–30% was obtained by applying a 1–2% solution of 2,4-D just below the tapping cut on the scratched bark two to three times a year, without damaging the tree.

Ethephon (2-chloroethyl phosphonic acid) is commonly used to regulate latex flow (Abraham *et al.*, 1968). The chemical is applied either close to or in the tapping cut. It dissolves slowly in the presence of water, releasing ethylene in the bark, which delays latex plugging and increases the duration of latex flow (Boatman, 1966). Stimulation with ethephon results in a 20–100% increase in latex yield (Varghese and Abrahams, 2005).

The puncture system of tapping is another advance in the tapping method. It involves one puncture each week on a scraped area of the bark treated with stimulant. Latex flows directly into a closed receptacle, safeguarding it from natural coagulants. The collection is made at 2- to 3-day intervals without causing latex coagulation. Puncture tapping saves in labour costs, but it is capital-intensive in equipment and chemicals and requires good management.

As laticifers are the major site of rubber biosynthesis in the rubber tree, their number affects rubber yields. Wu *et al.* (2002) found that application of jasmonates (JAs) to the bark of young rubber trees increased laticifer formation. Linolenic acid, a precursor molecule in JA synthesis, also increased secondary laticifer formation (Hao and Wu, 2000). The biosynthesis of NR is increased by tapping and endogenous accumulation or exogenous application of JA (Yu *et al.*, 2007). How the JA signal regulates rubber biosynthesis is not yet known.

Diseases and pests

South American leaf blight (SALB) caused by *M. ulei* is the most devastating disease of the rubber tree. This disease caused the demise of the rubber industry in South America and could pose an eventual global threat. The fungus responsible for the disease can be controlled by a number of fungicides. Some *Hevea* species possess resistance to the disease and some strains of susceptible species are also resistant. Other important leaf diseases include abnormal leaf fall caused by *Phytophthora* spp., leaf spot caused by *Corynespora cassicola*, and powdery mildew caused by *Oidium* spp. Unlike SALB, *C. cassicola* is prevalent in all rubber producing regions of the world and is capable of causing serious damage. Control involves the planting of resistant trees.

Pink disease caused by *Corticium salmonicolor* causes drying of the main stem of immature trees, and black stripe caused by *Phytophthora* damages the tapping panel. Root diseases deserving mention are white root disease by *Rigidoporus lignosus*, brown root disease by *Phellinus noxius* and red root disease by *Ganoderma philippii*. Tapping panel dryness is a physiological disorder in the bark tissue which can result from excessive tapping. It causes reduction and ultimate cessation of latex flow.

Insects are not a serious threat to the rubber tree. Notable among insects are termites, mites, cockchafers and thrips.

Research Needs

At the top of the research priority is the creation and selection of *Hevea* clones that produce high yields with low tapping frequency. Adaptation to marginal climates, tolerance to low temperatures and resistance to insects and pathogens are important research areas for sustainable cultivation. Rubber trees are prone to wind damage; thus, breeding for plant ediotypes suited for windy climates and development of management practices to shelter trees from high wind velocity impact are essential areas of research to reduce loss from this weather calamity. Improvements in latex quality for compatibility with technical processes and for properties important to the final rubber product such as the ratio of sol to gel rubber deserve research attention. Improving the quality and increasing the yield of lumber are becoming increasingly important to keep rubber plantations competitive with other plantation crops such as oil palm. Research aimed at developing timber-specific clones and cultural practices to optimize timber yields require acceleration to meet this objective. Honey is an important ancillary enterprise on rubber plantations. To increase honey output, research directed towards developing an apiary designed for rubber plantations and determination of the optimum bee population and tree density from multilocation trials will be necessary.

Much of the improvement in the rubber tree both in terms of latex as well as timber has been made using breeding followed by selection. Poor fruit set and non-synchronization of flowering among different cultivars are important constraints in *Hevea* breeding. Research into reproductive biology and pollen storage methods that can help alleviate this problem are under way, but deserve much concerted effort (Hamaz and Chan, 1996; Hamaz *et al.*, 1999, 2002).

Allergy to *Hevea* latex is threatening its use for intimate purposes. The International Union of Immunological Societies and the World Health Organization have identified 13 proteins responsible for latex allergy (Yeang, 2004). Research is in progress (Bernstein *et al.*, 2003; Yeang *et al.*, 2006) and deserves appropriate resources so that quantization of the concentration and understanding of the nature and behaviour of residual latex proteins can be accomplished at an early date. Once the allergens are pinpointed, diagnostics for latex allergy, immunotherapy to remove the effect of latex allergy and immunoassays to control the quality of the latex products can be developed. Breeders should be geared to developing clones devoid of allergen proteins, once known.

Breeding and Improvement

Full discussion of rubber tree breeding is beyond the scope of this chapter; readers are referred to Clement-Demange *et al.* (2007), Priyadarshan (2007) and Priyadarshan *et al.* (2009).

The main objective of rubber breeding has been improved yield, with attention to such parameters as the growth of the trunk during both the immature and mature phase, stability of the stand, resistance to stresses (such as tapping panel dryness, wind damage, moisture deficit, low temperature) and yield per tree over a specified period. Little attention has been paid to rootstock improvement.

Conventional breeding

Germplasm pool

The rubber tree germplasm pool can be separated broadly into the 'Wickham' population and the 'Amazonian' population. The Wickham collection, as mentioned previously, has its origin in the seeds collected by Henry Wickham in 1876. This population has been used extensively in rubber tree breeding. During 1981, with the initiative from the International Rubber Research and Development Board (IRRDB), a new collection of seeds and budwoods was obtained from Brazilian Amazonia.

Genetic diversity in both germplasm pools has been studied. Based on yield and various yield components, Mydin *et al.* (1992) grouped the Wickham clones into eight different clusters. Abraham (2000) studied 80 wild accessions and grouped them into nine different clusters. Thus, there is sufficient genetic diversity available on hand for breeders to find the genes needed for improvement of the desired traits.

Collection has also been made of allied species of *H. brasiliensis*. They are a potential gene pool, particularly for resistance to leaf diseases (Priyadarshan and Gonçalves, 2003).

Selection

Ortet or mother tree selection is the screening for outstanding genotypes derived from natural recombination. The clones selected in this way are called primary clones. A number of important early clones such as Tjir 1, PR 107, GT1, PB 28/59 and AVROS 255 were derived using this election procedure. At least 16 primary clones are considered prime progenitors of many modern clones (Priyadarshan, 2007). By 1930, an agreement had emerged that the primary clones had reached the yield plateau and the emphasis shifted from prime clones to recombinants produced out of controlled pollination (Tan, 1987). This led to the establishment of polyclonal seed gardens. The choice of this controlled open-pollination method was based on the knowledge that the two important characters of yield and trunk girth in rubber tree were additive in nature (Nga and Subramaniam, 1974; Tan, 1981) and thus selection based on general combining ability should be reliable and the best parent clones could be selected based on the evaluation of the seedling progenies. Simmonds (1986) emphasized that an optimum number of parents were crucial for seed gardens and suggested a layout involving nine clones with all hetero neighbours.

Hybridization and selection

Hybridization and selection has been used extensively for rubber tree improvement. Since the rubber tree is propagated by clones, once a desirable clone has been identified, it can be fixed easily through vegetative multiplication. Primary clones provided the parents with potential for use in the early hybridization programmes. Subsequent parental selection came from previously improved clones. Hybridization and selection continues to be an important technique for improving important rubber traits.

The steps in hybridization and selection consist of the production of full-sib families, followed in turn by seedling evaluation trial, small-scale clonal trial and large-scale clonal trial. Selection is practised at every step of the process. Most hybridization work has been carried out on Wickham material following generation-wise assortive mating (GAM) involving crossing of 'the best of the best', with strong emphasis on selection for high yield (Wycherley, 1976).

In order to shorten the breeding cycle and for early release of clones, efforts have been made to evaluate clones at the juvenile stage.

A modified Hamaker–Morri–Mann method, in which 2- to 3-year-old plants are tapped, has been widely used to quantify latex yields (Tan and Subramaniam, 1976). Early-age latex yield of high, medium, and low latex-yielding clones revealed that tapping tests of 2-year-old clones gave good indication of their future mature ability (Varghese *et al.*, 1993).

Molecular breeding

In vitro culture and genetic transformation

The duration of the breeding cycle and the height of the tree make rubber breeding very time-consuming. In addition, breeders have reached a yield plateau with conventional breeding methods, necessitating the use of biotechnology to induce, increase and exploit new genetic variation (Priyadarshan, 2007). The ability to produce plants by tissue culture is a prerequisite for direct gene transfer. Tissue culture with different parts of the rubber plant has been applied successfully to regenerate the whole plant.

ANTHER AND SEED CULTURE. Anther culture consists of three phases: (i) production of embryo from callus; (ii) maturity of embryo; and (iii) plant regeneration. Chen (1984) has described the suitable growth mediums for *Hevea* callus growth and embryo formation and for the differentiation of apical bud from the embryo. It takes nearly 50 days for embryo development from callus and another 2–3 months for the apical bud to develop from the embryo.

Venkatachalam *et al.* (2006) described an agrobacterium-mediated transformation protocol using immature anther-derived calli as initial explants used to generate genetically engineered plants from a high-yielding Indian clone, RRII 105. Calli were co-cultured with *Agrobacterim tumefaciens* harbouring a plasmid vector containing the Hb superoxide dismutase (SOD) gene and the reporter gene and selectable marker gene used were beta-glucuronidase (Gus) gene (uidA) and neomycin phosphotransferase (nptII), respectively. The germinated plantlets were GUS positive and the integration of uidA, nptII and HbSOD transgenes into *Hevea* genome was confirmed by polymerase chain reaction (PCR), as well as Southern blot analysis.

SOMATIC EMBRYOGENESIS AND MERISTEM CULTURE. There are four steps in somatic embryogenesis: (i) callogenesis, (ii) differentiation, (iii) multiplication and (iv) plant regeneration. The full *Hevea* regeneration procedure including the composition of the growth medium at different steps has been narrated by Carron *et al.* (1995). Significant genotype-medium (El Hadrami *et al.*, 1991) and tissue-medium interactions (Veisseire *et al.*, 1994) have been observed in this plant regeneration procedure. Instead of commonly used semisolid medium, Etienne *et al.* (1997) standardized a pulsed-air temporary immersion system in which embryonic calluses were immersed in an autoclavable filtration unit, which increased embryo production by 3–4 times, amounting to 400 embryos/g fresh weight and a lower number of abnormal embryos. Clones PR 107 and PB 260 regenerated this way were highly responsive (Carron *et al.*, 1995).

There are three steps in meristem culture: (i) primary culture, (ii) multiplication with rooting and (iii) acclimatization. Juvenile stem pieces are used for this purpose. The procedure of *Hevea* regeneration is given by Carron *et al.* (1995). The acclimatization of regenerated seedlings is very important to the establishment of plants in the soil. Clones RRII 105, PB 5/51, PB 235, IRCA 438, IRCA 440, IRCA 442, PR 107 and GT1 have been multiplied using this method (Carron *et al.*, 1995).

Molecular markers and QTLs

All marker systems, except SNPs (single nucleotide polymorphisms), have been applied in *Hevea* (Priyadarshan, 2007). Molecular studies showed the existence of three genetic groups and many alleles in the Amazonian populations (Seguin *et al.*, 1996). Lekawipat *et al.* (2003) studied diversity in 40 Wickham accessions and 68 Amazonian accessions. For this task, 170 alleles from 12 microsatellite markers spread among all genotypes were employed. An average of 14 alleles was available per locus. Amazonian accessions were clearly more variable than Wickham clones. The microsatellites of Amazonian accessions were more polymorphic than cultivated Wickham clones. The wild accessions could be divided into three clusters based on the three Amazonian regions from which they were collected. CIRAD (the Centre de Cooperation Internationale en Recherche Agronomique pour le Développement, Corsica, France) has established a set of microsatellite markers that can identify over 300 clones (Hayashi, 2009).

Lespinasse *et al.* (2000b) constructed a saturated genetic linkage map of *Hevea* species based on RFLP, AFLP, microsatellite and isozyme markers. Further, genetic determinant of the resistance source of SALB (F 4542), widely used in backcrossing programmes, were characterized by a genetic map (Lespinasse *et al.*, 2000a). Hayashi (2009) mapped QTLs for resistance to SALB using 195 F1 progeny derived from a cross between PB 260 (susceptible) and RO 38 (resistant) clones that could be used beneficially in marker-assisted breeding. Several promising SALB resistance clones have been selected by CIRAD (Guen *et al.*, 2008).

Genetics of rubber synthesis

Rubber elongation factor (REF) (Goyvaerts *et al.*, 1991), hydroxylmethylglutarylcoenzyme A reductase (Chye *et al.*, 1992) and small rubber particle protein (SRPP) (Oh *et al.*, 1999) are the three main genes expressed in latex. REF (6.1%) is the most abundantly expressed gene followed by SRPP (3.7%) (Han *et al.*, 2000). These expressed gene sequences when compared with public databases of identified genes, about 16% of the database matched ESTs encoding rubber biosynthesis-related protein. Expression of rubber synthesis genes could be a basis for selection and breeding for high rubber yields in future. Kush *et al.* (1990) suggested using transcript levels as molecular markers for early selection. The transcript levels of hydrolytic enzymes (polygalacturonase and cellulase) could indicate better laticifer development potential.

Sequence of the 3-hydroxy-3-methylglutaryl-CoA synthase (HMGS) (Sirinupong *et al.*, 2005) and 3-hydroxy-3-methylglutaryl-CoA reductase (HMGR) (Chye *et al.*, 1992) genes in the MVA pathway have been reported. Sando *et al.* (2008) obtained full-length cDNA of genes encoding all the enzymes that catalysed the six steps of the MVA pathway. Alignment analysis and polygenetic analysis revealed the presence of three acetyl-CoA acetyltransferase genes, two CHMGS genes and four HMGR genes. The MVA pathway genes in *H. brasiliensis* were found to complement MVA pathway deletion mutation in yeast. These findings are valuable to breeding for latex flow improvement in the rubber tree.

Natural Rubber Processing

Natural rubber is sold either as 'latex concentrates' or 'dry rubber'. Latex concentrate is made from freshly tapped, uncoagulated latex. On the other hand, dry rubber is made from coagulation of field latex or remilled rubber sheets. Natural rubber is processed into different types and grade based on end use. Latex concentrate is the basic constituent of contraceptives, surgical dipped goods and rubber threads. For making extra tough rubber for applications like tank liners, ribbed smoked sheets are used. Pale crepe is suited for medical sundries, footwear, cements and adhesives.

Latex concentrate

Latex concentrates are processed by centrifugation, evaporation, creaming, or electrodecantation. Commercial latex is produced mostly by centrifugation into two types: (i) HA (high ammonia) latex, preserved with 0.7% ammonia; and (ii) LA-TZ (low ammonia) latex, preserved with 0.2% ammonia + 0.04–0.05% lauric acid as ammonium laurate.

Dry rubber

Sheets

Rubber is shaped into sheets at the factory scale or cottage scale, depending on the

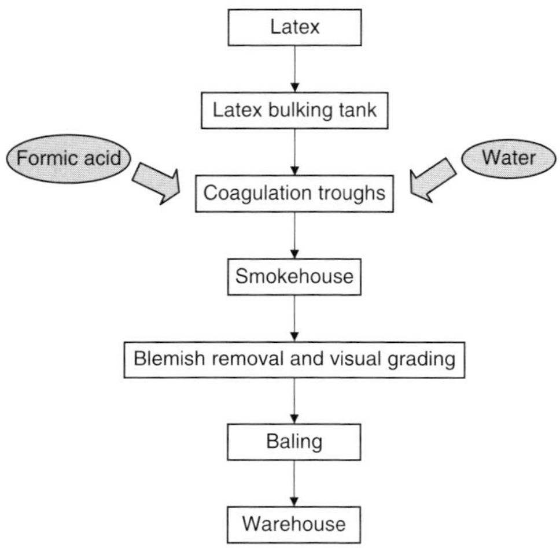

Fig. 17.6. Ribbed smoke sheet processing.

holding size (Fig. 17.6). It includes blending, coagulation, milling, drying and finishing. Rubber from different sources is blended to make the rubber uniform. The blend is diluted with water and formic acid or acetic acid is added to bring about coagulation. The coagulum is pressed between rollers to produce sheets of uniform thickness. The rollers are grooved appropriately to make 'ribs' on the sheets to expand the surface area for drying. The sheets are sized and hung on lines to dry in a smokehouse. Blemishes are removed manually and the sheets are graded visually. The dried sheets are baled, separating the sheets with talc to prevent adhesion.

Crepes

Crepe rubber is produced from latex (pale or sole crepe), cup lumps, scraps, or poor-quality sheet rubber (re-milled crepe). After blending the latex, coagulation is carried out by a procedure called 'fractional' coagulation. The first fraction is the unstable off-colour pale yellow that produces low-grade crepe, the remaining latex is a wither material. The coagulum is washed and pressed between rollers to shape rubber into thin crepes. The crepes are then dried in hot chambers or tunnels, or on drying floors.

Block rubber

This process produces technically graded rubber from latex and field coagulum (Fig. 17.7). Relatively sophisticated machinery is used to process large volumes of raw material at a time. Latex from different sources is first blended in large tanks and chemicals added to control viscosity and affect colour. The blend is then coagulated by adding formic acid in long coagulation troughs. The coagulum is pressed between rollers to form crepes, mashed into small pieces by a hammer mill and finally into crumbs by an extruder. An alternative method of making crumb is to add incompatible oil to coagulated latex and then feed it through a creper for transformation into crumbs. Crumbs are dried by hot air. The dry rubber is baled by means of a hydraulic press and wrapped in polythene to prevent adhesion between bales in the crate.

Quality control

Solid rubber

Grading of solid rubber based on measurable technical specifications originated in Malaysia in the 1960s and was adapted subsequently by other rubber producing countries. The grading of technically specified rubber (TSR) is based on dirt content, ash content, volatile matter content, nitrogen content, plasticity and colour measured using ISO (International Organization for Standardization) test methods.

TSR grades based on ISO specification (Table 17.2) are: (i) TSR L – high-quality and light-coloured rubber prepared from latex; (ii) TSR CV – viscosity-stabilized, high-quality latex rubber; (iii) TSR 5 – good-quality latex rubber, darker than TSR L; (iv) TSR 10 – good-quality grade derived from field coagulum, suitable for general purpose use; (v) TSR 20 – for general purpose use below TSR 10 in quality; and (vi) TSR 50 – up to 0.50% wt dirt content.

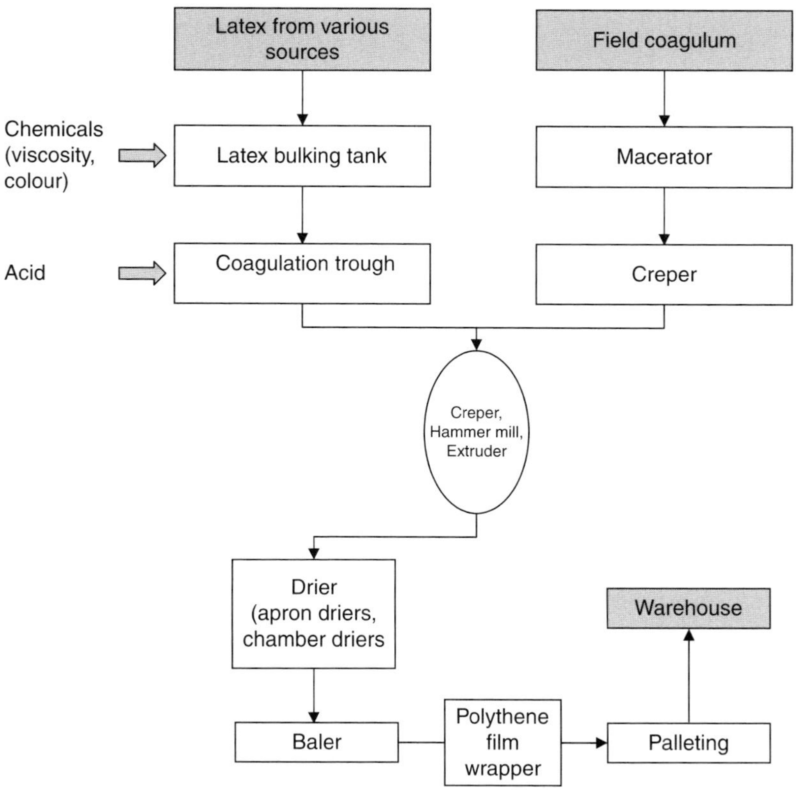

Fig. 17.7. Block rubber processing.

Latex concentrate

Latex concentrate is graded according to technical specifications established separately by the American Society for Testing Materials (ASTM) and the British Standards Institution (BSI).

Manufacturing of Rubber Goods

The main steps of rubber manufacturing are preparing, compounding, shaping and vulcanization (Fig. 17.8). Preparing involves the breaking down of rubber molecules (mastication) so that the rubber is plastic when hot.

Table 17.2. Technical specifications (TSR), (UNCTAD 2009, with permission).

	Grades					
Parameters	TSR CV	TS L	TSR 5	TSR 10	TSR 20	TSR 50
Dirt (max) % wt	0.05	0.05	0.05	0.10	0.20	0.50
Ash (max) % wt	0.60	0.60	0.50	0.75	1.00	1.50
Nitrogen (max) % wt	0.60	0.60	0.50	0.60	0.60	0.60
Volatile matter (max) % wt	0.80	0.80	0.80	0.80	0.80	0.80
Initial Wallace plasticity Po (min)		30	30	30	30	30
Plasticity retention index (PRI) (min)	60	60	60	50	40	30
Colour Lovibond scale (individual value, max)		6				

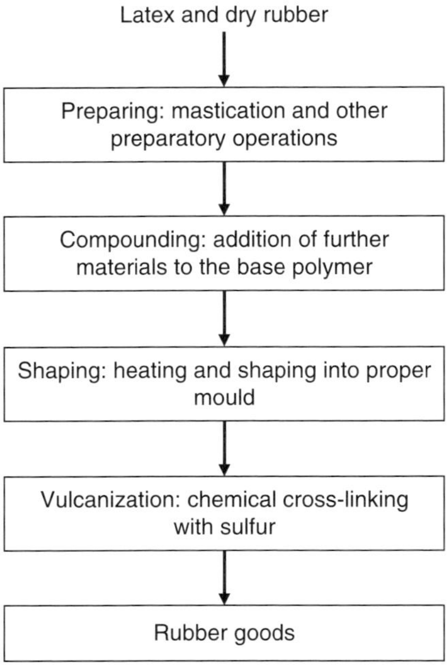

Fig. 17.8. Manufacturing of rubber goods.

Further materials to fit the specification of the final manufactured goods and sulfur and other vulcanization constituents are added to the masticated rubber. The rubber mix is then heated and shaped into the appropriate mould or fractioned on to a fabric. The final step is vulcanization to add strength to the manufactured goods. Sulfur used in vulcanization actuates cross-linking of polymer molecules by atomic bridges, making the rubber more durable and more resistant to chemical attack. In modern vulcanization, several additives as well as sulfur are used to achieve different ends. It is now possible to carry out the vulcanization reaction at temperatures of 140–180°C using accelerators that allow the reaction between rubber and sulfur to take place at lower temperature and in less time. Antioxidants extend the life of rubber products by preventing breakage of covalent bonds by oxygen or ozone and reinforcing agents to increase stiffness, tensile strength and resistance to abrasion.

Exceptions to the above manufacturing process are latex goods, which are produced using liquid technologies involving either dipping shape formers or entrapping air to produce foams.

Natural Rubber Commerce

The statistics on production, consumption and prices for NR are presented in Table 17.3. The area under rubber tree cultivation has increased significantly from an estimated 3.9 million ha in 1961 to 8.0 million ha in 2005 (FAO, 2007). The largest area of over 2.67 million ha of rubber planting is in Indonesia. World annual production increased from 2.1 million t in 1961 to 9.1 million t in 2005. Thailand is the largest NR producer in the world, followed by Indonesia. The five largest NR consuming countries in order are China, USA, Japan, India and Malaysia. The annual usage rate between 1998 and 2005 for NR increased by 4%, compared to 3% for synthetic rubber (SR).

The chain of trade from smallholders traditionally involved village and town dealers, wholesale dealers and exporters. In order to improve the return of smallholders, several rubber producing countries have organized them into groups so that they can process their rubber collectively and sell it in bulk to fetch a better price.

Two modules exist in the trading of rubber: (i) open trading and (ii) direct trading.

Open trading

The primary market serves the producer end and the terminal market meets the consumer end of the rubber trade (Fig. 17.9). Both spot and future trading is conducted in these markets. Kuala Lumpur and Singapore are the main primary markets and Hamburg, Shanghai, Amsterdam, Paris, Tokyo, New York and London are the terminal markets.

Direct trading

International tyre companies have their buying agents located in the main rubber producing companies to ensure uninterrupted raw material supply. Large rubber estates and smallholder organizations have established direct links with particular consumers in order to arrive at a mutually agreed price through negotiation and avoid market fluctuations.

Table 17.3. Statistical summary of the world rubber situation (adapted from IRG, 2009).

	Thousand tonnes		
	2006	2007	2008
Natural rubber production			
Latin America	202	228	241
Africa	423	445	443
Asia	9331	9386	9406
Total	*9698*	*9707*	*9876*
Natural rubber consumption			
North America	1148	1157	1179
Latin America	528	565	562
European Union	1302	1377	1189
Other Europe	252	273	255
Africa	120	118	112
Asia/Oceania	5961	6391	6453
Total	*9329*	*9884*	*9726*
World supply demand surplus/deficit	369	–177	150
World stocks	3676	3955	4201
Natural rubber prices			
Europe, TSR 20 €/t	1646	1619	2511
SICOM, RSS3, S$/t	3344	3444	260
New York, TSR20, US$/t	2113	2321	1702

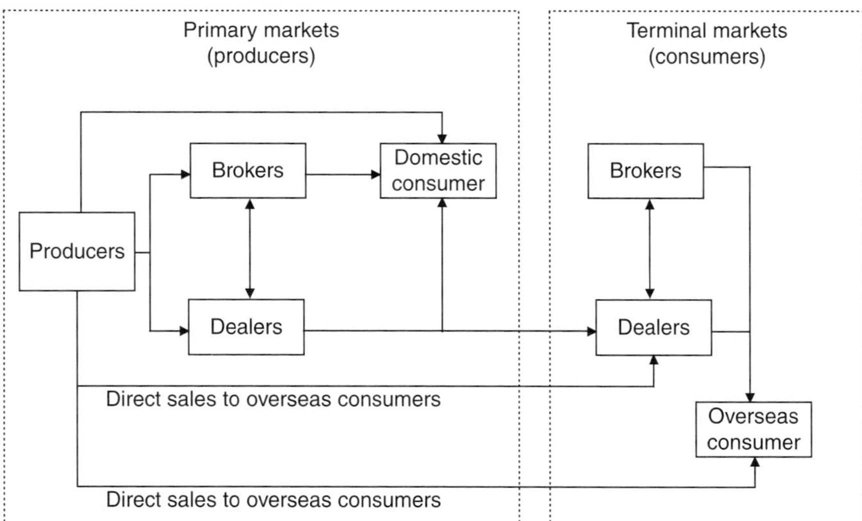

Fig. 17.9. Rubber marketing channels.

Rubber trading agreements and trading consortiums

The International Natural Rubber Agreement under the auspices of the UN Conference on Trade and Development (UNCTAD) was signed between rubber producing and rubber importing countries in 1979 to reduce excessive market price fluctuations and keep the price level remunerative to the producers. The market intervention mechanism triggered when the price reached +/− 15% of the reference price set in the agreement between the rubber exporting and importing countries. A joint fund, with contributions from both the rubber exporting and rubber importing counties, was set up to carry out this timely market intervention. The operational provisions of the 1979 Agreement were retained in the later Agreements reached in 1987 and 1995, but modifications were added in the periodicity of reviews of the reference price and the level of price fluctuation for triggering intervention. A disagreement between exporting and importing countries on the reference price resulted in the withdrawal of the rubber exporting countries from the agreement and the subsequent termination of the agreement itself.

In 2001, Indonesia, Malaysia and Thailand established an organization called the 'International Tripartite Rubber Organization' (ITRO) to manage the level of rubber production in the member countries to ensure a minimum trading price. They launched the International Rubber Company in October 2003 to pool their resources and use it to stabilize world NR prices.

Natural Rubber Products

Natural rubber is used in the manufacture of more than 40,000 products, including more than 400 medical products. Apart from minor uses as crepes for shoe soles and adhesives, raw rubber is almost never used itself in products; instead vulcanized rubber is the norm. Figure 17.10 lists the type of products made out of NR for different uses. Over 50% of NR is used in the manufacture of tyres and tyre products. Synthetic rubber cannot match NR in tear strength and heat resistance, making it the preferred material for high-performance tyres used on racing cars, aircraft, trucks and buses. There is very little potential to replace NR with SR for these

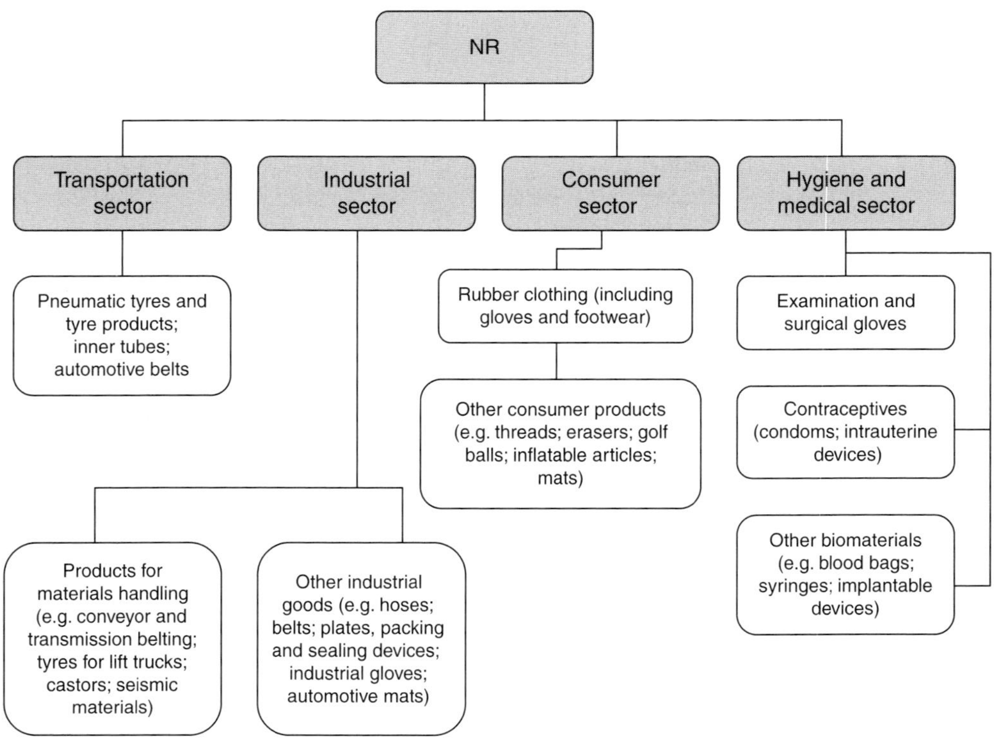

Fig. 17.10. Major end uses of natural rubber.

applications. In most tyres, both NR and SR are used (Table 17. 4). Generally, the larger the size of the tyre, the greater is the share of NR.

Technical factors limit the substitution of NR by SR for latex products. Because natural latex is waterproof whereas most synthetics absorb some water, NR latex is best suited for condoms and medical examination equipment. NR latex also offers the best protection available against pathogens such as HIV. The competition to *Hevea* latex will probably come from other NR sources because of the allergy problems associated with it.

Additional Uses of the Rubber Tree

The rubber tree is an important source of timber. One hectare of rubber plantation yields approximately 190 m^3 of timber (Arshad *et al.*, 1995). Rubber honey is another ancillary source of income. Because of the positive carbon footprint envisioned for trees, global carbon trading will bring in an income to plantation owners in the future.

The seed from the rubber tree has an average oil yield of 40% (Njoku *et al.*, 1996). This oil has not yet been exploited for economic purposes, except for some reported possible uses in soaps, alkyd resins and lubricants (Sthapitanonda *et al.*, 1981; Njoku *et al.*, 1996). The possible utilization of rubber seed for biodiesel production has been investigated (Quick, 1989; Ikwuagwu *et al.*, 2000). Quick (1989) found rubber seed oil methyl ester compared favourably to soybean, sunflower and linseed oils. Ikwuagwu *et al.* (2000) observed that *trans*-methylation improved the fuel properties of rubber oil and brought it closer to the fuel properties of commercial diesel. However, refining reduced the storage life of the oil. The decrease in oxidative stability of the refined oil and methyl ester resulted from the decrease in the viscosity of the parent rubber seed oil. The oxidative stability of commercial diesel fuels is increased by adding antioxidants and dispersants.

Table 17.4. Typical tyre composition by weight, (UNCTAD 2009, with permission).

Passenger tyre	Average weight: new ~ 11 kg, scrap 9 kg (%)
Natural rubber	14
Synthetic rubber	27
Carbon black	28
Steel	14–15
Fabric, fillers, accelerators, antiozonants, etc.	16–17
Truck tyre	Average weight: new ~ 54 kg, scrap 45 kg (%)
Natural rubber	27
Synthetic rubber	14
Carbon black	28
Steel	14–15
Fabric, fillers, accelerators, antiozonants, etc.	16–17

The rubber tree possesses traits that make it a suitable candidate for reforestation and land reclamation. It is a fast-growing tree developing canopy early to provide ground cover. *Hevea* is a surface feeder; its root system is near to the topsoil, binding soil particles to safeguard against erosion. The deciduous habit of the tree also enriches the soil by annually providing organic matter from the fallen leaves.

Since latex can be extracted easily from the rubber tree plant, it raises the possibility that the tree can be used for the production of commercially valuable proteins. Genes encoding proteins of commercial value such as for pharmaceuticals and personal care products can be inserted into the rubber tree through genetic transformation and the encoded protein collected continually with the latex. A transgenic rubber tree producing a bacterial enzyme (GUS), a mouse antibody and a human serum albumin has been reported by the Rubber Research Institute of Malaysia (Arokiaraj *et al.*, 2002; Yeang *et al.*, 2002). However, to realize this utilization outlet, much concerted effort will be needed.

Fate of Used Rubber Products

Tyres and rubber wastes have much greater significance as a disposal problem than as a potential source for recycling. Rubber articles more than a few millimeters in thickness do not degrade easily, even in harsh environments. Vulcanized rubbers cannot be reformed without chemical or mechanical degradation. Rubber products such as tyres that go through vulcanization are virtually indestructible and, because of their composite nature, are difficult to recycle and pose a major scrap disposal problem. Medical goods cannot be recovered because of health and safety problems and are dumped or incinerated. However, hybrid materials known as thermoplastic elastomers produced by mixing rubber and plastic materials or complex polymerization techniques do not require vulcanization and can be processed like plastic by shaping while hot and then cooling.

Secondary production of unadulterated thermoset rubber is not feasible as it is mixed with other chemicals or materials during manufacture. The recovery of rubber is also complicated by its frequent contamination with fluids or by its partial degradation from the combined effect of oxygen and heat. Therefore, recovered rubber can only be used either with virgin rubber in amounts small enough not to lower the quality of the finished articles seriously or as particulate fillers in lower-grade products such as solid tyres or sports surfaces. Scrap rubber tyres can be used for a number of purposes without processing, e.g. as ballast for plastic covering film in agriculture, boat fenders, artificial fish reefs, coastal defences and river and reservoir bank enforcement. The problem common to these reuses is that these scrap tyres are taken away from recycling and turn in time into an environmental problem.

Retreading can extend the life of tyres. Used tyres are inspected rigorously for integrity; thereafter, the remaining tread is removed by buffing. The buffed material is collected for use as rubber crumb. Approximately 80% of the original tyre is preserved in retreading and 20% of the material is new. Rubber crumb is used as a minor particulate filler for certain products and as a major constituent for carpet underlay and sports surfaces.

The reused rubber ultimately needs disposal. A variety of alternatives are being explored to find environmentally friendly and economical methods of scrap rubber disposal. Mixing rubber crumb with the asphalt (rubberized asphalt) prolongs the life of the resulting roads, but construction material costs are higher. Therefore, its use has been largely for roads that are not easily accessible or where wear rates are excessive. The US Congress passed the Intermodal Surface Transportation Efficiency Act in 1991 requiring progressively higher levels of tyre crumbs in asphalt for road surface to boost rubberized asphalt production.

The objective of thermal decomposition of scrap rubber is to reclaim its constituent parts. Depending on the reclamation process, it can yield gases, fuel oil, carbon black, zinc oxide, steel, sulfur and hydrocarbons. The thermal decomposition process involves prolysis, gasification, hydrogenation and catalytic extraction. However, the technology at present is not economically viable.

Increasingly, scrap rubber is being used to generate energy. Burning of scrap rubber generates approximately 32 MJ/kg, which is greater than the 29 MJ/kg of coal. Scrap rubber tyres are also fed to cement kilns, where they not only heat the kiln but also impart improved properties to the cement by adding iron from the tyre steel reinforcement.

Current Status and Future Prospects

Breeding and selection for improved clones has resulted in significant yield gains and the rubber tree is now grown in geographical areas beyond its natural climatic zone. The germplasm base of the rubber tree has been expanded by exploration, which could produce future gains. However, due to low yields of new germplasm and the long length of the tree-breeding process, benefits will be distributed over a lengthy period. Specific objective-targeted germplasm pre-breeding programmes could enhance gain by providing evaluated new parents for recombination breeding. Molecular marker-assisted selection will play a greater role in the future to assist breeding for traits like drought tolerance, cold tolerance and resistance to diseases. Tapping panel dryness continues to be a serious physiological disorder limiting latex yield and requires a better understanding of the underlying causes through physiological and biochemical investigations in order to prevent it from developing. Although effective control measures for serious diseases are in place in most rubber tree producing countries, the battle between plants and diseases is everlasting and so breeding for disease will always remain a priority in crop production. To alleviate the problem of variation among bud-grafted populations, commercialization of tissue culture for rubber tree propagation should receive special attention. Another way to produce true-to-mother plants is apomixis, but genes for apomixis have not yet been characterized for induction into the rubber tree.

Generation of income from ancillary products is essential for the long-term viability of rubber tree plantations. Ways to increase yields of quality timber and honey deserve attention. Selection and development of clones targeted to trees suited for timber should help decrease the inherent defects of rubber wood and increase the per unit area timber yield. Optimization of plantation conditions to suit bee visits and having apiaries on the plantations can provide additional income from the sale of honey.

The major drawback of latex production from *Hevea* is its heavy reliance on manual labour for tapping. Newer methods have been found that reduce tapping frequencies, but much more efficiency needs to be achieved in this area. However, with the possibility being remote that an annual crop could be improved to a level to match the rubber tree in latex volume and quality, its dominance in the vulcanized rubber industry will continue. Allergy problems associated with *Hevea* latex threaten its market for contact rubber products. Some research progress has been reported on the alleviation of this problem. However, other latex crops such as guayule may begin to share this market segment with the rubber tree.

Technological advances have lowered the energy cost of rubber tyre production and tyre durability. The quality of retread tyres has improved, and so has their acceptability. New ways of recycling used rubber products are being found and, as a result, ecological problems from tyre dumping have lessened. Since NR production has reached a plateau, ways to increase the service life of tyres will attract research attention. Preservation of latex, pre-vulcanization of latex, radiation vulcanization and deprotenization are some of the technological challenges for the future.

The industries based on NR are mature, so only a slow gradual increase in raw material demand is anticipated. Global NR production has increased 3.0%/year over the past 50 years, while demand has increased by 3.25%/year. Thus, a tight balance between demand and supply has always prevailed. The inelasticity of perennial crops, including the rubber tree, to control supply in the short run to adjust to market demand has subjected NR to frequent price fluctuations, sometimes at levels causing serious economic hardship for growers. Rubber producing countries have tried to put in some mechanism for price stability, but market forces have prevailed. The superiority of NR over SR ensures its steady demand in the manufacturing of different rubber products. All predictions of future cars see them using rubber tyres and thus the need for NR to produce high quality tyres. The current trend of greater reliance on non-fossil-based products should increase the demand for NR. Increased use of NR in buildings for earthquake damage control can be foreseen as more people realize that such systems save life and property.

References

Abraham, P.A., Wycherley, P.R. and Pakianathan, S.W. (1968) Stimulation of latex flow in *Hevea brasiliensis* by 4-amino-3, 5, 6-trichloropicolinic acid and 2-chloroethane phosphonic acid. *Journal of Rubber Research Institute of Malaya* 20, 291–305.

Abraham, S.T. (2000) Genetic parameters and divergence in certain wild genotypes of *Hevea brasiliensis* (Willd. ex Adr. de Juss) Muell. Ar. PhD thesis, Mahatma Gandhi University, Kottayam, Kerala, India.

Archer, B.L., Audley, B.G., Cockbain, E.G. and McSweeney, G.P. (1963) The biosynthesis of rubber. Incorporation of mevalonate and isopentenyl pyrophosphate into rubber by *Hevea brasiliensis*-latex fractions. *Biochemistry Journal* 89, 565–574.

Arokiaraj, P., Rueker, F., Obermayr, E., Shamsul Bahri, A.R., Jaafar, H., Carter, D.C. and Yeang, H.Y. (2002) Expression of human serum albumin in transgenic *Hevea brasiliensis*. *Journal of Rubber Research* 5, 157–166.

Arshad, N.L., Othman, R. and Yacob, A.R.W. (1995) *Hevea* wood availability in Pennisular Malaysia. *RRIM Planters Bulletin* 224–225, 73–83.

Bandurski, R.S. and Teas, H.J. (1957) Rubber biosynthesis in latex of *Hevea brasiliensis*. *Plant Physiology* 32, 643–648.

Baulkwill, W.J. (1989) The history of natural rubber production. In: Webster, C.C. and Baulkwill, W.J. (eds) *Rubber*. Longman Scientific and Technical, Essex, UK, pp. 1–56.

Bernstein, D.I., Biagini, R.E., Karnani, R., Hamilton, R., Murphy, K., Bernstein, C., Arif, S.A., Brendts, B. and Yeang, H.Y. (2003) *In vitro* sensitization to purified *Hevea brasiliensis* proteins in health-care workers sensitized to natural rubber latex. *Journal of Allergy and Clinical Immunology* 111, 610–616.

Boatman, S.G. (1966) Preliminary physiological studies on the promotion of latex flow by plant growth regulators. *Journal of the Rubber Research Institute of Malaya* 19, 243–258.

Carron, M.P., Etienne, H., Michaux-Ferriere, N. and Montoro, P. (1995) Somatic embryogenesis in rubber tree *Hevea brasiliensis* Muell. Arg. In: Bajaj, Y.P.S. (ed.) *Biotechnology in Agriculture and Forestry: Somatic Embryogenesis and Synthetic Seeds*, Volume 30. Springer, Berlin, pp. 353–369.

Chappell, J. (1995) Biochemistry and molecular biology of isoprenoid biosynthetic pathway in plants. *Annual Review of Plant Physiology and Plant Molecular Biology* 46, 521–547.

Chen, Z. (1984) Rubber (*Hevea*). In: Sharp, W.R., Evans, D.A., Ammirato, P.V. and Yamada, Y. (eds) *Handbook of Plant Cell Culture – Crop Species*, Volume II. Macmillan, UK, pp. 546–571.

Chow, K., Wan, K., Isa, M.N.M., Bahari, A., Tan, S., Harikrisna, K. and Yeang, H. (2007) Insight into rubber biosynthesis from transcriptome analysis of *Hevea brnsiliensis* latex. *Journal of Experimental Botany* 58, 2429–2440.

Chye, M.L., Tan, C.T. and Chua, N.H. (1992) Three gene encode 3-hydroxy-3-methyl glutaryl-coenzyme A reductase in *Hevea brasiliensis*. hmg1and hmg3 are differently expressed. *Plant Molecular Biology* 19, 473–484.

Clement-Demange, A., Legnate, H., Seguin, M., Carron, M.P., Le Guen, V. and Chapuset, T. (2000) Rubber tree. In Charrier, A., Jacquot, M., Harmon, S. and Nicolas, D. (eds) *Tropical Plant Breeding*. Collection Reperes. CIRAD-ORSTOM, Montpellier, France, 455–480.

Clement-Demange, A., Priyadarshan, P.M., Thuy Hoa, T.T. and Venkatachalam, P. (2007) *Hevea* rubber breeding and genetics. *Plant Breeding Review* 29, 177–283.

Cornish, K. (1993) The separate roles of plant *cis* and *trans* prenyl transferases in *cis*-1,4 polyisoprene biosynthesis. *European Journal of Biochemistry* 218, 267–271.

Cornish, K. (2001) Biochemistry of natural rubber, a vital raw material, emphasizing biosynthetic rate, molecular weight and compartmentalization, in evolutionarily divergent plant species. *Natural Products Report* 18, 182–189.

Cornish, K. and Siler, D.J. (1995) Effect of different allylic diphosphates on the initiation of new rubber molecules and on *cis*-1,4-polyisoprene biosynthesis in guayule (*Parthenium argentatum* Gray). *Journal of Plant Physiology* 147, 301–305.

Cornish, K. and Siler, D.J. (1996) Characterization of *cis*-prenyl transferase activity localized in a buoyant fraction of rubber particles from *Ficus elastica* latex. *Plant Physiology and Biochemistry* 34, 377–384.

Cornish, K., Siler, D.J., Grosjean, O. and Goodman, N. (1993) Fundamental similarities in the rubber particle architecture and function in three evolutionary divergent plant species. *Journal of Natural Rubber Research* 8, 275–285.

Cornish, K., Castillón, J. and Scott, D.J. (2000) Rubber molecular weight regulation, *in vitro*, in plant species that produce high and low molecular weights *in vivo*. *Biomacromolecules* 1, 632–641.

Dean, W. (1987) *Brazil and the Struggle for Rubber*. Cambridge University Press, New York.

Dennis, M.S. and Light, D.R. (1989) Rubber elongation factor from *Hevea brasiliensis*. *Journal of Biological Chemistry* 264, 18608–18617.
Dickenson, P.B. (1969) Electron microscopical studies of latex vessel of *Hevea brasiliensis*. *Journal Rubber Research Institute, Malaya* 21, 543–559.
Dijkman, M.J. (1951) *Hevea – Thirty Years of Research in the Far East*. University of Miami Press, Coral Gables, Florida.
El Hadrami, I., Carron, M.P., d'Auzac, J. (1991) Influence of exogenous hormones on somatic embryogenesis in *Hevea brasiliensis*. *Annals of Botany* 67, 511–515.
Etienne, H., Lartaud, M., Michaux-Ferriere, N., Carron, M.P., Berthouly, M. and Teisson, C. (1997) Improvement of somatic embryogenesis in *Hevea brasiliensis* (Muell. Arg.) using the temporary immersion technique. *In Vitro Cellular and Developmental Biology – Plant* 33, 81–87.
FAO (Food and Agriculture Organization of the United Nations) (2007) *FAOSTAT* (http://faostat.fao.org/default.aspx, accessed 12 February 2009).
George, C.M. (1962) Mature rubber manuring: effect of fertilizers on yield. *Rubber Board Bulletin* 5, 202.
Gonçalves, P. de S., Martin, A.L.M., Bortoletto, N. and Tanzini, M.R. (1996) Estimates of genetic parameters and correlations of juvenile characters based on open pollinated progenies of *Hevea*. *Brazilian Journal of Genetics* 19, 105–111.
Goyvaerts, E., Dennis, M., Light, D. and Chua, N.H. (1991) Cloning and sequencing of cDNA encoding the rubber elongation factor of *Hevea brasiliensis*. *Plant Physiology* 997, 317–321.
Guen, V.L., Guyot, J., Reis Mattos, C.R., Seguin, M. and Garcia, D. (2008) Long lasting rubber tree resistance to *Microcyclus ulei* characterized by reduced conidial emission and absence of telomorph. *Crop Protection* 27, 1498–1503.
Hamaz, S. and Chan, J.L. (1996) Pollen storage of *Hevea*. *Journal of Natural Rubber Research* 11, 115–124.
Hamaz, S., Chan, J.L. and Yeang, H.Y. (1999) *Hevea brasiliensis* seedlings from hand-pollination using cryostored pollen validated by isozymes. *Journal of Rubber Research* 2, 62–68.
Hamaz, S., Chan, J.L. and Yeang, H.Y. (2002) Pollen tube growth and fruit-set success in *Hevea brasiliensis* hand-pollination influenced by the choice of clone and female flower. *Euphytica* 123, 1–8.
Han, K., Shin, D.O., Yang, J., Kim, I.J., Oh, S.K. and Chow, K.S. (2000) Gene expressed in the latex of *Hevea brasiliensis*. *Tree Physiology* 20, 504–510.
Hao, B.Z. and Wu, J.L. (2000) Laticifer differentiation in *Hevea brasiliensis*: induction by exogenous jasmonic acid and linolenic acid. *Annals of Botany* 85, 37–43.
Hayashi, Y. (2009) Production of natural rubber from para rubber tree. *Plant Biotechnology* 26, 67–70.
Ikwuagwu, O.E., Ononogbu, I.C. and Njoku, O.U. (2000) Production of biodiesel using rubber [*Hevea brasiliensis* (Kunth. Muell.)] seed oil. *Industrial Crops and Products* 12, 57–62.
IRG (International Rubber Study Group) (2009) Rubber Statistical Bulletin April–June 2009 edition (http://www.rubberstudy.com/statistics-quarstat.aspx, accessed 8 August 2009).
Jessy, M.D., Mathew, M., Jacob, S. and Punnoose, K.I. (1994) Comparative evaluation of basin and drip system of irrigation in rubber. *Indian Journal of Natural Rubber Research* 7, 51–56.
Koyama, T., Wititsuwannakul, D., Asawatrerattanakul, K., Wititsuwannakul, R., Ohya, N., Tanaka, Y. and Ogura, K. (1996) Isopentenyl diphosphate isomerase in rubber latex. *Phytochemistry* 43, 769–773.
Kush, A.E., Goyvaerts, M.L. and Chye Chua, N.H. (1990) Laticifer specific gene expression in *Hevea brasiliensis* (rubber tree). *Proceedings of the National Academy of Sciences of the United States of America* 87, 1787–1790.
Lekawipat, N., Teerawatannasuk, K., Rodier-Goud, M., Seguin, M., Vanavichit, A., Toojinda, T. and Tragoonrung, S. (2003) Genetic diversity analysis of wild germplasm and cultivated clones of *Hevea brasiliensis* Muell. Arg. by using microsatellite markers. *Journal Rubber Research* 6, 36–47.
Lespinasse, D., Grivet, L., Troispoux, V., Rodier-Goud, M., Pinard, F. and Seguin, M. (2000a) Identification of QTLs involved in the resistance to South American Leaf Blight (*Microcyclus ulei*) in the rubber tree. *Theoretical and Applied Genetics* 100, 975–984.
Lespinasse, D., Rodier-Goud, M., Grivet, L., Leconte, A., Legnaté, H. and Seguin, M. (2000b) A saturated genetic linkage map of rubber tree (*Hevea* spp.) based on RFLP, AFLP, microsatellite and isozyme markers. *Theoretical and Applied Genetics* 100, 127–138.
Lynen, F. (1969) Biochemical problems of rubber synthesis. *Journal of Rubber Research Institute Malaysia* 21, 289–406.
Madhavan, S., Greenblatt, G.A., Foster, M.A., Benedict, C.R. (1989) Stimulation of isopentenyl pyrophosphate incorporation into polyisoprene in extracts from guayule plants (*Parthenium argentatum* Gray) by low temperature and 2-(3,4- dichlorophenoxy) triethylamine. *Plant Physiology* 89, 506–511.

Mydin, K.K., Nair, V.G., Sethuraj, M.R., Saraswathy, P. and Panikkar, A.O.N. (1992) Genetic divergence in *Hevea brasiliensis. Indian Journal of Natural Rubber Research* 5, 120–124.

Nga, B.H. and Subramaniam, S. (1974) Variation in *Hevea brasiliensis.* 1. Yield and girth data of the 1937 hand pollinated seedlings. *Journal of Rubber Research Institute Malaysia* 24, 69–74.

Njoku, O.U., Ononogbu, I.C. and Owusu, J.Y. (1996) An investigation of oil of rubber (*Hevea brasiliensis*). *Journal of Rubber Research Institute Sri Lanka* 78, 52–59.

Oh, S.K., Kang, H.S., Shin, D.S., Yang, J., Chow, K.S., Yeang, H.Y., Wagner, B., Breiteneder, H. and Han, K.H. (1999) Isolation, characterization, and functional analysis of a novel cDNA clone encoding a small rubber particle protein (SRPP) from *Hevea brasiliensis. Journal of Biological Chemistry* 274, 17132–17138.

Peng, S., Xu, J., Li, H. and Tian, W. (2009) Cloning and molecular characterization of *HbCOII* from *Hevea brasiliensis. Bioscience, Biotechnology, and Biochemistry* 73, 665–670.

Philip, V. (1997) Soil moisture and nutrient influence on growth and yield of *Hevea brasiliensis* in Tripura. PhD thesis, Indian Institute of Technology, Kharagpur, India.

Priyadarshan, P.M. (2003) Breeding *Hevea brasiliensis* for environmental constraints. *Advances in Agronomy* 79, 351–400.

Priyadarshan, P.M. (2007) Rubber. In: Kole, C. (ed.) *Technical Crops*. Springer-Verlag, Berlin, pp. 143–174.

Priyadarshan, P.M. and Crement-Demange, A. (2004) Breeding *Hevea* rubber: Formal and molecular genetics. In: Hall, J.C., Dunlap, J.C. and Friedmann, T. (eds) Advances in Genetics. Elsevier Academic Press, Amsterdam, pp. 51–115.

Priyadarshan, P.M. and Gonçalves, P.S. (2003) *Hevea* gene pool for breeding. *Genetic Resources and Crop Evolution* 50, 101–114.

Priyadarshan, P.M. and Gonçalves, P.S. and Omokhafe, K.O. (2009) Breeding *Hevea* rubber. In: Jain, S.M. and Priyadarshan, P.M. (eds) *Breeding Plantation Tree Crops*. Springer + Business Media, Berlin, pp. 469–522.

Quick, G.R. (1989) Oil seed as energy crops. In: Robbelen, G., Downey, R.K. and Ashri, A. (eds) Oil Crops of the World. McGraw-Hill, New York, pp. 118–131.

Rhines, C.E. (1958) Technical developments in natural rubber production. *Economic Botany* 12, 80–86.

Rohmer, M., Kani, M., Simonin, P., Sutter, B. and Sahm, H. (1993) Isoprenoid biosynthesis in bacteria: a novel pathway for early steps leading to isopentenyl diphosphate. *Biochemical Journal* 295, 517–524.

Sando, T., Takaoka, C., Mukai, Y., Yamashita, A., Hattori, M., Ogasawara, N., Fukusaki, E. and Kobayashi, A. (2008) Cloning and characterization of mevalonate pathway genes in a natural rubber producing plant *Hevea brasiliensis. Bioscience, Biotechnology, and Biochemistry* 72, 2049–2060.

Schoon, T.G.F. and van der Bie, G.J. (1955) Particle size distribution in brominated *Hevea* lattices. *Journal of Polymer Science* 16, 63–88.

Schwarz, M.K. (1994) Terpen biosynthesis in *Ginkgo biloba*. PhD thesis, Eidgenössische Technische Hochschule (ETH), Zürich, Switzerland.

Seguin, M., Besse, P., Lespinasse, D., Lebrun, P., Rodier-Goud, M. and Nicolas, D. (1996) *Hevea* molecular genetics. *Plantations, Recherché Développment* 3, 77–88.

Sekhar, B.C. (2004) Asia should tap new 'tyre rubber'. *Rubber Asia* 18, 49–52.

Simmonds, N.W. (1986) Theoretical aspects of synthetic/polycross populations of rubber seedlings. *Journal of Natural Rubber Research* 1, 1–15.

Sirinupong, N., Suwanmanee, P., Doolittle, R.F., Suvachitanont, W. (2005) Molecular cloning of a new cDNA and expression of 3-hydroxy-3-Methylglutaryl-CoA synthase gene from *Hevea brasiliensis*. *Planta* 221, 502–512.

Sthapitanonda, K. Vimolchalao, C., Munsakul, S. and Udomsakdhi, B. (1981) Rubber-seed oil for paint. *Journal of the National Research Council of Thailand* 13, 27–42.

Southorn, W.A. and Edwin, E.E. (1968) Latex flow studies. II. Influence of lutoids on the stability of *Hevea* latex. *Journal of Rubber Research Institute of Malaya* 20, 187–200.

Tan, H. (1981) Estimates of genetic parameters and their implications in *Hevea* breeding. In: Yap, T.C., Graham, K.M. and Sukanu, J. (eds) *Crop Improvement Research*. Society for the Advancement of Breeding Research in Asia and Oceania (SABRAO), Kuala Lumpur, pp. 439–446.

Tan, H. (1987) Strategies in rubber tree breeding. In: Abbott, A.J. and Atkin, R.K. (eds) *Improving Vegetatively Propagated Crops*. Academic Press, London, pp. 28–54.

Tan, H. and Subramaniam, S. (1976) A five-parent diallel cross analysis for certain characters of young *Hevea* seedlings. *Proceedings International Rubber Conference, Kuala Lumpur.* Rubber Research Institute of Malaysia, Kuala Lumpur, Malaysia, pp. 13–26.

Tan, H., Khoo, S.K. and Ong, S.H. (1996) Selection of advanced polycross progenies in *Hevea* improvement. *Journal of Natural Rubber Research* 11, 215–225.

Tanaka, Y. (1989) Structure and biosythesis mechanism of natural polyisoprene. *Progress in Polymer Science* 14, 339–371.

Tangpakdee, J. (1998) Structural characterization of natural rubber. PhD thesis, Tokyo University of Agriculture and Technologys, Tokyo.

Tangpakdee, J., Tanaka, Y., Ogura, K., Koyama, T., Wititsuwannakul, R. and Wititsuwannakul, D. (1997) Isopentenyl diphosphate isomerase and prenyl transferase activities in bottom fraction of C-serum from *Hevea* latex. *Phytochemistry* 45, 261–267.

UNCTAD (United Nations Conference on Trade and Development) (2009) Rubber (http://www.unctad.org/infocomm/anglais/rubber/sitemap.htm, accessed 12 July 2009).

Varghese, Y.A. and Abraham, S.T. (2005) Rubber. In: Chopra, V.L. and Peter, K.V. (eds) *Handbook of Industrial Crops*. Food Products Press, New York, pp. 403–458.

Varghese, Y.A., John, A., Premkumari, D., Panikkar, A.O.N. and Sethuraj, M.R. (1993) Early evaluation in *Hevea*: growth and yield at the juvenile phase. *Indian Journal of Natural Rubber Research* 6, 19–23.

Veisseire, P., Cailloux, F. and Coudret, A. (1994) Effect of conditioned media on the somatic embryogenesis of *Hevea brasiliensis*. *Plant Physiology and Biochemistry* 32, 571–676.

Venkatachalam, P., Jayashree, R., Rekha, K., Sushmakumari, S., Sobha, S., Kumari Jayshree, P., Kala, R.G. and Thulaseedharan, A. (2006) Rubber tree (*Hevea brasiliensis* Muell. Arg.). *Methods in Molecular Biology* 344, 153–164.

Vijayakumar, K.R., Dey, S.K., Chandrashekhar, T.R., Devakumar, A.S., Mohanakrishna, T., Rao, P.S. and Sethuraj, M.R. (1998) Irrigation requirement of rubber in the subhumid tropics. *Agricultural Water Management* 35, 245–259.

Webster, C.C. and Paardekooper, E.C. (1989) Botany of rubber tree. In: Webster C.C. and Baulkwill, W.J. (eds) *Rubber*. Longman Scientific and Technical, Essex, UK, pp. 57–84.

Wititsuwannakul, D., Wititsuwannakul, R. and Suwanmanee, P. (1990) 3-hydroxy-3-methylglutaryl coenzyme A reductase from the latex of *Hevea brasiliensis*. *Phytochemistry* 29, 1401–1403.

Wu, J.L., Hao, B.Z. and Tan, H.Y. (2002) Wound-induced differentiation in *Hevea brasiliensis* shoots mediated by jasmonic acid. *Journal of Rubber Research* 5, 53–63.

Wychereley, P.R. (1976) Rubber. In: Simmonds, N.W. (ed.) *Evolution of Crop Plants*. Longman, London, pp. 77–80.

Wycherley, P.R. (1992) The genus *Hevea:* botanical aspects. In: Sethuraj, M.R. and Mathew, N.M. (eds) *Natural Rubber: Biology, Cultivation, and Technology*. Elsevier, Amsterdam, pp. 50–66.

Yeang, H.Y. (2004) Natural rubber latex allergens: new developments. *Current Opinion in Allergy and Clinical Immunology* 4, 99–104.

Yeang, H.Y., Arokiaraj, P., Jaafar, H., Siti, A.M., Rajamanikram, S., Chan, J.L., Jafari, S., Leelavathy, R., Hamzah, S. and van Der Logt, C.P.E. (2002) Expression of a functional recombinant antibody fragment in the latex of transgenic *Hevea brasiliensis*. *Journal of Rubber Research* 5, 215–225.

Yeang, H.Y, Hamilton, R.G., Bernstein, D.I., Arif, S.A.M., Chow, K.S., Loke, Y.H., Raulf, H.M., Wagner, S., Breitender, H. and Biagini, R.E. (2006) Allergen concentration in natural rubber latex. *Clinical and Experimental Allergy* 36, 1078–1086.

Yu, J.H., Yang, S.G., Huang, M.J. and Tian, W.M. (2007) Effects of season, exploitation and exogenous jasmonic acid on the laticifer differentiation in mature rubber tree. *Chinese Journal of Tropical Crops* 28, 1–5.

18

Guayule: Culture, Breeding and Rubber Production

Dennis T. Ray, Michael A. Foster, Terry A. Coffelt
and Colleen McMahan

Introduction

Guayule (*Parthenium argentatum*, Compositae) is a perennial shrub (Fig. 18.1) native to the Chihuahuan Desert of north-central Mexico and the Trans Pecos of south-west Texas (Stockton Plateau and Big Bend Region) of the USA (NAS, 1977). Native populations are found on semiarid plateaus scattered throughout approximately 300,000 km^2 of rangeland and over a range of climatic conditions (Foster and Coffelt, 2005).

Guayule has long been known as a potential source of natural rubber, essentially identical to that from the rubber tree (Thompson and Ray, 1988; Siler and Cornish, 1994). The use of guayule rubber by native populations to make balls for games was first reported by the Spanish in the early 1500s (Thompson and Ray, 1988). Use of guayule rubber as a commercial source began in the late 1800s, when it was 'rediscovered' by a Mexican boundary survey party.

Guayule has been through four efforts to make it a commercial crop in North America (Ray, 1993). In the early 1900s, guayule was considered as an alternative source of natural rubber in the USA due to the high price of imported rubber from the Amazon region (Bonner, 1991). This initial attempt at commercialization started with the harvesting of wild guayule stands in Mexico and accounted for up to 24% of the total rubber imported into the USA by 1910 (Bonner, 1991). At this time, 20 extraction plants were either operational or under construction in Mexico. This production came to a halt in 1912, because of the Mexican Revolution, and the effort moved across the border into the USA, with plantings in Arizona and California (Ray, 1993). This first commercialization effort came to a halt in 1929 as a result of the Great Depression (Ray, 1993).

The second effort to use guayule as a source for natural rubber was the Emergency Rubber Project (ERP) of World War II. Natural rubber production had moved almost exclusively to large plantations of the Brazilian rubber tree grown in South-east Asia, and these sources were cut off at the beginning of the war (Ray, 1993). This project was very successful, generating the bulk of our knowledge about the basic biology of guayule, and developed the germplasm on which the current breeding programmes are based. The effort ended with the end of the war and the development of synthetic rubber.

Guayule was investigated seriously a third time, starting in the late 1970s, because of the quadrupling of crude oil prices. The fear was that if the oil supply could be manipulated, there might again be a shortage of natural rubber due to either natural disaster or political

Fig. 18.1. Guayule field plots.

unrest in South-east Asia. This led to the enactment of the Native Latex Commercialization and Economic Development Act of 1978 and the Critical Agricultural Materials Act of 1984 (Laws 95-592 and 98-284). Again, a tremendous amount of work was accomplished, resulting in significant yield increases and the refinement of cultural practices to fit modern mechanized agriculture (Whitworth and Whitehead, 1991; Ray, 1993; Foster and Coffelt, 2005). This third effort again showed that guayule could be planted, cultivated, harvested and processed as a source of natural rubber; however, as the political climate changed, this effort was also terminated.

The future for guayule appeared bleak until the report of latex allergy to *Hevea* rubber products in the general population (Ownby *et al.*, 1994). This made the development of an alternative, safe source of natural rubber imperative, and guayule proved to be a source of non-allergenic latex for those with latex allergy (Siler *et al.*, 1996). Guayule latex was found to contain many fewer proteins than *Hevea* latex, and in much lower quantities (Cornish, 1996, 1998; Cornish and Lytle, 1999).

The commercialization of non-allergenic guayule latex became a reality when the Yulex Corporation was granted the exclusive licence to US Patent No. 5580942 (Cornish, 1996) and to US Patent No. 5718050 (Cornish, 1998) on guayule latex processing and products, respectively. To date, Yulex has established a business organization, developed a financial base, increased seed of promising lines, built a processing plant, developed several products for entry into the high-value medical devices marketplace and is in the process of planting large acreages to support the industry.

Cultural Practices

Site selection and field preparation

The selection of suitable sites for guayule production hinges on soil type, temperature, annual precipitation and irrigation. The preferred soil is a well-drained type such as a sandy or gravelly loam. Production areas during the ERP included areas with limited below-freezing winter temperatures: the Texas Trans Pecos, south-west New Mexico, southern Arizona and the Salinas Valley of California (NAS, 1977). Shrubs in native populations tolerate freezing temperatures but may be susceptible under cultivated conditions. Recent studies have indicated that certain guayule lines established on the Southern High Plains near Plainview, Texas, can survive winter temperatures of –5°C (Foster *et al.*, 2008). Although guayule production will require added irrigation, precipitation will augment these water needs. Each production area will have its own irrigation scheme, whether it is furrow, flood, centre pivot, solid-set sprinkler, or drip. Successful establishment may require a combination of irrigation types (sprinkler for establishment and furrow for maintenance). Irrigation water salinity will exclude certain areas from production. Field preparation will involve practices common in selected production areas. Whether establishment is by direct seeding or transplanting, planting on raised beds spaced 1 m apart has been customary (Foster *et al.*, 2002b).

Plant establishment

Two methods of establishing field plantings have been used (Hammond and Polhamus, 1965). The most widely used technique involves direct seeding in nurseries or greenhouses for the production of seedlings for transplanting. Another method, used only on an experimental scale, involves direct field seeding. Today's production costs reflect the enormous expenses involved in the operation of nurseries and greenhouses to produce seedlings. Establishment costs could be reduced substantially with the development of successful direct-seeding techniques. Direct seeding could reduce the cost of establishment to below US$400/ha versus US$900–1200/ha for transplanting (Bucks *et al.*, 1986). Recent costs of transplanting guayule in Arizona are estimated to be US$1600/ha (D.W. Swiger, Arizona, 2003, personal communication). This figure includes both greenhouse and field transplanting costs.

Seed conditioning

Successful plant establishment in the field or greenhouse depends on good quality seed. Guayule is a prolific seed producer, blooming and setting seeds continuously throughout the growing season when moisture is available. The guayule seed is an achene with attached bracts and a pair of sterile florets (Chandra and Bucks, 1986). However, many of the seeds are either empty or not viable. The percentage of filled seed in collections made during the ERP varied tremendously (Taylor, 1946). Extremes of 0–70% were present, but percentages usually ranged from 10 to 45%.

Natural or primary dormancy in guayule seed persists in varying degrees in the mature, unthreshed condition. The delayed germination of guayule seed is attributed to two types of primary dormancy: an inner seed coat dormancy, which may last 12 months or longer, and an embryo dormancy of about 2 months (Benedict and Robinson, 1946; Emparan and Tysdal, 1957; Hammond, 1959). Research during the ERP revealed that: (i) the primary dormancy, in addition to physical barriers, could also be due to the presence of inhibitors that were partially removed by leaching during seed cleaning; and (ii) intensive oxidative treatments with calcium hypochlorite degraded the tissue surrounding the embryo and facilitated its germination (Roberts, 1946).

Seed treatments with polyethylene glycol (PEG), growth regulators and other chemicals can enhance the planting quality of many crop and vegetable seeds (Khan *et al.*, 1978). The beneficial effects of these treatments include rapid and uniform germination and improved seedling vigour. Partially delayed germination could be corrected by treating the seeds with a solution of calcium hypochlorite [$Ca(ClO)_2$)] or sodium hypochlorite (NaOCl) containing about 1.5% available chlorine (McCallum, 1929; Naqvi and Hanson, 1980, 1982; Tipton, 1981; Tipton *et al.*, 1981; Jorge *et al.*, 2002). Continuous exposure to light also overcame guayule seed dormancy (Hammond, 1959; Chandra and Bucks, 1986). Gibberellins substituted for light in completely breaking both embryo and inner seed coat dormancy and promoted seedling emergence of non-dormant achenes under a soil cover (Hammond, 1959; Naqvi and Hanson, 1980; Tipton *et al.*, 1981; Chandra and Bucks, 1986). This suggested that the action of gibberellins involved the hormonal enhancement of seed viability and vigour characteristics. Washing guayule seeds with water and treating with 0.5% NaOCl removes many phenolic compounds from the chaff that can cause a significant inhibition of germination and radicle growth (Naqvi and Hanson, 1982). Since guayule seed has more than one type of dormancy, each method of breaking dormancy acts on a different dormancy factor.

A seed treatment/conditioning procedure has been developed that is currently recommended for enhancing the planting quality of guayule seed (Bucks *et al.*, 1983; Chandra and Bucks, 1986). The procedure involves imbibing guayule seeds under aerobic conditions in a medium containing 25% PEG (MW 8000), 10^{-4} M gibberellic acid (GA), 0.05% potassium nitrate and 0.1% thiram fungicide adjusted to pH 8.0 with a saturated solution of calcium hydroxide. The seeds are treated for 3–4 days at 25°C in continuous light. Guayule seed conditioning enhances both germination and the development of normal seedlings over a broad temperature range.

The effects of nine preconditioning treatments on the percentage and rate of germination and emergence and seedling fresh weight have been studied recently (Jorge *et al.*, 2006). Three-year-old seed of lines AZ-1, AZ-3 and N9-3 were analysed using X-rays to determine the seed lot quality. There were differences in quality (seeds containing embryos and endosperm), but no differences were found for percentage and rate of germination and emergence and seedling fresh weight among the nine treatments. These data suggest that seed quality is of greater importance in determining germination and emergence in older seed lots rather than conditioning treatments.

Transplanting

Guayule has been established successfully using various transplanting techniques (Kelley *et al.*, 1946; Tingey and Clifford, 1946; Erickson and Smith, 1947; Hammond and Polhamus, 1965; Tipton, 1981; Gonzalez and Rektorik, 1986). Seedlings grown in nursery trays and transplanted into the field grow slowly, attaining only a few mm of top growth and about 100 mm of root growth during the first 4 weeks after transplanting (Miyamoto and Bucks, 1985). Frequent irrigation is required for this period because of the shallow root system of the transplants. The frequency under furrow irrigation may vary from 3 days to a week, depending on air temperature, bed, soil and water conditions. Weekly furrow irrigations of 20 mm were applied for 5 weeks to a plot at El Paso, Texas, and seedling survival was more than 95% (Miyamoto *et al.*, 1984c). Transplant survival was greater than 95% following spring plantings in a loam soil at Mesa, Arizona (Bucks *et al.*, 1984). The transplants were first furrow-irrigated with 110 mm of water immediately after planting and then were sprinkler-irrigated once or twice a week for 7 weeks at 18 mm per application. Transplants have been established during spring months with 100–250 mm of water having salinity of less than 4 dS/m (Miyamoto and Bucks, 1985).

Direct seeding

Guayule seed is small (about 1000–1500 seeds/g) and must be planted shallow for optimum emergence. Guayule seed germination begins 3–5 days after planting, followed by emergence about 10 days later (Miyamoto and Bucks, 1985). Seedlings grow slowly and produce about 10 mm of top growth and 60 mm of root growth by 2 weeks after emergence. Therefore, frequent irrigation is crucial during the first 3–4 weeks after planting to promote seed germination, prevent soil crusting, facilitate emergence and to protect young seedlings against desiccation (Foster and Moore, 1992; Foster *et al.*, 1999, 2002b). Rain showers which splash accumulated soluble salts on to seedling leaves compound seedling mortality. When primary leaves appear on the seedlings, the soil surface should then be allowed to dry to reduce risk from damping-off and other seedling diseases, as well as salt damage to the young seedlings. Recommended irrigation techniques for plant establishment by direct seeding include: (i) conventional and low rate sprinklers, (ii) drip, or (iii) alternate row watering with furrow irrigation.

Good seed quality, seedbed preparation and precision planting are essential for successful direct seeding (Whitworth, 1981a). Direct-seeding field studies during the ERP involved planting either dry or pregerminated seed at rates of 5–15 kg/ha (Tingey, 1943, 1945a,b, 1952; Cowley, 1945a,b; Hammond and Polhamus, 1965; McGinnies and Mills, 1980). Emergence was a major problem and varied from less than 10% to over 50%, with the greatest emergence occurring at the highest moisture level. Conditioned seed with germination rates varying from 56 to 92% have been used for direct seeding in Texas and Arizona and involved seeding 100 seeds/m or about 0.5 kg seed/ha, depending on seed size and weight (Foster and Moore, 1992; Foster *et al.*, 1993, 1999, 2002a,b).

Current recommendations for direct seeding guayule are that acceptable stands occur when seeds are: (i) conditioned with polyethylene glycol, gibberellic acid and light; (ii) planted accurately on the soil surface using fluid drilling or precision planting techniques; and (iii) irrigated precisely, being careful not to under or over irrigate (Bucks *et al.*, 1986; Foster and Moore, 1992; Foster *et al.*, 1999, 2002a,b). A planting rate of at least 40 seeds/m is required to obtain moderate plant populations when the initial seed germination rate is at least 60%. Seed quality, seedling vigour and salt tolerance are still the main problems to be solved before direct seeding can be recommended as the preferred establishment practice. Recent evidence also suggests that soilborne insects such as crickets can pose a threat to newly germinated seedlings within 1 week of emergence (L. Johnson, Arizona, 2009, personal communication). Attacks by insects at this stage have been observed to devastate stands of direct-seeded guayule completely. Stands with over 70% emergence have been reduced to zero overnight. Investigations on insecticides that could be used to control insects and protect the young seedlings are ongoing.

Cover crops and synthetic shade covers have been investigated for increasing seedling emergence and survival under sprinkler irrigation. Polyshade strips and cloth increased plant survival over no shading when conditioned seeds were sown with a precision planter (Bucks *et al.*, 1987). Shading by wheat increased plant survival in summer and fall (autumn) treatments, but not during spring plantings. Synthetic materials decreased soil, air and cotyledon temperatures, and increased plant water potential versus no shade.

Although it has usually been accepted that direct seeding could reduce establishment costs (Bucks *et al.*, 1986), it has never been determined whether the savings could be offset because the direct-seeded shrubs might not produce at the same level as transplants. Transplanted seedlings, grown in nursery containers in the greenhouse, are usually 7–15 weeks old, vary in height from 100 to 200 mm and have a shallow, fibrous root system (Carranza and Ramirez, 1981; Siddiqui *et al.*, 1982; Miyamoto and Bucks, 1985). These plants have a definite growth advantage over direct-seeded seedlings during the establishment season. Guayule has been direct seeded successfully in field plots, but transplanting is still the most reliable method of stand establishment.

Irrigation

Irrigation is a critical factor influencing guayule establishment and production. Little was known prior to 1943 about the irrigation of guayule. The irrigation amounts applied during 3 years (1943–1945) of the ERP averaged about 500 mm and formed the basis for water required for production (Roberts, 1946). Nakayama *et al.* (1991) have proposed higher requirements of 1000–1300 mm. Recent studies applying over 2000 mm of water showed that guayule used all of the water applied, with none penetrating below the root zone (D. Hunsaker, Arizona, 2009, personal communication).

Compared to other crops being grown in arid areas, the biggest advantage of guayule with respect to water use may be its drought tolerance, which can permit flexibility in irrigation scheduling. Guayule also does not have critical periods, such as flowering or seed set, when lack of irrigation can cause crop failure in other crops. However, compared to ERP standards, recent research has shown more supplemental water than previously thought must be applied to increase yields and to shorten the growth cycle.

Water quality

Water supply will determine the location of irrigated guayule production, but of more immediate concern is water quality. Salt tolerance of established guayule has been reported as higher than lucerne and almost as tolerant as pima (*Gossypium barbadense* L.) and upland (*G. hirsutum* L.) cotton (Miyamoto *et al.*, 1990). However, guayule is highly susceptible to salinity at emergence and seedling stages, and tolerance at these stages is lower than carrots, one of the most salt-sensitive crops currently grown in the south-western USA.

Irrigation with waters containing soluble salts presents a constant hazard to crop production (Longenecker and Lyerly, 1959). Those salts affect crop yields in two ways. First, excessive amounts of soluble salts in the soil limit plant growth by rendering the soil water less available to the plant. The plant roots cannot extract sufficient water for growth. Second, yields can be affected indirectly by large amounts of sodium in irrigation waters (Wadleigh and Gauch, 1944; Retzer and Mogen, 1946; Miyamoto *et al.*, 1984c; Maas *et al.*, 1986). Excess sodium affects growth directly by its toxic action in the plant and also by reducing soil structural properties indirectly.

Increased salinity affects guayule production by: (i) reducing dry matter production; (ii) decreasing rubber production; (iii) decreasing water-use efficiency (amount of rubber produced per unit of water applied); and (iv) not interacting with increased plant population to enhance rubber or resin production (Tingey, 1952; Moore and Murphy, 1979; Abrahams *et al.*, 1984; Miyamoto *et al.*, 1984b,c, 1990; Miyamoto and Bucks, 1985; Zittlosen and Fangmeier, 1986; Hoffman *et al.*, 1988; Foster and Moore, 1992; Fowler and Tinguely, 1993; Foster *et al.*, 1999, 2002b). These studies suggest that if reduced rates of growth are acceptable and salt accumulation is minimized, then water salinity of up to 1.0 dS/m could be used for plant establishment and up to 4.5 dS/m for plant growth without risking significant mortality. The critical effect of salinity seems to be on plant mortality, indicating that survival rather than growth reduction at high levels of salinity will be the limiting factor for guayule production. Off-centred or double-row planting combined with alternate row watering could help minimize salt accumulation when water of high salinity is used for irrigation.

Water stress

Early investigations confirmed that stress played an important role in rubber production. Approaches to controlling water stress focused on the theory that plant stress, caused by soil water deficits, could increase rubber production. Most literature indicates that decreasing irrigation results in increased rubber content, but also causes shrub biomass reductions. The effect of irrigation on rubber yield has not been consistent. Shrub biomass and rubber yields were greatest in a sandy loam soil at the higher irrigation levels, whereas the greatest rubber yields on a silty clay loam occurred at the lowest moisture levels (Hunter and Kelley, 1946; Bucks *et al.*, 1984, 1985c). Rubber content increased during periods of high stress (Benedict *et al.*, 1947). They indicated that rubber accumulation could be forced by alternating periods of low and high moisture stress. Water applied at 680 and 1230 mm in the second growing season produced equal amounts of shrub, and the highest rubber yield was obtained with no irrigation (Veihmeyer and Hendrickson, 1961).

Yield and crop water stress relationships have indicated that guayule grown in arid environments can be more sensitive to water stress that occurs late in the growing season as compared to earlier in the year (Bucks *et al.*, 1985a). A simple, reliable method was needed for following water stress so that the stress/rubber production interrelationship could be defined clearly. The crop water stress index (CWSI), developed for other economic crops, was applied to transplanted guayule by Nakayama and Bucks (1983). Shrub canopy temperature measurements based on remote infrared thermometric techniques and atmospheric vapour pressure deficits from meteorological parameters were used to relate plant water stress to the soil water status. It has been found that plant yields in a study at Mesa, Arizona, were related negatively and linearly to the computed CWSI range of 0.20 for low stress to 0.75 for high stress (Nakayama and Bucks, 1984). The CWSI was also related negatively to soil water content and evapotranspiration. An inverse correlation between rubber yield and seasonally averaged CWSI has also been reported (Garrot *et al.*, 1986). The CWSI has been used for scheduling water applications for maintaining irrigation treatments (Ray *et al.*, 1986).

Fertilization

Scientists in the ERP classified guayule as a low user of the major nutrients (McGinnies and Mills, 1980). Even at the high densities under which nursery seedlings were grown, the demand on the soil did not appear to be greater than for other field crops. More recent research has shown that the plant does not require high nutrient levels, except with high irrigation applications (Bucks *et al.*, 1985c). Coffelt and Williams (2009) found indirect evidence from analysing waste water from a latex extraction plant that guayule might require higher levels of K when grown for optimum yields than previously thought. Current recommendations are that fertilization requirements should be based on soil fertility and the general condition of the plants.

Transplants

Growth and rubber accumulation of transplants grown in outdoor gravel culture for 8.5 months were affected by nitrogen more than any of the other major nutrient elements (Bonner, 1944). Plants receiving most or all of their nitrogen as nitrate grew better and yielded more rubber than shrubs receiving their nitrogen as ammonium. The effect of plant spacing, fall irrigation and fertilization on rubber production was studied during the winter at Salinas, California, in 1-year-old guayule (Tingey and Foote, 1946). The rubber yield in shrubs fertilized in July and receiving an additional irrigation in September was 240 kg/ha versus 190 kg/ha in the non-fertilized shrubs.

Plant height and width were significantly greater in the nitrogen treatments than in the non-fertilized shrubs when split applications of nitrogen (4.5, 9.0, 18.0 kg/ha) were applied to transplanted guayule in California (Cannell and Younger, 1983). Urea, ammonium phosphate and calcium nitrate at 112 kg of actual nitrogen/ha were applied in a guayule seed increase field at Marana, Arizona (Rubis, 1983). Plant biomass in the fertilizer treatments was 10.9% greater than the control, and calcium nitrate treatments alone yielded 20% more than the control.

Although guayule can withstand extreme drought, Bucks *et al.* (1985c) found that water and nitrogen applications of 2850 mm and 210 kg/ha were required to achieve the greatest production with 2-year-old shrubs. A gravel culture technique was used to study the influence of nitrogen, phosphorus and potassium on guayule shrub growth and rubber accumulation (Thomas and Hickman, 1989). Increasing available nitrogen and potassium raised plant dry weight, but dry weight was not affected by changes in phosphorus levels.

Direct seeding

The effects of fertilization on direct-seeded guayule have involved the early practice of either sowing multiple rows on wide nursery beds at heavy rates and thinning, or thickly seeding on beds in the field to establish guayule as a row crop. It should be emphasized that heavy seeding rates were used in each instance. Nitrogen fertilizer side-dressed at 280 kg/ha proved to be beneficial in a trial conducted during 1943 in California to test the feasibility of growing guayule as a row crop by direct seeding (Hammond and Polhamus, 1965).

The effects of irrigation and fertilizer on the rubber production of thickly seeded guayule were investigated at four locations in California (Kelley *et al.*, 1946). Guayule was seeded in rows 180 mm apart in seven-row nursery beds. After thinning, 30 kg of nitrogen, 130 kg of phosphorus and 80 kg/ha of potassium were applied. An additional 100 kg/ha of nitrogen was applied in split applications of 50 kg each in early and late summer. The application of fertilizer did not increase rubber production. The results indicated that for maximum rubber production, dense guayule stands were required. The optimum irrigation and fertility treatments were related closely to soil and climatic conditions. A 3 month seedling establishment period would require at least 300 mm of water and 56 kg/ha of nitrogen (Bucks *et al.*, 1986).

Weed control

Weed control in guayule nurseries was one of the most expensive operations during the ERP. Nearly 3000 people were used to hand weed 223 ha of nurseries near Salinas, California (Mihail *et al.*, 1991). This labour requirement was reduced 90% by the use of close cultivation between the bands of plants and through the development of post-emergence petroleum oils. Oil sprays gave effective control of seedling grasses and some broadleaf weeds on a 472 ha plantation being cultivated for the Guayule Stockpiling Project near Crystal City, Texas, in 1951 (Hammond and Polhamus, 1965). More recent research has focused on the evaluation of modern herbicides (Mihail *et al.*, 1991; Foster and Coffelt, 2005).

Transplants

Guayule is a poor competitor against annual and perennial broadleaf and grass weeds. In a commercial situation, guayule will not be harvested for at least two growing seasons. Therefore, growers will need effective, economical weed control during establishment and up to the time when the shrubs have grown too large to permit mechanical cultivation.

Studies in Arizona and California showed that trifluralin was a promising pre-emergence treatment for controlling annual broadleaf weeds and grasses (Siddiqui *et al.*, 1982; Elder *et al.*, 1983; Kidd, 1983). Several pre-plant incorporated herbicides were tested in field plots in New Mexico (Whitworth, 1983). Guayule was most tolerant to DCPA (9.0 kg ai/ha), fluridone (0.5–1.8 kg ai/ha), metolachlor (1.1–2.2 kg ai/ha) and pendimethalin (0.6–1.1 kg ai/ha). The results of herbicide evaluation trials in Australia indicated that successful stand establishment would be achieved only by transplanting with pre-emergence weed control (Milthorpe *et al.*, 1991). A pre-plant-incorporated tank mix application of oxyfluorfen plus oryzalin at a rate of 2 + 2 kg ai/ha gave excellent weed control for a full spectrum of weeds for at least 6–8 months. Pre-emergence treatments of isoxaben, pendimethalin and trifluralin were tested at Fort Stockton, Texas, for weed control in Mexican bulk transplants. Shrub biomass, rubber and resin content and rubber and resin yield in the treatments were all equal to or greater than the control.

With the support of the preceding studies, a Special Local Needs registration has been issued for the use of pendimethalin (Prowl 3.3 EC) for pre-emergence control of most annual grasses and certain broadleaf weeds in transplanted guayule in Arizona (Arizona Department of Agriculture, 2003). Pendimethalin may be applied for short-term (4 months) or long-term (6–8 months) weed control at rates of 5.6 and 11.2 l/ha, respectively. The herbicide should be applied after transplanting as band or broadcast sprays, without allowing spray to contact guayule leaves, shoots, or buds. Pendimethalin can also be applied through chemigation systems as a supplemental weed control practice.

Post-emergence, over-the-top herbicide treatments are toxic to guayule transplants during periods of active growth. No herbicides are labelled for post-emergence weed control in guayule. Unshielded glyphosate applications resulted in high guayule mortality (Siddiqui *et al.*, 1982; Ferraris, 1986). The recommended treatments were paraquat and glyphosate applied as shielded sprays when the shrubs were dormant. Shielded band sprays of paraquat in spring to control winter weeds and glyphosate in fall to control summer weeds, supplemented by hand weeding, have been successful in Australia (Ferraris, 1986). Following harvest by clipping, it was proposed to band spray oryzalin over the beds and incorporate trifluralin in the furrows.

Recently, Williams *et al.* (2009) found that organic material from guayule plants could cause herbicides to be tied up in the soil. This indicates that for herbicide applications to regrowth, higher rates of these herbicides may be needed to achieve desirable weed control levels. Higher rates of herbicides may also be needed for crops following guayule if the organic material left in the soil is not dispersed prior to planting the succeeding crop.

Broadcast sprays for weed control in established guayule should be applied to dormant shrubs, and spot or individual plant treatments can be used for localized weed infestations. Glyphosate, oryzalin and oxyflurofen applied as directed sprays to dormant guayule shrubs controlled established weeds effectively and were not toxic to the shrubs (Ferraris, 1986; Foster *et al.*, 1986; Milthorpe *et al.*, 1991). Post-emergence broadcast sprays of glyphosate, 2,4-D and bromoxynil were applied to a dormant guayule stand with no shrub injury (Foster *et al.*, 1989).

Direct seeding

As mentioned previously, guayule is difficult to establish by direct seeding because the seedlings grow slowly and, at the same time, offer little competition to emerging weed seedlings. An effective pre-plant or pre-emergence herbicide is required for optimum stand establishment. Several pre-plant, soil incorporated herbicides have been evaluated in New Mexico (Whitworth, 1981b; Boyse *et al.*, 1983; Whitworth, 1983). Only bensulide, DCPA and pendimethalin demonstrated adequate selectivity on direct-seeded guayule.

Pre-emergence treatments of DCPA, pendimethalin and prodiamine were tested on a Dalby clay at Fort Stockton, Texas (Foster *et al.*, 1993). Guayule seedling establishment with DCPA (4.5 and 9.0 kg ai/ha), pendimethalin (0.3 kg ai/ha) and prodiamine (0.3 kg ai/ha) was not significantly different from the control in 1991. A study conducted on a Delnorte very gravelly loam at Fort Stockton confirmed the adequate selectivity of DCPA 9.0 (kg ai/ha) and pendimethalin (0.6 kg ai/ha) as pre-emergence weed control treatments for direct-seeded guayule (Foster and Coffelt, 2005).

DCPA (4.5, 9.0, 11.0 kg/ha), bensulide (2.2, 3.4, 4.5 kg/ha) and pendimethalin (0.6, 1.1 and 2.2 kg/ha) were applied as pre-plant incorporated treatments on a Casa Grande sandy clay loam before direct seeding at Maricopa, Arizona (Foster *et al.*, 2002a). Pendimethalin was safe for use in guayule direct seeding. This herbicide has a broad spectrum of weed control and is used widely by growers for weed control in other crops.

Harvesting

Rubber is located principally in the cortical parenchyma cells of the shrubs, with two-thirds in the stem and branches and the remainder in the roots (NAS, 1977). Previously, harvesting procedures centred on digging the whole plant to access rubber in the roots as well as the branches. Current methods involve clipping the plants at 100 mm above the soil surface and milling only the plant tops (Coffelt and Nakayama, 2007). Lloyd (1911) was the first to suggest harvesting guayule by clipping instead of digging, although the Intercontinental Rubber Company tried clipping, but without much success. Clipping was investigated during the ERP because this method offered the advantage of not having to replant after harvesting. Pollarding or allowing plants to regrow following the initial harvest is the current recommended harvesting method.

New germplasm, coupled with improved agronomic practices, have increased rubber yields significantly compared with the standard USDA or Mexican germplasm (Estilai and Dierig, 1996). The University of California has released seven cultivars (CAL-1 to CAL-7) (Tysdal *et al.*, 1983; Estilai, 1985, 1986). The Arizona Agricultural Experiment Station and the USDA-ARS jointly released six lines (AZ-1 to AZ-6), which were bred for their ability to regenerate after cutting to allow for multiple harvests, vigour of regrowth and bulk rubber yield (Ray *et al.*, 1999). Rubber yields of these lines were 58–101% greater than the old USDA lines. This newer germplasm is the basis for current commercialization efforts not only in the USA, but also in Australia, South Africa, Europe and other countries.

Whole plant yields

Early studies found that a maximum rubber yield of 1500–1900 kg/ha was obtained after about 3 years (Kelley *et al.*, 1946; Tingey, 1952). Recent studies have shown that shrub yield increases proportionately with increasing irrigation up to about 3000 mm for the first

2 years of growth (Bucks *et al.*, 1984, 1985b,c; Miyamoto *et al.*, 1984a). Thus, the quantity of water required depends on the yield desired. The maximum shrub yields increased with the length of growing season. The 2-year-old whole plant rubber yields were about 35% greater on a sandy soil (Yuma, Arizona) than on a loam (Mesa, Arizona). The higher yields on the sand versus the loam soil were attributed to better soil aeration, a longer effective growing season and more vigorous plant root development. Moderate to high irrigation amounts with low nitrogen applications gave the highest biomass production that resulted in the highest rubber yields.

When guayule established from transplants was grown for two seasons, 120–160 mm/ha of water were required to produce 1000 kg of dry shrub (Miyamoto and Bucks, 1985). The annual water use of guayule for maximum shrub production was similar to that of lucerne, but the requirement to produce 1000 kg of biomass was almost twice that of lucerne. Water-use efficiencies have ranged from 0.70–0.85 kg/m^3 for dry matter, 0.045–0.055 kg/m^3 for resin and 0.030–0.040 kg/m^3 for rubber production based on the evapotranspiration of the guayule stand (Bucks *et al.*, 1985a).

Rubber yields from direct-seeded shrubs have been compared to transplants (Tingey and Clifford, 1946; Foster *et al.*, 2002b). Rubber yields from most direct-seeded lines were not significantly different from transplants, and the growth period until harvest could be shortened by nearly a year by seeding directly in the field.

Clipped plant yields

Because a greater proportion of rubber yield is contained in the shrub branches, guayule may be harvested commercially by above-ground clipping. Clipping has been proposed to (i) increase rubber productivity per unit area, (ii) allow growers an early return on their investment and (iii) delay the cost of stand re-establishment by permitting a stand to remain productive longer (Ray *et al.*, 1986; Foster and Coffelt, 2005). Limited clipping research during the ERP showed that the quality of the rubber from clipped tops was equal to that from whole plants, and rubber in the roots of clipped plants did not decrease (McGinnies and Haase, 1975). Recent clipping studies have usually supported the clipping method of harvest by showing that: (i) rubber yield is greater in clipped than whole plants; (ii) sequential clipping several times possibly would increase yields compared with a single whole plant harvest at an older age; (iii) clipping 2-year-old plants results in no yield differences between direct-seeded shrubs and transplants; and (iv) salinity (especially for irrigation water above 6 dS/m) causes significant reductions in regrowth, survival, resin content and rubber production in plant shoots and roots after clipping.

Some guayule selections responded favourably to clipping at ground level and were capable of survival and vigorous regrowth (Estilai *et al.*, 1988). A study conducted at Marana, Arizona, evaluated regrowth of three new germplasm lines (AZ1, AZ2 and AZ3) and an unreleased breeding line (G-14) following the harvest of 1- and 2-year-old plants (Coffelt *et al.*, 2001). All lines had good regrowth, except AZ1. The results from these two studies demonstrated that at least two harvests from a single planting of new germplasm lines should be possible and that lines responded differentially for the ability to regrow.

Plants with regrowth potential are essential for multiple harvests by clipping (McGinnies and Haase, 1975; Garrot and Ray, 1983; Bucks *et al.*, 1985b; Estilai *et al.*, 1988; Hoffman *et al.*, 1988; Maas *et al.*, 1988; Ray *et al.*, 1997; Foster *et al.*, 1999, 2002b; Coffelt *et al.*, 2001). Clipping can distribute the cost of stand establishment across several harvest cycles since multiple clipping harvests would eliminate the need to re-establish shrub stands after each harvest. Results from these studies also show that lines vary in their ability to regrow following clipping, so line selection becomes important. Regrowth should be one of the critical traits evaluated by breeders and agronomists in future studies prior to releasing new germplasm.

Postharvest handling

Production and harvesting practices for guayule have been developed for bulk/solid rubber uses (Foster and Coffelt, 2005), especially for tyres. One of these practices is to field-cure the shrub for 10–45 days prior to processing, which dehydrates the shrub, thus reducing the weight of shrub material to be transported, and maximizes the amount of solid rubber recovered by the milling process (Taylor and Chubb, 1952). However, this is an unacceptable practice when harvesting guayule for latex because natural processes that occur in the plant following harvest, most notably dehydration, result in a rapid loss of extractable latex (Cornish *et al.*, 2000; McMahan *et al.*, 2006). Previous work has also shown that long-term storage under high temperatures can affect both rubber and latex yield and quality adversely (Black *et al.*, 1986; Schloman *et al.*, 1986; Estilai and Hamerstrand, 1989; Dierig *et al.*, 1990; Nakayama and Coates, 1996; Cornish *et al.*, 2000).

Because immediate processing of guayule shrub for latex on a commercial scale is not feasible, Coffelt *et al.* (2009a) developed a method for storing freshly harvested shrub prior to processing for latex extraction that maintained latex concentration and yield. Their method involved keeping harvested shrub moist to protect the latex fraction prior to dry chipping and subsequent wet-grind and quantification processes. Storing harvested shrub under moist conditions may allow the industry more flexible harvesting and processing schedules by extending the time period between harvesting and latex extraction processes without significant latex loss. Coffelt *et al.* (2009a) also noted that wet storage for 2–3 weeks increased latex extraction by over 100% in some cases compared to freshly harvested shrub. The reasons for this increase are yet to be determined. Total rubber and resin were not affected by any of the storage treatments (wet or dry) compared to freshly harvested shrub (Coffelt *et al.*, 2009a).

Two previous studies noted genotypic differences among the lines used in storage studies, indicating that improvements in storage behaviour might be possible using plant breeding and selection (Estilai and Hamerstrand, 1989; Dierig *et al.*, 1990). However, a third study (Coffelt *et al.*, 2009a) did not observe any genotypic differences. The differences between studies may be due to the storage methods being studied and/or the genotypes used in the studies.

Plant Breeding

Guayule yields were first increased by planting larger areas and improving cultivation techniques rather than through breeding. This is partly because guayule is a difficult species to work with from the breeder's point of view: (i) it is a perennial; (ii) it is physiologically immature for 1 year before the first harvest at year 2; (iii) reproduction is essentially asexual (asexual reproduction by apomixis); and (iv) because breeding plots must be maintained for many years to evaluate multiple harvests from regrowth. The first breeding successes were through either mass selection or the selection of individual high-yielding plants, even though the available germplasm from which selections have been made appears to be rather narrow (Thompson and Ray, 1988; Ray, 1993; Ray *et al.*, 2005).

There have been successes in guayule plant breeding, such as increases in rubber yield from 300 to 1000 kg/ha rubber (Estilai and Ray, 1991). However, guayule still contains many wild characteristics such as indeterminate flowering, seed shattering and both sexual and asexual reproduction occurring in the same plant, which will be desirable to change to facilitate commercialization.

Germplasm sources

The domestication and development of guayule as a crop was initiated in 1910 by W.B. McCallum, who was then employed by the Intercontinental Rubber Company. A breeding and selection programme was started in 1916 at Continental, Arizona, and transferred to Salinas, California, in 1925. A selection made by McCallum in Salinas, '593', was the principal cultivar used in production in the 1920s, 1930s and the ERP (Thompson and Ray, 1988). During the ERP, a major activity was germplasm and cultivar development, and the breeding material developed during this time became the basis for the research efforts started in the 1970s and continuing until today (Thompson and Ray, 1988).

The first collection of guayule germplasm was made by McCallum in 1912. When civil strife and revolution began in northern Mexico, McCallum gathered seeds from wild stands in order to move his cultural operations into the USA. The seeds were planted initially at Valley Center, California, and evaluation of germplasm was conducted subsequently at Continental, Arizona, and Salinas, California (Thompson and Ray, 1988).

There were two main germplasm collection expeditions during the ERP. Powers, McCallum and Olson collected 66 accessions from 24 locations in Mexico; and Powers and Federer collected 368 accessions from 21 locations in Texas. These accessions were then planted and evaluated at Salinas in 1943. In 1948, Hammond and Hinton collected an additional 184 accessions from 93 locations in Mexico (Thompson and Ray, 1988).

The USDA guayule breeding programme at Salinas, California, was terminated in 1959 and 24 germplasm lines, developed by H.M. Tysdal from the Powers, Hammond and Hinton collections, plus line '593' developed by McCallum, were selected for storage at the USDA National Seed Storage Laboratory (now the National Center for Genetic Resources Preservation) at Fort Collins, Colorado, in 1965. These 25 lines, selected on the basis of their rubber production and plant growth characteristics, were the only ones saved from the hundreds of selections, breeding lines and accessions stored at Shafter and Salinas, California. These 25 lines plus the line 'Bulk Richardson' (from D.D. Rubis; a bulk seed collection from Mexico made by Richardson) became what were commonly called the 26 USDA germplasm lines, from which the breeding programmes in the 1970s began (Thompson and Ray, 1988).

Interestingly, 21 of the 26 USDA lines came from the state of Durango, Mexico. The apparent narrow germplasm base is accentuated by the fact that 15 of the lines descended from the Powers, McCallum and Olson collection No. 4265, which was a bulked seed collection from five plants at one location. The original diploid material came from collection No. 4254, which was also bulked seed from five plants at one location (Thompson and Ray, 1988).

In 1976, R.C. Rollins made collections from 45 locations in Mexico. In 1977, C.T. Mason collected related *Parthenium* species throughout Mexico, Naqvi and Hanson collected guayule from 50 locations in Mexico (also in 1977) and in 1982 Tipton and Gregg collected seeds from 10 native populations in Texas. An extensive effort was mounted in 1982 by Mexican scientists, who collected 3000 accessions from 310 locations from six states (Thompson and Ray, 1988). Unfortunately, it is unclear where most of these accessions are today.

All breeding approaches depend on the existing genetic variability found in the available germplasm. However, even though this genetic base appears to be rather narrow, it has not been a hindrance to guayule breeding programmes. This is probably because the facultative nature of apomixis in polyploid guayule continually releases new variability with each seed harvest. In fact, with the limited scale of the present plant improvement programmes, this variability is created faster than it can be exploited by breeders (Thompson and Ray, 1988; Ray *et al.*, 2005).

Most guayule germplasm today consists of reproducing triploid ($3n = 54$) and tetraploid ($4n = 72$) accessions apomictically, because they received most of the attention in previous breeding programmes (Hammond and Polhamus, 1965; Thompson and Ray, 1988; Ray *et al.*, 2005). Sexually reproducing, largely self-incompatible diploids ($2n = 36$) have had only limited use in guayule breeding programmes.

At present, the USDA-ARS National Arid Land Plant Genetic Resources Unit in Parlier, California, has 144 *P. argentatum* accessions and five interspecific hybrids of different *Parthenium* species. Twenty-five of these accessions have PI numbers, with the remainder carrying western regional numbers, but unfortunately as many as 64 accessions may not have viable seed. This is an important problem that has been recognized by the National Plant Germplasm System. A collection trip by Coffelt, Foster and Stout was sponsored in Texas in 2005 to try to recollect some of the original collections. However, guayule could not be found at most of the original collection sites. Where guayule plants were found, little or no seed was present. Other species were found at many of these sites, especially *P. incanum*. It is possible that many of these original sites had been misidentified or that other species have replaced the original guayule populations. One site was found with viable guayule seed near Bakersfield, Texas. Seed has been collected at this site and added to the collection. This represents the only wild source of guayule collected since the early collection trips. Studies to determine the best methods for long-term storage of guayule seed to maintain viability have been initiated, but recommendations are not yet available.

Breeding methods

Selection in guayule has been aided significantly by the description of the components of yield and their relationships to rubber production (Thompson *et al.*, 1988; Dierig *et al.*, 1989). In general, rubber content (%) was not correlated with rubber yield, and in fact was often correlated negatively. Fresh and dry weights, as well as other characteristics related to biomass production, were correlated highly and consistently with rubber yield (Thompson *et al.*, 1988; Dierig *et al.*, 1989). The characteristics shown to be the best predictors of rubber content were plant fresh and dry weight, per cent dry weight and plant volume, and the best predictive model for rubber yield included plant height and width, volume and dry weight (Dierig *et al.*, 1989).

Ray *et al.* (1993) tested the relatedness of apomictic parents and their open-pollinated, half-sib progeny families for eight components of yield. Heritability estimates were made by measuring the components of yield in both the parents and progeny. The parent plants were all open-pollinated progeny of a single-plant selection made by D.D. Rubis, and measurements were made when the parent plants were 3 years old and the progeny plants 2 years old. For rubber yield, rubber content, resin content, fresh weight, dry weight, per cent dry weight, height and width, none of the parent–progeny regressions were significantly different from zero. For all characters, a large range of phenotypic variation was observed, and the range and standard deviation of the parents were greater than among the progeny. This was probably due to the compounding of environmental effects (the parent plants were a year older than the progeny plants), rather than a difference in genetic variability (Dierig *et al.*, 2001). Linear correlations were performed to study the relationship between rubber yield and the other seven characters, and fresh and dry weights were correlated highly and positively with rubber yield in all populations. Thompson *et al.* (1988) found significant correlations between rubber content and resin content that were higher than correlations of any other character with rubber content. This high correlation means that breeders should be able to create new lines that are higher in both rubber and resin than older lines. Because both rubber and resin are important characters in determining the value of guayule end products, breeding for simultaneous increases in these traits is important to ensure successful commercialization. Evidence that this is possible is found in the release of six new germplasm lines that are higher in rubber and resin than the older USDA lines (Ray *et al.*, 1999).

Biomass appears to be the best predictor of rubber yield (rubber yield = plant biomass × rubber concentration). Thus, plant growth or biomass production can be used as a primary selection index for rubber yield. However, selection for large plant size may be disadvantageous because of harvesting; its transportation and handling and the efficiency of rubber extraction in the processing plant are significant economic factors in the production of rubber from guayule. For this reason, selection of plants with higher rubber concentration in concert with adequate biomass production must receive primary attention. Such selection is difficult because there is often a negative correlation between rubber concentration and biomass (Thompson *et al.*, 1988).

Advances and potential for improvement

In many instances, the breeding of new and conventional crops is the same. The major differences are that in new crops: (i) the plant breeder starts with a different and frequently unique and exotic germplasm base from which to develop a crop; (ii) the breeder is often totally unfamiliar with the species, the germplasm and potential end products; (iii) the traits to be improved frequently have not been identified by researchers, industry, or growers; and (iv) there is often a paucity of previous research, including the appropriate technology for evaluating, selecting and breeding of the products and co-products sought. New crop breeders must be flexible in their approach to breeding where so much is unknown. The breeder must be innovative and able to change approaches and methodology rapidly to meet the opportunities and constraints as they are encountered.

The primary objective for all guayule breeding programmes has been to increase rubber yield. Secondary objectives have included improving rubber quality, resin yields, seedling and mature plant vigour, plant architecture, regeneration following harvest by clipping and tolerance to salinity, drought, diseases and pests (Thompson and Ray, 1988; Estilai and Ray, 1991; Ray *et al.*, 2005). However, because of the relatively few researchers involved in guayule breeding, the secondary objectives have not received much attention over time. The most extensively employed breeding approach in guayule has been single-plant selections from within apomictic polyploid populations. Selection of individual plants is usually the simplest and most rapid method when heritabilities for desired characters are high. If heritabilites are high, increases can be made in a short period of time, but the long-term potential is for only modest gains because apomixis restricts the release of new genetic combinations. Thus, the degree of success using this method depends first on the amount of heterogeneity in the population; second, whether or not the differences are genetic; and third, on the number of plants that can be screened (Thompson and Ray, 1988). This method increased annual rubber yields from approximately 300 to 1000 kg/ha by selecting for the components of yield described previously, but predominately by selecting simultaneously for rubber concentration (%) and dry matter or biomass production (Estilai and Ray, 1991).

When heritabilities are low, single-plant selection is not as effective as family selection. In family selection, families of progeny, either full-sibs or half-sibs, are used to evaluate the quality of the parent plants. Thus, parent plants are not selected on their own merits, but on those of their progeny. The disadvantage of family selection is that there is a lengthened generation interval. However, because guayule is a perennial plant with almost continuous flowering, many generations of progeny can be obtained from a single plant once it has been selected as a suitable parent.

Mass selection is one of the oldest plant breeding methods and significant gains can be achieved in a relatively short period of time because only the top-yielding plants in a population are selected to become the parents of the next generation. Today, mass selection is used to enhance germplasm and develop cultivars, especially in crops where there are few individuals involved and cross-pollination is the major mode of reproduction. Mass selection has been used in sexual diploid populations by Ray *et al.* (1995) in which, after three cycles of selection, a diploid line tolerant to *Verticillium dahliae* was developed.

Mass selection has never been used extensively in polyploid guayule because to enhance populations using this method, one must be able to screen many plants effectively (hundreds at minimum, thousands optimally), the resulting progeny must be uniform and cross-pollination must be the major mode of reproduction. Many of the important characters in guayule appear to be multigenic, in which heritability is low, and screening procedures for large numbers of samples have not been developed. In addition, polyploid guayule populations are fairly variable because of new genetic combinations resulting from facultative apomixis in highly heterozygous plants, and from environmental effects compounded over several years of growth. The apomictic nature of polyploid guayule greatly reduces the chance of attaining significant gains using mass selection.

Hybridization of apomictic polyploids is a method that has been suggested but has been used sparingly because of the problems of separating the offspring that arise from sexual reproduction from the apomicts. Plants expressing high levels of sexuality could be identified using the method of Keys *et al.* (2002), and crossed to produce new genetic combinations from which further selections could be made. Seed would be collected from the hybrid plants, planted and tested for apomictic potential. If the resulting progeny are predominately apomictic, seed from them would be placed in progeny trials and tested for possible release as new lines. If the plants are predominately sexual, they could be backcrossed to enhance certain characteristics, self-pollinated to produce a segregating population from which more selections could be made, or apply standard breeding strategies generally not used in guayule.

Interspecific hybridization has been applied on a limited scale only. None of the other *Parthenium* species produce an appreciable amount of rubber, although they should be considered as potential sources of vigour, increased resin content, increased biomass, disease and insect resistance, regrowth ability after clipping and cold tolerance.

The major disadvantage of interspecific hybrids is that it will take a large number of backcross generations to guayule to increase the rubber content as well as to keep the desirable new trait(s). The University of California-Riverside has released three germplasm lines (Cal-1, Cal-2 and Cal-5) that have been developed from interspecific crosses of guayule with three different *Parthenium* species. These three have increased vigour, biomass production and resistance to *Verticillium* wilt. AZ-101, a vigorous natural interspecific hybrid, is an open-pollinated cross between guayule and *P. tomentosum* var. *stramonium* (Thompson and Ray, 1988; Estilai and Ray, 1991).

Diploids have been used in guayule breeding because of their sexual (non-apomictic) reproduction, and thus the ability to use standard breeding methodologies. While there are problems in using diploids, such as significantly lower yields and increased susceptibility to root diseases, these yield and disease problems have been overcome by using modified recurrent selection schemes to increase yield and mass selection to develop *Verticillium*-tolerant lines (Estilai and Ray, 1991; Ray *et al.*, 1995). These improved diploid lines either can be crossed to apomictic polyploids or have their chromosome numbers doubled with colchicine. Diploids could also be used to release new genetic combinations by crossing them as the female parent to apomictic polyploids. The resulting apomictic progeny plants might contain new and useful combinations of genes, because meiosis in the microspore mother cells of the apomictic polyploid male plants is normal. Once high-yielding polyploids are identified, they could be crossed on to diploids, resulting in populations with enough variation from which to make selections (Thompson and Ray, 1988; Estilai and Ray, 1991).

A potential breeding method that can make the most of limited resources in guayule is the pedigreed natural crossing method (Hammons, 1964; Coffelt, 1989). Guayule meets the requirements for use of this method by having natural cross-pollination between potential parents (species or diploids or polyploids) and dominant markers to identify hybrids. The advantages of this method are that crossing is not dependent on limited time available for a single scientist or trained assistant to perform the cross; identification, harvesting and isolation of hybrids can be done by semi-skilled workers on land unsuitable for yield trials and other experiments; and it is more economical than making crosses in the greenhouse. The biggest disadvantages are that the pedigree of the hybrids is based on a parental line rather than a single plant and large amounts of land may be needed to identify hybrids. The advantages of this method of producing large numbers of hybrids with little effort should outweigh the disadvantage of individual parent plant identification. The higher out-crossing rate of guayule compared with self-pollinated species should result in a larger number of hybrids being identified with the same amount of land.

Yield trials have been used successfully to evaluate guayule germplasm lines under various environmental conditions (Estilai and Ray, 1991; Majeau *et al.*, 2003). However, this valuable tool has not been available to breeders consistently, due to a lack of continuous funding. More consistent funding is needed to carry these trials to completion and initiate new ones as new germplasm becomes available.

Another important aspect of yield trials is their use in estimating genotype, environment and genotype × environment interactions.

Coffelt *et al.* (2005) found that location, line and plant age were significant and the interactions not significant for all traits measured in a study at two locations with two plant ages. Environment accounted for over 50% of the variability observed in all traits, followed by plant age (16%) and line (10%). These results point to the tremendous impact that environment has on guayule plant growth, biomass and latex content. Coffelt *et al.* (2005) could not determine from these tests whether temperature, soil type, moisture, fertility, or a combination of these or other environmental factors, were responsible for this response. Some of the non-significant interactions may have been significant if a larger or wider germplasm base could have been evaluated. Dierig *et al.* (2001) also observed significant environmental effects even within a single field. Additional studies are needed to determine the environmental factor(s) responsible for the large environmental response observed in these studies.

It is important to breeders that the genotype × environment interaction is not significant since this means selection for superior lines can be done at one location. The superior lines should be superior at other locations where guayule is grown. Anecdotal evidence supports this conclusion, since the AZ lines (Ray *et al.*, 1999) tested by Coffelt *et al.* (2005) have been observed to give similar results when evaluated under diverse environments such as Australia, South Africa and China.

Progress in selection for rubber/latex traits has been hampered because of the difficulty in determining rubber and latex yield in single plants. The analyses for rubber and latex contents are labour-intensive and expensive, greatly limiting the number of samples that can be processed. The amount of leaves, the moisture content of the shrub and deterioration of the latex during processing can all interfere, and must be considered, in the analysis of rubber, and especially latex (Teetor *et al.*, 2009). In addition, morphological traits have not been identified that correlate consistently with rubber or latex content. Improvements in these areas could speed the breeding progress greatly.

Research is needed to establish the relationship between latex and rubber concentrations and yields. If rubber and latex concentrations and/or yields are closely related, then previous relationships established between rubber concentration/yield and the various yield components can be expected to be the same as their relationships with latex concentration/yield. However, if rubber concentration is not closely related to latex concentration, then studies will need to be conducted to establish the relationships between latex concentration and traits such as plant biomass, latex yield, rubber concentration and yield, resin concentration and yield, plant height and width, etc. Recent studies (Coffelt *et al.*, 2009a,b) have indicated inconsistent relationships between latex and rubber concentrations. In one study (Coffelt *et al.*, 2009a), latex concentration and yield varied with storage conditions prior to chipping, whereas rubber concentrations and yield did not. In another study (Coffelt *et al.*, 2009b) of the effects of plant population and planting dates over several harvest dates, rubber and latex concentrations were similar. These studies suggest that more research defining the relationships between latex concentration and yield and rubber concentration and yield will need to be done before meaningful breeding programmes can be started.

Guayule Products and By-products

Natural rubber, *cis*-1,4-polyisoprene, is the largest volume elastomer in commerce. Produced in tropical countries by the Brazilian rubber tree, its performance in critical applications such as aircraft tyres and medical devices has not been duplicated by synthetics. Prices continue to increase on a macro scale and supplies continue to tighten, especially with increased industrialization in China and India. Guayule is the only alternative rubber-producing crop ready to be scaled up to meet the burgeoning demand for natural rubber.

Latex

The Yulex Corporation (Maricopa, Arizona) has scaled a 'green' aqueous-based guayule latex extraction process successfully (Cornish, 1996, 1998) to over 2041 kg of fresh shrub per day capacity. Commercial guayule latex produced in the USA has been used to develop a multitude of products, primarily for medical applications, including examination gloves, dental dams and catheters. Guayule latex meets ASTM D1076-06 Category 4 latex requirements for low total protein and no detectable antigenic protein (McMahan *et al.*, 2007a). Products produced from this latex do not elicit the Type I allergic response of subjects sensitized to *Hevea* natural rubber (Siler and Cornish, 1994; Schloman *et al.*, 1996; Siler *et al.*, 1996; Cornish and Lytle, 1999; Cornish *et al.*, 2001).

Solid guayule rubber

Production of solid natural rubber bales from guayule took place during the three previous commercialization periods: early 20th century, during the ERP (Cooperrider and Culley, 1943) and in the late 1970–early 1980s (Velásquez *et al.*, 1978). The material produced commercially was used primarily for tyres, at various percentages in various components. When extracted by organic solvents, the rubber contained high levels of resin (low-molecular weight, acetone-extractable material) that diminished the physical properties of the raw rubber and limited utility. Few results have been published; however, lower treadwear was noted for 100% guayule tyres (Doering, 1934). During the ERP, 100% replacement of *Hevea* with guayule rubber was needed and tyres considered equal to those from *Hevea* were built and tested successfully (Hammond and Polhamus, 1965).

In a pilot-scale test, the Bridgestone/Firestone facility in Sacaton, Arizona, produced a total of 8.8 t of fractionated rubber product in 1988–1990 (Schloman, 2005). More than 3.3t (37%) of the rubber produced at the Sacaton prototype processing facility met specifications derived from those for TSR20 natural rubber (Grade 20 Technically Specified Rubber). The TSR20 rubber was tested as a replacement for natural rubber in aircraft tyres. However, these tyres did not pass all dynamic testing requirements. So, without further processing, the product was not considered a direct drop-in replacement for *Hevea* natural rubber (Lukich, 1992). However, light truck tyres built with guayule in place of *Hevea* did meet US Army proving grounds durability testing (Lucas, 1996).

When formulated into tyre compounds, guayule rubber exhibits lower cure rates and lower oxidative stability compared to *Hevea* natural rubber (Schloman *et al.*, 1996) and therefore requires adjustment in compounding and processing. Guayule rubber compounds with similar vulcanization kinetics and tensile strength to that of *Hevea* rubber can be produced by formulation adjustments (Hammond and Polhamus, 1965; Ramos de Valle and Aramburo, 1981; Schloman *et al.*, 1996; Cornish *et al.*, 2008). Development of a solid rubber bale from the currently used aqueous extraction process is now under development (K. Cornish, Arizona, 2009, personal communication).

Resin and its uses

Natural plant resins are insoluble in water, soluble in polar organic solvents and are used commercially in cosmetics, pharmaceuticals and as modifiers in synthetic rubber and plastics. They are viscous and can be solid or semi-solid, clear or translucent, yellowish or brownish in colour. Natural plant resin (acetone extract) is found in whole guayule shrub in the range of 8–10% (dry weight basis) and in similar concentrations in the leaf stream (Cornish and Schloman, 2003). Slightly lower concentrations, about 7%, have been measured in the lignocellulosic bagasse after latex extraction (water-based process) (McMahan *et al.*, 2007b). In addition to the bagasse, leaves comprise 30% (winter) to 60% (summer) of the harvested guayule shrub. The leaves are normally removed prior to processing because they do not contain high quality rubber; but, like the bagasse, they are rich in terpenoid resin.

Numerous studies of solvent-extracted resin from guayule bagasse have demonstrated its potential as an adhesive tackifier, a polymer modifier and as an additive in formulations of paints and coatings. Guayule resin adhesives have demonstrated the ability to bond to many substances, even under water, to numerous surfaces (Nakayama, 2005). Water- and abrasion-resistant coatings have been made using guayule solvent-extracted resin in coating formulations (Belmares *et al.*, 1980). Further, the solvent-extracted resin can be used as an adhesive modifier of amine-cured epoxy resin for making strippable coatings with good impact resistance and hardness (Thames and Kaleem, 1991; Thames and Wagner, 1991). Peelable coatings are important in temporary protection of commercial and military structures and vehicles, and epoxy-amine polymers can be formulated as low volatile organic compound (VOC) coatings with excellent chemical resistance, water resistance and corrosion resistance (Thames *et al.*, 1996).

Guayule resin-modified marine coatings inhibit surface fouling by barnacles and seagrass (Greenfield, 1992). Isolated resin fractions (solvent extraction) exhibit varying toxicity to shrimp and/or barnacles (Thames *et al.*, 1996), suggesting the natural products responsible for antifouling can be concentrated in controlled-release paints or plastics. Antifouling paints are important to the economic interests of the US military and industry. Other important coatings applications for which guayule resin use holds potential are: underwater paints, water-based architectural paints, UV-curable chemical agent resistant coating (CARC) resistant paints and UV-curable corrosion-resistant primers.

Guayule resin is the quintessential tackifier for natural rubber because of mutual compatibility. It is considered miscible with natural rubber, and with elastomers of similar polarity. As an additive to rubber compounds, it causes softening and tackifying in the green state, with plasticization effects in the cured state. When resin is added to guayule rubber, both adhesion and tack increase with concentration of resin. As expected, the shear properties decrease with increasing concentration of resin (Gumbs, 2004).

Finally, incorporation of guayule resin protects wood against termite, molluscan borer and fungal attacks in a persistent manner, an important feature of a wood preservative (Bultman *et al.*, 1991, 1998). Composites made from combinations of wood, plastic and guayule bagasse or resin yield termite- and fungus-resistant products (Nakayama *et al.*, 2004). Incorporation of guayule resin with existing wood coatings or adhesives may provide both insect control and modified adhesive properties (Nakayama, 2005).

Bagasse

Production of natural rubber from guayule leaves 80–90% of the biomass (dry weight) unused. This biomass co-product can be used as a source of raw materials. For instance, it has been used to fabricate high-density composite boards that have termite-resistant properties (Nakayama *et al.*, 2001, 2003, 2004; Chow *et al.*, 2008; Holt *et al.*, 2009), as a soil amendment (Schloman, unpublished data), as fuels (Nakayama *et al.*, 2003) and, after further treatment, as a source of speciality chemicals.

Guayule bagasse is a potential source of fermentable sugars. It is unusually high in pentosans (Touzinsky, 1980). However, sugar accessibility depends on the extent to which the structure of bagasse woody tissue can be modified. In the conversion of the woody fragment to fermentable sugars and a preliminary assessment of guayule, bagasse saccharification and fermentation has shown that hemicellulose hydrolysis with dilute H_2SO_4 yields 12% xylose and 5% arabinose based on bagasse charge (Schloman, 1983). Only 49% of the solubilized pentosans were hydrolysed. The residual lignocellulose was treated with 70% H_2SO_4 to disrupt the cellulose crystallite structure and then saccharified with dilute acid. Hexose yields were low: 6% glucose and 1% fructose, based on the original bagasse charge.

Guayule bagasse has important potential as a biofuel feedstock. It is in the form of a finely divided solid of which harvesting and grinding costs are borne by the rubber production process. It is a high-energy feedstock of over 20,000 kJ/kg high heat value (Boateng *et al.*, 2009). It is transportable, relatively high in density and produced 12 months a year. Guayule bagasse is similar to other hardwoods in composition (Chow *et al.*, 2008) and as a feedstock for ethanol production (K. Holtman, California, 2009, personal communication). It can be used to produce very viscous, high-energy content (~ 30 MJ/kg) pyrolysis liquid (bio-oil) in yields averaging over 60% without any catalyst (Boateng *et al.*, 2009). For guayule biorefinery operations, it may also be used to generate steam and electricity by combustion.

Summary

Guayule has a long history of use as a source of natural rubber. Although similar in quality to natural rubber from *Hevea brasiliensis*, guayule rubber was not competitive economically until the occurrence of latex allergy in the general population. Continued pressure on worldwide *Hevea* rubber supplies has contributed to renew interest in the use of guayule rubber in tyre applications.

The selection of suitable sites for guayule production depends on well-drained soils, mild winter temperatures and available irrigation. The ERP recommended areas in southwestern USA with limited below-freezing winter temperatures. However, recent studies have shown that certain guayule lines can survive winter temperatures of –5°C on the Southern High Plains near Plainview, Texas.

Transplanting has been and still is the most reliable method of guayule stand establishment. Direct seeding has been successful in research plots in Texas, New Mexico and Arizona using seed conditioning techniques and precision planting. Direct-seeded plants were shown to produce rubber yields comparable to transplants grown under the same field conditions.

Although guayule is a semi-arid, drought-tolerant shrub, it must be irrigated for maximum sustained production. It may require from 1000 to 1300 mm of applied water (irrigation plus rainfall) per year to attain maximum production. Rubber yields in transplants were shown to increase proportionally with increasing irrigation up to almost 3000 mm for two growing seasons.

Clipping in transplanted and direct-seeded stands can distribute the cost of stand establishment across several harvest cycles, since multiple clipping harvests would eliminate the need to re-establish shrub stands after each harvest. This translates into lower stand establishment costs. Results have shown that lines vary in their ability to regrow following clipping, so line selection becomes important if this type of harvesting system is to be used.

Increased salinity results in decreased biomass and rubber yields. Guayule transplants have been shown to maintain production at salinities up to 4.5–4.6 dS/m. The major salinity problem may be one of survival and establishment and not necessarily growth reduction. Direct-seeding guidelines to minimize salt damage to susceptible seedlings include sprinkler irrigation for germination and emergence, followed by furrow irrigation for plant maintenance and production. Drip irrigation can also be used to reduce water requirements and manage salt damage. Research has revealed that guayule does not require high nutrient levels, except with high irrigation applications.

Trifluralin and pendimethalin are safe for pre-emergence weed control in transplanted guayule. A Special Local Needs registration for pendimethalin has been granted for the pre-emergence control of broadleaf and grass weeds in Arizona. Post-emergence treatments have not been successful as over-the-top sprays, except during the guayule dormant period. No pre- or post-emergence treatments are currently labelled for use in direct seeding.

The major differences in breeding guayule and other new crops compared to traditional crops are: (i) the plant breeder starts with a different and frequently unique and exotic germplasm base from which to develop a crop; (ii) the breeder is often totally unfamiliar with the species, the germplasm and potential end products; (iii) the traits to be improved frequently have not been identified by researchers, industry and/or growers; and (iv) there is often a paucity of previous research, including the appropriate technology for evaluating, selecting and breeding of the products and co-products sought. New crop breeders must be flexible in their approach to breeding where so much is unknown. The breeder must be innovative and able to change approaches and methodology rapidly to meet the opportunities and constraints as they are encountered. Breeding guayule is difficult because of several factors, such as its perennial growth; long immature stage before the initial harvest at 2 years of age; facultative apomictic reproduction system; and necessity for evaluating multiple harvests. In spite of these difficulties, there have been successes through guayule plant breeding resulting in significant increases in yield per area. Rubber and resin yields have been increased by 300% in some lines.

These increases have been accomplished mainly through selection of high-yielding individual plants, but also through mass selection. Other breeding methods such as pedigreed natural selection, interspecific and intraspecific hybridization and family selection can also be used in breeding guayule. Utilizing these methods requires a long-term commitment for the programme to be successful. Genetically modifying guayule is another tool that might be used for improving it; however, initial experiments using this method have not yet proved successful.

For future progress in guayule breeding to be made, much work remains to be done. The relationship between solid rubber and latex rubber and the factors affecting this relationship need to be identified and understood. The inheritance of important traits needs to be determined, the genes involved identified and their location mapped to specific chromosomes. The large environmental effects on resin and rubber also need to be determined.

The available germplasm appears on first glance to be rather narrow but because of the facultative nature of apomictic reproduction in guayule, genetic variability is being released continually. Three collection trips, one in the early 1900s and two in the early 1940s account for most of the germplasm used in current breeding programmes and what is available in the National Plant Germplasm system today. A more recent collection trip in Texas found guayule was no longer growing at many of the older sites. Collection of guayule germplasm from its natural habitat in Mexico and the USA needs to be a top priority before it is all lost. In conjunction with collection, research needs to be done on storing seed properly for the long term. Much of the previously collected seed is no longer available because it has lost viability.

Guayule research and development priorities during the past decade have focused appropriately on variety development, agronomic studies and latex extraction processes. Rubber remains the primary driver in development. The installation of the first Yulex production plant in Maricopa, Arizona, in 2006, and its operation using guayule produced in the USA, represents the culmination of much of that research. The next priority is co-product development, as increasing quantities of guayule bagasse are produced.

Acknowledgement

We would like to thank Valerie Hunter Teetor for invaluable help in editing and formatting the manuscript, as well as performing many of the experiments reported herein.

References

Abrahams, J.E., Fangmeier, D.D. and Ray, D.T. (1984) Utilizing drip irrigation to establish direct seeded guayule. Arizona Agricultural Experiment Station. J. Paper No. 4001. Tucson, Arizona, pp. 7.

Arizona Department of Agriculture (2003) Issuance of SLN-24c registration for the use of pendimethalin (Prowl 3.3 EC) in guayule production in the state of Arizona. EPA SLN No. AZ-030007.

Belmares, H., Jimenez, L.L. and Ortega, M. (1980) New rubber peptizers and coatings derived from guayule resin (*Parthenium argentatum* Gray). *Industrial and Engineering Chemistry Product Research and Development* 19, 107–111.

Benedict, H.M. and Robinson, J. (1946) Studies on germination of guayule seed. USDA Technical Bulletin 921, 48 pp.

Benedict, H.M., McRary, W.L. and Slattery, M.C. (1947) Response of guayule to alternating periods of low and high moisture stresses. *Botanical Gazette* 108, 535–549.

Black, L.T., Swanson, C.L. and Hammerstrand, G.E. (1986) Effects of storage on the molecular weight of rubber contained in guayule. *Rubber Chemistry and Technology* 59, 123–129.

Boateng, A., Mullen, C.A., Goldberg, N., Hicks, K.B., McMahan, C., Cornish, K. and Whalen, M. (2009) Energy-dense liquid fuel intermediates by pyrolysis of guayule (*Parthenium argentatum*) shrub and bagasse. *Fuel* 88, 2207–2215.

Bonner, J. (1944) Effect of varying nutritional treatments on growth and rubber accumulation in guayule. *Botanical Gazette* 105, 352–364.

Bonner, J. (1991) The history of rubber. In: Whitworth, J.W. and Whitehead, E.E. (eds) *Guayule Natural Rubber*. Office of Arid Lands Studies, University of Arizona, Tucson, Arizona, pp. 1–6.

Boyse, L.G., Whitworth, J.W. and Clark, D.R. (1983) Herbicide effects on guayule stand and rubber content. In: *Abstracts of the Guayule Rubber Society Fourth Annual Meeting*. Riverside, California, pp. 59.

Bucks, D.A., Dierig, D.A., Chandra, G.R., Backhaus, R.A., Roth, R.L. and Alcorn, S.M. (1983) Progress in direct seeding of guayule. In: *Abstracts of the Guayule Rubber Society Fourth Annual Meeting*. Riverside, California, pp. 14.

Bucks, D.A., Nakayama, F.S. and French, O.F. (1984) Water management for guayule rubber production. *Transactions of the American Society of Agricultural Engineers* 27, 1863–1870.

Bucks, D.A., Nakayama, F.S., French, O.F., Legard, W.W. and Alexander, W.L. (1985a) Irrigated guayule-production and water use relationships. *Agricultural Water Management* 10, 95–102.

Bucks, D.A., Nakayama, F.S., French, O.F., Rasnick, B.A. and Alexander, W.L. (1985b) Irrigated guayule-plant growth and production. *Agricultural Water Management* 10, 81–93.

Bucks, D.A., Roth, R.L., Nakayama, F.S. and Gardner, B.R. (1985c) Irrigation water, nitrogen, and bioregulation for guayule production. *Transactions of the American Society of Agricultural Engineers* 28, 1196–1205.

Bucks, D.A., Roth, R.L., Powers, D.E. and Chandra, G.R. (1986) Direct seeding for economical guayule field establishment. In: Fangmeier, D.D. and Alcorn, S.M. (eds) *Proceedings of the Fourth International Guayule Research and Development Conference*. Guayule Rubber Society, Tucson, Arizona, pp. 77–87.

Bucks, D.A., Powers, D.E., Chandra, G.R., Allen, S.G. and Fink, D.H. (1987) Shading techniques for the direct seeding of guayule. In: *Abstracts of the Guayule Rubber Society Seventh Annual Meeting*. Annapolis, Maryland.

Bultman, J.D., Gilbertson, R.L., Adaskaveg, J., Amburgey, T.L., Parikh, S.V. and Bailey, C.A. (1991) The efficacy of guayule resin as a pesticide. *Bioresource Technology* 35, 1997–2001.

Bultman, J.D., Chen, S.L. and Schloman, W.W. (1998) Anti-termitic efficacy of the resin and rubber in fractionator overheads from a guayule extraction process. *Industrial Crops and Products* 8, 133–143.

Cannell, G.H. and Younger, V.B. (1983) Irrigation, nitrogen and water use in guayule. In: *Abstracts of the Guayule Rubber Society Fourth Annual Meeting*. Riverside, California, pp. 48.

Carranza, A. and Ramirez, L.E. (1981) Establishment of guayule as influenced by different techniques and times of transplanting under dryland conditions. In: *Abstracts of the Guayule Rubber Society Annual Meeting*. Phoenix, Arizona, p. 24.

Chandra, G.R. and Bucks, D.A. (1986) Improved quality of chemically treated guayule (*Parthenium argentatum* Gray) seeds. In: Fangmeier, D.D. and Alcorn, S.M. (eds) *Proceedings of the Fourth International Guayule Research and Development Conference*. Guayule Rubber Society, Tucson, Arizona, pp. 59–68.

Chow, P., Nakayama, F.S., Blahnik, B., Youngquist, J.A. and Coffelt, T.A. (2008) Chemical constituents and physical properties of guayule wood and bark. *Industrial Crops and Products* 28, 303–308.

Coffelt, T.A. (1989) Natural crossing of peanut in Virginia. *Peanut Science* 16, 46–48.

Coffelt, T.A. and Nakayama, F.S. (2007) A six step harvesting procedure for guayule small plots for laboratory analyses. In: Janick, J. and Whipkey, A. (eds) *Issues in New Crops and New Uses*. ASHS Press, Alexandria, Virginia, pp. 66–71.

Coffelt, T.A. and Williams, C.F. (2009) Characterization and recycling of waste water from guayule latex extraction. *Industrial Crops and Products* 29, 648–653.

Coffelt, T.A., Dierig, D.A., Ray, D.T. and Nakayama, F.S. (2001) Regrowth after harvest of one- and two-year old guayule plants. In: *Abstracts of the Fifth National Symposium – New Crops and New Uses*. Association for the Advancement of Industrial Crops, Phoenix, AZ, pp. 50.

Coffelt, T.A., Nakayama, F.S., Ray, D.T., Cornish, K. and McMahan, C.M. (2009a) Post-harvest storage effects on guayule latex, rubber, and resin contents and yields. *Industrial Crops and Products* 29, 326–335.

Coffelt, T.A., Nakayama, F.S., Ray, D.T., Cornish, K., McMahan, C.M. and Williams, C.F. (2009b) Plant population, planting date, and germplasm effects on guayule latex, rubber, and resin yields. *Industrial Crops and Products* 29, 255–260.

Cooperrider, C.K. and Culley, M.J. (1943) The wild guayule situation in Mexico: preliminary report, supplement I and supplement II. USDA Forest Service, Tucson, Arizona, pp. 35.
Cornish, K. (1996) Hypoallergenic natural rubber products from *Parthenium argentatum* (Gray) and other non-*Hevea brasiliensis* species. United States Patent No. 5,580,942.
Cornish, K. (1998) Hypoallergenic natural rubber products from *Parthenium argentatum* (Gray) and other non-*Hevea brasiliensis* species. United States Patent No. 5,718,050.
Cornish, K. and Lytle, C.D. (1999) Viral impermeability of hypoallergenic, low protein, guayule latex films. *Journal of Biomedical Materials Research Part A* 47, 434–437.
Cornish, K. and Schloman, W.W. Jr (2003) Guayule rubber. In: Kroschwitz, J.I. (ed.) *Encyclopedia of Polymer Science and Technology*, Volume II, 3rd edn. Wiley Interscience. Hoboken, New Jersey, pp. 670–698.
Cornish, K., Chapman, M.H., Brichta, J.L., Vinyard, S.H. and Nakayama, F.S. (2000) Post-harvest stability of latex in different sizes of guayule branches. *Industrial Crops and Products* 12, 25–32.
Cornish, K., Brichta, J.L., Yu, P., Wood, D.F., McGlothlin, M.W. and Martin, J.A. (2001) Guayule latex provides a solution for the critical demands of the non-allergenic medical products market. *Agro Food Industry Hi-Tech* 12, 27–31.
Cornish, K., Williams, J., Hall, J.L. and McCoy, R.G. III (2008) Production and properties of Yulex ® – the natural solution to latex allergy. *Rubber Chemistry and Technology* 81, 709–722.
Cowley, W.R. (1945a) Emergence and survival of guayule from pregerminated and dry seed planted during the hot season on irrigated land in the lower Rio Grande Valley. US Forest Service, Emergency Rubber Project, Los Angles, California, Final Report 4-45, pp. 9.
Cowley, W.R. (1945b) The effect of irrigation at time of planting on emergence and survival of pregerminated guayule seed planted in a water furrow. US Forest Service, Emergency Rubber Project, Los Angles, California, Final Report 3a-45, pp. 5.
Dierig, D.A., Thompson, A.E. and Ray, D.T. (1989) Relationship to morphological variables to rubber production in guayule. *Euphytica* 44, 259–264.
Dierig, D.A., Thompson, A.E. and Ray, D.T. (1990) Effects of field storage on guayule rubber quantity and quality. *Rubber Chemistry and Technology* 64, 211–218.
Dierig, D.A., Ray, D.T., Coffelt, T.A., Nakayama, F.S., Leake, G.S. and Lorenz, G. (2001) Heritability of height, width, resin, rubber, and latex in guayule (*Parthenium argentatum*). *Industrial Crops and Products* 13, 229–238.
Doering, J.H. (1934) Guayule rubber in tires and tubes. *Industrial and Engineering Chemistry* 26, 541–543.
Elder, N.G., Elmore, C.L. and Beaupre, C. (1983) Effect of preplant and preemergence herbicides on weed control, growth, and rubber content of transplanted guayule (*Parthenium argentatum*). In: Gregg, E.C., Tipton, J.L. and Huang, H.T. (eds) *Proceedings of the Third International Guayule Conference*. Guayule Rubber Society, Pasadena, California, pp. 582.
Emparan, P.R. and Tysdal, H.M. (1957) The effect of light and other factors on breaking the dormancy of guayule seed. *Agronomy Journal* 49, 15–19.
Erickson, L.C. and Smith, P.F. (1947) Studies on handling and transplanting guayule nursery stock. USDA Technical Bulletin 924, 57 pp.
Estilai, A. (1985) Registration of CAL-5 guayule germplasm. *Crop Science* 25, 369–370.
Estilai, A. (1986) Registration of CAL-6 and CAL-7 guayule germplasm. *Crop Science* 26, 1261–1262.
Estilai, A. and Dierig, D.A. (1996) Improvement in rubber and resin yields of guayule through plant breeding. In: Princen, L.H. and Rossi, C. (eds) *Proceedings of the Ninth International Conference on Jojoba and its Uses, and the Third International Conference on New Industrial Crops and Products*. Versa Press, East Peoria, Illinois, pp. 255–258.
Estilai, A. and Hamerstrand, G.E. (1989) Post-harvest degradation of guayule rubber. *Rubber Chemistry and Technology* 62, 635–642.
Estilai, A. and Ray, D.T. (1991) Genetics, cytogenetics, and breeding of guayule. In: Whitworth, J.W. and Whitehead, E.E. (eds) *Guayule Natural Rubber*. Office of Arid Lands Studies, University of Arizona, Tucson, Arizona, pp. 47–92.
Estilai, A., Naqvi, H.H. and Waines, J.G. (1988) Developing guayule as a domestic rubber crop. *California Agriculture* 43, 29–30.
Ferraris, R. (1986) Agronomic practices for the production of guayule. In: Steward, F.A. and Lucas, S.M. (eds) *Potential Production of Natural Rubber from Guayule in Australia*. Commonwealth Scientific and Industrial Research Organization, Melbourne, Australia, pp. 97–110.
Foster, M.A. and Coffelt, T.A. (2005) Guayule agronomics: establishment, irrigated production, and weed control. *Industrial Crops and Products* 22, 27–40.

Foster, M.A. and Moore, J. (1992) Direct-seeding techniques for guayule stand establishment in west Texas. *Journal of Production Agriculture* 5, 163–167.
Foster, M.A., Carrillo, T.D. and Moore, J. (1986) Weed control in established guayule stands. In: Fangmeier, D.D. and Alcorn, S.M. (eds) *Proceedings of the Fourth International Guayule Research and Development Conference*. Guayule Rubber Society, Tucson, Arizona, pp. 318–320.
Foster, M.A., Ranne, D.W. and Moore, J. (1989) Weed control in dormant guayule with bromoxynil, glyphosate, and 2,4-D. *El Guayulero* 11, 18–24.
Foster, M.A., Kleine, L.G. and Moore, J. (1993) Response of direct-seeded guayule to preemergence herbicides. *Journal of the American Oil Chemists Society* 70, 1239–1240.
Foster, M.A., Fowler, J.L., Kleine, L.G. and Puppala, N. (1999) Establishing new guayule lines by direct seeding. *Industrial Crops and Products* 9, 93–100.
Foster, M.A., Coffelt, T.A. and Majeau, G. (2002a) Effects of herbicides on direct-seeded guayule. In: *Abstracts of the Association for the Advancement of Industrial Crops Annual Meeting*. Saskatoon, Saskatchewan, Canada, p. 4.
Foster, M.A., Fowler, J.L., Kleine, L.G., Grote, M.M. and Puppala, N. (2002b) Performance of direct-seeded and transplanted guayule. *Industrial Crops and Products* 15, 23–31.
Foster, M.A., Coffelt, T.A., Petty, A.K. and Nesmith, D.M. (2008) Guayule production on the Southern High Plains. In: *Abstracts of Association for the Advancement of Industrial Crops 20th Annual Meeting*. College Station, Texas, pp. 26.
Fowler, J.L. and Tinguely, R.C. (1993) Direct seeding of guayule as influenced by polyacrylamide soil conditioner, bed shape and seed cover. *Industrial Crops and Products* 2, 13–21.
Garrot, D.J. and Ray, D.T. (1983) Regrowth in guayule in response to clipping: a final report. In: *Abstracts of the Guayule Rubber Society Fourth Annual Meeting*. Riverside, California, pp.18.
Garrot, D.J., Ray, D.T. and Fangmeier, D.D. (1986) Water and stress effects on growth and rubber accumulation in guayule as determined by the crop water stress index. In: Fangmeier, D.D. and Alcorn, S.M. (eds) *Proceedings of the Fourth International Guayule Research and Development Conference*. Guayule Rubber Society, Tucson, Arizona, pp. 413.
Gonzalez, C.L. and Rektorik, R.J. (1986) Establishment of guayule transplants under dryland conditions in south Texas. *El Guayulero* 8, 7–15.
Greenfield, D. (1992) Guayule: a resin alternative for anti-fouling? (anti-fouling coating). *Modern Paint and Coatings* 82, 8.
Gumbs, R.W. (2004) Low cost solventless guayule resin adhesives. Final Report to US Department of Agriculture on Grant No. 2004-33610-14337, pp. 1–34.
Hammond, B.L. (1959) Effect of gibberellin, sodium hypochlorite, light, and planting depth on germination of guayule seed. *Agronomy Journal* 51, 621–623.
Hammond, B.L. and Polhamus, L.G. (1965) Research on guayule (*Parthenium argentatum*): 1942–1959. USDA Technical Bulletin No. 1327, Washington, DC, 157 pp.
Hammons, R.O. (1964) Pedigreed natural crossing – a new technique. In: *Proceedings of the Third Nature Peanut Research Conference*. Auburn, Alabama, pp. 49–53.
Hoffman, G.J., Shannon, M.C., Maas, E.V. and Grass, L. (1988) Rubber production of salt-stressed guayule at various plant populations. *Irrigation Science* 9, 213–226.
Holt, G.A., Blodgett, T.L. and Nakayama, F.S. (2005) Characteristics of pellet fuel from cotton gin byproducts utilizing various process treatments. In: *Proceedings of the International Conference of Industrial Crops for Rural Development*. Pardo comunicacion, S.L., Murcia, Spain, pp. 89–103.
Holt, G.A., Coffelt, T.A., Chow, P. and Nakayama, F.S. (2009) Biobased composition boards made from cotton gin and guayule wastes: select physical and mechanical properties. *International Journal Materials and Product Technology* 36, 104–114.
Hunter, A.S. and Kelley, O.J. (1946) The growth and rubber content of guayule as affected by variations in soil moisture stresses. *Journal of the American Society of Agronomy* 38, 118–134.
Jorge, M.H.A., Ray, D.T. and Foster, M.A. (2002) Germination of guayule (*Parthenium argentatum* Gray) seed. In: *Abstracts of the Association for the Advancement of Industrial Crops Annual Meeting*. Saskatoon, Saskatchewan, Canada, pp. 3.
Jorge, M.H.A., Veatch-Blohm, M.E., Ray, D.T. and Foster, M.A. (2006) Guayule seed germination under different conditioning treatments. *Industrial Crops and Products* 24, 60–65.
Kelley, O.J., Haise, H.R., Markham, L.C. and Hunter, A.S. (1946) Increased rubber production from thickly seeded guayule. *Journal of the American Society of Agronomy* 38, 589–613.
Keys, R.N., Ray, D.T. and Dierig, D.A. (2002) Characterization of apomictic potential in guayule (*Parthenium argentatum*) *in vivo* and *in vitro*. *Journal of the American Society for Horticultural Science* 127, 404–408.

Khan, A.A., Tao, K.L., Knypl, J.S., Borkowska, B. and Powell, L.E. (1978) Osmotic conditioning of seeds: physiological and biochemical changes. *Acta Horticulture* 83, 267–278.
Kidd, B.E. (1983) Guayule tolerance to four herbicides. In: Gregg, E.C., Tipton, J.L. and Huang, H.T. (eds) *Proceedings of the Third International Guayule Conference*. Guayule Rubber Society, Pasadena, California, pp. 483–489.
Lloyd, F.E. (1911) Guayule (*Parthenium argentatum* Gray) – a rubber plant of the Chihuahuan Desert. Publication 139, Carnegie Institute, Washington, DC, 213 pp.
Longenecker, D.E. and Lyerly, P.J. (1959) Some relations among irrigation water quality, soil characteristics and management practices in the Trans-Pecos area. Miscellaneous Publication 373, Texas Agricultural Experiment Station, College Station, Texas, 18 pp.
Lucas, W. (1996) Summary test report for the guayule tire evaluation of the M1028, commercial utility cargo vehicle. US Army Proving Ground, Yuma, Arizona, pp. 1–5.
Lukich, L.T. (1992) Final report – US Navy/USDA Guayule Program. Goodyear Tire and Rubber Company, Akron, Ohio, pp. 13.
Maas, E.V., Donovan, T.J., Francois, L.E. and Hammerstrand, G.E. (1986) Salt tolerance of guayule. In: Fangmeier, D.D. and Alcorn, S.M. (eds) *Proceedings of the Fourth International Guayule Research and Development Conference*. Guayule Rubber Society, Tucson, Arizona, pp. 101–107.
Maas, E.V., Donovan, T.J. and Francois, L.E. (1988) Salt tolerance of irrigated guayule. *Irrigation Science* 9, 199–211.
McCallum, W.B. (1929) Method of treating and sowing guayule seed. US Patent No. 1,735,835.
McGinnies, W.G. and Haase, E.F. (1975) Guayule: a rubber-producing shrub for arid and semiarid regions. Arid Lands Resource Information Paper No. 7. Office of Arid Lands Studies, University of Arizona, Tucson, Arizona, pp. 267.
McGinnies, W.G. and Mills, J.L. (1980) *Guayule Rubber Production: The World War II Emergency Rubber Project*. Office of Arid Lands Studies, University of Arizona, Tucson, Arizona, pp. 199.
McMahan, C.M., Cornish, K., Coffelt, T.A., Nakayama, F.S., McCoy, R.G., Brichta, J.L. and Ray, D.T. (2006) Post-harvest storage effects on guayule latex quality from agronomic trials. *Industrial Crops and Products* 24, 321–328.
McMahan, C.M., Williams, J. and Cornish, K. (2007a) Guayule latex: physical and chemical properties. *Rubber and Plastics News* 37, 16–18.
McMahan, C.M., Cornish, K. and Whalen, M. (2007b) *Parthenium argentatum* as a source of bio-based products. Agricultural Biomass, Biobased Products, and Biofuels Symposium. American Chemical Society, Chicago Illinois.
Majeau, G.H., Ray, D.T., Coffelt, T.A. and Foster, M.A. (2003) New guayule (*Parthenium argentatum* Gray) variety trials. In: *Abstracts of the Association for the Advancement of Industrial Crops Annual Meeting*. Portland, Oregon, pp. 14.
Mihail, J.D., Alcorn, S.M. and Whitworth, J.W. (1991) Plant health: the interactions of guayule, microorganisms, arthropods, and weeds. In: Whitworth, J.W. and Whitehead, E.E. (eds) *Guayule Natural Rubber*. Office of Arid Lands Studies, University of Arizona, Tucson, Arizona, pp. 73–216.
Milthorpe, P., Paterson-Jones, J.C., Whitworth, J.W., Abel, G. and Miller, W.P. (1991) Case histories of guayule production in Australia, South Africa, and the United States. In: Whitworth, J.W. and Whitehead, E.E. (eds) *Guayule Natural Rubber*. Office of Arid Lands Studies, University of Arizona, Tucson, Arizona, pp. 367–385.
Miyamoto, S. and Bucks, D.A. (1985) Water quantity and quality requirements of guayule: current assessment. *Agricultural Water Management* 10, 205–219.
Miyamoto, S., Davis, J. and Piela, K. (1984a) Water use, growth and rubber yields of four guayule selections as related to irrigation regimes. *Irrigation Science* 5, 95–103.
Miyamoto, S., Piela, K., Davis, J. and Fenn, L.B. (1984b) Salt effects on emergence and seedling mortality of guayule. *Agronomy Journal* 76, 295–300.
Miyamoto, S., Piela, K. and Gobran, G.R. (1984c) Salt effects on transplant mortality, growth and rubber yields of guayule. *Irrigation Science* 5, 275–284.
Miyamoto, S., Davis, J. and Madrid, L. (1990) Salt tolerance of guayule. Bulletin 1651. Texas Agricultural Experiment Station, Texas A&M University, College Station, Texas, pp. 18.
Moore, J. and Murphy, J.M. (1979) Sprinkler irrigation with saline water. In: Miyamoto, S. (ed.) *Proceedings of the Inter-American Conference on Salinity and Water Management Technology*. Texas A&M University Agricultural Research Center, El Paso, Texas, pp. 213–221.
Nakayama, F.S. (2005) Guayule future development. *Industrial Crops and Products* 22, 3–13.

Nakayama, F.S. and Bucks, D.A. (1983) Application of a foliage temperature based crop water stress index to guayule. *Journal of Arid Environments* 6, 269–276.
Nakayama, F.S. and Bucks, D.A. (1984) Crop water stress index, soil water, and rubber yield relations for the guayule plant. *Agronomy Journal* 76, 791–794.
Nakayama, F.S. and Coates, W. (1996) Storage effects on rubber content of laboratory- and field-prepared guayule shrub. In: Princen, L.H. and Rossi, C. (eds) *Third International Conference on New Industrial Crops and Products*, Catamarca, Argentina. pp. 243–246.
Nakayama, F.S., Bucks, D.A., Gonzalez, C.L. and Foster, M.A. (1991) Water and nutrient requirements of guayule under irrigated and dryland production. In: Whitworth, J.W. and Whitehead, E.E. (eds) *Guayule Natural Rubber*. Office of Arid Lands Studies, University of Arizona, Tucson, Arizona, pp. 145–182.
Nakayama, F.S., Vinyard, S.H., Chow, P., Bajwa, D.S., Youngquist, J.A., Muehl, J.H. and Krzysik, A.M. (2001) Guayule as a wood preservative. *Industrial Crops and Products* 14, 105–111.
Nakayama, F.S., Holt, G.A., Coffelt, T.A., Vinyard, S.H. and Faber, A.L. (2003) Pellet fuel from guayule plant material. In: *Abstracts of the Association for the Advancement of Industrial Crops Annual Meeting*. Portland, Oregon, pp. 62.
Nakayama, F.S., Youngquist, J.A. and Chow, P. (2004) Composites comprising plant material from *Parthenium* spp. and plastic. United States Patent No. 6,787,590.
Naqvi, H.H. and Hanson, G.P. (1980) Recent advances in guayule seed germination procedures. *Crop Science* 20, 501–504.
Naqvi, H.H. and Hanson, G.P. (1982) Germination and growth inhibitors in guayule (*Parthenium argentatum* Gray) chaff and their possible influence in seed dormancy. *American Journal of Botany* 69, 985–989.
NAS (National Academy of Sciences) (1977) *Guayule: An Alternative Source of Natural Rubber*. Washington, DC, pp. 80.
Ownby, D.R., Ownby, H.E., McCullough, J.A. and Shafer, A.W. (1994) The prevalence of anti-latex IgE antibodies in 1000 volunteer blood donors. *Journal of Allergy and Clinical Immunology* 97, 1188–1192.
Ramos de Valle, L.F. and Aramburo, F. (1981) Rheological characterization of guayule rubber I. *Journal of Rheology* 25, 379–403.
Ray, D.T. (1993) Guayule: a source of natural rubber. In: Janick, J. and Simon, J.E. (eds) *New Crops*. John Wiley and Sons Inc, New York, pp. 338–342.
Ray, D.T., Garrot, D.J., Fangmeier, D.D. and Coates, W. (1986) Clipping as an agronomic practice in guayule. In: Fangmeier, D.D. and Alcorn, S.M. (eds) *Proceedings of the Fourth International Guayule Research and Development Conference*. Guayule Rubber Society, Tucson, Arizona, pp. 185–191.
Ray, D.T., Dierig, D.A., Thompson, A.E. and Diallo, M.M. (1993) Parent–offspring relationships in apomictic guayule. *Journal of the American Oil Chemists Society* 70, 1235–1237.
Ray, D.T., Orum, T.V., Bigelow, D.M. and Alcorn, S.M. (1995) Selection of diploid and screening of polyploidy guayule lines for *Verticillium* tolerance. *Industrial Crops and Products* 4, 303–310.
Ray, D.T., Dierig, D.A., Coffelt, T.A. and Stumpf, D.K. (1997) Latex production in guayule germplasm. In: *Abstracts of the Association for the Advancement of Industrial Crops Annual Meeting*. Saltillo, Mexico, pp. 76.
Ray, D.T., Dierig, D.A., Thompson, A.E. and Coffelt, T.A. (1999) Registration of six guayule (*Parthenium argentatum* Gray) germplasms with high yielding ability. *Crop Science* 39, 300.
Ray, D.T., Coffelt, T.A. and Dierig, D.A. (2005) Breeding guayule for commercial production. *Industrial Crops and Products* 22, 15–25.
Retzer, J.L. and Mogen, C.A. (1946) The salt tolerance of guayule. *Journal of the American Society of Agronomy* 38, 728–742.
Roberts, P.H. (1946) Final report, the Emergency Rubber Project: a report on our wartime guayule program. US Forest Service, Emergency Rubber Project, Los Angeles, California, pp. 234.
Rubis, D.D. (1983) Influence of irrigation and fertilizer treatments on guayule. In: *Abstracts of the Guayule Rubber Society Fourth Annual Meeting*. Riverside, California, USA, pp. 59.
Schloman, W.W. Jr (1983) Guayule byproduct evaluation II. Fermentable sugars from bagasse. In: *Abstracts of the Guayule Rubber Society Fourth Annual Meeting*, Riverside, California, p. 40.
Schloman, W.W. Jr (2005) Processing guayule for latex and bulk rubber. *Industrial Crops and Products* 22, 41–47.
Schloman, W.W. Jr, Garrot, D.J. and Ray, D.T. (1986) Water stress and seasonal effects on rubber quality in irrigated guayule. *Journal of Agricultural and Food Chemistry* 34, 683–685.
Schloman, W.W. Jr, Wyzgoski, F., McIntyre, D., Cornish, K. and Siler, D.J. (1996) Characterization and performance testing of guayule latex. *Rubber Chemistry and Technology* 69, 215–222.

Siddiqui, I.A., Connell, J.L. and Loctov, P. (1982) Report on the feasibility of commercial development of guayule in California. California Department of Food and Agriculture, Sacramento, California, pp. 51.
Siler, D.J. and Cornish, K. (1994) Hypoallergenicity of guayule rubber particle proteins compared to *Hevea* latex proteins. *Industrial Crops and Products* 2, 307–313.
Siler, D.J., Cornish, K. and Hamilton, R.G. (1996) Absence of cross-reactivity of IgE antibodies from subjects allergic to *Hevea brasiliensis* latex with a new source of natural rubber latex from guayule (*Parthenium argentatum*). *Journal of Allergy and Clinical Immunology* 98, 895–902.
Taylor, C.A. (1946) The propagation of guayule: studies covering seed, nursery, and direct-seeding practices. US Forest Service, Emergency Rubber Project, Los Angeles, California, pp. 84.
Taylor, K.W. and Chubb, R.L. (1952) Rubber recovery from freshly harvested guayule. *Industrial and Engineering Chemistry* 44, 879–882.
Teetor, V.H., Ray, D.T. and Schloman, W.W. Jr (2009) Evaluating chemical indices of guayule rubber content: guayulins A and B. *Industrial Crops and Products* 29, 590–598.
Thames, S.F. and Kaleem, K. (1991) Guayule resin in amine-epoxy strippable coatings. *Bioresource Technology* 35, 185–190.
Thames, S.F. and Wagner, J.P. (1991) Recent advances in guayule coproduct research and development. In: Whitworth, J.W. and Whitehead, E.E. (eds) *Guayule Natural Rubber*. Office of Arid Lands Studies, University of Arizona, Tucson, Arizona, pp. 311–350.
Thames, S.F., Schuman, T.P., Reichel, L.W., Purvis, W.A. and Poole, P.W. (1996) *Guayule CoProducts: Emerging Technology in Industrial, Marine, and Peelable Coatings*. 41st International SAMPE Symposium, ACS Symposium Series, pp. 223–239.
Thomas, J.R. and Hickman, M.V. (1989) Effect of N, P, and K on growth, mineral composition, resin, and rubber accumulation in guayule (*Parthenium argentatum* Gray). *El Guayulero* 11, 25–35.
Thompson, A.E. and Ray, D.T. (1988) Breeding guayule. *Plant Breeding Reviews* 6, 93–165.
Thompson, A.E., Ray, D.T., Livingston, M. and Dierig, D.A. (1988) Variability of rubber and plant growth characteristics among single plant selections from a diverse guayule breeding population. *Journal of the American Society for Horticultural Science* 113, 608–611.
Tingey, D.C. (1943) The comparative value of dry, pregerminated, and threshed guayule seed in direct field planting. Final Report 28. US Forest Service, Emergency Rubber Project, Los Angeles, California. pp. 9.
Tingey, D.C. (1945a) The effect of depth of seeding and frequency of irrigation on the number of seedlings emerging in guayule as compared with lettuce and celery. Final Report 29. US Forest Service, Emergency Rubber Project, Los Angeles, California, pp. 12.
Tingey, D.C. (1945b) The effect of seeding in dry and moist soil at various depths, with and without seeder press wheels, and frequency of irrigation on emergence in guayule, lettuce, and celery. Final Report 30-31. US Forest Service, Emergency Rubber Project, Los Angeles, California, pp. 12.
Tingey, D.C. (1952) Effect of spacing, irrigation, and fertilization on rubber production in guayule sown directly in the field. *Journal of the American Society of Agronomy* 44, 298–302.
Tingey, D.C. and Clifford, E.D. (1946) Comparative yields of rubber from seeding guayule directly in the field and transplanting nursery stock. *Journal of the American Society of Agronomy* 38, 1068–1072.
Tingey, D.C. and Foote, W.H. (1946) Effect of irrigation on the resumption of growth of guayule transplants. *Journal of the American Society of Agronomy* 38, 896–904.
Tipton, J.L. (1981) Seed handling and transplant production. In: *Abstracts of the Guayule Rubber Society Annual Meeting*. Phoenix, Arizona, pp. 7.
Tipton, J.L., Craver, J.L. and Blackwell, J. (1981) A method for harvesting, cleaning, and treating achenes of guayule (*Parthenium argentatum* Gray). *Horticultural Science* 16, 296–298.
Touzinsky, G.F. (1980) Guayule as a source of papermaking fiber. In: *Abstracts of the Fifth International Dissolving Pulps Conference*. Technical Association of the Pulp and Paper Industry, Vienna, pp. 311–315.
Tysdal, H.M., Estilai, A., Siddiqui, I.A. and Knowles, P.F. (1983) Registration of four guayule germplasms. *Crop Science* 23, 189.
Veihmeyer, F.J. and Hendrickson, A.H. (1961) Response of a plant to soil-moisture changes as shown by guayule. *Hilgardia* 30, 621–637.
Velásquez, M.A., Martínez, O. and Aguirre, J. (1978) Revisión histórica de la producción de hule de guayule en México de 1903 a 1951. In: Campos Lopez, E. (ed.) *Guayule: Reencuentro en el Desierto*. CIQA, México, DF, pp. 27–70.

Wadleigh, C.H. and Gauch, H.G. (1944) The influence of high concentrations of sodium sulfate, sodium chloride, and magnesium chloride on the growth of guayule in sand cultures. *Soil Science* 58, 399–403.

Whitworth, J.W. (1981a) Direct seeding of guayule. In: *Abstracts of the Guayule Rubber Society Annual Meeting*. Phoenix, Arizona, pp. 5.

Whitworth, J.W. (1981b) Herbicides for weed control in guayule. *El Guayulero* 3, 13.

Whitworth, J.W. (1983) Guayule research in New Mexico. In: Gregg, E.C., Tipton, J.L. and Huang, H.T. (eds) *Proceedings of the Third International Guayule Conference*. Guayule Rubber Society, Pasadena, California, pp. 105–110.

Whitworth, J.W. and Whitehead, E.E. (1991) *Guayule Natural Rubber: A Technical Publication with Emphasis on Recent Findings*. Guayule Administrative Management Committee and USDA Cooperative Research Service, Office of Arid Lands Studies, University of Arizona, Tucson, Arizona.

Williams, C.F., Coffelt, T.A. and Watson, J.E. (2009) Increased soil sorption of pendimethalin due to deposition of guayule derived detritus. *Soil Science Society of America Journal* 73, 1952–1957.

Zittlosen, R.H. and Fangmeier, D.D. (1986) Sprinkler irrigation for establishing directly seeded guayule. In: Fangmeier, D.D. and Alcorn, S.M. (eds) *Proceedings of the Fourth International Guayule Research and Development Conference*. Guayule Rubber Society, Tucson, Arizona, pp. 433–440.

19

Gums, Resins and Waxes

Enrico Casadei and Ben Chikamai

Introduction

Gums, resins and waxes represent important natural resources for many countries in different regions of the world due to their demand in international trade. These products are employed in a wide range of food and pharmaceutical products, and in several other technical applications. They form an important group of non-wood forest products (NWFPs) and are the basis of a multi-billion dollar industry, and this fact is indicative of the potential of NWFPs for value addition at various stages from harvesting of raw materials to end uses.

Gums of plant origin consist of mixtures of polysaccharides that are either water-soluble or absorb water and swell to form gel, which on hydrolysis yield simple sugars. Plant gums are of different types, the major ones being exudate, seed and seaweed. A concise coverage of all plant gums would constitute a book on its own. This chapter has focused on exudate gums to demonstrate the diversity of the products and their applications. Exudate gums are generally produced following an injury to the tree caused by either extreme desiccation, which results in the cracking of the bark (natural exudation), damage by animals (insect borer, browsing or breakage by wild/domestic animals) or deliberate injury (tapping). The three major exudate gums – gum arabic, gum tragacanth and gum karaya possess a unique range of functionalities (Phillips and Williams, 2001). They have been important items of international trade in the food, pharmaceutical, adhesive, paper, textile and other industries for centuries.

The term 'gum arabic' is used with varying degrees of precision by different groups of people. In the context of its use as a food additive and in the pharmaceutical industry, gum arabic of commerce is defined by regulatory authorities (FAO/WHO) as a dried exudate obtained from the stems and branches of *Acacia senegal* or *A. seyal* (fam. Leguminosae). Gum arabic consists mainly of high-molecular weight polysaccharides and their calcium, magnesium and potassium salts, which on hydrolysis yield arabinose, galactose, rhamnose and glucuronic acid.

Apart from very small amounts which originate outside the continent, Africa can boast to be the world's sole source of gum arabic, a commodity which involves the labour of many tens of thousands of people in the producing countries and which is consumed or used by many millions of others worldwide. The *Acacia* trees which yield the gum occur in a wide belt of semi-arid land across sub-Saharan Africa.

Except for Sudan, and to some extent Nigeria, Chad, Senegal, Niger and Mali, where

Fig. 19.1. Gum arabic exudation.

initiatives have been undertaken to introduce plantations of *A. senegal*, the bulk of gum arabic and acacia gum is derived from natural stands and by natural exudation (Fig. 19.1).

Gum tragacanth is obtained from the dried sap of several species of Middle Eastern legumes of the genus *Astragalus*, including *A. adscendens*, *A. gummifer* and *A. tragacanthus*. Iran is the biggest producer of the best quality of this gum. Gum tragacanth is a viscous, odourless, tasteless, water-soluble mixture of polysaccharides obtained from sap which is drained from the root of the plant and dried. The gum seeps from the plant in twisted ribbons or flakes, which can be powdered. When added to water, tragacanth absorbs water and becomes a gel that can be stirred into a paste. The gum is used in veg-tanned leatherworking as an edge slicking and burnishing compound and occasionally is used as a stiffener in textiles. It contains an alkaloid that historically has been used as a herbal remedy for such conditions as cough and diarrhoea. As a mucilage or paste, it has been used as a topical treatment for burns. It is used in pharmaceuticals and foods as an emulsifier, thickener, stabilizer and texturing additive (code E413). Also, it is the traditional binder used in the making of artist's pastels, as it does not adhere to itself the same way as other gums (such as gum arabic) do when dry.

Gum karaya, sometimes known as sterculia gum, is the dried exudation from species in the genus *Sterculia*. In India, it is produced from S. *urens*, while in Africa the major species is *S. setigera*. Senegal is the world's leading producer, followed by India, which, however, leads in world consumption. The highest grade sorts of gum karaya are white, translucent and almost free of bark. The lower grades vary from light yellow to brown and may contain as much as 3% of insoluble impurities. Powdered gum karaya is white to greyish white.

Gum karaya is a complex polysaccharide of high-molecular weight. A molecular weight as high as 9,500,000 has been reported. On hydrolysis, it yields galactose, rhamnose and galacturonic acid. Gum karaya occurs as a partially acetylated derivative. The acid number has been found to vary from 13.4 to 22.7. The variation in acid number is influenced not only by the source of the sample, but also by its age. The gum has a peculiar property of splitting off free acetic acid and this loss is correlated loosely with the particle size. Trimethylamine has also been identified in the hydrolysis products. Gum karaya contains 12–14% moisture and less than 1% acid insoluble ash.

The food industry consumes less than 50% of all the gum karaya. Its main function is to impart stability through binding and emulsifying properties. At 0.2–0.4%, karaya prevents the bleeding of free water and the formation of large ice crystals in ice pops, water ices and sherbets. A large part of the karaya is used in the pharmaceutical industry as laxative. The coarse particles absorbing water swell enormously, forming a discontinuous type of mucilage that is very effective as a laxative. It is also used as a denture adhesive in which the finely powdered gum is dusted on the dental plate and swells when it touches the moist surface of the gums. This gives a comfortable and tight fit of the plate.

Gum resins of plant origin are produced by plants in the families Burseraceae, Leguminosae and Pinaceae. In this chapter, focus will be given to gums from species in the Burseraceae family. Burseraceae is a family of 17 genera with some 560 species, which are widespread in the tropics, especially in Africa, Malaysia and South America. These are trees or shrubs characterized by aromatic resins from the bark used even in biblical times as cosmetics and perfumes. The main gum resins – frankincense, myrrh and opoponax are the hardened, resinous exudates obtained from trees of certain *Boswellia* and *Commiphora* species, which are common in the hot drylands (Fig. 19.2). The resins, particularly frankincense, are used in unprocessed form for both fragrance and flavour purposes, but there is some production of distilled essential oils and extracts.

The word 'frankincense' comes from an old French word *fraunk-encens*, which means true or real incense. It is commonly called olibanum, which comes from an Arabic word *al-luban* meaning 'the milk' of the true incense. Both terms are often used interchangeably, though the term olibanum is more common in commerce. Myrrh is an Arabic word meaning bitter. It is an aromatic resin obtained from trees and shrubs from *C. myrrha*. Opoponax is a resinous exudate obtained from two species of *Commiphora*, namely *C. holtziana* (syn. *C. erythraea*), which produces the medicinal type, and *C. guidottii*, which produces the sweet-scented type known as sweet myrrh. In Somali, opoponax is known as hagar.

Waxes is a general term used to refer to the mixture of long-chain apolar lipids forming a protective coating on plant leaves and fruits, but also in animals, algae, fungi and bacteria. Waxes may be natural secretions of plants or animals, produced artificially by purification from natural petroleum, or completely synthetic. In this chapter, only vegetable waxes will be considered, which are obtained from different plants such as *Myrica faya*, *Euphorbia cerifera* and *E. antisyphilitica*, *Copernica cerifera*, *Simmondsia chinensis* and others (Fig. 19.3).

The various materials named waxes do not form a chemically homogeneous group. All waxes are water-resistant materials made up of various substances, including hydrocarbons (normal and branched alkanes and alkenes), ketones, diketones, primary and secondary alcohols, aldehydes, etc.

Chemically, a wax is a type of lipid that may contain a wide variety of long-chain primary alcohols and fatty acids. They are usually distinguished from fats by the lack of triglyceride esters of glycerin and three fatty acids.

A number of waxes are produced commercially in large amounts for use in cosmetics, lubricants, polishes, surface coatings, inks and many other applications.

Waxes are extremely versatile because of their deformability and traditional application as candles. They have properties that are common with oils, fats, gums, resins and pitches in that they dissolve easily in fat solvents based on temperature. They can

Fig. 19.2. Collection of gum resins, Wajir, Kenya.

Fig. 19.3. Different types of commercial waxes (Museum Services Corporation, http://www.museumservicescorporation.com).

serve effectively as functional vehicles for pigments since they can wet and disperse them uniformly. Capability of being emulsified with water makes them valuable for the preparation of self-lustreing floor and furniture polishes, and also for more exacting aqueous creams such as pharmaceuticals and cosmetics.

Botanical Sources and Production Areas

Gums

Arabic

There are close to 900 *Acacia* species capable of producing gum. These are located primarily in tropical climates, with about 130 of them located specifically on the African continent. Close to 80% of gum arabic is produced by *A. senegal* and the remainder is produced by either *A. laeta* or *A. seyal*, with each species contributing 10% to the total supply of gum. The gum produced by *A. senegal* is commonly referred to as 'hard gum' and the gum from *A. seyal* as 'flaky gum'. *A. senegal* is native to the hot and dry regions of Africa and parts of the Middle East. Its distribution in nature shows varietal preferences. Var. *senegal* is the most widespread, being present in the African Sahel from Senegal and Mauritania in the west to the Horn of Africa and southwards to Tanzania. In Asia, it is found in Pakistan and the drier parts of north-eastern India. Var. *kerensis* is restricted to the Horn of Africa, var. *rostrata* to southern Africa, while var. *leiorachis* has discontinuous range in eastern and southern Africa. *A. seyal* has two varieties, with var. *seyal* being the more extensive. It is present from Senegal to the Horn of Africa and southwards to Tanzania. Var. *fistula* is confined to eastern and central Africa. *A. laeta* is confined to the central and West African Sahel from Chad to Senegal.

Tragacanth

Tragacanth is a natural polysaccharide obtained from around 20 species of Leguminosae of the genus *Astragalus*, and in particular *A. adscendens*, *A. creticus*, *A. gummifer* and *A. tragacantha*, originated from south-east Europe and south-west Asia. The main producer is Iran, but the product is also present in Syria and Turkey. The gum has antiseptic, emollient, anorexic and laxative effects. It seems that the product has an effect of delaying the growth of cancer cells.

Karaya

This gum is an extract of *Sterculia* trees. The majority of commercial gum karaya is obtained from *S. urens*, which is a soft-wooded tree that grows to approximately 10 m. It is native to India and Pakistan, where it is found on the dry, rocky hills and plateaus; it grows there almost exclusively, where it is cultivated for karaya production. It has also been reported in Africa – eastern and southern Africa (northern Kenya, Malawi and Zambia). However, the main karaya producing species is *S. setigera*, which is found in the west, central and eastern African Sahel from Senegal to Sudan. A common method of production is to make blazes on the tree trunk with some sharp-edged tool. The gum begins to exude as soon as the blaze is made and exudation continues for many days. Exudation is better in hot months from March to May/June. No tapping is done in the rainy season. The yield of gum varies from 1 to 5 kg/tree/season, depending on many factors (Iqbol, 1993). Due to indiscriminate and unscientific tapping, the trees often die. Although an improved tapping method has been developed, it has not been put into practice. The collected gum is washed and dried. The gum is then graded. The highest grade sorts of gum karaya are white, translucent and almost free of bark. The lower grades vary from light yellow to brown and may contain as much as 3% of insoluble impurities. Powdered gum karaya is white to greyish white.

Gum resins

Frankincense or gum olibanum

Frankincense or gum olibanum is obtained from species in the genus *Boswellia*, which belongs to the Burseraceae family. The genus is composed of 20 or so species extending from the Ivory Coast to the Horn of Africa and southwards to Madagascar. It is also found in the Middle East and India. The main sources of commercial incense in Africa are *B. frereana* (endemic to northern Somalia, i.e. Somaliland and Puntland) and is said to produce the best quality and sought after incense (commercially called meydi); *B. sacra* (Somaliland, Puntland, South Yemen and Oman) and commercially called 'beeyo'; *B. papyrifera* (Djibouti, Eritrea, Ethiopia and Sudan); and *B. neglecta* (Ethiopia, Kenya, Somalia, Tanzania and Uganda). In India, the main source is *B. serrata*, which occurs in the drier parts of northern India.

Myrrh and opoponax

The source of myrrh and opoponax is from *Commiphora* species, which are small trees or shrubs with short, thorny branches. True myrrh is produced by *C. myrrha* (Nees) Engl. (syn. *C. molmol* Engl.), a variable species found in southern Arabia and the Horn of Africa – Somalia, south-eastern Ethiopia and north-east Kenya. Numerous other *Commiphora* spp. yield resin and it is not clear to what extent these enter commerce, either as adulterants or as inferior types of myrrh. These include *C. abyssinica*, *C foliacea*, *C. playfairii*, *C. serrulata*, *C. africana*, *C. schimperi*, etc. The various pharmacopoeias refer to myrrh as being obtained from '*C. molmol* and other species of *Commiphora*'. Indian false myrrh (bdellium) comes from *C. mukul*.

The name 'opoponax' is derived from that of its original source, *Opoponax chironium*, but production today is entirely from *Commiphora* spp., namely *C. guidottii* and *C. holtizziana* (syn. *C. erythraea*). *C. guidotti* is endemic in Somalia, while *C. holtizziana* is found in Somalia, south-eastern Ethiopia and northern Kenya.

Vegetable waxes

Bayberry

This is an aromatic, green vegetable wax. Bayberry wax is removed from the surface of the fruit of the bayberry (wax-myrtle) shrub by boiling the berries in water and skimming the wax from the surface of the water. It is made up primarily of esters of lauric, myristic and palmitic acid. Bayberry wax is used primarily in the manufacture of candles and other products where distinctive fragrance is desirable.

Candelilla

Candelilla is derived from the leaves of the small candelilla shrub native to northern Mexico and south-western USA, *Euphorbia cerifera* and *E. antisyphilitica*, from the family Euphorbiaceae. It is yellowish-brown, hard, brittle and opaque to translucent. The wax is obtained by boiling the leaves and stems with diluted sulfuric acid, the surface is skimmed and allowed to solidify. It is refined by further treatment with sulfuric acid and subsequent passage through filter presses. Its melting point is 67–79°C. Candelilla wax is used mostly mixed with other waxes to harden them without raising their melting point.

Carnauba

Carnauba is obtained from the leaves of the carnauba palm, *C. prunifera*, a plant native to and grown only in the north-eastern Brazilian states of Piaui, Ceará and Rio Grande do Norte. It is known as 'queen of waxes' and usually comes in the form of hard yellow-brown flakes. It is obtained from the leaves of the carnauba palm by collecting them, beating them to loosen the wax, then refining and bleaching the wax.

Castor

Also called hydrogenated castor oil, castor is a hard, brittle vegetable wax produced by the hydrogenation (chemical combination with hydrogen) of pure castor oil, in the presence of a nickel catalyst. The wax is odourless and insoluble in water.

Esparto

Esparto is a by-product in the artisanal preparation of paper. Two coarse grass species, *Lygeum spartum* and *Stipa tenacissima*, native to southern Spain and northern Africa, are the source of this wax. The yellowish-white wax is obtained from waste liquors in the preparation of esparto (grass) pulp for papermaking.

Japan

Japan is a pale-yellow, waxy, water-insoluble solid with a gummy feel, obtained from the berries of certain sumacs native to Japan and China, such as *Rhus verniciflua* (Japanese sumac tree) and *R. succedanea* (Japanese wax tree). It is not a true wax but could be considered a vegetable tallow that contains 10–15% palmitin, stearin and olein with about 1% japanic acid. Japan wax is sold in flat squares or disks and has a rancid odour. It is extracted by expression and heat, or by the action of solvents.

Jojoba

This liquid wax is produced from the seed of the jojoba (*Simmondsia chinesis*) plant, a shrub native to southern Arizona, southern California and north-western Mexico. The oil makes up approximately 50% of the jojoba seed by weight. Unrefined jojoba oil appears as a clear golden liquid at room temperature with a slightly fatty odour. Refined jojoba oil is colourless and odourless.

Ouricury

Ouricury is a brown wax obtained from the leaves of a Brazilian feather palm, *Syagrus coronata* or *Cocos coronata*, by scraping the leaf surface. Harvesting ouricury wax is more difficult than harvesting carnauba wax, as ouricury wax does not flake off the surface of the leaves.

Rice bran

Derived from rice bran, the main components are aliphatic acids (wax acids) and higher alcohol esters. The aliphatic acids consist of palmitic acid (C-16), behenic acid (C-22), lignoceric acid (C-24), other wax acids (C-26), etc. The higher alcohol esters consist of ceryl alcohol (C-26), melyssil alcohol (C-30), etc. Rice bran wax also contains unsaponifiable constituents such as free fatty acids (palmitic acid), squalene and phospholipids.

Soy

This wax, produced from the soybean plant, is quite a recent invention. It has been developed as a natural, cheaper alternative for beeswax. The demand for natural wax candles has been growing since the last decade, but beeswax is about ten times more expensive than paraffin. Testing different natural plant waxes, it has been possible to produce a vegetable wax made with partially hydrogenated soy oil, coconut oil and palm oil. It has also been possible to reduce the cost of an economical natural wax candle by blending beeswax with soy wax.

Status of Management

Gums and resins

Rural and urban people in Africa and other parts of the developing world are heavily dependent on NWFPs for a wide range of needs, including food, medicine and construction materials. Many of these NWFPs are important sources of income and employment at the local level, with some being traded at the international level.

Gums and resins represent important products generating income at the level of local population involved in their production, processing and trading. Local perspectives on use and management of gum and resin producing trees in different producing countries in Africa are related strictly to how local communities interact with forest vegetation. Gums and resins represent natural resources for populations living in drylands; however, poverty has forced local communities to overexploit their forests resources. Most of the producing areas are arid to semi-arid, with marginal or no agricultural potential. Overexploitation of natural resources is common and the degradation of natural resources increases food insecurity further and reduces local livelihood options.

This negative trend may be further intensified due to climate change (e.g. drought). Consequently, there is an urgent need for improved land-use strategies that will make the vast arid and semi-arid land resources contribute optimally to the livelihoods of local people and national development goals.

The perspective of local communities maintaining forest resources on the basis of their forest-related values and indigenous knowledge systems is becoming more and more a commitment of governments and the private sector. The concepts of resource degradation and sustainable use of resources should therefore be considered as a development including interactions among processes of degradation, resource conservation and resource enrichment.

In many gum and resin producing countries, programmes have been launched since the 1990s for the plantation and protection of trees, strengthening management and controlled exploitation of resources, applying rehabilitation of arid and semi-arid lands and providing training to local communities. A regional master plan for training was published in 2005 by the Network for Natural Gums and Resins in Africa (NGARA), developing seven modules that are considered basic to create awareness at the level of the decision makers and stakeholders of the sector for sustainable development of resources, increase income to the poorest people of the sector, add value to exported products and better control of desertification, which are considered the key issues of a long-term programme. In this modality, the Regional Master Plan for Training is seen as a plan that must be implemented at national and regional levels to go step-by-step with the strategic approach of the long-term programme to combat desertification, alleviate poverty in rural areas and increase food security in the region. The seven modules cover the areas of production, collection, market, technology and quality control of gums and resins, including environmental conservation and management.

The future of the gums and resins market is affected by instability considering the large fluctuation of demand and offer. This situation has affected many of the producing countries of the region and their economy with negative effects to the different segments of the gums and resins sector, from producers to processors and exporters. The main competitors in the market are represented by synthetic substitutes that can jeopardize the gums and resins market and sustainable development of the resources, particularly when market demand is increasing and the offer cannot keep pace with the market request. To overcome the fluctuation in the supply of gums and resins, the producing countries have started to discuss the organization of a system to stabilize the offer, creating a stock of products in different countries of the region that will be used under specific regional agreements.

Waxes

As petroleum-based wax supplies continue to erode on a global scale, major wax end-users are treating the transition to vegetable-based and synthetic waxes as a marketing opportunity.

The candle business consumes nearly 50% of the global wax supply, despite slow growth in this mature sector of the market. Candle companies are taking advantage of the opportunity to market candles made from vegetables and other plant-based waxes as cleaner burning and more environmentally friendly. Studies conducted by Kline and

Company (a consulting company, Little Falls, New Jersey, USA) indicate that vegetable, palm and synthetic waxes are all set to expand over the next 15 years as declining petroleum wax supplies and speciality applications make room for alternatives. The actual trend reflects a significant growth in the market over the next 10–15 years coming from non-petroleum waxes, which is expected to grow to nearly one-fourth of the total supply by 2020.

The Kline study provides a detailed analysis of supply, demand, pricing, competition and trends in the global wax industry, with a focus on both conventional and synthetic waxes. The report also includes profiles for the top 20 global and regional suppliers, including technology vendors, as well as major global wax end-users.

Considering the increasing market demand for vegetable waxes and possible use not only in the candle industry but also in other sectors, including the food industry and cosmetic and pharmaceutical areas, timely investment should facilitate development of the appropriate technology for the elaboration of new products, originating from a large variety of plants deriving from Asia, Africa or North America and currently expanded worldwide due to their ample distribution in this past century. A correct management of natural resources and elaboration of derived synthetic products will represent a growing business activity in this sector.

Composition and Properties

Gums

Arabic

This gum is a complex high-molecular weight polysaccharide. The precise chemical nature differs according to the botanical origin of the gum, and these differences are reflected in the analytical properties as well as the functional properties and uses to which gum arabic is put. On hydrolysis, gum arabic yields D-galactose, L-arabinose, L-rhamnose and D-glucuronic acids and its 4-0 methyl ether. It also contains small amounts of protein (2.5–3.5%) comprising a total of 18 amino acids and minerals (cations), which exist as part neutralized salt of acidic polysaccharides. A total of 14 cations have been identified with calcium, potassium, sodium and magnesium being most abundant. The relative proportion of cations varies depending on the relative abundance of the water-soluble cations in the soil. The presence of a carbohydrate component and protein in gum arabic makes it belong to a member of the arabino-galactan-protein [AGP] family.

The values and relative proportions of some of the above analytical parameters of gum arabic vary according to the botanical origin and geographical differences. Gum arabic from *A. senegal*, for example, has negative specific rotation, higher values of nitrogen and hence protein, viscosity and rhamnose content, compared to that from *A. seyal*. Typical values of gum arabic from *A. senegal* and *A. seyal* are given in Table 19.1.

The functional properties of gum arabic (especially emulsification and stabilization) are determined more by its molecular nature. Gum arabic is a member of the arabinogalactan protein family comprising a main component (arabinogalactan that represents about 88.4% of the total gum) with a molecular mass of 1.27×10^5, an AGP complex (10.4%° of the total) with a molecular mass of 1.45×10^5 and containing about 50% of the total protein and a glycoprotein (GP) with a molecular mass of 2.5×10^5 and containing 25% of the protein. It is the AGP that is responsible for the functional properties of emulsification and stabilization and this component is well represented in both species.

Tragacanth

The physical characteristics of the gum shows that when added to water, the soluble tragacanthin dissolves to give a colloidal hydrosol solution, while the insoluble bassorin component swells to a gel-like state. When a small proportion of water is used, a soft, adhesive paste is formed. If more water is added, the paste forms a uniform mixture that separates in 1 or 2 days, with most of the dissolved tragacanthin in the upper layer and the greater portion of insoluble bassorin in the lower layer. The ability of tragacanth to swell in water to give thick, viscous dispersions or pastes has accounted for many of its uses in the pharmaceutical and food industries.

Table 19.1. Physicochemical and carbohydrate data of gum arabic from four African countries (Chikamai *et al.*, 1996).

	A. senegal Ethiopia		*A. senegal* Kenya		*A. senegal* Nigeria		*A. senegal* Sudan		*A. seyal* Sudan
	A	C	A	C	A	C	A	C	A
	n = 3	*n* = 2	*n* = 4	*n* = 7	*n* = 3	*n* = 1	*n* = 3	*n* = 3	*n* = 2
Moisture content (%)	13	12.8	13.6	12.2	10.6	10	11.7	11.6	10.4
Ash (%)	3.9	4.5	4.6	4.2	4.4	3.9	3.1	3.5	2.2
Specific rotation (deg)	–25	–30	–29	–31	–29	–32	–25	–27	+52
Nitrogen content (%)	0.35	0.26	0.51	0.49	0.4	0.37	0.32	0.32	0.16
Protein content	2.2	1.7	3.3	3.2	2.9	2.2	2.1	2.1	1.1
Intrinsic viscosity (ml/g)	18	17	17	17	15	17	16	17	11
Emulsion stability (%)	97	97	90	91	97	93	99	96	97
Tannin (25% sol.)	–	–	–	–	–	–	–	–	–
Gel (25% sol.)	+	+	+	+	+	+	+	+	–
Sugar composition									
Glucuronic acid	23	16	16	26	16	20	25	21	11
Galactose	46	22	17	35	43	42	42	48	36
Arabinose	29	18	25	30	29	28	35	30	40
Rhamnose	12	10	14	14	9	10	12	9	

A = Authentic; C = Commercial.

As with most other gums, viscosity is the most important property of the gum tragacanth solution. The viscosity of a 1% solution of the highest grade of gum is about 3400 cps. A 2–4% concentration gives a thick paste when properly hydrated. In a cold preparation, the maximum viscosity is usually reached after 24 h, but it can be obtained in about 2 h by raising the temperature of the solution to about 50°C.

Gum tragacanth solutions are acidic, usually in the pH range of 5–6. Its maximum initial viscosity is at pH 8. However, its maximum stable viscosity is around pH 5 and is said to decrease at pH below 4 or above 6. Compared to other gums, however, tragacanth is fairly stable over a wide pH range down to extremely acidic conditions at about pH 2. For this reason, it has been widely used in food products such as salad dressings, where stable viscosities at low pHs are required.

The best quality tragacanth gum is tasteless, whitish, yellowish or pale-brown in colour and translucent in appearance. The lower grades are generally more strongly coloured than the higher grades. The gum is obtained in two basic physical forms, namely ribbons (superior quality) and flakes (inferior quality). These two forms are obtained from different subspecies of the shrub. Both types of the shrubs normally do not grow in the same locality (Iqbol, 1993). The best type of gum is obtained from artificial incisions rather than from natural exudations. Abundant rainfall prior to the tapping season, and dry conditions during the harvesting season, constitute optimum conditions for gum production.

Karaya

This gum is a complex, partially acetylated polysaccharide obtained as a calcium and magnesium salt. The polysaccharide component of gum karaya has a high-molecular weight and is composed of galacturonic acid, beta-D-galactose, glucuronic acid, L-rhamnose and other residues. It absorbs water very rapidly to form viscous mucilage at low concentrations, although it is one of the least soluble of the gum exudates.

The quality of gum karaya depends on the thoroughness of impurity removal. Food-grade gum is usually a white to pinkish grey powder with a slight vinegar odour. Pharmaceutical grades of karaya may be almost clear or translucent.

Gum karaya is the least soluble of the commercial plant exudates, but it absorbs water rapidly and swells to form viscous colloidal solutions, even at low concentrations (1%). The swelling behaviour of gum karaya is dependent on the presence of acetyl groups in its structure. Deacetylation through alkali treatment results in a water-soluble gum. When used in higher concentrations in water (up to 4%), karaya forms gels or pastes. Unlike other gums, karaya swells in 60% alcohol, but remains insoluble in other organic solvents. Karaya may absorb up to 100 times its weight in water. Because the gum is partially acetylated, it may release acetic acid during storage.

Gum resins

Myrrh and frankincense

These are oleo-gum-resins, i.e. the crude material contains an essential oil, a water-soluble gum and alcohol-soluble resin. Myrrh contains about 3–8% essential oil, 30–60% water-soluble gum and 25–40% alcohol-soluble resin (Tucker, 1986). Frankincense contains about 5–9% essential oil, 65–85% alcohol-soluble resin and the remainder as water-soluble gums.

Detailed chemical analysis of myrrh, hagar and olibanum has not been carried out. However, preliminary results show that an authentic myrrh sample has two major components – about 50% furanoudesma -1,3,-diene and 15% lindestrene (Baser *et al.*, 2003). This is particularly true for wild harvest and late stage tapped myrrh. Other compounds in a descending order of relative concentration are furanodiene (8.8%), germacrene B (6.6%), isofuranogermacrene (6.1%), 2 methoxy-furanodiene (4.6%) and β-elemene (3.8%).

Hagar

Hagar has four main components – germacrene D (23%), furanogermacra 1, ten (15)-diene-6-one (13.4%) and two unidentified furanosequiterpenes (18% and 11.4%). Other important components are curzerenone (10.5%), germacrene B (7.4%), β-selinene (7%), furanosesquiterpine (6.2%), unknown (6.0%), β-elemene (5%) and 2 methoxyfuranogermacrane (3.1%).

B. papyrifera has octyl acetate as the main constituent (46.8%). Other main compounds are α-pinene (6.1%), 1-octanol (5.9%), limonene (4.8%) and linalool (3.6%). Analysis of the essential oil of olibanum from *B. neglecta* shows the following main components: α-pinene (27%), α-thujene 25%, terpinen-4-ol (23%) and α-cymene (4.3%).

Waxes

The range of lipid types in plant waxes is highly variable, both in nature and in composition. In addition, there may be hydroxy-β-diketones, oxo-β-diketones, alkenes, branched alkanes, acids, esters, acetates and benzoates of aliphatic alcohols, methyl, phenylethyl and triterpenoid esters, and many more. The amount of each lipid class and the nature and proportions of the various molecular species within each class vary greatly according to the plant species and the site of wax deposition.

Carnauba

Carnauba contains mainly esters of fatty acids (80–85%), fatty alcohols (10–16%), acids (3–6%) and hydrocarbons (1–3%). Specific for carnauba wax is the content of esterified fatty diols (about 20%), hydroxylated fatty acids (about 6%) and cinnamic acid (about 10%). Cinnamic acid, an antioxidant, may be hydroxylated or methoxylated. It also contains free acids (straight-chain acids with even-numbered carbon chains from C-24 to C-28), free alcohols (straight-chain alcohols with even-numbered carbon chains from C-30 to C-34), hydrocarbons (straight-chain odd-numbered carbon chains from C-27 to C-31) and resins.

Candelilla

This wax consists primarily of odd-numbered *n*-alkanes (C-29 to C-33), together with esters of acids and alcohols with even-numbered carbon chains (C-28 to C-34). Free acids, free alcohols, sterols, neutral resins and mineral matter are also present.

Jojoba

Jojoba is a fluid (melting point about 7°C) formed quite exclusively of alcohols esterified with long-chain fatty acids (more than 98%) with a total of 38–44 carbon atoms. The fatty acids are 18:1*n*-9 (about 10%), 20:1*n*-9 (about 70%) and 22:1*n*-9 (15–20%), while the fatty alcohols have predominantly 20 and 22 carbon atoms and one double bond. The iodine value is approximately 80. Jojoba oil is relatively shelf-stable when compared with other vegetable oils. It has an oxidative stability index of approximately 60, which means that it is more shelf-stable than oils of safflower, canola, almond or squalene, but less than castor, macadamia nut and coconut.

Ouricury

Ouricury resembles carnauba wax for the physical properties, so it can be used as a substitute where light colour is not required, e.g. in carbon paper inks, moulding lubricants and polishes. Its melting point is 81–84°C.

Esparto

This melts at 73°C. While its composition is highly variable, it contains hydrocarbons, esters, alcohol (C-28) and triterpenoids.

Japan

Japan is not a true wax but is more like a vegetable tallow found in the kernel and outer skin of the berries of *Rhus* and *Toxicodendron* species, including those yielding Japanese lacquer. It contains a high amount of palmitic acid triglycerides (93–97%), long-chain dicarboxylic acids including C-22 and C-23 chains (4–5.5%) and free alcohols (12–16%). Its melting point is 45–53°C.

Rice bran

Rice bran wax contains esters of fatty acids (C-26 to C-30) and long-chain alcohols (C-26 to C-30) and a large amount of unsaponifiable matter (55–67%).

Uses

Gums

Arabic

This is used as a stabilizing, flavour-fixing, gelling, thickening, film-forming encapsulating and binding agent in many industries like the food industry, pharmaceutical industry, cosmetic industry, printing industry and textile industry (Table 19.2).

Its non-food applications include pharmaceutical, cosmetic, lithographic and offset preparations. It was used extensively as an adhesive, but this use has been taken over almost entirely by synthetics. Similarly, its uses as a sizing and finishing material in the textile industry have also given way mostly to modern substitutes. However, small quantities continue to be used in papermaking. The gum is also used to a limited extent in polishes, contact insecticides and pesticides, photographic emulsions and pharmaceuticals.

Tragacanth

Tragacanth is one of the oldest gums known and its use has dated back from pre-Christian times. It is widely used in pharmaceuticals and cosmetics, as a thickening agent in syrups, salad dressings and sauces, in textile

Table 19.2. Gum arabic (414) is a food additive that may be used in the following foods under the conditions of good manufacturing practices (GMP), as outlined in the Preamble of the Codex General Standard of Food Additives (Codex Stan 192–1995 updated online).

Number	Food category	Number	Food category
01.1.2	Dairy-based drinks, flavoured and/or fermented (e.g. chocolate milk, cocoa, eggnog, drinking yoghurt, whey-based drinks)	08.3	Processed comminuted meat, poultry and game products
01.3	Condensed milk and analogues (plain)	08.4	Edible casing (e.g. sausage casings)
01.4.3	Clotted cream (plain)	09.3	Semi-preserved fish and fish products, including molluscs, crustaceans and echinoderms
01.4.4	Cream analogues	09.4	Fully preserved, including canned or fermented fish and fish products, including molluscs, crustaceans and echinoderms
01.5	Milk powder and cream powder and powder analogues (plain)	10.2.3	Dried and/or heat coagulated egg products
01.6.1	Unripened cheese	10.3	Preserved eggs, including alkaline, salted and canned eggs
01.6.2	Ripened cheese	10.4	Egg-based desserts (e.g. custard)
01.6.4	Processed cheese	11.6	Table-top sweeteners, including those containing high-intensity sweeteners
01.6.5	Cheese analogues	12.2.2	Seasonings and condiments
01.7	Dairy-based desserts (e.g. pudding, fruit or flavoured yoghurt)	12.3	Vinegars
01.8.1	Liquid whey and whey products, excluding whey cheeses	12.4	Mustards
02.2.2	Fat spreads, dairy fat spreads and blended spreads	12.5	Soups and broths
02.3	Fat emulsion mainly of type oil-in-water, including mixed and/or flavoured products based on fat emulsions	12.6	Sauces and like-products
02.4	Fat-based desserts excluding dairy-based dessert products of food category 01.7	12.7	Salads (e.g. macaroni salad, potato salad) and sandwich spreads, excluding cocoa- and nut-based spreads of food categories 04.2.2.5 and 05.1.3
03.0	Edible ices, including sherbet and sorbet	12.8	Yeast and like-products
04.1.2	Processed fruit	12.9	Soybean-based seasonings and condiments
04.2.2.2	Dried vegetables (including mushrooms and fungi, roots and tubers, pulses and legumes, and aloe vera) and seaweeds in vinegar, oil, brine or soybean sauce	12.10	Protein products other than from soybeans
04.2.2.3	Vegetables (including mushrooms and fungi, roots and tubers, pulses and legumes, and aloe vera) and seaweeds in vinegar, oil, brine or soybean sauce	13.3	Dietetic foods intended for special medical purposes (excluding products of food category 13.1)

Table 19.2.

Number	Food category	Number	Food category
04.2.2.4	Canned or bottled (pasteurized) or retort pouch vegetables (including mushrooms and fungi, roots and tubers, pulses and legumes, and aloe vera) and seaweeds	13.4	Dietetic formulae for slimming purposes and weight reduction
04.2.2.5	Vegetables (including mushrooms and fungi, roots and tubers, pulses and legumes, and aloe vera), seaweeds and nut and seed purees and spreads (e.g. peanut butter)	13.5	Dietetic foods (e.g. supplementary foods for dietary use), excluding products of food categories 13.1–13.4 and 13.6
04.2.2.6	Vegetables (including mushrooms and fungi, roots and tubers, pulses and legumes, and aloe vera), seaweed and nut and seed pulps and preparations (e.g. vegetable desserts and sauces, candied vegetables) other than food category 04.2.2.5	13.6	Food supplements
04.2.2.8	Cooked or fried vegetables (including mushrooms and fungi, roots and tubers, pulses and legumes, and aloe vera) and seaweeds	14.1.4	Water-based flavoured drinks, including 'sport', 'energy' or 'electrolyte' drinks and particulated drinks
05.0	Confectionery	14.2.1	Beer and malt beverages
06.3	Breakfast cereals, including rolled oats	14.2.2	Cider and perry
06.4.3	Pre-cooked pastas and noodles and like products	14.2.4	Wines (other than grape)
06.5	Cereal and starch-based desserts (e.g. rice pudding, tapioca pudding)	14.2.5	Mead
06.6	Batters (e.g. for breading or batters for fish or poultry)	14.2.6	Distilled spirituous beverages containing more than 15% alcohol
06.7	Pre-cooked or processed rice products, including rice cakes (oriental type only)	14.2.7	Aromatized alcoholic beverages (e.g. beer, wine and spirituous cooler-type beverages, low alcoholic refreshers)
06.8	Soybean products (excluding soybean-based seasonings and condiments of food category 12.9)	15.0	Ready-to-eat savouries
07.0	Bakery wares	16.0	Composite foods – foods that could not be placed in categories 01–15
08.2	Processed meat, poultry and game products in whole pieces or cuts		

sizing and as an adhesive. It dissolves readily in cold water to give a solution of very high viscosity, which additionally is highly resistant to strong acidic conditions, and the gum is therefore used primarily as a stabilizer and thickener in acid preparations.

Karaya

Karaya is used as a food additive (INS number 416) and as an emulsifier, stabilizer and thickening agent. In food, gum karaya is a food additive that may be used in most food categories, as indicated for gum arabic (Table 19.1) under the conditions of good manufacturing practices (GMP) as outlined in the Preamble of the Codex General Standard on Food Additives.

Of all the gum karaya produced, only 10% is used as a food additive; the remainder goes into pharmaceutical products (Anderson, 1993).

The three most important uses of karaya in pharmaceutical products are as a dental adhesive for false teeth, in the manufacture of colostomy bag fixing, and as a bulk laxative. Colostomy bag fixing is the most common use of gum karaya, in which the gum's qualities are difficult to equal. Its use in dental fixatives started declining when research showed that habitual use of acidic gum karaya had an adverse effect on any remaining natural teeth. In the first two applications, there has been some substitution of karaya by cheaper carboxymethylcellulose derivatives, although recent American trade reports have suggested that some of these substitutes are not as effective as karaya. In France and the USA, 95% of imported karaya is used in pharmaceutical products.

Gum karaya is also used to adulterate gum tragacanth, due to their similar physical characteristics.

Gum resins

Frankincense and myrrh

These are important natural plant products used in several industries that include pharmacology, food, flavour, liqueur and beverage, cosmetics, perfumery and others. Moreover, frankincense and myrrh have several local applications in medicinal, hygienic and insecticide areas that could be developed through research. They are widely used in the traditional medicines of several countries for treatment of a wide variety of ailments from embalming to cancer, leprosy, bronchitis, diarrhoea, dysentery, typhoid, mouth ulcers, inflammatory complaints, viral hepatitis, female disorders, infections/wounds, coughs, tumours and others.

Today, Western medicine does not promote/validate any of the historical or current Eastern medicinal practices. However, practitioners of aromatherapy believe in its power to reduce anxiety or stress. It is also promoted as an aid in meditation and prayer – a throwback to the times when it was the primary scent in the temple. In the East, it is widely used as a medicine.

Frankincense is still a main ingredient in many different types of incense. It is also popular in commercial incense mixtures – and the raw 'tears' are readily available to burn directly on hot coals, just as in ancient times.

It is also important in the perfumery industry as a scent and as a fixative. Oil from frankincense can take up to 6 h to evaporate, making it an important ingredient in many perfumes. The current potpourri market has also found a niche for the 'tears' and oil.

Myrrh has been used since several millennia as a medicine, as well as for ceremonial and religious purposes. In many cultures in Europe, Asia and Africa, myrrh has enjoyed various traditional and industrial uses and applications. A recent study conducted in Saudi Arabia on the prevalence and pattern of use of alternative medicine, based on interviews of 1408 individuals, revealed the most frequently used medicines were honey (40%), black seed (39%) and myrrh (35%) (Al-Faris *et al.*, 2008). This traditional medicinal use of myrrh extends to several countries, where it is used for treatments of a wide variety of ailments from embalming to cancer, leprosy, bronchitis, diarrhoea, dysentery, typhoid, mouth ulcers, inflammation, viral hepatitis, female disorders, wounds, coughs, tumours, etc. The British Herbal Pharmacopoeia (1996) indicates myrrh tincture as a mouthwash for gingivitis and ulcers. Myrrh is also an important drug in Chinese traditional medicine (Yen, 1992). In Somalia and Ethiopia, decoction of myrrh resin is used to treat stomachache; it is mixed with powdered charcoal to make ink for writing on parchments and burnt in houses and in the bush to chase away snakes (Dekebo *et al.*, 2002). Modern uses include flavouring foods, drinks and confectionary items and as an additive of products for personal use such as perfumes, deodorants, shampoos, bath lotions, toilet soaps, toothpastes, mouthwashes, air fresheners, etc.

Opoponax

Opoponax oil (CAS No. 8021-36-1) is currently used worldwide at levels below 100 ppm in selected cigarette brands. A consumable resin can be extracted from opoponax by cutting the plant at the base of the stem and sun-drying the juice that flows out. Though people often find the taste acrid and bitter, the highly flammable resin can be burned as incense to produce a scent somewhat like balsam or lavender. The resin has been used in the treatment of spasm and, before that, as an emmenagogue in the treatment of asthma, chronic visceral infections, hysteria and hypochondria. Opoponax resin is sold most frequently in dried irregular pieces, though tear-shaped gems are not uncommon. Opoponax is also used in the production of certain perfumes and is the fragrance of special scented candles.

Waxes

Waxes are a primary finishing material used since the 18h century in the manufacture of candles, in cosmetics, in the food industry, as well as automobile waxes, shoe polishes, instrument polishes, and floor and furniture polishes.

Bayberry

Bayberry is used primarily in the manufacture of candles and other products where a distinctive fragrance is desirable. Bayberry wax is a delicate and rare wax. Bayberry candles are made from authentic bayberry wax and just the right amount of beeswax to create the perfect, all natural bayberry candle.

Candelilla

This wax finds use in the cosmetics industry, as a component of lip balms and lotion bars. One of its major uses was as a binder for chewing gums. Candelilla wax can be used as a substitute for carnauba wax and beeswax. It is also used for making varnish. As a food additive, candelilla wax has the number E902 and is used as a glazing agent (Table 19.3).

Carnauba

Carnauba wax can produce a glossy finish and as such is used in automobile waxes, shoe polishes, instrument polishes, and floor and furniture polishes, especially when

Table 19.3. Candelilla wax (E902) is a food additive that may be used in the following foods under the conditions of good manufacturing practices (GMP), as outlined in the Preamble of the Codex General Standard of Food Additives (Codex Stan 192–1995 updated online).

Number	Food category	Number	Food category
05.3	Chewing gum	05.1.4	Cocoa and chocolate products
14.1.5	Coffee, coffee substitutes, tea, herbal infusions and other hot cereal and grain beverages, excluding cocoa	05.2	Confectionery including hard and soft candy, nougats, etc., other than food categories 05.1, 05.3 and 05.4
05.4	Decorations (e.g. for fine bakery wares), toppings (non-fruit) and sweet sauces	07.2	Fine bakery wares (sweet, salty, savoury) and mixes
13.6	Food supplements	05.1.5	Imitation chocolate, chocolate substitute products
15.0	Ready-to-eat savouries	04.1.1.2	Surface-treated fresh fruit
04.2.1.2	Surface-treated fresh vegetables (including mushrooms and fungi, roots and tubers, pulses and legumes, and aloe vera), seaweeds and nuts and seeds	14.1.4	Water-based flavoured drinks, including 'sport', 'energy' or 'electrolyte' drinks and particulated drinks

mixed with beeswax. It is used as a coating on dental floss. Use for paper coatings is the most common application in the USA. It is the main ingredient in surfboard wax, combined with coconut oil. Because of its hypoallergenic and emollient properties, as well as its shine, carnauba wax appears as an ingredient in many cosmetics formulas where it is used to thicken lipstick, eyeliner, mascara, eye shadow, foundation, deodorant, various skincare preparations, suncare preparations, etc. It is the finish of choice for most briar tobacco or smoking pipes. It produces a high-gloss finish when buffed on to wood. This finish dulls with time rather than flaking off (as is the case with most other finishes used.) Though too brittle to be used by itself, carnauba wax is often combined with other waxes (principally beeswax) to treat and waterproof many leather products, where it provides a high-gloss finish and increases the leather's hardness and durability. It is also used in the pharmaceutical industry as a tablet-coating agent.

In foods, it is used as a formulation aid, lubricant, release agent, anti-caking agent and surface finishing agent in baked foods and mixes, chewing gum, confections, frostings, fresh fruits and juices, gravies, sauces, processed fruits and juices, soft candy, etc. (Table 19.4).

When used as a mould release, carnauba, unlike silicone or PTFE, is suitable for use with liquid epoxy, epoxy moulding compounds (EMC) and some other plastic types. Carnauba wax is compatible with epoxies and generally enhances its properties, along with those of most other engineering plastics. As a mould release agent it is used for hair, latex, polyurethanes, epoxies, polyester resins, and silicone rubbers.

An aerosol mould release is formed by suspending carnauba wax in a solvent. This aerosol version is used extensively in moulds for semiconductor devices. Semiconductor manufacturers also use chunks of carnauba wax to break in new epoxy moulds or to release the plunger when it sticks. Carnauba

Table 19.4. Carnauba wax (E903) is a food additive that may be used in the following foods under the conditions of good manufacturing practices (GMP), as outlined in the Preamble of the Codex General Standard of Food Additives (Codex Stan 192–1995 updated online).

Number	Food category	Max level
07.0	Bakery wares	GMP
05.3	Chewing gum	1200 mg/kg
05.1.4	Cocoa and chocolate products	5000 mg/kg
14.1.5	Coffee, coffee substitutes, tea, herbal infusions and other hot cereal and grain beverages, excluding cocoa	200 mg/kg
05.2	Confectionery including hard and soft candy, nougats, etc., other than food categories 05.1, 05.3 and 05.4	5000 mg/kg
05.4	Decorations (e.g. for fine bakery wares), toppings (non-fruit) and sweet sauces	4000 mg/kg
13.6	Food supplements	5000 mg/kg
05.1.5	Imitation chocolate, chocolate substitute products	5000 mg/kg
04.1.2	Processed fruits	400 mg/kg
15.0	Ready-to-eat savouries	200 mg/kg
04.1.1.2	Surface-treated fresh fruit	400 mg/kg
04.2.1.2	Surface-treated fresh vegetables (including mushrooms and fungi, roots and tubers, pulses and legumes, and aloe vera), seaweeds and nuts and seeds	400 mg/kg
14.1.4	Water-based flavoured drinks, including 'sport', 'energy', or 'electrolyte' drinks and particulated drinks.	200 mg/kg

is used in melt/castable explosives to produce an insensitive explosive formula such as Composition B, which is a blend of RDX and TNT.

Castor

Castor is used in polishes, cosmetics, electrical capacitors, carbon paper, lubrication and coatings and greases where resistance to moisture, oils and other petrochemical products is required. Castor wax is also useful in polyurethane coating formulations, as it contains three secondary hydroxyl groups. These coating compositions are useful as a topcoat varnish for leather, wood and rubber.

Esparto

Because of its hard, well-blending, easy-to-emulsify properties, esparto wax is often used to give smoothness to polishes. It is used mainly as a substitute for carnauba wax, being cheaper and easily available on the market.

Japan

Japan is used chiefly in the manufacture of candles, furniture polishes, floor waxes, wax matches, soaps, food packaging, pharmaceuticals and as a substitute for beeswax.

Jojoba

Most jojoba oil is consumed as an ingredient in cosmetics and personal care products, especially skincare and haircare.

Ouricury

This has properties similar to carnauba wax and is thus used as a cheaper substitute in products such as floor and furniture polish. The wax, however, has a high resin content and, for that reason, is regarded as inferior. Only about 7000 t are produced annually.

Rice bran

Rice bran is used in paper coating, textiles, fruit and vegetable coatings, pharmaceuticals, candles, moulded novelties, electric insulation, textile and leather sizing, waterproofing, carbon paper, typewriter ribbons, printing inks, lubricants, crayons, adhesives, chewing gum and cosmetics. In cosmetics, rice bran wax is used as an emollient and is the basic material for some exfoliation particles. It may also serve as a substitute for carnauba wax.

Soy

This wax is used mainly in the candle industry because soy wax candles do not emit the soot and fumes that come from regular paraffin wax candles. Candle wax is preferred by many people as a cheap alternative for beeswax. Vegetable wax is made with partially hydrogenated soy oil, coconut oil and palm oil.

Production, Trade and Marketing

Gums

Arabic

Africa is the principal producer of gum arabic, accounting for about 98% of global production (ITC, 2008). Asia is the other region outside Africa that contributes the difference. A total of 15 countries from Africa are known to produce true gum arabic from either *A. senegal* or *A. seyal*. Based on 2006 trade statistics, Table 19.5 gives an overview of producing and exporting countries. From the table, it is clear that Sudan, Chad and Nigeria are the principal players, followed by Senegal, Mali, Tanzania, Ethiopia and Mauritania. About nine other countries contribute a total of less that 1%.

Between 1992 and 2006, the export volume of gum arabic increased by 117% from just under 30,000 t (1993) to above 70,000 t in 2005. Table 19.6 shows the trend in production over this period. There is generally a 5- to 10-year cycle in the burst and boom production trend, which is influenced much more by climatic conditions (especially drought). This explains the production figures of 1992/3 and 2003. Overall, however, export volumes

Table 19.5. Major African producing and exporting countries, (ITC [TradeMap] based on COMTRADE statistical data).

Country		%
Major producing countries	Sudan	43
	Chad	35
	Nigeria	18
Minor producing countries	Senegal	1.1
	Mali	0.8
	Tanzania	0.5
	Ethiopia	0.4
	Mauritania	0.4
	Cameroon	0.3
Others	Ten African countries	0.3

Table 19.6. Raw gum arabic exports (tonne) 1992–2006, (ITC [TradeMap] based on COTONNERADE statistical data).

Year	Sudan	Chad	Nigeria	Africa/ others	Subtotal/ others	Asia	Total
1992	17,061	2,450	8,385	3,073	30,942	726	31,668
1993	13,475	3,701	7,042	2,243	26,461	756	27,217
1994	23,341	4,558	9,822	3,751	41,472	684	42,156
1995	18,143	7,001	9,914	2,821	37,879	814	38,693
1996	17,671	7,365	12,164	3,349	40,549	435	40,984
1997	17,342	8,527	10,199	5,301	41,369	696	42,065
1998	25,053	12,584	8,166	2,296	48,099	384	48,483
1999	19,305	11,312	8,598	3,399	42,614	912	43,526
2000	21,916	11,682	8,239	4,009	45,846	2,251	48,097
2001	26,105	12,881	8,747	2,137	49,870	471	50,341
2002	34,162	10,664	6,556	2,724	54,106	258	54,364
2003	13,217	9,672	50	3,097	26,036	835	26,871
2004	27,444	12,044	15,407	2,393	57,288	762	58,050
2005	33,078	14,186	19,313	3,930	70,507	1,879	72,386
2006	23,149	17,812	21,231	3,474	65,666	709	66,375

have increased following various initiatives, including improved production, quality control practices in major producing countries like Nigeria and Senegal, opening up of opportunities in new producing countries like Chad and Tanzania, as well as initiating programmes in countries like Niger.

Sudan, Chad and Nigeria contributed 92–94% of the global market during the period 2005/6, with the other African countries contributing about 5%, while the Asian countries contributed 1–3%. Among the African countries, Sudan is the leader and accounted for 50% in 2002 and 35% in 2006. Chad has shown the greatest progress in gum export, rising from a mere 2000t in 2002 to 17,000t in 2006, a 750% rise (ITC, 2008). Nigeria has also greatly increased its export volumes, which made it the second leading producer in 2005/06.

The EU is the largest market for gum arabic imports, with about 200,000t being imported between 2003 and 2007. France is the leading importing country in Europe, whose share of the global market over the period 1993–2007 was 30%, followed by Britain with 10% and Germany/Italy with a combined volume of 10%. The USA is a major partner of the EU and imports mostly processed gum arabic. However, in terms of global volumes, the USA is second, representing 17% of the global market. India and Japan import 13% on average, with the former having significantly increased imported volumes over the past few years. An important phenomenon is the increasing consumption of gum arabic among the emerging countries of Eastern Europe, as well as in South America and others in Asia, which are now having a combined market share of about 20% of the imports.

The prices of gum arabic between and among producer countries vary greatly and are generally not yet easy to obtain. However, recent collaboration between ITC and NGARA is helping resolve this difficulty (ITC, 2008). Additionally, the newly established Africa Gum Arabic Association (AGAA) is helping to come up with harmonized pricing for the two main sources of gum arabic, i.e. *A. senegal* gum (Hashab, Chadian Kitir or Nigerian grade 1) and *A. seyal* gum (Sudanese talha, French friable and Nigerian grade 2). Generally, prices are influenced by production trends in the producing countries and have varied from as low as US$1500/t to as high as US$6000/t for hard gum, i.e. *A. senegal*, and from US$750–1500/t for friable gum, i.e. from *A. seyal* gum (Muller and Okoro, 2004). Sometimes, price fluctuations have varied sharply in succeeding years, influenced by supply conditions, thus affecting users greatly. This does not bode well for the gum arabic market and efforts are being made to address the situation. For example, in the current season (2008/9), the AGAA has announced harmonized prices of US$3000 FOB for *A. senegal* gum (Hashab, Kitir, Nigeria grade 1) and US$2000 for *A. seyal* gum (talha, friable or Nigerian grade 2).

As regards market access, there are no significant barriers to the development of gum arabic and no specific quotas or taxes on imports. Some countries ask for a certificate of origin only, while others like the USA request a fumigation certificate to ensure that the packaging meets phytosanitary standards. The only challenge to the gum arabic market is in stabilizing the supply of goods and consistent quality.

Karaya

World production of gum karaya is currently at about 5500 t/annum and is on a declining trend. India is the only regular producer, overwhelmingly dominating international trade in the gum. From the end of the 1960s until the mid-1980s, the annual export averaged 4000–6000 t (FAO, 1995). Total gum production was about three times that amount, since most Indian gum was, and probably still is, consumed domestically. At the end of the 1980s, gum supply collapsed, reaching an all-time low of 570 t in 1991–1992. The reason for this dramatic decrease was the switch to a government-controlled market. Recently, government controls have been relaxed and export has increased again, but is still much below the level of the mid-1980s. It is far from certain that India can regain its original dominance in the gum karaya market, since increasing amounts are being exported by African countries. Senegal is the biggest African producer and exports annually around 1000 t. Sudan has the potential to become an important supplier in the future, if the same efforts are applied for gum karaya as for gum arabic.

The USA consumes roughly one-half of gum karaya and Western Europe around 30%. There is a significant re-export trade from European ports. Consumption in the USA is roughly stable, whereas in Europe there is evidence of some decline. Exports are controlled largely by a large number of long-established merchants based in Bombay. During 1991/92, India exported 573.6 t of gum karaya, mainly to Japan, France, USA, West Germany, UK, Belgium, Italy, UAE and the Netherlands, in that order. The price of Indian gum karaya varies between US$2250/t and US$6000/t, depending on the grade.

Tragacanth

Commercial gum tragacanth supplies are dominated by Iran, which is by far the world's largest producer. Gum production is labour-intensive and is carried out from remote hostile areas of Iran and Turkey. After collection, the gum is cleaned and selected by hand into five 'ribbon' (superior quality) and five flake (inferior quality) grades, which may range in price from US$10–62/kg (Anderson, 1993). The best quality gum is of Iranian origin. Iran and Turkey are the main producing countries, about 70% of the supplies originating from Iran alone. Iran's average annual production potential has been estimated at 400 t. It is also known that the gum is produced in Afghanistan and Syria, but export consignments are very rare.

Tragacanth finds markets in many different countries, but the European Community, the USA, Japan and the former Soviet Union are the major importing regions. At present, the world market for gum tragacanth is estimated to be no more than 500 t/year, about 300 t of which is exported by Iran (FAO, 1995). The price of ribbons is US$3000–4000/t for the lower grade and up to US$22,000/t for the highest grade.

Gum resins

Gum resins (frankincense, myrrh and opoponax), unlike gum arabic, are not usually recorded as such in trade statistics, making it difficult to quantify international trade. The information reported here is therefore derived from an examination of data on 'other natural gums and resins' from importing countries to ensure that gum arabic and other gums, e.g. karaya or tragacanth, are excluded. On the basis of such analysis, Africa remains the principal producer of this group of gum resins. When all the trade data are consolidated from around the world, the following inferences become apparent:

- World demand is estimated at around 5000 t.
- Ethiopia, Kenya and Somalia are the principal producers and exporters (Coppen, 2005). Export volumes from Kenya from 2005–2008 varied between 1200 and 1500 t, with Ethiopia exporting about two and a half times more than the amount coming from Somalia (especially Puntland and Somaliland). Kenya is emerging as the principal supplier of medicinal type opoponax (hagar), while Ethiopia remains the principal exporter of frankincense and myrrh.
- China and Ethiopia are the largest markets, though the Middle East, North Africa and the USA also import significant amounts from source. Within the EU, Germany is the biggest importer (and re-exporter) of the resins.
- Prices generally vary depending on the type of gum resin. Meydi (i.e. frankincense from *B. frereana*) from Somalia is the most highly priced. For example, in 2005 grade 1 type opoponax (hagar) sold for prices ranging between US$2.50 and US$4.00/kg, while meydi was sold for US$20/kg.

Waxes

The global waxes market in 2005 was estimated at a value of more US$7.2 billion and it is expected to grow with an annual rate of nearly 6% to reach US$10.9 billion by 2010. Most of the wax used globally today is derived from petroleum and is known as mineral or paraffin wax. Paraffin waxes occupy 75–77% of the global wax production.

There are more than 100 companies active in the global waxes market that produce and/or process various types of waxes, primarily mineral or paraffin waxes, synthetic waxes and natural waxes. Mergers and acquisitions increasingly are taking place in Europe and the USA as smaller organizations face strong competition from cheaper Asian imports. The USA was the largest market, with a 34.1% share. Europe (24.0%) and Asia-Pacific (23.6%) have similar market shares.

The main driver for growth within the natural waxes market is enhanced demand from end-user industries; particularly perfumed and coloured candles, and a wide variety of consumer products that are wax-based.

Conclusions

Gums and resins are used largely in western countries in many applications. This demand generates an important commercial flux from the African Sahelian countries to the rest of the world and thus represents an important resource for populations that are among the poorest on the planet. It is clear that development of the demand for these products would benefit the producing countries, which urgently need to meet an economic development.

Thanks to its unique physical properties, acacia gum has been used in various foods and pharmaceutical products. More recently, western countries have discovered that gum arabic is also a dietary fibre with very interesting nutritional properties. In view of the increasing consumer awareness of the connection between health and diet, the food industry is increasingly developing foods with enhanced nutritional properties, so-called 'functional foods', 'nutraceutics' or other explicit denominations. The use of acacia gum as a nutritional ingredient clearly represents a real opportunity to develop demand, and consequently a potential resource for the producing countries.

The demand for gum arabic has increased tremendously over the past few years, reaching the export volumes (about 70,000 t) that were witnessed in the 1960s and 1970s. Various factors, including improved production, quality control practices and opening up of opportunities in new producing countries, partly explain this current increase. Further incentives through the NGARA and AGAA in collaboration with the Association for International Promotion of Gums (AIPG) should see production increase to about 100,000 t/year.

World demand for gum karaya is generally on the decline, going by trade volumes. India was once the main producer and exporter, but recently Senegal has emerged as the leading exporter, accounting for about 1000 t/year. However, it is believed that India consumes most of the gum karaya produced. Opportunities exist in Africa for increasing production from other countries in the Sahel if awareness is raised and communities are trained in sound harvesting and postharvest handling.

Demand for gum tragacanth fell greatly during the 1980s, from several thousand tonnes to 200–300 t/year, for several reasons. The Iran/Iraq war made supplies erratic. The Iranian government tried to fix the price, which made the gum non-competitive, and high inflation rates in Turkey had the same effect. Iran's recent recovery in the tragacanth export market suggests that with a correct understanding of the world tragacanth market and the supply of premium product, there are huge prospects for securing a bigger and better market for gum tragacanth. Deliberate cultivation of gum producing *Astragalus* species may be rewarding in appropriate regions. However, little is known about the genetic factors that enhance gum yield and quality.

Information regarding gum resins (frankincense, myrrh and opoponax) has been generally scanty, thereby affecting their commercial opportunities, despite having been articles for international trade since ancient times. However, considering the pharmacological and industrial applications of these valuable resources, it is expected that the enormous economic opportunity that these resources have could provide an expansion of demand, at both national and international levels. Hagar is being used increasingly in both herbal and conventional medicine, which provides opportunities for expanded production. The NGARA is currently working with several member countries in the Horn of Africa to improve the production and marketing systems, which should see the sector recover to reclaim some of the market, while at the same time improving the livelihoods of the communities and economies of member countries.

Renewable alternatives to wax are enjoying new opportunities as a result of a raft of factors. Among these are dwindling supplies of petroleum-based waxes, skyrocketing paraffin wax prices and growing consumer preference for greener products. In the candle market, where 50% of the global wax supply is being consumed, companies are already making the move to vegetable-based materials such as soybean oil, palm oil, beeswax and even tallow.

References

Al-Faris, E.A., Al-Rowais, N., Mohamed, A.G., Al-Rukban, M.O., Al- Kurdi, A., Al-Noor, M.A.B., Al-Harby, S. and Sheikh, A. (2008) Prevalence and pattern of alternative medicine use: the results of a household survey. *Annuals of Saudi Medicine* 28, 4–10.

Anderson, D.M.W. (1993) Some factors influencing the demand of gum arabic (*Acacia senegal* (L.) Wild) and other water-soluble tree exudates. *Forest Ecology and Management* 58, 1–18.
Baser, K.H.C., Demirci, B., Dekebo, A., Dagne, E. (2003) Essential oils of some Boswellia spp. Myrrh and opopanax. *Flavour Frag Journal* 18, 153–156.
Chikamai, B.N., Phillips, G.O. and Casadei, E. (1996) The characterization and specification of gum Arabic. Project No. TCP/RAF/4557, FAO, Rome.
Coppen, J.J.W. (2005) Overview of international trade and markets. In: *Production and Marketing of Gum Resins: Frankincense, Myrrh and Opoponax*. NGARA Publication Series No. 5, Nairobi, Kenya.
Dekebo, A., Dague, E. and Sterner, O. (2002) Furanosesquiterpenes from *Commiphora sphaerocarpa* and related adulterants of true myrrh. *Fitoterapia* 73(1), 48–55.
FAO (1995) Gum resins and latexes of plant origin. (Non-wood-forest-products 6.) FAO, Rome.
FAO/WHO Joint Expert Committee on Food Additives (1998) Published specifications on gum arabic FNP 52 Add 6; republished in FNP 52 Add 7 (1999) to include editorial changes.
Iqbol, M. (1993) International trade in non-wood-forest-products: an overview. Food and Agriculture Organization of the United Nations, Rome.
ITC (International Trade Centre) (2008) Gum arabic. *Market News Service*, September edn.
Kline & Company Inc., (2007) Global opportunities and threats in the wax business, 2006–2020. Report #Y635, 2007.
Muller, D. and Okoro, C. (2004) *Production and Marketing of Gum Arabic*. NGARA Publication Series No. 2, Nairobi, Kenya.
Phillips, G.O. and Williams, P.A. (2001) Tree exudate gums: natural and versatile food additives and ingredients. *Food Ingredients and Analysis International* 23, 26–28.
Tucker, A.O. (1986) Frankincense and myrrh. *Economic Botany* 40(4), 425–433.
Yen, K.Y. (1992) *The Illustrated Chinese Materia Medica Crude and Prepared*. SMC Publishing, Inc, Taipei, Taiwan.

20

Botanical Insecticides, Deterrents, Repellents and Oils

Murray B. Isman

Introduction

Botanicals have been used for pest control for at least two millennia. Plant materials or extracts containing pyrethrins, rotenoids or nicotine have held a central place in the arsenal of products used by farmers to protect their crops from the ravages of insects and related pests. However, following the discovery of the insecticidal properties of DDT in 1939 by Paul Muller and subsequent development and commercialization of other chemicals, synthetic insecticides have dominated pest management. The outstanding efficacy and ease in use of conventional chemical insecticides have led to their overwhelming adoption in crop and forest protection, structural and urban pest control, management of stored product pests and protection against blood-feeding insects and ectoparasites of humans and animals.

Well-documented examples of negative environmental impacts (viz. reduced populations of fish, birds and wildlife; disruption of natural biological control and pollination; widespread groundwater contamination) and emerging concerns for harmful effects on human health through long-term low-level exposure have resulted in increasingly stringent regulations for the use of conventional synthetic pesticides in most developed countries (Isman, 2006). However, while the general public embraces the notion that 'natural' products are inherently better for their health and the environment, there are sufficient examples of natural products with demonstrated toxicity to humans to warrant some level of regulatory scrutiny of botanical products, as well as those from other natural sources (Trumble, 2002).

Given the past history of humans using plant materials for pest control, the widespread use of plants and plant derivatives as medicines, culinary herbs and colouring agents, and the enhanced worldwide awareness of human impacts on the environment, it might seem surprising that so few botanical insecticides are currently used in commercial pest management practice. Table 20.1 lists the major botanical insecticides used at present. These include two traditional products (pyrethrum, rotenone), one product commercially developed in the 1980s (neem) and one developed in the past decade (essential oils).

Traditional Botanical Insecticides

The traditional botanical insecticide still in widespread use

Among the eight to ten botanical insecticides that have seen commercial use as crop protectants in the past 50 years, pyrethrum

Table 20.1. Major botanical insecticides: basic properties (modified from Isman, 2005).

	Pyrethrum	Rotenone	Neem	Essential oils
Country of origin	Kenya, Australia	SE Asia, Venezuela	India	Mediterranean region, SE Asia, China
Active ingredients	Pyrethrins (esters)	Rotenoids (isoflavonoids)	Azadirachtin (limonoids)	Monoterpenes, simple phenols
Actives in technical product (%)	20–50	5–7	10–20	50–95
Approx. price (US$/ kg)	80–90	3–5	150–200	7–22
Mode of action	Neurotoxic (axonic poison)	Cytotoxic (mitochondrial poison)	Moult disruptor (ecdysone antagonist)	Neurotoxic (octopamine agonist)
Action on pests	Contact/ knockdown	Stomach poison	IGR[a] /antifeedant	Contact/ knockdown
Target pests	Broad spectrum, including ectoparasites	Primarily used as fish poison; garden pests	Broad spectrum, especially chewing insects	Primarily softbodied insects; urban pests, mosquitoes
Mammalian toxicity	Minimal	Moderate; very toxic to fish	Non-toxic	Non-toxic

[a]IGR, Insect Growth Regulator.

remains alone among the traditional products still enjoying widespread use (Table 20.1). It is probably the most widely and heavily used botanical insecticide worldwide. Pyrethrum refers to the oleoresin obtained by solvent extraction of the flowers of *Tanacetum cinerariaefolium*, a daisy native to south-eastern Europe. As recently as a decade ago, the majority (> 75%) of the world's supply of pyrethrum was produced in Kenya and Tanzania. However, production began in Tasmania (Australia) in 1996 and has grown to the point where it now produces almost one-half of the world supply. Whereas the traditional method of preparation generated a resin containing 20–25% of the insecticidal constituents (pyrethrins I and II, esters of chrysanthemic acid; Fig. 20.1), newer processing methods in Australia generate a technical grade material comprising 50% pyrethrins by weight.

Pyrethrins are classic axonic poisons against insects, disrupting nerve transmission through a biochemical action not dissimilar from that of DDT. What characterizes pyrethrin toxicity in insects best is their rapid knockdown, notably in flying insects, some of which are immobilized within 1 s. Pyrethrins are common active ingredients in consumer (household) insecticide products, where apparent speed-of-kill is often more important to the user than absolute efficacy based on the administered dose. Natural pyrethrins are relatively unstable in the presence of the UV component of sunlight, greatly reducing their persistence under field conditions and therefore limiting their agricultural use. Indeed, solving the problem of persistence led to the development of the synthetic pyrethroids, a class of chemicals that dominated world insecticide use from the 1980s to the start of the current century.

In the State of California, USA, almost 73,000 t of pyrethrins were sold in 2005, but this number dropped to 14,000 Mt in 2007 (Cal DPR, 2007). Of the total amount applied in 2007, approximately 50% was used in public health and a further 25% for structural pest control; as such, the quantities used in agriculture are rather modest compared to other natural pesticides used in agriculture such as *Bacillus thuringiensis* (82,000 t) and spinosad (42,000 t).

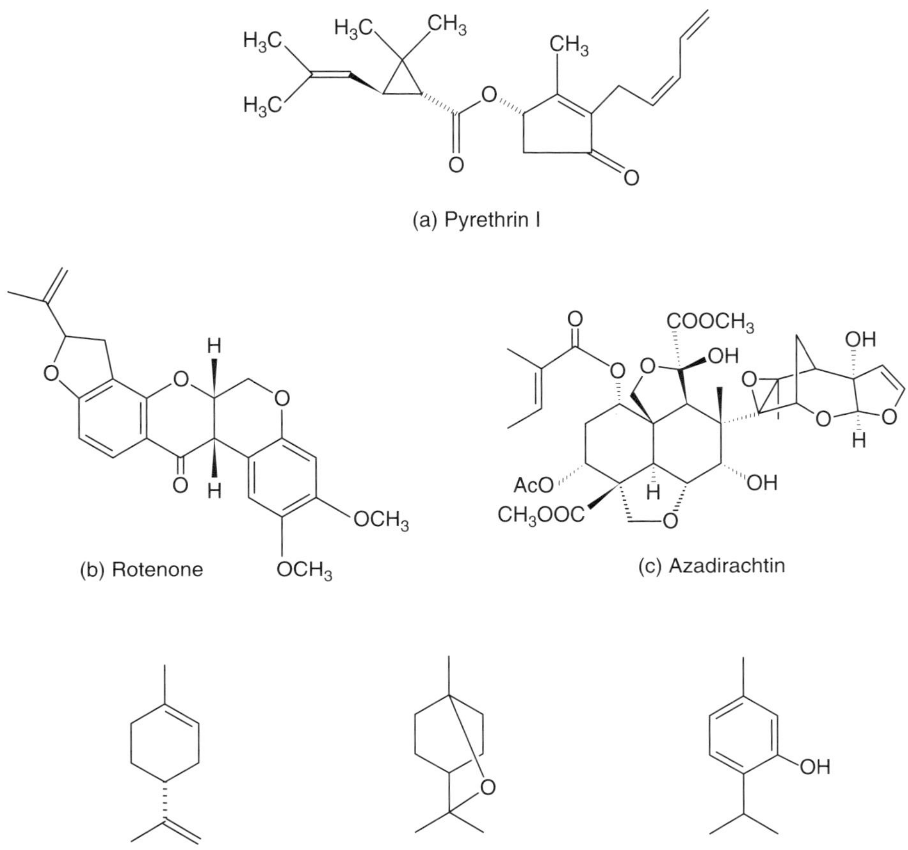

(a) Pyrethrin I (b) Rotenone (c) Azadirachtin (d) *d*-Limonene (e) 1,8-Cineole (f) Thymol

Fig. 20.1. Active ingredients of major botanical insecticides.

Traditional botanical insecticides in declining use

Rotenone

Rotenone (Fig. 20.1) is an isoflavonoid obtained from the roots or rhizomes of tropical legumes in the genera *Derris*, *Lonchocarpus* and *Tephrosia*. The term is also applied loosely to the dried powder obtained by grinding the roots, or to the resin resulting from extraction, containing a number of closely related isoflavonoids. The material has been used for over 350 years in South-east Asia as a fish poison. Indeed, the outstanding toxicity of rotenone to fish explains the fact that over 90% of the rotenone used in North America is for management of 'trash' fish species in the freshwater. Rotenone sees little use today in agriculture, although it continues to be sold for home and garden use. Pure rotenone is relatively toxic to mammals; the oral acute LD_{50} in rats is 132 mg/kg. Domestic products seldom contain more than 1% active ingredient, accounting for its relative safety. As an inhibitor of the electron transport chain in mitochondria, rotenone is a broad-spectrum cytotoxin.

Nicotine

Nicotine and the related alkaloids nornicotine and anabasine are highly insecticidal constituents obtained from aqueous extracts of tobacco (*Nicotiana* spp.; Solanaceae) and *Anabasis aphylla* (Chenopodiaceae). As agonists of the neurotransmitter acetylcholine, nicotine-type alkaloids are synaptic poisons toxic to both insects and mammals. Nicotine is a good example of a plant natural product whose safety cannot be assumed – in fact, it is quite toxic to humans through ingestion, dermal exposure or inhalation. An oral human lethal dose has been estimated to be as low as 60 mg. As a result, it is seldom used today in North America or Europe, although it continues to be used in China and crude tobacco extracts are used in Africa (Morse *et al.*, 2002).

Other traditional botanicals

A handful of other plant materials have seen limited commercial use as insecticides, but these uses are in decline, at least in industrialized countries. One of these is sabadilla, a powder based on the ground seeds of the South American plant *Schoenocaulon officinale* (Liliaceae). In purity, the active principles, steroidal-type alkaloids, are quite toxic to mammals but, as with rotenone, the concentration of alkaloids in the powdered seeds is quite low, providing a margin of safety to the user. Wood of the Caribbean tree *Ryania speciosa* (Flacourtiaceae) contains a suite of unique alkaloids that block neuromuscular junctions. The ground stem wood, containing < 1% ryanodine and its alkaloid analogues, has been used in organic fruit production. *Quassia amara* (Simaroubaceae), a small tree from Brazil, contains bitter triterpenoids in its wood and bark. Woodchips and ground bark of this species have been used traditionally as an insecticide, as have plant parts from the related tree, *Ailanthus altissima*.

Newer Botanical Insecticides

Neem

It is ironic to refer to neem as a 'newer' insecticide, as the medicinal use of this tree in its native India, and likely its use to protect stored products, dates back at least 2000 years. However, extensive research into the insecticidal properties of the Indian neem tree, *Azadirachta indica* (Meliaceae), in Germany, the UK and the USA in the 1970s and 1980s laid the foundation for the introduction of botanical insecticides based on refined neem seed extracts in the 1990s. Neem seeds contain complex triterpenoids, among which the most important is azadirachtin (Fig. 20.1). This substance interferes with synthesis of the insect moulting hormone, α-ecdysone, as well as other physiologically active neuropeptides in insects, producing a wide range of physiological and behavioural effects, such as anorexia. In addition, azadirachtin is the most potent natural feeding deterrent for insects discovered to date. Exactly which of these actions account for field efficacy of neem insecticides is not clearly understood. Importantly, azadirachtin and botanical preparations based on neem seed extracts are virtually non-toxic to mammals and wildlife, making them among the safest of all insecticides. Refined neem products have been expensive to produce, which may account for their limited penetration into the marketplace thus far (Isman, 2004).

Plant essential oils

Aromatic oils obtained through steam distillation of many plants, most notably those of the mint family (Lamiaceae), are widely used in the fragrance and flavour industries. Although there have been long traditions of using aromatic plants to protect stored grain in some tropical regions (Belmain and Stevenson, 2001), the acute insecticidal action of certain essential oils or their major constituents have been reported only recently and commercial exploitation of these for pest management begun (Isman, 2000). Oils obtained from clove, rosemary, thyme, eucalyptus and various mint species have demonstrated contact and fumigant toxicity to a wide spectrum of insects, including human head lice (Toloza *et al.*, 2008), as well as prophylactic action against some plant pathogenic fungi (Isman and Machial, 2006). The active principles are monoterpenes, sesquiterpenes and their biogenically related phenols (Fig. 20.1). In addition to direct toxicity to insects, many of these substances are deterrents or repellents. Given the longstanding and widespread use of most of these plants as culinary herbs and/or indigenous medicines by numerous societies, they are generally recognized as safe to humans and animals, although a small number of the essential oils in isolation are moderately toxic to mammals. Several commercial insecticides based on plant essential oils intended for professional, agricultural, veterinary and consumer applications have been introduced in the past decade.

Search for New Insecticides from Plants

Terrestrial plants produce a staggering array of 'secondary' plant metabolites – i.e. those not involved in primary plant metabolism – and it is widely acknowledged that many of these likely function in chemical defence against herbivory. Scientists have long viewed the plant kingdom, particularly tropical species, as fertile ground for the discovery of novel substances with potential value as human therapeutics. This has also been the case for the discovery of natural insecticides of plant origin. Over the past three decades, the scientific literature has swelled with hundreds (if not thousands) of reports of bioactivity of plant extracts or isolated natural products in insects based on laboratory bioassays. Few of these discoveries have led to the commercial development of new botanical insecticides, for reasons discussed later in this chapter.

Strategies for discovery

While it is certainly true that discovery of pharmaceuticals or insecticides from plants has occurred occasionally through purely random screening of regional flora, the probability of discovery can be heightened if certain strategies are deployed. A taxonomic approach has proven useful in many cases (e.g. Satasook *et al.*, 1994). Congeners of plant species known to contain bioactive principles often contain suites of chemical analogues, some of which could be as or more active than the first discovered principle. Based on the past 30 years, there is wide agreement that particular plant families are especially fruitful in the search for new insecticidal constituents, among which are the mint (Lamiaceae), mahogany (Meliaceae), sunflower (Asteraceae) and custard apple (Annonaceae) families (Jacobson, 1989).

A second productive strategy is one based on anthropology – investigation of plants and plant materials traditionally used by indigenous human populations for pest control or as natural medicines. For example, the traditional use of dried, powdered flowers of yellow azalea (*Rhododendron molle*; Ericaceae) in China led to the isolation of grayanoid diterpenes with insecticidal properties. Endemic medicinal plants warrant investigation as natural insecticides – many outstanding natural insecticides have been discovered previously or concurrently in pharmaceutical screening of plant extracts (e.g. McLaughlin, 2008).

One limitation in the potential production of a plant-based insecticide is the relative availability of the plant resource. If the plant in question cannot be cultivated readily, harvest of natural populations may not be adequate to satisfy the need for starting material on a commercial scale. Therefore, an alternative search strategy involves screening waste or by-products of plant-based industries. For example, extraction and screening of sawdust from tropical timber operations led to the discovery of potent insecticides in the stemwood of the Malaysian tree, *Azadirachta excelsa*, a close relative of the Indian neem tree (Isman, 1997; Schmutterer *et al.*, 2002).

Methods of discovery

While there are thousands of reports of the bioactivity of plant extracts or isolated compounds from plants against insects, the wide variety of extraction methods employed, bioassay endpoints examined and species of insects tested make comparisons between individual studies extremely tenuous. Scientific investigations of this type continue to be commonplace. At the very least, where insecticidal action is a major endpoint, a positive control (botanical insecticide in current use) should be included in all bioassays. Without such a 'yardstick' no assessment of 'relative' bioactivity can be rendered.

Successful screening depends on a reliable and robust bioassay. Criteria that distinguish good bioassays are: (i) reproducibility; (ii) linearity over a reasonable dose or concentration range; and (iii) predetermined endpoints. A simple and low cost bioassay with considerable throughput is the brine shrimp (*Artemia salina*) bioassay, as employed by McLaughlin and colleagues (Alkofahi *et al.*, 1989). While this bioassay successfully identified certain annonaceous acetogenins as natural insecticides (as well as several anticancer drugs), it is basically a measure of general cytotoxicity and may not necessarily be a good predictor of bioactivity against agricultural pests. One inescapable difficulty in searching for natural insecticides is the fact of interspecific differences in susceptibility – a substance or extract that is toxic to chewing pests may not be effective against sucking pests, subterranean pests, stored product pests, etc. Thus, screening based on a particular type of insect may overlook valuable bioactivity against other insects completely. Because plant secondary compounds often have more than one physiological or behavioural effect on insects, it is important to decide at the outset precisely what endpoint (outcome) is to be measured in the bioassay. The most obvious (and often easiest to quantify) is mortality over a short time interval, but various sublethal effects (inhibition of growth, interference with development, feeding or oviposition deterrence) should not be discounted, as they can make valuable contributions to field 'efficacy'.

Bioactive Principles from Plants with Commercial Potential

Among the thousands of plant substances that have demonstrated bioactivity against insects, three types appear to be the best candidates for commercialization as pesticides. These are the acetogenins from the custard apple family (Annonaceae), isobutylamides from the pepper family (Piperaceae) and limonoids from the chinaberry tree (*Melia azedarach* and *M. toosendanin*).

Acetogenins are cytotoxic constituents found in the seeds of certain edible fruits from tropical species in the genus *Annona* (e.g. custard apple, sweetsop) or from bark of the temperate pawpaw (Alkofahi *et al.*, 1989). Patents for their use as insecticides were issued over 20 years ago, both in Germany and the USA. Toxicologically, they share their mode-of-action, as mitochondrian electron transport chain inhibitors, with rotenone. Consequently, certain acetogenins in purity are quite toxic to mammals, although, as with other natural insecticides, the risk of human poisonings is greatly mitigated through their use in botanical preparations where the active principles are in low concentrations. Recent studies suggest their suitability for use as pesticides against agricultural pests (Leatemia and Isman, 2004; Alvarez Colom *et al.*, 2007). None the less, pesticide registration seems to be the greatest impediment to the commercial development of these materials as insecticides.

Over 40 years ago, Miyakado and colleagues isolated isobutylamides as the insecticidal principles in the culinary spice, black pepper. Numerous studies of these substances in this species and the neotropical *Piper tuberculatum* have indicated their suitability for insect control against a wide range of pests (Scott *et al.*, 2003; Debonsi *et al.*, 2009). The isobutylamides are axonic poisons in insects, possibly acting through a mechanism comparable to that of the pyrethrins. The same compounds are also efficacious against certain plant pathogenic fungi (Navickiene *et al.*, 2000).

Limonoids from the seeds of the chinaberry tree, seeds of the African *M. volkensii* and from bark of the closely related Chinese tree, *M. toosendan*, have well-documented feeding deterrent and growth inhibitory properties against several pest species (Schmutterer, 2002). Though not as effective an insect growth regulator and antifeedant as azadirachtin found in seeds of the Indian neem tree, they have shown potential for use as insecticides under field conditions. Extracts of *M. toosendan*, rich in the limonoid toosendanin and analogues, has been used in commercial insecticides sold in China, in mixtures with conventional insecticides, or with nicotine, another natural product. Extracts of these *Melia* species tend to be relatively non-toxic to mammals. Instead, it is their chemical complexity that makes regulatory approval in industrialized countries a significant challenge.

Powdered flowers of the yellow azalea, *R. molle* (Ericaceae), contain grayanoid diterpenes, among which rhodojaponin-III is especially active against agricultural pests. A botanical insecticide based on this compound has been commercialized recently in China (Zhong *et al.*, 2006).

Potential Modes of Action

Physiological

Acute toxicity

Conventional screening of novel compounds for insecticidal action has long focused on those compounds that are the most potent and rapid acting, a reflection of the fact that the vast majority of synthetic insecticides target the insect nervous system. However, natural insect–plant chemical interactions are generally subtler; plants have typically evolved to discourage herbivory, rather than kill attacking insects. As such, the search for acutely insecticidal substances produced by plants is a search for exceptions to the rule. While there certainly are some plant compounds that through neurotoxic or cytotoxic action kill insects at low doses – viz. pyrethrins, acetogenins, isobutylamides – these are in the distinct minority among insect bioactive principles from plants. Many bioassays typically enumerate insect mortality after short time intervals, e.g. 24 or 48 h, following direct topical administration or through exposure to residues on inert surfaces (e.g. glass) or on plant surfaces. However, restricting the observation period to such a short duration would completely miss slow-acting lethal toxicity of plant substances. One outstanding example of the latter is azadirachtin; as an insect growth regulator that interferes with moulting, the lethal consequences often take 3 days or more to be manifested (Isman, 1997). Similarly, a toxin that acts through inhibition of protein synthesis such as the natural insecticide rocaglamide from *Aglaia odorata* (Meliaceae) can take 3 days or longer for appreciable toxicity to be observed (Satasook *et al.*, 1993).

Insect growth regulation

Insect growth regulators are defined as those substances that interfere with development and moulting in insects. Interference can result from substances that serve as agonists or antagonists of insect juvenile hormones or moulting hormones, or that prevent the synthesis of chitin, a key component of the insect exoskeleton. A handful of plant natural products have been discovered that affect insects through these types of mechanisms, the best example of which is azadirachtin from seeds of the Indian neem tree. Others, such as the precocenes for *Ageratum houstonianum* (Bowers, 1982), have shown little if any commercial potential. The scientific literature in this area is confused by numerous reports of plant extracts or isolated substances that cause malformed insects during pupation or the development of adult insects in laboratory bioassays, when in fact many substances that interfere with insect nutrition can produce similar effects.

Larval growth inhibition

The vast majority of plant substances that affect insects in laboratory bioassays do so by inhibiting growth of immature insects. Such inhibition can result from reduced consumption, which in turn can be a consequence of direct feeding deterrence or from post-ingestion malaise. Such growth inhibitory substances or their extracts could be efficacious as crop protectants (Wheeler and Isman, 2001; Leatemia and Isman, 2004). Under field conditions, a product that slowed the growth rate of pests would broaden the window of opportunity for natural enemies (predators and parasites) or abiotic factors (weather) to act as mortality factors. In practice though, a product that does not kill insects directly in a reasonable time frame would not be popular with growers who are accustomed to conventional insecticides that typically suppress pest populations within hours.

Insect behaviour modification

The use of insect pheromones for mating confusion, mass trapping or bait-and-kill applications demonstrates the utility of insect behaviour-modifying substances in pest management. Based on the observation that numerous plant compounds deter insect feeding or oviposition (egg-laying), or repel insects, the concept of using deterrents or repellents as non-toxic crop protectants has long been touted.

Feeding (deterrents)

It is probable that more plant compounds deter insect feeding than affect insects in any other behavioural or physiological manner. There is abundant scientific literature documenting insect feeding deterrence from plant extracts or from isolated substances thereof. In spite of this, there is not a single commercial antifeedant used in crop protection (Isman, 2002). Some people will argue that neem insecticides, containing the outstanding antifeedant and insect growth regulator, azadirachtin, constitute an example of a commercial antifeedant. However, it is the insect growth regulating activities of azadirachtin that are more consistent and reliable under field conditions, at least for the majority of pest species against which neem products are used (Isman, 2004).

There are two key reasons why antifeedants have not proven successful for crop protection to date. First, insects quickly habituate to substances that deter feeding. The vast majority of scientific studies reporting antifeedant effects of plant compounds against insects base their conclusions on experiments where naïve insects are exposed a single time to the test compound. But, when the same insects are subsequently exposed to the same antifeedant, or continually exposed, the deterrent activity wanes. This has been demonstrated clearly using azadirachtin and the Asian armyworm, *Spodoptera litura* (Bomford and Isman, 1996). Translated to a field situation, an antifeedant would appear to lose efficacy rapidly within days, if not hours of application. This phenomenon of habituation in caterpillars has been shown to extend to a wide range of chemicals, as well as to complex mixtures (i.e. plant extracts) (Akhtar *et al.*, 2003). Second, insect species, even closely related ones, vary widely in their response to a specific antifeedant; a suite of pests on a crop cannot all be expected to respond to an antifeedant in the same manner. In contrast, most insecticides enjoy a relatively broad spectrum of activity against pests, enhancing their utility and value as crop protectants. This should not preclude the possible existence of a potent antifeedant effective against a key pest (e.g. Colorado potato beetle) with the potential for commercial exploitation. Insect pests with restricted host plant ranges such as the Colorado potato beetle (*Leptinotarsa decemlineata*) tend to show greater sensitivity to antifeedants derived from non-host plants. As the focal point for potato pest management, the market for a product that could deter feeding by this pest alone could be sufficient to warrant commercialization.

Repellents

By definition, a repellent is a substance or stimulus that causes an organism to make an oriented movement away from the source. The term 'insect repellent' has become synonymous with personal protectants against biting flies, mosquitoes and other blood-feeding insects and arthropods, even though it remains unclear as to whether most of these products actually 'repel' pests or simply mask the attractiveness of the host, resulting in non-acceptance by the pest. At present, DEET (*N,N*-diethyl *m*-toluamide) is the most widely used insect repellent, but biologists and chemists are actively engaged in finding safe and effective alternatives or replacements. Several plant oils or their constituents have been commercialized as insect 'repellents' in the past decade. These include oils obtained from soybean, lemongrass, cinnamon and the compounds 3,8-*p*-menthane-diol (from lemon eucalyptus), citronellal (from lemongrass) and 2-phenethylpropionate (from groundnut). All of these materials appear effective against mosquitoes based on short-term tests with humans, although their duration of effect has been the subject of some debate (Fradin and Day, 2002). Flea and tick control products for companion animals based on *d*-limonene, a constituent of citrus peel oil, or oils of peppermint, cinnamon, clove, thyme and lemongrass, have been introduced recently.

d-Limonene is heavily used as an alternative to conventional pesticides for structural pest control (i.e. termite management) in California, and other plant oils (clove, peppermint, etc.) are used in the USA by professional pest control operators as 'flushing agents' for cockroach control and for 'perimeter treatments' of homes against ants and termites, suggesting that repellence makes a strong contribution to the efficacy of these products. Repellent activity may also underlie the use of these oils in the long-term protection of foods and food products through their incorporation into packaging materials (Wong *et al.*, 2005).

Product Development and Commercialization

For any pesticide, the road from the laboratory bench to commercial application can be a long one, with many hurdles to clear along the route. Even a natural product with outstanding bioactivity against insects in the laboratory may not reach the marketplace unless a long list of practical criteria is met. For the most part, insecticidal plant natural products are considerably less efficacious overall than comparable conventional (synthetic) products. Therefore, proper formulation – providing good coverage and delivering the active ingredient to the target pest – can make the difference between product success and failure. Botanical insecticides have some properties not shared with conventional products, e.g. complex mixtures of active principles, geographic and temporal chemical variation. The different factors that influence the development and commercialization of botanicals for insect control are discussed in this section.

Resource availability

Production of a botanical insecticide requires continuous availability of raw plant material on a commercial scale. To achieve this scale requires intentional cultivation of the source plant, as is the case with pyrethrum, rotenone, neem and many of the plants from which essential oils are obtained. Alternatively, biomass that is a waste product of another industry, e.g. citrus peel (*Citrus x sinensis*) from orange juice production as the starting material for extraction of *d*-limonene (Fig. 20.1d), can be an economical and efficient source. Harvest of plant material from natural populations can be sufficient at the outset of product development, but if a botanical insecticide is successful in the marketplace on a greater than merely local scale, then this practice may not be sufficient to meet demand for long.

The Indian neem tree, originally limited to the Indian subcontinent, has been introduced to Africa, the Americas and Australia, to the extent that it is now almost pantropical. In Africa, it is used as a windbreak and as a source of firewood, among other uses, whereas neem plantations in Australia, Mexico and other countries were established in the 1990s primarily with a view to production of natural insecticides from the seed. The most spectacular success in the cultivation of plants for the production of botanical insecticides has been seen in Tasmania, Australia, which in little more than a decade has grown to become the world's largest single supplier of pyrethrum.

Processing and standardization

An important consideration in the production of botanical insecticides is the degree to which the raw plant material must be processed to concentrate the active principles, and the cost of that process. The relative high price for pyrethrum and neem result from both the cost of the raw commodity and that of processing required to produce a technical grade active ingredient for manufacture. Processing may be required not only to concentrate active ingredients, but also to enhance the stability of actives, extending the shelf life of the concentrate. The ability to store concentrates for long periods allows the production of the end-use products to be uncoupled in time and space from the cultivation and harvest of the raw plant starting material. A further advantage for a stable concentrate is that it allows batches of material to be blended as a means of achieving product consistency, facilitating quality control. Plant secondary chemistry is notorious for geographical and temporal variation. Botanical insecticides typically contain suites of closely related compounds, often with two or more key active principles, for example the two pyrethrin analogues in pyrethrum or the two major azadirachtin analogues in neem seed extracts. Stability of these actives, an analytical method for their quantification and the opportunity to mix batches of extracts can make it possible to produce a consistent product.

Toxicology and regulatory approval

With few exceptions, regulatory approval remains the most formidable barrier to the commercialization of botanical insecticides. In part, this situation is a consequence of a regulatory system that emerged and has evolved in response to the rise of synthetic agrochemicals following World War II. Assessing potential environmental and human health risks of pesticides based on laboratory animal tests has long been controversial, even when chemicals are considered on an individual basis. Add to this the chemical complexity and variability inherent in botanical insecticides and it is no wonder that many regulatory agencies are unwilling to relax their data requirements for these alternative products. In defence of those who assume the responsibility for protecting the environment and public health, it is clearly naïve to assume that natural products are likely to be safer than synthetic chemicals, as there are several examples of natural products that are quite toxic to humans and other mammals (e.g. nicotine, strychnine). One interesting exception to this is the exemption from registration afforded to specific plant materials by the US Environmental Protection Agency. This list includes several essential oils, or their major constituents, obtained from culinary herbs and thus deemed 'generally regarded as safe' based on the long-use history of these materials by humans. Some jurisdictions provide similar exemption for products not applied directly to food plants; however, many other jurisdictions fail to provide any distinction between natural products and synthetic chemicals.

Uses of Botanical Insecticides and Repellents

Prospects for developed countries

Worldwide, botanical insecticides likely command at most 1% of the US$700 million per annum insecticide market. The demand for these products is expected to grow most in developed countries (USA, Japan, EU members) where increasing public awareness of environmental sustainability and the linkage between chronic health effects and environmental pollutants is mirrored in the growth of demand for organic produce and 'natural' alternatives for urban pest problems. The demand for these alternative products (of which botanicals are an important group) is highest in situations where non-occupational human exposures are greatest – urban pest control in the home and in institutional/commercial settings, viz. schools, hospitals, restaurants, warehouses. For these markets, human safety (or at least the perception of safety) prevails over absolute efficacy, and premium prices can be commanded. This is also true in the case of pest control for companion animals (dogs, cats) and other highly valued animals (e.g. horses). Sales of organic products in the USA in 2005 represented nearly 2.5% of total food sales, which grew in real terms from US$4 billion to US$18 billion in the past decade. However, sales of botanical insecticides apparently have not tracked this spectacular growth in organic farming. Sales of botanical insecticides in California remained stable or even contracted between 2004 and 2007 (Cal DPR, 2007), consistent with trends

Table 20.2. Botanical insecticides used in California (USA), 2007 (data from Cal DPR, 2007).

Insecticide	Kg applied	Non-agricultural uses
d-Limonene	31,340	89% used for structural pest control
Pyrethrins	7,790	51% for public health, 24% for structural pest control
Rotenone	3,350	98% for fish management
Azadirachtin (neem)	1,010	Primarily for agriculture
Strychnine	480	Primarily for agriculture
Sabadilla	95	Primarily for agriculture

for other biopesicides (e.g. microbials such as *B. thuringiensis* and spinosad, both of which are approved for organic farming). Quantities of botanical insecticides applied, and their main uses in California in 2007, are shown in Table 20.2. In short, botanicals will continue to show growth in diverse niche markets, but are not expected to challenge conventional pesticides for the lucrative agricultural market in the near term.

Prospects for developing countries

Farmers in developing countries, especially in tropical and subtropical regions, stand to benefit the most from botanical insecticides (Isman, 2008). The majority of fatalities resulting from accidental poisonings attributed to synthetic insecticides (95% of 220,000 worldwide annually) occur in developing countries. This is a consequence of limited literacy and farmer training, and a lack of safety equipment. Crude botanical preparations may be far less efficacious than conventional pesticides, but the risk of poisoning among applicators and farm workers is generally far less as well. Due to high cost, most farmers in developing countries cannot afford conventional pesticides unless these receive heavy government subsidies. On the other hand, many such countries have long established traditions of using plant materials or crude plant extracts for pest management; the plants used are often readily available at no cost apart from the labour required for their collection and preparation. In some countries, the opportunity might even exist to develop local cottage industries focused on preparing insecticides based on endemic plants used by indigenous peoples. Finally, the regulation of pesticides, and its enforcement, is less strenuous in the poorest countries, at least in regard to domestic food production.

In conclusion, it is unrealistic to expect botanical insecticides to displace conventional (synthetic) insecticides for the protection of the world's major agronomic crops (cotton, maize, soybean, rice, oilseeds). Instead, botanical insecticides hold the most promise in the developed world in urban applications where a premium is placed on human and animal safety and in developing countries where they constitute an affordable tool for crop protection.

References

Akhtar, Y., Rankin, C.H. and Isman, M.B. (2003) Decreased response to feeding deterrents following prolonged exposure in the larvae of a generalist herbivore, *Trichoplusia ni* (Lepidoptera: Noctuidae). *Journal of Insect Behavior* 16, 811–831.

Alkofahi, A., Rupprecht, J.K., Anderson, J.E., McLaughlin, J.L., Mikolajczak, K.L. and Scott, B.A. (1989) Search for new pesticides from higher plants. In: Arnason, J.T., Philogene, B.J.R. and Morand, P. (eds) *Insecticides of Plant Origin*. American Chemical Society, Washington, DC, pp. 25–43.

Alvarez Colom, O., Neske, A., Popich, S. and Bardon, A. (2007) Toxic effects of annonaceous acetogenins from *Annona cherimolia* (Magnoliales: Annonaceae) on *Spodoptera frugiperda* (Lepidoptera: Noctuidae). *Journal of Pest Science* 80, 63–67.

Belmain, S. and Stevenson, P. (2001) Ethnobotanicals in Ghana: reviving and modernizing age-old farmer practice. *Pesticide Outlook* 12, 233–238.

Bomford, M.K. and Isman, M.B. (1996) Desensitization of fifth instar *Spodoptera litura* (Lepidoptera: Noctuidae) to azadirachtin and neem. *Entomologia experimentalis et applicata* 81, 307–313.

Bowers, W.S. (1982) Toxicology of the precocenes. In: Coats, J.R. (ed.) *Insecticide Mode of Action*. Academic Press, New York, pp. 403–427.

Cal DPR (California Department of Pesticide Regulation) (2007) Summary of pesticide use report data 2007, indexed by chemical (http://www.cdpr.ca.gov/docs/pur/pur07rep/chmrpt07.pdf, accessed 24 March 2009).

Debonsi, H.M., Miranda, J.E., Murata, A.T., Bortoli, S.A. de, Kato, M.J., Bolzani, V.S. and Furlan, M. (2009) Isobutyl amides – potent compounds for controlling *Diatraea saccharalis*. *Pest Management Science* 65, 47–51.

Fradin, M.S. and Day, J.F. (2002) Comparative efficacy of insect repellents against mosquito bites. *New England Journal of Medicine* 347, 13–18.

Isman, M.B. (1997) Neem insecticides. *Pesticide Outlook* 8, 32–38.

Isman, M.B. (2000) Plant essential oils for pest and disease management. *Crop Protection* 19, 603–608.

Isman, M.B. (2002) Insect antifeedants. *Pesticide Outlook* 13, 152–157.

Isman, M.B. (2004) Factors limiting commercial success of neem insecticides in North America and Western Europe. In: Koul, O. and Wahab, S. (eds) *Neem: Today and in the New Millennium*. Kluwer Academic, Dordrecht, The Netherlands, pp. 33–41.

Isman, M.B. (2005) Problems and opportunities for the commercialization of botanical insecticides. In: Regnault-Roger, C., Philogene, B.J.R. and Vincent, C. (eds) *Biopesticides of Plant Origin*. Lavoisier Tech et Doc., Paris, pp. 283–291.

Isman, M.B. (2006) The role of botanical insecticides, deterrents and repellents in modern agriculture and an increasingly regulated world. *Annual Review of Entomology* 51, 45–66.

Isman, M.B. (2008) Botanical insecticides: for richer, for poor. *Pest Management Science* 64, 8–11.

Isman, M.B. and Machial, C.M. (2006) Pesticides based on plant essential oils: from traditional practice to commercialization. In: Rai, M. and Carpinella, M.C. (eds) *Naturally Occurring Bioactive Compounds*. Elsevier BV, Amsterdam, pp. 29–44

Isman, M.B., Gunning, P.J. and Spollen, K.M. (1997) Tropical timber species as sources of botanical insecticides. In: Hedin, P.A., Hollingworth, R.M., Masler, E.P., Miyamoto, J. and Thompson, D.G. (eds) *Phytochemicals for Pest Control*. American Chemical Society, Washington, DC, pp. 27–37.

Jacobson, M. (1989) Botanical pesticides: past, present and future. In: Arnason, J.T., Philogene, B.J.R. and Morand, P. (eds) *Insecticides of Plant Origin*. American Chemical Society, Washington, DC, pp. 1–10.

Leatemia, J.A. and Isman, M.B. (2004) Efficacy of crude seed extracts of *Annona squamosa* against diamondback moth, *Plutella xylostella* L. in the greenhouse. *International Journal of Pest Management* 50, 129–133.

McLaughlin, J.L. (2008) Paw paw and cancer: annonaceous acetogenins from discovery to commercial products. *Journal of Natural Products* 71, 1311–1321.

Morse, S., Ward, A., McNamara, N. and Denholm, I. (2002) Exploring the factors that influence the uptake of botanical insecticides by farmers: a case study of tobacco-based products in Nigeria. *Experimental Agriculture* 38, 469–479.

Navickiene, H.M.D., Alecio, A.C., Kato, M.J., Bolzani, V.S., Young, M.C.M. and Cavalheiro, A.J. (2000) Antifungal amides from *Piper hispidum* and *Piper tuberculatum*. *Phytochemistry* 55, 621–626.

Satasook, C., Isman, M.B. and Wiriyachitra, P. (1993) Toxicity of rocaglamide, an insecticidal natural product, to the variegated cutworm, *Peridroma saucia* (Lepidoptera: Noctuidae). *Pesticide Science* 36, 53–58.

Satasook, C., Isman, M.B., Ishibashi, F., Medbury, S., Wiriyachitra, P. and Towers, G.H.N. (1994) Insecticidal bioactivity of crude extracts of *Aglaia* species (Meliaceae). *Biochemical Systematics and Ecology* 22, 121–127.

Schmutterer, H. (ed.) (2002) *The Neem Tree*, 2nd edn. Neem Foundation, Mumbai, India.

Schmutterer, H., Ermel, K. and Isman, M.B. (2002) The tiam, sentang or marrango tree: *Azadirachta excelsa* (Jack). In: Schmutterer, H. (ed.) *The Neem Tree*, 2nd edn. Neem Foundation, Mumbai, India, pp. 760–769.

Scott, I.M., Jensen, H., Scott, J.G., Isman, M.B., Arnason, J.T. and Philogene, B.J.R. (2003) Botanical insecticides for controlling agricultural pests: piperamides and the Colorado potato beetle *Leptinotarsa decemlineata* Say (Coleoptera: Chrysomelidae). *Archives of Insect Biochemistry and Physiology* 54, 212–225.

Toloza, A.C., Vassena, C. and Picollo, M.I. (2008) Ovicidal and adulticidal effects of monoterpenoids against permethrin-resistant human head lice, *Pediculus humanus capitis. Medical and Veterinary Entomology* 22, 335–339.

Trumble, J.T. (2002) Caveat emptor: safety considerations for natural products used in arthropod control. *American Entomologist* 48, 7–13.

Wheeler, D.A. and Isman, M.B. (2001) Antifeedant and toxic activity of *Trichilia americana* extract against the larvae of *Spodoptera litura. Entomologia experimentalis et applicata* 98, 9–16.

Wong, K.K.Y., Signal, F.A., Campion, S.H. and Motion, R.L. (2005) Citronella as an insect repellent in food packaging. *Journal of Agricultural and Food Chemistry* 53, 4633–4636.

Zhong, G., Liu, J., Weng, Q., Hu, M. and Luo, J. (2006) Laboratory and field evaluations of rhodojaponin-III against the imported cabbage worm *Pieris rapae* (L.) (Lepidoptera: Pieridae). *Pest Management Science* 62, 976–981.

21

Principles of Plant-based Remediation of Contaminated Soils

Peter M. Kopittke, Enzo Lombi, Neal W. Menzies and Ravi Naidu

Introduction – Nature and Magnitude of the Problem

Soil contamination is defined as the accumulation of hazardous compounds within the soil environment. While human activities have contaminated the environment since the beginning of civilization, the rate of this process has escalated dramatically since the start of the Industrial Revolution. As a result, enormous financial investments are now required in order to rehabilitate the soil to a productive and non-environmentally-damaging endpoint. To this end, an ever-increasing number of technologies have been developed (Lombi and Hamon, 2004). Among these technologies, plant-based approaches have attracted attention for several years in view of their potential benefits in terms of cost, ability to improve soil conditions and public acceptance. This chapter reviews their application, potential and drawbacks in the context of various contamination scenarios.

The contamination of soil represents a serious environmental and economic challenge. Although the magnitude of the soil contamination problem is difficult to quantify accurately, it is estimated that, globally, hundreds of thousands of sites are contaminated and require remediation. While emerging economies in South America and Asia (such as China) are expected to increase demand in the medium term, current demand is driven by the USA, which remains the largest market for soil remediation technologies. Indeed, the USA contains an estimated 500,000 contaminated sites, with the cost of remediation currently valued at US$10 billion/year, increasing to approximately US$650 billion over the next 30–35 years (CEI, 2005). Demand for remediation technologies is also high in Europe, with 250,000 sites identified as requiring clean-up; a number which is expected to increase a further 50% by 2025 (EEA, 2007). Current expenditure on the management of these sites in Europe corresponds to an average of 0.07% of the national GDP, of which an average of 35% comes from public budgets (EEA, 2007).

A variety of compounds may accumulate as contaminants in soil; these may be either inorganic (including trace elements and nutrient contaminants) or organic (including petroleum hydrocarbons, PCBs and pesticides). The contribution of different types of contaminants to the total global contaminant load remains difficult to quantify. In Europe, the main soil contaminants (estimated as the frequency that the contaminant was identified as the most important contaminant on a site) are heavy metals (37%), mineral oil (34%), polycyclic aromatic hydrocarbons (PAH, 13%) and aromatic hydrocarbons (6%) (EEA, 2007).

Of the inorganic contaminants, trace elements are generally of the greatest concern and include both trace metals (such as Ni, Cu, Pb and Cd) and metalloids (such as As). Contamination of soil with trace elements may occur through a variety of activities, including the use of roads, mining, the use of fertilizers, metal processing and the disposal of biosolids and industrial wastes to land. In contrast to organic contaminants, trace elements do not undergo microbial or chemical degradation. As a result, trace elements (and their toxic effects) persist in the environment indefinitely. Remediation of sites contaminated by trace elements typically involves either the removal of the contaminant (for example, excavation, leaching, or phytoextraction) or stabilization of the contaminant (for example, solidification or phytostabilization). In addition to trace elements, the application of nutrients (and in particular, N and P) to the soil in excess of the rate of nutrient removal can have detrimental impacts on both the soil environment and surrounding water bodies. The movement of nutrients into waterways is concerning due to their adverse impacts on the health of both humans (for example, causing methemoglobinaemia) and the ecosystem. The major causes of excessive nutrient application to soil are wastewater effluent, septic systems, runoff/infiltration from animal feedlots and agricultural fertilization. Generally, nutrient contamination (eutrophication) is 'remediated' by reducing nutrient inputs (for example, optimizing fertilizer inputs).

Petroleum hydrocarbons are the most commonly encountered organic contaminants in urban areas and may be released from a variety of sources including distribution depots, oil refineries, gasworks and waste processing and disposal facilities. In addition to the techniques used for inorganic contaminants, organic compounds can potentially be remediated using volatilization and biodegradation. However, the behaviour of hydrocarbon contaminants in soil depends on both their structure (for example, aliphatic or aromatic) and size. Pesticides are also important organic contaminants and have been used in agricultural systems for centuries. While pesticides include both inorganic salts (such as salts of lead and copper) and synthetic organic compounds, the latter dominate current usage. Once within the soil environment, pesticides may be retained (adsorbed by the soil particles), transformed by abiotic or biochemical processes, or transported to the atmosphere or water bodies.

The release of contaminants to soil may occur through a variety of activities. In urban areas, industrial activities and municipal wastes are the major causes of soil contamination, while in rural areas, the use of agrochemicals and biosolids can result in widespread contamination across the broader landscape. Although mining activities generally affect only a small proportion of the total landscape, they can potentially have comparatively large environmental impacts. In Europe, the EEA (2007) identified the main sources of soil contamination (estimated as the proportion of sites where preliminary investigations had been completed) as being industrial production (41%), municipal waste treatment (15%), oil industry (14%) and industrial waste treatment and disposal (7.3%).

Plant-based Soil Remediation Approaches

The term 'phytoremediation' encompasses all plant-based techniques that are used to remediate environmental problems. For the remediation of soil contamination, a number of plant-based approaches have been developed and are commonly divided into four categories: (i) phytoextraction, (ii) phytostabilization, (iii) phytovolatilization and (iv) plant-assisted biodegradation (sometimes termed phytodegradation).

When considering the suitability of plant-based approaches for the remediation of contaminated sites, it is first necessary to consider the management and environmental issues associated with the site (Pierzynski *et al.*, 2002). With the exception of phytostabilization, all of the approaches listed above are clean-up technologies. Phytostabilization, as discussed below, is a containment approach that does not aim to reduce the contaminant concentration in the soil, but rather, the availability and movement of contaminants. This aspect has to be considered in terms of the legislative framework in which a remediation effort is conducted. Phytostabilization, even if possibly the most promising and cost-effective approach, may not comply with the requirements of legislation based purely on total contaminant concentrations. Furthermore, given that plant-based approaches take longer to remediate a site than conventional systems (see below), the time frame available for treatment should also be identified. Consideration should also be given to the future land use, ecological risk issues (including the potential movement of contaminants to receptor organisms) and the likelihood that the selected approach will succeed (given the site characteristics) and achieve the desired results (Pierzynski *et al.*, 2002).

Phytoextraction

Phytoextraction refers to the use of plants to extract metals from the soil. The metals which are accumulated in the harvestable parts of the plants are therefore removed at harvest. Two different approaches have been pursued to this end: natural hyperaccumulation and chemically enhanced phytoextraction.

Hyperaccumulator plants

Hyperaccumulation refers to the use of plants naturally able to accumulate large amounts of metals in their shoots. According to McGrath *et al.* (2002), hyperaccumulator plants are those that:

- accumulate 100- to 1000-fold more metals than normal plants growing on soils with background metal concentrations;
- are characterized by an enhanced root–shoot translocation so that shoot–root metal concentration ratios are > 1; and
- are hypertolerant to the metal they accumulate.

These simple characteristics represent a robust framework for the identification of hyperaccumulator plants. This is necessary because several studies have claimed hyperaccumulation simply on the basis of total metal concentrations in plants or as a result of short-term hydroponic studies where metal concentrations used were not environmentally relevant and tolerance was not assessed.

To date, over 400 species have been identified as hyperaccumulators, with this number increasing continuously as new species are identified (Baker *et al.*, 2000). For instance, no As hyperaccumulator plants were reported prior to 2001 when *Pteris vittata* L. (Chinese brake fern) was first reported to accumulate remarkable concentrations of this metalloid (Ma *et al.*, 2001). Since then, the list of As hyperaccumulators has increased substantially, with several other ferns having been reported to share the same characteristics. However, most known hyperaccumulators are Ni accumulators, with some reported to accumulate other metals such as Zn and Cd. The situation remains uncertain in relation to the hyperaccumulation of Cu and Pb. While Brooks (1998) reported 37 taxa of Cu hyperaccumulators, mainly from central Africa, convincing evidence for Cu hyperaccumulation (that is, evidence which satisfies the requirements listed above) is still lacking (McGrath *et al.*, 2002). For example, *Elsholtzia splendens* Nakai ex F. Maek. has been reported by several authors to be a Cu hyperaccumulator. However, Song *et al.* (2004) demonstrated that this species was not a hyperaccumulator, but rather, a tolerant excluder. Similarly, while several authors claimed to have identified Pb hyperaccumulators (based on the total concentrations of Pb in the shoot), a critical review of the published data revealed that the majority of these studies were either conducted in hydroponic conditions, or reported shoot Pb concentrations much lower than those in the soil. For instance, several authors have reported *Sedum alfredii* as a Pb hyperaccumulator simply on the basis of Pb concentrations (for example, Liu *et al.*, 2007, and Grubor, 2008). However, Long *et al.* (2009) demonstrated that, when growing in contaminated soils, Pb was retained in roots, thus indicating exclusion as a tolerance strategy for Pb. Therefore elevated concentrations of a metal in plant tissues do not necessarily indicate hyperaccumulation. Although Arshad *et al.* (2008) reported large concentrations (over

100 mg/kg) of Pb in *Pelargonium* spp., these results were obtained when the plant was grown in a soil containing almost 40,000 mg Pb/kg. Thus, in this case, phytoextraction would be quite unreasonable as the volume of Pb-contaminated plant biomass would exceed the mass of the soil.

In the case of Pb (and other elements strongly partitioning on the soil solid phase), a specific problem in terms of phytoextraction relates to the fact that this process is bioavailability limited. Thus, for this reason, the concept of chemically enhanced phytoextraction was introduced in the late 1990s.

Chemically enhanced phytoextraction

Chemically enhanced phytoextraction was introduced by Blaylock *et al.* (1997) to overcome the limitations of phytoextraction by natural hyperaccumulators. These limitations are related to (i) the typically small biomass of hyperaccumulator plants (although exceptions do exist); (ii) the unavailability of natural hyperaccumulators for important contaminants such as Cu and Pb; and (iii) the low phytoavailability of some contaminants (such as Pb), all of which result in a lower rate of uptake and accumulation in plants. With this approach, the bioavailability limitation is overcome by the application of a mobilizing agent and the use of high biomass plants to accumulate the mobilized metal. This also offers the advantage that established agricultural practices can be used, as the plants employed are common crops such as canola (*Brassica napus*) or maize (*Zea mays*). Furthermore, the application of some mobilizing agents, such as EDTA (ethylenedinitrilotetraacetic acid), has been found to enhance the root–shoot translocation of metals, thereby favouring the accumulation of the contaminants in the harvestable part of the plant.

In this approach, a high biomass-yielding crop is grown in the contaminated soils and when the plant reaches maximum vegetative growth, a chemical agent (which is able to mobilize the metal of interest) is added to the soil, thereby increasing the concentration of the metal rapidly in soil pore water. Concomitantly, the root system is damaged by the mobilizing agent and/or the metal, with a resulting lessening of the plant's physiological barriers responsible for metal exclusion. To this extent, the use of herbicides or other compounds (such as surfactants) which are able to reduce the integrity of the root plasma membranes has also been proposed (for example, see Maier *et al.*, 2001). Thereafter, the plant simply acts as a 'straw', whereby the mobilized metal is transported to the shoot through the transpiration stream. As a consequence of the damage to the root system and the accumulation of the toxic metals, the plant is often significantly damaged and is therefore harvested soon after the application of the chemical to the soil.

Following the pioneering study of Blaylock *et al.* (1997), a significant body of work has been conducted with a review of the current literature on the subject approaching 250 articles in international journals. The performance of different plant species in chemically assisted phytoextraction has varied considerably as a function of the contamination scenario, the metal involved and the mobilizing agent employed. Ideally, plants utilized for this approach should have fast and vigorous growth, a high biomass production, an extensive root system and a tolerance to metals. Saifullah *et al.* (2009) recently reviewed the literature for EDTA-aided Pb phytoextraction and suggested that monocotyledonous plants were generally more tolerant of metals and were therefore well suited for this approach. However, dicotyledonous plants tend to accumulate larger amounts of metals in their shoots, probably because of the more pronounced damage of EDTA to the plant's physiological barriers, which would normally restrict the uptake of toxic metals.

In addition to EDTA (which remains the most studied mobilizing agent for chemically assisted phytoextraction), a series of other chemical compounds has been examined. Among these, chelating agents (such as DTPA, CDTA, NTA, HBED and EDDS) have been studied extensively for their ability to form stable complexes with metals. Various drivers have underpinned this research. First of all, EDTA is not ideal as a mobilizing agent as it is has a low specificity for many of the target metals. For instance, Wu *et al.* (1999) stated that Fe(III) might outcompete Pb for chelation by EDTA. Furthermore, given that the metal–EDTA complexes are generally negatively charged (and hence are not attracted to the soil's cation exchange capacity), they are prone to leaching. This problem is compounded by the high stability and slow degradation of the metal–EDTA complexes (Lombi *et al.*, 2001; Meers *et al.*, 2005b). The mobility and the toxicity of the mobilized metals (and the chelating agent itself) have been reported to persist for several months, with the risk that successive crops can experience phytotoxicity and that leaching may occur. Several authors have reported that the increase in metal solubility caused by EDTA is much larger than the increase in metal uptake (Lombi *et al.*, 2001; Schmidt, 2003; Meers *et al.*, 2005a). In some instances, only approximately 1% of the mobilized Pb is taken up by the plant, with the remainder potentially available for leaching. These issues have prompted warning of the danger of using EDTA in chemically assisted phytoextraction (Saifullah *et al.*, 2009). This warning can be extended to include all chelating agents that are recalcitrant in soil and maintain large concentrations of toxic metals in soil pore water for extended periods of time. As a consequence, substantial effort has been directed toward the study of less persistent mobilizing agents, ranging from low-molecular weight organic acids to EDDS and other comparatively easily degradable aminopolycarboxylic acids. The results obtained thus far, however, have shown that while able to reduce the risk of metal leaching, these mobilizing compounds do not perform as well as EDTA in terms of enhancing metal accumulation (Evangelou *et al.*, 2008).

Feasibility of phytoextraction

As mentioned above, the feasibility of phytoextraction is conditioned by a series of issues. Clearly, as it is based on the use of plants to remediate contaminated soils, this approach is potentially only applicable under conditions that allow a plant cover to be established, and it is limited to situations in which the contaminants are in the rooting zone of the plants. In cases where the contaminants are likely to be toxic, this approach also requires plants to be tolerant to all contaminants present. For instance, *Thlaspi caerulescens* can hyperaccumulate and is hypertolerant to Cd and Zn, but is as sensitive as other species to elevated concentrations of Cu (Lombi *et al.*, 2001). Furthermore, contaminated soils are often unbalanced in terms of pH, nutrients and water availability. These conditions need to be improved to enhance plant growth conditions if phytoextraction is to be successful.

Another limitation to phytoextraction using hyperaccumulator plants is given by the challenge of growing these plants successfully. Many hyperaccumulating species are slow growing and compete only poorly with weeds under non-phytotoxic conditions. In addition, the seeds of these plants generally are obtained from wild populations and exhibit large differences in terms of growing patterns and plant size (Schwartz *et al.*, 2006). Therefore, a concerted effort is required to select homogeneous (and well-performing) lines and to develop appropriate agronomic practices for the farming of hyperaccumulator plants. Progress has already been made toward the development of commercially viable phytoextraction of Ni by *Alyssum* spp. Chaney *et al.* (2007) reviewed the literature on this matter, covering the optimization of soil factors such as pH and fertilization, as well as other agronomic practices (such as the selection of agronomically improved lines, weed control, plant density and harvesting).

Even though a considerable body of literature is available on phytoextraction, almost all of these studies have been conducted either in the laboratory or in the glasshouse; very few studies have reported field trials. Among these, perhaps the most complete is that of McGrath *et al.* (2006), which examines several years of field trials. This work focused on phytoextraction of Zn and Cd by *T. caerulescens* and *Arabidopsis halleri* L. over 3 years. In an extended growing season (14 months), *T. caerulescens* extracted up to 21% of the total Cd and 4% of the total Zn. This study suggests that if growth is optimized, phytoextraction could potentially be feasible for the clean-up of moderately Cd-contaminated land.

Several authors have reported theoretical calculations based on known accumulation rates and assumed plant yields. The following example, based on a high-biomass plant and a low-biomass hyperaccumulator plant, can be used to assess the conditions under which phytoextraction could potentially represent a feasible option. Given that no convincing alternatives to EDTA have been reported (and this chelating agent has been considered too risky due to its leaching potential), chemically enhanced phytoextraction has not been considered in this example. The assumptions of the example are as follows:

- Initial metal concentration in soil (mg/kg): 10 Cd, 1500 Zn
- Target metal concentrations (mg/kg): 3 Cd, 300 Zn
- Metal concentration in plants (mg/kg): *T. caerulescens* Zn 10,000, Cd 700; maize Zn 1000, Cd 10
- Plant yield (t/ha/year): *T. caerulescens* 5, maize 25
- Soil density (g/cm^3): 1.3
- Soil depth (m): 0.2
- Plant metal concentrations are considered constant (i.e. independent from soil metal concentrations).

The results indicate that by using maize, the clean-up goals will be achieved in 125 years for Zn and in 73 years for Cd. The time required for phytoremediation decreases with *T. caerulescens* to 62 years for Zn and 5 years for Cd. This simple calculation is sufficient to identify some key issues in phytoextraction. Firstly, a small biomass is not necessarily a limiting factor if the concentration of metals in the plant is sufficiently large. McGrath *et al.* (2006) reported an average biomass of *T. caerulescens* in the field of approximately 1 t/ha, while Robinson *et al.* (1998) estimated a biomass of 2.6 t/ha for natural populations in southern France. Given that these yields were achieved without agronomic optimization or selection of seed lines with increasing performance, the 5 t/ha used in the example above is not unreasonable. In other words, the large biomass of conventional crops does not offset the difference in metal concentrations between non-hyperaccumulator and hyperaccumulator plants. Secondly, phytoextraction is likely to be feasible only when the contamination is not severe and the time available for clean-up is in the order of several years.

The main criticism of such calculations relates to uncertainties in the metal concentration that can be expected in the plant, as this is likely to change over time as the metal of interest is removed from the soil. To overcome this issue, Zhao *et al.* (2003) compiled a number of studies conducted using *T. caerulescens* to derive a relationship between metal concentration in the soil and in the plant. Using this relationship, they developed a simple model able to describe the phytoextraction potential of this plant. Their results also suggested that phytoextraction of Cd from moderately contaminated soils could be feasible.

Phytostabilization: *In situ* Immobilization and Revegetation

The accumulation of contaminants in soil can lead to reduced plant growth and, in some cases, to complete death of the vegetative cover. Apart from a direct impact on the immediate environment, these bare soils pose a danger to surrounding human populations and ecosystems due to contaminant transfer, particularly through dust inhalation and ingestion. In such circumstances, it is desirable to lower the mobility of the contaminants and form a permanent vegetative cover to stabilize the contaminants *in situ*. The use of plants to stabilize a site in order to minimize the movement of contaminants is termed 'phytostabilization'. Phytostabilization aims to reduce contaminant mobility by limiting leaching, reducing erosion, decreasing contaminant bioavailability in the food chain and improving the site's aesthetics. The goal of phytostabilization is to have a long-term succession of the plant community, promote soil development processes, microbial diversity and, finally, to restore soil ecosystem functions to a state of self-sustainability (Mendez and Maier, 2008).

In contrast to phytoextraction, phytostabilization focuses on sequestering pollutants in soil rather than plant tissues. As phytoextraction aims to maximize the movement of the contaminant into the shoot, plants selected for phytostabilization should minimize their shoot accumulation in order to reduce the risk of transferring the contaminant into fauna that consume the plants. Generally, the term 'phytostabilization' encompasses both the use of amendments to immobilize the contaminant (*in situ* immobilization) and the use of plants to stabilize the soil, both of which are considered here. For heavily contaminated areas, *in situ* immobilization and revegetation can be an effective and cost-efficient remediation technique. The application of amendments to stabilize the contaminant often reduces the mobility and bioavailability of the contaminant, thereby reducing its movement through the food chain. Furthermore, the subsequent revegetation of the site stabilizes the soil, reduces erosion (both wind and water) and improves aesthetics.

Phytostabilization (revegetation) is a widely used and accepted method for the remediation of disturbed sites. Compared to other approaches, phytostabilization has several advantages. In addition to stabilizing the soil and reducing erosion, the vegetation reduces deep drainage (by increasing transpiration), thereby limiting leaching of contaminants into groundwater. Revegetation also potentially provides a permanent, self-sustaining remediation option. Phytostabilization is aesthetically pleasing (a successfully remediated site contains a 'green cover'), requires little maintenance once established and is a 'green technology', which generally has public acceptance. Furthermore, the successful use of a plant-based system is often substantially cheaper than traditional techniques (Pierzynski *et al.*, 2002). However, the process of remediating a site using phytostabilization is comparatively slow. Furthermore, given that the contaminant remains on site, long-term monitoring is required to ensure the effectiveness of remediation.

While phytostabilization may be used to remediate any contaminated site, it is most commonly used on sites of comparatively low economic value where it is not economically feasible or practical to remove/immobilize the contaminant using more expensive techniques (such as excavation, extraction, or thermal treatment). In addition to the broader considerations required for plant-based remediation systems, there are two key factors to consider when examining phytostabilization: the *in situ* immobilization of contaminants and the selection of plant species. Given that phytostabilization is most commonly used on sites contaminated by trace metals (rather than organic contaminants), the following discussion will focus primarily on metal-contaminated sites.

In situ immobilization of contaminants

The aim of *in situ* immobilization is to reduce contaminant solubility and hence reduce the likelihood that it will be transferred through the food chain by being (i) taken up by plants, (ii) leached into the groundwater and/or (iii) available to soil organisms. Studies have used a range of parameters to measure the effectiveness of immobilizing amendments, including changes in soil fractionation (including the water-soluble, exchangeable, organically bound and residual fractions), tissue concentrations in plants, plant yield/biomass and ecotoxicological assays.

The effectiveness of any given immobilizing amendment will depend on both the properties of the contaminant and the properties of the soil. Hence, careful consideration of the underlying chemical process occurring in contaminated soil is required to ensure that the desired outcome is achieved. Some amendments immobilize contaminants through multiple processes and thus may be more effective than other compounds which act only through a single mechanism. For example, the incorporation of red mud (a highly alkaline waste material) into a contaminated soil is more effective at reducing acid extractability of metals than lime ($CaCO_3$), as red mud not only increases pH (as does the lime) but also increases the capacity of the soil to adsorb metals specifically due to its high sesquioxide content (Lombi *et al.*, 2002). However, care must be taken to ensure that the addition of amendments does not have negative impacts on contaminant mobility/availability. For example, while the addition of hydroxyapatite ($Ca_5(PO_4)_3OH$) (up to 5%

w/w) was found to reduce solubility of Al, Ba, Cd, Co, Mn, Ni, Pb and U, its application increased the solubility of oxyanion contaminants (As and Cr) simultaneously due to increased competition from phosphate for specific adsorption sites (Seaman *et al.*, 2001). Similarly, the addition of organic amendments may increase the mobility of some contaminants due to increased competition for specific adsorption sites or due to increased complexation with dissolved organic matter (DOM), which thereby increases the soluble metal concentration.

In situ immobilization is most suited to moderately contaminated soils, as the bioavailability of the contaminant is reduced, but the total concentration is not altered. For highly contaminated materials, *in situ* immobilization and revegetation of the site may be insufficient to reduce the risk of off-site movement of material by leaching and erosion, or the incidental, or deliberate, ingestion of the contaminated soil by animals (Beyer *et al.*, 2007). For highly contaminated materials, capping of the site with non-contaminated material may be the most appropriate remediation (Garbaciak *et al.*, 1997; Ashley *et al.*, 2004; Farmer *et al.*, 2006). In designing vegetated capping systems, consideration must be given to the supply of nutrients and water to the vegetation (Wehr *et al.*, 2005) and to the isolation of phytotoxic contaminants below the rooting zone of the plants (Menzies and Mulligan, 2000).

pH (liming)

Lime ($CaCO_3$) has been used in agricultural systems for centuries and remains one of the most commonly used soil amendments. Compared to other *in situ* immobilization methods, the application of lime (typically at a rate of 1–10 t/ha) is comparatively cheap and effective. As well as reducing contaminant solubility, the addition of lime may improve plant growth (and hence promote stabilization of the site) by overcoming the limitations imposed on plant growth in acidic soils, such as Al toxicity or Ca deficiency. Repeated applications of lime may be necessary as lime may show a decrease in effectiveness with time (Chlopecka and Adriano, 1996). Consideration should also be given to the liming material selected; $Ca(OH)_2$ has been reported to be less effective than KOH in immobilizing Cd due to competition between Ca and Cd for binding sites (Bolan *et al.*, 2003a).

In contaminated soils, the influence of pH on metal solubility will depend on the species (form) in which the metal is present. For cationic trace metals (such as Cu^{2+}, Pb^{2+}, Zn^{2+}, Ni^{2+} and Cd^{2+}), an increase in pH generally reduces the solubility of metal contaminants due to precipitation of insoluble metal hydroxides (Lindsay, 1979), co-precipitation with carbonates and increased specific adsorption (for soils containing variable charge minerals [see below]). For example, Lombi *et al.* (2002) investigated the influence of liming on the fractionation of Cd, Cu, Ni, Pb and Zn in contaminated soils from France (pH 5.5, contaminated by industrial activities) and the UK (pH 6.4, contaminated by sewage sludge applications). The authors reported that the addition of lime (at a rate sufficient to increase pH to *c.* 7.5) resulted in significant decreases in the soluble and exchangeable metals and significant increases in both the carbonate-bound fraction and the Fe- and Mn-oxide-bound fraction (i.e. specifically adsorbed metals). Generally, a decrease in the size of the labile pool of metals (such as that reported by Lombi *et al.*, 2002) reduces plant uptake. Using soil from a decommissioned Zn/Pb smelter in the UK, Gray *et al.* (2006) reported that the addition of lime significantly decreased soluble concentrations of Zn, Cd, Pb, Ni and Cu, decreased metal concentrations in the tissues of *Festuca rubra* L. and improved plant biomass yield. Similarly, the addition of lime to an acidic sandy soil (pH 4.3) reduced concentrations of $Ca(NO_3)_2$-extractable Pb, reduced shoot tissue Pb concentrations and reduced the phototoxic effects of Pb in lettuce (Geebelen *et al.*, 2002).

While an increase in soil pH may reduce metal solubility, some studies have reported that reduced solubility may not necessarily result in reduced toxicity. For example, using data from more than 500 chronic toxicity tests in experimentally aged soils and in field contaminated soils, Smolders *et al.* (2009) reported that soil pH was generally a good predictor of metal solubility but a poor predictor of metal toxicity across soils. This apparent discrepancy possibly reflects the formation of organo-metal complexes at near-neutral and alkaline pH values. Sauve *et al.* (1998) found that while the solubility of Pb decreased linearly from pH 3 to 6.5 (and was independent of soil organic matter within that pH range), from pH 6.5 to 8.0, the higher pH promoted the formation and dissolution of organo-Pb complexes, thereby increasing Pb solubility.

For anionic trace metals (such as arsenate [AsO_4^{3-}] and chromate [CrO_4^{2-}]), their retention in soil is controlled predominantly by variable charge minerals (through specific adsorption) and organic matter (through complexation/chelation). While for cationic metals, an increase in pH is expected to increase specific adsorption and for anionic metals, an increase in pH is expected to increase solubility due to a decrease in specific adsorption (see 'specific adsorption' below). In addition, an increase in pH may increase soluble As concentrations due to a pH-induced mobilization of humic acids (Warwick *et al.*, 2005; Klitzke and Lang, 2009).

Precipitation

The addition of some amendments to soil results in the precipitation of the contaminant due to the formation of low-solubility compounds. The main compounds which have been investigated are P-containing minerals, due to their ability to react with a range of trace metals, metalloids and radionuclides to form secondary precipitates that are stable over a wide range of environmental conditions. Phosphate minerals decrease solubility of trace metals due to precipitation and/or sorption. Hydroxyapatite ($Ca_5(PO_4)_3OH$) is the most commonly used P compound as it is comparatively abundant and inexpensive, although studies have also investigated the use of other compounds such as superphosphate fertilizers. Most studies investigating P compounds have tended to focus on Pb-contaminated soils due to the very low solubility of Pb–orthophosphate complexes. Pyromorphites (such as $Pb_5(PO_4)_3Cl$, $Pb_5(PO_4)_3OH$ and $Pb_5(PO_4)_3F$) are the most stable Pb-phosphate minerals which form under normal environmental conditions (Ryan *et al.*, 2001), with chloropyromorphite ($Pb_5(PO_4)_3Cl$) the least soluble of the common pyromorphites (log K: $Pb_5(PO_4)_3Cl$ = −84.4, $Pb_5(PO_4)_3OH$ = −62.8 and $Pb_3(PO_4)_2$ = −44.5; Parkhurst, 2006).

In a study using a Pb-contaminated soil from adjacent to a smelter, Ryan *et al.* (2001) reported that the addition of hydroxyapatite resulted in a decrease in the exchangeable, carbonate, Fe/Mn-associated and organic/sulfide-associated Pb fractions and an increase in the residual Pb fraction due to the precipitation of chloropyromorphite. Similarly, in a long-term field trial, P amendments were found to increase the residual Pb fraction due to precipitation of pyromorphite-like minerals (Chen *et al.*, 2003). This P-induced increase in residual Pb typically results in a decrease in the phytoavailability of Pb. Using soil contaminated from a smelter, Zhu *et al.* (2004) reported that the addition of P-containing compounds (hydroxyapatite, phosphate rock and single superphosphate) at rates of up to 0.5% w/w decreased concentrations of Pb in the shoot and root tissue by up to *c.* 70%, due to a decrease in the more labile Pb fractions (water soluble, exchangeable, carbonate, Fe-Mn oxide and organic-bound), and an increase in the residual Pb fraction. Some studies have demonstrated that the *in situ* immobilization of Pb is enhanced by pre-acidification of the soil, in order to increase the solubility of the Pb and thereby enhance its ability to react with the added P. Hettiarachchi *et al.* (2001) reported that the pre-acidification of a Pb-contaminated soil significantly lowered bioavailable Pb following the addition of P due to the increased precipitation of hydroxypyromorphite ($Pb_5(PO_4)_3OH$).

While Pb is the most commonly studied contaminant for precipitation using phosphates, other metals, such as Cd and Zn, can also form highly insoluble phosphates (Nriagu, 1974). Application of apatite (0.4% w/w) has been reported to decrease exchangeable concentrations of Cd and Pb, and thereby reduce plant uptake (Chlopecka and Adriano, 1997). Similarly, the application of hydroxyapatite (up to 5% w/w) reduced solubility of Al, Ba, Cd, Co, Mn, Ni, Pb and U due to transfer from labile forms (in particular, water soluble and exchangeable fractions) to secondary phosphate precipitates (Seaman *et al.*, 2001). However, these authors also noted that the application of P increased the solubility of oxyanion contaminants (As and Cr) due to increased competition from phosphate for specific adsorption sites (see below). Similarly, while hydroxyapatite reduced availability of Cd, Cu, Ni, Pb and Zn, it increased availability of As due to phosphate–arsenate competition (Boisson *et al.*, 1999b). Therefore, care must be taken when applying P compounds to ensure that the mobility of other contaminants is not increased. In addition, high rates of hydroxyapatite application may reduce plant growth by decreasing the availability of essential nutrients such as Mg and Mn (Boisson *et al.*, 1999a).

Ion exchange

The application of compounds with an ion exchange capacity (typically a cation exchange capacity, CEC) may reduce contaminant uptake by plants by reducing their concentration in the soil solution. In their review of 500 chronic toxicity tests, Smolders *et al.* (2009) reported that the toxicity threshold (based on total soil metal concentration) rose proportionally to the charge of the soil, thereby demonstrating the importance of ion exchange capacity. Several classes of compounds have been suggested for increasing the soil CEC, including zeolites and other clay minerals such as montmorillonite.

Zeolites are crystalline, hydrated aluminosilicates of alkali and alkaline earth cations that possess an infinite three-dimensional crystal structure. More than 170 zeolites have been identified, including more than 40 naturally occurring species. The permanent charge of zeolites (often in the order of 100–500 $cmol_{(c)}$/kg) results from the isomorphous substitution of Al^{3+} for Si^{4+} in the silicate tetrahedral structure. Zeolites are also known as 'molecular sieves' due to their ability to sort molecules selectively based on their size (which also gives rise to 'selective' cation exchange). This property arises from their internal structure, which consists of a regular pore structure of specific dimensions. The ability of zeolites to reduce soluble concentrations of metals has been demonstrated in both the glasshouse and field. Indeed, zeolites have a high affinity for many metals, including Pb, Cd, Zn and Cu (Ouki and Kavannagh, 1997). Castaldi *et al.* (2005) showed that the addition of zeolite to a contaminated soil generally resulted in a decrease in the soluble, $Ca(NO_3)_2$-extractable and EDTA-extractable fractions of Pb, Zn and Cd, which resulted in decreased metal uptake in white lupin (*Lupinus albus*) and increased shoot and root growth. Similarly, Chlopecka and Adriano (1996) reported that the addition of zeolite (1.5% w/w) reduced the exchangeable Zn concentration, reduced the shoot Zn concentration and improved plant growth and yield. However, while zeolites may reduce the free ionic metal concentration, the increase in pH associated with the addition of the zeolite may result in an increase in metal leaching due to an increase in the DOM fraction. Oste *et al.* (2002) reported that while the addition of zeolites decreased free ionic Cd and Zn substantially, the increase in soil pH (from pH 6 to 8) resulted in an increase in the dissolved organic matter concentration and an increased leaching of metal–DOM complexes. Furthermore, in some instances, despite a reduction in the soluble metal concentration, the addition of zeolite (0.5% w/w) has been reported to reduce plant growth due to adverse effects on soil structure (Geebelen *et al.*, 2002).

Clay minerals (and in particular, smectitic minerals such as montmorillonite) have also been investigated for their ability to reduce soil solution concentrations of trace metals. The addition of 1% w/w bentonite (montmorillonite) was found to reduce $Ca(NO_3)_2$-extractable Pb significantly, although it was less effective than zeolite (0.5% w/w) and lime (1% w/w) (Geebelen *et al.*, 2002). Similarly, the addition of bentonite to a soil amended with sewage sludge reduced the water-extractable (14–75% reduction) and exchangeable (12–42% reduction) fractions of Zn, Cd, Cu and Ni, which in turn reduced plant uptake (Usman *et al.*, 2004; Usman *et al.*, 2005). Delaminated kaolinite (in which the accessible surface area is greatly increased) has also been reported to be effective in reducing the concentration of metals in the soil solution; Menzies *et al.* (2009) reported that 2.5 t/ha delaminated kaolinite reduced plant uptake of Cd by up to 64% due to a decrease in the soil solution concentration.

While the addition of compounds with a high ion exchange capacity may reduce the concentration of metals in the soil solution, they are also likely to reduce the soil solution concentrations of some essential plant nutrients (in particular, Ca, Cu, Mg, Mn and Zn). The application of a delaminated kaolinite (with a CEC of 200 $cmol_{(c)}$/kg) at a rate of 2.5 t/ha was found to decrease shoot tissue concentrations of Ca, P, Mg, Mn and Zn significantly (Menzies *et al.*, 2009). Therefore, when applying ion exchange amendments, care must be taken to ensure that growth is not limited by deficiencies of the other nutrients for which the amendment also has a high affinity.

Specific adsorption

The addition of amendments containing Fe-/Mn-/Al-oxides (variable charge minerals) may reduce contaminant concentration by specifically adsorbing the contaminants. The ability of the amendment to adsorb contaminants specifically depends on both the contaminant and the properties of the soil. Of particular importance is the pH of the soil in relation to the point of zero charge (PZC); soils are negatively charged at pH values > PZC and positively charged at pH values < PZC (see Uehara and Gillman, 1981, for more information).

For cationic trace metals, the rate of specific adsorption (and hence removal from the soil solution) increases as pH increases. This increased adsorption results from both an increase in the negative charge of the surface and a disturbance of the water sheath surrounding the metal ion, which makes it more prone to react with the surface. Thus, for cations, the rate of specific adsorption generally approximates the order of the first hydrolysis constant (pK_a) (i.e. Cu > Pb > Zn > Co > Ni > Cd > Mn), with maximum adsorption occurring in higher pH soils where the minerals are negatively charged (for example, see McKenzie, 1980). The addition of bauxite residue (and Fe oxide-rich waste material) to two contaminated soils was found to result in a substantial decrease in soluble and exchangeable concentrations of Cd, Cu, Ni, Pb and Zn, and an increase in the Fe–Mn oxide-associated metals (Lombi *et al.*, 2002). While this decrease in availability is partly attributable to an increase in pH (bauxite residue has a pH of 10.5), the adsorption of metals by the Fe–Mn oxide fraction decreased the acid extractability of the metals, suggesting that the risk of metal re-mobilization is less when amending using red mud rather than lime. Similarly, the addition of red mud (5% w/w) to contaminated soils reduced labile (1 M NH_4NO_3-extractable) Cd by up to 91%, Zn by up to 94%, Ni by up to 71% and Pb by up to 83% (Friesl *et al.*, 2004). This reduction in labile metals reduced shoot tissue concentrations of both Cd and Zn, but not Pb. Steelshot (a waste product high in Fe oxides) has also been investigated for its ability to immobilize contaminants. Geebelen *et al.* (2002) reported that the application of 1% steelshot reduced $Ca(NO_3)_2$-extractable Pb by *c.*50%, but was not as effective as 1% hydroxyapatite or 1% lime. However, the authors concluded that while the addition of steelshot reduced extractable Pb concentrations, it reduced plant growth due to Fe or Mn phytotoxicity.

For anions, maximum adsorption occurs at the pK_a of its conjugate acid. At pH values lower than the pK_a, less than half of the acid has dissociated (and hence it is not in an anionic form), while at pH values higher than the pK_a, the variable charge surface becomes more negative and hence less able to react with anions. Garcia-Sanchez *et al.* (2002) reported that the addition of Fe/Al oxyhydroxides to two polluted mining soils reduced water-extractable As (present as $H_2AsO_4^-$ and $HAsO_4^{2-}$ under typical soil conditions) by *c.* 50–100% due to specific adsorption, although the capacity of the amendments to adsorb As was pH dependent with a maximum at *c.* pH 4.

Thus, for both cations and anions, ions with pK_a values in the range of soil pH values tend to behave as specifically adsorbing ions, while the ions at the extremes of the pK_a range are generally indifferent and only participate in exchange reactions.

The rate at which an ion specifically adsorbs will also be influenced by competition from other specifically adsorbing ions. For example, Codling and Dao (2007) reported that while the addition of P compounds to a contaminated soil decreased Pb availability, it increased As availability due to competition between phosphate (pK_a values of 2, 7 and 12) and arsenate (pK_a values of 2, 7 and 11.5). Similarly, the application of hydroxyapatite increased the solubility of oxyanion contaminants (As and Cr) due to increased competition from phosphate for specific adsorption sites (Seaman *et al.*, 2001).

Organic amendments

Organic amendments may ameliorate metal toxicity to plants by redistributing metals to less available fractions. Organic materials contain a mixture of functional groups, including sulfonic, carbonyl, carboxyl and amine groups. Thus, the addition of organic materials may interact with metals by complexation, exchange, adsorption/desorption and/or precipitation/dissolution, thereby influencing their solubility, availability and mobility. Generally, the preference of humic acids and peat for divalent cations follows: Cu > Pb > Fe > Ni = Co = Zn > Mn = Ca (Adriano, 2001).

Typically, the addition of organic amendments results in a decrease in metal mobility due to adsorption and the formation of stable complexes, thereby resulting in a shift from labile fractions to organically bound forms. Bolan *et al.* (2003b) reported that the addition of biosolids reduced the phytotoxicity of Cd by decreasing both the soluble and exchangeable Cd and increasing the organic-bound Cd, presumably due to complexation of the Cd by the organic matter. Using an industrially contaminated soil, Ruttens *et al.* (2006a) found that the addition of a commercial compost (5% w/w) reduced the $Ca(NO_3)_2$-extractable concentrations of Zn, Cd, Cu and Pb and reduced the tissue concentrations of these metals. Similarly, the addition of composted biosolids to a Cd-contaminated soil was reported to result in a decrease in both the phytotoxicity and tissue concentrations of Cd due to a redistribution of Cd from the water-soluble and exchangeable fractions to the organic fraction (Shuman *et al.*, 2002).

For some contaminants, the addition of organic amendments may increase, rather than decrease, mobility. For anionic contaminants (such as arsenate and chromate), competition between DOM and the anionic contaminant for sorption sites on variable charged minerals is an important process, with DOM potentially displacing the sorbed contaminant and hence increasing the soluble concentration (see Redman *et al.*, 2002). Indeed, while Mench *et al.* (2003) reported that the addition of a municipal compost reduced 0.1 M $Ca(NO_3)_2$-extractable concentrations of Cd, Zn and Cu (due to an increase in pH and an increase in sorption), water-extractable concentrations of As increased substantially due to competition between the negatively charged organic acids and arsenate for the soil's binding sites. The addition of organic amendments (and an increase in DOM) may also increase contaminant solubility and mobility due to an increase in the formation of soluble complexes. This is particularly problematic for Cu and Pb, which have a very high affinity for organic matter. For example, Hsu and Lo (2000) reported that the mobility of Cu increased in the presence of DOM due to the formation of complexes with humic and fulvic acids. Similarly, Ruttens *et al.* (2006b) reported that while the addition of compost (5% w/w) to a smelter-contaminated soil from Belgium reduced leaching of both Zn and Cd by *c.* 80%, leaching of Cu increased by *c.* 17-fold and Pb by *c.* 30-fold.

Thus, the influence of organic amendments on mobility and availability depends on the properties of the organic compound, with factors favouring the formation of soluble organo-metallic complexes (such as an increase in pH or an increase in the degree of humification) potentially increasing metal solubility due to either competition for adsorption sites or an increase in the formation of soluble metal complexes. Consequently, while the addition of organic amendments often results in an initial decrease in metal availability, the long-term decomposition of the organic material (and hence an increase in DOM) may actually enhance metal mobility.

Selection of plant species

The selection of appropriate plant species is of critical importance to the success of the overall phytostabilization process; the selection of inappropriate species potentially will result in system failure. A range of plant species may be used, including commercially available plants, native species or metal-tolerant species; although ecosystem composition (or end use) may be determined at least in part by site requirements or public opinion. The inclusion of legumes provides the potential for the fixation of atmospheric N, although legumes are often more sensitive to contaminants than other species and hence may be more difficult to establish.

While many studies have investigated the tolerance of various plant species to a range of inorganic and organic contaminants, most have been conducted using agricultural crops rather than species more commonly used for phytostabilization. Thus, only limited information is available comparing the suitability of plant species for use in phytostabilization systems (Table 21.1).

When selecting plant species for phytostabilization, several criteria should be considered to ensure that the most suitable species are chosen. In particular, plants should:

1. Be tolerant of the contaminant;
2. Minimize the movement of the contaminant into the shoots (i.e. the plant should not be a hyperaccumulator);
3. Be tolerant of other site-specific factors likely to influence plant growth (for example, drought, salinity and acidity); and
4. Be fast growing (but not invasive) and self-propagating.

Table 21.1. Examples of studies comparing the suitability of plant species for use in phytostabilization systems.

Plants	Metal	Plant parameters measured[1]	Experimental system	Example of suitable species	Reference
Four Australian trees	As	Plant height, shoot tissue As	Contaminated soil	*Eucalyptus cladocalyx*	King *et al.*, 2008
Native Mexican plants	As	Shoot tissue As, BCF	Contaminated soil, plant survey	*Baccharis neglecta*	Flores-Tavizon *et al.*, 2003
Mexican plants	Cd, Cu, Pb, Zn	Shoot tissue concentrations, BCF	Contaminated soil, plant survey	*Asphodelus fistulosus*	Gonzalez and Gonzalez-Chavez, 2006
Native plants from Spain	Cu, Pb, Zn	Shoot tissue concentrations, BCF	Contaminated soil, plant survey	*Lygeum spartum*	Conesa *et al.*, 2006, 2007
Five grasses from western USA	Cu	Biomass, shoot tissue Ni, EC_{50}, PT_{50}	Sand culture	*Agrostis gigantea*	Paschke and Redente, 2002
Seven perennial grasses	Cu	Biomass, shoot tissue Cu, EC_{50}, PT_{50}	Sand culture	*Urochloa mosambicensis*	Kopittke *et al.*, 2009b
Twelve tropical grasses	Cu	Biomass and shoot tissue Cu, PT_{50}	Contaminated soil	*Pennisetum americanum*	Plenderleith and Bell, 1990
Four Australian trees	Cu	Biomass and tissue Cu, PT_{50}	Solution culture	*Eucalyptus camaldulensis*	Reichman *et al.*, 2006
Four Australian trees	Mn	Biomass and tissue Mn, PT_{50}	Solution culture	*Melaleuca leucadendra*	Reichman *et al.*, 2004
Five grasses from western USA		Biomass, shoot tissue Ni, EC_{50}, PT_{50}	Sand culture	*Agrostis gigantea*	Paschke *et al.*, 2005
Seven tropical grasses	Mn	Biomass and shoot tissue Mn	Soil and sand culture	*Cenchrus ciliaris* cv. Biloela	Smith, 1979
Seven perennial grasses	Ni	Biomass, shoot tissue Ni, EC_{50}, PT_{50}	Sand culture	*Urochloa mosambicensis*	Kopittke *et al.*, 2010
Three Australian trees	Zn	Biomass and tissue Zn, PT_{50}	Solution culture	*Eucalyptus camaldulensis*	Reichman *et al.*, 2001
A range of grasses and forbs	Zn	Biomass, shoot tissue Ni, EC_{50}, PT_{50}	Sand culture	*Linum perenne*	Paschke *et al.*, 2000, 2006

[1] BCF = bioconcentration factor; EC_{50} = the half-maximal effective concentration; PT_{50} = the 50% phytotoxicity threshold.

Tolerant of the contaminant

Ideally, plants used to stabilize a contaminated site should be tolerant of that contaminant. For most sites where phytostabilization is used, the objective is to achieve satisfactory groundcover rather than near-maximum plant growth, and hence 50% maximum growth is often considered satisfactory. For this purpose, the 'EC_{50}' (the half-maximal effective concentration) is useful as it identifies the concentration of the contaminant which causes a 50% reduction in plant growth. Therefore, for a particular contaminant, assessment of the EC_{50} allows comparison of the tolerance of a range of plants to that contaminant; higher EC_{50} values indicating a higher tolerance to the contaminant.

Movement of contaminant into the shoots

For phytostabilization, the rate of contaminant transfer into the shoot should be minimized to reduce the risk of contaminant transfer through the ecosystem. For contaminants where translocation of the contaminant from the root to the shoot is of concern, examination of the contaminant concentration in plant shoots (i) provides an estimate of the likelihood that toxic levels of that contaminant will move through the ecosystem (due to consumption of plant material by animals) and (ii) enables comparisons of the extent to which different plant species accumulate contaminants in their shoots. In addition, the measurement of shoot concentrations allows the calculation of critical tissue concentrations for toxicity (which can aid in the assessment of the long-term potential for plant establishment).

A common method of expressing accumulation in the shoots is the 'bioconcentration factor' (BCF), where BCF = total element concentration in shoot tissue ÷ total element concentration in soil. Contaminant-accumulating plants are defined as species in which the tissue concentration is higher than the concentration found in the soil (i.e. BCF > 1) (McGrath and Zhao, 2003). Therefore, for phytostabilization (where the aim is to minimize contaminant transfer), suitable plant species should have a BCF < 1, and ideally << 1.

To provide an estimate of the potential movement of a contaminant through the ecosystem, two factors should be considered. Firstly, it is necessary to know the concentration of the contaminant which can be tolerated in the diet of animals. Typically, this is taken as the 'MTL' (maximum tolerable level), which is defined as the dietary level that, when fed for a defined period of time, will not impair animal health and/or performance (National Research Council (USA), 2005). Secondly, it is necessary to estimate the concentration of the contaminant which could potentially accumulate in the shoots. This can be estimated as the 'PT_{50}' (the 50% phytotoxicity threshold), which is defined as the concentration of contaminant in the shoots corresponding to a 50% reduction in shoot growth. If a plant has a low PT_{50} value, growth of the plant will be reduced before high concentrations of the contaminant accumulate in the shoot. In contrast, if a plant has a high PT_{50} value, plant growth will be reduced only after high levels of the contaminant have accumulated in the shoot. Although the PT_{50} does not represent the maximum concentration possible in the shoots, at concentrations higher than the PT_{50}, plant growth will be relatively poor and hence the mass of shoots available for consumption by animals will be low.

Thus, for any given species, it is possible to compare the concentration of contaminant which can potentially accumulate in the shoots (PT_{50}) to the concentration of contaminant which can be tolerated in the diet of animals (MTL). Such a comparison estimates the likelihood (risk) that toxic levels of the contaminant will be transferred through the food chain (Table 21.2).

While the comparison of the PT_{50} and the MTL is useful for the selection of specific plants, comparatively few data are available for species suitable for phytostabilization. Therefore, it is useful to compare the MTL to 'generalized' PT_{50} values which have been estimated from a range of plants (Table 21.3). These general PT_{50} values enable broad assumptions to be made regarding which of the contaminants are likely to be more toxic to plants and which are likely to be more toxic to animals consuming the plant shoots; however, they do not take into account differences between plant species.

Table 21.2. Use and interpretation of the 'PT_{50}' (50% phytotoxicity threshold) and 'MTL' (maximum tolerable level) for the selection of plants for the phytostabilization of contaminated sites.

Criterion	Interpretation
PT_{50} < MTL	*Lower risk of transfer of the contaminant through the food chain – the contaminant is likely to be more toxic to plants than to animals which consume the plant shoots.* Plant growth is likely to be reduced at contaminant concentrations lower than that which would result in the accumulation of the contaminant in the shoots at concentrations of concern to animals. The greater the magnitude of the difference between the PT_{50} and the MTL, the lower the risk.
$PT_{50} \approx$ MTL	*Moderate risk of transfer of the contaminant through the food chain – the contaminant is likely to be approximately equally toxic to plants and the animals consuming the plant shoots.* Plant growth is likely to be reduced at contaminant concentrations approximately equal to those which result in the accumulation of the contaminant in the shoots at concentrations of concern to animals.
PT_{50} > MTL	*Higher risk of transfer of the contaminant through the food chain – the contaminant is likely to be more toxic to animals consuming the plants than to the plants themselves.* The plants are likely to continue growing and producing substantial biomass even when the contaminant is accumulating in the shoots at levels which are of concern to animals consuming the plant. The greater the magnitude of the difference between the PT_{50} and the MTL, the greater the risk.

Table 21.3. Approximate MTL (maximum tolerable level) and general PT (phytotoxicity threshold) for a range of trace metals.

	PT – plant toxicity[a] (mg/kg)	MTL – fauna toxicity[b] (mg/kg)	Toxicity more likely to animals or plants	Likelihood that toxic levels of contaminant will be transferred into wider ecosystem
As	5–20	30	Plants	Low
Cd	10–100	5–10	Animals	High
Cu	15–30	40	Plants	Low
Mn	200–2000	2,000	Both	Moderate
Ni	25–100	100	Both	Moderate
Pb	50–100	100	Both	Moderate
Zn	100–1000	500	Both	Moderate

[a]Macnicol and Beckett, 1985, and Kabata-Pendias and Pendias, 2001.
[b]National Research Council (USA), 2005.

Thus, using generalized values, it would appear that the risk of toxic concentrations being transferred through the ecosystem (due to consumption of plant material) is greatest for Cd-contaminated sites and least for As- and Cu-contaminated sites (Table 21.3).

Tolerant to other site-specific factors

For the revegetation and stabilization of contaminated sites, it is important to identify all factors limiting plant growth. In addition to being tolerant to the contaminant, plants selected for the stabilization of a site should be tolerant to other site-specific factors, such as low soil fertility, salinity, drought and extreme pH (acidity or alkalinity). In some instances, it may be possible to address some of the limitations by covering the contaminated material with a relatively thin layer of clean soil. Unlike a capping layer, which limits access of the plant roots to the waste, in this situation the clean soil provides a portion of the root zone where plants can obtain nutrients, while the waste acts as a subsoil. This approach has been used to revegetate alkaline bauxite residue, the thin layer of clean soil permitting the plant to obtain nutrients such as Fe and Mn (Wehr *et al.*, 2005). However, this is an expensive process and it may not be possible to obtain sufficient soil to cover large sites. Alternatively, it may be possible to amend the existing (contaminated) soil sufficiently to allow plant growth. The application of appropriate amendments may improve the likelihood of establishing vegetation on the contaminated site and may include fertilizers (to overcome nutrient deficiencies), lime (to overcome soil acidity), or organic materials (to add nutrients or improve the physical properties of the soil).

Fast growing, non-invasive and self-propagating

Perennial grasses provide a quick groundcover to assist in limiting erosion (both water and wind). Furthermore, grasses are generally more tolerant of many contaminants than are broadleaf species (for example, Kukier and Chaney, 2004). Trees are slower growing but over the longer term they provide a canopy cover and a deep root network to help stabilize the soil. The native species offer the advantage of being well adapted to climatic conditions of the site, though introduced species may be faster growing and/or more tolerant of the hostile conditions commonly found in contaminated sites. However, if using introduced species, care must be taken to ensure the plant is not invasive and thereby decreases regional biodiversity. Furthermore, for the long-term effectiveness of a revegetation programme, it is essential to ensure that the species selected will self-propagate.

Phytovolatilization

Phytovolatilization refers to the use of plants to enhance the volatilization of contaminants and applies to both organic and inorganic pollutants. The latter are restricted to those elements that form methyl- and hydride-derivatives (as well as the elemental form in the case of Hg). Plants can be used to enhance the volatilization process through several mechanisms. For organic contaminants, the role of the plant is generally that of increasing the flux of contaminants toward the atmosphere through the transpiration process. For instance, Marr *et al.* (2006) reported that poplar trees could increase the vertical concentration gradient of naphthalene in the groundwater, thereby increasing the upward diffusive flux of the contaminant and enhancing direct volatilization into the atmosphere.

The phytovolatilization process has been studied in detail in the case of Se. This element is released in the atmosphere as dimethylselenide (DMSe) and dimethyldiselenide (DMDSe). The Se volatilization process is performed by both plants and microorganisms. Firstly, selenate is reduced to selenite and/or selenide. These reduced Se species are then assimilated into organic forms such as selenomethionine and selenocystein, which are subsequently methylated to non-volatile dimethylselenonium compounds. Finally, these methylated compounds are converted into the DMSe and DMDSe (Zhang and Frankenberger, 2000). Since both microorganism and plants can volatilize Se, it is difficult to distinguish the relative proportion of these two components. It is, however, assumed that the rhizosphere plays an important role in the volatilization of Se, with the plant supplying organic C sources to the microorganisms. This plant-induced Se volatilization has been studied by Dungan *et al.* (2002) and de Souza *et al.* (1999), who reported that while microbial volatilization was enhanced by root exudates, microorganisms appeared to stimulate Se uptake and volatilization by plants. It should be noted that volatile Se species are less toxic than inorganic forms (Wilber, 1980). Furthermore, the volatilized Se could be beneficial if it was redeposited in areas that were Se deficient. This appears to be the case in California, and in particular in the San Joaquin Valley, where the majority of the Se volatilization studies have taken place (Lin *et al.*, 2000).

Another element that has been investigated in the frame of phytovolatilization is Hg. Plants are not able to reduce this element to Hg^0, which is the volatile form. However, the mechanism responsible for the volatilization of Hg by bacteria is well characterized. Transgenic plants, expressing the bacterial *mer* operon, have been shown to volatilize Hg

(Heaton *et al.*, 2003). However, public acceptance has been low since Hg is not an essential element and there are concerns regarding the redeposition of Hg. To overcome this problem, the approach has been modified and an effort is being made to enhance phytochelatine synthesis in plants in order to store the sorbed Hg rather than volatilizing it (Meagher and Heaton, 2005).

Plant-assisted Biodegradation

The subject of biodegradation of organic contaminants is too vast to be included in this chapter comprehensively. Plants can contribute directly to the degradation of organic contaminants through various metabolic processes. For instance, the release of enzymes into the soil has been reported to cause degradation of contaminants. The degradation of trinitrotoluene by plant-derived nitroreductases and laccases has been demonstrated in the laboratory (Wolfe *et al.*, 1993; Boyajian and Carreira, 1997). However, the role of plants in biodegradation is more often related to their ability to stimulate microbial degradation. Therefore, we aim to discuss here the fundamental principle that underpins the role of plants in the biodegradation of organic contaminants by microorganisms. To this end, plants exert an important role by providing a suitable environment in the rhizosphere. These processes have been reviewed thoroughly and recently by Wenzel (2009), who identified three key factors controlling plant-assisted biodegradation: (i) the health of the plant–microbial community; (ii) the presence of degradation capability within the system; and (iii) the bioavailability of the contaminants.

Central to the success of plant-induced biodegradation is the stimulation of the microbial community exercised by the release of nutrients, and especially C substrates into the rhizosphere. These compounds have multiple effects; enhancing the size and activity of the microbial community in the rhizosphere, providing co-metabolites in the biodegradation process and stimulating selection of the appropriate degradation capabilities. This latter aspect is related to the similarity of some of the material released by the plants, such as acetylene, biphenyl, *p*-coumaric acid, morin, or palmitic acid, which resemble the structure of organic contaminants (Siciliano and Germida, 1997). This mechanism may explain the evolution of degrading microbial communities in the rhizosphere of plants.

The role of co-metabolites is important in the not uncommon cases where the contaminant of interest cannot be utilized as the sole C source for the microorganism. As the pattern of root exudation is largely species specific, the ability of different plants to stimulate the desired microbial community will vary from plant to plant.

Another parameter that influences microbial degradation is the bioavailability of the contaminant. Plant roots exert a substantial influence in the control and modification of key soil parameters linked to contaminant bioavailability such as pH, redox potential, abundance and quality of organic matter and organic compounds, and mineralogy (Lombi *et al.*, 2001). An aspect of root activities that has a direct impact on the enhancement of the bioavailability of organic contaminants is the release of biosurfactants (Read *et al.*, 2003).

Potential Reuse of the Biomass

Common to all aspects of phytotechnologies for soil remediation is the issue related to the use of the plant biomass produced during the clean-up or stabilization process. This will, of course, depend on the phytoremediation approach used, the plant species used, the nature of the contaminant and its presence in the plant biomass.

Phytoextraction will produce a biomass in which the contaminant is accumulated in the plant material. The end use of this biomass is therefore dependent on the nature of the accumulated metal. For valuable metals, recovery (i.e. phytomining) may be economically feasible and hence may be possible for metals such as Ni, Zn and Tl. For example, Chaney *et al.* (2007) reported that while cropping Ni-enriched soils cost US$250–500/ha, assuming 400 kg Ni/ha is extracted by plants and that Ni has a value of US$40/kg, the authors estimate the value of the crop to be US$16,000/ha. A similar situation may be possible for the phytoextraction of Tl, which is used increasingly in the manufacture of crystals, dyes, pigments, electric and electronic equipments, semiconductors, optical and infrared systems and fibreglass cables. Due to this increased demand, the price of Tl has increased steadily since the late 1980s by approximately 15%/year (USGS, 2003). The current price for Tl oxide (approximately US$1250/kg) makes Tl one of the most expensive metals after Pt, Au and Pl. Therefore, there is a significant opportunity to recycle Tl-loaded biomass. In the case of Tl, a high-yielding plant, *Iberis intermedia*, could make phytoextraction feasible, with its biomass reaching up to 10 t/ha. Anderson *et al.* (1999) reported that *I. intermedia* could hyperaccumulate Tl up to 4000 mg/kg when grown in a soil containing only 16 mg/kg, and suggested that the phytomining of Tl from moderately contaminated soils using *I. intermedia* would produce twice the return from a crop of wheat and at the same time would clean up the soil. A similar scenario may be possible for Zn, even though the biomass of Zn hyperaccumulators is smaller than in the case of Ni and Tl hyperaccumulators. In contrast, phytoextraction of toxic elements with little or no values such as Pb, Cd and As may generate secondary wastes. In these instances, however, if the biomass contained a larger concentration of metal than the soil, a reduction in the volume of material that needed to be disposed in secure landfills could be achieved.

Another possible avenue in the case of Se is the potential use of Se-enriched plant materials as feed supplements (Banuelos and Mayland, 2000). The importance of Se compounds as anticancer agents is becoming increasingly clear. Simultaneously, the occurrence of Se deficiency in human nutrition is being increasingly documented. The combination of these issues may generate new avenues for the use of Se-enriched plants in counteracting micronutrient, and in particular Se, deficiencies. A similar use of B-rich plants following phytoextraction has been suggested by Robinson *et al.* (2007).

Various options have also been suggested for the reuse of the plant biomass obtained during the phytostabilization processes. In this case, the contaminant concentration in the plants may not be relevant in terms of metal recycling, but rather, the biomass may be suitable for use as a biofuel, as traditional industrial products such as solid wood and reconstructed products, or for the production of plant derivates. For example, Robinson *et al.* (2007) suggested that poplar used for the remediation of B-contaminated soil could be used for the production of bioenergy. A similar solution was suggested by Licht and Isebrands (2002). In these cases, the limiting factor is likely to be the availability of suitable incinerators. Also, the final disposal of ashes derived from the process may require an assessment of the risk caused by the potential presence of contaminants.

Finally, the use of medicinal plants has been suggested as a potentially viable option in phytostabilization. For instance, Zheljazkov *et al.* (2008) suggested the cultivation of medicinal plants in an area impacted by a Pb–Zn smelter near Plovdiv, Bulgaria. They reported that cultivation of conventional crop in the area might pose a hazard due to the potential for the accumulation of toxic metals in agricultural products. This study assessed the characteristics of essential oils produced by *Marrubium vulgare*, *Melissa officinalis* and *Origanum heracleoticum* and reported that metals in these products were low, thereby suggesting a potentially viable alternative to conventional crops.

Potential, Limitations and Way Forward

Phytoremediation offers a 'green', 'natural' and aesthetically pleasing method for the remediation of contaminated sites. In addition, phytoremediation may offer an economic benefit compared to conventional techniques (Cunningham and Berti, 2000; Pierzynski *et al.*, 2002). For these reasons, interest in the use of plant-based techniques to remediate contaminated sites continues to grow. However, consideration must also be given to the potential limitations of plant-based approaches. Although specific limitations to the various approaches are outlined in the appropriate sections above, there are also some broad limitations which apply to plant-based approaches in general. Of particular importance, plant-based techniques require longer time frames to remediate contaminated sites than do convention methods such as landfilling. Therefore, plant-based approaches are likely to be of limited use for high-value land where it is normally desirable to remediate the site as quickly as possible. Similarly, a prolonged remediation process may be viewed as undesirable by the managers of contaminated land who continue to maintain legal liability for the site throughout the process. Site managers (and to a lesser degree, regulatory authorities) often prefer to have the contaminant removed from the site rather than having it stabilized *in situ*; once a contaminant has been removed, it clearly no longer represents any potential hazard (for that specific site at least). For sites remediated by phytostabilization, it is also more difficult to demonstrate a reduction in risk than for those sites where the contaminant has been removed. A particular difficulty is that phytostabilization may not comply with the requirements of legislation, which are often based on total contaminant concentration, although there has been a growing trend towards a risk-based approach which fundamentally embodies the axiom that, for chemical risk, 'the dose makes the poison'.[1] That is that there has to be exposure before harm from the exposure can be realized. Exposure is related to the bioavailable fraction rather than total metal loading.

The successful implementation of plant-based approaches requires a true multidisciplinary effort, with collaboration between soil scientists, agronomists, hydrologists, ecotoxicologists and economists. Due to the difficulties currently facing phytoextraction (particularly chemically assisted phytoextraction), phytostabilization is perhaps a more widely achievable solution at this time. Nevertheless, as plant-based techniques are refined and improved, these technologies will contribute increasingly to the global effort to remediate our soil; the substance on which our very existence depends.

Notes

[1] 'All substances are poisons; there is none which is not a poison. The right dose differentiates a poison…' (Paracelsus, 1493–1541).

References

Adriano, D.C. (2001) *Trace Elements in Terrestrial Environments: Biogeochemistry, Bioavailability and Risks of Metals*. Springer-Verlag, New York.

Anderson, C.W.N., Brooks, R.R., Chiarucci, A., LaCoste, C.J., Leblanc, M., Robinson, B.H., Simcock, R. and Stewart, R.B. (1999) Phytomining for nickel, thallium and gold. *Journal of Geochemical Exploration* 67, 407–415.

Arshad, M., Silvestre, J., Pinelli, E., Kallerhoff, J., Kaemmerer, M., Tarigo, A., Shahid, A., Guiresse, M., Pradere, P. and Dumat, C. (2008) A field study of lead phytoextraction by various scented *Pelargonium* cultivars. *Chemosphere* 71, 2187–2192.

Ashley, P.M., Lottermoser, B.G., Collins, A.J. and Grant, C.D. (2004) Environmental geochemistry of the derelict Webbs Consols mine, New South Wales, Australia. *Environmental Geology* 46, 591–604.

Baker, A.J.M., McGrath, S.P., Reeves, R.D. and Smith, J.A.C. (2000) Metal hyperaccumulator plants: a review of the ecology and physiology of a biological resource for phytoremediation of metal-polluted soils. In: Terry, N., Banuelos, G. and Vangronsveld, J. (eds) *Phytoremediation of Contaminated Soil and Water*. Lewis Publisher, Boca Raton, Florida, pp. 85–107.

Banuelos, G.S. and Mayland, H.F. (2000) Absorption and distribution of selenium in animals consuming canola grown for selenium phytoremediation. *Ecotoxicology and Environmental Safety* 46, 322–328.

Beyer, W.N., Gaston, G., Brazzle, R., O'Connell, A.F. and Audet, D.J. (2007) Deer exposed to exceptionally high concentrations of lead near the Continental Mine in Idaho, USA. *Environmental Toxicology and Chemistry* 26, 1040–1046.

Blaylock, M.J., Salt, D.E., Dushenkov, S., Zakharova, O., Gussman, C., Kapulnik, Y., Ensley, B.D. and Raskin, I. (1997) Enhanced accumulation of Pb in Indian mustard by soil-applied chelating agents. *Environmental Science and Technology* 31, 860–865.

Boisson, J., Ruttens, A., Mench, M. and Vangronsveld, J. (1999a) Evaluation of hydroxyapatite as a metal immobilizing soil additive for the remediation of polluted soils. Part 1. Influence of hydroxyapatite on metal exchangeability in soil, plant growth and plant metal accumulation. *Environmental Pollution* 104, 225–233.

Boisson, J., Mench, M., Vangronsveld, J., Ruttens, A., Kopponen, P. and De Koe, T. (1999b) Immobilization of trace metals and arsenic by different soil additives: evaluation by means of chemical extractions. *Communications in Soil Science and Plant Analysis* 30, 365–387.

Bolan, N.S., Adriano, D.C., Mani, P.A. and Duraisamy, A. (2003a) Immobilization and phytoavailability of cadmium in variable charge soils. II. Effect of lime addition. *Plant and Soil* 251, 187–198.

Bolan, N.S., Adriano, D.C., Duraisamy, P. and Mani, A. (2003b) Immobilization and phytoavailability of cadmium in variable charge soils. III. Effect of biosolid compost addition. *Plant and Soil* 256, 231–241.

Boyajian, G.E. and Carreira, L.H. (1997) Phytoremediation: a clean transition from laboratory to marketplace? *Nature Biotechnology* 15, 127–128.

Brooks, R.R. (1998) *Plants that Hyperaccumulate Heavy Metals*. CAB International, Wallingford, UK.

Castaldi, P., Santona, L. and Melis, P. (2005) Heavy metal immobilization by chemical amendments in a polluted soil and influence on white lupin growth. *Chemosphere* 60, 365–371.

CEI (Canadian Environmental Industries) (2005) Soil remediation technologies: assessment, clean-up, decommissioning, rehabilitation. Canadian Environmental Industries (Energy and Environmental Industries Branch) (http://www.ic.gc.ca/eic/site/ea-ae.nsf/eng/ea02201.html, accessed June 2009).

Chaney, R.L., Angle, J.S., Broadhurst, C.L., Peters, C.A., Tappero, R.V. and Sparks, D.L. (2007) Improved understanding of hyperaccumulation yields commercial phytoextraction and phytomining technologies. *Journal of Environmental Quality* 36, 1429–1443.

Chen, M., Ma, L.Q., Singh, S.P., Cao, R.X. and Melamed, R. (2003) Field demonstration of *in situ* immobilization of soil Pb using P amendments. *Advances in Environmental Research* 8, 93–102.

Chlopecka, A. and Adriano, D.C. (1996) Mimicked *in situ* stabilization of metals in a cropped soil: bioavailability and chemical form of zinc. *Environmental Science and Technology* 30, 3294–3303.

Chlopecka, A. and Adriano, D.C. (1997) Influence of zeolite, apatite and Fe-oxide on Cd and Pb uptake by crops. *Science of the Total Environment* 207, 195–206.

Codling, E.E. and Dao, T.H. (2007) Short-term effect of lime, phosphorus, and iron amendments on water-extractable lead and arsenic in orchard soils. *Communications in Soil Science and Plant Analysis* 38, 903–919.

Conesa, H.M., Faz, A. and Arnaldos, R. (2006) Heavy metal accumulation and tolerance in plants from mine tailings of the semiarid Cartagena-La Union mining district (SE Spain). *Science of the Total Environment* 366, 1–11.

Conesa, H.M., Faz, Á. and Arnaldos, R. (2007) Initial studies for the phytostabilization of a mine tailing from the Cartagena-La Union Mining District (SE Spain). *Chemosphere* 66, 38–44.

Cunningham, S.D. and Berti, W.R. (2000) Phytoextraction and phytostabilization: technical, economic, and regulatory considerations of the soil–lead issue. In: Terry, N. and Banuelos, G. (eds) *Phytoremediation of Contaminated Soil and Water*. Lewis Publishers, New York, pp. 359–376.

Dungan, R.S., Yates, S.R. and Frankenberger, W.T. (2002) Volatilization and degradation of soil-applied dimethylselenide. *Journal of Environmental Quality* 31, 2045–2050.

EEA (European Environment Agency) (2007) Progress in management of contaminated sites – assessment (CSI 015). European Environment Agency (ISpecification20041007131746) (http://themes.eea.europa.eu/IMS/ISpecs/ISpecification20041007131746/, accessed June 2009).

Evangelou, M.W.H., Ebel, M., Hommes, G. and Schaeffer, A. (2008) Biodegradation: the reason for the inefficiency of small organic acids in chelant-assisted phytoextraction. *Water Air and Soil Pollution* 195, 177–188.

Farmer, J.G., Paterson, E., Bewley, R.J.F., Geelhoed, J.S., Hillier, S., Meeussen, J.C.L., Lumsdon, D.G., Thomas, R.P. and Graham, M.C. (2006) The implications of integrated assessment and modelling studies for the future remediation of chromite ore processing residue disposal sites. *Science of the Total Environment* 360, 90–97.

Flores-Tavizon, E., Alarcon-Herrera, M.T., Gonzalez-Elizondo, S. and Olguin, E.J. (2003) Arsenic tolerating plants from mine sites and hot springs in the semi-arid region of Chihuahua, Mexico. *Acta Biotechnologica* 23, 113–119.
Friesl, W., Horak, O. and Wenzel, W.W. (2004) Immobilization of heavy metals in soils by the application of bauxite residues: pot experiments under field conditions. *Journal of Plant Nutrition and Soil Science* 167, 54–59.
Garbaciak, S., Spadaro, P., Thornburg, T. and Fox, R. (1997) Sequential risk mitigation and the role of natural recovery in contaminated sediment projects. *Water Science and Technology* 37, 331–336.
Garcia-Sanchez, A., Alvarez-Ayuso, E. and Rodriguez-Martin, F. (2002) Sorption of As(V) by some oxyhydroxides and clay minerals. Application to its immobilization in two polluted mining soils. *Clay Minerals* 37, 187–194.
Geebelen, W., Vangronsveld, J., Adriano, D.C., Carleer, R. and Clijsters, H. (2002) Amendment-induced immobilization of lead in a lead-spiked soil: evidence from phytotoxicity studies. *Water, Air, and Soil Pollution* 140, 261–277.
Gonzalez, R.C. and Gonzalez-Chavez, M.C.A. (2006) Metal accumulation in wild plants suffounding mining wastes. *Environmental Pollution* 144, 84–92.
Gray, C.W., Dunham, S.J., Dennis, P.G., Zhao, F.J. and McGrath, S.P. (2006) Field evaluation of *in situ* remediation of a heavy metal contaminated soil using lime and red-mud. *Environmental Pollution* 142, 530–539.
Grubor, M. (2008) Lead uptake, tolerance, and accumulation exhibited by the plants *Urtica dioica* and *Sedum spectabile* in contaminated soil without additives. *Archives of Biological Sciences* 60, 239–244.
Heaton, A.C.P., Rugh, C.L., Kim, T., Wang, N.J. and Meagher, R.B. (2003) Toward detoxifying mercury-polluted aquatic sediments with rice genetically engineered for mercury resistance. *Environmental Toxicology and Chemistry* 22, 2940–2947.
Hettiarachchi, G.M., Pierzynski, G.M. and Ransom, M.D. (2001) *In situ* stabilization of soil lead using phosphorus. *Journal of Environmental Quality* 30, 1214–1221.
Hsu, J.H. and Lo, S.L. (2000) Characterization and extractability of copper, manganese, and zinc in swine manure composts. *Journal of Environmental Quality* 29, 447–453.
Kabata-Pendias, A. and Pendias, H. (2001) *Trace Elements in Soils and Plants*, 3rd edn. CRC Press, Boca Raton, Florida.
King, D.J., Doronila, A.I., Feenstra, C., Baker, A.J.M. and Woodrow, I.E. (2008) Phytostabilisation of arsenical gold mine tailings using four Eucalyptus species: growth, arsenic uptake and availability after five years. *Science of the Total Environment* 406, 35–42.
Klitzke, S. and Lang, F. (2009) Mobilization of soluble and dispersible lead, arsenic, and antimony in a polluted, organic-rich soil – effects of pH increase and counterion valency. *Journal of Environmental Quality* 38, 933–939.
Kopittke, P.M., Blamey, F.P.C., Kopittke, R.A. and Menzies, N.W. (2010) Tolerance of seven perennial grasses to high concentrations of nickel in sand culture. *Environmental Chemistry* 7, DOI: 10.1071/EN09100.
Kopittke, P.M., Blamey, F.P.C., Sheldon, A.R. and Menzies, N.W. (2009) Tolerance of perennial grasses to high copper in sand culture. *Environmental Chemistry* 6, 253–259.
Kukier, U. and Chaney, R.L. (2004) *In situ* remediation of nickel phytotoxicity for different plant species. *Journal of Plant Nutrition* 27, 465–495.
Licht, L.A. and Isebrands, J.G. (2002) Linking phytoremediated pollutant removal to biomass economic opportunities. *Biomass and Bioenergy* 28, 203–218.
Lin, Z.Q., Schemenauer, R.S., Cervinka, V., Zayed, A., Lee, A. and Terry, N. (2000) Selenium volatilization from a soil–plant system for the remediation of contaminated water and soil in the San Joaquin Valley. *Journal of Environmental Quality* 29, 1048–1056.
Lindsay, W.L. (1979) *Chemical Equilibria in Soils*. John Wiley and Sons, New York.
Liu, D., Li, T., Yang, X., Islam, E., Jin, X. and Mahmood, Q. (2007) Enhancement of lead uptake by hyperaccumulator plant species *Sedum alfredii* Hance using EDTA and IAA. *Bulletin of Environmental Contamination and Toxicology* 78, 280–283.
Lombi, E. and Hamon, R.E. (2004) Remediation of polluted soils. In: Hillel, D. (ed.) *Encyclopedia of Soils in the Environment*. Academic Press, New York, pp. 379–385.
Lombi, E., Zhao, F.J., Dunham, S.J. and McGrath, S.P. (2001) Phytoremediation of heavy metal-contaminated soils: natural hyperaccumulation versus chemically enhanced phytoextraction. *Journal of Environmental Quality* 30, 1919–1926.
Lombi, E., Zhao, F.J., Zhang, G., Sun, B., Fitz, W., Zhang, H. and McGrath, S.P. (2002) *In situ* fixation of metals in soils using bauxite residue: chemical assessment. *Environmental Pollution* 118, 435–443.

Long, X.X., Zhang, Y.G., Jun, D. and Zhou, Q.X. (2009) Zinc, cadmium and lead accumulation and characteristics of rhizosphere microbial population associated with hyperaccumulator *Sedum alfredii* Hance under natural conditions. *Bulletin of Environmental Contamination and Toxicology* 82, 460–467.
Ma, L.Q., Komar, K.M., Tu, C., W.H., Z., Cai, Y. and Kennelley, E.D. (2001) A fern that hyperaccumulates arsenic – a hardy, versatile, fast-growing plant helps to remove arsenic from contaminated soils. *Nature (London)* 409, 579.
McGrath, S.P. and Zhao, F.J. (2003) Phytoextraction of metals and metalloids from contaminated soils. *Current Opinion in Biotechnology* 14, 277–282.
McGrath, S.P., Zhao, F.J. and Lombi, E. (2002) Advances in phytoremediation of metals, metalloids and radionuclides. *Advances in Agronomy* 75, 1–56.
McGrath, S.P., Lombi, E., Gray, C.W., Caille, N., Dunham, S.J. and Zhao, F.J. (2006) Field evaluation of Cd and Zn phytoextraction potential by the hyperaccumulators *Thlaspi caerulescens* and *Arabidopsis halleri*. *Environmental Pollution* 141, 115–125.
McKenzie, R.M. (1980) The adsorption of lead and other heavy metals on oxides of manganese and iron. *Australian Journal of Soil Research* 18, 61–73.
Macnicol, R.D. and Beckett, P.H.T. (1985) Critical tissue concentrations of potentially toxic elements. *Plant and Soil* 85, 107–129.
Maier, R.M., Neilson, J.W., Artiola, J.F., Jordan, F.L., Glenn, E.P. and Descher, S.M. (2001) Remediation of metal-contaminated soil and sludge using biosurfactant technology. *International Journal of Occupational Medicine and Environmental Health* 14, 241–248.
Marr, L.C., Booth, E.C., Andersen, R.G., Widdowson, M.A. and Novak, J.T. (2006) Direct volatilization of naphthalene to the atmosphere at a phytoremediation site. *Environmental Science and Technology* 40, 5560–5566.
Meagher, R. and Heaton, A. (2005) Strategies for the engineered phytoremediation of toxic element pollution: mercury and arsenic. *Journal of Industrial Microbiology and Biotechnology* 32, 502–513.
Meers, E., Ruttens, A., Hopgood, M., Lesage, E. and Tack, F.M.G. (2005a) Potential of *Brassic rapa*, *Cannabis sativa*, *Helianthus annuus* and *Zea mays* for phytoextraction of heavy metals from calcareous dredged sediment derived soils. *Chemosphere* 61, 561–572.
Meers, E., Ruttens, A., Hopgood, M.J., Samson, D. and Tack, F.M.G. (2005b) Comparison of EDTA and EDDS as potential soil amendments for enhanced phytoextraction of heavy metals. *Chemosphere* 58, 1011–1022.
Mench, M., Bussière, S., Boisson, J., Castaing, E., Vangronsveld, J., Ruttens, A., De Koe, T., Bleeker, P., Assunção, A. and Manceau, A. (2003) Progress in remediation and revegetation of the barren Jales gold mine spoil after *in situ* treatments. *Plant and Soil* 249, 187–202.
Mendez, M.O. and Maier, R.M. (2008) Phytostabilization of mine tailings in arid and semiarid environments – an emerging remediation technology. *Environmental Health Perspectives* 116, 278–283.
Menzies, N.W. and Mulligan, D.R. (2000) Vegetation dieback on clay-capped pyritic mine waste. *Journal of Environmental Quality* 29, 437–442.
Menzies, N.W., Snars, K.E., Kopittke, G.R. and Kopittke, P.M. (2009) Amelioration of cadmium contaminated soils using cation exchangers. *Journal of Plant Nutrition*, 32, 1321–1335.
National Research Council (USA) (2005) *Mineral Tolerance of Animals*, second edn. National Academy of Sciences, Washington, DC.
Nriagu, J.O. (1974) Lead orthophosphates. 4. Formation and stability in environment. *Geochimica et Cosmochimica Acta* 38, 887–898.
Oste, L.A., Lexmond, T.M. and Van Riemsdijk, W.H. (2002) Metal immobilization in soils using synthetic zeolites. *Journal of Environmental Quality* 31, 813–821.
Ouki, S.K. and Kavannagh, M. (1997) Performance of natural zeolites for the treatment of mixed metal-contaminated effluents. *Waste Management and Research* 15, 383–394.
Parkhurst, D. (2006) PhreeqcI v2.12.5, United States Geological Survey (http://water.usgs.gov/owq/software.html, accessed March 2006).
Paschke, M.W. and Redente, E.F. (2002) Copper toxicity thresholds for important restoration grass species of the western United States. *Environmental Toxicology and Chemistry* 21, 2692–2697.
Paschke, M.W., Redente, E.F. and Levy, D.B. (2000) Zinc toxicity thresholds for important reclamation grass species of the western United States. *Environmental Toxicology and Chemistry* 19, 2751–2756.
Paschke, M.W., Valdecantos, A. and Redente, E.F. (2005) Manganese toxicity thresholds for restoration grass species. *Environmental Pollution* 135, 313–322.
Paschke, M.W., Perry, L.G. and Redente, E.F. (2006) Zinc toxicity thresholds for reclamation forb species. *Water Air and Soil Pollution* 170, 317–330.

Pierzynski, G., Kulakow, P., Erickson, L. and Jackson, L. (2002) Plant system technologies for environmental management of metals in soils: educational materials. *Journal of Natural Resources and Life Sciences Education* 31, 31–37.
Plenderleith, R.W. and Bell, L.C. (1990) Tolerance of twelve tropical grasses to high soil concentrations of copper. *Tropical Grasslands* 24, 103–110.
Read, D.B., Bengough, A.G., Gregory, P.J., Crawford, J.W., Robinson, D., Scrimgeour, C.M., Young, I.M., Zhang, K. and Zhang, X. (2003) Plant roots release phospholipid surfactants that modify the physical and chemical properties of soil. *New Phytologist* 157, 315–326.
Redman, A.D., Macalady, D.L. and Ahmann, D. (2002) Natural organic matter affects arsenic speciation and sorption onto hematite. *Environmental Science and Technology* 36, 2889–2896.
Reichman, S.M., Asher, C.J., Mulligan, D. and Menzies, N.W. (2001) Seedling responses of three Australian tree species to toxic concentrations of zinc in solution culture. *Plant and Soil* 235, 151–158.
Reichman, S.M., Menzies, N.W., Asher, C.J. and Mulligan, D. (2004) Seedling responses of four Australian tree species to toxic concentrations of manganese in solution culture. *Plant and Soil* 258, 341–350.
Reichman, S.M., Menzies, N.W., Asher, C.J. and Mulligan, D.R. (2006) Responses of four Australian tree species to toxic concentrations of copper in solution culture. *Journal of Plant Nutrition* 29, 1127–1141.
Robinson, B.H., Leblanc, M., Petit, D., Brooks, R.R., Kirkman, J.H. and Gregg, P.E.H. (1998) The potential of *Thlaspi caerulescens* for phytoremediation of contaminated soils. *Plant and Soil* 203, 47–56.
Robinson, B.H., Green, S.R., Chancerel, B., Mills, T.M. and Clothier, B.E. (2007) Poplar for the phytomanagement of boron contaminated sites. *Environmental Pollution* 150, 225–233.
Ruttens, A., Mench, M., Colpaert, J.V., Boisson, J., Carleer, R. and Vangronsveld, J. (2006a) Phytostabilization of a metal contaminated sandy soil. I: Influence of compost and/or inorganic metal immobilizing soil amendments on phytotoxicity and plant availability of metals. *Environmental Pollution* 144, 524–532.
Ruttens, A., Colpaert, J.V., Mench, M., Boisson, J., Carleer, R. and Vangronsveld, J. (2006b) Phytostabilization of a metal contaminated sandy soil. II: Influence of compost and/or inorganic metal immobilizing soil amendments on metal leaching. *Environmental Pollution* 144, 533–539.
Ryan, J.A., Zhang, P.C., Hesterberg, D., Chou, J. and Sayers, D.E. (2001) Formation of chloropyromorphite in a lead-contaminated soil amended with hydroxyapatite. *Environmental Science and Technology* 35, 3798–3803.
Saifullah, M.E., Qadir, M., Caritat, P. de, Tack, F.M.G., Du Laing, G. and Zia, M.H. (2009) EDTA-assisted Pb phytoextraction. *Chemosphere* 74, 1279–1291.
Sauve, S., McBride, M. and Hendershot, W. (1998) Soil solution speciation of lead (II): effects of organic matter and pH. *Soil Science Society of America Journal* 62, 618–621.
Schmidt, U. (2003) Enhancing phytoextraction: the effect of chemical soil manipulation on mobility, plant accumulation, and leaching of heavy metals. *Journal of Environmental Quality* 32, 1939–1954.
Schwartz, C., Sirguey, C., Peronny, S., Reeves, R.D., Bourgaud, F. and Morel, J.L. (2006) Testing of outstanding individuals of *Thlaspi caerulescens* for cadmium phytoextraction. *International Journal of Phytoremediation* 8, 339–357.
Seaman, J.C., Arey, J.S. and Bertsch, P.M. (2001) Immobilization of nickel and other metals in contaminated sediments by hydroxyapatite addition. *Journal of Environmental Quality* 30, 460–469.
Shuman, L.M., Dudka, S. and Das, K. (2002) Cadmium forms and plant availability in compost-amended soil. *Communications in Soil Science and Plant Analysis* 33, 737–748.
Siciliano, S.D. and Germida, J.J. (1997) Mechanisms of phytoremediation: biochemical and ecological interactions between plants and bacteria. *Environmental Reviews* 6, 65–79.
Smith, F.W. (1979) Tolerance of seven tropical pasture grasses to excess manganese. *Communications in Soil Science and Plant Analysis* 10, 853–867.
Smolders, E., Oorts, K., van Sprang, P., Schoeters, I., Janssen, C.R., McGrath, S.P. and McLaughlin, M.J. (2009) Toxicity of trace metals in soil as affected by soil type and aging after contamination: using calibrated bioavailability models to set ecological soil standards. *Environmental Toxicology and Chemistry*, 28, 1633–1642.
Song, J., Zhao, F.-J., Luo, Y.-M., McGrath, S.P. and Zhang, H. (2004) Copper uptake by *Elsholtzia splendens* and *Silene vulgaris* and assessment of copper phytoavailability in contaminated soils. *Environmental Pollution* 128, 307–315.
Souza, M.P. de, Chu, D., Zhao, M., Zayed, A.M., Ruzin, S.E., Schichnes, D. and Terry, N. (1999) Rhizosphere bacteria enhance selenium accumulation and volatilization by Indian mustard. *Plant Physiology* 119, 565–573.

Uehara, G. and Gillman, G. (1981) *The Mineralogy, Chemistry, and Physics of Tropical Soils with Variable Charge Clays*. Westview Press, Boulder, Colorado.
USGS (United States Geological Survey) (2003) Mineral commodity summary. United States Geological Survey (http://minerals.usgs.gov/minerals/pubs/mcs/, accessed June 2009).
Usman, A.R.A., Kuzyakov, Y. and Stahr, K. (2004) Effect of clay minerals on extractability of heavy metals and sewage sludge mineralization in soil. *Chemistry and Ecology* 20, 123–135.
Usman, A.R.A., Kuzyakov, Y. and Stahr, K. (2005) Effect of immobilizing substances and salinity on heavy metals availability to wheat grown on sewage sludge-contaminated soil. *Soil and Sediment Contamination* 14, 329–344.
Warwick, P., Inam, E. and Evans, N. (2005) Arsenic's interaction with humic acid. *Environmental Chemistry* 2, 119–124.
Wehr, J.B., So, H.B., Menzies, N.W. and Fulton, I. (2005) Hydraulic properties of layered soils influence survival of Rhodes grass (*Chloris gayana* Kunth.) during water stress. *Plant and Soil* 270, 287–297.
Wenzel, W.W. (2009) Rhizosphere processes and management in plant-assisted bioremediation (phytoremediation) of soils. *Plant and Soil* 321, 385–408.
Wilber, C.G. (1980) Toxicology of selenium – a review. *Clinical Toxicology* 17, 171–230.
Wolfe, N.L., Ou, T.Y. and Carreira, L.H. (1993) Biochemical remediation of TNT contaminated soils. US Army Corps Eng., Washington, D.C.
Wu, J., Hsu, F.C. and Cunningham, S.D. (1999) Chelate-assisted Pb phytoextraction: Pb availability, uptake, and translocation constraints. *Environmental Science and Technology* 33, 1898–1904.
Zhang, Y.Q. and Frankenberger, W.T. (2000) Formation of dimethylselenonium compounds in soil. *Environmental Science and Technology* 34, 776–783.
Zhao, F.J., Lombi, E. and McGrath, S.P. (2003) Assessing the potential for zinc and cadmium phytoremediation with the hyperaccumulator *Thlaspi caerulescens*. *Plant and Soil* 249, 37–43.
Zheljazkov, V.D., Jeliazkova, E.A., Kovacheva, N. and Dzhurmanski, A. (2008) Metal uptake by medicinal plant species grown in soils contaminated by a smelter. *Environmental and Experimental Botany* 64, 207–216.
Zhu, Y.G., Chen, S.B. and Yang, J.C. (2004) Effects of soil amendments on lead uptake by two vegetable crops from a lead-contaminated soil from Anhui, China. *Environment International* 30, 351–356.

22

Phytoremedial Crops and Current Research

Priya Padmanabhan and Shivendra V. Sahi

Introduction

Contamination of soil and water with toxic metals poses serious threats to humans, animals and the ecosystem. Toxic metals and xenobiotics also affect crop yields, soil biomass and fertility. Heavy metal pollution of the environment has resulted from the burning of fossil fuels, residues from mines and smelting industries, municipal wastes, fertilizers, pesticides and sewage. Metals such as Cd, Pb or Hg are considered non-essential and are potentially highly toxic due to reactivity with S and N atoms. Even metals essential for the growth of living organisms can be toxic at higher concentrations. A metal or metalloid species may be considered a contaminant if it occurs where it is unwanted, or in a form or concentration that is detrimental to living organisms.

The use of plants to remediate soils contaminated with hazardous metals is a promising technique as it is more cost-effective compared to other alternate remediation strategies involving soil washing, excavation, or disposal of contaminated soils (Chaney *et al.*, 1997). It is also appealing as this can be carried out *in situ*, minimizing human exposure to hazardous metal residues and is aesthetically pleasing to the public. Phytoremediation refers to a group of techniques that involves the use of plants to extract, remove, concentrate, degrade or immobilize toxic contaminants from water or soil.

Metal Hyperaccumulator Plants

Plants with the ability to grow in the presence of a high concentration of metals are categorized into hyperaccumulator, indicator or tolerant. Hyperaccumulators collect very high concentrations of metal in aboveground plant parts in the natural habitat during their normal growth and development. Indicators are plants that normally grow above underground ore deposits. They are particularly important in biogeochemical prospecting for ores. A tolerant plant species can tolerate and grow in the presence of a particular metal that is toxic to most other plants. While indicators and hyperaccumulator plants are also tolerant, tolerant species may not necessarily be indicators or hyperaccumulators.

To date, more than 400 plant species belonging to families such as Asteraceace, Brassicaceae, Caryophyllaceae, Poaceae, Violaceae and Fabaceae are reported as hyperaccumulators. The Brassicaceae is the best represented among these metal hyperaccumulator families, with 87 *Brassica* species classified as metal hyperaccumulators. Plants that accumulate $> 0.1\%$ of dry weight of Ni, Co, Cu and Pb, $> 1\%$ of Zn and $> 0.01\%$ of Cd are placed in the category of hyperaccumulators (Clemens, 2001). These plants are adapted to the environmental conditions in their habitat and the metal accumulation property assists them in their defence against herbivores and fungal infections (Martens and Boyd, 2002).

Natural hyperaccumulators are low biomass producing plants and very often are selective for an individual metal (Clemens *et al.*, 2002). Almost 75% of the identified hyperaccumulators accumulate Ni and are termed as nickelophilous plants (Baker and Brooks, 1989). Among the earliest known natural metal hyperaccumulators are a group of small, weedy alpine plants called Alpine pennycress (*Thlaspi* spp.), which lack the standard pathogen defence mechanism. *Thlaspi* spp. exhibit large interspecific and intraspecific variations, which make them important plants for studying hyperaccumulation.

Mechanisms of Metal Accumulation in Plants

The mechanism of metal hyperaccumulation in plants is very complex. The main processes involved in metal homeostasis and resistance are uptake of metals, translocation and sequestration.

For uptake, metal ions should be bioavailable to plants. The bioavailability of metals to the plants may be affected by various factors like metal content in the soil, pH, presence of organic substances, as well as plant and/or microbial activities and their interactions (Pilon-Smits and Pilon, 2002). Plants have various mechanisms that make the metals bioavailable for uptake. Many plants and bacteria are known to release metal chelators such as siderophores, organic acids and phenolics that affect the solubility and bioavailability of metals for plant uptake (Ross, 1994). Many graminaceous plants secrete mugineic acid and avenic acid into the rhizosphere as phytosiderophores (Fan *et al.*, 1993; Kinnersely, 1993). Phytosiderophores secreted by cereals under Fe deficiency have been shown to enhance the mobilization of Fe, Zn, Cu and Mn in the soil (Romheld and Marschner, 1986). Many metal-chelating proteins related to metallothioneins and phytochelatins also function as siderophores in plants (Robinson *et al.*, 1993; Rauser, 1999).

Some plants can mobilize soil-bound metal ions by specific plasma membrane-bound metal reductases (Salt and Wagner, 1993). Plant roots also extrude H^+ via ATPases which replace cations, thereby acidifying the soil to help solubilization of metals (Taiz and Zeiger, 2002). Cieslinski *et al.* (1998) reported the presence of many low-molecular weight organic acids such as acetic and succinic acid only in the rhizosphere of the Cd-accumulating genotype Kyle of wheat, and not in non-accumulator genotype Arcola. In *Allysium*, transport and bioaccumulation of Ni were greatly enhanced by the amino acid histidine (Kramer *et al.*, 1996). Von Wiren *et al.* (1999) showed that nicotianamine is an effective chelator of Fe^{3+} and it has an important role in scavenging Fe. Buckwheat secretes oxalic acid from the roots in response to Al and accumulates non-toxic Al-oxalate in the leaves (Ma *et al.*, 1997).

The uptake of solubilized metals is a cellular phenomenon governed by the rules of membrane transport. Metal transporters export metal ions out of the root symplast into the xylem apoplast for root–shoot translocation. In the xylem, translocation is an energy-driven process (Salt and Rauser, 1995). It has been suggested that several chelators (e.g. malate, citrate, histidine or nicotianamine) are involved in the translocation of different metal cations through the xylem (Stephan *et al.*, 1996; Von Wiren *et al.*, 1999).

Sequestration is the final step in metal accumulation. Metals are usually sequestered into the vacuole. Storage may also be in the apoplast, or in specialized cells like epidermal cells and trichomes (Salt and Kramer, 2000). The metal ions are bound by chelators and chaperones. Chelators help in the metal detoxification process by buffering cytosolic metal concentrations. Functions of chaperones are to deliver metal ions specifically to organelles and metal-requiring proteins. Citrate, malate and oxalate have been found to be associated with a range of processes that involve differential metal tolerance, metal transport through the xylem and vacuolar metal sequestration (Rauser, 1999).

Molecular Mechanisms of Metal Tolerance in Plants

Heat shock protein

Heat shock proteins (HSPs) are a group of proteins whose high-level production is triggered by elevated temperatures or other stresses. Increase in HSP expression has also been reported in plants exposed to heavy metals (Lewis *et al.*, 1999). Hall (2002) suggested the role of HSPs in controlling metal damage was in protecting cell membranes from metal injury. Heckathorn *et al.* (2004) observed production of small HSPs in the chloroplast of maize as an early response to heavy metal accumulation in leaves, which could be to limit damage to photosynthesis. Tseng *et al.* (1993) found that in rice both heat and heavy metal stress enhanced the expression of small HSPs (16–20 kDa). Neumann *et al.* (1995) obtained increased levels of HSP17 in roots of *Armeria maritima* grown on Cu-rich soil. Wollgiehn and Neumann (1999) observed that although HSP17 was enhanced in cell cultures of *Silene vulgaris* exposed to a range of heavy metals, zero or very low amounts of HSPs were found in plants growing on metalliferous soils, suggesting that HSPs were not responsible for heritable metal tolerance. Large HSP (HSP70) has also been found to increase in response to Cd stress (Neumann *et al.*, 1994). HSP70 was localized in the nucleus, cytoplasm and plasma membrane, leading to inference that HSP70 could be involved in the protection of membranes against Cd stress.

Metal-binding molecules

Metallothionein

Metallothioneins (MTs) are cysteine-rich, low-molecular weight proteins which bind metal ions and are found throughout the animal and plant kingdom. MTs are generally categorized into two classes: MT1 and MT2, based on the alignment of cysteine residues (Prasad, 1999). The first plant MT was reported from wheat and it was shown to bind Zn (Lane *et al.*, 1987). More than 50 MT-like sequences have been identified from plants (Rauser, 1999). They show great variation from mammalian and fungal MTs. Metallothioneins play a role in detoxification of heavy metals and in homeostasis of intracellular metal ions (Cobbett and Goldsbrough, 2002).

There are several reports on the introduction of MTs from animal sources into plants to reduce metal accumulation in shoots by trapping the metal in the roots. Transfer of a mammalian MT in tobacco under the control of a constitutive promoter was able to reduce the translocation of Cd into the shoot (Maiti *et al.*, 1989; Elmayan and Tepfer, 1994). Zimeri *et al.* (2005) reported that the MT1 knockdown *Arabidopsis* plant lines were all hypersensitive to Cd and accumulated several-fold lower levels of As, Cd and Zn than wild type, while Cu and Fe levels were unaffected.

Understanding the role of MT genes in plants has been hampered by the lack of protein data. This is due to the unsuccessful attempts to purify MT proteins from plants, except in the case of Ec (early cysteine labelled) protein from wheat and *Arabidopsis* MT1 and MT2 (Murphy *et al.*, 1997). Animal MTs were among the first genes expressed in transgenic plants (Misra and Gedamu, 1989); however, little is known about the functions of the various plant MT sequences themselves or why there is so much sequence diversity. Unravelling the exact functions of these proteins and the metals that bind to these proteins is a future challenge. Data suggest that they could function as antioxidants, although there is a lack of conclusive evidence (Dietz *et al.*, 1999), while a role in plasma membrane instability is another suggestion (Salt *et al.*, 1998). Other proposed functions include detoxification of metals (Cu), cytosolic Zn buffering, scavenging of metals during leaf senescence and metal secretion through trichomes (Rauser, 1999)

Phytochelatins

Phytochelatins (PCs) are a family of small, metal-binding peptides that were first identified in fission yeast (*Schizosaccharomyces pombe*) (Murasugi *et al.*, 1981). Later Grill *et al.* (1985) reported the ubiquitous occurrence of the same peptides in various cells of plants exposed to Cd. Unlike MTs which are gene-encoded, PCs are enzymatically synthesized. PC synthesis is induced rapidly in cells and tissues following exposure to a variety of metals or metalloids such as Cd, Ni, Zn, Ag, Hg, Pb, As and Se (Rauser, 1995). PCs have been identified in a wide variety of plant species including monocots, dicots, gymnosperms, algae, etc. (Gekeler *et al.*, 1989). They are synthesized non-translationally from reduced glutathione (GSH) in a transpeptidation reaction (Grill *et al.*, 1989) catalysed by the enzyme PC synthase. γ-GluCys dipeptidyl transpeptidase (EC 2.3.2.15), named PC synthase, catalyses the synthesis of PCs by transferring a γ-GluCys moiety of GSH to GSH or to other PCs (Zenk, 1996).

The role of PCs in heavy metal tolerance has been well-studied and characterized in Cd-sensitive mutants of *Arabidopsis*, *cad1* and *cad2* (Howden and Cobbett, 1992; Howden *et al.*, 1995a,b). These mutants are deficient in PC production due to mutations in γ-glutamyl-cys synthetase in the case of *cad2* mutants, or in PC synthase in the case of *cad1* mutants. Howden *et al.* (1995a,b) observed that the Cd-sensitive mutants of *Arabidopsis* varied in their ability to accumulate PCs, and the amount of PCs accumulated by the mutants correlated with the degree of sensitivity to Cd. Ha *et al.* (1999) found PCs to be involved in the detoxification of Cd and arsenate in *Arabidopsis* and *S. pombe*, but they played no role in the detoxification of Zn, Ni and Se. Cd accumulation in Indian mustard is accompanied by a rapid induction of PC synthesis (Haag-Kerwer *et al.*, 1999). The role of PC synthase in Cd tolerance has been shown in studies using azuki beans (Inouhe *et al.*, 2000). According to Xiang and Oliver (1998), exposure of *Arabidopsis* plants to Cd and Cu resulted in increased transcription of the genes involved in GSH biosynthetic pathway and of GSH reductase. Studies of transgenic Indian mustard plants, in which the expression of the enzymes of the GSH biosynthetic pathway was increased, noted simultaneous increases in PC biosynthesis and Cd tolerance (Zhu *et al.*, 1999). Wild-type Indian mustard plants respond to exposure to Cd with increased levels of expression of genes involved in PC biosynthesis (Schafer *et al.*, 1998).

PC synthase genes were isolated from higher plants including *Arabidopsis* and wheat (Clemens *et al.*, 1999; Vatamaniuk *et al.*, 1999). The studies conducted in *Arabidopsis* have been interpreted to indicate a role for PCs in the homeostasis of essential metal ion metabolism (Zenk, 1996).

Salt *et al.* (1989) suggested a possible role of PCs in Cu tolerance in copper-tolerant *Mimulus guttatus*. The As-tolerant *Holcus lanatus* plants were found to contain higher levels of PCs and As than the wild-type plants when grown in the presence of high levels of As (Hartley-Whitaker *et al.*, 2001). However, there are also some conflicting reports on the role of PCs in metal tolerance (Steffens, 1990; Ernst *et al.*, 1992). De Knecht *et al.* (1992) showed that differential Cd in *S. vulgaris* was not due to the differential production of PCs.

In addition to detoxification functions, PCs also play other important roles in the cell including heavy metal homeostasis, S metabolism or as antioxidants (Dietz *et al.*, 1999). Steffens (1990) suggested that their involvement in the detoxification process of heavy metals might be a consequence of these other functions.

Metal transporters

Metal transporters are specialized proteins that are essential for maintaining intracellular metal homeostasis. These proteins mediate metal uptake by root cells and metal transfer between cells and organs. Metal transporters are involved in metal detoxification by mediating the transport of metal cations or metal chelates from the cytosol to the vacuolar compartment (Salt and Wagner, 1993). Different classes of metal transporters are discussed below.

Cation diffusion facilitator family protein

Cation diffusion facilitators (CDFs) are a family of heavy metal transporters implicated in the transport of Zn, Cd and Co from the cytoplasm to the outside of cells, or into organellar compartments. The CDF family of metal transporters was first described in bacteria (Nies, 1992) and subsequently also identified in yeast, animals and plants (Paulsen and Saier, 1997; van der Zaal *et al.*, 1999). Some members of the CDF family function in heavy metal uptake while others catalyse efflux. They are found in plasma membranes and also in intracellular membranes.

A plant CDF transporter gene designated *ZAT* (zinc transporter) was first cloned from *Arabidopsis* (van der Zaal *et al.*, 1999). Overexpression of *ZAT* in transgenic plants resulted in a significant increase in Zn resistance and a strongly increased Zn content under high Zn exposure than wild-type plants. Bloss *et al.* (2002) found that proteoliposomes containing the reconstituted heterologously expressed ZAT1 transporter from *Arabidopsis* could accumulate Zn, and subcellular localization studies confirmed its presence in the vacuolar membrane (Kobae *et al.*, 2004). *ZTP1*, a *ZAT* gene highly similar to the *Arabidopsis ZAT* gene, has also been cloned from the Zn hyperaccumulator, *T. caerulescens* (Assuncao *et al.*, 2001). Maser *et al.* (2001) suggested that the CDF family be renamed as the cation efflux family (CE) since there was lack of basic information about the energetics of the CDF transporters but the efflux function of the characterized family members was known. In addition, ZAT-related proteins should be called metal-tolerance proteins (MTPs) to allow more systematic naming as the plant CE family expands and to include plant CE family members that may have different metal transport characteristics than ZAT.

Persans *et al.* (2001) have characterized a CDF transporter (*TgMTP1*) from *T. goesingense* that is thought to be involved in the accumulation of metal ions in the shoot vacuoles. The *T. goesingense MTP1* conferred tolerance to a broad spectrum of heavy metals including Ni, Cd and Co based on the expression studies in yeast. Kobae *et al.* (2004) isolated a mutant of *Arabidopsis* lacking *AtMTP1* that was more sensitive to increased Zn levels in the medium, confirming that AtMTP1 is an important factor in keeping the cytosol free of excess Zn. They proposed that AtMTP1 was localized in the vacuolar membrane and involved in sequestration of excess Zn in the cytoplasm into vacuoles to maintain Zn homeostasis. A CDF transporter, *ShMTP1*, has also been characterized from *Stylosanthes hamata*, a tropical legume that can grow in acidic, high Mn^{2+} soils (Delhaize *et al.*, 2003). It imparts Mn^{2+} tolerance when expressed in yeast and *Arabidopsis* through sequestration into internal organelles and probably functions as a proton/Mn^{2+} antiporter. A search of the *Arabidopsis* genome shows the existence of eight genes for proteins with homology to the CDF family (Maser *et al.*, 2001).

Zn-regulated transporter and Fe-regulated transporter protein

Zn-regulated transporter and Fe-regulated transporter protein (ZIP) is named after zinc-regulated transporter ZRT1 of *Saccharomyces cerevisiae* and Fe-regulated transporter IRT1 of *A. thaliana* (Eide et al., 1996; Zhao and Eide 1996), two members of the ZIP family proteins identified initially. IRT1 appears to be involved in Fe accumulation and is expressed solely in roots of Fe limited plants (Maser et al., 2001). Later studies showed that IRT1 could also transport Zn^{2+}, Mn^{2+} and Cd^{2+} (Korshunova et al., 1999). It was suggested that the involvement of IRT1 in the accumulation of metals other than Fe was only under iron-limiting conditions based on the observation that iron-limited plants accumulated higher levels of other metal ions such as Zn, Mn and Cd (Cohen et al., 1998). About 85 ZIP family members have now been identified from bacteria, archaea and all types of eukaryotes, and grouped into four main subfamilies (Gaither and Eide, 2001). Various members of ZIPs have been shown to facilitate transport of Fe, Zn, Mn and Cd in plants (Guerinot, 2000).

A plant species may possess several ZIP genes. The *Arabidopsis* genome contains 18 such genes. This could be due to the fact that different ZIP proteins are involved in different phases of metal transport or may have different substrate specificity. Grotz *et al.* (1998) reported cloning of genes from *A. thaliana* encoding ZIP proteins, *ZIP1*, *ZIP2* and *ZIP3*. Expression in yeast of these closely related genes conferred Zn uptake activities. Based on competition assays, ZIP2 (but not ZIP1 or ZIP3) was strongly inhibited by Cd^{2+} and Cu^{2+}. *ZIP1* and *ZIP3* were closely related to *IRT1*, while *ZIP2* was more distantly related.

ZIP1 and *ZIP3* were induced in the roots in response to Zn deficiency. *ZIP2* mRNA could not be detected in plants under Zn-sufficient or -deficient conditions. A fourth *A. thaliana* ZIP family gene, *ZIP4*, was identified in the course of genome sequencing, which was also induced in both roots and shoots of Zn-deficient plants. No Zn or Fe uptake activity was detected in yeast expressing *ZIP4*. The ZIP1 and ZIP3 proteins located on plasma membrane were suggested to play a role in the uptake of Zn from the rhizosphere, whereas ZIP4, which contained a potential chloroplast targeting sequence, was thought to mediate transport of Zn into plastids.

Pence *et al.* (2000) isolated a Zn transporter which is also a member of the *ZIP* gene family cDNA (*ZNT1*) from the Zn/Cd-hyperaccumulating plant *T. caerulescens*. This transporter is homologous to *Arabidopsis* ZIP4 (Grotz *et al.*, 1998). *ZNT1* is highly expressed in roots and shoots of *T. caerulescens*, both under conditions of Zn deficiency and at normal nutritional Zn supply. But in the non-hyperaccumulator species, *T. arvense*, it is expressed under Zn-deficient conditions and also strongly downregulated at normal Zn supply (Assuncao *et al.*, 2001). It was suggested that the high expression of Zn transporters in *T. caerulescens*, irrespective of Zn availability, might be the major reason for the enhanced Zn uptake of this species. In general, alterations of the patterns of Zn-responsive transcriptional regulation of Zn transporters might play a pivotal role in Zn hyperaccumulation (Lasat *et al.*, 2000; Pence *et al.*, 2000).

Natural resistance associated macrophage protein

Natural resistance associated macrophage proteins (Nramps) comprise a novel family of highly conserved integral membrane proteins that are implicated in the transport of divalent metal ions in a wide range of organisms, including bacteria, fungi, plants and animals. Three Nramp gene homologues, *OsNramp1*, *OsNramp2* and a partial length of *OsNramp3*, were first reported in rice (Belouchi *et al.*, 1997). Currently, six members of the *Arabidopsis Nramp* gene family have been partially characterized (Williams *et al.*, 2000; Maser *et al.*, 2001). *AtNramp1*, *AtNramp3* and *AtNramp4* are upregulated in *Arabidopsis* roots under Fe-limiting conditions and confer tolerance at toxic concentrations of external Fe(II) (Curie *et al.*, 2000; Thomine *et al.*, 2000). *AtNramp3* is also linked to Cd uptake and sensitivity and its expression in the vascular bundles throughout the plant (roots, stems and leaves) suggest that it may function in long-distance metal transport (Thomine *et al.*, 2000). Bereczky *et al.* (2003) observed that *LeNramp1* expressed in tomato roots (*Lycopersicon esculentum*) was upregulated by Fe deficiency and opined that it may function in mobilizing Fe in the vascular parenchyma under limiting conditions. *Nramp* transcript from barley (*Hordeum vulgare*) is downregulated in the presence of Cd at adequate nitrogen supply, but strongly upregulated by Cd under N-deficiency (Finkemeier *et al.*, 2003).

In soybean, homologues of the *Nramp* family have been identified and proposed to participate in Fe(II) transport and Fe homeostasis in the nodule to support symbiotic N_2 fixation (Kaiser *et al.*, 2003). *Nramps* are also shown to be involved in the uptake of other metal ions such as Cu in yeast. Oomen *et al.* (2008) identified *TcNRAMP3* and *TcNRAMP4*, the closest homologues to *AtNRAMP3* and *AtNRAMP4* in *T. caerulescens* and characterized them by expression analysis and heterologous expression in yeast and *A. thaliana*. When expressed in yeast, TcNRAMP3 and TcNRAMP4 transport the same metals as their respective *A. thaliana* orthologues: Fe, Mn and Cd but not Zn for Nramp3; Fe, Mn, Cd and Zn for NRAMP4. Also, the inactivation of *AtNramp3* and *AtNramp4* in *A. thaliana* resulted in strong Cd and Zn hypersensitivity, which was fully rescued by *TcNramp3* or *TcNramp4* expression. Recently, Wei *et al.* (2009) reported that TcNRAMP3 was predominantly expressed in roots of *T. caerulescens* and the expression of TcNRAMP3 was induced by iron starvation and by the heavy metals Cd and Ni in roots. Furthermore, overexpression of TcNRAMP3 in tobacco resulted in slight Cd sensitivity of root growth and did not influence Ni resistance, suggesting their role in metal homeostasis.

Heavy metal ATPase

The heavy metal ATPases (P-type ATPases) form a super family of transporters that function to pump a variety of cations across cell membranes (Axelsen and Palmgren, 1998). P1B-ATPases1 are a subgroup of P-ATPases that are implicated in the transport of heavy metals (Cu^{+}, Cu^{2+}, Zn^{2+}, Co^{2+}) across biological membranes (Williams and Mills, 2005). This group of transporters has also been described as the CPx-ATPases because they contain a conserved intramembranous cysteine–proline cysteine/histidine/serine sequence (Solioz and Vulpe, 1996).

A. thaliana has eight members of the type 1B subfamily (Cobbett *et al.*, 2003), an unusually high number compared with non-plant species analysed to date. Baxter *et al.* (2003) designated these members from HMA1 to HMA8, while two of them, HMA6 and HMA7, were previously designated as PAA1 and RAN1, respectively. In *Arabidopsis*, HMA1, HMA2, HMA3 and HMA4 are likely to be Zn21/Co21/Cd21/Pb21 ATPases, whereas RAN1, PAA1 and HMA5 are candidate Cu21/Ag21 ATPases. Only two of the *Arabidopsis* HMAs have been assigned transport specificity to date. RAN1 (AtHMA7) falls into the Cu^{+}/Ag^{+} cluster and is important in the delivery of Cu ions across post-Golgi membranes to create functional ethylene receptors (Woeste and Kieber, 2000). PAA1 (AtHMA6) also falls into the Cu^{+}/Ag^{+} cluster and studies involving paa1 mutants suggest that it is responsible for the delivery of Cu to the plastid, particularly the Cu-dependent proteins plastocyanin and Cu/Zn superoxide dismutase in the plastid (Shikanai *et al.*, 2003).

ABC transporter

The ABC-transporter super family is one of the largest protein families (Henikoff *et al.*, 1997). The best-characterized among these are the 'full-size transporters'. In plants, knowledge about this group of proteins is still scarce. There are 53 genes in *Arabidopsis* encoding for full-size ABC proteins. They can be divided into two groups based on the topology: multidrug-resistance proteins (MDRs, also known as PGPs for P-glycoproteins) and multidrug-resistance-related proteins (MRPs). The role of ABC transporters in the uptake of Cd into the vacuole in the form of heavy metal chelates was shown by Salt and Rauser (1995), who found that the transport of a phytochelatin–Cd across the oat root tonoplast was energized by MgATP. The MRPs are considered to be involved in the transport of phytochelatin–Cd or GS–Cd complexes across the tonoplast (Rea *et al.*, 1998)

CAtion eXchanger

Cation exchangers (CAXs) are a group of proteins that export cations of the cytosol to maintain optimal ionic concentrations in the cell (Shigaki *et al.*, 2006). They are energized by the pH gradient established by proton pumps, such as H^{+} ATPase or H^{+} pyrophosphatase (Kamiya and Maeshima, 2004). The term 'CAX' was first used to describe the cation antiporters CAX1 and CAX2 from the model plant *A. thaliana* isolated and characterized by Hirschi *et al.* (1996). Different isoforms of *Arabidopsis* Ca^{2+} ATPases (ACA) and Na^{+}/H^{+} (NHX) antiporters are localized to different cellular locations (Yokoi *et al.*, 2002; Schiott *et al.*, 2004). Two plant CAXs from rice and soybean localized to the plasma membrane have been reported (Luo *et al.*, 2005; Qi *et al.*, 2005). Hirschi *et al.* (2000) attributed expression of CAX2 to the increased accumulation of Ca^{2+}, Cd^{2+} and Mn^{2+} and improved Mn^{2+} tolerance in tobacco plants. A mung bean (*V. radiata*) CAX gene, *VCAX1*, encoding a high capacity Ca^{2+} transporter, was able to suppress the Ca^{2+} sensitivity of yeast mutant K665 (Ueoka-Nakanishi *et al.*, 2000). However, the functions of many other CAX family genes in plants are unknown (Maser *et al.*, 2001).

Copper transporter family

Copper transporter family (CTR) members are responsible for high-affinity Cu transport in the plasma membrane and tonoplast (Lee *et al.*, 2002). A putative copper transporter, *A. thaliana* Cu transporter (COPT1) has been identified for the ability of its cDNA to complement functionally an *S. cerevisiae* mutant defective in high-affinity Cu uptake (Kampfenkel *et al.*, 1995). Sancenon *et al.* (2003)

isolated a five-member family (COPT1-5) of potential Cu transporters from *Arabidopsis*, including the previously identified COPT1. In addition, they studied the expression pattern of *COPT1* in transgenic plants expressing a reporter gene under the control of the *COPT1* promoter. *CaMV35S::COPT1* antisense transgenic plants showed a decrease in *COPT1* expression and the plants also displayed dramatically increased root length, which was completely reversed by the addition of copper.

Crops Used for Metal Accumulation

T. caerulescens is one of the best-known metal hyperaccumulators. It grows successfully on Ni-contaminated sites and accumulates 3% of its dry mass as metal. This plant also has a remarkable capacity to accumulate extremely high levels of non-labile Zn and Cd in its shoots (Lasat, 2002). *T. caerulescens* has been shown to accumulate Zn concentrations up to 26,000 ppm without exhibiting any toxic symptoms (Brown *et al.*, 1995a,b). Escarre *et al.* (2000) found that *T. caerulescens* extracted 22% of soil exchangeable Cd from contaminated sites and showed considerable ability to tolerate Cd. Several species of *Thlaspi* are known to hyperaccumulate more than one metal. *T. caerulescens* accumulates Cd, Ni, Pb and Zn; *T. goesingense* and *T. ochroleucum* accumulate Ni and Zn; and *T. rotundifolium* accumulates Ni, Pb and Zn (Kramer *et al.*, 1996; Prasad, 2005). *T. rotundifolium* can accumulate 0.13–8.2 g Pb/kg (dry mass) in their leaves (Reeves and Brooks, 1983).

Arabis gemmifera is also reported as a hyperaccumulator of Cd and Zn and its phytoextraction capacitiy almost equals *T. caerulescens* (Kubota and Takenaka, 2003). Studies with *Pista stratiotes* show that it can extract different metals like Ag, Cd, Cr, Cu, Hg, Ni, Pb and Zn at different rates, although concentrations higher than 5 mM result in growth inhibition (Odjegba and Fasidi, 2004). *A. halleri*, a close relative of the model plant *A. thaliana*, is reported as a Cd hypertolerant and Zn hyperaccumulator. Long *et al.* (2002) reported that *Sedum alfredii*, a high-biomass plant, could hyperaccumulate Zn. Wang *et al.* (2003) found that *Polygonum hydropiper* and *Rumex acetosa* had the potential for phytoremediating Zn-polluted soils. Chen *et al.* (2000) reported that vetiver grass could be used to remediate Cd-polluted soil. Ianneli *et al.* (2002) observed that *Phragmites australis* plants concentrated Cd in roots followed by shoots.

Pitchel *et al.* (2000) demonstrated the capacity of *Taraxacum officinale* and *Ambrosia artimisiifolia* to remove Pb and Cd successfully from soils in growth chamber studies. *B. juncea* has been identified to remove Pb from soil (Clemens, 2001). *Sesbania drummondii*, a leguminous fast-growing, high-biomass plant distributed in southern coastal regions of the USA can accumulate > 4% Pb in its shoot dry mass at 1000 mg/l $Pb(NO_3)_2$ under hydroponic conditions (Sahi *et al.*, 2002). According to Boonyapookana *et al.* (2005), *Helianthus annuus* can concentrate Pb in the leaf and stem and it could be used in the restoration of abandoned mines and factory sites. Chandrasekhar *et al.* (2005) noted that *Hemidesmus indicus* hyperaccumulated Pb mainly in its roots and shoots. Wei *et al.* (2004) recorded high concentration of Cd in *Solanum nigrum* and *Conyza canadensis* and tolerance to the combined action of Cu and Zn.

Tian *et al.* (2004) showed that *Spartina alterniflora* plants could tolerate Hg and that they possessed the ability to transform organic Hg into inorganic form. This inorganic Hg accumulates in the underground parts of the plants and is then transferred back to the soil by diffusion and permeation. The water fern (*Azolla caroliniana*) is capable of purifying water polluted by Hg and Cr by accumulating them into its tissues (Bennicelli *et al.*, 2004).

Lemna gibba has been found suitable for As phytoremediation of mine tailing waters because of its high accumulation potential (Mkandawire and Dudel, 2005). Ma *et al.* (2001) reported that Chinese brake fern (*Pteris vittata*) was able to hyperaccumulate high concentrations of As in its fronds. Arsenic transportation from roots to shoots in *P. vittata* was reported by Doucleff and Terry (2002). Ferns like *P. cretica*, *P. longofolia* and *P. umbrosa* are able to extract As from contaminated soils (Zhao *et al.*, 2002; Caille *et al.*, 2004).

Alyssium is proficient in storing Ni in the leaf epidermal cell vacuoles or in the basal portions of the trichomes and, as a result, is useful for Ni extraction from contaminated soils (Broadhurst *et al.*, 2004). *Stanleya pinnata* is known to accumulate Se (Parker *et al.*, 2003). *B. juncea* can extract and store nearly 2g Se/kg (dry mass) in the leaves (Orser *et al.*, 1999). *Austromyrtus bidwilli* is identified as an Mn hyperaccumulator species (Bidwell *et al.*, 2002). Another plant, *Phytolacca actinosa*, has also been shown as an Mn accumulator that grows rapidly with high biomass, making it suitable for the phytoremediation of Mn-contaminated soils (Xue *et al.*, 2004). *B. oleracea* var. *acephala* and *Iberis intermedia* are reported as hyperaccumulators of Tl (Al Najar *et al.*, 2005). *S. drummondii* can accumulate high Cu and Hg content in its shoots and roots under hydroponic conditions (Israr *et al.* 2006; Sahi *et al.*, 2007). A list of plant species having phytoremediation potential is presented in Table 22.1.

The phytoremediation efficiency of plants can be further improved through various approaches such as optimization of agronomic practices, amendments of soil through addition of organic acids and synthetic chelators like EDTA, and improvement in their biomass and metal accumulation ability through breeding.

Improving Plants for Phytoremediation Through Genetic Engineering

The process of metal uptake and translocation is monitored by several genes. Developing transgenic plants that extract more metal from the soil and accumulate it at their root surface or in their aboveground biomass can enhance plant metal remediation capability. This goal can be arrived at by more efficient sequestration of metals in plant storage compartments, overproduction of metal-chelating molecules, or by increasing activity of stress-responsive enzymes (Pilon-Smits and Pilon, 2002).

Table 22.1. List of metal hyperaccumulator plants.

Species	Metal	Reference
Miconia lutescens	Al	Bech *et al.*, 1997
Melastoma malabathricum	Al	Watanabe *et al.*, 1998
Pteris vittata L.	As	Ma *et al.*, 2001
Thlaspi caerulescens	Cd	Madico *et al.*, 1992
Sedum alfredii	Cd	Deng *et al.*, 2006
Haumaniastrum robertii	Co	Baker and Brooks, 1989
Leersia hexandra Swartz	Cr	Zhang *et al.*, 2007
Aeolanthus biformifolius	Cu	Morrison *et al.*, 1979
Ipomea alpina	Cu	Baker and Walker, 1990
Phytolacca acinosa Roxb	Mn	Xue *et al.*, 2004
Thlaspi goesingense	Ni	Reeves and Brooks, 1983
Alyssum bertholoni	Ni	Brooks and Radford, 1978
Alyssum argenteum All	Ni	McCutcheon and Schnoor, 2003
Berkheya codii	Ni	Brooks, 1998
Psychotria douarrei	Ni	Baker *et al.*, 1985
Brassica juncea	Pb	Lombi *et al.*, 2001
Thlaspi rotundifolium	Pb	Reeves and Brooks, 1983
Minuartia verna	Pb	Ernst, 1974
Sesbania drummondii	Pb	Sahi *et al.*, 2002
Astragalus racemosus	Se	Beath *et al.*, 1937
Arabidopsis halleri	Zn	Ernst, 1968
Thlaspi caerulescens	Zn	Ernst, 1982
Sedum alfredii	Zn	Deng *et al.*, 2006

A transgenic approach has been employed to bioengineer plants capable of remediating methylmercury-contaminated soils. Transgenic *A. thaliana* plants expressing bacterial *merA* and *merB* were capable of growing on media containing 50-fold higher methylmercury concentrations than the wild-type plant (Bizily *et al.*, 2000). Tomato plants engineered to express the bacterial gene 1-aminocyclopropane-1-carboxylic acid (ACC) deaminase have shown enhanced metal tolerance and accumulated greater metal levels (Cd, Co, Cu, Mg, Ni, Pb and Zn) than the non-transgenic plants (Grichko *et al.*, 2000). Evans *et al.* (1992) generated transgenic *A. thaliana* plants expressing *PsMTA* gene encoding for the metallothionein-like compounds from pea.

Transgenic *A. thaliana* plants overexpressing YCF1 (yeast protein which detoxifies Cd) showed enhanced tolerance and accumulated greater amounts of Cd and Pb (Song *et al.*, 2003). Transgenic Indian mustard plants overexpressing the γ-glutamylcysteine synthase or glutathione synthetase and adenosine triphosphate sulfurylase were found to be more efficient in removing a greater number of metals from the contaminated soil collected from a USEPA (US Environmental Protection Agency) superfund site compared with wild-type plants (Bennett *et al.*, 2003).

Brassica plants genetically modified to express *E. coli gsh I* gene coding for γ-glutamylcysteine synthetase targeted to the plastids showed increased tolerance to Cd, higher concentrations of phytochelatins, γ-GluCys and glutathione, and total non-protein thiols compared with wild-type seedlings (Zhu *et al.*, 1999). *A. thaliana* plants genetically modified with two genes, *AsrC* (arsenate reductase) and *γ-ECS* (γ-glutamylcysteine synthetase) from *E. coli*, obtained 4- to 17-fold greater fresh shoot weight and accumulated 2- to 3-fold more As/g of tissue than wild-type or plants expressing *AsrC* or γ-ECS alone (Dhanker *et al.*, 2002).

Grichko *et al.* (2000) generated transgenic tomato plants expressing the bacterial gene 1-aminocyclopropane-1-carboxylic acid (ACC) deaminase. Transgenic plants were found to accumulate more metals like Cd, Co, Cu, Mg, Ni, Pb or Zn in their tissues than the non-transgenic plants.

Indian mustard plants engineered to overexpress ATP sulfurylase or cystathionine-γ synthase (CGS) have shown enhanced Se tolerance than wild-type plants with Se concentrated in their shoots (Van Huysen *et al.*, 2004). Transgenic *B. juncea* overexpressing the selenocysteine methyltransferase (SMT) gene from *Astragalus bisulcatus* demonstrated increased tolerance to Se compounds, in particular selenate (LeDuc *et al.*, 2004).

Limitations of Phytoremediation and Future Prospects

Phytoremediation is a promising and emerging field with great potential for future applications. It has gained acceptance as an effective green technology to clean up metal-contaminated sites. But, phytoremediation techniques have several limitations. This approach is slower with respect to the time needed to achieve the anticipated result compared to conventional approaches. Factors like soil properties, extent of metal contamination and level of metals in the soil, as well as climatic conditions, determine the degree of success of the clean-up operation. The bioavailability of the pollutant also limits the use of phytoremediation.

In the past few years, substantial progress has been made in understanding the mechanisms underlying the processes involved in the phytoextraction of metals, especially those involving the uptake and hyperaccumulation of metals in plants. The insights gained in this area, however, are still incomplete, awaiting further elucidation of the regulatory processes involved in the pathways that lead to metal accumulation and tolerance in plants. Wide-scale commercial applications of phytoremediation will depend on the success achieved in overcoming currently limiting factors through a multidisciplinary approach integrating areas such as breeding, agronomy, soil biology, environmental biology and microbiology.

References

Al-Najar, H., Kaschl, A., Schulz, R. and Romheld, V. (2005) Effects of thallium fractions in the soil and pollution origin in thallium uptake by hyperaccumulator plants: a key factor for assessment of phytoextraction. *International Journal of Phytoremediation* 7, 55–67.

Assuncao, A.G.L., Costa Martins, P.D.A., De Folter, S., Vooijs, R., Schat, H. and Aarts, M.G.M. (2001) Elevated expression of metal transporter genes in three accessions of the metal hyperaccumulator *Thlaspi caerulescens*. *Plant Cell and Environment* 24, 217–226.

Axelsen, K.B. and Palmgren, M.G. (1998) Evolution of substrate specificities in the P-type ATPase superfamily. *Journal of Molecular Evolution* 46, 84–101.

Baker, A.J.M. and Brooks, R.R. (1989) Terrestrial higher plants which hyperaccumulate metallic-elements: a review of their distribution, ecology and phytochemistry. *Biorecovery* 1, 81–126.

Baker, A.J.M. and Walker, P.L. (1990) Ecophysiology of metal uptake by tolerant plant. In: Shaw, A.J. (ed.) *Heavy Metal Tolerance in Plants: Evolutionary Aspects*. CRC Press, Boca Raton, Florida, pp. 155–177.

Baker, A.J.M., Brooks, R.R. and Kersten, W.J. (1985) Accumulation of nickel by *Psychotria* species from the Pacific Basin. *Taxonomy* 34, 89–95.

Baxter, I., Tchieu, J., Sussman, M.R., Boutry, M., Palmgren, M.G., Gribskov, M., Harper, J.F. and Axelsen, K.B. (2003) Genomic comparison of P-type ATPase ion pumps in *Arabidopsis* and rice. *Plant Physiology* 132, 618–628.

Beath, O.A., Eppsom, H.F. and Gillbert, G.S. (1937) Selenium distribution in and seasonal variation of vegetation type occurring on seleniferous soils. *Journal of the American Pharmacists Association* 26, 394–405.

Bech, J., Poschenrieder, C., Llugany, M., Barcelo, J., Tume, P., Tobias, F.J., Barranzuela, J.L. and Vasquez E.R. (1997) Arsenic and heavy metal contamination of soil and vegetation around a copper mine in Northern Peru. *Science of the Total Environment* 203, 83–91.

Belouchi, A., Kwan, T. and Gros, P. (1997) Cloning and characterization of the OsNramp family from *Oryza sativa*, a new family of membrane proteins possibly implicated in the transport of metal ions. *Plant Molecular Biology* 33, 1085–1092.

Bennett, L.E., Burkhead, J.L., Hale, K.L., Terry, N., Pilon, M. and Pilon-Smits, E.A. (2003) Analysis of transgenic Indian mustard plants for phytoremediation of metal-contaminated mine tailings. *Journal of Environmental Quality* 32, 432–440.

Bennicelli, R., Stepniewska, Z., Banach, A., Szajnocha, K. and Ostrowski, J. (2004) The ability of *Azolla caroliniana* to remove heavy metals (Hg (II), Cr (III), Cr (VI) from municipal waste water. *Chemosphere* 55, 141–146.

Bereczky, Z., Wang, H.-Y., Schubert, V., Ganal, M. and Bauer, P. (2003) Differential regulation of *nramp* and *irt* metal transporter genes in wild type and iron uptake mutants of tomato. *The Journal of Biological Chemistry* 278, 24697–24704.

Bidwell, S.D., Batianoff, G.N., Woodrow, J.E. and Sommer-Knudson, J. (2002) Hyperacumulation of manganese in the rainforest tree *Austromyrtus bidwilli* (Myrtaceae) from Queensland, Australia. *Functional Plant Biology* 29, 899–905.

Bizily, S.P., Rugh, C.L. and Meagher, R.B. (2000) Phytodetoxification of hazardous organomercurials by genetically engineered plants. *Nature Biotechnology* 18, 213–217.

Bloss, T., Clemens, S. and Nies, D.H. (2002) Characterization of the ZAT1p zinc transporter from *Arabidopsis thaliana* in microbial model organisms and reconstituted proteoliposomes. *Planta* 214, 783–791.

Boonyapookana, B., Parkplan, P., Techapinyawat, S., DeLaune, R.D. and Jugusujinda, A. (2005) Phytoaccumulation of lead by sunflower (*Helianthus annuus*), tobacco (*Nicotiana tabacum*), and vetiver (*Vetiver zizanioides*). *Journal of Environmental Science and Health* A 40, 117–137.

Brooks, R.R. (1998) *Plants that Hyperaccumulate Heavy Metals*. CAB International, Wallingford, UK, p. 380.

Brooks, R.R. and Radford, C.C. (1978) Nickel accumulation by European species of the genus Alyssum. *Proceedings of the Royal Society (London)* Sect. B203, 387–403.

Broadhurst, C.L., Chaney, R.L., Angle, J.S., Maugel, T.K., Erbe, E.F. and Murphy C.A. (2004) Simultaneous hyperaccumulation of nickel, manganese, and calcium in *Alyssium* leaf trichomes. *Environmental Science and Technology* 38, 5797–5802.

Brown, S.L., Chaney, R.L., Angle, J.S. and Baker, A.J.M. (1995a) Zinc and cadmium uptake by hyperaccumulator *Thlaspi caerulescens* grown in nutrient solution. *Soil Science Society of America Journal* 59, 125–133.

Brown, S.L., Chaney, R.L., Angle, J.S. and Baker, A.J.M. (1995b) Zinc and cadmium uptake by hyperaccumulator *Thlaspi caerulescens* and metal tolerant *Silene vulgaris* grown on sludge-amended soils. *Environmental Science and Technology* 29, 1581–1585.

Caille, N., Swanwick, S., Zhao, F.J. and McGrath, S.P. (2004) Arsenic hyperaccumulation by *Pteris vittata* from arsenic contaminated soils and the effect of liming and phosphate fertilization. *Environmental Pollution* 132, 113–120.

Chandrasekhar, K., Kamala, C.T., Chary, N.S., Balaram, V. and Garcia, G. (2005) Potential of *Hemidesmus indicus* for phytoextraction of lead from industrially contaminated soils. *Chemosphere* 58, 507–514.

Chaney, R.L., Mallik, M., Li, Y.M., Brown, S.L., Brewer, E.P., Angle, J.S. and Baker, A.J.M. (1997) Phytoremediation of soil metals. *Current Opinion in Biotechnology* 8, 279–84.

Chen, H.M., Zheng, C.R., Tu, C. and Shen, Z.G. (2000) Chemical methods and phytoremediation of soil contaminated with heavy metals. *Chemosphere* 41, 229–234.

Cieslinski, G.K.C., Van Rees, J., Szmigielska, A.M., Krishnamurti, G.S.R. and Huang, P.M. (1998) Low-molecular-weight organic acids in rhizosphere soils of durum wheat and their effect on cadmium bioaccumulation. *Plant and Soil* 203, 109–117.

Clemens, S. (2001) Molecular mechanisms of plant metal homeostasis and tolerance. *Planta* 212, 475–486.

Clemens, S., Kim, E.J., Neumann, D. and Schroeder, J.I. (1999) Tolerance to toxic metals by a gene family of phytochelatin synthases from plant and yeast. *The European Molecular Biology Organization Journal* 18, 3325–3333.

Clemens, S., Palmgren, M.G. and Kramer, U. (2002) A long way ahead: understanding and engineering plant metal accumulation. *Trends in Plant Science* 7, 309–315.

Cobbett, C. and Goldsbrough, P. (2002) Phytochelatins and metallothioneins: roles in heavy metal detoxification and homeostasis. *Annual Review of Plant Biology* 53, 159–182.

Cobbett, C.S., Hussain, D. and Haydon, M.J. (2003) Structural and functional relationships between type 1B heavy metal transporting P-type ATPases in *Arabidopsis*. *New Phytologist* 159, 315–321.

Cohen, C.K., Fox, T.C., Gravin, D.F. and Kochian, C.V. (1998) The role of iron-deficiency stress responses in stimulating heavy-metal transport in plants. *Plant Physiology* 116, 1063–1072.

Curie, C., Alonso, J.M., Le Jean, M., Ecker, J.R. and Briat, J.F. (2000) Involvement of Nramp1 from *Arabidopsis thaliana* in iron transport. *Biochemical Journal* 347, 749–755.

De Knecht, J.A., Koevoets, P.L.M., Verkleij, J.A.C. and Ernst, W.H.O. (1992) Evidence against a role for phytochelatins in naturally selected increased cadmium tolerance in *Silene vulgaris* (Moench) Garcke. *New Phytologist* 122, 681–688.

Delhaize, E., Kataoka, T., Hebb, D.M., White, R.G. and Ryan, P.R. (2003) Genes encoding proteins of the cation diffusion facilitator family that confer manganese tolerance. *The Plant Cell* 15, 1131–1142.

Deng, D.M., Shu, W.S., Zhang, J., Zou, H.L., Lin, Z., Ye, Z.H. and Wong, M.H. (2006) Zinc and cadmium accumulation and tolerance in populations of *Sedum alfredii*. *Environmental Pollution* 147, 381–386.

Dhanker, O.P., Li, Y.J., Rosen, B.P., Shi, J., Salt, D., Senecoff, J.F., Sashti, N.A. and Meagher, R.B. (2002) Engineering tolerance and hyperaccumulation of arsenic in plants by combining arsenate reductase and g-glutamylcysteine synthase expression. *Nature Biotechnology* 20, 1140–1145.

Dietz, K.-J., Baier, M. and Kramer, U. (1999) Free radicals and reactive oxygen species as mediators of heavy metal toxicity in plants. In: Prasad, M.N.V. and Hagemeyer, J. (eds) *Heavy Metal Stress in Plants: From Molecules to Ecosystems*. Springer Verlag, Berlin, pp. 73–97.

Doucleff, M. and Terry, N. (2002) Pumping out the arsenic. *Nature Biotechnology* 20, 1094–1095.

Eide, D., Broderius, M., Fett, J. and Guerinot, M.L. (1996) A novel iron-regulated metal transporter from plants identified by functional expression in yeast. *Proceedings of the National Academy of Sciences of the United States of America* 93, 5624–5628.

Elmayan, T. and Tepfer, M. (1994) Synthesis of a bifunctional metallothionein/-glucuronidase fusion protein in transgenic tobacco plants as a means of reducing leaf cadmium levels. *The Plant Journal* 6, 433–440.

Ernst, W.H.O. (1968) Zur Kenntnis der Soziologia und Okologie der Schwermetallvegetation Grossbritaniens. *Berichte der Deutschen Botanischen Gesellschaft*. 81, 116–124.

Ernst, W.H.O. (1974) *Schwermetallvegetation Grossbritaniens*. Gustav Fischer Verlag, Stuttgart, Germany.

Ernst, W.H.O. (1982) Schwermetallvegetation. In: Kinzel, H. (ed.) *Pflanzenokologie und Mineralstoffwechsel*. Verlag Eugen Ulmer, Stuttgart, Germany, pp. 472–506.

Ernst, W.H.O., Verkleij, J.A.C. and Schat, H. (1992) Metal tolerance in plants. *Acta Botanica Neerlandica* 41, 229–248.

Escarre, J., Lefebre, C., Gruber, W., LeBlanc, M., Lepart, J., Riviere, Y. and Delay. B. (2000) Zinc and cadmium hyperaccumulation by *Thlaspi caerulescens* from metalliferous and non-metalliferous sites in Meditteranean area: implications for phytoremediation. *New Phytologist* 145, 429–437.

Evans, K.M., Gatehouse, J.A., Linsay, W.P., Shi, J., Tommey, A.M. and Robinson, N.J. (1992) Expression of the pea metallothionein-like gene *PsMTA* in *Escherichia coli* and *Arabidopsis thaliana* and analysis of trace metal ion accumulation: implications for *PsMTA* function. *Plant Molecular Biology* 20, 1019–1028.

Fan, T.W., Colmer, T.D., Lane, A.N. and Higashi, R.M. (1993) Determination of metabolites by 'H NMR and GC: analysis for organic osmolytes in crude tissue extracts. *Analytical Biochemistry* 214, 260–271.

Finkemeier, I., Kluge, C., Metwally, A., Georgi, M., Grotjohann, N. and Dietz, K.-J. (2003) Alterations in Cd-induced gene expression under nitrogen deficiency in *Hordeum vulgare. Plant, Cell and Environment* 26, 821–833.

Gaither, L.A. and Eide, D.J. (2001) Eukaryotic zinc transporters and their regulation. *Biometals* 14, 251–270.

Gekeler, W., Winnaker, E.-L. and Zenk, M.H. (1989) Survey of the plant kingdom for the ability to bind heavy metals through phytochelatins. *Z Naturforsch Teil C* 44, 361–369.

Grichko, V.P., Filby, B. and Glick, B.R. (2000) Increased ability of transgenic plants expressing the bacterial enzyme ACC deaminase to accumulate Cd, Co, Cu, Ni, Pb, and Zn. *Journal of Biotechnology* 81, 45–53.

Grill, E., Winnaker, E.-L. and Zenk, M.H. (1985) Phytochelatins: the principal heavy metal-complexing peptides of higher plants. *Science* 230, 674–676.

Grill, E., Loffler, S., Winnaker, E.-L. and Zenk, M.H. (1989) Phtychelatins, the heavy-metal-binding peptides of plants, are synthesized from glutathione by a specific gamma-glutamylcysteine dipeptidyl transpeptidase (phytochelatin synthase). *Proceedings of the National Academy of Sciences of the United States of America* 86, 6838–6842.

Grotz, N., Fox, T., Connolly, E., Park, W., Guerinot, M.L. and Eide, D. (1998) Identification of a family of zinc transporter genes from *Arabidopsis* that respond to zinc deficiency. *Proceedings of the National Academy of Sciences of the United States of America* 95, 7220–7224.

Guerinot, M.L. (2000) The Zip family of metal transporters. *Biochimica et Biophysica Acta* 1465, 190–198.

Ha, S.-B., Smith, A.P., Howden, R., Dietrich, W.M., Bugg, S., O'Connell, M.J., Goldsbrough, P.B. and Cobbett, C.S. (1999) Phytochelatin synthase genes from *Arabidopsis* and the yeast, *Schizosaccharomyces pombe. The Plant Cell* 11, 1153–1164.

Haag-Kerwer, A., Schafer, H.J., Heiss, S., Walter, C. and Rausch, T. (1999) Cadmium exposure in *Brassica juncea* causes a decline in transpiration rate and leaf expansion without effect on photosynthesis. *Journal of Experimental Botany* 50, 1827–1835.

Hall, J.L. (2002) Cellular mechanisms for heavy metal detoxification and tolerance. *Journal of Experimental Botany* 53, 1–11.

Hartley-Whitaker, J., Ainsworth, G., Vooijs, R., Ten Bookum, W., Schat, H. and Meharg, A.A. (2001) Phytochelatins are involved in differential arsenate tolerance in *Holcus lanatus. Plant Physiology* 126, 299–306.

Heckathorn, S.A., Mueller, J.K., LaGuidice, S., Zhu, B., Barrett, T., Blair, B. and Dong, Y. (2004) Chloroplast small heat-shock proteins protect photosynthesis during heavy metal stress. *American Journal of Botany* 91, 1312–1318.

Henikoff, S., Greene, E.A., Pietrokovski, S., Bork, P., Attwood, T.E. and Hood, L. (1997) Gene families: the taxonomy of protein paralogs and chimeras. *Science* 278, 609–614.

Hirschi, K., Zhen, R., Cunningham, K.W., Rea, P.A. and Fink, G.R. (1996) CAX1, a H+/Ca2+ antiporter from *Arabidopsis. Proceedings of the National Academy of Sciences of the United States of America* 93, 8782–8786.

Hirschi, K.D., Korenkov, V.D., Wilganowski, N.L. and Wagner, G.J. (2000) Expression of *Arabidopsis* CAX2 in tobacco. Altered metal accumulation and increased manganese tolerance. *Plant Physiology* 124, 125–133.

Howden, R. and Cobbett, C.S. (1992) Cadmium-sensitive mutants of *Arabidopsis thaliana. Plant Physiology* 100, 100–107.

Howden, R., Goldsbrough, P.B., Andersen, C.R. and Cobbett, C.S. (1995a) Cadmium-sensitive, cad1 mutants of *Arabidopsis thaliana* are phytochelatin deficient. *Plant Physiology* 107, 1059–1066.

Howden, R., Goldsbrough, P.B., Anderson, C.R. and Cobbett, C.S. (1995b) A cadmium-sensitive, glutathione-deficient mutant of *Arabidopsis thaliana. Plant Physiology* 107, 1067–1073.

Ianneli, M.A., Pietrini, F., Fiore, L., Petrilli, L. and Massacci, A. (2002) Antioxidant response in *Phragmites australis* plants. *Plant Physiology* 107, 1059–1066.
Inouhe, M., Ito, R., Ito, S., Sasada, N., Tohoyama, H. and Joho, M. (2000) Azuki bean cells are hypersensitive to cadmium and do not synthesize phytochelatins. *Plant Physiology* 123, 1029–1036.
Israr, M., Sahi, S.V., Datta, R. and Dibyendu, Sarkar, D. (2006) Bioaccumulation and physiological effects of mercury in *Sesbania drummondii. Chemosphere* 65, 591–598.
Kaiser, B.N., Moreau, S., Castelli, J., Thomson, R., Lambert, A., Bogliolo, S., Puppo, A. and Day, D.A. (2003) The soybean NRAMP homologue, GmDMT1 is a symbiotic divalent metal transporter capable of ferrous iron transport. *Plant Journal* 35, 295–304.
Kamiya, T. and Maeshima, M. (2004) Residues in internal repeats of the rice cation/H+ exchanger are involved in the transport and selection of cations. *The Journal of Biological Chemistry* 279, 812–819.
Kampfenkel, K., Kushnir, S., Babiychuk, E., Inze, D. and Van Montagu, M. (1995) Molecular characterization of a putative *Arabidopsis thaliana* copper transporter and its yeast homologue. *The Journal of Biological Chemistry* 270, 28479–28486.
Kinnersely, A.M. (1993) The role of phytochelates in plant growth and productivity. *Plant Growth Regulation* 12, 207–217.
Kobae, Y., Uemura, T., Sato, M.H., Ohnishi, M., Mimura, T. and Maeshima, M. (2004) Zinc transporter of *Arabidopsis thaliana* AtMTP1 is localized to vacuolar membranes and implicated in zinc homeostasis. *Plant and Cell Physiology* 45, 1749–1758.
Korshunova, Y.O., Eide, D., Clark, W.G., Guerinot, M.L. and Pakrasi, H.B. (1999) The IRT1 protein from *Arabidopsis thaliana* is a metal transporter with a broad substrate range. *Plant Molecular Biology* 40, 37–44.
Kramer, U., Cotter-Howells, J.D., Charnock, J.M., Baker, A.J.M. and Smith, J.A.C. (1996) Free histidine as a metal chelator in plants that accumulate nickel. *Nature* 379, 635–638.
Kubota, H. and Takenaka, C. (2003) *Arabis gemmifera* is a hyperaccumulator of Cd and Zn. *International Journal of Hazardous Substance* 5, 197–220.
Lane, B., Kajiota, R. and Kennedy, T. (1987) The wheat-germ Ec protein is a zinc-containing metallothionein. *Biochemistry and Cell Biology* 65, 1001–1005.
Lasat, M.M. (2002) Phytoextraction of toxic metals: a review of biological mechanisms. *Journal of Environmental Quality* 31, 109–120.
Lasat, M.M., Pence, N.S., Garvin, D.F., Ebbs, S.D. and Kochian L.V. (2000) Molecular physiology of zinc transport in the Zn hyperaccumulator *Thlaspi caerulescens. Journal of Experimental Botany* 51, 71–79.
LeDuc, D.L., Tarun, A.S., Montes-Bayon, M., Meija, J., Malit, M.F., Wu, C.P., AbdelSamie, M., Chiang, C.Y., Tagmount, A., DeSouza, M., Neuhierl, B., Bock, A., Caruso, J. and Terry N. (2004) Overexpression of selenocysteine methyltransferase in *Arabidopsis* and Indian mustard increases selenium tolerance and accumulation. *Plant Physiology* 135, 377–383.
Lee, J., Pena, M.M.O., Nose, Y. and Thiele, D.J. (2002) Biochemical characterization of the human copper transporter Ctr1. *The Journal of Biological Chemistry* 277, 4380–4387.
Lewis, S., Handy, R.D., Cordi, B., Billinghurst, Z. and Depledge, M.H. (1999) Stress proteins (HSPs): methods of detection and their use as an environmental biomarker. *Ecotoxicology* 8, 351–368.
Lombi, E., Zhao, F.J., Dunham, S.J. and McGrath, S.P. (2001) Phytoremediation of heavy metal contaminated soils: natural hyperaccumulation versus chemically-enhanced phytoextraction. *Journal of Environmental Quality* 30, 1919–1926.
Long, X.X., Yang, X.E., Ye, Z.Q., Ni, W.Z. and Shi, W.Y. (2002) Differences of uptake and accumulations of zinc in four species of *Sedum. Acta Botanica Sinica* 44, 152–157.
Luo, G.Z., Wang, H.W., Huang, J., Tian, A.G., Wang, Y.J., Zhang, J.S. and Chen, S.Y. (2005) A putative plasma membrane cation/proton antiporter from soybean confers salt tolerance in *Arabidopsis. Plant Molecular Biology* 59, 809–820.
Ma, J.F., Zheng, H.J. and Matsumoto, H. (1997) Detoxifying aluminum with buckwheat. *Nature* 390, 569–570.
Ma, L.Q., Komar, K.M., Tu, C., Zhang, W., Cai, Y. and Kennelley, E.D. (2001) A fern that hyperaccumulates arsenic. *Nature* 409, 579.
McCutcheon, S.C. and Schnoor, J.L. (2003) *Phytoremediation*. John Wiley and Sons, New Jersey.
Madico, J., Poschenrieder, C., Vazquez, M.D. and Barcelo, J. (1992) Effects of high zinc and cadmium concentrations on the metallophyte *Thlaspi caerulescens* (Brassicaceae). *Planta* 2, 495–504.
Maiti, I.B., Wagner, G.J., Yeargan, R. and Hunt, A.G. (1989) Inheritance and expression of the mouse metallothionein gene in tobacco. *Plant Physiology* 91, 1020–1024.

Martens, S.N. and Boyd, R.S. (2002) The defensive role of Ni hyperaccumulation by plants: a field experiment. *American Journal of Botany* 89, 998–1003.
Maser, P., Thomine, S., Schroeder, J.I., Ward, J.M., Hirschi, K., Sze, H., Talke, I.N., Amtmann, A., Maathuis, F.J., Sanders, D., Harper, J.F., Tchieu, J., Gribskov, M., Persans, M.W., Salt, D.E., Kim, S.A. and Guerinot, M.L. (2001) Phylogenetic relationships within cation transporter families of *Arabidopsis*. *Plant Physiology* 126, 1646–1667.
Misra, S. and Gedamu, L. (1989) Heavy metal tolerant transgenic *Brassica napus* L. and *Nicotiana tabacum* L. plants. *Theoretical and Applied Genetics* 78, 161–168.
Mkandawire, M. and Dudel, E.G. (2005) Accumulation of arsenic in *Lemna gibba* L. (duckweed) in tailing waters of two abandoned uranium mining sites in Saxony, Germany. *Science of the Total Environment* 336, 81–89.
Morrison, R.S., Brooks, R.R., Reeves, R.D. and Malaisse, F. (1979) Copper and cobalt uptake by metallophytes from Zaïre. *Plant and Soil* 53, 535–539.
Murasugi, A., Wada, C. and Hayashi, Y. (1981) Cadmium-binding peptide induced in fission yeast, *Schizosacharomyces pombe*. *Journal of Biochemistry* 90, 1561–1564.
Murphy, A., Zhou, J., Goldsbrough, P.B. and Taiz, L. (1997) Purification and immunological identification of metallothioneins 1 and 2 from *Arabidopsis thaliana*. *Plant Physiology* 113, 1293–1301.
Neumann, D., Lichtenberger, O., Gunther, D., Tschiersch, K. and Nover, L. (1994) Heat shock proteins induce heavy metal tolerance in higher plants. *Planta* 194, 360–367.
Neumann, D., Nieden, U.Z., Lichtenberger, O. and Leopold, I. (1995) How does *Armeria maritime* tolerate high heavy metal concentrations? *Journal of Plant Physiology* 146, 704–717.
Nies, D.H. (1992) CzcR and CzcD, gene products affecting regulation of cobalt, zinc and cadmium resistance (czc) in *Alcaligenes eutrophus*. *Journal of Bacteriology* 174, 8102–8110.
Odjegba, V.J. and Fasidi, I.O. (2004) Accumulation of trace elements by *Pistia stratiotes*: implications for phytoremediation. *Ecotoxicology* 13, 637–646.
Oomen, R.J.F.J., Wu, J., Lelievre, F., Blanchet, S., Richaud, P., Barbier-Brygoo, H., Aarts, M.G.M. and Thomine, S. (2008) Functional characterization of NRAMP3 and NRAMP4 from the metal hyperaccumulator *Thlaspi caerulescens*. *New Phytologist* 181, 637–650.
Orser, C.S., Salt, D.E., Pickering, I.I., Epstein, A. and Ensley, B.D. (1999) *Brassica* plants to provide enhanced mineral nutrition: selenium phytoenrichment and metabolic transformation. *Journal of Medicinal Food* 1, 253–261.
Parker, D.R., Feist, L.J., Varvel, T.W., Thomason, D.N. and Zhang, Y.Q. (2003) Selenium phytoremediation potential of *Stanleya pinnata*. *Plant and Soil* 249, 157–165.
Paulsen, I.T. and Saier, M.H. Jr (1997) A novel family of ubiquitous heavy metal ion transport proteins. *Journal of Membrane Biology* 156, 99–103.
Pence, N.S., Larsen, P.B., Ebbs, S.D., Letham, D.L.D., Lasat, M.M., Garvin, D.F., Eide, D. and Kochian, L.V. (2000) The molecular physiology of heavy metal transport in the Zn/Cd hyperaccumulator *Thlaspi caerulescens*. *Proceedings of the National Academy of Sciences of the United States of America* 97, 4956–4960.
Persans, M.W., Nieman, K. and Salt, D.E. (2001) Functional activity and role of cation-efflux family members in Ni hyperaccumulation in *Thlaspi goesingense*. *Proceedings of the National Academy of Sciences of the United States of America* 98, 9995–10000.
Pilon-Smits, E. and Pilon, M. (2002) Phytoremediation of metals using transgenic plants. *Critical Reviews in Plant Science* 21, 439–456.
Pitchel, J., Kuroiwa, K. and Sawyerr, H.T. (2000) Distribution of Pb, Cd and Ba in soils and plants of two contaminated soils. *Environmental Pollution* 110, 171–178.
Prasad, M.N.V. (1999) Metallothioneins and metal binding complexes in plants. In: Prasad, M.N.V. and Hagemeyer, J. (ed.) *Heavy Metal Stress in Plants: From Molecules to Ecosystems*. Springer Verlag, Berlin, pp. 51–72.
Prasad, M.N.V. (2005) Nickelophilous plants and their significance in phytotechnologies. *Brazilian Journal of Plant Physiology* 17, 113–128.
Qi, B.S., Li, C.G., Chen, Y.M., Lu, P.L., Hao, F.S., Shen, G.M., Chen, J. and Wang, X.C. (2005) Functional analysis of rice Ca^{2+}/H^+ antiporter OsCAX3 in yeast and its subcellular localization in plant. *Progress in Biochemistry Biophysics* 32, 876–881 [in Chinese].
Rauser, W.E. (1995) Phytochelatins and related peptides: structure, biosynthesis, and function. *Plant Physiology* 109, 1141–1149.

Rauser, W.E. (1999) Structure and function of metal chelators produced by plants – the case of organic acids, amino acids, phytin and metallothioneins. *Cell Biochemistry and Biophysics* 31, 19–48.

Rea, P.A., Li, Z.-S., Lu, Y.-P. and Drozdowicz, Y.M. (1998) From vacuolar GS-X pumps to multispecific ABC transporters. *Annual Review of Plant Physiology and Plant Molecular Biology* 49, 727–760.

Reeves, R.D. and Brooks, R.R. (1983) Hyperaccumulation of lead and zinc by two metallophytes from mining areas of Central Europe. *Environmental Pollution Series A, Ecological and Biological* 31, 277–285.

Robinson, N.J., Tommey, A.M., Kuske, C. and Jackson, P.J. (1993) Plant metallothioneins. *Biochemistry Journal* 295, 1–10.

Romheld, V. and Marschner, H. (1986) Mobilization of iron in the rhizosphere of different plant species. *Advances in Plant Nutrition* 2, 155–204.

Ross, S.M. (1994) *Toxic Metals in Soil–Plant Systems.* Wiley, Chichester, UK.

Sahi, S.V., Bryant, N.L., Sharma, N.C. and Singh, S.R. (2002) Characterization of a lead hyperaccumulator shrub, *Sesbania drummondii. Environmental Science and Technology* 36, 4676–4680.

Sahi, S.V., Israr, M., Srivastava, A.K., Gardea-Torresdey, J.L. and Parsons, J.G. (2007) Accumulation, speciation and cellular localization of copper in *Sesbania drummondii. Chemosphere* 67, 2257–2266.

Salt, D.E. and Kramer, U. (2000) Mechanisms of metal hyperaccumulation in plants. In: Raskin, I. and Ensely, B. (eds) *Phytoremediation of Toxic Metals.* John Wiley and Sons, Inc, New York, pp. 231–246.

Salt, D.E. and Rauser, W.E. (1995) MgATP-dependent transport of phytochelatins across the tonoplast of oat roots. *Plant Physiology* 107, 1293–1301.

Salt, D.E. and Wagner, G.J. (1993) Cadmium transport across tonoplast of vesicles from oat roots. Evidence for a CdC2/HC antiport activity. *Journal of Biological Chemistry* 268, 12297–12302.

Salt, D.E., Thurman, D.A., Tomsett, A.B. and Sewell, A.K. (1989) Copper phytochelatins of *Mimulus guttatus. Proceedings of Royal Society, B, London* 236, 79–89.

Salt, D.E., Smith, R.D. and Raskin, I. (1998) Phytoremediation. *Annual Review of Plant Physiology and Plant Molecular Biology* 49, 643–668.

Sancenon, V., Puig, S., Mira, H., Thiele, D.J. and Penarrubia, L. (2003) Identification of a copper transporter family in *Arabidopsis thaliana. Plant Molecular Biology* 51, 577–587.

Schafer, H.J., Haag-Kerwer, A. and Rausch, T. (1998) cDNA cloning and expression analysis of genes encoding GSH synthesis in roots of the heavy metal accumulator *Brassica juncea* L.: evidence or Cd-induction of a putative mitochondrial g-glutamylcysteine synthetase isoform. *Plant Molecular Biology* 37, 87–97.

Schiott, M., Romanowsky, S.M., Baekgaard, L., Jakobsen, M.K., Palmgren, M.G. and Harper, J.F. (2004) A plant plasma membrane Ca2+ pump is required for normal pollen tube growth and fertilization. *Proceedings of the National Academy of Sciences of the United States of America* 101, 9502–9507.

Shigaki, T., Rees, I., Nakhleh, L. and Hirschi K.D. (2006) Identification of three distinct phylogenetic groups of CAX cation/protonantiporters. *Journal of Molecular Evolution* 63, 815–825.

Shikanai, T., Muller-Moule, P., Munekage, Y., Niyogi, K.K. and Pilon, M. (2003) PAA1, a P-type ATPase of *Arabidopsis*, functions in copper transport in chloroplasts. *The Plant Cell* 15, 1333–1346.

Solioz, M. and Vulpe, C. (1996) CPx-type ATPases: a class of P-type ATPases that pump heavy metals. *Trends in Biochemical Sciences* 21, 237–241.

Song, W.Y., Sohn, E.J., Martinoia, E., Lee, Y.J., Yang, Y.Y., Jasinski, M., Forestier, C., Hwang, I. and Lee, Y. (2003) Engineering tolerance and accumulation of lead and cadmium in transgenic plants. *Nature Biotechnology* 21, 914–919.

Steffens, J.C. (1990) The heavy metal-binding peptides of plants. *Annual Review of Plant Physiology and Molecular Biology* 41, 553–575.

Stephan, U.W., Schmidke, I., Stephan, V.W. and Scholz, G. (1996) The nicotianamine molecule is made-to-measure for complexation of metal micronutrients in plants. *Biometals* 9, 84–90.

Taiz, L. and Zeiger, E. (2002) *Plant Physiology*, 3rd edn. Sinauer Associates, Sunderland, Massachusetts.

Thomine, S., Wang, R., Ward, J.M., Crawford, N.M. and Schroeder, J.I. (2000). Cadmium and iron transport by members of a plant transporter gene family in *Arabidopsis* with homology to NRAMP genes. *Proceedings of the National Academy of Sciences of the United States of America* 97, 4991–4996.

Tian, J.L., Zhu, H.T., Yang, Y.A. and He, Y.K. (2004) Organic mercury tolerance, absorption and transformation in *Spartina* plants. *Journal of Plant Physiology and Molecular Biology* (China) 30, 577–582.

Tseng, T.S., Tzeng, S.S., Yeh, C.H., Chang, F.C., Chen, Y.M. and Lin C.Y. (1993) The heat-shock response in rice seedlings – isolation and expression of cDNAs that encode class-I low molecular-weight heat shock proteins. *Plant and Cell Physiology* 34, 165–168.

Ueoka-Nakanishi, H., Tsuchiya, T., Sasaki, M., Nakanishi, Y., Cunningham, K.W. and Maeshima, M. (2000) Functional expression of mung bean Ca^{2+}/H^{+} antiporter in yeast and its intracellular localization in the hypocotyl and tobacco cells. *European Journal of Biochemistry* 267, 3090–3098.
van der Zaal, B.J., Neuteboom, L.W., Pinas, J.E., Chardonnens, A.N., Schat, H., Verkleij, J.A.C. and Hooykaas, P.J.J. (1999) Over-expression of a novel *Arabidopsis* gene related to putative zinc-transporter genes from animals can lead to enhanced zinc resistance and accumulation. *Plant Physiology* 119, 1047–1055.
Van Huysen, T., Terry, N. and Pilon-Smits, E.A.H. (2004) Exploring the selenium phytoremediation potential of transgenic Indian mustard overexpressing ATP sulfurylase or cystathionine-γ-synthase. *International Journal of Phytoremediation* 6, 1–8.
Vatamaniuk, O.K., Mari, S., Lu, Y.P. and Rea, P.A. (1999) AtPCS1, a phytochelatin synthase from *Arabidopsis*: isolation and *in vitro* reconstitution. *Proceedings of the National Academy of Sciences of the United States of America* 96, 7110–7115.
Von Wiren, N., Klair, S., Bansal, S., Briat, J.F., Khodr, H., Shiori, T., Leigh, R.A. and Hider, R.C. (1999) Nicotianamine chelates both FeIII and FeII. Implications for metal transport in plants. *Plant Physiology* 119, 1107–1114.
Wang, Q.R., Cui, Y.S., Liu, X.M., Dong, Y.T. and Christie, P. (2003) Soil contamination and plant uptake of heavy metals at polluted sites in China. *Journal of Environmental Science Health* 38, 823–838.
Watanabe, T., Osaki, M., Yoshihara, T. and Tadano, T. (1998) Distribution and chemical speciation of aluminum in the Al accumulator plant, *Melastoma malabathricum* L. *Plant and Soil* 201, 165–173.
Wei, S.H., Zhou, Q.X., Wang, X., Cao, W., Ren, L.P. and Song, Y.F. (2004) Potential of weed species applied to remediation od soils contaminated with heavy metals. *Journal of Environmental Science* (China) 16, 868–873.
Wei, W., Chai, T., Zang, Y., Han, L., Xu, J. and Guan, Z. (2009) The *Thlaspi caerulescens* NRAMP homologue TcNRAMP3 is capable of divalent cation transport. *Molecular Biotechnology* 41, 15–21.
Williams, L.E. and Mills, R.F. (2005) P1B-ATPases–an ancient family of transition metal pumps with diverse functions in plants. *Trends in Plant Science* 10, 491–502.
Williams, L.E., Pittman, J.K. and Hall, J.L. (2000) Emerging mechanisms for heavy metal transport in plants. *Biochimica et Biophysica Acta* 1465, 104–126.
Woeste, K.E. and Kieber, J.J. (2000) A strong loss-of-function mutation in RAN1 results in constitutive activation of the ethylene response pathway as well as a rosette lethal phenotype. *The Plant Cell* 12, 443–455.
Wollgiehn, R. and Newmann, D. (1999) Metal stress response and tolerance of cultured cells from *Silene vulgaris* and *Lycopersicon peruvianum*: role of heat stress proteins. *Journal of Plant Physiology* 154, 547–553.
Xiang, C. and Oliver, D.J. (1998) Glutathione metabolic genes coordinately respond to heavy metal and jasmonic acid in *Arabidopsis*. *The Plant Cell* 10, 1539–1550.
Xue, S.G., Chen, Y.X., Reeves, R.D., Baker, A.J., Lin, Q. and Fernando, D.R. (2004) Manganese uptake and accumulation by the hyperaccumulator plant *Phytolacca actinosa* Roxb. (Phytolaccaceae). *Environmental Pollution* 131, 393–399.
Yokoi, S., Quintero, F.J., Cubero, B., Ruiz, M.T., Bressan, R.A. and Hasegawa, P.M. (2002) Differential expression and function of *Arabidopsis* thaliana NHX Na+/H+ antiporters in the salt stress response. *The Plant Journal* 30, 529–539.
Zenk, M.H. (1996) Heavy metal detoxification in higher plants: a review. *Gene* 179, 21–30.
Zhang, X.H., Liu, J., Huang, H.T., Chen, J., Zhu, Y.N. and Wang, D.O. (2007) Chromium accumulation by the hyperaccumulator plant *Leersia hexandra* Swartz. *Chemosphere* 67, 1138–1143.
Zhao, F.J., Dunham, S.J. and McGrath, S.P. (2002) Arsenic hyperaccumulation by different fern species. *New Phytologist* 15, 27–31.
Zhao, H. and Eide, D. (1996) The yeast ZRT1 gene encodes the zinc transporter of a high affinity uptake system induced by zinc limitation. *Proceedings of the National Academy of Sciences of the United States of America* 93, 2454–2458.
Zhu, Y.L., Pilon-Smits, E.A.H., Tarun, A.S., Weber, S.U., Jouanin, L. and Terry, N. (1999) Cadmium tolerance and accumulation in Indian mustard is enhanced by overexpressing-glutamylcysteine synthetase. *Plant Physiology* 121, 1169–1177.
Zimeri, A.M., Dhanker, O.P., McCaig, B. and Meagher, R.B. (2005) The plant MT1 metallothioneins are stabilized by binding cadmium and are required for cadmium tolerance and accumulation. *Plant Molecular Biology* 58, 839–855.

Appendix: Botanical name of plants mentioned in the text by common name

Common name	Botanical name
Abaca	*Musa textilis*
Acacia	*Acacia* spp.
Alder buckthorn	*Rhamnus frangula*
Almond	*Prunus dulcis*
Aloe	*Aloe vera*
Annatto	*Bixa orellena*
Apricot	*Prunus armeniaca*
Avocado	*Persea americana*
Azalea, yellow	*Rhododendron molle*
Babassu	*Orbignya oleifera*
Bamboo	*Bambusa vulgaris*
Barley	*Hordeum vulgare*
Bean, azuki	*Vigna angularis*
Bermudagrass	*Cynodon* spp.
Betel nut laurel	*Litsea stocksii*
Big bluestem	*Andropogon gerardii*
Borage	*Borago officianalis*
Buckwheat	*Fagopyrum esculentum*
Calendula	*Calendula officinalis*
California Bay	*Umbelluria californica*
Camelina	*Camelina sativa*
Canola	*Brassica napus*
Carnation	*Dianthus caryophyllus*
Carrot	*Daucus carota*
Cascara sagrada	*Rhamnus purshiana*
Cassava	*Manihot esculenta*
Castor	*Ricinus communis*
Catechu	*Acacia catechu*
Celery	*Apium graveolens*
Chinaberry	*Melia* spp.
Chinese vegetable tallow	*Triadica sebifera*
Cinnamon	*Cinnamomum verum*

Clove	*Syzygium aromaticum*
Cocoa	*Theobroma cacao*
Coconut	*Cocos nucifera*
Coriander	*Coriandrum sativum*
Cosmos	*Cosmos bipinnatus*
Cotton	*Gossypium hirsutum*
Cow Cockle	*Saponaria vaccaria*
Cowpea	*Vigna unguiculata*
Crambe	*Crambe abyssinica*
Cuphea	*Cuphea* spp.
Custard apple	*Annona cherimola*
Dhupa	*Vateria indica*
Eastern gamagrass	*Tripsicum dactyloides*
Ethiopian mahogany	*Trichilia emetica*
Eucalyptus	*Eucalyptus globus*
Fern, Chinese brake	*Pteris vittata*
Fern, water	*Azolla caroliniana*
Flame of the forest	*Butea monosperma*
Flax	*Linum usitatissimum*
Grape	*Vitis vinifera*
Groundnut	*Arachis hypogaea*
Guayule	*Parthenium argentatum*
Hemp	*Cannabis sativa*
Henna	*Lawsonia inermis*
Honesty	*Lunaria biennis*
Indiangrass	*Sorghastrum nutans*
Indigo	*Indigofera tinctoria*
Irish potato	*Solanum tuberosum*
Ironweed	*Vernonia galamensis*
Jatropha	*Jatropha curcas*
Jojoba	*Simmondsia chinensis*
Jute	*Corchorus* spp.
Kamala	*Mallotus philippensis*
Karanja	*Pongamia pinnata*
Kenaf	*Hibiscus cannabinus*
Khakan	*Salvadora oleoides*
Kokum	*Garcinia* spp.
Kudzu	*Pueraria lobata*
Kusum	*Schleichera oleosa*
Lemon eucalyptus	*Corymbia citriodora*
Lemongrass	*Cymbopogon* spp.
Lesquerella	*Lesquerella filiformis*
Lettuce	*Lactuca sativa*
Linseed	*Linum usitatissimum*
Lucerne	*Medicago sativa*
Macadamia nut	*Macadamia integrifolia*
Madder	*Rubia tinctorum*
Madder, Indian	*Oldenlandia umbellata*
Mahua	*Madhuca* spp.
Maize	*Zea mays*
Malabar nut	*Adhatoda vasica*
Mango	*Mangifera indica*

Mangosteen	*Garcinia cornea*
Marigold	*Calendula officinalis*
Meadowfoam	*Limnanthes* spp.
Milkweed	*Asclepias syrica*
Mint	*Mentha* spp.
Miscanthus	*Miscanthus* spp.
Morinda, great	*Morinda citrifolia*
Mung bean	*Vigna radiata*
Mustard, Indian	*Brassica juncea*
Mustard, Asian	*Brassica tournefortii*
Nahor	*Mesua ferrea*
Napiergrass	*Pennisetum purpureum*
Neem	*Azadirachta indica*
Nettle	*Urtica dioica*
Niger	*Guizotia abyssinica*
Oat	*Avena sativa*
Oil palm	*Elaeis guineensis*
Oiticica	*Licania rigida*
Olive	*Olea europaea*
Pangola grass	*Digitaria* spp.
Pawpaw	*Asimina triloba*
Pea	*Pisum sativum*
Pennycress	*Thlaspi arvense*
Pepper, black	*Piper nigrum*
Pepper, red	*Capsicum annuum*
Perilla	*Perilla frutescens*
Pine	*Pinus* spp.
Pineapple	*Ananas comosus*
Pine, Loblolly	*Pinus taeda*
Pine, Longleaf	*Pinus palustris*
Pine, Shortleaf	*Pinus echinata*
Pine, Slash	*Pinus elliotti*
Pisa	*Actinodaphne hookeri*
Pomegranate	*Punica granatum*
Poplar	*Populus* spp.
Poplar, Bigtooth aspen	*Populus grandidentata*
Poplar, European Aspen	*Populus tremula*
Poplar, Eastern Cottonwood	*Populus deltoides*
Poplar, Trembling Aspen	*Populus tremuloides*
Poppy	*Papaver spp.*
Potato	*Solanum tuberosum*
Rambutan	*Nephelium lappaceum*
Ramie	*Bohemeria nivea*
Rapeseed	*Brassica napus*
Reed canarygrass	*Phalaris arundinacea*
Rhubarb	*Rheum palmatum*
Rice	*Oryza sativa*
Rosemary	*Rosmarinus officinale*
Rubber tree	*Hevea brasiliensis*
Rubber plant	*Ficus elastica*
Russian dandelion	*Taraxacum kok-saghyz*
Rye	*Secale cereale*

Rye, wild	*Leymus* spp.
Safflower	*Carthamus tinctorius*
Saffron	*Crocus sativus*
Sal	*Shorea* spp.
Sandalwood, red	*Pterocarpus santalinus*
Sea buckthorn	*Hippophae rhamnoides*
Senna	*Cassia angustifolia* or *C. acutifolia*
Sesame	*Sesamum indicum*
Shea	*Butyrospermum parkii*
Snapdragon	*Antirrhinum majus*
Sorghum	*Sorghum bicolor*
Soybean	*Glycine max*
Spartina	*Spartina alterniflora*
Sugarbeet	*Beta vulgaris*
Sugarcane/energy cane	*Saccharum officinarum*
Sunflower	*Helianthus annuus*
Sweet gum	*Liquidambar styraciflua*
Sweet potato	*Ipomoea batatas*
Sweetsop	*Annona squamosa*
Switchgrass	*Panicum virgatum*
Sycamore	*Acer pseudoplatanus*
Thyme	*Thymus vulgaris*
Tobacco	*Nicotiana tabacum*
Tomato	*Lycopersicon esculentum*
Tung	*Vernicia fordii*
Turmeric	*Curcuma longa*
Undi	*Calophyllum inophyllum*
Vetiver grass	*Chrysopogon zizanioides*
Walnut	*Juglans regia*
Watermelon	*Citrullus lanatus*
Wheat	*Triticum aestivum*
Willow	*Salix* spp.
Yellow dock	*Rumex crispus*

Index

Note: page numbers in *italics* refer to figures and tables.